AF352791

Brain Injury and Neurodegeneration: Molecular, Functional, and Translational Approach

Brain Injury and Neurodegeneration: Molecular, Functional, and Translational Approach

Editors

Kumar Vaibhav
Meenakshi Ahluwalia
Pankaj Gaur

MDPI • Basel • Beijing • Wuhan • Barcelona • Belgrade • Manchester • Tokyo • Cluj • Tianjin

Editors

Kumar Vaibhav
Augusta University
Augusta
GA
USA

Meenakshi Ahluwalia
Augusta University
Augusta
GA
USA

Pankaj Gaur
Georgetown University
Medical Center
Washington
DC
USA

Editorial Office
MDPI
St. Alban-Anlage 66
4052 Basel, Switzerland

This is a reprint of articles from the Special Issue published online in the open access journal *Biomedicines* (ISSN 2227-9059) (available at: https://www.mdpi.com/journal/biomedicines/special_issues/Brain_Injury_Neurodegeneration).

For citation purposes, cite each article independently as indicated on the article page online and as indicated below:

LastName, A.A.; LastName, B.B.; LastName, C.C. Article Title. *Journal Name* **Year**, *Volume Number*, Page Range.

ISBN 978-3-0365-8384-6 (Hbk)
ISBN 978-3-0365-8385-3 (PDF)

Cover image courtesy of Kumar Vaibhav

© 2023 by the authors. Articles in this book are Open Access and distributed under the Creative Commons Attribution (CC BY) license, which allows users to download, copy and build upon published articles, as long as the author and publisher are properly credited, which ensures maximum dissemination and a wider impact of our publications.

The book as a whole is distributed by MDPI under the terms and conditions of the Creative Commons license CC BY-NC-ND.

Contents

About the Editors

Kumar Vaibhav

Kumar Vaibhav (MS, PhD, Asst. Prof.) is a Neuroscientist, Educator, Mentor, and Director of the Brain injury, Senescence and Translational Neuroscience laboratory in the Department of Neurosurgery, Medical College of Georgia at Augusta University, Augusta, GA, USA. Dr. Vaibhav is also affiliated with the Department of Oral Biology and Diagnostic Sciences, Center for Excellence in Research, Scholarship and Innovation, Dental College of Georgia, Augusta University, Augusta, GA and the Transdisciplinary Research Initiative in Inflammaging and Brain Aging (TRIBA), Augusta University, Augusta, GA. His main research focus is on the outcomes of traumatic brain injury and intracerebral hemorrhages with emphasis on immune interaction, the systemic influence of diseases, and alteration in the endocannabinoid system in acute and chronic pathology. His lab is further investigating pathology related to brain injury-induced senescence, neurodegeneration, and dementia.

Meenakshi Ahluwalia

Meenakshi Ahluwalia (MS, PhD) is a senior post-doctoral fellow in the Department of Pathology, Medical College of Georgia, and is currently working in the Georgia Cancer Center at Augusta University, Augusta, GA. She has expertise in traumatic brain injury, hemorrhagic stroke, glioma, and genetics. Her research interest is to develop non-invasive therapy for brain insult and oncolytic treatment for the repression of glioma.

Pankaj Gaur

Pankaj Gaur (MS, PhD—Neuroscience) is a highly accomplished researcher and scientist who has made significant contributions to the fields of neuroscience, immunology, and cancer research. Dr. Gaur is affiliated as a Research Instructor (Faculty) at Lombardy Cancer Center, Georgetown Medical College, Georgetown University, Washington DC. Dr. Gaur's primary research interest revolves around identifying innovative strategies to enhance specific immune responses against cancer, autoimmune diseases (such as multiple sclerosis and systemic lupus erythematosus), as well as various brain disorders including Parkinson's disease, epilepsy, traumatic brain injury, and neurodegeneration. His aim is to unravel the intricate mechanisms underlying immune responses, ultimately leading to the development of novel therapeutic approaches.

Preface to "Brain Injury and Neurodegeneration: Molecular, Functional, and Translational Approach"

We are glad to introduce this Reprint of the Special Issue "Brain Injury and Neurodegeneration: Molecular, Functional, and Translational Approach" for the readers. This Special Issue includes 19 articles, including 1 editorial, 10 research articles, and 8 reviews on current topics of brain injury, pathology, and neurodegeneration. This collection has shed light on many aspects of brain pathologies, related meta-analysis, and alternative medicines. This Issue will be a good read for scientists, medical professionals, students, and for the general public.

We would like to acknowledge our gratitude to the authors for publishing in our Issue, and the reviewers for doing a wonderful job of promptly assessing the articles. We are also thankful to our Editor in Chief, Prof. Dr. Shaker A. Mousa, our Assistant Editor, Ms. Dora Xie, and the entire *Biomedicines* editorial staff for the support we received throughout this journey. Last but not least, I would like to applaud our team of guest editors, Drs. Meenakshi Ahluwalia and Pankaj Gaur, who helped and supported the cause.

Kumar Vaibhav, Meenakshi Ahluwalia, and Pankaj Gaur
Editors

 *biomedicines*

Editorial

Brain Injury and Neurodegeneration: Molecular, Functional, and Translational Approach

Meenakshi Ahluwalia [1,2,3], **Pankaj Gaur** [4] **and Kumar Vaibhav** [3,5,6,*]

[1] Department of Pathology, Medical College of Georgia, Augusta University, Augusta, GA 30912, USA; mahluwalia@augusta.edu
[2] Georgia Cancer Center, Medical College of Georgia, Augusta University, Augusta, GA 30912, USA
[3] Department of Neurosurgery, Medical College of Georgia, Augusta University, Augusta, GA 30912, USA
[4] Lombardi Comprehensive Cancer Center, Georgetown University Medical Center, Washington, DC 20057, USA; pg684@georgetown.edu
[5] Department of Oral Biology and Diagnostic Sciences, Center for Excellence in Research, Scholarship and Innovation, Dental College of Georgia, Augusta University, Augusta, GA 30912, USA
[6] Transdisciplinary Research Initiative in Inflammaging and Brain Aging (TRIBA), Augusta University, Augusta, GA 30912, USA
* Correspondence: kvaibhav@augusta.edu; Tel.: +1-(706)-721-4691; Fax: +1-(706)-721-7619

Citation: Ahluwalia, M.; Gaur, P.; Vaibhav, K. Brain Injury and Neurodegeneration: Molecular, Functional, and Translational Approach. *Biomedicines* **2023**, *11*, 1947. https://doi.org/10.3390/biomedicines11071947

Received: 4 July 2023
Accepted: 6 July 2023
Published: 10 July 2023

Copyright: © 2023 by the authors. Licensee MDPI, Basel, Switzerland. This article is an open access article distributed under the terms and conditions of the Creative Commons Attribution (CC BY) license (https://creativecommons.org/licenses/by/4.0/).

Recently, we have achieved substantial progress in our understanding of brain injury and neurodegeneration. We have enhanced our knowledge about different brain pathologies or insults (accidental or non-accidental), such as hemorrhages, traumatic brain injury (TBI), ischemia, hypoxic/hypobaric insults, and neurological disorders such as Parkinson's disease and Alzheimer's disease (PD and AD). Brain pathology is multifactorial, encompassing a cascade of inflammation, necrotic, and apoptotic pathways. It is well known that brain insults or injuries to the brain may lead to neurological disorders over time, and genetic or environmental factors play essential roles in the progression of a brain disease. The absence of a specific cure to limit injury progression after an insult has persuaded the scientific community to study the mechanisms behind brain injury and degenerative cascade and to explore different therapeutic strategies.

This Special Issue, entitled "Brain Injury and Neurodegeneration: Molecular, Functional, and Translational Approach", has addressed various important aspects of brain injury and neurodegeneration, such as TBI, cerebral hypoxia, epilepsy, AD, and SARS-CoV-2-mediated brain damage. This Special Issue has received an enthusiastic response globally, resulting in the publication of 18 peer-reviewed articles, including 10 research articles and 8 reviews.

The recent SARS-CoV-2 pandemic caused more than 3 million deaths globally [1]. Similarly to past coronavirus outbreaks, including SARS and MERS, SARS-CoV-2 infections were associated with fever, dry cough, pneumonia, fatigue, and acute respiratory distress syndrome (ARDS) [2]. However, each host–pathogen interaction leaves a footprint on the health of different organs [3]. Clinical studies on COVID-19 patients have revealed atypical symptoms and neurological signs, including headaches, anosmia, nausea, dysgeusia, damage to respiratory centers, and cerebral infarction [2,4]. Extreme cytokines release (cytokine storm) occurs due to aberrant immune pathways, and microglial activation propagates CNS damage in COVID-19 patients [1–4]. Moreover, elderly with neurological problems such as PD and AD showed a higher incidence of COVID-19-associated complications [4]. In this Special Issue, an Editor's Choice article by Donegani et al. [5] provided proof of CNS damage via an [18F]-FDG PET scan. Twenty-two SARS-CoV-2 patients underwent whole-body [18F]-FDG PET examination, including a dedicated brain acquisition between May and December 2020 after their recovery from SARS-CoV-2 infection, and fourteen patients were found to have persistent hyposmia in bilateral fusiform gyri and parahippocampal, and in left insula, as compared to the controls [5].

Our Special Issue has a large number of TBI-related articles, which includes five research articles and one review. TBI is a global health concern as it results in substantial death and disability [6–11]. In light of this, our group [12] has provided a comprehensive review on TBI. In this Editor's Choice review, we have provided a broad understanding of TBI pathology, mechanisms, inflammation, and immune interactions. Understanding these mechanisms and exploring potential targets for neuroprotective treatments are crucial for advancing new therapies. This review delves into the molecular events that occur following TBI, encompassing inflammation and programmed cell death, and offers an overview of the current literature and therapeutic approaches, contributing to a deeper understanding of secondary injuries caused by TBI. Cheng et al. [13] have shown that stretch injury (an in vitro model of TBI) in SH-SY5Y neuroblastoma cells altered the mitochondrial membrane potential and triggered oxidative DNA damage at 24 h. Stretch injury increased neuronal stress via reducing brain-derived neurotrophic factors (BDNFs) and increasing amyloid-β. Mechanistically, neuronal injury was exaggerated through the loss of the insulin pathway and via increased glycogen synthase kinase 3β (GSK-3β)S9/p-Tau protein levels [13]. In addition, Puhakka et al. [14] reported that small non-coding RNAs (sncRNAs) play a crucial role in modulating post-TBI neuroinflammation. Further, increased expression of the miR-146a profile and of 3′tRF-IleAAT and 3′tRF-LysTTT was found to be associated with behavioral deficits in animals with chronic TBI-induced neuroinflammation [14]. Another study by Vorn et al. [15] profiled the plasma exosomal microRNAs from young adults with mild TBI and from healthy individuals, and identified 25 dysregulated exosomal miRNAs in the chronic mTBI group 4.48 mean years after injury. These miRNAs are associated with pathways of neurological disease, organismal injury and abnormalities, and psychological disease, and can be useful to be diagnostic markers for chronic mTBI [15].

A higher percentage of patients with TBI die from secondary pathological processes despite the application of preventative measures and the provision of medical supervision [8–11]. Post-traumatic epilepsy (PTE) is one of the most common debilitating implications of TBI [16]. Post-traumatic seizures (PTS) are experienced in increasing TBI individuals who also display resistance to traditional anti-seizure medications (ASMs). Hentig et al. [17] identified an upregulated sonic hedgehog (Shh) signaling pathway in zebrafish after CNS injury that helps in regeneration. Shh signaling increases excitatory amino acid transporters (Eaat)2a to inhibit TBI-induced glutamate excitotoxicity and subsequent seizure sequelae [17]. Further, Ghosh et al. [18] provided an account for epilepsy, the roles played by various neurotransmitters and their corresponding receptors in the pathophysiology of epilepsy. One of the forefront areas of epilepsy research is drug-resistant epilepsy (DRE), which is the focus of this review. The authors mentioned that drug-resistant epilepsy (DRE) remains a prominent focus of research due to its link to psychosocial complications and premature mortality, and to a high prevalence among epileptic patients [18]. The review examines various hypotheses relating to DRE and explores unconventional therapeutic strategies and combination therapy. Additionally, recent studies supporting modern treatment approaches for epilepsy are discussed, with specific emphasis on the mTOR pathway, blood–brain barrier breakdown, and inflammatory pathways.

There are other factors, CNS insults and diseases, such as hypoxia, alcohol, viral infections, ischemia, and neurodegeneration, which contribute towards acute and chronic deficits in CNS function. Our collection also highlights the interesting findings in these particular areas. Both chronic alcoholism and human herpesvirus-6 (HHV-6) infection may cause movement-related disorders and promote neuroinflammation. Jain et al. [19] observed decreased perivascular CD68$^+$/Iba1$^+$ microglia in the postmortem brain from alcoholic individuals as compared to the dominant CD68$^+$/Iba1$^-$ microglial subpopulation in the control brains. All the control brains were HHV-6 negative. Further, HHV-6 infection in alcoholics elevated microglial dystrophic changes with higher Iba1$^+$ cells and compounded the microglia-mediated neuroinflammation. Another research article by Baltanas et al. [20] explored the rare, biallelic variants of the AGTPBP1 gene that caused its loss of function, and led to childhood-onset neurodegeneration with cerebellar atrophy

(CONDCA). Mutations in AGTPBP1 led to the substantial loss of cerebellar Purkinje cells in the mouse model of cerebellar ataxia, and might be used for CONDCA modeling in mice. Asik et al. [21] provided a review of Alzheimer's disease in relation to Amyloid-β (Aβ). They have provided a comprehensive view of Amyloid-β (Aβ)-related pathology and mechanisms and of the current clinical status of anti-amyloid therapy. They emphasized that the relationship between dysfunctional mitochondrial and the progression of AD required further research. Yoshida et al. [22] reported that higher oxidative stress in hippocampal mitochondria leads to cognitive impairment in a 5xFAD mice model of Alzheimer's disease. They further mentioned that age can be a vital factor in elevated oxidative stress in AD pathology, and preventing mitochondrial oxidative damage may be important to protect cognitive function.

Hypoxia as a result of the deprivation of oxygen, temporary or chronic, can either be adaptive or pathological. A systematic review by Stoica et al. [23] utilized the "Preferred Reporting Items for Systematic Reviews and Meta-Analyses (PRISMA)" filtering method, and probed five internationally renowned medical databases. Using this method, they identified 45 eligible papers and provided information on pathophysiology, mechanisms, and consequent clinical conditions following hypoxic episodes. Ischemia occurs as a result of transient or permanent interruption blood supply in a given region, which leads to poor oxygenation, inflammation, and oxidative stress. Fatty acid-binding proteins (FABPs) mediate lipid metabolism and regulate the dynamics of fatty acids. Following transient MCAO, the levels of FABP3, FABP5, and FABP7 were found to be upregulated in the brain. Guo et al. [24] reported that the FABP inhibitor, i.e., FABP ligand 6 [4-(2-(5-(2-chlorophenyl)-1-(4-isopropylphenyl)-1H-pyrazol-3-yl)-4-fluorophenoxy)butanoic acid] (referred to here as MF6), minimized the prostaglandin E2 (PGE2)-mediated inflammation in ischemic brain. Flavonoids, icariin (ICA), and icaritin (ICT) derived from *Herba epimedii* have been identified as neuroprotective phytochemicals. Wu et al. [25] reported that both ICA and ICT treatment improved neuronal cell apoptosis, minimized oxidative stress, and countered extracellular matrix (ECM) accumulation in mice brains post-acute cerebral ischemia. Demyanenko et al. [26] reviewed the role of histone deacetylases and their inhibitors in ischemic stroke.. The authors showed that ischemic stroke generally reduces gene expression via suppression of the acetylation of histones H3 and H4. Inhibitors to histone deacetylases promoted functional recovery post-cerebral ischemia by inducing neurogenesis and angiogenesis in the injured areas of brain. This review aimed to explore neuroprotective activities of histone deacetylase inhibitors in ischemic stroke. In line with neuroprotection, Haider et al. [27] investigated the role of mitoquinone in chronic neuroprotection post-TBI, and found that mitoquinone reduced gliosis, decreased oxidative stress, limited neuroinflamamtion, and improved axonal integrity and neuronal survival in an open-head CCI mouse model of moderate TBI.

With the increase in the expectancy of the life span of humans, incidences of accidental injury and neurodegenerative diseases (NDs) have risen and have imposed a considerable burden on the family, society, and the nation. The review by Khan et al. [28] explores the mechanisms of action of the phytochemicals and nutraceuticals available to date for various NDs. The group has reviewed clinical and pre-clinical studies involving phytochemicals in neurodegeneration. Despite phytochemicals showing a robust effect in animal studies, mixed results were observed in several clinical trials, and therefore, the authors stressed a need to reassess their efficacies in more robust clinical studies [28]. While traditional medicines and chemical inhibitors are being actively studied in pre-clinical and clinical settings, non-invasive methods such as exercise [29,30], whole-body vibration [31], or ischemic conditioning [32–34] can enhance endogenous protection against many diseases. Nhu et al. [35] reviewed treadmill exercise (TE) on neural mitochondria in PD. Parkinson's disease is the second most common neurodegenerative disorder [36,37], and TE has been widely applied in its rehabilitation [30,38]. For this systematic review [35], the CAMARADES checklist was used to assess the methodological quality of the studies. The review findings supported the hypothesis that treadmill exercise could attenuate neuronal mitochondrial dysregulation and respiratory deficiency in PD and could slow down the progression of PD.

In conclusion, this Special Issue represents a novel and exciting perspective on brain injury, hypoxia, ischemia, and neurodegeneration, and it represents a comprehensive research resource for readers on disease-related pathology, mechanisms, and translational approach. However, we acknowledge that this field of neuroscience is under active research, and several new findings are being made daily. Therefore, this Special Issue of articles, along with new discoveries, will be an interesting and substantial read for scholars in this field.

Funding: Authors' research has been supported by grants from the National Institutes of Neurological Diseases and Stroke (NS114560), National Institutes of Child Health and Development (HD094606), and AURI Research support to K.V.

Conflicts of Interest: The authors declare no financial or competing conflicts of interest. The funders had no role in the design of the study; in the collection, analyses, or interpretation of data; in the writing of the manuscript; in the decision to publish the results; or in the editing of articles.

References

1. Ahluwalia, P.; Vaibhav, K.; Ahluwalia, M.; Mondal, A.K.; Sahajpal, N.; Rojiani, A.M.; Kolhe, R. Infection and Immune Memory: Variables in Robust Protection by Vaccines Against SARS-CoV-2. *Front. Immunol.* **2021**, *12*, 660019. [CrossRef] [PubMed]
2. Jarrahi, A.; Ahluwalia, M.; Khodadadi, H.; da Silva Lopes Salles, E.; Kolhe, R.; Hess, D.C.; Vale, F.; Kumar, M.; Baban, B.; Vaibhav, K.; et al. Neurological consequences of COVID-19: What have we learned and where do we go from here? *J. Neuroinflamm.* **2020**, *17*, 286. [CrossRef] [PubMed]
3. Ahluwalia, P.; Ahluwalia, M.; Vaibhav, K.; Mondal, A.; Sahajpal, N.; Islam, S.; Fulzele, S.; Kota, V.; Dhandapani, K.; Baban, B.; et al. Infections of the lung: A predictive, preventive and personalized perspective through the lens of evolution, the emergence of SARS-CoV-2 and its pathogenesis. *EPMA J.* **2020**, *11*, 581–601. [CrossRef]
4. Hsu, P.C.; Shahed-Al-Mahmud, M. SARS-CoV-2 mediated neurological disorders in COVID-19: Measuring the pathophysiology and immune response. *Life Sci.* **2022**, *308*, 120981. [CrossRef] [PubMed]
5. Donegani, M.I.; Miceli, A.; Pardini, M.; Bauckneht, M.; Chiola, S.; Pennone, M.; Marini, C.; Massa, F.; Raffa, S.; Ferrarazzo, G.; et al. Brain Metabolic Correlates of Persistent Olfactory Dysfunction after SARS-Cov2 Infection. *Biomedicines* **2021**, *9*, 287. [CrossRef]
6. Coronado, V.G.; Xu, L.; Basavaraju, S.V.; McGuire, L.C.; Wald, M.M.; Faul, M.D.; Guzman, B.R.; Hemphill, J.D. Surveillance for traumatic brain injury-related deaths—United States, 1997–2007. *Morb. Mortal. Wkly. Rep. Surveill. Summ.* **2011**, *60*, 1–32.
7. Stevens, J.A.; Corso, P.S.; Finkelstein, E.A.; Miller, T.R. The costs of fatal and non-fatal falls among older adults. *Inj. Prev. J. Int. Soc. Child Adolesc. Inj. Prev.* **2006**, *12*, 290–295. [CrossRef]
8. Bramlett, H.M.; Dietrich, W.D. Quantitative structural changes in white and gray matter 1 year following traumatic brain injury in rats. *Acta Neuropathol.* **2002**, *103*, 607–614. [CrossRef]
9. Bramlett, H.M.; Dietrich, W.D. Progressive damage after brain and spinal cord injury: Pathomechanisms and treatment strategies. *Prog. Brain Res.* **2007**, *161*, 125–141. [CrossRef]
10. Taylor, C.A.; Bell, J.M.; Breiding, M.J.; Xu, L. Traumatic Brain Injury-Related Emergency Department Visits, Hospitalizations, and Deaths—United States, 2007 and 2013. *Morb. Mortal. Wkly. Rep. Surveill. Summ.* **2017**, *66*, 1–16. [CrossRef]
11. Destounis, A.; Tountas, C.; Theodosis-Georgilas, A.; Zahos, P.; Kasinos, N.; Palios, J.; Beldekos, D. An unusual case of double-chambered left ventricle: A case of double-chambered left ventricle communicated with right ventricle through a ventricular septal defect presented during only in diastole and a concomitant mitral valve prolapse. *J. Echocardiogr.* **2019**, *17*, 167–168. [CrossRef]
12. Jarrahi, A.; Braun, M.; Ahluwalia, M.; Gupta, R.V.; Wilson, M.; Munie, S.; Ahluwalia, P.; Vender, J.R.; Vale, F.L.; Dhandapani, K.M.; et al. Revisiting Traumatic Brain Injury: From Molecular Mechanisms to Therapeutic Interventions. *Biomedicines* **2020**, *8*, 389. [CrossRef]
13. Cheng, P.W.; Wu, Y.C.; Wong, T.Y.; Sun, G.C.; Tseng, C.J. Mechanical Stretching-Induced Traumatic Brain Injury Is Mediated by the Formation of GSK-3β-Tau Complex to Impair Insulin Signaling Transduction. *Biomedicines* **2021**, *9*, 1650. [CrossRef]
14. Puhakka, N.; Das Gupta, S.; Vuokila, N.; Pitkänen, A. Transfer RNA-Derived Fragments and isomiRs Are Novel Components of Chronic TBI-Induced Neuropathology. *Biomedicines* **2022**, *10*, 136. [CrossRef]
15. Vorn, R.; Suarez, M.; White, J.C.; Martin, C.A.; Kim, H.S.; Lai, C.; Yun, S.J.; Gill, J.M.; Lee, H. Exosomal microRNA Differential Expression in Plasma of Young Adults with Chronic Mild Traumatic Brain Injury and Healthy Control. *Biomedicines* **2021**, *10*, 36. [CrossRef]
16. Yu, T.; Liu, X.; Sun, L.; Wu, J.; Wang, Q. Clinical characteristics of post-traumatic epilepsy and the factors affecting the latency of PTE. *BMC Neurol.* **2021**, *21*, 301. [CrossRef]
17. Hentig, J.; Campbell, L.J.; Cloghessy, K.; Lee, M.; Boggess, W.; Hyde, D.R. Prophylactic Activation of Shh Signaling Attenuates TBI-Induced Seizures in Zebrafish by Modulating Glutamate Excitotoxicity through Eaat2a. *Biomedicines* **2021**, *10*, 32. [CrossRef]

18. Ghosh, S.; Sinha, J.K.; Khan, T.; Devaraju, K.S.; Singh, P.; Vaibhav, K.; Gaur, P. Pharmacological and Therapeutic Approaches in the Treatment of Epilepsy. *Biomedicines* **2021**, *9*, 470. [CrossRef]

19. Jain, N.; Smirnovs, M.; Strojeva, S.; Murovska, M.; Skuja, S. Chronic Alcoholism and HHV-6 Infection Synergistically Promote Neuroinflammatory Microglial Phenotypes in the Substantia Nigra of the Adult Human Brain. *Biomedicines* **2021**, *9*, 1216. [CrossRef]

20. Baltanás, F.C.; Berciano, M.T.; Santos, E.; Lafarga, M. The Childhood-Onset Neurodegeneration with Cerebellar Atrophy (CONDCA) Disease Caused by AGTPBP1 Gene Mutations: The Purkinje Cell Degeneration Mouse as an Animal Model for the Study of this Human Disease. *Biomedicines* **2021**, *9*, 1157. [CrossRef]

21. Mohamed Asik, R.; Suganthy, N.; Aarifa, M.A.; Kumar, A.; Szigeti, K.; Mathe, D.; Gulyás, B.; Archunan, G.; Padmanabhan, P. Alzheimer's Disease: A Molecular View of β-Amyloid Induced Morbific Events. *Biomedicines* **2021**, *9*, 1126. [CrossRef] [PubMed]

22. Yoshida, N.; Kato, Y.; Takatsu, H.; Fukui, K. Relationship between Cognitive Dysfunction and Age-Related Variability in Oxidative Markers in Isolated Mitochondria of Alzheimer's Disease Transgenic Mouse Brains. *Biomedicines* **2022**, *10*, 281. [CrossRef] [PubMed]

23. Stoica, S.I.; Bleotu, C.; Ciobanu, V.; Ionescu, A.M.; Albadi, I.; Onose, G.; Munteanu, C. Considerations about Hypoxic Changes in Neuraxis Tissue Injuries and Recovery. *Biomedicines* **2022**, *10*, 481. [CrossRef] [PubMed]

24. Guo, Q.; Kawahata, I.; Degawa, T.; Ikeda-Matsuo, Y.; Sun, M.; Han, F.; Fukunaga, K. Fatty Acid-Binding Proteins Aggravate Cerebral Ischemia-Reperfusion Injury in Mice. *Biomedicines* **2021**, *9*, 529. [CrossRef] [PubMed]

25. Wu, C.T.; Chen, M.C.; Liu, S.H.; Yang, T.H.; Long, L.H.; Guan, S.S.; Chen, C.M. Bioactive Flavonoids Icaritin and Icariin Protect against Cerebral Ischemia-Reperfusion-Associated Apoptosis and Extracellular Matrix Accumulation in an Ischemic Stroke Mouse Model. *Biomedicines* **2021**, *9*, 1719. [CrossRef] [PubMed]

26. Demyanenko, S.; Dzreyan, V.; Sharifulina, S. Histone Deacetylases and Their Isoform-Specific Inhibitors in Ischemic Stroke. *Biomedicines* **2021**, *9*, 1445. [CrossRef]

27. Haidar, M.A.; Shakkour, Z.; Barsa, C.; Tabet, M.; Mekhjian, S.; Darwish, H.; Goli, M.; Shear, D.; Pandya, J.D.; Mechref, Y.; et al. Mitoquinone Helps Combat the Neurological, Cognitive, and Molecular Consequences of Open Head Traumatic Brain Injury at Chronic Time Point. *Biomedicines* **2022**, *10*, 250. [CrossRef]

28. Khan, A.; Jahan, S.; Imtiyaz, Z.; Alshahrani, S.; Antar Makeen, H.; Mohammed Alshehri, B.; Kumar, A.; Arafah, A.; Rehman, M.U. Neuroprotection: Targeting Multiple Pathways by Naturally Occurring Phytochemicals. *Biomedicines* **2020**, *8*, 284. [CrossRef]

29. Khan, M.B.; Alam, H.; Siddiqui, S.; Shaikh, M.F.; Sharma, A.; Rehman, A.; Baban, B.; Arbab, A.S.; Hess, D.C. Exercise Improves Cerebral Blood Flow and Functional Outcomes in an Experimental Mouse Model of Vascular Cognitive Impairment and Dementia (VCID). *Transl. Stroke Res.* **2023**. *ePub ahead of print*. [CrossRef]

30. Wang, R.; Tian, H.; Guo, D.; Tian, Q.; Yao, T.; Kong, X. Impacts of exercise intervention on various diseases in rats. *J. Sport Health Sci.* **2020**, *9*, 211–227. [CrossRef]

31. Yin, H.; Berdel, H.O.; Moore, D.; Davis, F.; Liu, J.; Mozaffari, M.; Yu, J.C.; Baban, B. Whole body vibration therapy: A novel potential treatment for type 2 diabetes mellitus. *Springerplus* **2015**, *4*, 578. [CrossRef]

32. Khan, M.B.; Hoda, M.N.; Vaibhav, K.; Giri, S.; Wang, P.; Waller, J.L.; Ergul, A.; Dhandapani, K.M.; Fagan, S.C.; Hess, D.C. Remote ischemic postconditioning: Harnessing endogenous protection in a murine model of vascular cognitive impairment. *Transl. Stroke Res.* **2015**, *6*, 69–77. [CrossRef]

33. Vaibhav, K.; Braun, M.; Khan, M.B.; Fatima, S.; Saad, N.; Shankar, A.; Khan, Z.T.; Harris, R.B.S.; Yang, Q.; Huo, Y.; et al. Remote ischemic post-conditioning promotes hematoma resolution via AMPK-dependent immune regulation. *J. Exp. Med.* **2018**, *215*, 2636–2654. [CrossRef]

34. Hess, D.C.; Khan, M.B.; Kamat, P.; Vaibhav, K.; Dhandapani, K.M.; Baban, B.; Waller, J.L.; Hoda, M.N.; Blauenfeldt, R.A.; Andersen, G. Conditioning medicine for ischemic and hemorrhagic stroke. *Cond. Med.* **2021**, *4*, 124–129.

35. Nhu, N.T.; Cheng, Y.J.; Lee, S.D. Effects of Treadmill Exercise on Neural Mitochondrial Functions in Parkinson's Disease: A Systematic Review of Animal Studies. *Biomedicines* **2021**, *9*, 1011. [CrossRef]

36. Kalia, L.V.; Lang, A.E. Parkinson's disease. *Lancet* **2015**, *386*, 896–912. [CrossRef]

37. Sveinbjornsdottir, S. The clinical symptoms of Parkinson's disease. *J. Neurochem.* **2016**, *139*, 318–324. [CrossRef]

38. Mehrholz, J.; Kugler, J.; Storch, A.; Pohl, M.; Elsner, B.; Hirsch, K. Treadmill training for patients with Parkinson's disease. *Cochrane Database Syst. Rev.* **2015**, CD007830. [CrossRef]

Disclaimer/Publisher's Note: The statements, opinions and data contained in all publications are solely those of the individual author(s) and contributor(s) and not of MDPI and/or the editor(s). MDPI and/or the editor(s) disclaim responsibility for any injury to people or property resulting from any ideas, methods, instructions or products referred to in the content.

Article

Brain Metabolic Correlates of Persistent Olfactory Dysfunction after SARS-Cov2 Infection

Maria Isabella Donegani [1,2], Alberto Miceli [1,2], Matteo Pardini [1,3], Matteo Bauckneht [1,2,*], Silvia Chiola [4,5], Michele Pennone [1,2], Cecilia Marini [1,6], Federico Massa [1,3], Stefano Raffa [1,4], Giulia Ferrarazzo [1,4], Dario Arnaldi [1,4], Gianmario Sambuceti [1,4], Flavio Nobili [1,3] and Silvia Morbelli [1,2]

[1] IRCCS Ospedale Policlinico San Martino, 16131 Genova, Italy; isabella.donegani@gmail.com (M.I.D.); albertomiceli23@gmail.com (A.M.); matteo.pardini@unige.it (M.P.); michele.pennone@hsanmartino.it (M.P.); cecilia.marini@unige.it (C.M.); fedemassa88@gmail.com (F.M.); Stefanoraffa@gmail.com (S.R.); giulia.ferrarazzo@gmail.com (G.F.); dario.arnaldi@unige.it (D.A.); sambuceti@unige.it (G.S.); flaviomariano.nobili@hsanmartino.it (F.N.); silviadaniela.morbelli@hsanmartino.it (S.M.)

[2] Nuclear Medicine Unit, Department of Health Sciences, University of Genoa, 516126 Genoa, Italy

[3] Department of Neuroscience (DINOGMI), University of Genoa, 516126 Genoa, Italy

[4] Humanitas Clinical and Research Center–IRCCS, Via Manzoni 56, Rozzano, 20089 Milan, Italy; silvia.chiola@cancercenter.humanitas.it

[5] Department of Biomedical Sciences, Humanitas University, Via Rita Levi Montalcini 4, Pieve Emanuele, 20090 Milan, Italy

[6] CNR Institute of Molecular Bioimaging and Physiology (IBFM), 20090 Milano, Italy

[*] Correspondence: matteo.bauckneht@gmail.com

Citation: Donegani, M.I.; Miceli, A.; Pardini, M.; Bauckneht, M.; Chiola, S.; Pennone, M.; Marini, C.; Massa, F.; Raffa, S.; Ferrarazzo, G.; et al. Brain Metabolic Correlates of Persistent Olfactory Dysfunction after SARS-Cov2 Infection. *Biomedicines* **2021**, *9*, 287. https://doi.org/10.3390/biomedicines9030287

Academic Editor: Kumar Vaibhav

Received: 13 January 2021
Accepted: 9 March 2021
Published: 12 March 2021

Publisher's Note: MDPI stays neutral with regard to jurisdictional claims in published maps and institutional affiliations.

Copyright: © 2021 by the authors. Licensee MDPI, Basel, Switzerland. This article is an open access article distributed under the terms and conditions of the Creative Commons Attribution (CC BY) license (https://creativecommons.org/licenses/by/4.0/).

Abstract: We aimed to evaluate the brain hypometabolic signature of persistent isolated olfactory dysfunction after SARS-CoV-2 infection. Twenty-two patients underwent whole-body [18F]-FDG PET, including a dedicated brain acquisition at our institution between May and December 2020 following their recovery after SARS-Cov2 infection. Fourteen of these patients presented isolated persistent hyposmia (smell diskettes olfaction test was used). A voxel-wise analysis (using Statistical Parametric Mapping software version 8 (SPM8)) was performed to identify brain regions of relative hypometabolism in patients with hyposmia with respect to controls. Structural connectivity of these regions was assessed (BCB toolkit). Relative hypometabolism was demonstrated in bilateral parahippocampal and fusiform gyri and in left insula in patients with respect to controls. Structural connectivity maps highlighted the involvement of bilateral longitudinal fasciculi. This study provides evidence of cortical hypometabolism in patients with isolated persistent hyposmia after SARS-Cov2 infection. [18F]-FDG PET may play a role in the identification of long-term brain functional sequelae of COVID-19.

Keywords: 18F-FDG PET; anosmia; COVID-19; SARS-CoV-2; olfactory dysfunction

1. Introduction

Coronavirus disease 2019 (COVID-19) due to SARS-CoV-2 infection was initially thought to be mainly restricted to the respiratory system, but it has become evident that this disease also involves multiple other organs, including the central and peripheral nervous system. Indeed, neurological complications such as stroke, encephalopathy, delirium, meningitis, seizures, and cranial nerve deficits have been reported in patients with COVID-19 [1]. Besides these more severe manifestations and complications, other frequent symptoms of COVID-19 are loss of smell (anosmia) and taste (ageusia) which can occur as first symptoms of infection or in the absence of any other clinical features [1]. In a European study including more than 400 COVID-19 patients, olfactory dysfunction and ageusia were reported in 86% and 82% of patients, respectively [2]. Different underlying mechanisms have been advocated to explain the presence of anosmia in patients with COVID-19. These include olfactory cleft syndrome, direct damage of olfactory sensory neurons, postviral anosmia syndrome, cytokine storm, and/or impairment of the olfactory perception centers in the brain [3]. Indeed the olfactory bulb might represent a potential route of entry of SARS-CoV-2 in the CNS, and the investigation of the pathophysiology of olfactory dysfunction might help to further understand the pathogenesis and long-term implications of CNS involvement in COVID-19 [4]. In this framework, the availability of sensitive biomarkers tracking disease substrates might speed up the investigation of CNS involvement in patients with SARS-CoV-2 infection and might be used to monitor and better predict the risk of long-term effects. To date, there are few, partially conflicting, results on magnetic resonance imaging (MRI) abnormalities in patients with COVID-related anosmia [5,6]. Indeed normal, transiently increased, and even reduced volume of the olfactory bulb has been reported in patients with isolated or persistent anosmia [5,6]. Similarly, [^{18}F]-Fluorodeoxyglucose ([^{18}F]-FDG) PET data of COVID-19 patients with anosmia have to-date been made available only through case reports and small case series of patients with a self-reported reduction in smell [7–11]. This very preliminary evidence has been mainly acquired at the time of viral infection or just after recovery in patients affected by moderate to severe disease, thus complicating data interpretation [11].

[^{18}F]-FDG PET may represent a sensitive tool to further confirm SARS-CoV-2 neurotropism through the olfactory pathway. Furthermore, given the potential functional and cognitive sequelae of COVID-19 and the established role of PET to support differential diagnosis of cognitive impairment, [^{18}F]-FDG PET can represent a suitable tool to identify the concomitant involvement of cortical structures potentially relevant for subsequent persistent cognitive, sensory or emotion disturbances [8]. Moreover, an increasing number of patients showing persistent symptoms (such as fatigue, dyspnea, anosmia/dysgeusia, memory impairment, and pain) have been described after recovery from SARS-CoV-2, defining an emerging chronic syndrome, so-called Long Covid [12,13]. Given these premises, we aimed to evaluate the presence of regional brain hypometabolism in patients with persistent isolated and objectively-assessed olfactory dysfunction after recovery from SARS-CoV-2 infection.

2. Material and Methods

2.1. Patients

Patients with anosmia after SARS-CoV-2 infection were recruited among subjects who underwent whole-body [^{18}F]-FDG PET, including a dedicated brain acquisition for clinical reasons other than SARS-CoV-2 infection in our institution between May and December 2020 (following their recovery after infection). The main inclusion criteria were previous SARS-CoV-2 infection, confirmed by polymerase chain reaction (PCR) at the time of initial symptoms, PET examinations performed during the recovery phase of SARS-CoV-2 infection, and an olfactory test still demonstrating olfactory dysfunction. The recovery phase was defined as when at least one negative swab test after infection was available. Exclusion criteria were demonstration of brain lesions on MRI, previous diagnosis of encephalopathy/encephalitis or cerebrovascular disorders due to or concomitant with the SARS-CoV-2 infection, or any other previous or current neurological or psychiatric disease. Patients that previously required mechanical ventilation or showed severe respiratory distress syndrome due to SARS-CoV-2 infection were also excluded, given the potential independent effect of these clinical scenarios on brain metabolism. Patients with a history of anosmia before SARS-CoV-2 infection, as well as those treated with chemotherapy in the last 3 months or previous radiotherapy in the head and neck district for oncological reasons, were also excluded. The study was approved by the Regional Ethical Committee (CER Liguria code 671/2020), all procedures and informed consent collection were in accordance with the ethical standards of the 1964 Helsinki declaration.

2.2. Olfactory Test

Olfaction was assessed by means of the Smell diskettes olfaction test [14] on the same day of PET examination. In fact, while self-reported newly onset loss of smell is important from an infection control perspective, self-reporting may result in misdiagnosis. The test was based on reusable diskettes as applicators of 8 different odorants. Using a questionnaire with illustrations, the test was designed as a triple forced multiple-choice test resulting in a score of 0 to 8 correct answers. Hyposmia was defined as making at least 2 mistakes on the questionnaire; the number of correct answers was recorded.

2.3. [^{18}F]-FDG Brain PET Acquisition and Image Processing

A dedicated [^{18}F]-FDG Brain PET acquisition was performed in all recruited patients according to the European Association of Nuclear Medicine (EANM) guidelines on two Siemens Biograph PET/CT systems (16 and mCT Flow 40, respectively) in the same center [15]. Images preprocessing was conducted using Statistical Parametric Mapping software version 8 (SPM8; Wellcome TrustCenter for Neuroimaging, London, UK) [16]. See Supplementary Materials for further details.

2.4. Voxel-Wise Analysis of Hypometabolic Signature of Olfactory Dysfunction after SARS-CoV-2 Infection

After preprocessing, smoothed images underwent a whole-brain voxel-wise group analysis to identify regions of relative hypometabolism with respect to a control group of 61 subjects consisting of 48 healthy controls acquired on the Biograph 16 system and previously recruited in our laboratory without any neurologic or psychiatric disease as detailed elsewhere [17] and thirteen subjects with smoldering multiple myeloma with both normal body and brain scans acquired on Biograph mCT Flow 40 PET/CT system (age 61.1 ± 11.1; 10 males). Patients with smoldering myeloma had no present or previous history of neurologic or psychiatric diseases and were never submitted to chemotherapy. Age, gender, and scanner type were included as nuisance variables in the analysis. We set a height threshold of family-wise error (FWE)-corrected $p < 0.05$ for multiple comparisons, at both the peak and cluster levels. Details on SPM analysis are included in the Supplementary Materials.

2.5. Structural Connectivity of Regions of Hypometabolism in Patients with Olfactory Dysfunction

The hypometabolic clusters in patients with hyposmia with respect to controls (hyposmia clusters) which had been obtained by means of the whole brain voxel-based analysis in SPM8, were saved as a volumetric region of interest (VOI). To assess the structural connectivity of metabolic correlates of hypo/anosmia after SARS-CoV-2 infection, we used the "Brain Connectivity and Behaviour" (BCB_ toolkit (18, http://www.toolkit.bcblab.com (accessed on 10 December 2020)), which included diffusion MRI data from healthy control subjects. Moreover, using the disconnectome pipeline in the BCB toolkit we computed structural connection maps of all voxels included in the hyposmia clusters by tracking fibers passing through them to identify their structural connectivity with other brain areas [18,19]. Briefly, the hypometabolic clusters present in patients with hyposmia with respect to controls (hyposmia clusters) and obtained by means of the whole brain voxel-based analysis in SPM8 were saved as VOI [16]. First, using the Tractotron pipeline, we evaluated the probability of the major white matter tracts crossing the hyposmia clusters and considered as significant only those voxels with a probability of at least 0.5. Moreover, using the disconnectome pipeline of the BCB toolkit [18,19], we computed the structural connection maps of all voxels included in the hyposmia clusters by tracking fibers passing through them to identify their structural connectivity with other brain areas as previously described [20]. Details about this procedure are detailed elsewhere [21].

3. Results

3.1. Patients

Twenty-two consecutive patients (12 males and 10 females; mean age 64 ± 10.5 years, range 35–79) in the recovery phase of SARS-CoV-2 infection were submitted to whole-body [^{18}F]-FDG PET in our center from 1 May and 1 December 2020. [^{18}F]-FDG PET was performed between 4 and 12 weeks after the first positive RT-PCR nasopharyngeal swab for SARS-CoV-2. Only nineteen of these patients met our inclusion criteria and were submitted to Smell diskettes olfaction test, which indicated the presence of hyposmia in fourteen of them who have been ultimately included in the present analyses. Figure 1 and its notes report the steps that narrowed the final study group. Further details on reasons for patients' exclusion are reported in the Supplementary Materials. Characteristics of the 14 analyzed patients are detailed in Table 1.

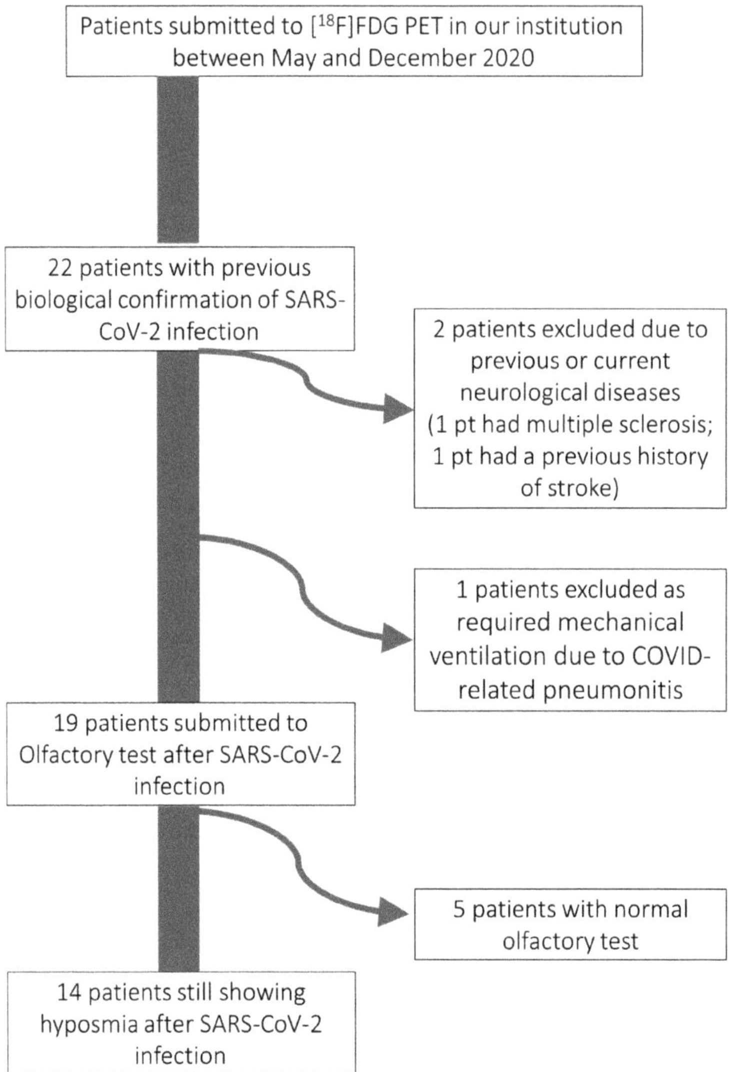

Figure 1. Flow-chart reporting steps that narrowed the final study group to fourteen patients still presenting with hyposmia during early recovery after SARS-CoV-2 infection. Pt, patients.

Table 1. Patients' Characteristics.

Characteristics	SARS-CoV-2 Patients with Hyposmia ($n = 14$) *
Age (years)	64.4 ± 10.9 (range 51–79)
Sex	
Male	7/14
Female	7/14
Time Since Diagnosis of SARS-CoV-2 infection (weeks) †	8.3 ± 2.1 (range 4–14)
Time Since first negative swab after proven SARS-CoV-2 infection (weeks)	4.0 ± 1.9 (range 1–7)
Olfactory test (number of correct answers)	
6/8	2
5/8	2
4/8	5
3/8	2
2/8	3

Values are shown as mean ± standard deviation (range). * None of these patients was complaining of other known possible sequelae of COVID-19 such as fatigue, chest pain, dyspnea, or reported any other focal neurological signs both at the time of SARS-CoV-2 infection and at the time of PET. † None of the patients had proven previous COVID-related lung involvement or previously received steroids, hydroxychloroquine, or other medication specifically aimed to support patients' response to COVID-19 (other than paracetamol).

3.2. Hypometabolism in Patients with Isolated Persistent Hyposmia after SARS-CoV-2 Infection

With respect to the controls, patients with hyposmia after SARS-CoV-2 infection were characterized by relative hypometabolism in parahippocampal (Brodmann area (BA) 36), fusiform (BA 20 and 37) gyri in both hemispheres and in the insula in the left hemisphere (BA 13). Clusters of significant hypometabolism in patients with hyposmia after SARS-CoV-2 are reported in Figure 2. Details on coordinates and z-score are reported in Table 2.

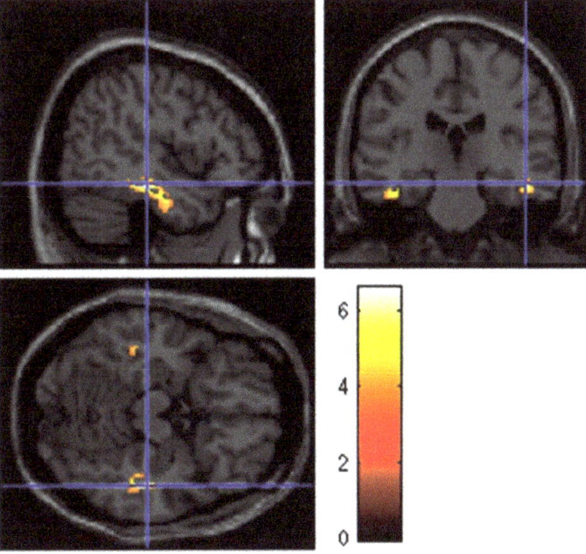

Figure 2. Hypometabolism with respect to controls in patients still presenting with hyposmia during early recovery after SARS-CoV-2 infection was highlighted in parahippocampal and fusiform gyri in both hemispheres (BA 20, 36, 37) and in the insula in the left hemisphere (BA 13). Height threshold of significance was set at $p < 0.05$ FWE-corrected at the cluster level. Regions of significant difference are shown color-graded in terms of Z values. Talairach coordinates and further details are available in Table 2.

Table 2. Whole-brain mapping of relative hypometabolism in patients with persistent olfactory dysfunction after SARS-CoV-2 infection with respect to controls.

	Cluster Level		Peak Level					
Cluster Extent	Corrected p-Value	Cortical Region	Maximum Zscore	Talairach Coordinates			Cortical Region	BA
260	0.032							
		R-limbic	5.68	45	−26	−9	Parahippocampal Gyrus	36
		R-Temporal	3.45	45	−22	−11	Fusiform Gyrus	20
		R-Temporal	3.41	45	−33	−9	Fusiform Gyrus	37
155	0.034	L-Limbic	5.15	−39	−29	−13	Parahippocampal Gyrus	36
		L-sublobar	3.36	−44	−37	17	Insula	13

$p < 0.05$, corrected for multiple comparisons with the Family-Wise error option both at peak and cluster level were accepted as statistically significant. In the 'cluster level' section on the left, the corrected p-value and the brain lobe with hypometabolism are reported. In the 'peak level' section on the right, the Z score and peak coordinates, the corresponding cortical region, and Brodmann area (BA) are reported. L, left; R, right.

3.3. Hyposmia Clusters Tractography and Connectivity

The hyposmia cluster was found to be included in the bilateral longitudinal fasciculi (ILF) with a probability 0.82 and 1 for the left and right ILF, respectively. The tractography results for the hyposmia cluster are shown in Figure 3.

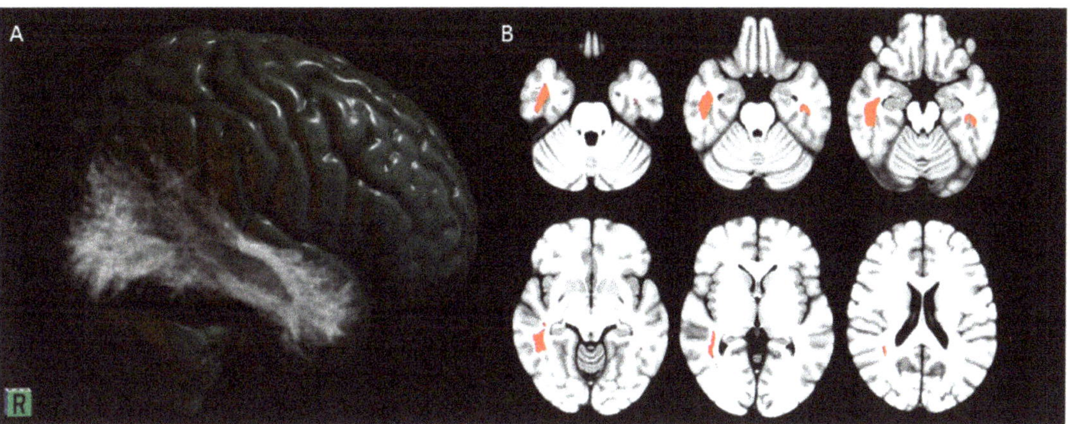

Figure 3. Structural connectivity of regions of hypometabolism in patients with olfactory dysfunction generated through the Brain Connectivity and Behaviour (BCB) toolkit (http://www.toolkit.bcblab.com (accessed on 10 December 2020)), which includes diffusion MRI data from healthy control subjects. Panel (**A**): The connectome map indicated a significant probability of connection of the hyposmia cluster with the inferior longitudinal fasciculus; Panel (**B**): tractography results of the hyposmia cluster.

4. Discussion

The present brief communication provides a demonstration of brain hypometabolism, namely in the bilateral limbic cortex, in a group of patients with isolated persistent hyposmia proven by olfactory test more than four weeks after SARS-CoV-2 infection. The highlighted area of hypometabolism also encompassed the insula in the left hemisphere and included the bilateral ILF.

One of the ongoing hypotheses to explain the anosmia of patients with COVID-19 (in the absence of nasal congestion) is that the virus enters the CNS through the first neurons of the olfactory pathway located in the olfactory mucosa [1]. Post-infectious olfactory dysfunction is thought to be caused by damage to the olfactory epithelium or central olfactory processing pathways [22]. The present evidence of hypometabolism in two symmetric, similar regions within the limbic cortex may support the occurrence of a distal involvement of the olfactory pathway. Moreover, the bilateral involvement of key cortical structures is sound from the pathophysiological point of view. In fact, hyposmia might not be subjectively perceived in case of unilateral involvement of the olfactory pathway [23]. To-date one case report provided [^{18}F]-FDG PET data in a patient with mild COVID-19 and isolated and persistent anosmia (in absence of any other COVID-related symptom) [10]. In fact, Karimi-Galougahi and colleagues reported the case of a 27-year-old woman with persistent anosmia for six weeks presenting hypometabolism of the left orbitofrontal cortex but with preserved metabolism in temporal cortex [10]. However, the evaluation of images was mainly based on visual inspection without observed-independent analysis or comparison with a control database which might have helped to more accurately evaluate a small region such as the medial temporal lobe. Evidence about the involvement of the limbic cortex after recovery from COVID-19 was also provided by two well-documented cases of patients submitted to whole body PET to assess metabolic activity of residual lung lesions just after COVID-related pneumonia [8]. Hypometabolism of the olfactory/rectus gyrus was present in both patients with additional hypometabolims within the amygdala, hippocampus, parahippocampus, cingulate cortex, pre-/post-central gyrus, thalamus/hypothalamus, cerebellum, pons, and the medulla in only one of them (who was not reporting anosmia). However, at the time of infection both these patients required hospitalization in intensive care unit and in one case, mechanical ventilation was needed. These more severe presentations and in particular mechanical ventilation may, at least in theory, have played a role on [^{18}F]-FDG PET hypometabolic regions especially during early recovery [9,24,25]. However, this limitation does not apply to our patients' group. Recently, Guedj and colleagues provided the first brain FDG PET data in Long COVID patients and again demonstrated bilateral hypometabolism in the bilateral rectal/orbital gyrus, amygdala and the hippocampus, brainstem and bilateral cerebellum. Thus, our findings largely confirm the topography of brain hypometabolism in patients in long COVID patients although associated with persistent hyposmia or with other persistent functional complaints [26]. The presence and the topography of hypometabolism after recovery in all the mentioned case reports, in previous small group studies and in our hyposmic group repetitively highlighted an involvement of limbic regions and might point to the risk of developing long-term neurological (possibly cognitive) sequelae, a hypothesis requiring studies in patients with a much longer recovery from infection [8,27,28]. Indeed, olfactory cortical area feeds into multimodal integration relevant for cognition control and the hippocampal regions is known to exchange input for storage of olfactory memory (also relevant for working memory [29]). Almeria et al. evaluated the impact of COVID-19 on neurocognitive performance in thirty-five patients with confirmed COVID-19 infection and found that the presence of anosmia and dysgeusia at the time of infection were among the main risk factors for cognitive impairment related with attention, memory and executive function [28].

Of note, the present group of patients showed hypometabolism also in the insula in the left hemisphere. The insula is densely interconnected with orbitofrontal and anterior cingulate cortices, amygdala, and hippocampus [30] and plays a key role in processing self-awareness. Indeed, olfaction aims to provide critical information about the environment subsequently directed at the cortical level also for multisensory integration.

Regarding the connectivity data, the involvement of the ILF is in line with observations of its role in hyposmia in Parkinson's Disease [31]. Interestingly, the ILF has been shown to be affected early on in viral infections, such as in HIV [32] and hepatitis C virus (HCV) [33].

Finally, it should be noted that MRI cortical signal has been evaluated in patients with COVID-19 and anosmia in few small studies. However, abnormalities have been substantially reported in the very early phase of infection [5,6,34,35]. In this framework, while the presence of hypometabolism at [^{18}F]-FDG PET cannot prove the direct spread of the virus along the olfactory pathway and cortex, the high sensitivity of FDG PET for cortical deafferentation may act as a measurable biomarker of persistent impairment of the transmission along the olfactory pathway [36]. Further investigation might help to understand if [^{18}F]-FDG PET data could be used to predict the prognosis of olfactory function recovery also at the single patient level [34]. The present study has some limitations, mainly related to its naturalistic observational nature and to the small group of patients being submitted to [^{18}F]-FDG PET for other clinical reasons, including the suspect or follow-up of oncological diseases. Brain lesions were radiologically excluded in all patients, and to reduce the potential confounding effect of comorbidities, we also excluded patients submitted to chemotherapy in the last three months or who underwent radiotherapy in the head and neck district. Indeed despite the small number of included patients, the present study provides a group analysis on brain metabolism of patients with persisting olfactory dysfunction after SARS-CoV-2 infection for the first time proven by olfactory test. The demonstration of hyposmia by means of an olfactory test (that was not possible in larger epidemiologic studies) and the exclusion of patients who suffered from COVID-related pneumonia, requested COVID-oriented treatment or mechanic ventilation is a further strength of the present study [37].

5. Conclusions

The COVID-19 outbreak has impacted clinical neurology in previous months, and other challenges might come in the next future. For several reasons, a not negligible number of neurological and cognitive complaints might emerge once the acute phase of the pandemic crisis is overcome. It will be of great scientific and clinical relevance to describe COVID-19 related cognitive symptoms (likely to be reversible or in any case not progressive) and to identify and characterize biomarkers that will help us to support clinical differential diagnosis with respect to cognitive impairment caused by neurodegenerative disease. $[^{18}F]$-FDG PET might play a role in this clinical setting. To this aim, we will need to be aware of the confounding effect of subtle sequelae of SARS-COV2 infection and on their reflection on PET and other biomarkers as demonstrated in the present study.

Supplementary Materials: Descriptive Details on patients' selection and further details on image preprocessing and analysis are available online at https://www.mdpi.com/2227-9059/9/3/287/s1.

Author Contributions: Conceptualization, S.M. and M.I.D.; methodology, S.M., M.I.D., A.M., and M.P. (Matteo Pardini); software, S.M., M.P. (Matteo Pardini); validation and formal analysis, S.M., M.I.D., and M.P. (Matteo Pardini); investigation, all authors.; data curation, all authors; writing—original draft preparation, S.M., M.I.D., and M.P. (Matteo Pardini); writing—review and editing, all authors; visualization, all authors; funding acquisition, S.M. and M.B.; All authors have read and agreed to the published version of the manuscript.

Funding: This work was supported by a grant from the Italian Ministry of Health—Rete Italiana di Neuroscienze.

Institutional Review Board Statement: The study was conducted according to the guidelines of the Declaration of Helsinki and approved by the Institutional Review Board.

Informed Consent Statement: Informed consent was obtained from all subjects involved in the study. Written informed consent for publication was obtained from all participants.

Data Availability Statement: The datasets used and/or analyzed during the current study are available from the corresponding author on reasonable request.

Acknowledgments: Authors are grateful to radiologic technologists, nurses, and administrative personal of the Nuclear Medicine Unit of the IRCCS Policlinico San Martino for their contribution to the acquisition of brain PET data of the patients included in this study. We also would like to thank them for their constant presence and support in the assistance of patients who underwent $[^{18}F]$-FDG PET for oncologic and neurologic clinical reasons at our institution during the pandemic emergency. This work was supported by a grant from the Italian Ministry of Health—Rete Italiana di Neuroscienze.

Conflicts of Interest: Silvia Morbelli and Flavio Nobili have received speaker honoraria from G.E. Healthcare. All other authors declare no conflict of interest.

References

1. Ellul, M.A.; Benjamin, L.; Singh, B.; Lant, S.; Michael, B.D.; Easton, A.; Kneen, R.; Defres, S.; Sejvar, J.; Solomon, T. Neurological associations of COVID-19. *Lancet Neurol.* **2020**, *19*, 767–783. [CrossRef]
2. Lechien, J.R.; Chiesa-Estomba, C.M.; De Siati, D.R.; Horoi, M.; Le Bon, S.D.; Rodriguez, A.; Dequanter, D.; Blecic, S.; El Afia, F.; Distinguin, L.; et al. Olfactory and gustatory dysfunctions as a clinical presentation of mild-to-moderate forms of the coronavirus disease (COVID-19): A multicenter European study. *Eur. Arch. Otorhinolaryngol.* **2020**, *277*, 2251–2261. [CrossRef]
3. Saussez, S.; Lechien, J.R.; Hopkins, C. Anosmia: An evolution of our understanding of its importance in COVID-19 and what questions remain to be answered. *Eur. Arch. Otorhinolaryngol.* **2020**, 1–5. [CrossRef]
4. Fotuhi, M.; Mian, A.; Meysami, S.; Raji, C.A. Neurobiology of COVID-19. *J. Alzheimers. Dis.* **2020**, *76*, 3–19. [CrossRef] [PubMed]
5. Galougahi, M.K.; Ghorbani, J.; Bakhshayeshkaram, M.; Naeini, A.S.; Haseli, S. Olfactory Bulb Magnetic Resonance Imaging in SARS-CoV-2-Induced Anosmia: The First Report. *Acad Radiol.* **2020**, *27*, 892–893. [CrossRef] [PubMed]
6. Politi, L.S.; Salsano, E.; Grimaldi, M. Magnetic Resonance Imaging Alteration of the Brain in a Patient With Coronavirus Disease 2019 (COVID-19) and Anosmia. *JAMA Neurol.* **2020**, *77*, 1028–1029. [CrossRef] [PubMed]

7. Annunziata, S.; Bauckneht, M.; Albano, D.; Argiroffi, G.; Calabrò, D.; Abenavoli, E.; Linguanti, F.; Laudicella, R.; Young Committee of the Italian Association of Nuclear Medicine (AIMN). Impact of the COVID-19 pandemic in nuclear medicine departments: Preliminary report of the first international survey. *Eur. J. Nucl. Med. Mol. Imaging* **2020**, *47*, 2090–2099. [CrossRef] [PubMed]
8. Guedj, E.; Million, M.; Dudouet, P.; Tissot-Dupont, H.; Bregeon, F.; Cammilleri, S.; Raoult, D. 18F-FDG brain PET hypometabolism in post-SARS-CoV-2 infection: Substrate for persistent/delayed disorders? *Eur. J. Nucl. Med. Mol. Imaging* **2020**, 1–4. [CrossRef]
9. Delorme, C.; Paccoud, O.; Kas, A.; Hesters, A.; Bombois, S.; Shambrook, P.; Boullet, A.; Doukhi, D.; Le Guennec, L.; Godefroy, N.; et al. COVID-19-related encephalopathy: A case series with brain FDG-positron-emission tomography/computed tomography findings. *Eur. J. Neurol.* **2020**, *27*, 2651–2657. [CrossRef]
10. Karimi-Galougahi, M.; Yousefi-Koma, A.; Bakhshayeshkaram, M.; Raad, N.; Haseli, S. 18FDG PET/CT Scan Reveals Hypoactive Orbitofrontal Cortex in Anosmia of COVID-19. *Acad Radiol.* **2020**, *27*, 1042–1043. [CrossRef] [PubMed]
11. Morbelli, S.; Ekmekcioglu, O.; Barthel, H.; Albert, N.L.; Boellaard, R.; Cecchin, D.; Guedj, E.; Lammertsma, A.A.; Law, I.; Penuelas, I.; et al. EANM Neuroimaging Committee. COVID-19 and the brain: Impact on nuclear medicine in neurology. *Eur. J. Nucl. Med. Mol. Imaging* **2020**, *47*, 2487–2492. [CrossRef] [PubMed]
12. Meeting the challenge of long COVID. *Nat. Med.* **2020**, *26*, 1803. [CrossRef] [PubMed]
13. The Lancet. Facing up to long COVID. *Lancet* **2020**, *396*, 1861. [CrossRef]
14. Briner, H.R.; Simmen, D. Smell diskettes as screening test of olfaction. *Rhinology* **1999**, *37*, 145–148. [PubMed]
15. Varrone, A.; Asenbaum, S.; Vander Borght, T.; Booij, J.; Nobili, F.; Någren, K.; Darcourt, J.; Kapucu, O.L.; Tatsch, K.; Bartenstein, P.; et al. EANM procedure guidelines for PET brain imaging using [18F]FDG, version 2. *Eur. J. Nucl. Med. Mol. Imaging* **2009**, *36*, 2103–2110. [CrossRef]
16. Friston, K.J.; Holmes, A.P.; Worsley, K.J.; Poline, J.P.; Frith, C.D.; Frackowiak, R.S. Statistical parametric maps in functional imaging: A general linear approach. *Hum. Brain Mapp.* **1994**, *2*, 189–210. [CrossRef]
17. Morbelli, S.; Bauckneht, M.; Arnaldi, D.; Picco, A.; Pardini, M.; Brugnolo, A.; Buschiazzo, A.; Pagani, M.; Girtler, N.; Nieri, A.; et al. 18F-FDG PET diagnostic and prognostic patterns do not overlap in Alzheimer's disease (AD) patients at the mild cognitive impairment (MCI) stage. *Eur. J. Nucl. Med. Mol. Imaging* **2017**, *44*, 2073–2083. [CrossRef]
18. Foulon, C.; Cerliani, L.; Kinkingnehun, S.; Levy, R.; Rosso, C.; Urbanski, M.; Volle, E.; Thiebaut de Schotten, M. Advanced lesion symptom mapping analyses and implementation as BCBtoolkit. *Gigascience* **2018**, *7*, 1–17. [CrossRef]
19. Rojkova, K.; Volle, E.; Urbanski, M.; Humbert, F.; Dell'Acqua, F.; Thiebaut de Schotten, M. Atlasing the frontal lobe connections and their variability due to age and education: A spherical deconvolution tractography study. *Brain Struct. Funct.* **2016**, *221*, 1751–1766. [CrossRef]
20. Thiebaut de Schotten, M.; Dell'Acqua, F.; Ratiu, P.; Leslie, A.; Howells, H.; Cabanis, E.; Iba-Zizen, M.T.; Plaisant, O.; Simmons, A.; Dronkers, N.F.; et al. From Phineas Gage and Monsieur Leborgne to H.M.: Revisiting Disconnection Syndromes. *Cereb. Cortex.* **2015**, *25*, 4812–4827. [CrossRef]
21. Massa, F.; Grisanti, S.; Brugnolo, A.; Doglione, E.; Orso, B.; Morbelli, S.; Bauckneht, M.; Origone, P.; Filippi, L.; Arnaldi, D.; et al. The role of anterior prefrontal cortex in prospective memory: An exploratory FDG-PET study in early Alzheimer's disease. *Neurobiol. Aging* **2020**, *96*, 117–127. [CrossRef]
22. Meng, X.; Deng, Y.; Dai, Z.; Meng, Z. COVID-19 and anosmia: A review based on up-to-date knowledge. *Am. J. Otolaryngol.* **2020**, *41*, 102581. [CrossRef]
23. Diodato, A.; De Brimont, M.R.; Yim, Y.S.; Derian, N.; Perrin, S.; Pouch, J.; Klatzmann, D.; Garel, S.; Choi, G.B.; Fleischmann, A. Molecular signatures of neural connectivity in the olfactory cortex. *Nat. Commun.* **2016**, *7*, 12238. [CrossRef]
24. Antczak, J.; Popp, R.; Hajak, G.; Zulley, J.; Marienhagen, J.; Geisler, P. Positron emission tomography findings in obstructive sleep apnea patients with residual sleepiness treated with continuous positive airway pressure. *J. Physiol. Pharmacol.* **2007**, *5* (Suppl. 5), 25–35.
25. Kas, A.; Soret, M.; Pyatigoskaya, N.; Habert, M.O.; Hesters, A.; Le Guennec, L.; Paccoud, O.; Bombois, S.; Delorme, C.; on the behalf of CoCo-Neurosciences study group and COVID SMIT PSL study group. The cerebral network of COVID-19-related encephalopathy: A longitudinal voxel-based 18F-FDG-PET study. *Eur. J. Nucl. Med. Mol. Imaging* **2021**, *15*, 1–15.
26. Guedj, E.; Campion, J.Y.; Dudouet, P.; Kaphan, E.; Bregeon, F.; Tissot-Dupont, H.; Guis, S.; Barthelemy, F.; Habert, P.; Ceccaldi, M.; et al. 18 F-FDG brain PET hypometabolism in patients with long COVID. *Eur. J. Nucl. Med. Mol. Imaging* **2021**, *26*, 1–11.
27. Sollini, M.; Morbelli, S.; Ciccarelli, M.; Cecconi, M.; Aghemo, A.; Morelli, P.; Chiola, S.; Gelardi, F.; Chiti, A. Long Covid hallmarks on [18F]FDG-PET/CT: A case-control study. *Eur. J. Nucl. Med. Mol. Imaging* **2021**, *7*, 1–11. [CrossRef]
28. Almeria, M.; Cejudo, J.C.; Sotoca, J.; Deus, J.; Krupinski, J. Cognitive profile following COVID-19 infection: Clinical predictors leading to neuropsychological impairment. *Brain Behav. Immun. Health* **2020**, 100163. [CrossRef] [PubMed]
29. Aqrabawi, A.J.; Kim, J.C. Olfactory memory representations are stored in the anterior olfactory nucleus. *Nat. Commun.* **2020**, *11*, 1246. [CrossRef] [PubMed]
30. Christopher, L.; Koshimori, Y.; Lang, A.E.; Criaud, M.; Strafella, A.P. Uncovering the role of the insula in non-motor symptoms of Parkinson's disease. *Brain* **2014**, *137*, 2143–2154. [CrossRef]
31. Haghshomar, M.; Dolatshahi, M.; Ghazi Sherbaf, F.; Sanjari Moghaddam, H.; Shirin Shandiz, M.; Aarabi, M.H. Disruption of Inferior Longitudinal Fasciculus Microstructure in Parkinson's Disease: A Systematic Review of Diffusion Tensor Imaging Studies. *Front. Neurol.* **2018**, *26*, 598. [CrossRef]

32. Gongvatana, A.; Schweinsburg, B.C.; Taylor, M.J.; Theilmann, R.J.; Letendre, S.L.; Alhassoon, O.M.; Jacobus, J.; Woods, S.P.; Jernigan, T.L.; Ellis, R.J.; et al. White matter tract injury and cognitive impairment in human immunodeficiency virus-infected individuals. *J. Neurovirol.* **2009**, *15*, 187–195. [CrossRef] [PubMed]

33. Bladowska, J.; Zimny, A.; Knysz, B.; Małyszczak, K.; Kołtowska, A.; Szewczyk, P.; Gąsiorowski, J.; Furdal, M.; Sąsiadek, M.J. Evaluation of early cerebral metabolic, perfusion and microstructural changes in HCV-positive patients: A pilot study. *J. Hepatol.* **2013**, *59*, 651–657. [CrossRef] [PubMed]

34. Kandemirli, S.G.; Altundag, A.; Yildirim, D.; Tekcan Sanli, D.E.; Saatci, O. Olfactory Bulb MRI and Paranasal Sinus CT Findings in Persistent COVID-19 Anosmia. *Acad Radiol.* **2021**, *28*, 28–35. [CrossRef] [PubMed]

35. Laurendon, T.; Radulesco, T.; Mugnier, J.; Gérault, M.; Chagnaud, C.; El Ahmadi, A.A.; Varoquaux, A. Bilateral transient olfactory bulb edema during COVID-19-related anosmia. *Neurology* **2020**, *95*, 224–225. [CrossRef] [PubMed]

36. Magistretti, P.J.; Pellerin, L. Astrocytes Couple Synaptic Activity to Glucose Utilization in the Brain. *News Physiol. Sci.* **1999**, *14*, 177–182. [CrossRef]

37. Passali, G.C.; Bentivoglio, A.R. Comment to the article "Olfactory and gustatory dysfunctions as a clinical presentation of mild-to-moderate forms of the coronavirus disease (COVID-19): A multicenter European study". *Eur. Arch. Otorhinolaryngol.* **2020**, *277*, 2391–2392. [CrossRef] [PubMed]

Review

Revisiting Traumatic Brain Injury: From Molecular Mechanisms to Therapeutic Interventions

Abbas Jarrahi [1,†], Molly Braun [1,2,3,†], Meenakshi Ahluwalia [4], Rohan V. Gupta [1], Michael Wilson [1,5], Stephanie Munie [1,6], Pankaj Ahluwalia [4], John R. Vender [1], Fernando L. Vale [1], Krishnan M. Dhandapani [1] and Kumar Vaibhav [1,*]

[1] Department of Neurosurgery, Medical College of Georgia, Augusta University, Augusta, GA 30912, USA; ajarrahi@augusta.edu (A.J.); mobraun@uw.edu (M.B.); Rgupta@augusta.edu (R.V.G.); wilsonma@evms.edu (M.W.); munie@musc.edu (S.M.); jvender@augusta.edu (J.R.V.); fvalediaz@augusta.edu (F.L.V.); kdhandapani@augusta.edu (K.M.D.)

[2] Department of Psychiatry and Behavioral Sciences, University of Washington School of Medicine, Seattle, WA 98195, USA

[3] VISN 20 Northwest Mental Illness Research, Education and Clinical Center (NW MIRECC), VA Puget Sound Health Care System, Seattle, WA 98108, USA

[4] Department of Pathology, Medical College of Georgia, Augusta University, Augusta, GA 30912, USA; mahluwalia@augusta.edu (M.A.); pahluwalia@augusta.edu (P.A.)

[5] School of Medicine, Eastern Virginia Medical School, Norfolk, VA 23501, USA

[6] College of Medicine, Medical University of South Carolina, Charleston, SC 29425, USA

[*] Correspondence: kvaibhav@augusta.edu; Tel.: +1-(706)-721-4691

[†] These authors contributed equally to this study.

Received: 8 September 2020; Accepted: 26 September 2020; Published: 29 September 2020

Abstract: Studying the complex molecular mechanisms involved in traumatic brain injury (TBI) is crucial for developing new therapies for TBI. Current treatments for TBI are primarily focused on patient stabilization and symptom mitigation. However, the field lacks defined therapies to prevent cell death, oxidative stress, and inflammatory cascades which lead to chronic pathology. Little can be done to treat the mechanical damage that occurs during the primary insult of a TBI; however, secondary injury mechanisms, such as inflammation, blood-brain barrier (BBB) breakdown, edema formation, excitotoxicity, oxidative stress, and cell death, can be targeted by therapeutic interventions. Elucidating the many mechanisms underlying secondary injury and studying targets of neuroprotective therapeutic agents is critical for developing new treatments. Therefore, we present a review on the molecular events following TBI from inflammation to programmed cell death and discuss current research and the latest therapeutic strategies to help understand TBI-mediated secondary injury.

Keywords: neurotrauma; neuroinflammation; excitotoxicity; oxidative stress; apoptosis; edema; brain injury; therapeutic strategies

---

1. Introduction

Traumatic brain injury (TBI), a leading cause of death and disability, is an international public health concern. An estimated 53–69 million individuals worldwide sustain a TBI annually [1], and up to 2 percent of the population lives with neurological disabilities caused by a TBI [2,3]. TBI occurs when an external mechanical force causes a disruption in normal brain functioning. While commonly discussed as a single clinical entity, TBI embodies a complex and heterogeneous pathology (Figures 1 and 2). As such, comprehensive knowledge of the cellular and molecular events post-TBI remains a long-standing goal of preclinical research, with the hope that this knowledge will spur the expansion of novel therapeutics.

TBI is categorized according to pathophysiology, etiology, and severity, as assessed by neuroimaging and physiological responses. The Glasgow Coma Scale (GCS) is most commonly utilized to define the severity of brain injury in clinical settings, where patients are assessed following initial resuscitation and within 48 h post-injury [4]. A GCS score of 13–15 is classified as mild injury, a score of 9–12 is classified as moderate injury, and a score of <9 is classified as severe injury. Another assessment tool similar to the GCS is the Full Outline of Unresponsiveness (FOUR) score, which can be used in intubated patients and includes an assessment of brainstem function [5].

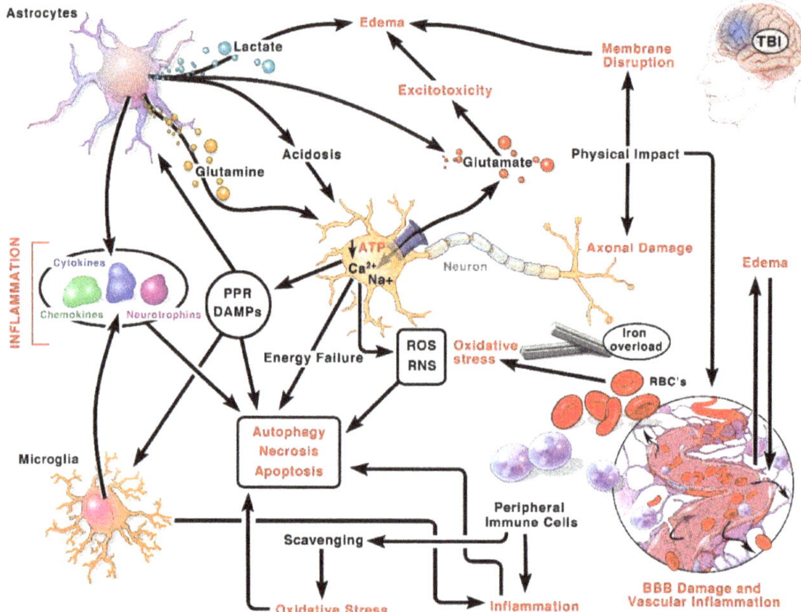

Figure 1. Pathophysiology of TBI. A schematic flow chart of the pathological changes after TBI that lead to acute and chronic neurovascular damage and immune activation. Immediately after the insult neurovascular damage occurs, and large amounts of DAMPs are released causing gliosis and peripheral immune cell infiltration. The initial function of these immune cells is to contain the injury and remove debris and dead cells. However, unregulated immune cells cause enhanced inflammation and injury progression. Furthermore, energy failure, oxidative stress, prolonged inflammation, and excitotoxicity lead to progressive injury with white matter damage and chronic behavioral deficits. Abbreviations: DAMP: Damage associated molecular patterns; PRR: Pattern recognition receptors; ROS: Reactive oxygen species; RNS: Reactive nitrogen species; RBC: Red blood cells; Na^+: Sodium ion; Ca^{2+}: Calcium ion; ATP: Adenosine triphosphate; TBI: Traumatic brain injury.

The pathogenesis of TBI may be divided into two injury-mechanisms: primary and secondary injury. Primary injury entails the direct brain damage that occurs immediately after the impact. The initial injury mechanisms could give rise to extraparenchymal hemorrhages (epidural hematoma, subdural hematoma, subarachnoid hemorrhage, and intraventricular hemorrhage); focal contusions and intraparenchymal hemorrhages; traumatic axonal (focal or diffuse) injury (TAI) due to shearing of WM tracts; and cerebral edema (Figures 1 and 2). Secondary injury mechanisms are also initiated at the moment of the traumatic incident but are believed to continue for many years through a series of cellular, physiological and molecular processes impacting all kinds of cells in the brain. Blood-brain barrier (BBB)-disturbance, excitotoxicity, mitochondrial dysfunction, oxidative stress, inflammation, and cell loss are the principal identified mechanisms orchestrating secondary injury mechanisms [6]. Thus, a pathophysiological and anatomical based classification system for TBI that links the precise pathological mechanisms with the appropriate therapeutic interventions would enhance the translation of therapies from bench to bedside [7]. Therefore, this review provides a synopsis of the main mechanisms of secondary brain damage, along with targeted current and potential neuroprotective therapeutic interventions in preclinical and clinical settings. In the following sections, we will present the historical context for targeting a number of secondary injury pathways. We will discuss the rationale, preclinical evidence, and where appropriate, the translational data in humans. This will provide a segue to why we discuss all these topics, show where we have failed, and perhaps gives a clue why some targets were unsuccessful. This can guide both experimental studies and clinical trial design as we seek efficacy treatments.

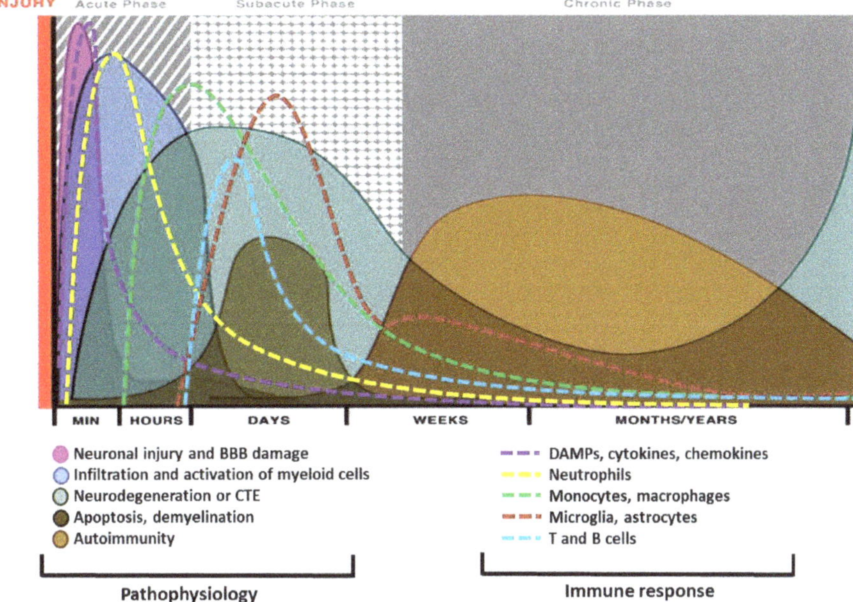

Figure 2. Different phases of traumatic brain injury (TBI) pathophysiology and relative immune response. Mechanical insult leads to acute neuronal injury and blood-brain barrier (BBB) damage, which initiates gliosis and glial injury minutes after TBI and continues for days after injury. Necrotic and apoptotic cell death start immediately after the insult and peak within h to days. Axonal shearing is another event that leads to demyelination and white matter injury. Neurodegeneration, traumatic encephalopathy, and axonal injury may sustain for years after a single TBI. Acute insult and neurovascular damage lead to myeloid accumulation and recruitment of T-cells that last for years and may cause chronic neurodegeneration and neuropathology. Immune cells respond to trauma in a timely manner and a differential pattern of activations has been observed by various studies. An impact to the head leads to cellular damage and results in the rapid release of damage-associated molecular patterns (DAMPs). DAMPs stimulate local cells to release inflammatory mediators such as cytokines and chemokines. These mediators recruit myeloid cells specifically neutrophils as first responders, which phagocytize debris and damaged cells promoting the containment of the injury site. As neutrophil numbers begin to decline, infiltrated monocytes and glia get activated and accumulate around the site of injury to perform further phagocytic or repair functions. Depending on the severity of the brain injury, myeloid cells can recruit T and B cells. T and B cells appear at the sites of brain pathology at later time points in the response (3–7 days post-injury) and may persist for weeks to months. Other abbreviation is as CTE: Chronic traumatic encephalopathy.

2. Excitotoxicity

Excitotoxicity is a pathological process where accumulation of excitatory neurotransmitters, usually glutamate and over-activation of their receptors (NMDAR), causes BBB damage, loss of neuronal membrane integrity, edema, and cell loss after TBI [8]. Studies involving the administration of membrane-resealing polymers following controlled cortical impact (CCI) reported reduced BBB permeability and cerebral damage, and improved functional recovery [9,10] but failed to rescue degenerating cells [11]. This may suggest that resealing of these membranes prevents further alterations of the membrane but did not rescue degenerating cells that were already damaged by the initial episode of TBI-induced excitotoxicity. Similarly, persistent elevated glutamate in cerebral tissue and CSF link with TBI severity in patients [12,13]. Although NMDAR antagonism mitigated TBI-induced neurological damage in rodents [14], global NMDAR antagonists showed side effects and were associated with poor therapeutic windows [15]. Thus, revelation of the mechanisms linking glutamate excitotoxicity, NMDAR activation, and consequent neurological damage, may offer a roadmap to improve neurological outcomes without any adverse effect.

2.1. Glutamate

Glutamate, the principal excitatory neurotransmitter, is essential for normal brain function; however, metabolic perturbations occurring immediately after neurotrauma result in loss of ATP production and subsequent failure of neuronal Na^+-K^+ ATPases. These changes disrupt the homeostatic balance of the electrochemical gradient, causing intracellular sodium accumulation and neuronal depolarization to exacerbate release of synaptic glutamate. In addition, mechanical stretching of neuronal membranes induces micropore formation to aggravate intracellular sodium influx. This ionic shift exacerbates the opening of voltage-gated calcium channels and neuronal depolarization to further the excessive synaptic release of glutamate. Microdialysis studies have shown that increased levels of extracellular glutamate post-TBI correlated with the severity of injury, while elevated glutamate levels in CSF and brain tissues correlated with worse outcomes after clinical TBI [12,16–19].

2.2. Glutamate Receptors

Glutamate receptors are present on membranes of neurons and glial cells both at synaptic and extra-synaptic regions. There are two types of glutamate receptors: (GluR)-ionotropic (iGluR) and metabotropic (mGluR). When glutamate binds to iGluRs (ligand-gated nonselective cation channels), it activates the ion channels, and when it binds to mGluRs (G protein-coupled receptors), it either upregulates or downregulates signal transduction pathways. iGluRs are in turn categorized into four subtypes, including N-methyl-D-aspartate receptors (NMDAR), kainate receptors (KAR), α-amino-3-hydroxy-5-methyl-4-isoxazolepropionic acid receptors (AMPAR), and delta receptors. Glutamate increases intracellular calcium primarily through activation of postsynaptic ionotropic receptors, such as NMDARs [20,21]; however, KARs and calcium-permeable AMPARs also may add to elevated intracellular calcium. Excessive inflow of calcium activates phospholipases, endonucleases, and proteases (calpains), which then precipitate neuronal cell loss via a process deemed excitotoxicity. In addition, GluRs, which are particularly sensitive to mechanical injury, may also contribute to delayed depolarization and injury [22].

2.2.1. Synaptic and Extrasynaptic NMDARs

Based on their location, NMDARs can exert opposing effects. Although stimulation of synaptic NMDAR upregulates brain-derived neurotrophic factor (BDNF) and cAMP response element-binding protein (CREB) activity to promote neuronal survival, extrasynaptic NMDAR activation leads to excitotoxicity and neuronal cell death. This occurs via promotion of a CREB shut-off pathway and inhibition of BDNF gene expression. Specific targeting of NMDARs based on their location could pave the way towards more effective neuroprotective therapies [23–25].

2.2.2. NMDAR Subunits

A more comprehensive analysis of NMDARs shows that they can be made up of seven variable subunits, including a GluNR1 subunit, four GluNR2 subunits (GluN2A, GluN2B, GluN2C, or GluN2D), and two GluNR3 subunits (GluNR3A and GluNR3B). NMDARs are heterotetramers with a strictly regulated subunit composition [26]. Further, extra-synaptic and synaptic NMDARs have different subunit compositions which could be a reason for their opposing cellular effects. The majority of extrasynaptic NMDARs are composed of NR1/NR2B subunits, while synaptic receptors also contain NR2A subunits. Moreover, NMDARs at immature sites markedly contain NR1/NR2B subunits, while NMDAR composition is more diverse at mature sites and switches from NR2B to NR2A on synaptic maturation, termed as NR2B to NR2A developmental switch [27]. Recently, it has been reported that higher expression of the NMDAR subunit NR2A protected neuronal connectivity in the injured brain, while activation of NR2B-containing NMDARs contributed towards loss of connectivity, suggesting a potential role for NMDARs in the restructuring of the neuronal network post-TBI [28].

2.2.3. Therapeutic Strategies Targeting NMDARs

Many therapeutic interventions have been designed based on targeting NMDARs, including NMDAR antagonists, NMDAR subunit inhibitors, and partial agonists of glycine/NMDAR.

NMDAR Antagonists

Administration of NMDAR antagonists (e.g., MK-801) improved outcomes after experimental TBI [29–32]; yet clinical trials exploring the efficacy of NMDAR antagonists were halted due to poor efficacy, major side effects, poor drug efficacy, restricted therapeutic windows, and interference with normal synaptic transmission. These disappointing results, which suggest limited utility of broad-spectrum NMDAR antagonists after acute brain injury, illustrate the translational challenges involved in limiting the detrimental effects of glutamate and suggest the need for alternative approaches [15,33–35].

MK-801: Rats treated with MK-801 prior to injury demonstrated enhanced cognition and axo-dendritic integrity [30], and co-application of MK-801 with other NMDAR antagonists had additive neuroprotective effects [31]. Treatment with MK-801 prior to injury, significantly attenuated neurological deficits; however, MK-801 had minimal influence on neurologic scores when was administered post-injury [32].

Memantine: Intraperitoneal treatment of memantine (10 and 20 mg/kg), a non-competitive NMDAR antagonist, immediately after TBI inhibits neuronal death in rats [36]. Memantine (1–10 μM), when applied to rat hippocampal neurons in vitro, inhibited extrasynaptic NMDAR-induced currents while largely sparing synaptic activity [37]. In a randomized controlled trial of moderate TBI patients, enteral administration of memantine (30 mg twice daily for 7 days) post-injury resulted in substantial improvement in GCS scores at 3 days and significant reduction in neuronal damage at 7 days, as evident from reduced serum levels of neuron-specific enolase (NSE) [38].

Ketamine: In a moderate TBI rat model, administration of ketamine, a non-competitive NMDAR antagonist, at a sub-anesthetic dose (10 mg/kg daily) for 7 days resulted in protection of neuronal dendrites and spines, attenuation of post-traumatic neuroinflammation, and thus, improved neurobehavioral outcomes [39]. Similarly, a significant association between ketamine treatment and reduced spreading depolarization events have been reported in TBI, SAH, and hemispheric stroke patients. Spreading depolarizations are linked with neuronal damage and poor outcome [40].

Magnesium: Magnesium can bind to NMDARs, blocking the passage of ions through the channel, and can therefore be utilized to augment neuroprotection. A study in rats reported that magnesium deficiency worsened TBI outcomes while magnesium administration improved neurological outcomes and mortality after TBI [41]. A Cochrane systematic review including three randomized controlled trials (RCTs) published in 2008, found no beneficial role for magnesium treatment in acute TBI patients in terms of improving neurological outcomes or mortality and therefore did not support its use [42]. Another methodical review of the use of magnesium sulfate in acute TBI management, including eight RCTs published in 2015, also found no significant improvement in mortality but did report a nonsignificant improvement in GOS and GCS scores with magnesium therapy [43].

NMDAR Subunit Inhibitors

Ifenprodil: Ifenprodil, a selective inhibitor of NR2B subunit, can differentially suppress extrasynaptic NMDARs, and therefore could attenuate cell death cascades [44]. Similarly, Ifenprodil treatment mitigated brain edema and reduced injury volume in a CCI model of rats [45]. In an in vitro model of TBI, Ifenprodil reduced NMDA-activated currents but failed to limit fluid shear stress-induced Ca^{2+} influx in primary rat astrocytes [46].

Ro25-6981: TBI modifies NMDAR expression and functioning. For example, moderate TBI in rats caused rapid recruitment of NR2B to membrane rafts successively inducing autophagy. Ro25-6981, a selective inhibitor of NR2B, markedly mitigated autophagy [47]. Activation of NR2B-containing NMDARs may be further involved in the insertion of calcium-permeable AMPARs (CP-AMPARs) in an in vitro model of TBI, further worsening the intracellular calcium overload [48]. Stretch injury of cultured cortical neurons resulted in enhanced extrasynaptic current transmission facilitated by NR2B-units of NMDARs, with no marked variations in synaptic transmission through NMDAR [49]. In addition, either Ro25-6981 or memantine treatment barred this injury-induced increase in CP-AMPAR activity [49]. Further, administration of Ro25-6981 (6 mg/kg, i.p.), attenuated edema post-TBI in mice and limited the NMDA-induced excitotoxicity and release of HMGB1 from injured cortical neurons in vitro [14].

Traxoprodil (CP-101,606): Intravenous (IV) infusion of traxoprodil in patients, another NR2B antagonist, for up to 72 h post-injury was reported to be safe and well-tolerated in all cases of TBI [50,51]. Although in an RCT where severe TBI patients were given a 72-h infusion of traxoprodil within 8 h post-injury, improvements in the dichotomized Glasgow Outcome Scale (dGOS) at 6 months and mortality rate were noticed; however, this improvement was not significant [52].

Temporal alteration of NMDARs and the partial agonist D-cycloserine (DCS): TBI in rodent models, leads to dynamic changes in NMDARs with early hyperactivation followed by weeks of functional loss. Subacute administration (24 to 72 h post-TBI) of DCS, a partial agonist of glycine/NMDARs, upregulated BDNF expression, restored impaired hippocampal long-term potentiation, and enhanced recovery of neurobehavioral and cognitive functions in mice [53]. The complex role of NMDARs in TBI pathophysiology adds to the translational challenges faced by therapeutic interventions targeting this mechanism.

2.3. Postsynaptic Density Protein 95 (PSD-95) and PSD-95 Inhibitors

PSD-95 is a membrane-associated guanylate kinase (MAGUK) that interacts with NMDARs, AMPARs, and potassium channels and plays a role in synaptic plasticity [54]. PSD-95 combines the NR2B subunit of NMDARs with neuronal nitric oxide synthase (nNOS), to form a complex NMDAR/PSD-95/nNOS, and adds to neurotoxicity [55]. Interestingly, inhibition of nNOS by disturbing NMDAR-PSD-95 interactions prevented nitration of protein and cell death in vitro [56]. ZL006, an inhibitor of nNOS-PSD95 interaction, prevented neuronal apoptosis and improved sensorimotor and cognitive outcomes in rodents [57]. Similarly, disruption of NMDAR-PSD-95 interaction by a synthetic peptide (now known as NA-1) blocked excitotoxicity in cultured neurons, limited ischemic cerebral damage, and improved neurological function in rats without altering calcium influx or synaptic activity [58]. The strategy appears to be very promising as inhibition of this interaction between PSD-95 and NMDAR-mediated neurotoxic signaling pathways has demonstrated reduced infarct size and improved outcomes after stroke in macaques [59]. Further, NMDAR-PSD-95-nNOS complex inhibitor, known as NA-1, has been shown to treat ruptured cerebral aneurism, reduce ischemic brain damage, and improve neurological scores, and has become the first stroke therapy to demonstrate efficacy in humans (NCT00728182) after initial results in primates [60–62]. In addition, two phase III clinical trials (NCT02315443; NCT02930018) have been completed in acute stroke and awaiting publication. Interestingly, TBI activated endoplasmic reticulum-associated PKR-like ER kinase (PERK) which phosphorylates CAMP response element-binding protein (CREB) and PSD-95, resulting in reduced brain-derived neurotropic factor (BDNF) and PSD-95 in the injured cortex. However, either PERK inhibitor GSK2656157 or overexpression of kinase-dead mutant of PERK (PERK-K618A) in primary neurons rescued loss of dendrites and memory in mice [63]. Further, a rodent model of CCI showed loss of PSD-95 with loss of neuronal NeuN in contused cortex and directly correlated with behavioral abnormalities [64]. UCCB01-147 [also known as Tat-NPEG4(IETDV)(2), (Tat-N-dimer)], a dimeric PSD-95 inhibitor, was observed to be neuroprotective in an experimental stroke model [65] but failed to demonstrate those beneficial effects in experimental TBI [66]. Therefore, it can be argued that PSD-95 alone or with other effector proteins may provide a potential clinical therapeutic target to improve memory and learning deficits but must be translated carefully for better results post-TBI.

2.4. mGluR2 Receptors and Gap Junctions

Recently, it was reported that mGluR2 receptors in a fluid percussion TBI model once activated, upregulated gap junction protein expression, suggesting a possible role for mGluR2 and gap junctions in secondary brain injury [67]. Gap junctions play critical roles in neuronal differentiation and circuit formation in the developing CNS and allow for the passage of ions, small molecules, and secondary messengers (Ca, IP3, cAMP, etc.) in the adult brain [68,69]. However, upregulated gap junctions in insulted brain may enhance the secretion of pro-apoptotic factors and secondary messengers such as Ca^{2+} to add to cell death. Therefore, mefloquine, a gap-junction blocker may be a valuable therapeutic tool in mitigating TBI-induced excitotoxicity.

2.5. Glutamate Transporters

The solute carrier 1 (SLC1) family of neurotransmitter transporters includes a five-member family of excitatory amino acid transporters (EAAT) that mediate the rapid uptake of synaptic glutamate via a process coupled to ion gradients [70]. Amongst the EAAT, EAAT1 [Glutamate Aspartate Transporter (GLAST)] (human/rodent homolog) and EAAT2 [Glutamate Transporter 1 (GLT-1)] are essential for glutamate clearance related to neurotransmission, whereas EAAT3, EAAT4, and EAAT5 exhibit less prominent parts in regulating neuronal excitability [70]. In particular, EAAT2 is expressed in glia and mediates 95 percent of glutamate uptake in CNS [71]. Notably, when ionic gradients are lost, sodium-glutamate transporters can reverse the transport direction to secrete a high amount of glutamate [72]. Given that excessive glutamate is associated with excitotoxicity, targeted enhancement of EAAT2 may circumvent the issues associated with administration of NMDAR antagonists. Postmortem analysis of human brains obtained after TBI showed lower expression of the glial glutamate transporter EAAT2, which might have impaired reuptake of extracellular glutamate and thus, have led to excitotoxicity [73,74]. Similarly, lowered EAAT2 expression inversely correlated with CSF glutamate levels up to 7 days post-injury in CCI model of rodents [75,76].

MS-153 (GLT-1 activator): Functionally, administration of GLT-1 antisense oligodeoxynucleotides exacerbated hippocampal injury and increased mortality after TBI [76], whereas acute administration of (R)-(-)-5-methyl-1-nicotinoyl-2-pyrazoline (MS-153), a GLT-1 activator, decreased neurodegeneration and attenuated calpain activation in cortical and hippocampal tissue for up to two weeks after fluid percussion injury. While these latter findings warrant further exploration of GLT-1 activators, MS-153 upregulated GLT-1 in the naïve brain but not after brain injury in rats, suggesting that mechanisms independent from GLT-1 may mediate the observed beneficial effects [77]. Future studies incorporating more selective pharmacological activators and genetic overexpression approaches will provide clarity regarding the translational potential of targeting GLT-1 after TBI.

2.6. Blood Glutamate Scavengers

Glutamate transporters located on brain capillary endothelial cells facilitate brain-to-blood efflux of glutamate and play a role in glutamate homeostasis. When glutamate concentrations are kept low in blood, this results in a larger concentration gradient of glutamate and enhances its brain-to-blood efflux [78,79]. Glutamate levels in the blood might be reduced via two enzymes: glutamate-oxaloacetate transaminase (GOT or AST) (L-glutamate + oxaloacetate $\rightleftharpoons$ α-ketoglutarate + L-aspartate) and glutamate-pyruvate transaminase (GPT or ALT) (L-glutamate + pyruvate $\rightleftharpoons$ α-ketoglutarate + L-alanine). The two serum enzymes (SGOT and SGPT) and co-substrates (oxaloacetate and pyruvate) may potentially act as glutamate scavengers. In agreement, treatment with co-substrate, oxaloacetate, and pyruvate resulted in a reduction of glutamate in blood and enhanced neuronal survival and neurological outcomes in experimental TBI studies [80,81]. In addition, recombinant GOT1 has also shown promising results in TBI [82], and ischemic stroke [83,84]. Of note, riboflavin (Vitamin B2) was found to be a potent scavenger of blood glutamate, resulting in reduced infarct size after ischemia [85]. Similarly, Hoane and colleagues found that riboflavin significantly reduced sensorimotor and cognitive impairment, reduced edema, and astrogliosis after TBI [86] Furthermore, a double-blind, randomized phase IIb clinical trial in stroke patients also demonstrated riboflavin significantly scavenged glutamate in patient blood [85]. Taken together, these studies suggest that therapies utilizing blood glutamate scavenging may be promising therapeutic avenues for reducing glutamate-induced excitotoxicity.

2.7. GABAergic Excitotoxicity

Although glutamate is a major player in excitotoxicity, elevated concentrations of other neurotransmitters have also been detected in the extracellular space of injured brains. These other neurotransmitters, such as GABA, may also aid in excitotoxicity in both specific and distant cell populations. A microdialysis study in experimental open-skull weight drop TBI, showed elevated GABA in the cortical extracellular space [87]. Similarly, hippocampal cell loss has been reported in multiple experimental models of TBI and may correlate with neurocognitive deficits that occur in TBI patients [88,89]. Immature neurons in the adult hippocampal sub-granular zone express voltage-gated channels similar to their embryonic equivalents and may be depolarized by GABAergic activity via chloride gradient reversal [90,91]. Since selective necrotic cell loss among immature adult-born neurons has been reported after CCI injury; therefore, the properties of these immature neurons may be relevant to pathophysiology in TBI [92–94]. However, the mechanism behind this particular cellular selectivity for necrosis among immature cells is still known. GABAergic excitotoxicity also contributes to neuronal loss as exposure of isoflurane to the immature pyramidal cells in culture increased intracellular calcium and led to cellular death [95], suggesting that intracellular calcium overload may add to immature neuronal death. These studies, coupled with the increased extracellular GABA concentrations as observed following TBI [87], may implicate an unexplored role for GABAergic excitotoxicity post-TBI. Understanding the mechanisms of GABAergic excitotoxicity in the post-traumatic brain may be valuable therapeutically, as several GABA antagonists, approved by the U.S. Food and Drug Administration (FDA), are available. In terms of treating more canonical glutamatergic excitotoxicity, it may be argued that antagonists to GluR or inhibitors that block the release of glutamate inhibitors (e.g., lamotrigine) exert beneficial effects not only through inhibition of either GluR or glutamate, but also by minimizing the neuronal metabolism. However, given the contradictory findings of clinical trials utilizing magnesium sulfate as an NMDAR antagonist in acute stroke patients [96,97], a more comprehensive knowledge of excitotoxicity following TBI, including non-canonical mechanisms (such as GABAergic excitotoxicity), is critical for developing effective therapeutic strategies. Further, excitotoxicity and as a result, influx of excessive calcium into cells lead to oxidative stress, mitochondrial dysfunction, activation of Nox family member, and oxidation of cellular biomolecules such as lipids, proteins, and DNA. Furthermore, excessive amount of intracellular calcium activates several proteases and phospholipases, and thus, mediates degradation of cellular proteins and lipids, and enhances ROS production, and contributes significantly to secondary injury post-TBI [98,99].

3. Oxidative Stress

TBI results in cerebral circulation dysfunction, microvascular impairment, and moderate hypoxia [100,101]. Although primary traumatic injury occurs as a result of the physical impact, tissue injury is amplified by secondary injury mechanisms. Oxidative damage is unambiguously one of the most confirmed "secondary injury" pathways observed in TBI. The brain is very sensitive to free radical-mediated damage because of the presence of abundant polyunsaturated lipids and a high rate of endogenous oxidative metabolism. Therefore, a balance between oxidant production and antioxidant machinery is essential for normal functioning of the brain.

3.1. Oxidant-Antioxidant Balance

The reperfusion of blood circulation after trauma ensures the survival of neurons but also elevates the generation of free radicals and reactive oxygen species (ROS) [102–104]. The generated free radicals, such as hydrogen peroxide, superoxides, nitric oxide (NO), etc., also cause excitotoxicity and impair the metabolic activity of cells. Further, superoxide radicals generated due to catalytic activity after TBI react with NO to form another potent oxidant peroxynitrite, which impairs cerebrovascular function [105,106]. The ROS possesses an unpaired electron and thus, readily binds to different macromolecules such as protein, nucleic acid, or lipid to cause damage. The endogenous antioxidant system comprises of glutathione (GSH), and various enzymes (i.e., glutathione reductase (GR), glutathione-S-transferase (GST), glutathione peroxidase (GPx), catalase (CAT), superoxide dismutase (SOD), and uric acid) neutralizes these ROS, preventing the oxidation of macromolecules. In a normal brain, oxidants and antioxidants exist in equilibrium; however, the unwarranted production of ROS following brain injury overburdens the efficiency of the endogenous antioxidants and shifts the equilibrium towards oxidants by depleting endogenous antioxidants. This disrupted balance increases membranous lipid peroxidation, oxidation of proteins, DNA break, and inhibition of the mitochondrial respiration, which ultimately throws cells into apoptosis or necrosis [107]. Overall, increased oxidants and reduced activity of antioxidant defense systems may contribute toward the pathogenesis post-TBI.

3.2. Superoxide Radicals and Superoxide Scavengers

Kontos and colleagues demonstrated an acute increase in brain microvascular superoxide radical ($O_2\bullet^-$) contents as a result of compromised autoregulatory function after fluid percussion injury [108,109]. Within an injured brain, several possible sources of $O_2\bullet^-$ may be operating from the very first minute of impact to a few h post-injury, including the arachidonic acid-prostaglandin cascade, oxidation of leaked hemoglobin and biogenic amines (e.g., norepinephrine, dopamine, 5-hydroxytryptamine), xanthine oxidase activity, and mitochondrial leakage. At later time points, activated microglia and infiltrating neutrophils and macrophages also provide additional sources of $O_2\bullet^-$ Superoxide $O_2\bullet^-$ is less reactive to biological substrates than hydrogen peroxide (H_2O_2). Once formed, $O_2\bullet^-$ undergoes dismutation to form H_2O_2 in a reaction that is catalyzed by SOD: $O_2\bullet^- + O_2^- + 2H^+ \rightarrow H_2O_2 + O_2$ [110]. The H_2O_2 formed is altered to water by peroxidases, such as Gpx and peroxiredoxin, or is dismuted to water and oxygen by CAT. Both CAT and GPx enzymes are abundant in the brain, though the latter has a sevenfold greater activity [111]. In the absence of the antioxidant system, as in TBI, $O_2\bullet^-$ actually exists in equilibrium with hydroperoxyl radicals ($HO_2\bullet$): $O_2\bullet^- + H^+ \rightarrow HO_2\bullet$, which is a considerably more powerful oxidizing or reducing agent [112]. $O_2\bullet^-/HO_2\bullet$ cause the pH to fall into acidic ranges (i.e., tissue acidosis), fueling an equilibrium shift in favor of $HO_2\bullet$, which is much more reactive than $O_2\bullet^-$, particularly toward lipids.

SOD and polyethylene glycol (PEG)-conjugated SOD (PEG-SOD): In humans, the three forms of SOD are SOD1 (Cu/Zn-SOD), SOD2 (Mn-SOD), and SOD3 (Cu/Zn-SOD), which are respectively sited in the cytoplasm, mitochondria, and outside the cell. In cats, the administration of SOD reverses the microvascular dysfunction post-TBI [113]. Transgenic mice overexpressing human Cu/Zn SOD activity have reduced acute injuries, prevented brain edema, and inhibited BBB permeability following TBI [114,115]. Studies using both Cu/Zn-SOD and Mn-SOD transgenic and knockout mice have further solidified the protective role of these enzymes against head trauma [115–119]. In a randomized controlled phase-II trial with more metabolically stable PEG-SOD (2000–10,000 U/kg intravenously administered 4 h post-TBI), the higher doses (5000 and 10,000 U/kg) decreased the duration when ICP > 20 mm of Hg. There was a statistically significant improvement in patient outcomes measured using the Glasgow Outcome Scale (GOS) at 3 and 6 months post-injury between patients who received placebo and those who received 10,000 U/kg PEG-SOD [120]. However, a subsequent larger phase-III randomized trial with higher doses of PEG-SOD (10,000 or 20,000 U/kg within 8 h post-TBI failed to show significant improvement in neurological outcomes or patient survival [121]. Nevertheless, it is imperative to shed light on a 4-h difference in the time of administration of PEG-SOD post-injury between these two trials.

OPC-14117: The superoxide radical scavenger "OPC-14117" reduced cortical damage and improved neuronal survival and cognitive functions following CCI in rats [122]. A controlled randomized trial assessing the safety of OPC-14117 (240 mg per day) in HIV-associated cognitive impairment found it to be tolerable and resulted in improvement of clinical global impression scale scores and nonsignificant enhancement of cognitive test scores [123]. In light of the neuroprotective effects of OPC-14117 in this preclinical TBI study and the safety of use in humans, the potential benefits of OPC-14117 treatment after TBI are worth investigating.

3.3. Iron, Hydroxyl Radicals, and Iron Chelators

The abundance of iron in the CNS makes it vulnerable to oxidative insult [124]. Under normal circumstances, iron is maintained in a non-catalytic state by plasma transferrin and intracellular ferritin [110]. However, in the event of tissue acidosis, when pH falls below 6, both proteins readily release their iron into the traumatized brain parenchyma. Further, hemorrhage occurs as a result of mechanical impact provides an obvious pool of iron released from hemoglobin via interaction with H_2O_2 or lipid hydroperoxides (LOOH) at acidic pH [125,126]. Once released into the brain parenchyma, iron actively catalyzes Fenton's reaction to generate ROS [110]. Free iron or iron compounds participate in production of ROS in two ways: First, autoxidation of Fe^{2+} produces $O_2 \bullet^-$ [110]: $Fe^{2+} + O_2 \rightarrow Fe^{3+} + O_2 \bullet^-$ and/or secondly, oxidation of Fe^{2+} by H_2O_2 gives hydroxyl radicals ($\bullet OH$) via Fenton's reaction: $Fe^{2+} + H_2O_2 \rightarrow Fe^{3+} + \bullet OH + OH^-$.

Deferoxamine and dextran-conjugated deferoxamine: Experimental TBI studies of deferoxamine (iron chelator) treatment in rodents have shown neuroprotective effects [127,128]. However, IV infusion of deferoxamine may cause profound hypotension, but binding of deferoxamine to dextran may alleviate this effect as low dose dextran-conjugated deferoxamine following TBI improved neurological outcomes as compared to the deferoxamine group alone [129].

N,N'-Di(2-hydroxybenzyl)ethylenediamine-N,N'-diacetic acid monohydrochloride (HBED): HBED is an iron chelator, can cross the BBB, and has a relatively longer half-life compared to that of deferoxamine [130]. HBED treatment resulted in reduced cortical damage and restored neurological functions in mice post-TBI [131].

3.4. Nitric oxide Synthase (NOS) and NOS Inhibitors

NOS catalyzes L-arginine to give NO and citrulline at the expense of NADPH. The three NOS isoforms [neuronal (nNOS or NOS1), inducible (iNOS or NOS2), and endothelial (eNOS or NOS3)] are acutely upregulated in rats following TBI, with levels peaking at 6 to 12 h post-injury and then declining. eNOS is expressed solely in endothelial cells, nNOS predominantly in neurons but also in polymorphonuclear cells, and iNOS in immune cells such as myeloid cells [132]. Subsequent to injury, an upsurge of NO occurs, possibly because of the hyperactivity of iNOS, and the inhibition of iNOS could be neuroprotective [133]. Clinically, NO levels can be assessed indirectly through CSF measurement of the end products of NO metabolism, and a peak concentration is found at 1–3 days following TBI [134–136].

nNOS, NG-nitro L-arginine methyl ester (L-NAME), and 7-nitro indazole (7-NI): Both L-NAME (nonselective NOS inhibitor) and 7-NI (nNOS inhibitor) were reported to exert neuroprotection only when administered within an hour of injury in mice, indicating a narrow therapeutic window post-TBI. L-arginine, the physiological precursor of NO, when co-administered with NOS inhibitors reverses their protective effects [137]. Pretreatment with L-NAME or 7-NI reduced NOS activity after FPI in rat, while 7-NI also improved neurobehavioral outcomes post-TBI [138]. However, another FPI model of TBI study in rats found no beneficial role of L-NAME administration post-TBI with regards to mortality, and pretreatment leads to prolonged hypertensive episodes and increased mortality [139].

eNOS and L-arginine: TBI-induced upregulation of eNOS plays a beneficial role by maintaining cerebral blood flow (CBF) post-head injury. Administration of L-arginine post-TBI in rats activated eNOS, improved CBF, and attenuated neurological deficits without altering cerebral perfusion pressure (CPP) [140]. Similarly, eNOS-deficient mice subjected to CCI have shown lower CBF at the cortical injury site compared to wild-type mice, and L-arginine did not improve the CBF nor the contusion volume in eNOS-deficient mice [141]. In humans, microdialysates from severe TBI patients showed elevated levels of NO metabolites in the first 24 h post-injury followed by a gradual decline over 5 days. There was also a significant direct relationship between the concentration of NO metabolites and regional CBF [142]. In severe TBI patients, L-arginine administration at 48 h post-injury had a better response in improving internal carotid artery flow volume than at 12 h following brain injury [143]. The above findings further emphasize the importance of determining the effective therapeutic window for drug administration following brain injury. In humans, the NOS3 (eNOS) gene, located at 7q35–36, has several allelic variations and patients having the −786C allele showed lower CBF values than other severe TBI patients. Thus, genetic makeup may be a potential contributing factor to the variability in TBI patient outcomes [144].

Statins: Statins are HMG-CoA reductase inhibitors with proven neuroprotective activity in experimental TBI studies through targeting of multiple secondary injuries [103]. More precisely, statins upregulate eNOS, have a palliative effect on cerebral autoregulation (CA), and improve stroke outcomes [145].

iNOS and iNOS inhibitors (Aminoguanidine (AG), L-NIL and 1400W): iNOS is an inducible type of NOS which is stimulated by injury-induced stimuli, such as inflammatory modulators, ROS, etc. [146–149]. In rats exposed to FPI, intraperitoneal injection of 100 mg/kg aminoguanidine (AG) twice daily for 3 days reduced the total cortical necrotic neuron counts but not the contusion volume [150]. Blast induced-TBI in rats showed that those receiving AG either prophylactically or after the injury performed better on neurobehavioral tests and had reduced cortical neuron degeneration compared with those receiving saline injection [151]. In an FPI model of brain trauma in rats at 6 h after injury, the following 3 iNOS inhibitors were given at 6 h post-injury: aminoguanidine (AG; 100 mg/kg intraperitoneally), L-N-iminoethyl-lysine (L-NIL; 20 mg/kg intraperitoneally), or N-[3-(aminomethyl)benzyl]acetamide (1400W; 20 mg/kg subcutaneously). All three improved neurofunctional outcomes, but AG also reduced brain edema [152].

Ronopterin (also termed 4-amino-tetrahydrobiopterin or VAS203): In a phase IIa RCT with moderate to severe TBI subjects (NO Synthase inhibition in traumatic brain injury, NOSTRA), patients were given various IV infusion doses (15–30 mg/kg) of ronopterin, a NOS inhibitor. Ronopterin treatment showed no marked alteration in ICP, CPP, or brain metabolism. Other than a transitory acute kidney injury in half of the patients receiving the highest dose, no other toxic side effects were reported. Additionally, ronopterin had a neuroprotective role shown by significant improvement of extended Glasgow Outcome Scores (eGOS) at 6 months [153]. These positive results lead NOSTRA trial into phase III which is current and whose study protocol has been published [154].

3.5. Peroxynitrite and Peroxynitrite Scavengers

Peroxynitrite is formed by the reaction of superoxide and NO radicals, which contributes to impaired cerebral vascular reactivity after TBI [105,106]. Peroxynitrite interacts with DNA, proteins, and lipids via oxidizing or radical-mediated mechanisms. In addition, it reacts with tyrosine residues of proteins to yield nitrotyrosine and impairs activity [155].

Penicillamine and penicillamine methyl ester: The thiol-containing compound penicillamine and the more brain-permeable penicillamine methyl ester are sulfhydryl-based scavengers of peroxynitrite. Both of these compounds have improved early neurological recovery in TBI mice. Although penicillamine remains mainly within the cerebral microvasculature, it showed greater neurological recovery than highly penetrable penicillamine methyl ester, highlighting the significance of early scavenging of intravascular peroxynitrite [156].

Tempol and α-phenyl-N-tert-butyl-nitrone (PBN): The peroxynitrite radical scavengers tempol [157–159] and PBN [160] have demonstrated neuroprotective activity in experimental TBI and could be good candidates to minimize nitrosative stress.

3.6. Lipid Peroxidation (LP) and LP Inhibitors

Free radical-mediated LP is an extensively studied mechanism of oxidative injury in TBI [161]. The brain cell membrane is abundant in polyunsaturated fatty acids e.g., arachidonic acid (AA), which is extremely susceptible to •OH-induced peroxidation. LP starts when a radical species, such as •OH, extracts hydrogen from an allyl group (AA + R• → AA• + RH), converting this allylic carbon into an "alkyl" radical (AA•). During the propagation stage, the resulting alkyl radical binds with a molecule of oxygen to form a lipid peroxyl radical (AA-OO•; AA• + O2 → AA-OO•). The peroxyl radical then reacts with a neighboring AA within the membrane and gains its electron to generate a lipid hydroperoxide (AA-OOH) and a resultant alkyl radical (AA•; AA-OO• + AA → AA-OOH + AA•). This cycle of generation of alkyl radicals is continuous and compromises cellular and sub-cellular membranous integrity. Finally, in the termination step of the LP, lipid radials react with another radical, giving rise to highly reactive, potentially neurotoxic aldehydes known as carbonyls. The neurotoxic aldehydes, 4-hydroxynonenal (4-HNE), and 2-propenal (acrolein) bind to basic amino acids (arginine, histidine, and lysine) and sulfhydryl-containing cysteine residues in cellular proteins, and alter their conformation and function.

Tirilazad (lazaroid, 21-aminosteroid, a LP inhibitor): Previously, tirilazad has been shown to enhance neurological recovery and survival in experimental TBI [162–164]. In a multicenter phase III study, moderate-severely injured patients, treated with tirilazard mesylate (10 mg/kg intravenous), starting within 4 h post-injury and repeated for every 6 h up to 5 days, did not show significant GOS on recovery/survival at 6 months. However, a post hoc analysis of data discovered that tirilazad lowered mortality rates in male TBI patients with accompanying traumatic subarachnoid hemorrhage (tSAH) [165].

U83836E: U83836E is a very effective second-generation lazaroid with a unique structure giving it the ability to scavenge lipid peroxyl and to inhibit LP. Further, U83836E also showed the ability to preserve mitochondrial respiratory function in rodents post-TBI [159].

LP carbonyl (4-HNE or acrolein) scavengers

Hydralazine: Despite being shown to have a good ability to scavenge LP carbonyls, hydralazine is a powerful vasodilator exacerbating hypotension in TBI and therefore, is not recommended post-TBI [166,167].

Phenelzine: Phenelzine, a monoamine oxidase inhibitor (MAO-I), is also a good scavenger of LP carbonyls because of the presence of its hydrazine functional group. Phenelzine inhibited oxidative damage and mitigated mitochondrial dysfunction in isolated rat brain mitochondria when subjected to exogenous acrolein or 4-HNE. Further, rats subjected to CCI and given a single dose of 10 mg/kg phenelzine subcutaneously 15 min post-TBI, protected cortical tissue from injury at 2 weeks [168]. Additionally, repeated doses of phenelzine (an initial dose of 10 mg/kg subcutaneous 15 min post-TBI followed by a repeated dose of 5 mg/kg subcutaneous every 12 h up to 60 h post-TBI) protected cortical tissue loss and attenuated mitochondrial impairment in CCI rats [169].

β-Phenylethylidenehydrazine (PEH): PEH is an active metabolite of phenelzine (β-phenylethylhydrazine). Both phenelzine and PEH possess a hydrazine functional group, and therefore, react with LP carbonyls and other LP aldehydes to form hydrazones. Because the compounds have different impacts on MAO inhibition, use of PEH may avoid the drug interactions seen with the use of phenelzine and certain sympathomimetics (tyramine) [170].

3.7. Nuclear Factor Erythroid 2-Related Factor 2 (Nrf2)-Antioxidant Response Element (ARE) Pathway

Kelch-like ECH-associated protein 1 (Keap1)-Nrf2-ARE signaling regulates endogenous antioxidant defense system and thus, plays an important role in protecting cells from intrinsic and extrinsic oxidants and electrophiles. Nrf2 heterodimerizes with other transcription factors and binds to ARE leading to expression of ARE-regulated genes that enhance cell survival. Kelch ECH associating protein 1 (Keap1) is a cytosolic repressor of Nrf2, which enhances its proteasomal degradation by binding with it [171]. Following TBI, Nrf2-knockout mice exhibited exacerbated brain injury with increased expression of inflammatory cytokines tumor necrotic factor-α (TNF-α), and interleukins (IL-1β and IL-6), and decreased activity of antioxidant enzymes NADPH: quinone oxidoreductase-1 (NQO-1) and glutathione S-transferase alpha-1 (GST-alpha1) [172].

Nrf2 activators: Sulforaphane (SFN), an Nrf2 activator, resulted in upregulation of the antioxidant enzymes heme oxygenase 1 (HO-1) and NQO-1 and lead to significant reduction in neurological dysfunction, injury volume, and neuronal death in rodent CCI models [173]. Tetra-butylhydroquinone (tBHQ), another Nrf2 activator, reduced nuclear factor kappa B (NF-κB) activation and inflammatory cytokines (TNF-α, IL-1β, IL-6) and attenuated cortical injury and brain edema in closed head-injured mice [174]. In another closed-head mouse model, tBHQ activated Nrf2 and attenuated NADPH oxidase (NOX2) in order to reduce cerebral edema and neurologic deficits [175]. Carnosic acid (CA) activates Nrf2 and in turn upregulates cytoprotective (ARE) genes and inhibits pro-inflammatory genes (through suppression of NF-κB). Early (15 min) and delayed (8 h) administration of CA after CCI in mice, resulted in restored mitochondrial respiration, and reduced neuronal cytoskeletal breakdown [167,176]. In mice subjected to repetitive mild TBI, CA administration improved cognitive and motor functions [177].

Melatonin and N-acetylserotonin (NAS): Melatonin (N-acetyl 5-methoxytryptamine) is a neurohormone with multiple physiological functions and has demonstrated to have anti-inflammatory, antioxidant, and antiapoptotic properties [178–180]. NAS, a precursor of melatonin, is a melatonin receptor 1C (MT3) agonist that is also shown to exhibit neuroprotection against TBI in preclinical studies. It exhibits antioxidant properties by directly scavenging oxidants and indirectly acting through antioxidant enzymes [181]. Melatonin might act through Nrf2-ARE signaling, as melatonin treatment in rodents upregulated antioxidant enzymes (HO-1 and NQO-1) downstream to Nrf2, while knockout of Nrf2 partially reversed its neuroprotective effects [182]. Moreover, a double-blinded randomized placebo-control clinical trial is investigating the sublingual melatonin in the treatment of post-concussion syndrome following mild pediatric TBI [183].

N-acetylcysteine (NAC) and N-acetylcysteine amide (NACA): Administration of both NAC and its more BBB-permeable form, NACA, reduced cortical damage in rodents after experimental TBI [184]. NACA treatment, following TBI in rats, activated the Nrf2-ARE pathway, attenuated oxidative stress, and inhibited neuronal degeneration [185]. A systematic review including twenty animal studies and three human trials concluded that although there is sufficient preclinical evidence for neuroprotective effects of NAC/NACA after TBI, well-designed clinical studies are lacking [186].

3.8. Endothelial Targeted Antioxidant Enzyme Therapy

Following TBI, the damaged endothelium is a key site for oxidative stress, and the damaged endothelial cells upregulate the expression of cell adhesion molecules, such as Intercellular Adhesion Molecule 1 (ICAM-1). A novel approach is the use of targeted endothelial nanomedicine/antibodies.

For example, anti-ICAM-1/CAT is an anti-ICAM-1 antibody conjugated to the antioxidant enzyme CAT. In an experimental model of TBI, anti-ICAM-1/CAT treatment reduced oxidative stress at the BBB and attenuated neuropathological outcomes [187].

4. Inflammation

Post-traumatic cerebral inflammation starts within minutes of injury and is characterized by upregulation and secretion of mediators (such as DAMPs, cytokines, and chemokines), infiltration of neutrophils and other myeloid cells, and subsequent glial activation and leukocyte recruitment (Figure 2) [188]. BBB impairment during the acute post-traumatic period allows for the entry of circulating neutrophils, monocytes, and lymphocytes to the injury site and directly influences neuronal survival and death [188–190]. The accumulated peripheral and resident immune cells in the brain parenchyma release inflammatory mediators, including but not limited to DAMPs, cytokine, chemokines, ROS, prostaglandins, and complement factors [20], which further potentiates inflammation in the injured brain by recruiting more immune cells to the injury site [191]. However, over time, subsequent production of anti-inflammatory mediators and endogenous protectants suppress both humoral and cellular immune activation (Figure 2). In addition to the infiltrating peripheral blood cells, the resident microglia are activated. These activated microglia help in clearing cell debris and promote tissue remodeling. However, activated microglia also release various neurotoxic substances, such as ROS, RNS, and excitatory neurotransmitters, such as glutamate, that may exacerbate neuronal death [192]. In addition, proliferation and migration of reactive astrocytes, and the development of a glial scar after brain trauma impair axonal regrowth. Overall, complex astrogliosis and the trafficking of immune cells to the injury site can promote tissue repair and neurogenesis via the release of neurotrophic factors, [29] or can exacerbate tissue damage through increased inflammatory mechanisms as well (Figure 2).

4.1. Inflammatory Mediators

Within minutes to hours after the injury, damaged cells release many intracellular components, such as heat shock proteins (HSP 60 and 70), nucleic acids, and high mobility group protein B1 (HMGB1) into circulation and the extracellular space [14,193,194]. These released intracellular components act as damage-associated molecular patterns (DAMPs) and activate pattern recognition receptors (PRR) for downstream cell signaling [194–196]. In response, astrocytes, microglia, and neurons at the injury site begin secreting cytokines and chemokines [191,197]. In addition to their contribution in immune processes and roles in homeostasis, cytokines also function as messengers of intracellular communication [198], while the chemotactic cytokines, or chemokines, regulate leukocyte activation and migration [199]. These inflammatory signals activate microglia and astrocytes, recruit peripheral immune cells, and increase migration to the site of injury. Once inside the brain, peripheral immune cells secrete large amounts of inflammatory mediators, which add to further tissue damage and remodeling [200–203]. Many studies have reported upregulated expression of IL-1β, TNF-α, IL-6, CCL2, CCL3, CXCL1, CXCL2, CXCL8/IL-8, CXCL10, CCR2, CCR5, CXCR4, and CX3CR1 within 6 h of TBI [204–206]. Similar to animal models of TBI, the levels of many cytokines and chemokines peak at 4 h post-injury in patients [207,208]. Collectively, the previous reports suggest that early upregulation of inflammatory mediators is a robust response to injury that also adds to the subsequent secondary injury and chronic neuropathological processes.

4.2. Cellular (Innate and Adaptive) Responses

The inflammatory mediators released post-TBI not only alter the residential CNS cells, but also recruit peripheral cells into the brain. These immune cells polarize toward pro-or anti-inflammatory phenotypes due to surrounding signals from the tissue microenvironment [194,205,209,210]. In cortical impact TBI models, neutrophils are among the first cell types to respond to injury. Upregulated adhesion molecules on vascular endothelium mediate neutrophil entry into the traumatized brain during the early h of the first day post-injury [195,211,212]. Despite the essential role of neutrophil recruitment during peripheral infections and damage, they release huge amounts of ROS and RNS in traumatized brain causing oxidative cellular damage. The presence of neutrophils in the injured brain becomes greatly reduced by 3–5 days post-TBI, and mononuclear leukocytes begin to predominate [195,211,213]. These infiltrated cells are mostly CD45^{hi}CCR2$^+$Ly6C$^+$ monocytes, with a small number of dendritic cells (DCs), T lymphocytes, and natural killer (NK) cells. DCs and T cells infiltrate the brain in lower numbers in a similar fashion as monocytes, and perform specific functions depending on the subpopulations of cells present. DCs are categorized into two- T-cell stimulating conventional dendritic cells (cDCs), and interferon-α secreting plasmacytoid (pDCs) [213–215]. Further, T cells are categorized into four sub-types- T helper (T$_H$), memory T, cytotoxic T (T$_C$), and NK cells, each serve distinct functions. Besides infiltrating peripheral cells, residential microglia and astrocytes simultaneously get activated during early immune response. Activated microglia, along with infiltrating macrophages, phagocytose cellular debris, secrete inflammatory mediators, and add to the local inflammation [211,216–218].

Meningeal lymphatic vessels are specialized to facilitate drainage of immune cells and macromolecules into the deep cervical lymph nodes and act as sites for immune surveillance of the CNS [219,220]. Notably, following TBI, activated macrophages may drain into cervical lymph nodes and mediate long-term adaptive response through MHC Class II-dependent antigen presentation [213,221,222]. In agreement, T-lymphocytes get activated and recruited within deep cervical lymph nodes by antigen-presenting cells rather than at the site of CNS injury [223]. Interestingly, HLA-DR, an MHC Class II antigen expressed on macrophages, initiated adaptive immune responses by binding myelin basic protein (MBP) [224,225]. Thus, myelin-loaded macrophages may initiate white matter injury (WMI) post-TBI in a similar way to the autoimmune-mediated demyelination in multiple sclerosis [226]. The fact that pharmacological inhibition of MHC Class II reduced neurodegeneration after TBI [227], therefore, activated macrophages perhaps be a connected link between TBI and chronic adaptive immune responses.

T-lymphocytes do not routinely cross the BBB [228,229] but are functionally diverse subsets that mediate the specific adaptive responses against a presented antigen. In particular, T$_H$ cells stimulate antibody production and release cytokines that potentiate activation of macrophages and T$_C$ cells. Thus, infiltrating macrophages and T$_H$ cells cause neurodegeneration as evident in traumatized brain tissue from animals and TBI patients [227,230,231]. Naïve T$_H$ cells differentiate into three T$_H$ subtypes, such as T$_H$1, T$_H$2, and T$_H$17 on the basis of clues obtained from secreted cytokines by activated macrophages [209,232–234].

Further, T$_H$17 cells promoted microglial polarization after experimental autoimmune encephalomyelitis (EAE), while, in vitro, myelin-specific T-lymphocytes induced a pro-inflammatory phenotype in microglia via IL-17 [235,236]. Furthermore, inhibition of T$_H$17 influx into the brain protected WM and prevented chronic neurological deficits post-neonatal hypoxia-ischemic injury [237]. In agreement, myelin-reactive T$_H$17 cells were found to induce demyelination and to compromise remyelination in animal models of WMI and in multiple sclerosis patients [238–241]. Moreover, curcumin, an anti-inflammatory compound in the curry spice turmeric, improved TBI outcomes [242] and attenuated activation of T$_H$17 in ovalbumin-sensitized mice and in acute graft versus host disease [243,244]. Curcumin further mitigated RORγT-mediated T$_H$17 differentiation, decreased MBP-reactive T cells, and attenuated IL-17 secretion by activated T$_H$17 cells in EAE [245]. Therefore, therapeutic strategies aiming at T$_H$17 production/activity and/or MHC-II inhibition may provide potential possibilities to improve chronic functional outcomes following TBI.

By 10–14 days post-injury, most of the circulating immune cells are largely absent from the injury site. However, F4/80$^+$ macrophages and glial fibrillary acidic protein (GFAP)+ astrocytes have been detected at distant sites far from the primary damage. In addition, injured thalamic neurons and WMI have been seen many months after the initial impact and thus indicate an effect of chronic diffuse injury [217,246,247]. More detailed literature on the neuroimmunology of TBI and various components of the post-TBI immune response can be obtained from the recent reviews by McKee and Lukens and Jassam et al. [248,249].

Contrary to previous dogma, cerebral inflammation is now considered to have both injurious and beneficial roles in TBI resolution. The traumatized brain can benefit from inflammation if regulated; otherwise, excessive and chronic inflammatory cascades can take over and contribute to numerous neuropathologies [250]. Therefore, many therapies targeting either specific immune cells or inflammation are gaining interests scientifically and clinically.

4.3. Therapies Targeting Inflammation in TBI

Glucocorticoids: Glucocorticoids have broad anti-inflammatory actions. A randomized placebo-controlled trial (Corticosteroid Randomisation After Significant Head injury, CRASH) investigated the effects of an IV corticosteroid (methylprednisolone) infusion within 8 h of TBI in adult patients with a GCS score of 14 or less. They reported higher mortality rates at 2 and 24 weeks in the glucocorticoid-treated patients [251,252].

TNF-α inhibitors: In neuroinflammation, TNF-α induces microglial and astrocytic activation, and influences BBB permeability, glutamatergic transmission, and synaptic plasticity [253]. Treatment with intraperitoneal etanercept, a TNF blocker (repeated every 12 h up to 3 days starting immediately following injury) lowered neuronal and glial apoptosis, attenuated microglial and astrocytic activation, reduced the cerebral damage, and improved cognition and motor ability [254]. Additionally, systemic entanercept adminstration was able to penetrate the contused brain tissue reducing the brain contents of TNF-α, and also stimulated newly formed neuronogenesis [255,256]. Clinically, perispinal administration of etanercept (PSE) has shown early promising results in several studies investigating its effects in neurological recovery and chronic pain post-brain injury [257–260]. An observational study investigating the role of PSE in chronic management of TBI and stroke patients reported significant improvements of chronic neurological dysfunctions. This beneficial effect was observed irrespective of the duration of ailment with improvements noted in patients treated more than a decade after stroke and TBI [261]. The wide therapeutic window of etanercept makes it a valuable therapeutic tool in the management of patients after TBI. 3,6′-dithiothalidomide, a TNF-α synthesis inhibitor, has also shown neuroprotective effects in experimental mild TBI (mTBI) studies when given up to 12 h after injury [262,263]. However, mice lacking TNF-α showed increased post-traumatic mortality without altering the sequele of TBI pathophysiology, and number of infiltrating cells, suggesting a protective role after TBI [264] and could be modulated wisely to extract better outcomes.

IL-1 inhibitors: IL-1 receptor antagonist (IL-1ra) (Anakinra, recombinant human IL-1ra, rhIL-1ra): Both IL-1α and IL-1β bind to IL-1 receptor type 1 (IL-1r1) and initiate signaling. IL-1ra is an endogenous antagonist of the IL-1r1 and blocks receptor activation by IL-1 [265]. IL-1ra overexpressing transgenic mice showed improved neurological functions with a delayed secretion of pro-inflammatory cytokines in a closed-head injury (CHI) model [266]. A review examined previous experimental studies using anakinra in TBI, elucidated that anakinra has a narrow early therapeutic window with less neuroprotective effects when given at 2 h compared to 5 min or 15 min post-injury [267]. In a randomized controlled phase II trial, 100 mg anakinra was administered subcutaneously once daily for 5 days to severe TBI patients and was observed to be safe, penetrated into plasma and brain extracellular fluid, and modified the neuroinflammatory response. However, only twenty patients were recruited, and so the therapeutic effect of anakinra could not be concluded in the study. Furthermore, considering the evidence from experimental studies regarding the narrow acute therapeutic window for administration of anakinra, it would have been advisable to administer the medication at an earlier time point which could be a limitation of the study due to ethical committee requirements [268].

IL-1β neutralizing antibody (anti-IL-1β antibody, IgG2a/k): Intraventricular infusion of anti-IL-1β antibody (starting 5 min post-injury up to 14 days) reduced cortical microglial activation, minimized neutrophil and T cell cortical infiltration, diminished lesion volume, and improved cognitive function [269]. Further, intraperitoneal anti-IL-1β antibody at 30 min and 7 days post-CCI brain injury in mice reduced the ipsilateral hemispheric edema [270]. It has been reported that TBI leads to activation of NOX2 and subsequent NOD-like receptor family pyrin domain-containing 3 (NLRP3) inflammasome. The activation of NLRP3 inflammasome may lead to recruitment of IL-1 and caspase-1 following TBI [271,272]. Clinically, expressions of IL-18, IL-1β, caspase-1, and apoptosis-associated speck-like protein containing a caspase recruitment domain (ASC) were found to be consistent with poor outcomes after TBI [273], and targeting inflammasome through inhibiting ASC could show a promising result in curbing inflammation in mice [274].

IL-1α inhibitors: IL-1α is another important early mediator of inflammation following acute TBI and its inhibition may be neuroprotective [275]. An experimental research study has investigated the selective and combined inhibition of both IL-1 subtypes in TBI. They investigated the effect of inhibiting IL-1α (IL-1α-deficient mice), IL-1β (IL-1β-deficient mice), and IL-1r1 [IL-1r1-deficient mice and anakinra (IL-1ra)] in cortical FPI. IL-1r1 blockade caused a greater reduction in diffuse cytokine expression compared to individual ablation of IL-1α or IL-1β. Both genetic (IL-1r1-deficient mice) and pharmacological (anakinra) blockade of IL-1r1 protected mice from cognitive dysfunction, which was not seen with selective ablation of both IL-1 subtypes [276]. Therefore, broad targeting of IL-1r1 may be a more effective neuroprotective approach.

IL-6 inhibitors: In experimental mild TBI models, IL-6 and keratinocyte-derived chemokine (KC) were elevated 90 min post-injury, and IL-6 levels correlated with injury levels. This makes serum IL-6 a possible biomarker of TBI severity [277]. In a severe injury, serum IL-6 levels were used as a valid prognostic marker of ICP elevation only in TBI patients but could not be used in patients having both polytrauma and TBI [278]. In one study, mice were subjected to mild TBI, treated with rat monoclonal anti-IL-6 antibodies 10 min after injury, and then immediately exposed to hypoxia for 30 min. Anti-IL-6 treatment reduced brain inflammation and neuronal injury. Additionally, anti-IL-6 antibody administration abrogated the motor incoordination prompted by mTBI and hypoxia [279].

Clinically, there are many medications available for IL-6 inhibition, some of them are monoclonal anti-IL-6 antibodies (siltuximab) or anti-IL6-receptor antibodies (tocilizumab, sarilumab). Thus, IL-6 inhibition has a protective effect following TBI and may be the focus of future extensive TBI studies. Given that IL-1 inhibition improved cognitive but not motor functions and that IL-6 inhibition improved motor coordination, collective inhibition of both IL-1 and IL-6 could be a promising therapeutic avenue for further research.

HMGB1 inhibitors

TLR4 inhibitor: HMGB1 is a DAMP and binds to PRRs such as TLR2, TLR4, or receptors for advanced glycation end products (RAGE). As neuronal HMGB1 leads TLR4-mediated secretion of IL-6 from activated microglial/macrophages, either pharmacological inhibition of TLR4 by VGX-1027 or genetic mutation limited post-traumatic edema and inflammation in mice post-TBI [14].

Glycyrrhizin (Gly): Gly is a natural triterpenoid, which possesses antiviral and anti-inflammatory activities and suppresses HMGB1 activities via binding with its HMG box [280]. Previously, glycyrrhizic acid (600 mg/kg, i.p.) was reported to reduce ipsilateral brain edema 24 h post-TBI when administered 15 min prior to injury [14]. Similarly, IV administration of Gly 30 min post-injury downregulates HMGB1-(TLR4/RAGE)-NF-κB inflammatory pathway, resulting in attenuation of brain edema and improvement of motor function in TBI rats [281].

HMGB1 A-box: HMGB1 is comprised of an acidic C-terminus and the DNA-binding domains A-box (contains binding sites) and B-box (pro-inflammatory domain). Therefore, recombinant HMGB1 A-box fragment may compete with HMGB1 for binding to corresponding receptor, and thus, may be exert neuroprotection. In a CCI model of TBI in mice, IV administration of HMGB1 A-box daily for 3 days lowered pro-inflammatory cytokine levels, attenuated BBB breakdown, reduced cerebral edema, and enhanced neurobehavioral outcomes [282].

Other anti-inflammatory interventions

Natural anti-inflammatory compounds [curcumin and epigallocatechin-3-gallate (EGCG)]: Curcumin, the active compound in the spice turmeric, decreased glial activation, reduced cerebral edema, and improved neurological functions after CCI injury in mice, possibly by suppressing IL-1β, inhibiting NFκB, and downregulating AQP4 [242]. Natural compounds such as curcumin, which have broad anti-inflammatory effects, may prove to be effective treatments for brain injury since they likely target multiple inflammatory pathways to prevent cell death. For example, EGCG, an antioxidant in green tea, reduced cerebral edema and microglial activation following TBI by lowering expression of AQP4 and GFAP. EGCG further decreased oxidative stress by inhibiting NADPH oxidase activation and increasing SOD activity [283].

Cannabinoids: A fast-growing field of therapeutic intervention in different brain disorders is to modulate the endocannabinoid (eCB) system to harness favorable outcomes [205,284]. Cannabis-based research has come a long way from its introduction in 1838 by William O'Shaughnessy [285] for the treatment of migraines [286,287] and neuropathic pain [285,288,289]. The current cannabis preparations available clinically are Cesamet (nabilone), Marinol (dronabinol: Δ9-tetrahydrocannabinol [Δ9-THC]), and Sativex also branded as Nabiximols (Δ9-THC with cannabidiol) [290]. Sativex has been approved in thirty countries for multiple sclerosis-associated spasticity and central neuropathic pain [291], and for opioid-resistant cancer pain as well [292]. The eCB system is an endogenous system that gets activated by natural cannabinoids or cannabinoid-mimicking substances. The eCB system consists of two main cannabinoid receptors (CB1 and CB2), their endogenous ligands 2-arachidonyl glycerol [(2-AG) and N-arachidonoylethanolamine (anandamide)], and ligand-synthesizing and metabolizing enzymes regulating the secreted ligands [293,294]. The eCB system is essential for cellular homeostasis and physiology, and may have an important contribution in repair processes either after injury or during disease [294–298]. However, various inhibitors to AEA-metabolizing enzyme fatty acid amide hydrolase (FAAH) did not advance to phase III clinical trials as neurological therapeutics [299]. Additionally, an IV cannabinoid analog, dexanabinol (HU-211), has failed to show protective effect after head trauma in phase III clinical trials [300], but the scientific quest continues in the hope for other therapeutic preparations for the treatment of stroke and other brain pathologies [301–303]. Treatment with exogenous 2-AG attenuated inflammation, cerebrovascular injury, and subsequent neurological deficits after TBI [304]. Recently, we reported that selective activation of CB2R helped to reduce inflammatory macrophages and thus, protected CBF and behavioral function [205]. Synthetic CB2R agonists such as HU-910 and HU-914 showed enhanced recovery in rodents after closed-head injury [304]. Thus, eCB possesses potential targets for the modulation in diverse TBI pathologies. Exogenous compounds, such as the plant-derived phytocannabinoids or synthetic cannabinoids are being highly incorporated in basic research in TBI therapy. However, full characterization of the eCB system in the settings of TBI and other brain injuries is not fully revealed, but efforts are being made to understand its important part in brain homeostasis.

Remote Ischemic Conditioning (RIC): Most studies involving TBI therapies focus on various drugs that can target secondary injury mechanisms; however, a novel treatment known as RIC involves a non-pharmacological approach [305]. This treatment involves applying a blood pressure-type cuff on the arm, which tightens and loosens, subjecting the limb to short cycles of alternate ischemia and subsequent reperfusion to protect distant organs such as the brain, from any injury [306]. Previously, RIC has been stated to improve CBF [307], possibly through the secretion of humoral factors, such as endothelial NOS, NO, and/or nitrites [308], and anti-inflammatory factors that activate protective pathways against ischemia. These mechanisms may also activate mitochondrial ATP-sensitive potassium channels [309] that can restore the mitochondrial membrane potential and can suppress apoptosis following ischemia-reperfusion [310]. Recently, we reported that RIC improved hematoma resolution in a murine model of intracerebral hemorrhage (ICH) via modulation of AMP-activated protein kinase (AMPK) in macrophages [210]. RIC is gaining popularity, and more and more research has started focusing on this method in an effort to enhance endogenous protection [311–315]. A recent clinical trial on RIC following severe TBI found that RIC significantly decreased levels of blood biomarkers after TBI [313]. However, this clinical trial was relatively small and only looked at biomarkers in patient blood at 0, 6, and 24 h after RIC. Further clinical trials on RIC as a treatment for TBI are desired to evaluate its effectiveness, particularly in regard to patient outcomes, cognition, and standards of living. RIC appears to be a promising and non-invasive treatment for numerous conditions and can be easily combined with other treatments. For example, RIC combined with minocycline was proved to be a safe and low-cost intervention in a mouse thromboembolic stroke model [307]. Moreover, exposing rodents to intermittent and sub-acute RIC upregulated endogenous protection mechanisms limiting secondary injury following TBI [311,312]. Non-invasive approaches, such as RIC, will be better tools to provide neuroprotection after TBI but will still require further testing in clinical settings.

5. Programmed Cell Death (PCD)

PCD is another major cause of neuronal cell loss that can continue for days following TBI and is associated with poor prognosis [316]. Important PCD processes include cell cycle activation-dependent cell death, cell death mediated by caspases and pro-apoptotic members of Bcl-2 family, PARP/AIF-dependent death, and calpain/cathepsis-dependent death [317,318]. Understanding these many mechanisms of PCD is essential for developing a therapeutic intervention since multiple mechanisms of cell death are often simultaneously and excessively activated in response to injury. Blocking any one individual cell death mechanism may not be beneficial as other mechanisms can still compensate and lead to cell loss.

5.1. Cell Cycle Activation-Dependent Neuronal Cell Death

In several neurodegenerative disorders, markers of cell cycle reentry can be detected long before actual neuronal death, suggesting that mature neurons may re-enter the cell cycle and that these cell cycle events (CCE) may be upstream of neuronal cell death pathways [319]. TBI activates neuronal apoptosis, upregulating cell cycle markers (e.g., Cyclin D1, CDK4, E2F5, c-myc, and PCNA), while downregulating various endogenous cell cycle inhibitors [320–322]. Pharmacological inhibitors of cyclin-dependent kinases (CDKs), which regulate the cell cycle, have been stated to attenuate neuronal cell death and significantly improve outcomes after TBI in rodent models [321,323].

5.2. Caspase-Dependent Cell Death

Morphological cell changes, such as nuclear fragmentation, chromatic condensation, and membrane budding following TBI, occur via caspase-dependent cleavage of specific apoptotic substrates. Caspase-3 activation can follow either the extrinsic pathway, involving TNF and FAS receptors, or the intrinsic pathway, involving mitochondrial outer membrane permeabilization (MOMP). MOMP induces the release of cytochrome c (cyt c), an inner mitochondrial membranous protein, into the cytosol. There, cyt c binds with apoptosis-inducing factor (Apaf-1) to form an ATP-dependent complex, which in turn activates caspase-9 and caspase-3 [324].

Caspase-3 is an important effector caspase, which plays an important role in injury-induced neuronal loss after TBI [324]. Many studies have reported the link between activation of caspases and neuronal apoptosis in both clinical and pre-clinical TBI [325–328]. Treatment with various caspase inhibitors improves outcomes after experimental TBI, is effective therapeutically, and has a broad safety window [327,329]. Caspase-12 can be activated by endoplasmic reticulum (ER) stress, triggering apoptosis and resulting in neuronal cell death [328,330]. Bcl-2 and Bcl-xL are anti-apoptotic proteins that either inhibit MOMP directly or inhibit pro-apoptotic proteins to regulate MOMP and caspases. Pro-apoptotic proteins, consist of three subtypes; the first subtype consists of BAX and BAK, which can directly permeabilize mitochondrial membrane; the second subtype includes BID and BIM, which activate the first subtype; and the third type includes BAD, PUMA, and NOXA which can inactivate Bcl-2 and Bcl-xL [318]. The balance between the activities of pro- and anti-apoptotic members of Bcl-2 family is a major determinant of apoptosis after TBI. Increased Bcl-2 and Bcl-xL expression leads to survival of cells [331] while upregulation of BAX, BAD, or BIM promotes cell demise in the post-TBI brain [316].

5.3. Caspase-Independent Cell Death Pathways

Following MOMP, mitochondrial inner membrane proteins, such as cyt c, Smac/DIABLO, AIF, and endonuclease G (endoG), may be released into the cytosol and modulate cell death. Calpain I, a Ca^{2+}-dependent cysteine protease, is activated following TBI [332–334]. Activated calpain I cleaves the death-promoting Bcl-2 family members BID [335,336] and BAX [337], which then translocate to mitochondrial membranes. This results in the release of truncated apoptosis-inducing factor (tAIF) [335], cyt c [338], and endoG [336] in the case of BID or cyt c in the case of BAX [337,338]. These released proteins cause damage to nucleic acids and potentiate the release of inner membrane proteins [339–341]. While endoG cleaves internucleosomal (180 base pair) DNA [342], tAIF causes DNA cleavage on a large scale via interaction with phosphorylated histone H2AX (γH2AX) and cyclophilin A after translocation into nuclei [343,344]. Further, activated calpain I cleaves the Na^+-Ca^{2+} exchanger, which leads to accumulation of intracellular Ca^{2+} [345]. Cyt c has also been shown to translocate to the nucleus and is linked with cytosolic translocation of acetylated histone H2A in irradiated HeLa cells [346–348].

Most studies have confirmed that AIF mediates cell death, independent of caspase, Apaf-1, or cyt c [349,350]. PAR polymerase-1 (PARP-1), Cyclophilin A, and HSP-70 are key regulators of AIF and are responsible for translocation of AIF from mitochondria into the nucleus, which is partly mediated by activation of PARP-1 [351,352]. PARP-1 activation causes depletion of cytosolic NAD+ and subsequent mitochondrial dysfunction, which mediates the release of AIF from mitochondria [353,354]. The end-product of PARP-1 activation, poly (ADP-ribose) (PAR) polymer, also cause direct or calpain-mediated mitochondrial impairment and MOMP [355]. PARP-1 is activated in response to DNA damage and forms PAR polymers to repair DNA nicks. However, when DNA loss is extensive, PAR starts building up in the nucleus, and eventually translocates to mitochondria, and causes release of AIF [356–358]. Further, nuclear PAR glycohydrolase (PARG) hydrolyzes excessive PAR into ADP-ribose. These ADP-ribose translocate into the plasma membrane to stimulate melastatin-like transient receptor potential 2 (TRPM-2) channels to cause excessive Ca^{2+} influx into neurons [359]. In the main region of injury following cerebral ischemia, where bioenergetics conditions are compromised, cellular death takes place via AIF- and PARP-1-mediated processes rather than caspase-mediated cell death [349]. In fact, in PARP-1-dependent cell death, mitochondria release AIF and cyt c; however, caspases do not become activated because of depleted ATP [353,355].

5.4. Therapies Targeting Cell Death Pathways

HSP70: HSP70 is an interesting molecule that is extremely important in neuronal cell survival [360,361]. This mechanism of neuroprotection includes binding of HSP70 to Apaf-1 and AIF, thereby blocking the creation of apoptosome complexes, and subsequent activation of caspase-3 [362,363] and attenuating nuclear translocation of AIF [361,364,365]. In addition, deletion of HSP70 or HSP110 caused increased cell death with upregulated expression of ROS-induced P53-target genes, such as pig1, pig8, and pig12, while HSP70/110 boosting drug celastrol improved behavioral outcomes and protected brain cells from secondary injury following TBI [366]. It must be noted that stress-induced cellular death likely involves multiple pathways [318,367,368]. A key determinant in cell death dynamics in TBI is likely the cellular bioenergetics of the brain. When bioenergetic processes are preserved, caspase-dependent cell death mechanisms predominate, while under deficient bioenergetic conditions, when caspase is not activated, AIF may facilitate cell death. As AIF and caspases act through parallel pathways in apoptosis, targeting both pathways would have potentially additive therapeutic effects [369].

CDK inhibitors: Selective CDK inhibitors inhibit the cell cycle that leads to glial activation and neuronal apoptosis in TBI [370]. These CDK inhibitors are toxic when given chronically, as is done in cancer treatments; however, short-term treatments, such as could be the case for acute TBI, would pose less of an issue. In addition, roscovitine and CR-8 show strong neuroprotective effects when administered as a single dose at a clinically relevant delayed time point [370]. Another TBI study found that CDK inhibitors decreased neuronal death, lesion volume, astroglial scarring, microglial activation, and improved motor and cognition functions [321].

Minocycline: Minocycline is a second-generation tetracycline and has been reported to inhibit microglial activation and subsequent excitotoxicity in TBI [371–374]. Minocycline also inhibited caspase-dependent and independent mitochondrial cell death pathways by preventing release of cyt c in a chronic neurodegeneration model [375]. However, one study found only transient neuroprotection and no change in apoptosis [189]. Minocycline is already FDA approved as an antibiotic, has a long half-life, can readily cross the BBB, and is well tolerated in high doses [376]. In addition, clinical trials of minocycline treatment in acute spinal cord injury found improvements in neurological outcomes and no significant adverse effects [377]. Therefore, animal studies and clinical trials warrant further investigation of minocycline as a possible therapeutic intervention in TBI.

Progesterone: Progesterone treatment leads to reduced edema, neuroinflammation, neuronal excitotoxicity, and apoptosis after TBI in both animal studies and initial clinical trials [378]. Progesterone modulated AQP4 expression on astrocytes and decreased cerebral edema in rats after TBI [379]. Progesterone may also reduce LP and oxidative stress by upregulating antioxidant enzymes, such as SOD [380], and ROS scavengers, such as mitochondrial glutathione [381]. Progesterone also attenuated neuronal excitoxicity by inhibiting voltage-gated calcium channels [382]. Studies have also shown that progesterone may inhibit activity of cyt c and caspase-3, and upregulate anti-apoptotic Bcl-2 proteins [382]. Despite the widespread success in experimental TBI [378], stage III clinical trials found no significant improvements in progesterone-treated TBI patients [383]. Similar to progesterone studies, many seemingly promising TBI treatments also failed in later clinical trials [384]. For example, erythropoietin restored mitochondrial function in TBI, and thereby reduced oxidative stress and inflammation [385] but failed to show significant improvements in TBI patients in clinical trials [386]. These failures may be attributed to many different factors including mechanistic and physiological differences in animal systems, heterogeneity of the injuries in patients, and/or issues with dosage and durations of treatment [387]. For example, no drug optimization studies were done prior to the phase III progsterone trials even though pre-clinical trials found many parameters that were critical for treatment effectiveness [384]. Studies have stated that growth hormone replacement therapy, in patients with post-traumatic hypopituitarism, partially reversed cognitive impairment, and improved processing speed and memory after TBI [388,389]. Incidence of endocrine insufficiency/failure after TBI is quite high but its cognitive symptoms are often mistaken for signs of residual injury [390]. These hormone deficiencies may be easily treated with hormone replacement therapy, and clinical symptoms respond well to treatment. Greater awareness of hypopituitarism and adrenal and endocrine failure following TBI is desirable in order to better manage the chronic effects of TBI.

microRNAs (miRNAs): Widespread research on therapies utilizing miRNAs, small non-coding RNA molecules that regulate gene expression, is performed in many different disease research fields [391]. miRNAs have demonstrated great therapeutic potential; however, research on these therapies is still in its infancy and the part of miRNAs in secondary injury in TBI remains largely unexplored [392]. Recently, Sabirzhanov et al. found that upregulation of miR-711 in TBI coincided with downregulation of the pro-survival protein Akt and subsequent activation of apoptotic PUMA and BIM, and cytosolic translocation of cytochrome c and AIF [392]. Inhibitor of miR-711 decreased apoptosis, restored Akt, and attenuated long-term neurological deficits after TBI. Another miRNA, miR-21 repressed apoptosis and supported angiogenesis by increasing Bcl-2 expression, inhibiting BAX and caspase-3, and activating PTEN-Akt signaling [393,394]. Going forward, miRNA therapies may be a promising future direction for the development of novel interventions in TBI by enabling direct targeting and inhibition of cell loss and concurrent targeting of multiple effectors.

6. Conclusions

In conclusion, a comprehensive understanding of TBI pathophysiology will allow for the development of effective drugs or drug combinations that target multiple secondary injury mechanisms and can be administered during optimal therapeutic windows. This approach could augment the bench to bedside translation of neuroprotective treatments. A few noteworthy treatment strategies have made it to the clinical trial stage for the treatment of brain injury, including tranexamic acid, CDP-choline, methylphenidate, NA-1, CBD, and non-invasive RIC. However, there is still a lack of effective therapeutics for TBI, with treatment mainly consisting of emergency surgeries, maintaining ICP, and rehabilitation therapies. The heterogeneous nature and complex pathophysiology of TBI necessitates a combination of therapies to ameliorate secondary injury and post-traumatic deficits. The fact that neurons and other supporting cells, such as astrocytes, microglia, oligodendrocytes, and the brain vasculature can all undergo degeneration quickly after trauma in the injured brain, further complicates its management. Astrogliosis and neuroinflammation are key secondary injury events that contribute to neurological deficits and even to chronic neurodegeneration after TBI. Moreover, death of non-neuronal cells can compromise recovery and hinder neurotransmission. Future therapeutic strategies should be focused on secondary injury, aiming to minimize detrimental events such as neuroinflammation and create optimal conditions for regeneration and repair post-injury. Additionally, the classification of patients according to GCS score alone may not be an effective method for patient inclusion in clinical trials. Advanced analysis of brain imaging has emerged as a vital tool for identifying progression of disease and efficacy of treatment clinically. Therefore, utilization of novel imaging methods and biomarkers, in addition to GCS scores, may be more accurate criteria for patient recruitment in clinical trials of neuroprotective medications.

Author Contributions: A.J., M.B., M.A., K.M.D., and K.V. drafted the manuscript. R.V.G., M.W., S.M., P.A., J.R.V., and F.L.V. provided intellectual input. All authors have read and agreed to the published version of the manuscript.

Funding: Authors' research is supported by grants from the National Institutes of Neurological Diseases and Stroke (NS114560 to KV, NS065172, NS097825 to KMD and NS110378 to BB/KMD), National Institutes of Child Health and Development (HD094606 to KV), AURI (MCGFD08343 to KV), and American Heart Association (GRNT33700286 to KMD).

Acknowledgments: The authors thank Colby Zahn for illustration provided in this review.

Conflicts of Interest: Authors declare no financial or competing conflict of interest. The funders had no role in the design of the study; in the collection, analyses, or interpretation of data; in the writing of the manuscript, or in the decision to publish the results.

References

1. Dewan, M.C.; Rattani, A.; Gupta, S.; Baticulon, R.E.; Hung, Y.C.; Punchak, M.; Agrawal, A.; Adeleye, A.O.; Shrime, M.G.; Rubiano, A.M.; et al. Estimating the global incidence of traumatic brain injury. *J. Neurosurg.* **2018**. [CrossRef] [PubMed]

2. Hutchinson, P.J.; O'Connell, M.T.; Rothwell, N.J.; Hopkins, S.J.; Nortje, J.; Carpenter, K.L.; Timofeev, I.; Al-Rawi, P.G.; Menon, D.K.; Pickard, J.D. Inflammation in human brain injury: Intracerebral concentrations of IL-1alpha, IL-1beta, and their endogenous inhibitor IL-1ra. *J. Neurotrauma* **2007**, *24*, 1545–1557. [CrossRef] [PubMed]

3. Langlois, J.A.; Rutland-Brown, W.; Thomas, K.E. Traumatic Brain Injury in the United States: Emergency Department Visits, Hospitalizations, and Deaths. Available online: http://www.ncdsv.org/images/CDC_TBIintheUSEDVisitsHospitalizationsAndDeaths_2006.pdf (accessed on 1 September 2020).

4. Teasdale, G.; Jennett, B. Assessment of coma and impaired consciousness. A practical scale. *Lancet (Lond. Engl.)* **1974**, *2*, 81–84. [CrossRef]

5. Wijdicks, E.F.; Bamlet, W.R.; Maramattom, B.V.; Manno, E.M.; McClelland, R.L. Validation of a new coma scale: The FOUR score. *Ann. Neurol.* **2005**, *58*, 585–593. [CrossRef] [PubMed]

6. Loane, D.J.; Faden, A.I. Neuroprotection for traumatic brain injury: Translational challenges and emerging therapeutic strategies. *Trends Pharmacol. Sci.* **2010**, *31*, 596–604. [CrossRef] [PubMed]

7. Saatman, K.E.; Duhaime, A.C.; Bullock, R.; Maas, A.I.; Valadka, A.; Manley, G.T.; Workshop Scientific, T.; Advisory Panel, M. Classification of traumatic brain injury for targeted therapies. *J. Neurotrauma* **2008**, *25*, 719–738. [CrossRef]

8. Ankarcrona, M.; Dypbukt, J.M.; Bonfoco, E.; Zhivotovsky, B.; Orrenius, S.; Lipton, S.A.; Nicotera, P. Glutamate-induced neuronal death: A succession of necrosis or apoptosis depending on mitochondrial function. *Neuron* **1995**, *15*, 961–973. [CrossRef]

9. Whalen, M.J.; Dalkara, T.; You, Z.; Qiu, J.; Bermpohl, D.; Mehta, N.; Suter, B.; Bhide, P.G.; Lo, E.H.; Ericsson, M.; et al. Acute plasmalemma permeability and protracted clearance of injured cells after controlled cortical impact in mice. *J. Cereb. Blood Flow Metab.* **2008**, *28*, 490–505. [CrossRef]

10. Mbye, L.H.; Keles, E.; Tao, L.; Zhang, J.; Chung, J.; Larvie, M.; Koppula, R.; Lo, E.H.; Whalen, M.J. Kollidon VA64, a membrane-resealing agent, reduces histopathology and improves functional outcome after controlled cortical impact in mice. *J. Cereb. Blood Flow Metab.* **2012**, *32*, 515–524. [CrossRef]

11. Miller, B.F.; Keles, E.; Tien, L.; Zhang, J.; Kaplan, D.; Lo, E.H.; Whalen, M.J. The pharmacokinetics and pharmacodynamics of Kollidon VA64 dissociate its protective effects from membrane resealing after controlled cortical impact in mice. *J. Cereb. Blood Flow Metab.* **2014**, *34*, 1347–1353. [CrossRef]

12. Faden, A.I.; Demediuk, P.; Panter, S.S.; Vink, R. The role of excitatory amino acids and NMDA receptors in traumatic brain injury. *Science* **1989**, *244*, 798–800. [CrossRef] [PubMed]

13. Narayan, R.K.; Michel, M.E.; Ansell, B.; Baethmann, A.; Biegon, A.; Bracken, M.B.; Bullock, M.R.; Choi, S.C.; Clifton, G.L.; Contant, C.F. Clinical trials in head injury. *J. Neurotrauma* **2002**, *19*, 503–557. [CrossRef]

14. Laird, M.D.; Shields, J.S.; Sukumari-Ramesh, S.; Kimbler, D.E.; Fessler, R.D.; Shakir, B.; Youssef, P.; Yanasak, N.; Vender, J.R.; Dhandapani, K.M. High mobility group box protein-1 promotes cerebral edema after traumatic brain injury via activation of toll-like receptor 4. *Glia* **2014**, *62*, 26–38. [CrossRef] [PubMed]

15. Ikonomidou, C.; Turski, L. Why did NMDA receptor antagonists fail clinical trials for stroke and traumatic brain injury? *Lancet Neurol.* **2002**, *1*, 383–386. [CrossRef]

16. Palmer, A.M.; Marion, D.W.; Botscheller, M.L.; Bowen, D.M.; DeKosky, S.T. Increased transmitter amino acid concentration in human ventricular CSF after brain trauma. *Neuroreport* **1994**, *6*, 153–156. [CrossRef] [PubMed]

17. Hong, Z.; Xinding, Z.; Tianlin, Z.; Liren, C. Excitatory Amino Acids in Cerebrospinal Fluid of Patients with Acute Head Injuries. *Clin. Chem.* **2001**, *47*, 1458. [CrossRef]

18. Baker, A.J.; Moulton, R.J.; MacMillan, V.H.; Shedden, P.M. Excitatory amino acids in cerebrospinal fluid following traumatic brain injury in humans. *J. Neurosurg.* **1993**, *79*, 369–372. [CrossRef]

19. Chamoun, R.; Suki, D.; Gopinath, S.P.; Goodman, J.C.; Robertson, C. Role of extracellular glutamate measured by cerebral microdialysis in severe traumatic brain injury. *J. Neurosurg.* **2010**, *113*, 564–570. [CrossRef]

20. Werner, C.; Engelhard, K. Pathophysiology of traumatic brain injury. *Br. J. Anaesth.* **2007**, *99*, 4–9. [CrossRef]

21. Greve, M.W.; Zink, B.J. Pathophysiology of traumatic brain injury. *Mt. Sinai J. Med. N.Y.* **2009**, *76*, 97–104. [CrossRef]

22. Tavalin, S.J.; Ellis, E.F.; Satin, L.S. Mechanical perturbation of cultured cortical neurons reveals a stretch-induced delayed depolarization. *J. Neurophysiol.* **1995**, *74*, 2767–2773. [CrossRef] [PubMed]

23. Parsons, M.P.; Raymond, L.A. Extrasynaptic NMDA receptor involvement in central nervous system disorders. *Neuron* **2014**, *82*, 279–293. [CrossRef] [PubMed]

24. Hardingham, G.E.; Fukunaga, Y.; Bading, H. Extrasynaptic NMDARs oppose synaptic NMDARs by triggering CREB shut-off and cell death pathways. *Nat. Neurosci.* **2002**, *5*, 405–414. [CrossRef] [PubMed]

25. Hardingham, G.E.; Bading, H. Synaptic versus extrasynaptic NMDA receptor signalling: Implications for neurodegenerative disorders. *Nat. Rev. Neurosci.* **2010**, *11*, 682–696. [CrossRef] [PubMed]

26. Sanz-Clemente, A.; Nicoll, R.A.; Roche, K.W. Diversity in NMDA receptor composition: Many regulators, many consequences. *Neuroscientist* **2013**, *19*, 62–75. [CrossRef]

27. Tovar, K.R.; Westbrook, G.L. The incorporation of NMDA receptors with a distinct subunit composition at nascent hippocampal synapses in vitro. *J. Neurosci.* **1999**, *19*, 4180–4188. [CrossRef]

28. Patel, T.P.; Ventre, S.C.; Geddes-Klein, D.; Singh, P.K.; Meaney, D.F. Single-neuron NMDA receptor phenotype influences neuronal rewiring and reintegration following traumatic injury. *J. Neurosci.* **2014**, *34*, 4200–4213. [CrossRef]

29. Bush, T.G.; Puvanachandra, N.; Horner, C.H.; Polito, A.; Ostenfeld, T.; Svendsen, C.N.; Mucke, L.; Johnson, M.H.; Sofroniew, M.V. Leukocyte infiltration, neuronal degeneration, and neurite outgrowth after ablation of scar-forming, reactive astrocytes in adult transgenic mice. *Neuron* **1999**, *23*, 297–308. [CrossRef]

30. Phillips, L.L.; Lyeth, B.G.; Hamm, R.J.; Reeves, T.M.; Povlishock, J.T. Glutamate antagonism during secondary deafferentation enhances cognition and axo-dendritic integrity after traumatic brain injury. *Hippocampus* **1998**, *8*, 390–401. [CrossRef]

31. Faden, A.I.; O'Leary, D.M.; Fan, L.; Bao, W.; Mullins, P.G.; Movsesyan, V.A. Selective blockade of the mGluR1 receptor reduces traumatic neuronal injury in vitro and improvesoOutcome after brain trauma. *Exp. Neurol.* **2001**, *167*, 435–444. [CrossRef]

32. McIntosh, T.K.; Vink, R.; Soares, H.; Hayes, R.; Simon, R. Effects of the N-methyl-D-aspartate receptor blocker MK-801 on neurologic function after experimental brain injury. *J. Neurotrauma* **1989**, *6*, 247–259. [CrossRef]

33. Muir, K.W. Glutamate-based therapeutic approaches: Clinical trials with NMDA antagonists. *Curr. Opin. Pharm.* **2006**, *6*, 53–60. [CrossRef] [PubMed]

34. Roesler, R.; Quevedo, J.; Schroder, N. Is it time to conclude that NMDA antagonists have failed? *Lancet Neurol.* **2003**, *2*, 13. [CrossRef]

35. Hoyte, L.; Barber, P.A.; Buchan, A.M.; Hill, M.D. The rise and fall of NMDA antagonists for ischemic stroke. *Curr. Mol. Med.* **2004**, *4*, 131–136. [CrossRef] [PubMed]

36. Rao, V.L.; Dogan, A.; Todd, K.G.; Bowen, K.K.; Dempsey, R.J. Neuroprotection by memantine, a non-competitive NMDA receptor antagonist after traumatic brain injury in rats. *Brain Res.* **2001**, *911*, 96–100. [CrossRef] [PubMed]

37. Xia, P.; Chen, H.S.; Zhang, D.; Lipton, S.A. Memantine preferentially blocks extrasynaptic over synaptic NMDA receptor currents in hippocampal autapses. *J. Neurosci.* **2010**, *30*, 11246–11250. [CrossRef] [PubMed]

38. Mokhtari, M.; Nayeb-Aghaei, H.; Kouchek, M.; Miri, M.M.; Goharani, R.; Amoozandeh, A.; Akhavan Salamat, S.; Sistanizad, M. Effect of Memantine on Serum Levels of Neuron-Specific Enolase and on the Glasgow Coma Scale in Patients With Moderate Traumatic Brain Injury. *J. Clin. Pharm.* **2018**, *58*, 42–47. [CrossRef]

39. Wang, C.Q.; Ye, Y.; Chen, F.; Han, W.C.; Sun, J.M.; Lu, X.; Guo, R.; Cao, K.; Zheng, M.J.; Liao, L.C. Posttraumatic administration of a sub-anesthetic dose of ketamine exerts neuroprotection via attenuating inflammation and autophagy. *Neuroscience* **2017**, *343*, 30–38. [CrossRef]

40. Hertle, D.N.; Dreier, J.P.; Woitzik, J.; Hartings, J.A.; Bullock, R.; Okonkwo, D.O.; Shutter, L.A.; Vidgeon, S.; Strong, A.J.; Kowoll, C.; et al. Effect of analgesics and sedatives on the occurrence of spreading depolarizations accompanying acute brain injury. *Brain* **2012**, *135*, 2390–2398. [CrossRef]

41. McIntosh, T.K.; Faden, A.I.; Yamakami, I.; Vink, R. Magnesium deficiency exacerbates and pretreatment improves outcome following traumatic brain injury in rats: 31P magnetic resonance spectroscopy and behavioral studies. *J. Neurotrauma* **1988**, *5*, 17–31. [CrossRef]

42. Arango, M.F.; Bainbridge, D. Magnesium for acute traumatic brain injury. *Cochrane Database Syst. Rev.* **2008**. [CrossRef] [PubMed]

43. Li, W.; Bai, Y.A.; Li, Y.J.; Liu, K.G.; Wang, M.D.; Xu, G.Z.; Shang, H.L.; Li, Y.F. Magnesium sulfate for acute traumatic brain injury. *J. Craniofac. Surg.* **2015**, *26*, 393–398. [CrossRef] [PubMed]

44. Williams, K. Ifenprodil discriminates subtypes of the N-methyl-D-aspartate receptor: Selectivity and mechanisms at recombinant heteromeric receptors. *Mol. Pharm.* **1993**, *44*, 851–859.

45. Dempsey, R.J.; Başkaya, M.K.; Doğan, A. Attenuation of brain edema, blood-brain barrier breakdown, and injury volume by ifenprodil, a polyamine-site N-methyl-D-aspartate receptor antagonist, after experimental traumatic brain injury in rats. *Neurosurgery* **2000**, *47*, 399–404; discussion 404–406. [CrossRef] [PubMed]

46. Maneshi, M.M.; Maki, B.; Gnanasambandam, R.; Belin, S.; Popescu, G.K.; Sachs, F.; Hua, S.Z. Mechanical stress activates NMDA receptors in the absence of agonists. *Sci. Rep.* **2017**, *7*, 39610. [CrossRef]

47. Bigford, G.E.; Alonso, O.F.; Dietrich, D.; Keane, R.W. A novel protein complex in membrane rafts linking the NR2B glutamate receptor and autophagy is disrupted following traumatic brain injury. *J. Neurotrauma* **2009**, *26*, 703–720. [CrossRef]

48. Spaethling, J.; Le, L.; Meaney, D.F. NMDA receptor mediated phosphorylation of GluR1 subunits contributes to the appearance of calcium-permeable AMPA receptors after mechanical stretch injury. *Neurobiol. Dis.* **2012**, *46*, 646–654. [CrossRef]

49. Ferrario, C.R.; Ndukwe, B.O.; Ren, J.; Satin, L.S.; Goforth, P.B. Stretch injury selectively enhances extrasynaptic, GluN2B-containing NMDA receptor function in cortical neurons. *J. Neurophysiol.* **2013**, *110*, 131–140. [CrossRef]

50. Merchant, R.E.; Bullock, M.R.; Carmack, C.A.; Shah, A.K.; Wilner, K.D.; Ko, G.; Williams, S.A. A double-blind, placebo-controlled study of the safety, tolerability and pharmacokinetics of CP-101,606 in patients with a mild or moderate traumatic brain injury. *Ann. N. Y. Acad. Sci.* **1999**, *890*, 42–50. [CrossRef]

51. Bullock, M.R.; Merchant, R.E.; Carmack, C.A.; Doppenberg, E.; Shah, A.K.; Wilner, K.D.; Ko, G.; Williams, S.A. An open-label study of CP-101,606 in subjects with a severe traumatic head injury or spontaneous intracerebral hemorrhage. *Ann. N. Y. Acad. Sci.* **1999**, *890*, 51–58. [CrossRef]

52. Yurkewicz, L.; Weaver, J.; Bullock, M.R.; Marshall, L.F. The effect of the selective NMDA receptor antagonist traxoprodil in the treatment of traumatic brain injury. *J. Neurotrauma* **2005**, *22*, 1428–1443. [CrossRef] [PubMed]

53. Shohami, E.; Biegon, A. Novel approach to the role of NMDA receptors in traumatic brain injury. *CNS Neurol. Disord. Drug Targets* **2014**, *13*, 567–573. [CrossRef] [PubMed]

54. Funke, L.; Dakoji, S.; Bredt, D.S. Membrane-associated guanylate kinases regulate adhesion and plasticity at cell junctions. *Annu Rev. Biochem.* **2005**, *74*, 219–245. [CrossRef] [PubMed]

55. Christopherson, K.S.; Hillier, B.J.; Lim, W.A.; Bredt, D.S. PSD-95 assembles a ternary complex with the N-methyl-D-aspartic acid receptor and a bivalent neuronal NO synthase PDZ domain. *J. Biol. Chem.* **1999**, *274*, 27467–27473. [CrossRef]

56. Arundine, M.; Aarts, M.; Lau, A.; Tymianski, M. Vulnerability of Central Neurons to Secondary Insults after *In Vitro* Mechanical Stretch. *J. Neurosci.* **2004**, *24*, 8106–8123. [CrossRef]

57. Qu, W.; Liu, N.K.; Wu, X.; Wang, Y.; Xia, Y.; Sun, Y.; Lai, Y.; Li, R.; Shekhar, A.; Xu, X.M. Disrupting nNOS-PSD95 Interaction Improves Neurological and Cognitive Recoveries after Traumatic Brain Injury. *Cereb. Cortex* **2020**, *30*, 3859–3871. [CrossRef]

58. Aarts, M.; Liu, Y.; Liu, L.; Besshoh, S.; Arundine, M.; Gurd, J.W.; Wang, Y.-T.; Salter, M.W.; Tymianski, M. Treatment of Ischemic Brain Damage by Perturbing NMDA Receptor—PSD-95 Protein Interactions. *Science* **2002**, *298*, 846–850. [CrossRef]

59. Cook, D.J.; Teves, L.; Tymianski, M. Treatment of stroke with a PSD-95 inhibitor in the gyrencephalic primate brain. *Nature* **2012**, *483*, 213–217. [CrossRef]

60. Hill, M.D.; Martin, R.H.; Mikulis, D.; Wong, J.H.; Silver, F.L.; Terbrugge, K.G.; Milot, G.; Clark, W.M.; Macdonald, R.L.; Kelly, M.E.; et al. Safety and efficacy of NA-1 in patients with iatrogenic stroke after endovascular aneurysm repair (ENACT): A phase 2, randomised, double-blind, placebo-controlled trial. *Lancet Neurol.* **2012**, *11*, 942–950. [CrossRef]

61. Ballarin, B.; Tymianski, M. Discovery and development of NA-1 for the treatment of acute ischemic stroke. *Acta Pharm. Sin.* **2018**, *39*, 661–668. [CrossRef]

62. Wu, Q.J.; Tymianski, M. Targeting NMDA receptors in stroke: New hope in neuroprotection. *Mol. Brain* **2018**, *11*, 15. [CrossRef]

63. Sen, T.; Gupta, R.; Kaiser, H.; Sen, N. Activation of PERK Elicits Memory Impairment through Inactivation of CREB and Downregulation of PSD95 After Traumatic Brain Injury. *J. Neurosci.* **2017**, *37*, 5900–5911. [CrossRef]

64. Wakade, C.; Sukumari-Ramesh, S.; Laird, M.D.; Dhandapani, K.M.; Vender, J.R. Delayed reduction in hippocampal postsynaptic density protein-95 expression temporally correlates with cognitive dysfunction following controlled cortical impact in mice. *J. Neurosurg.* **2010**, *113*, 1195–1201. [CrossRef] [PubMed]

65. Bach, A.; Clausen, B.H.; Moller, M.; Vestergaard, B.; Chi, C.N.; Round, A.; Sorensen, P.L.; Nissen, K.B.; Kastrup, J.S.; Gajhede, M.; et al. A high-affinity, dimeric inhibitor of PSD-95 bivalently interacts with PDZ1-2 and protects against ischemic brain damage. *Proc. Natl. Acad. Sci. USA* **2012**, *109*, 3317–3322. [CrossRef]

66. Sommer, J.B.; Bach, A.; Mala, H.; Stromgaard, K.; Mogensen, J.; Pickering, D.S. In vitro and in vivo effects of a novel dimeric inhibitor of PSD-95 on excitotoxicity and functional recovery after experimental traumatic brain injury. *Eur. J. Neurosci.* **2017**, *45*, 238–248. [CrossRef] [PubMed]

67. Wang, Y.; Song, J.H.; Denisova, J.V.; Park, W.M.; Fontes, J.D.; Belousov, A.B. Neuronal gap junction coupling is regulated by glutamate and plays critical role in cell death during neuronal injury. *J. Neurosci.* **2012**, *32*, 713–725. [CrossRef]

68. Hartfield, E.M.; Rinaldi, F.; Glover, C.P.; Wong, L.F.; Caldwell, M.A.; Uney, J.B. Connexin 36 expression regulates neuronal differentiation from neural progenitor cells. *Plos ONE* **2011**, *6*, e14746. [CrossRef]

69. Todd, K.L.; Kristan, W.B., Jr.; French, K.A. Gap junction expression is required for normal chemical synapse formation. *J. Neurosci.* **2010**, *30*, 15277–15285. [CrossRef]

70. Amara, S.G.; Fontana, A.C. Excitatory amino acid transporters: Keeping up with glutamate. *Neurochem. Int.* **2002**, *41*, 313–318. [CrossRef]

71. Suchak, S.K.; Baloyianni, N.V.; Perkinton, M.S.; Williams, R.J.; Meldrum, B.S.; Rattray, M. The 'glial' glutamate transporter, EAAT2 (Glt-1) accounts for high affinity glutamate uptake into adult rodent nerve endings. *J. Neurochem.* **2003**, *84*, 522–532. [CrossRef]

72. Li, S.; Stys, P.K. Na(+)-K(+)-ATPase inhibition and depolarization induce glutamate release via reverse Na(+)-dependent transport in spinal cord white matter. *Neuroscience* **2001**, *107*, 675–683. [CrossRef]

73. Beschorner, R.; Dietz, K.; Schauer, N.; Mittelbronn, M.; Schluesener, H.J.; Trautmann, K.; Meyermann, R.; Simon, P. Expression of EAAT1 reflects a possible neuroprotective function of reactive astrocytes and activated microglia following human traumatic brain injury. *Histol. Histopathol.* **2007**, *22*, 515–526. [PubMed]

74. van Landeghem, F.K.; Weiss, T.; Oehmichen, M.; von Deimling, A. Decreased expression of glutamate transporters in astrocytes after human traumatic brain injury. *J. Neurotrauma* **2006**, *23*, 1518–1528. [CrossRef]

75. van Landeghem, F.K.; Stover, J.F.; Bechmann, I.; Bruck, W.; Unterberg, A.; Buhrer, C.; von Deimling, A. Early expression of glutamate transporter proteins in ramified microglia after controlled cortical impact injury in the rat. *Glia* **2001**, *35*, 167–179. [CrossRef] [PubMed]

76. Rao, V.L.; Dogan, A.; Bowen, K.K.; Todd, K.G.; Dempsey, R.J. Antisense knockdown of the glial glutamate transporter GLT-1 exacerbates hippocampal neuronal damage following traumatic injury to rat brain. *Eur. J. Neurosci.* **2001**, *13*, 119–128.

77. Fontana, A.C.; Fox, D.P.; Zoubroulis, A.; Mortensen, O.V.; Raghupathi, R. Neuroprotective Effects of the Glutamate Transporter Activator (R)-(-)-5-methyl-1-nicotinoyl-2-pyrazoline (MS-153) following Traumatic Brain Injury in the Adult Rat. *J. Neurotrauma* **2016**, *33*, 1073–1083. [CrossRef]

78. Gottlieb, M.; Wang, Y.; Teichberg, V.I. Blood-mediated scavenging of cerebrospinal fluid glutamate. *J. Neurochem.* **2003**, *87*, 119–126. [CrossRef]

79. Helms, H.C.C.; Nielsen, C.U.; Waagepetersen, H.S.; Brodin, B. Glutamate Transporters in the Blood-Brain Barrier. *Adv. Neurobiol.* **2017**, *16*, 297–314. [CrossRef]

80. Zlotnik, A.; Gruenbaum, S.E.; Artru, A.A.; Rozet, I.; Dubilet, M.; Tkachov, S.; Brotfain, E.; Klin, Y.; Shapira, Y.; Teichberg, V.I. The neuroprotective effects of oxaloacetate in closed head injury in rats is mediated by its blood glutamate scavenging activity: Evidence from the use of maleate. *J. Neurosurg. Anesth.* **2009**, *21*, 235–241. [CrossRef]

81. Zlotnik, A.; Sinelnikov, I.; Gruenbaum, B.F.; Gruenbaum, S.E.; Dubilet, M.; Dubilet, E.; Leibowitz, A.; Ohayon, S.; Regev, A.; Boyko, M.; et al. Effect of glutamate and blood glutamate scavengers oxaloacetate and pyruvate on neurological outcome and pathohistology of the hippocampus after traumatic brain injury in rats. *Anesthesiology* **2012**, *116*, 73–83. [CrossRef] [PubMed]

82. Boyko, M.; Gruenbaum, S.E.; Gruenbaum, B.F.; Shapira, Y.; Zlotnik, A. Brain to blood glutamate scavenging as a novel therapeutic modality: A review. *J. Neural. Transm. (Vienna)* **2014**, *121*, 971–979. [CrossRef] [PubMed]

83. Campos, F.; Sobrino, T.; Ramos-Cabrer, P.; Argibay, B.; Agulla, J.; Pérez-Mato, M.; Rodríguez-González, R.; Brea, D.; Castillo, J. Neuroprotection by glutamate oxaloacetate transaminase in ischemic stroke: An experimental study. *J. Cereb. Blood Flow Metab.* **2011**, *31*, 1378–1386. [CrossRef] [PubMed]

84. Pérez-Mato, M.; Ramos-Cabrer, P.; Sobrino, T.; Blanco, M.; Ruban, A.; Mirelman, D.; Menendez, P.; Castillo, J.; Campos, F. Human recombinant glutamate oxaloacetate transaminase 1 (GOT1) supplemented with oxaloacetate induces a protective effect after cerebral ischemia. *Cell Death Dis.* **2014**, *5*, e992. [CrossRef] [PubMed]

85. Da Silva-Candal, A.; Pérez-Díaz, A.; Santamaría, M.; Correa-Paz, C.; Rodríguez-Yáñez, M.; Ardá, A.; Pérez-Mato, M.; Iglesias-Rey, R.; Brea, J.; Azuaje, J.; et al. Clinical validation of blood/brain glutamate grabbing in acute ischemic stroke. *Ann. Neurol.* **2018**, *84*, 260–273. [CrossRef]

86. Hoane, M.R.; Wolyniak, J.G.; Akstulewicz, S.L. Administration of riboflavin improves behavioral outcome and reduces edema formation and glial fibrillary acidic protein expression after traumatic brain injury. *J. Neurotrauma* **2005**, *22*, 1112–1122. [CrossRef]

87. Nilsson, P.; Hillered, L.; Ponten, U.; Ungerstedt, U. Changes in cortical extracellular levels of energy-related metabolites and amino acids following concussive brain injury in rats. *J. Cereb. Blood Flow Metab.* **1990**, *10*, 631–637. [CrossRef]

88. Anderson, K.J.; Miller, K.M.; Fugaccia, I.; Scheff, S.W. Regional distribution of fluoro-jade B staining in the hippocampus following traumatic brain injury. *Exp. Neurol.* **2005**, *193*, 125–130. [CrossRef]

89. Sato, M.; Chang, E.; Igarashi, T.; Noble, L.J. Neuronal injury and loss after traumatic brain injury: Time course and regional variability. *Brain Res.* **2001**, *917*, 45–54.

90. Deng, W.; Aimone, J.B.; Gage, F.H. New neurons and new memories: How does adult hippocampal neurogenesis affect learning and memory? *Nat. Reviews. Neurosci.* **2010**, *11*, 339–350. [CrossRef]

91. Mongiat, L.A.; Schinder, A.F. Adult neurogenesis and the plasticity of the dentate gyrus network. *Eur. J. Neurosci.* **2011**, *33*, 1055–1061. [CrossRef]

92. Gao, X.; Deng-Bryant, Y.; Cho, W.; Carrico, K.M.; Hall, E.D.; Chen, J. Selective death of newborn neurons in hippocampal dentate gyrus following moderate experimental traumatic brain injury. *J. Neurosci. Res.* **2008**, *86*, 2258–2270. [CrossRef]

93. Gao, X.; Chen, J. Conditional knockout of brain-derived neurotrophic factor in the hippocampus increases death of adult-born immature neurons following traumatic brain injury. *J. Neurotrauma* **2009**, *26*, 1325–1335. [CrossRef]

94. Zhou, H.; Chen, L.; Gao, X.; Luo, B.; Chen, J. Moderate traumatic brain injury triggers rapid necrotic death of immature neurons in the hippocampus. *J. Neuropathol. Exp. Neurol.* **2012**, *71*, 348–359. [CrossRef] [PubMed]

95. Zhao, Y.L.; Xiang, Q.; Shi, Q.Y.; Li, S.Y.; Tan, L.; Wang, J.T.; Jin, X.G.; Luo, A.L. GABAergic excitotoxicity injury of the immature hippocampal pyramidal neurons' exposure to isoflurane. *Anesth. Analg.* **2011**, *113*, 1152–1160. [CrossRef] [PubMed]

96. Afshari, D.; Moradian, N.; Rezaei, M. Evaluation of the intravenous magnesium sulfate effect in clinical improvement of patients with acute ischemic stroke. *Clin. Neurol. Neurosurg.* **2013**, *115*, 400–404. [CrossRef] [PubMed]

97. Saver, J.L.; Starkman, S.; Eckstein, M.; Stratton, S.J.; Pratt, F.D.; Hamilton, S.; Conwit, R.; Liebeskind, D.S.; Sung, G.; Kramer, I.; et al. Prehospital use of magnesium sulfate as neuroprotection in acute stroke. *N. Engl. J. Med.* **2015**, *372*, 528–536. [CrossRef] [PubMed]

98. Khatri, N.; Thakur, M.; Pareek, V.; Kumar, S.; Sharma, S.; Datusalia, A.K. Oxidative Stress: Major Threat in Traumatic Brain Injury. *CNS Neurol. Disord. Drug Targets* **2018**, *17*, 689–695. [CrossRef] [PubMed]

99. Angeloni, C.; Prata, C.; Vieceli Dalla Sega, F.; Piperno, R.; Hrelia, S. Traumatic Brain Injury and NADPH Oxidase: A Deep Relationship. *Oxidative Med. Cell. Longev.* **2015**, *2015*, 370312. [CrossRef] [PubMed]

100. Toth, P.; Szarka, N.; Farkas, E.; Ezer, E.; Czeiter, E.; Amrein, K.; Ungvari, Z.I.; Hartings, J.A.; Buki, A.; Koller, A. Traumatic brain injury-induced autoregulatory dysfunction and spreading depression-related neurovascular uncoupling: Pathomechanism and therapeutic implications. *Am. J. Physiol. Heart Circ. Physiol.* **2016**. [CrossRef]

101. Veenith, T.V.; Carter, E.L.; Geeraerts, T.; Grossac, J.; Newcombe, V.F.; Outtrim, J.; Gee, G.S.; Lupson, V.; Smith, R.; Aigbirhio, F.I.; et al. Pathophysiologic Mechanisms of Cerebral Ischemia and Diffusion Hypoxia in Traumatic Brain Injury. *JAMA Neurol.* **2016**, *73*, 542–550. [CrossRef]

102. Ansari, M.A.; Roberts, K.N.; Scheff, S.W. Oxidative stress and modification of synaptic proteins in hippocampus after traumatic brain injury. *Free Radic. Biol. Med.* **2008**, *45*, 443–452. [CrossRef] [PubMed]

103. Cornelius, C.; Crupi, R.; Calabrese, V.; Graziano, A.; Milone, P.; Pennisi, G.; Radak, Z.; Calabrese, E.J.; Cuzzocrea, S. Traumatic brain injury: Oxidative stress and neuroprotection. *Antioxid. Redox Signal.* **2013**, *19*, 836–853. [CrossRef] [PubMed]

104. Readnower, R.D.; Chavko, M.; Adeeb, S.; Conroy, M.D.; Pauly, J.R.; McCarron, R.M.; Sullivan, P.G. Increase in blood-brain barrier permeability, oxidative stress, and activated microglia in a rat model of blast-induced traumatic brain injury. *J. Neurosci. Res.* **2010**, *88*, 3530–3539. [CrossRef] [PubMed]

105. DeWitt, D.S.; Prough, D.S. Blast-induced brain injury and posttraumatic hypotension and hypoxemia. *J. Neurotrauma* **2009**, *26*, 877–887. [CrossRef]

106. Vuceljic, M.; Zunic, G.; Romic, P.; Jevtic, M. Relation between both oxidative and metabolic-osmotic cell damages and initial injury severity in bombing casualties. *Vojnosanit. Pregl.* **2006**, *63*, 545–551. [CrossRef]

107. Tran, L.V. Understanding the pathophysiology of traumatic brain injury and the mechanisms of action of neuroprotective interventions. *J. Trauma Nurs.* **2014**, *21*, 30–35. [CrossRef]

108. Povlishock, J.T.; Kontos, H.A. Continuing axonal and vascular change following experimental brain trauma. *Cent. Nerv. Syst. Trauma* **1985**, *2*, 285–298. [CrossRef]

109. Kontos, H.A.; Wei, E.P. Superoxide production in experimental brain injury. *J. Neurosurg.* **1986**, *64*, 803–807. [CrossRef]

110. Halliwell, B.; Gutteridge, J. *Free Radicals in Biology and Medicine*; Oxford University Press: Oxford, UK, 2007.

111. Marklund, S.L.; Westman, N.G.; Lundgren, E.; Roos, G. Copper- and zinc-containing superoxide dismutase, manganese-containing superoxide dismutase, catalase, and glutathione peroxidase in normal and neoplastic human cell lines and normal human tissues. *Cancer Res.* **1982**, *42*, 1955–1961.

112. Smith, S.L.; Andrus, P.K.; Zhang, J.R.; Hall, E.D. Direct measurement of hydroxyl radicals, lipid peroxidation, and blood-brain barrier disruption following unilateral cortical impact head injury in the rat. *J. Neurotrauma* **1994**, *11*, 393–404. [CrossRef]

113. Kontos, H.A.; Povlishock, J.T. Oxygen radicals in brain injury. *Cent. Nerv. Syst. Trauma* **1986**, *3*, 257–263. [CrossRef] [PubMed]

114. Chan, P.H.; Epstein, C.J.; Li, Y.; Huang, T.T.; Carlson, E.; Kinouchi, H.; Yang, G.; Kamii, H.; Mikawa, S.; Kondo, T.; et al. Transgenic mice and knockout mutants in the study of oxidative stress in brain injury. *J. Neurotrauma* **1995**, *12*, 815–824. [CrossRef] [PubMed]

115. Mikawa, S.; Kinouchi, H.; Kamii, H.; Gobbel, G.T.; Chen, S.F.; Carlson, E.; Epstein, C.J.; Chan, P.H. Attenuation of acute and chronic damage following traumatic brain injury in copper, zinc-superoxide dismutase transgenic mice. *J. Neurosurg.* **1996**, *85*, 885–891. [CrossRef]

116. Lewen, A.; Fujimura, M.; Sugawara, T.; Matz, P.; Copin, J.C.; Chan, P.H. Oxidative stress-dependent release of mitochondrial cytochrome c after traumatic brain injury. *J. Cereb. Blood Flow Metab.* **2001**, *21*, 914–920. [CrossRef] [PubMed]

117. Lewen, A.; Sugawara, T.; Gasche, Y.; Fujimura, M.; Chan, P.H. Oxidative cellular damage and the reduction of APE/Ref-1 expression after experimental traumatic brain injury. *Neurobiol. Dis.* **2001**, *8*, 380–390. [CrossRef]

118. Xiong, Y.; Shie, F.S.; Zhang, J.; Lee, C.P.; Ho, Y.S. Prevention of mitochondrial dysfunction in post-traumatic mouse brain by superoxide dismutase. *J. Neurochem.* **2005**, *95*, 732–744. [CrossRef]

119. Pineda, J.A.; Aono, M.; Sheng, H.; Lynch, J.; Wellons, J.C.; Laskowitz, D.T.; Pearlstein, R.D.; Bowler, R.; Crapo, J.; Warner, D.S. Extracellular superoxide dismutase overexpression improves behavioral outcome from closed head injury in the mouse. *J. Neurotrauma* **2001**, *18*, 625–634. [CrossRef]

120. Muizelaar, J.P.; Marmarou, A.; Young, H.F.; Choi, S.C.; Wolf, A.; Schneider, R.L.; Kontos, H.A. Improving the outcome of severe head injury with the oxygen radical scavenger polyethylene glycol-conjugated superoxide dismutase: A phase II trial. *J. Neurosurg.* **1993**, *78*, 375–382. [CrossRef]

121. Muizelaar, J.P.; Kupiec, J.W.; Rapp, L.A. PEG-SOD after head injury. *J. Neurosurg.* **1995**, *83*, 942. [CrossRef]

122. Aoyama, N.; Katayama, Y.; Kawamata, T.; Maeda, T.; Mori, T.; Yamamoto, T.; Kikuchi, T.; Uwahodo, Y. Effects of antioxidant, OPC-14117, on secondary cellular damage and behavioral deficits following cortical contusion in the rat. *Brain Res.* **2002**, *934*, 117–124. [CrossRef]

123. The Dana Consortium. Safety and tolerability of the antioxidant OPC-14117 in HIV-associated cognitive impairment. The Dana Consortium on the Therapy of HIV Dementia and Related Cognitive Disorders. *Neurology* **1997**, *49*, 142–146. [CrossRef] [PubMed]

124. Zaleska, M.M.; Floyd, R.A. Regional lipid peroxidation in rat brain in vitro: Possible role of endogenous iron. *Neurochem. Res.* **1985**, *10*, 397–410. [CrossRef]

125. Sadrzadeh, S.M.; Graf, E.; Panter, S.S.; Hallaway, P.E.; Eaton, J.W. Hemoglobin. A biologic fenton reagent. *J. Biol. Chem.* **1984**, *259*, 14354–14356. [PubMed]

126. Sadrzadeh, S.M.; Eaton, J.W. Hemoglobin-mediated oxidant damage to the central nervous system requires endogenous ascorbate. *J. Clin. Investig.* **1988**, *82*, 1510–1515. [CrossRef]

127. Long, D.A.; Ghosh, K.; Moore, A.N.; Dixon, C.E.; Dash, P.K. Deferoxamine improves spatial memory performance following experimental brain injury in rats. *Brain Res.* **1996**, *717*, 109–117. [CrossRef]

128. Zhang, L.; Hu, R.; Li, M.; Li, F.; Meng, H.; Zhu, G.; Lin, J.; Feng, H. Deferoxamine attenuates iron-induced long-term neurotoxicity in rats with traumatic brain injury. *Neurol. Sci.* **2013**, *34*, 639–645. [CrossRef]

129. Panter, S.S.; Braughler, J.M.; Hall, E.D. Dextran-coupled deferoxamine improves outcome in a murine model of head injury. *J. Neurotrauma* **1992**, *9*, 47–53. [CrossRef]

130. Daglas, M.; Adlard, P.A. The Involvement of Iron in Traumatic Brain Injury and Neurodegenerative Disease. *Front. Neurosci.* **2018**, *12*, 981. [CrossRef]

131. Khalaf, S.; Ahmad, A.S.; Chamara, K.; Dore, S. Unique Properties Associated with the Brain Penetrant Iron Chelator HBED Reveal Remarkable Beneficial Effects after Brain Trauma. *J. Neurotrauma* **2018**. [CrossRef]

132. Gahm, C.; Holmin, S.; Mathiesen, T. Temporal profiles and cellular sources of three nitric oxide synthase isoforms in the brain after experimental contusion. *Neurosurgery* **2000**, *46*, 169–177. [CrossRef]
133. Cherian, L.; Hlatky, R.; Robertson, C.S. Nitric oxide in traumatic brain injury. *Brain Pathol. (Zur. Switz.)* **2004**, *14*, 195–201. [CrossRef]
134. Toczylowska, B.; Chalimoniuk, M.; Wodowska, M.; Mayzner-Zawadzk, E. Changes in concentration of cerebrospinal fluid components in patients with traumatic brain injury. *Brain Res.* **2006**, *1104*, 183–189. [CrossRef] [PubMed]
135. Clark, R.S.; Kochanek, P.M.; Obrist, W.D.; Wong, H.R.; Billiar, T.R.; Wisniewski, S.R.; Marion, D.W. Cerebrospinal fluid and plasma nitrite and nitrate concentrations after head injury in humans. *Crit. Care Med.* **1996**, *24*, 1243–1251. [CrossRef] [PubMed]
136. Uzan, M.; Tanriover, N.; Bozkus, H.; Gumustas, K.; Guzel, O.; Kuday, C. Nitric oxide (NO) metabolism in the cerebrospinal fluid of patients with severe head injury. Inflammation as a possible cause of elevated no metabolites. *Surg. Neurol.* **2001**, *56*, 350–356. [CrossRef]
137. Mesenge, C.; Verrecchia, C.; Allix, M.; Boulu, R.R.; Plotkine, M. Reduction of the neurological deficit in mice with traumatic brain injury by nitric oxide synthase inhibitors. *J. Neurotrauma* **1996**, *13*, 11–16. [CrossRef] [PubMed]
138. Wada, K.; Chatzipanteli, K.; Busto, R.; Dietrich, W.D. Effects of L-NAME and 7-NI on NOS catalytic activity and behavioral outcome after traumatic brain injury in the rat. *J. Neurotrauma* **1999**, *16*, 203–212. [CrossRef]
139. Lu, Y.C.; Liu, S.; Gong, Q.Z.; Hamm, R.J.; Lyeth, B.G. Inhibition of nitric oxide synthase potentiates hypertension and increases mortality in traumatically brain-injured rats. *Mol. Chem. Neuropathol.* **1997**, *30*, 125–137. [CrossRef]
140. Cherian, L.; Chacko, G.; Goodman, J.C.; Robertson, C.S. Cerebral hemodynamic effects of phenylephrine and L-arginine after cortical impact injury. *Crit. Care Med.* **1999**, *27*, 2512–2517. [CrossRef]
141. Hlatky, R.; Lui, H.; Cherian, L.; Goodman, J.C.; O'Brien, W.E.; Contant, C.F.; Robertson, C.S. The role of endothelial nitric oxide synthase in the cerebral hemodynamics after controlled cortical impact injury in mice. *J. Neurotrauma* **2003**, *20*, 995–1006. [CrossRef]
142. Hlatky, R.; Goodman, J.C.; Valadka, A.B.; Robertson, C.S. Role of nitric oxide in cerebral blood flow abnormalities after traumatic brain injury. *J. Cereb. Blood Flow Metab.* **2003**, *23*, 582–588. [CrossRef]
143. Rangel-Castilla, L.; Ahmed, O.; Goodman, J.C.; Gopinath, S.; Valadka, A.; Robertson, C. L-arginine reactivity in cerebral vessels after severe traumatic brain injury. *Neurol. Res.* **2010**, *32*, 1033–1040. [CrossRef] [PubMed]
144. Robertson, C.S.; Gopinath, S.P.; Valadka, A.B.; Van, M.; Swank, P.R.; Goodman, J.C. Variants of the endothelial nitric oxide gene and cerebral blood flow after severe traumatic brain injury. *J. Neurotrauma* **2011**, *28*, 727–737. [CrossRef]
145. Giannopoulos, S.; Katsanos, A.H.; Tsivgoulis, G.; Marshall, R.S. Statins and cerebral hemodynamics. *J. Cereb. Blood Flow Metab.* **2012**, *32*, 1973–1976. [CrossRef] [PubMed]
146. Cho, H.; Yun, C.W.; Park, W.K.; Kong, J.Y.; Kim, K.S.; Park, Y.; Lee, S.; Kim, B.K. Modulation of the activity of pro-inflammatory enzymes, COX-2 and iNOS, by chrysin derivatives. *Pharmacol. Res.* **2004**, *49*, 37–43. [CrossRef]
147. Khan, A.; Vaibhav, K.; Javed, H.; Tabassum, R.; Ahmed, M.E.; Khan, M.M.; Khan, M.B.; Shrivastava, P.; Islam, F.; Siddiqui, M.S.; et al. 1,8-cineole (eucalyptol) mitigates inflammation in amyloid Beta toxicated PC12 cells: Relevance to Alzheimer's disease. *Neurochem. Res.* **2014**, *39*, 344–352. [CrossRef] [PubMed]
148. Tabassum, R.; Vaibhav, K.; Shrivastava, P.; Khan, A.; Ahmed, M.E.; Ashafaq, M.; Khan, M.B.; Islam, F.; Safhi, M.M.; Islam, F. Perillyl alcohol improves functional and histological outcomes against ischemia-reperfusion injury by attenuation of oxidative stress and repression of COX-2, NOS-2 and NF-kappaB in middle cerebral artery occlusion rats. *Eur. J. Pharmacol.* **2015**, *747*, 190–199. [CrossRef]
149. Vaibhav, K.; Shrivastava, P.; Javed, H.; Khan, A.; Ahmed, M.E.; Tabassum, R.; Khan, M.M.; Khuwaja, G.; Islam, F.; Siddiqui, M.S.; et al. Piperine suppresses cerebral ischemia-reperfusion-induced inflammation through the repression of COX-2, NOS-2, and NF-kappaB in middle cerebral artery occlusion rat model. *Mol. Cell. Biochem.* **2012**, *367*, 73–84. [CrossRef]
150. Wada, K.; Chatzipanteli, K.; Kraydieh, S.; Busto, R.; Dietrich, W.D. Inducible nitric oxide synthase expression after traumatic brain injury and neuroprotection with aminoguanidine treatment in rats. *Neurosurgery* **1998**, *43*, 1427–1436. [CrossRef]

151. Moochhala, S.M.; Md, S.; Lu, J.; Teng, C.H.; Greengrass, C. Neuroprotective role of aminoguanidine in behavioral changes after blast injury. *J. Trauma* **2004**, *56*, 393–403. [CrossRef]

152. Louin, G.; Marchand-Verrecchia, C.; Palmier, B.; Plotkine, M.; Jafarian-Tehrani, M. Selective inhibition of inducible nitric oxide synthase reduces neurological deficit but not cerebral edema following traumatic brain injury. *Neuropharmacology* **2006**, *50*, 182–190. [CrossRef]

153. Stover, J.F.; Belli, A.; Boret, H.; Bulters, D.; Sahuquillo, J.; Schmutzhard, E.; Zavala, E.; Ungerstedt, U.; Schinzel, R.; Tegtmeier, F.; et al. Nitric oxide synthase inhibition with the antipterin VAS203 improves outcome in moderate and severe traumatic brain injury: A placebo-controlled randomized Phase IIa trial (NOSTRA). *J. Neurotrauma* **2014**, *31*, 1599–1606. [CrossRef] [PubMed]

154. Tegtmeier, F.; Schinzel, R.; Beer, R.; Bulters, D.; LeFrant, J.Y.; Sahuquillo, J.; Unterberg, A.; Andrews, P.; Belli, A.; Ibanez, J.; et al. Efficacy of Ronopterin (VAS203) in Patients with Moderate and Severe Traumatic Brain Injury (NOSTRA phase III trial): Study protocol of a confirmatory, placebo-controlled, randomised, double blind, multi-centre study. *Trials* **2020**, *21*, 80. [CrossRef] [PubMed]

155. Pacher, P.; Beckman, J.S.; Liaudet, L. Nitric oxide and peroxynitrite in health and disease. *Physiol. Rev.* **2007**, *87*, 315–424. [CrossRef]

156. Hall, E.D.; Kupina, N.C.; Althaus, J.S. Peroxynitrite scavengers for the acute treatment of traumatic brain injury. *Ann. N. Y. Acad. Sci.* **1999**, *890*, 462–468. [CrossRef] [PubMed]

157. Zhang, R.; Shohami, E.; Beit-Yannai, E.; Bass, R.; Trembovler, V.; Samuni, A. Mechanism of brain protection by nitroxide radicals in experimental model of closed-head injury. *Free Radic. Biol. Med.* **1998**, *24*, 332–340. [CrossRef]

158. Bonini, M.G.; Mason, R.P.; Augusto, O. The Mechanism by which 4-hydroxy-2,2,6,6-tetramethylpiperidene-1-oxyl (tempol) diverts peroxynitrite decomposition from nitrating to nitrosating species. *Chem. Res. Toxicol.* **2002**, *15*, 506–511. [CrossRef]

159. Hall, E.D.; Vaishnav, R.A.; Mustafa, A.G. Antioxidant therapies for traumatic brain injury. *Neurother. J. Am. Soc. Exp. Neurother.* **2010**, *7*, 51–61. [CrossRef]

160. Marklund, N.; Clausen, F.; Lewen, A.; Hovda, D.A.; Olsson, Y.; Hillered, L. alpha-Phenyl-tert-N-butyl nitrone (PBN) improves functional and morphological outcome after cortical contusion injury in the rat. *Acta Neurochir.* **2001**, *143*, 73–81. [CrossRef]

161. Gutteridge, J.M. Lipid peroxidation and antioxidants as biomarkers of tissue damage. *Clin. Chem.* **1995**, *41*, 1819–1828. [CrossRef]

162. Hall, E.D.; Yonkers, P.A.; McCall, J.M.; Braughler, J.M. Effects of the 21-aminosteroid U74006F on experimental head injury in mice. *J. Neurosurg.* **1988**, *68*, 456–461. [CrossRef]

163. McIntosh, T.K.; Thomas, M.; Smith, D.; Banbury, M. The novel 21-aminosteroid U74006F attenuates cerebral edema and improves survival after brain injury in the rat. *J. Neurotrauma* **1992**, *9*, 33–46. [CrossRef] [PubMed]

164. Dimlich, R.V.; Tornheim, P.A.; Kindel, R.M.; Hall, E.D.; Braughler, J.M.; McCall, J.M. Effects of a 21-aminosteroid (U-74006F) on cerebral metabolites and edema after severe experimental head trauma. *Adv. Neurol.* **1990**, *52*, 365–375.

165. Marshall, L.F.; Maas, A.I.; Marshall, S.B.; Bricolo, A.; Fearnside, M.; Iannotti, F.; Klauber, M.R.; Lagarrigue, J.; Lobato, R.; Persson, L.; et al. A multicenter trial on the efficacy of using tirilazad mesylate in cases of head injury. *J. Neurosurg.* **1998**, *89*, 519–525. [CrossRef] [PubMed]

166. Galvani, S.; Coatrieux, C.; Elbaz, M.; Grazide, M.H.; Thiers, J.C.; Parini, A.; Uchida, K.; Kamar, N.; Rostaing, L.; Baltas, M.; et al. Carbonyl scavenger and antiatherogenic effects of hydrazine derivatives. *Free Radic. Biol. Med.* **2008**, *45*, 1457–1467. [CrossRef] [PubMed]

167. Hall, E.D.; Wang, J.A.; Miller, D.M.; Cebak, J.E.; Hill, R.L. Newer pharmacological approaches for antioxidant neuroprotection in traumatic brain injury. *Neuropharmacology* **2019**, *145*, 247–258. [CrossRef] [PubMed]

168. Singh, I.N.; Gilmer, L.K.; Miller, D.M.; Cebak, J.E.; Wang, J.A.; Hall, E.D. Phenelzine mitochondrial functional preservation and neuroprotection after traumatic brain injury related to scavenging of the lipid peroxidation-derived aldehyde 4-hydroxy-2-nonenal. *J. Cereb. Blood Flow Metab.* **2013**, *33*, 593–599. [CrossRef] [PubMed]

169. Cebak, J.E.; Singh, I.N.; Hill, R.L.; Wang, J.A.; Hall, E.D. Phenelzine Protects Brain Mitochondrial Function In Vitro and In Vivo following Traumatic Brain Injury by Scavenging the Reactive Carbonyls 4-Hydroxynonenal and Acrolein Leading to Cortical Histological Neuroprotection. *J. Neurotrauma* **2017**, *34*, 1302–1317. [CrossRef]

170. Baker, G.; Matveychuk, D.; MacKenzie, E.M.; Holt, A.; Wang, Y.; Kar, S. Attenuation of the effects of oxidative stress by the MAO-inhibiting antidepressant and carbonyl scavenger phenelzine. *Chem. Biol. Interact.* **2019**, *304*, 139–147. [CrossRef]

171. Kensler, T.W.; Wakabayashi, N.; Biswal, S. Cell survival responses to environmental stresses via the Keap1-Nrf2-ARE pathway. *Annu. Rev. Pharm. Toxicol.* **2007**, *47*, 89–116. [CrossRef]

172. Jin, W.; Wang, H.; Yan, W.; Zhu, L.; Hu, Z.; Ding, Y.; Tang, K. Role of Nrf2 in protection against traumatic brain injury in mice. *J. Neurotrauma* **2009**, *26*, 131–139. [CrossRef]

173. Hong, Y.; Yan, W.; Chen, S.; Sun, C.R.; Zhang, J.M. The role of Nrf2 signaling in the regulation of antioxidants and detoxifying enzymes after traumatic brain injury in rats and mice. *Acta Pharm. Sin.* **2010**, *31*, 1421–1430. [CrossRef] [PubMed]

174. Jin, W.; Kong, J.; Wang, H.; Wu, J.; Lu, T.; Jiang, J.; Ni, H.; Liang, W. Protective effect of tert-butylhydroquinone on cerebral inflammatory response following traumatic brain injury in mice. *Injury* **2011**, *42*, 714–718. [CrossRef] [PubMed]

175. Lu, X.Y.; Wang, H.D.; Xu, J.G.; Ding, K.; Li, T. Pretreatment with tert-butylhydroquinone attenuates cerebral oxidative stress in mice after traumatic brain injury. *J. Surg. Res.* **2014**, *188*, 206–212. [CrossRef]

176. Miller, D.M.; Singh, I.N.; Wang, J.A.; Hall, E.D. Nrf2-ARE activator carnosic acid decreases mitochondrial dysfunction, oxidative damage and neuronal cytoskeletal degradation following traumatic brain injury in mice. *Exp. Neurol.* **2015**, *264*, 103–110. [CrossRef]

177. Maynard, M.E.; Underwood, E.L.; Redell, J.B.; Zhao, J.; Kobori, N.; Hood, K.N.; Moore, A.N.; Dash, P.K. Carnosic Acid Improves Outcome after Repetitive Mild Traumatic Brain Injury. *J. Neurotrauma* **2019**, *36*, 2147–2152. [CrossRef] [PubMed]

178. Lin, C.; Chao, H.; Li, Z.; Xu, X.; Liu, Y.; Hou, L.; Liu, N.; Ji, J. Melatonin attenuates traumatic brain injury-induced inflammation: A possible role for mitophagy. *J. Pineal Res.* **2016**, *61*, 177–186. [CrossRef]

179. Ding, K.; Xu, J.; Wang, H.; Zhang, L.; Wu, Y.; Li, T. Melatonin protects the brain from apoptosis by enhancement of autophagy after traumatic brain injury in mice. *Neurochem. Int.* **2015**, *91*, 46–54. [CrossRef] [PubMed]

180. Wu, H.; Shao, A.; Zhao, M.; Chen, S.; Yu, J.; Zhou, J.; Liang, F.; Shi, L.; Dixon, B.J.; Wang, Z.; et al. Melatonin attenuates neuronal apoptosis through up-regulation of K($^+$)-Cl($^-$) cotransporter KCC2 expression following traumatic brain injury in rats. *J. Pineal Res.* **2016**, *61*, 241–250. [CrossRef]

181. Luo, C.; Yang, Q.; Liu, Y.; Zhou, S.; Jiang, J.; Reiter, R.J.; Bhattacharya, P.; Cui, Y.; Yang, H.; Ma, H.; et al. The multiple protective roles and molecular mechanisms of melatonin and its precursor N-acetylserotonin in targeting brain injury and liver damage and in maintaining bone health. *Free Radic. Biol. Med.* **2019**, *130*, 215–233. [CrossRef]

182. Ding, K.; Wang, H.; Xu, J.; Li, T.; Zhang, L.; Ding, Y.; Zhu, L.; He, J.; Zhou, M. Melatonin stimulates antioxidant enzymes and reduces oxidative stress in experimental traumatic brain injury: The Nrf2-ARE signaling pathway as a potential mechanism. *Free Radic. Biol. Med.* **2014**, *73*, 1–11. [CrossRef]

183. Barlow, K.M.; Brooks, B.L.; MacMaster, F.P.; Kirton, A.; Seeger, T.; Esser, M.; Crawford, S.; Nettel-Aguirre, A.; Zemek, R.; Angelo, M.; et al. A double-blind, placebo-controlled intervention trial of 3 and 10 mg sublingual melatonin for post-concussion syndrome in youths (PLAYGAME): Study protocol for a randomized controlled trial. *Trials* **2014**, *15*, 271. [CrossRef]

184. Pandya, J.D.; Readnower, R.D.; Patel, S.P.; Yonutas, H.M.; Pauly, J.R.; Goldstein, G.A.; Rabchevsky, A.G.; Sullivan, P.G. N-acetylcysteine amide confers neuroprotection, improves bioenergetics and behavioral outcome following TBI. *Exp. Neurol.* **2014**, *257*, 106–113. [CrossRef]

185. Zhou, Y.; Wang, H.D.; Zhou, X.M.; Fang, J.; Zhu, L.; Ding, K. N-acetylcysteine amide provides neuroprotection via Nrf2-ARE pathway in a mouse model of traumatic brain injury. *Drug Des. Devel.* **2018**, *12*, 4117–4127. [CrossRef] [PubMed]

186. Bhatti, J.; Nascimento, B.; Akhtar, U.; Rhind, S.G.; Tien, H.; Nathens, A.; da Luz, L.T. Systematic Review of Human and Animal Studies Examining the Efficacy and Safety of N-Acetylcysteine (NAC) and N-Acetylcysteine Amide (NACA) in Traumatic Brain Injury: Impact on Neurofunctional Outcome and Biomarkers of Oxidative Stress and Inflammation. *Front. Neurol.* **2017**, *8*, 744. [CrossRef] [PubMed]

187. Lutton, E.M.; Farney, S.K.; Andrews, A.M.; Shuvaev, V.V.; Chuang, G.Y.; Muzykantov, V.R.; Ramirez, S.H. Endothelial Targeted Strategies to Combat Oxidative Stress: Improving Outcomes in Traumatic Brain Injury. *Front. Neurol.* **2019**, *10*, 582. [CrossRef]

188. Morganti-Kossmann, M.C.; Rancan, M.; Otto, V.I.; Stahel, P.F.; Kossmann, T. Role of cerebral inflammation after traumatic brain injury: A revisited concept. *Shock (Augustaga.)* **2001**, *16*, 165–177. [CrossRef]

189. Bye, N.; Habgood, M.D.; Callaway, J.K.; Malakooti, N.; Potter, A.; Kossmann, T.; Morganti-Kossmann, M.C. Transient neuroprotection by minocycline following traumatic brain injury is associated with attenuated microglial activation but no changes in cell apoptosis or neutrophil infiltration. *Exp. Neurol.* **2007**, *204*, 220–233. [CrossRef]

190. Kubes, P.; Ward, P.A. Leukocyte recruitment and the acute inflammatory response. *Brain Pathol. (Zur. Switz.)* **2000**, *10*, 127–135. [CrossRef]

191. Lucas, S.M.; Rothwell, N.J.; Gibson, R.M. The role of inflammation in CNS injury and disease. *Br. J. Pharmacol.* **2006**, *147* (Suppl. 1), S232–S240. [CrossRef]

192. Kreutzberg, G.W. Microglia: A sensor for pathological events in the CNS. *Trends Neurosci.* **1996**, *19*, 312–318. [CrossRef]

193. King, M.D.; Laird, M.D.; Ramesh, S.S.; Youssef, P.; Shakir, B.; Vender, J.R.; Alleyne, C.H.; Dhandapani, K.M. Elucidating novel mechanisms of brain injury following subarachnoid hemorrhage: An emerging role for neuroproteomics. *Neurosurg. Focus* **2010**, *28*, E10. [CrossRef] [PubMed]

194. Braun, M.; Vaibhav, K.; Saad, N.M.; Fatima, S.; Vender, J.R.; Baban, B.; Hoda, M.N.; Dhandapani, K.M. White matter damage after traumatic brain injury: A role for damage associated molecular patterns. *Biochim. Et Biophys. Acta. Mol. Basis Dis.* **2017**, *1863*, 2614–2626. [CrossRef] [PubMed]

195. Rhodes, J. Peripheral immune cells in the pathology of traumatic brain injury? *Curr. Opin. Crit. Care* **2011**, *17*, 122–130. [CrossRef]

196. Kigerl, K.A.; de Rivero Vaccari, J.P.; Dietrich, W.D.; Popovich, P.G.; Keane, R.W. Pattern recognition receptors and central nervous system repair. *Exp. Neurol.* **2014**, *258*, 5–16. [CrossRef] [PubMed]

197. Rothwell, N.J. Annual review prize lecture cytokines - killers in the brain? *J. Physiol.* **1999**, *514 Pt 1*, 3–17. [CrossRef] [PubMed]

198. Wang, C.X.; Shuaib, A. Involvement of inflammatory cytokines in central nervous system injury. *Prog. Neurobiol.* **2002**, *67*, 161–172. [CrossRef]

199. Lu, W.; Gersting, J.A.; Maheshwari, A.; Christensen, R.D.; Calhoun, D.A. Developmental expression of chemokine receptor genes in the human fetus. *Early Hum. Dev.* **2005**, *81*, 489–496. [CrossRef]

200. Dalgard, C.L.; Cole, J.T.; Kean, W.S.; Lucky, J.J.; Sukumar, G.; McMullen, D.C.; Pollard, H.B.; Watson, W.D. The cytokine temporal profile in rat cortex after controlled cortical impact. *Front. Mol. Neurosci.* **2012**, *5*, 6. [CrossRef]

201. Shein, S.L.; Shellington, D.K.; Exo, J.L.; Jackson, T.C.; Wisniewski, S.R.; Jackson, E.K.; Vagni, V.A.; Bayir, H.; Clark, R.S.; Dixon, C.E.; et al. Hemorrhagic shock shifts the serum cytokine profile from pro- to anti-inflammatory after experimental traumatic brain injury in mice. *J. Neurotrauma* **2014**, *31*, 1386–1395. [CrossRef]

202. Redell, J.B.; Moore, A.N.; Grill, R.J.; Johnson, D.; Zhao, J.; Liu, Y.; Dash, P.K. Analysis of functional pathways altered after mild traumatic brain injury. *J. Neurotrauma* **2013**, *30*, 752–764. [CrossRef]

203. White, T.E.; Ford, G.D.; Surles-Zeigler, M.C.; Gates, A.S.; Laplaca, M.C.; Ford, B.D. Gene expression patterns following unilateral traumatic brain injury reveals a local pro-inflammatory and remote anti-inflammatory response. *BMC Genom.* **2013**, *14*, 282. [CrossRef] [PubMed]

204. Vaibhav, K.; Braun, M.; Alverson, K.; Khodadadi, H.; Kutiyanawalla, A.; Ward, A.; Banerjee, C.; Sparks, T.; Malik, A.; Rashid, M.H.; et al. Neutrophil extracellular traps exacerbate neurological deficts after traumatic brain injury. *Sci. Adv.* **2020**. [CrossRef]

205. Braun, M.; Khan, Z.T.; Khan, M.B.; Kumar, M.; Ward, A.; Achyut, B.R.; Arbab, A.S.; Hess, D.C.; Hoda, M.N.; Baban, B.; et al. Selective activation of cannabinoid receptor-2 reduces neuroinflammation after traumatic brain injury via alternative macrophage polarization. *Brain Behav. Immun.* **2018**, *68*, 224–237. [CrossRef] [PubMed]

206. Tweedie, D.; Karnati, H.K.; Mullins, R.; Pick, C.G.; Hoffer, B.J.; Goetzl, E.J.; Kapogiannis, D.; Greig, N.H. Time-dependent cytokine and chemokine changes in mouse cerebral cortex following a mild traumatic brain injury. *ELife* **2020**, *9*. [CrossRef] [PubMed]

207. Helmy, A.; Carpenter, K.L.; Menon, D.K.; Pickard, J.D.; Hutchinson, P.J. The cytokine response to human traumatic brain injury: Temporal profiles and evidence for cerebral parenchymal production. *J. Cereb. Blood Flow Metab.* **2011**, *31*, 658–670. [CrossRef] [PubMed]

208. Helmy, A.; Antoniades, C.A.; Guilfoyle, M.R.; Carpenter, K.L.; Hutchinson, P.J. Principal component analysis of the cytokine and chemokine response to human traumatic brain injury. *Plos ONE* **2012**, *7*, e39677. [CrossRef]

209. Braun, M.; Vaibhav, K.; Saad, N.; Fatima, S.; Brann, D.W.; Vender, J.R.; Wang, L.P.; Hoda, M.N.; Baban, B.; Dhandapani, K.M. Activation of Myeloid TLR4 Mediates T Lymphocyte Polarization after Traumatic Brain Injury. *J. Immunol.* **2017**, *198*, 3615–3626. [CrossRef]

210. Vaibhav, K.; Braun, M.; Khan, M.B.; Fatima, S.; Saad, N.; Shankar, A.; Khan, Z.T.; Harris, R.B.S.; Yang, Q.; Huo, Y.; et al. Remote ischemic post-conditioning promotes hematoma resolution via AMPK-dependent immune regulation. *J. Exp. Med.* **2018**. [CrossRef]

211. Soares, H.D.; Hicks, R.R.; Smith, D.; McIntosh, T.K. Inflammatory leukocytic recruitment and diffuse neuronal degeneration are separate pathological processes resulting from traumatic brain injury. *J. Neurosci.* **1995**, *15*, 8223–8233. [CrossRef]

212. Carlos, T.M.; Clark, R.S.; Franicola-Higgins, D.; Schiding, J.K.; Kochanek, P.M. Expression of endothelial adhesion molecules and recruitment of neutrophils after traumatic brain injury in rats. *J. Leukoc. Biol.* **1997**, *61*, 279–285. [CrossRef]

213. Holmin, S.; Mathiesen, T.; Shetye, J.; Biberfeld, P. Intracerebral inflammatory response to experimental brain contusion. *Acta Neurochir.* **1995**, *132*, 110–119. [CrossRef] [PubMed]

214. Hausmann, R.; Kaiser, A.; Lang, C.; Bohnert, M.; Betz, P. A quantitative immunohistochemical study on the time-dependent course of acute inflammatory cellular response to human brain injury. *Int. J. Leg. Med.* **1999**, *112*, 227–232.

215. Hsieh, C.L.; Kim, C.C.; Ryba, B.E.; Niemi, E.C.; Bando, J.K.; Locksley, R.M.; Liu, J.; Nakamura, M.C.; Seaman, W.E. Traumatic brain injury induces macrophage subsets in the brain. *Eur. J. Immunol.* **2013**, *43*, 2010–2022. [CrossRef]

216. Kelley, B.J.; Lifshitz, J.; Povlishock, J.T. Neuroinflammatory responses after experimental diffuse traumatic brain injury. *J. Neuropathol. Exp. Neurol.* **2007**, *66*, 989–1001. [CrossRef]

217. Cao, T.; Thomas, T.C.; Ziebell, J.M.; Pauly, J.R.; Lifshitz, J. Morphological and genetic activation of microglia after diffuse traumatic brain injury in the rat. *Neuroscience* **2012**, *225*, 65–75. [CrossRef]

218. Semple, B.D.; Kossmann, T.; Morganti-Kossmann, M.C. Role of chemokines in CNS health and pathology: A focus on the CCL2/CCR2 and CXCL8/CXCR2 networks. *J. Cereb. Blood Flow Metab.* **2010**, *30*, 459–473. [CrossRef]

219. Louveau, A.; Smirnov, I.; Keyes, T.J.; Eccles, J.D.; Rouhani, S.J.; Peske, J.D.; Derecki, N.C.; Castle, D.; Mandell, J.W.; Lee, K.S.; et al. Structural and functional features of central nervous system lymphatic vessels. *Nature* **2015**, *523*, 337. [CrossRef]

220. Aspelund, A.; Antila, S.; Proulx, S.T.; Karlsen, T.V.; Karaman, S.; Detmar, M.; Wiig, H.; Alitalo, K. A dural lymphatic vascular system that drains brain interstitial fluid and macromolecules. *J. Exp. Med.* **2015**, *212*, 991–999. [CrossRef]

221. Holmin, S.; Schalling, M.; Hojeberg, B.; Nordqvist, A.C.; Skeftruna, A.K.; Mathiesen, T. Delayed cytokine expression in rat brain following experimental contusion. *J. Neurosurg.* **1997**, *86*, 493–504. [CrossRef]

222. Oehmichen, M.; Jakob, S.; Mann, S.; Saternus, K.S.; Pedal, I.; Meissner, C. Macrophage subsets in mechanical brain injury (MBI)–a contribution to timing of MBI based on immunohistochemical methods: A pilot study. *Leg. Med.* **2009**, *11*, 118–124. [CrossRef]

223. Walsh, J.T.; Zheng, J.; Smirnov, I.; Lorenz, U.; Tung, K.; Kipnis, J. Regulatory T Cells in Central Nervous System Injury: A Double-Edged Sword. *J. Immunol.* **2014**, *193*, 5013–5022. [CrossRef] [PubMed]

224. Pizzolla, A.; Gelderman, K.A.; Hultqvist, M.; Vestberg, M.; Gustafsson, K.; Mattsson, R.; Holmdahl, R. CD68-expressing cells can prime T cells and initiate autoimmune arthritis in the absence of reactive oxygen species. *Eur. J. Immunol.* **2011**, *41*, 403–412. [CrossRef] [PubMed]

225. Vergelli, M.; Pinet, V.; Vogt, A.B.; Kalbus, M.; Malnati, M.; Riccio, P.; Long, E.O.; Martin, R. HLA-DR-restricted presentation of purified myelin basic protein is independent of intracellular processing. *Eur. J. Immunol.* **1997**, *27*, 941–951. [CrossRef] [PubMed]

226. Berger, T.; Rubner, P.; Schautzer, F.; Egg, R.; Ulmer, H.; Mayringer, I.; Dilitz, E.; Deisenhammer, F.; Reindl, M. Antimyelin antibodies as a predictor of clinically definite multiple sclerosis after a first demyelinating event. *N. Engl. J. Med.* **2003**, *349*, 139–145. [CrossRef]

227. Tobin, R.P.; Mukherjee, S.; Kain, J.M.; Rogers, S.K.; Henderson, S.K.; Motal, H.L.; Newell Rogers, M.K.; Shapiro, L.A. Traumatic brain injury causes selective, CD74-dependent peripheral lymphocyte activation that exacerbates neurodegeneration. *Acta Neuropathol. Commun.* **2014**, *2*, 143. [CrossRef]

228. Mosley, R.L.; Hutter-Saunders, J.A.; Stone, D.K.; Gendelman, H.E. Inflammation and Adaptive Immunity in Parkinson's Disease. *Cold Spring Harb. Perspect. Med.* **2012**, *2*. [CrossRef]

229. Hickey, W.F.; Hsu, B.L.; Kimura, H. T-lymphocyte entry into the central nervous system. *J. Neurosci. Res.* **1991**, *28*, 254–260. [CrossRef]

230. Holmin, S.; Söderlund, J.; Biberfeld, P.; Mathiesen, T. Intracerebral Inflammation after Human Brain Contusion. *Neurosurgery* **1998**, *42*, 291–298. [CrossRef]

231. Hua, R.; Mao, S.S.; Zhang, Y.M.; Chen, F.X.; Zhou, Z.H.; Liu, J.Q. Effects of pituitary adenylate cyclase activating polypeptide on CD4($^+$)/CD8($^+$) T cell levels after traumatic brain injury in a rat model. *World J. Emerg. Med.* **2012**, *3*, 294–298. [CrossRef]

232. Gutcher, I.; Becher, B. APC-derived cytokines and T cell polarization in autoimmune inflammation. *J. Clin. Investig.* **2007**, *117*, 1119–1127. [CrossRef]

233. Kabelitz, D.; Medzhitov, R. Innate immunity-cross-talk with adaptive immunity through pattern recognition receptors and cytokines. *Curr. Opin. Immunol.* **2007**, *19*, 1–3. [CrossRef] [PubMed]

234. Fu, H.; Wang, A.; Mauro, C.; Marelli-Berg, F. T lymphocyte trafficking: Molecules and mechanisms. *Front. Biosci.* **2013**, *18*, 422–440.

235. Murphy, A.C.; Lalor, S.J.; Lynch, M.A.; Mills, K.H. Infiltration of Th1 and Th17 cells and activation of microglia in the CNS during the course of experimental autoimmune encephalomyelitis. *Brain Behav. Immun.* **2010**, *24*, 641–651. [CrossRef]

236. Rostami, A.; Ciric, B. Role of Th17 cells in the pathogenesis of CNS inflammatory demyelination. *J. Neurol. Sci.* **2013**, *333*, 76–87. [CrossRef] [PubMed]

237. Yang, D.; Sun, Y.-Y.; Bhaumik, S.K.; Li, Y.; Baumann, J.M.; Lin, X.; Zhang, Y.; Lin, S.-H.; Dunn, R.S.; Liu, C.-Y.; et al. Blocking Lymphocyte Trafficking with FTY720 Prevents Inflammation-Sensitized Hypoxic–Ischemic Brain Injury in Newborns. *J. Neurosci.* **2014**, *34*, 16467–16481. [CrossRef]

238. Baxi, E.G.; DeBruin, J.; Tosi, D.M.; Grishkan, I.V.; Smith, M.D.; Kirby, L.A.; Strasburger, H.J.; Fairchild, A.N.; Calabresi, P.A.; Gocke, A.R. Transfer of myelin-reactive th17 cells impairs endogenous remyelination in the central nervous system of cuprizone-fed mice. *J. Neurosci.* **2015**, *35*, 8626–8639. [CrossRef]

239. Cao, Y.; Goods, B.A.; Raddassi, K.; Nepom, G.T.; Kwok, W.W.; Love, J.C.; Hafler, D.A. Functional inflammatory profiles distinguish myelin-reactive T cells from patients with multiple sclerosis. *Sci. Transl. Med.* **2015**, *7*, 287ra274. [CrossRef]

240. Becher, B.; Segal, B.M. T(H)17 cytokines in autoimmune neuro-inflammation. *Curr. Opin. Immunol.* **2011**, *23*, 707–712. [CrossRef]

241. Grifka-Walk, H.M.; Lalor, S.J.; Segal, B.M. Highly polarized Th17 cells induce EAE via a T-bet independent mechanism. *Eur. J. Immunol.* **2013**, *43*, 2824–2831. [CrossRef]

242. Laird, M.D.; Sukumari-Ramesh, S.; Swift, A.E.; Meiler, S.E.; Vender, J.R.; Dhandapani, K.M. Curcumin attenuates cerebral edema following traumatic brain injury in mice: A possible role for aquaporin-4? *J. Neurochem.* **2010**, *113*, 637–648. [CrossRef]

243. Ma, C.; Ma, Z.; Fu, Q.; Ma, S. Curcumin attenuates allergic airway inflammation by regulation of CD4+CD25+ regulatory T cells (Tregs)/Th17 balance in ovalbumin-sensitized mice. *Fitoterapia* **2013**, *87*, 57–64. [CrossRef] [PubMed]

244. Park, M.J.; Moon, S.J.; Lee, S.H.; Yang, E.J.; Min, J.K.; Cho, S.G.; Yang, C.W.; Park, S.H.; Kim, H.Y.; Cho, M.L. Curcumin attenuates acute graft-versus-host disease severity via in vivo regulations on Th1, Th17 and regulatory T cells. *Plos ONE* **2013**, *8*, e67171. [CrossRef] [PubMed]

245. Xie, L.; Li, X.K.; Funeshima-Fuji, N.; Kimura, H.; Matsumoto, Y.; Isaka, Y.; Takahara, S. Amelioration of experimental autoimmune encephalomyelitis by curcumin treatment through inhibition of IL-17 production. *Int. Immunopharmacol.* **2009**, *9*, 575–581. [CrossRef]

246. Hernandez-Ontiveros, D.G.; Tajiri, N.; Acosta, S.; Giunta, B.; Tan, J.; Borlongan, C.V. Microglia activation as a biomarker for traumatic brain injury. *Front. Neurol.* **2013**, *4*, 30. [CrossRef]

247. Loane, D.J.; Kumar, A.; Stoica, B.A.; Cabatbat, R.; Faden, A.I. Progressive Neurodegeneration after Experimental Brain Trauma: Association with Chronic Microglial Activation. *J. Neuropathol. Exp. Neurol.* **2014**, *73*, 14–29. [CrossRef]

248. McKee, C.A.; Lukens, J.R. Emerging Roles for the Immune System in Traumatic Brain Injury. *Front. Immunol.* **2016**, *7*, 556. [CrossRef]

249. Jassam, Y.N.; Izzy, S.; Whalen, M.; McGavern, D.B.; El Khoury, J. Neuroimmunology of Traumatic Brain Injury: Time for a Paradigm Shift. *Neuron* **2017**, *95*, 1246–1265. [CrossRef]

250. Correale, J.; Villa, A. The neuroprotective role of inflammation in nervous system injuries. *J. Neurol.* **2004**, *251*, 1304–1316. [CrossRef]

251. Roberts, I.; Yates, D.; Sandercock, P.; Farrell, B.; Wasserberg, J.; Lomas, G.; Cottingham, R.; Svoboda, P.; Brayley, N.; Mazairac, G.; et al. Effect of intravenous corticosteroids on death within 14 days in 10008 adults with clinically significant head injury (MRC CRASH trial): Randomised placebo-controlled trial. *Lancet (Lond. Engl.)* **2004**, *364*, 1321–1328. [CrossRef]

252. Edwards, P.; Arango, M.; Balica, L.; Cottingham, R.; El-Sayed, H.; Farrell, B.; Fernandes, J.; Gogichaisvili, T.; Golden, N.; Hartzenberg, B.; et al. Final results of MRC CRASH, a randomised placebo-controlled trial of intravenous corticosteroid in adults with head injury-outcomes at 6 months. *Lancet (Lond. Engl.)* **2005**, *365*, 1957–1959. [CrossRef]

253. Tuttolomondo, A.; Pecoraro, R.; Pinto, A. Studies of selective TNF inhibitors in the treatment of brain injury from stroke and trauma: A review of the evidence to date. *Drug Des. Dev.* **2014**, *8*, 2221–2238. [CrossRef] [PubMed]

254. Chio, C.C.; Lin, J.W.; Chang, M.W.; Wang, C.C.; Kuo, J.R.; Yang, C.Z.; Chang, C.P. Therapeutic evaluation of etanercept in a model of traumatic brain injury. *J. Neurochem.* **2010**, *115*, 921–929. [CrossRef] [PubMed]

255. Cheong, C.U.; Chang, C.P.; Chao, C.M.; Cheng, B.C.; Yang, C.Z.; Chio, C.C. Etanercept attenuates traumatic brain injury in rats by reducing brain TNF- alpha contents and by stimulating newly formed neurogenesis. *Mediat. Inflamm.* **2013**, *2013*, 620837. [CrossRef]

256. Chio, C.C.; Chang, C.H.; Wang, C.C.; Cheong, C.U.; Chao, C.M.; Cheng, B.C.; Yang, C.Z.; Chang, C.P. Etanercept attenuates traumatic brain injury in rats by reducing early microglial expression of tumor necrosis factor-alpha. *BMC Neurosci.* **2013**, *14*, 33. [CrossRef]

257. Tobinick, E.; Rodriguez-Romanacce, H.; Levine, A.; Ignatowski, T.A.; Spengler, R.N. Immediate neurological recovery following perispinal etanercept years after brain injury. *Clin. Drug Investig.* **2014**, *34*, 361–366. [CrossRef] [PubMed]

258. Tobinick, E. Immediate Resolution of Hemispatial Neglect and Central Post-Stroke Pain After Perispinal Etanercept: Case Report. *Clin. Drug Investig.* **2020**, *40*, 93–97. [CrossRef]

259. Ralph, S.J.; Weissenberger, A.; Bonev, V.; King, L.D.; Bonham, M.D.; Ferguson, S.; Smith, A.D.; Goodman-Jones, A.A.; Espinet, A.J. Phase I/II parallel double-blind randomized controlled clinical trial of perispinal etanercept for chronic stroke: Improved mobility and pain alleviation. *Expert Opin. Investig. Drugs* **2020**, *29*, 311–326. [CrossRef]

260. Ignatowski, T.A.; Spengler, R.N.; Dhandapani, K.M.; Folkersma, H.; Butterworth, R.F.; Tobinick, E. Perispinal etanercept for post-stroke neurological and cognitive dysfunction: Scientific rationale and current evidence. *CNS Drugs* **2014**, *28*, 679–697. [CrossRef]

261. Tobinick, E.; Kim, N.M.; Reyzin, G.; Rodriguez-Romanacce, H.; DePuy, V. Selective TNF inhibition for chronic stroke and traumatic brain injury: An observational study involving 629 consecutive patients treated with perispinal etanercept. *CNS Drugs* **2012**, *26*, 1051–1070. [CrossRef]

262. Baratz, R.; Tweedie, D.; Rubovitch, V.; Luo, W.; Yoon, J.S.; Hoffer, B.J.; Greig, N.H.; Pick, C.G. Tumor necrosis factor-alpha synthesis inhibitor, 3,6′-dithiothalidomide, reverses behavioral impairments induced by minimal traumatic brain injury in mice. *J. Neurochem.* **2011**, *118*, 1032–1042. [CrossRef]

263. Baratz, R.; Tweedie, D.; Wang, J.Y.; Rubovitch, V.; Luo, W.; Hoffer, B.J.; Greig, N.H.; Pick, C.G. Transiently lowering tumor necrosis factor-alpha synthesis ameliorates neuronal cell loss and cognitive impairments induced by minimal traumatic brain injury in mice. *J. Neuroinflamm.* **2015**, *12*, 45. [CrossRef] [PubMed]

264. Stahel, P.F.; Shohami, E.; Younis, F.M.; Kariya, K.; Otto, V.I.; Lenzlinger, P.M.; Grosjean, M.B.; Eugster, H.P.; Trentz, O.; Kossmann, T.; et al. Experimental closed head injury: Analysis of neurological outcome, blood-brain barrier dysfunction, intracranial neutrophil infiltration, and neuronal cell death in mice deficient in genes for pro-inflammatory cytokines. *J. Cereb. Blood Flow Metab.* **2000**, *20*, 369–380. [CrossRef] [PubMed]

265. Thome, J.G.; Reeder, E.L.; Collins, S.M.; Gopalan, P.; Robson, M.J. Contributions of Interleukin-1 Receptor Signaling in Traumatic Brain Injury. *Front. Behav. Neurosci.* **2019**, *13*, 287. [CrossRef] [PubMed]

266. Tehranian, R.; Andell-Jonsson, S.; Beni, S.M.; Yatsiv, I.; Shohami, E.; Bartfai, T.; Lundkvist, J.; Iverfeldt, K. Improved recovery and delayed cytokine induction after closed head injury in mice with central overexpression of the secreted isoform of the interleukin-1 receptor antagonist. *J. Neurotrauma* **2002**, *19*, 939–951. [CrossRef]

267. Bergold, P.J. Treatment of traumatic brain injury with anti-inflammatory drugs. *Exp. Neurol.* **2016**, *275 Pt. 3*, 367–380. [CrossRef]

268. Helmy, A.; Guilfoyle, M.R.; Carpenter, K.L.; Pickard, J.D.; Menon, D.K.; Hutchinson, P.J. Recombinant human interleukin-1 receptor antagonist in severe traumatic brain injury: A phase II randomized control trial. *J. Cereb. Blood Flow Metab.* **2014**, *34*, 845–851. [CrossRef] [PubMed]

269. Clausen, F.; Hanell, A.; Bjork, M.; Hillered, L.; Mir, A.K.; Gram, H.; Marklund, N. Neutralization of interleukin-1beta modifies the inflammatory response and improves histological and cognitive outcome following traumatic brain injury in mice. *Eur. J. Neurosci.* **2009**, *30*, 385–396. [CrossRef] [PubMed]

270. Clausen, F.; Hanell, A.; Israelsson, C.; Hedin, J.; Ebendal, T.; Mir, A.K.; Gram, H.; Marklund, N. Neutralization of interleukin-1beta reduces cerebral edema and tissue loss and improves late cognitive outcome following traumatic brain injury in mice. *Eur. J. Neurosci.* **2011**, *34*, 110–123. [CrossRef]

271. Ma, M.W.; Wang, J.; Dhandapani, K.M.; Brann, D.W. NADPH Oxidase 2 Regulates NLRP3 Inflammasome Activation in the Brain after Traumatic Brain Injury. *Oxidative Med. Cell. Longev.* **2017**, *2017*, 6057609. [CrossRef]

272. Ismael, S.; Ahmed, H.A.; Adris, T.; Parveen, K.; Thakor, P.; Ishrat, T. The NLRP3 inflammasome: A potential therapeutic target for traumatic brain injury. *Neural Regen. Res.* **2020**, *16*, 49–57. [CrossRef]

273. Kerr, N.; Lee, S.W.; Perez-Barcena, J.; Crespi, C.; Ibañez, J.; Bullock, M.R.; Dietrich, W.D.; Keane, R.W.; de Rivero Vaccari, J.P. Inflammasome proteins as biomarkers of traumatic brain injury. *Plos ONE* **2018**, *13*, e0210128. [CrossRef] [PubMed]

274. Desu, H.L.; Plastini, M.; Illiano, P.; Bramlett, H.M.; Dietrich, W.D.; de Rivero Vaccari, J.P.; Brambilla, R.; Keane, R.W. IC100: A novel anti-ASC monoclonal antibody improves functional outcomes in an animal model of multiple sclerosis. *J. Neuroinflamm.* **2020**, *17*, 143. [CrossRef] [PubMed]

275. Brough, D.; Denes, A. Interleukin-1alpha and brain inflammation. *IUBMB Life* **2015**, *67*, 323–330. [CrossRef] [PubMed]

276. Newell, E.A.; Todd, B.P.; Mahoney, J.; Pieper, A.A.; Ferguson, P.J.; Bassuk, A.G. Combined Blockade of Interleukin-1alpha and -1beta Signaling Protects Mice from Cognitive Dysfunction after Traumatic Brain Injury. *ENeuro* **2018**, *5*. [CrossRef]

277. Yang, S.H.; Gustafson, J.; Gangidine, M.; Stepien, D.; Schuster, R.; Pritts, T.A.; Goodman, M.D.; Remick, D.G.; Lentsch, A.B. A murine model of mild traumatic brain injury exhibiting cognitive and motor deficits. *J. Surg. Res.* **2013**, *184*, 981–988. [CrossRef] [PubMed]

278. Hergenroeder, G.W.; Moore, A.N.; McCoy, J.P., Jr.; Samsel, L.; Ward, N.H., 3rd; Clifton, G.L.; Dash, P.K. Serum IL-6: A candidate biomarker for intracranial pressure elevation following isolated traumatic brain injury. *J. Neuroinflamm.* **2010**, *7*, 19. [CrossRef]

279. Yang, S.H.; Gangidine, M.; Pritts, T.A.; Goodman, M.D.; Lentsch, A.B. Interleukin 6 mediates neuroinflammation and motor coordination deficits after mild traumatic brain injury and brief hypoxia in mice. *Shock (Augustaga.)* **2013**, *40*, 471–475. [CrossRef]

280. Mollica, L.; De Marchis, F.; Spitaleri, A.; Dallacosta, C.; Pennacchini, D.; Zamai, M.; Agresti, A.; Trisciuoglio, L.; Musco, G.; Bianchi, M.E. Glycyrrhizin binds to high-mobility group box 1 protein and inhibits its cytokine activities. *Chem. Biol.* **2007**, *14*, 431–441. [CrossRef]

281. Gu, X.J.; Xu, J.; Ma, B.Y.; Chen, G.; Gu, P.Y.; Wei, D.; Hu, W.X. Effect of glycyrrhizin on traumatic brain injury in rats and its mechanism. *Chin. J. Traumatol.* **2014**, *17*, 1–7.

282. Yang, L.; Wang, F.; Yang, L.; Yuan, Y.; Chen, Y.; Zhang, G.; Fan, Z. HMGB1 a-Box Reverses Brain Edema and Deterioration of Neurological Function in a Traumatic Brain Injury Mouse Model. *Cell. Physiol. Biochem.* **2018**, *46*, 2532–2542. [CrossRef]

283. Zhang, B.; Wang, B.; Cao, S.; Wang, Y. Epigallocatechin-3-Gallate (EGCG) Attenuates Traumatic Brain Injury by Inhibition of Edema Formation and Oxidative Stress. *Korean J. Physiol. Pharmacol.* **2015**, *19*, 491–497. [CrossRef] [PubMed]

284. Reddy, V.; Grogan, D.; Ahluwalia, M.; Salles, É.L.; Ahluwalia, P.; Khodadadi, H.; Alverson, K.; Nguyen, A.; Raju, S.P.; Gaur, P.; et al. Targeting the endocannabinoid system: A predictive, preventive, and personalized medicine-directed approach to the management of brain pathologies. *EPMA J.* **2020**. [CrossRef] [PubMed]

285. Russo, E.B. History of cannabis as medicine: Nineteenth century irish physicians and correlations of their observations to modern research. In *Cannabis Sativa L.: Botany and Biotechnology*; Chanda, S., Lata, H., Elsohly, M.., Eds.; Springer International Publishing: Cham, Switzerland, 2017; pp. 63–78.

286. Russo, E.B. Clinical Endocannabinoid Deficiency Reconsidered: Current Research Supports the Theory in Migraine, Fibromyalgia, Irritable Bowel, and Other Treatment-Resistant Syndromes. *Cannabis Cannabinoid Res.* **2016**, *1*, 154–165. [CrossRef]

287. Rhyne, D.N.; Anderson, S.L.; Gedde, M.; Borgelt, L.M. Effects of Medical Marijuana on Migraine Headache Frequency in an Adult Population. *Pharmacotherapy* **2016**, *36*, 505–510. [CrossRef]

288. Russo, E.B.; Hohmann, A.G. Role of cannabinoids in pain management. In *Comprehensive Treatment of Chronic Pain by Medical, Interventional and Behavioral Approaches*; Deer, T., Gordin, V., Eds.; Springer: New York, NY, USA, 2013; pp. 181–197.

289. Serpell, M.; Ratcliffe, S.; Hovorka, J.; Schofield, M.; Taylor, L.; Lauder, H.; Ehler, E. A double-blind, randomized, placebo-controlled, parallel group study of THC/CBD spray in peripheral neuropathic pain treatment. *Eur. J. Pain (Lond. Engl.)* **2014**, *18*, 999–1012. [CrossRef]

290. Chen, D.J.; Gao, M.; Gao, F.F.; Su, Q.X.; Wu, J. Brain cannabinoid receptor 2: Expression, function and modulation. *Acta Pharm. Sin.* **2017**, *38*, 312–316. [CrossRef]

291. Rog, D.J.; Nurmikko, T.J.; Friede, T.; Young, C.A. Randomized, controlled trial of cannabis-based medicine in central pain in multiple sclerosis. *Neurology* **2005**, *65*, 812–819. [CrossRef]

292. Johnson, J.R.; Burnell-Nugent, M.; Lossignol, D.; Ganae-Motan, E.D.; Potts, R.; Fallon, M.T. Multicenter, double-blind, randomized, placebo-controlled, parallel-group study of the efficacy, safety, and tolerability of THC:CBD extract and THC extract in patients with intractable cancer-related pain. *J. Pain Symptom Manag.* **2010**, *39*, 167–179. [CrossRef]

293. Benyo, Z.; Ruisanchez, E.; Leszl-Ishiguro, M.; Sandor, P.; Pacher, P. Endocannabinoids in cerebrovascular regulation. *Am. J. Physiol. Heart Circ. Physiol.* **2016**, *310*, H785–H801. [CrossRef]

294. Schurman, L.D.; Lichtman, A.H. Endocannabinoids: A Promising Impact for Traumatic Brain Injury. *Front. Pharmacol.* **2017**, *8*, 69. [CrossRef]

295. Paloczi, J.; Varga, Z.V.; Hasko, G.; Pacher, P. Neuroprotection in Oxidative Stress-Related Neurodegenerative Diseases: Role of Endocannabinoid System Modulation. *Antioxid. Redox Signal.* **2018**, *29*, 75–108. [CrossRef]

296. Fernandez-Ruiz, J.; Moro, M.A.; Martinez-Orgado, J. Cannabinoids in Neurodegenerative Disorders and Stroke/Brain Trauma: From Preclinical Models to Clinical Applications. *Neurother. J. Am. Soc. Exp. Neurother.* **2015**, *12*, 793–806. [CrossRef]

297. Habib, A.; Chokr, D.; Wan, J.; Hegde, P.; Mabire, M.; Siebert, M.; Ribeiro-Parenti, L.; Le Gall, M.; Letteron, P.; Pilard, N.; et al. Inhibition of monoacylglycerol lipase, an anti-inflammatory and antifibrogenic strategy in the liver. *Gut* **2018**. [CrossRef]

298. Kho, D.T.; Glass, M.; Graham, E.S. Is the Cannabinoid CB2 Receptor a Major Regulator of the Neuroinflammatory Axis of the Neurovascular Unit in Humans? *Adv. Pharmacol. (San Diegocalif.)* **2017**, *80*, 367–396. [CrossRef]

299. Nozaki, C.; Markert, A.; Zimmer, A. Inhibition of FAAH reduces nitroglycerin-induced migraine-like pain and trigeminal neuronal hyperactivity in mice. *Eur. Neuropsychopharmacol.* **2015**, *25*, 1388–1396. [CrossRef]

300. Maas, A.I.; Murray, G.; Henney, H., 3rd; Kassem, N.; Legrand, V.; Mangelus, M.; Muizelaar, J.P.; Stocchetti, N.; Knoller, N. Efficacy and safety of dexanabinol in severe traumatic brain injury: Results of a phase III randomised, placebo-controlled, clinical trial. *Lancet Neurol.* **2006**, *5*, 38–45. [CrossRef]

301. Latorre, J.G.; Schmidt, E.B. Cannabis, Cannabinoids, and Cerebral Metabolism: Potential Applications in Stroke and Disorders of the Central Nervous System. *Curr. Cardiol. Rep.* **2015**, *17*, 627. [CrossRef]

302. Russo, E.B. Synthetic and natural cannabinoids: The cardiovascular risk. *Br. J. Cardiol.* **2015**, *22*, 7–9.

303. Pacher, P.; Steffens, S.; Hasko, G.; Schindler, T.H.; Kunos, G. Cardiovascular effects of marijuana and synthetic cannabinoids: The good, the bad, and the ugly. *Nat. Rev. Cardiol.* **2018**, *15*, 151–166. [CrossRef]

304. Magid, L.; Heymann, S.; Elgali, M.; Avram, L.; Cohen, Y.; Liraz-Zaltsman, S.; Mechoulam, R.; Shohami, E. Role of CB2 Receptor in the Recovery of Mice after Traumatic Brain Injury. *J. Neurotrauma* **2019**, *36*, 1836–1846. [CrossRef]

305. Hess, D.C.; Blauenfeldt, R.A.; Andersen, G.; Hougaard, K.D.; Hoda, M.N.; Ding, Y.; Ji, X. Remote ischaemic conditioning-a new paradigm of self-protection in the brain. *Nat. Rev. Neurol.* **2015**, *11*, 698–710. [CrossRef]

306. Saxena, P.; Newman, M.A.; Shehatha, J.S.; Redington, A.N.; Konstantinov, I.E. Remote ischemic conditioning: Evolution of the concept, mechanisms, and clinical application. *J. Card. Surg.* **2010**, *25*, 127–134. [CrossRef]

307. Hoda, M.N.; Fagan, S.C.; Khan, M.B.; Vaibhav, K.; Chaudhary, A.; Wang, P.; Dhandapani, K.M.; Waller, J.L.; Hess, D.C. A 2 × 2 factorial design for the combination therapy of minocycline and remote ischemic perconditioning: Efficacy in a preclinical trial in murine thromboembolic stroke model. *Exp. Transl. Stroke Med.* **2014**, *6*, 10. [CrossRef]

308. Hess, D.C.; Hoda, M.N.; Khan, M.B. Humoral Mediators of Remote Ischemic Conditioning: Important Role of eNOS/NO/Nitrite. *Acta Neurochir. Suppl.* **2016**, *121*, 45–48. [CrossRef]

309. Loukogeorgakis, S.P.; Williams, R.; Panagiotidou, A.T.; Kolvekar, S.K.; Donald, A.; Cole, T.J.; Yellon, D.M.; Deanfield, J.E.; MacAllister, R.J. Transient Limb Ischemia Induces Remote Preconditioning and Remote Postconditioning in Humans by a KATP Channel–Dependent Mechanism. *Circulation* **2007**, *116*, 1386–1395. [CrossRef]

310. Xu, M.; Wang, Y.; Ayub, A.; Ashraf, M. Mitochondrial K(ATP) channel activation reduces anoxic injury by restoring mitochondrial membrane potential. *Am. J. Physiol. Heart Circ. Physiol.* **2001**, *281*, H1295–H1303. [CrossRef]

311. Vaibhav, K.; Baban, B.; Khan, M.B.; Liu, J.Y.; Huo, Y.; Hess, D.C.; Dhandapani, K.M.; Hoda, M.N. Remote ischemic preconditioning protects from traumatic brain injury (TBI). *J. Neurotrauma* **2014**, *31*, A87.

312. Vaibhav, K.; Baban, B.; Wang, P.; Khan, M.B.; Pandya, C.; Ahmed, H.; Chaudhary, A.; Ergul, A.; Heger, I.; Hess, D.C.; et al. Remote Ischemic Conditioning (RIC) Attenuates Post-TBI Ischemic Injury and Improves Behavioral Outcomes. *Stroke A J. Cereb. Circ.* **2015**, *46*, ATP92.

313. Joseph, B.; Pandit, V.; Zangbar, B.; Kulvatunyou, N.; Khalil, M.; Tang, A.; O'Keeffe, T.; Gries, L.; Vercruysse, G.; Friese, R.S.; et al. Secondary brain injury in trauma patients: The effects of remote ischemic conditioning. *J. Trauma Acute Care Surg.* **2015**, *78*, 698–703; discussion 703–705. [CrossRef]

314. Pandit, V.; Khan, M.; Zakaria, E.R.; Largent-Milnes, T.M.; Hamidi, M.; O'Keeffe, T.; Vanderah, T.W.; Joseph, B. Continuous remote ischemic conditioning attenuates cognitive and motor deficits from moderate traumatic brain injury. *J. Trauma Acute Care Surg.* **2018**, *85*, 48–53. [CrossRef]

315. Sandweiss, A.J.; Azim, A.; Ibraheem, K.; Largent-Milnes, T.M.; Rhee, P.; Vanderah, T.W.; Joseph, B. Remote ischemic conditioning preserves cognition and motor coordination in a mouse model of traumatic brain injury. *J. Trauma Acute Care Surg.* **2017**, *83*, 1074–1081. [CrossRef] [PubMed]

316. Minambres, E.; Ballesteros, M.A.; Mayorga, M.; Marin, M.J.; Munoz, P.; Figols, J.; Lopez-Hoyos, M. Cerebral apoptosis in severe traumatic brain injury patients: An in vitro, in vivo, and postmortem study. *J. Neurotrauma* **2008**, *25*, 581–591. [CrossRef]

317. Bredesen, D.E. Key note lecture: Toward a mechanistic taxonomy for cell death programs. *Stroke A J. Cereb. Circ.* **2007**, *38*, 652–660. [CrossRef] [PubMed]

318. Bredesen, D.E. Programmed cell death mechanisms in neurological disease. *Curr. Mol. Med.* **2008**, *8*, 173–186. [CrossRef]

319. Stoica, B.A.; Byrnes, K.R.; Faden, A.I. Cell cycle activation and CNS injury. *Neurotox. Res.* **2009**, *16*, 221–237. [CrossRef] [PubMed]

320. Di Giovanni, S.; Movsesyan, V.; Ahmed, F.; Cernak, I.; Schinelli, S.; Stoica, B.; Faden, A.I. Cell cycle inhibition provides neuroprotection and reduces glial proliferation and scar formation after traumatic brain injury. *Proc. Natl. Acad. Sci. USA* **2005**, *102*, 8333–8338. [CrossRef] [PubMed]

321. Cernak, I.; Stoica, B.; Byrnes, K.R.; Di Giovanni, S.; Faden, A.I. Role of the cell cycle in the pathobiology of central nervous system trauma. *Cell Cycle (Georget. Tex.)* **2005**, *4*, 1286–1293. [CrossRef]

322. Stoica, B.; Byrnes, K.; Faden, A.I. Multifunctional Drug Treatment in Neurotrauma. *Neurother. J. Am. Soc. Exp. Neurother.* **2009**, *6*, 14–27. [CrossRef]

323. Hilton, G.D.; Stoica, B.A.; Byrnes, K.R.; Faden, A.I. Roscovitine reduces neuronal loss, glial activation, and neurologic deficits after brain trauma. *J. Cereb. Blood Flow Metab.* **2008**, *28*, 1845–1859. [CrossRef]

324. Yakovlev, A.G.; Faden, A.I. Caspase-dependent apoptotic pathways in CNS injury. *Mol. Neurobiol.* **2001**, *24*, 131–144. [CrossRef]

325. Wan, J.; Wang, J.; Cheng, H.; Yu, Y.; Xing, G.; Oiu, Z.; Qian, X.; He, F. Proteomic analysis of apoptosis initiation induced by all-trans retinoic acid in human acute promyelocytic leukemia cells. *Electrophoresis* **2001**, *22*, 3026–3037. [CrossRef]

326. Zhang, X.; Alber, S.; Watkins, S.C.; Kochanek, P.M.; Marion, D.W.; Graham, S.H.; Clark, R.S. Proteolysis consistent with activation of caspase-7 after severe traumatic brain injury in humans. *J. Neurotrauma* **2006**, *23*, 1583–1590. [CrossRef] [PubMed]

327. Yakovlev, A.G.; Knoblach, S.M.; Fan, L.; Fox, G.B.; Goodnight, R.; Faden, A.I. Activation of CPP32-like caspases contributes to neuronal apoptosis and neurological dysfunction after traumatic brain injury. *J. Neurosci.* **1997**, *17*, 7415–7424. [CrossRef]

328. Larner, S.F.; Hayes, R.L.; McKinsey, D.M.; Pike, B.R.; Wang, K.K. Increased expression and processing of caspase-12 after traumatic brain injury in rats. *J. Neurochem.* **2004**, *88*, 78–90. [CrossRef]

329. Knoblach, S.M.; Nikolaeva, M.; Huang, X.; Fan, L.; Krajewski, S.; Reed, J.C.; Faden, A.I. Multiple caspases are activated after traumatic brain injury: Evidence for involvement in functional outcome. *J. Neurotrauma* **2002**, *19*, 1155–1170. [CrossRef] [PubMed]

330. Nakagawa, T.; Zhu, H.; Morishima, N.; Li, E.; Xu, J.; Yankner, B.A.; Yuan, J. Caspase-12 mediates endoplasmic-reticulum-specific apoptosis and cytotoxicity by amyloid-beta. *Nature* **2000**, *403*, 98–103. [CrossRef] [PubMed]

331. Nathoo, N.; Narotam, P.K.; Agrawal, D.K.; Connolly, C.A.; van Dellen, J.R.; Barnett, G.H.; Chetty, R. Influence of apoptosis on neurological outcome following traumatic cerebral contusion. *J. Neurosurg.* **2004**, *101*, 233–240. [CrossRef]

332. Brophy, G.M.; Pineda, J.A.; Papa, L.; Lewis, S.B.; Valadka, A.B.; Hannay, H.J.; Heaton, S.C.; Demery, J.A.; Liu, M.C.; Tepas, J.J., 3rd; et al. alphaII-Spectrin breakdown product cerebrospinal fluid exposure metrics suggest differences in cellular injury mechanisms after severe traumatic brain injury. *J. Neurotrauma* **2009**, *26*, 471–479. [CrossRef]

333. McGinn, M.J.; Kelley, B.J.; Akinyi, L.; Oli, M.W.; Liu, M.C.; Hayes, R.L.; Wang, K.K.; Povlishock, J.T. Biochemical, structural, and biomarker evidence for calpain-mediated cytoskeletal change after diffuse brain injury uncomplicated by contusion. *J. Neuropathol. Exp. Neurol.* **2009**, *68*, 241–249. [CrossRef]

334. Mondello, S.; Robicsek, S.A.; Gabrielli, A.; Brophy, G.M.; Papa, L.; Tepas, J.; Robertson, C.; Buki, A.; Scharf, D.; Jixiang, M.; et al. alphaII-spectrin breakdown products (SBDPs): Diagnosis and outcome in severe traumatic brain injury patients. *J. Neurotrauma* **2010**, *27*, 1203–1213. [CrossRef]

335. Polster, B.M.; Basanez, G.; Etxebarria, A.; Hardwick, J.M.; Nicholls, D.G. Calpain I induces cleavage and release of apoptosis-inducing factor from isolated mitochondria. *J. Biol. Chem.* **2005**, *280*, 6447–6454. [CrossRef]

336. Takano, J.; Tomioka, M.; Tsubuki, S.; Higuchi, M.; Iwata, N.; Itohara, S.; Maki, M.; Saido, T.C. Calpain mediates excitotoxic DNA fragmentation via mitochondrial pathways in adult brains: Evidence from calpastatin mutant mice. *J. Biol. Chem.* **2005**, *280*, 16175–16184. [CrossRef]

337. Gao, G.; Dou, Q.P. N-terminal cleavage of bax by calpain generates a potent proapoptotic 18-kDa fragment that promotes bcl-2-independent cytochrome C release and apoptotic cell death. *J. Cell. Biochem.* **2000**, *80*, 53–72. [CrossRef]

338. Mandic, A.; Viktorsson, K.; Strandberg, L.; Heiden, T.; Hansson, J.; Linder, S.; Shoshan, M.C. Calpain-mediated Bid cleavage and calpain-independent Bak modulation: Two separate pathways in cisplatin-induced apoptosis. *Mol. Cell. Biol.* **2002**, *22*, 3003–3013. [CrossRef]

339. van Loo, G.; Saelens, X.; van Gurp, M.; MacFarlane, M.; Martin, S.J.; Vandenabeele, P. The role of mitochondrial factors in apoptosis: A Russian roulette with more than one bullet. *Cell Death Differ.* **2002**, *9*, 1031–1042. [CrossRef]

340. Daugas, E.; Nochy, D.; Ravagnan, L.; Loeffler, M.; Susin, S.A.; Zamzami, N.; Kroemer, G. Apoptosis-inducing factor (AIF): A ubiquitous mitochondrial oxidoreductase involved in apoptosis. *FEBS Lett.* **2000**, *476*, 118–123. [CrossRef]

341. Li, L.Y.; Luo, X.; Wang, X. Endonuclease G is an apoptotic DNase when released from mitochondria. *Nature* **2001**, *412*, 95–99. [CrossRef]

342. Ishihara, Y.; Shimamoto, N. Involvement of endonuclease G in nucleosomal DNA fragmentation under sustained endogenous oxidative stress. *J. Biol. Chem.* **2006**, *281*, 6726–6733. [CrossRef]

343. Artus, C.; Boujrad, H.; Bouharrour, A.; Brunelle, M.N.; Hoos, S.; Yuste, V.J.; Lenormand, P.; Rousselle, J.C.; Namane, A.; England, P.; et al. AIF promotes chromatinolysis and caspase-independent programmed necrosis by interacting with histone H2AX. *Embo J.* **2010**, *29*, 1585–1599. [CrossRef]

344. Zhang, X.; Chen, J.; Graham, S.H.; Du, L.; Kochanek, P.M.; Draviam, R.; Guo, F.; Nathaniel, P.D.; Szabo, C.; Watkins, S.C.; et al. Intranuclear localization of apoptosis-inducing factor (AIF) and large scale DNA fragmentation after traumatic brain injury in rats and in neuronal cultures exposed to peroxynitrite. *J. Neurochem.* **2002**, *82*, 181–191. [CrossRef]

345. Bano, D.; Munarriz, E.; Chen, H.L.; Ziviani, E.; Lippi, G.; Young, K.W.; Nicotera, P. The plasma membrane Na^+/Ca^{2+} exchanger is cleaved by distinct protease families in neuronal cell death. *Ann. N. Y. Acad. Sci.* **2007**, *1099*, 451–455. [CrossRef]

346. Nur, E.K.A.; Gross, S.R.; Pan, Z.; Balklava, Z.; Ma, J.; Liu, L.F. Nuclear translocation of cytochrome c during apoptosis. *J. Biol. Chem.* **2004**, *279*, 24911–24914. [CrossRef]

347. Zhao, S.; Aviles, E.R., Jr.; Fujikawa, D.G. Nuclear translocation of mitochondrial cytochrome c, lysosomal cathepsins B and D, and three other death-promoting proteins within the first 60 min of generalized seizures. *J. Neurosci. Res.* **2010**, *88*, 1727–1737. [CrossRef]

348. Fujikawa, D.G.; Shinmei, S.S.; Cai, B. Kainic acid-induced seizures produce necrotic, not apoptotic, neurons with internucleosomal DNA cleavage: Implications for programmed cell death mechanisms. *Neuroscience* **2000**, *98*, 41–53. [CrossRef]

349. Cregan, S.P.; Dawson, V.L.; Slack, R.S. Role of AIF in caspase-dependent and caspase-independent cell death. *Oncogene* **2004**, *23*, 2785–2796. [CrossRef]

350. Cande, C.; Vahsen, N.; Garrido, C.; Kroemer, G. Apoptosis-inducing factor (AIF): Caspase-independent after all. *Cell Death Differ.* **2004**, *11*, 591–595. [CrossRef]

351. Hong, S.J.; Dawson, T.M.; Dawson, V.L. Nuclear and mitochondrial conversations in cell death: PARP-1 and AIF signaling. *Trends Pharmacol. Sci.* **2004**, *25*, 259–264. [CrossRef]

352. Whalen, M.J.; Clark, R.S.; Dixon, C.E.; Robichaud, P.; Marion, D.W.; Vagni, V.; Graham, S.H.; Virag, L.; Hasko, G.; Stachlewitz, R.; et al. Reduction of cognitive and motor deficits after traumatic brain injury in mice deficient in poly(ADP-ribose) polymerase. *J. Cereb. Blood Flow Metab.* **1999**, *19*, 835–842. [CrossRef]

353. Alano, C.C.; Ying, W.; Swanson, R.A. Poly(ADP-ribose) polymerase-1-mediated cell death in astrocytes requires NAD+ depletion and mitochondrial permeability transition. *J. Biol. Chem.* **2004**, *279*, 18895–18902. [CrossRef]

354. Ying, W.; Alano, C.C.; Garnier, P.; Swanson, R.A. NAD+ as a metabolic link between DNA damage and cell death. *J. Neurosci. Res.* **2005**, *79*, 216–223. [CrossRef]

355. Moubarak, R.S.; Yuste, V.J.; Artus, C.; Bouharrour, A.; Greer, P.A.; Menissier-de Murcia, J.; Susin, S.A. Sequential activation of poly(ADP-ribose) polymerase 1, calpains, and Bax is essential in apoptosis-inducing factor-mediated programmed necrosis. *Mol. Cell. Biol.* **2007**, *27*, 4844–4862. [CrossRef] [PubMed]

356. Andrabi, S.A.; Kim, N.S.; Yu, S.W.; Wang, H.; Koh, D.W.; Sasaki, M.; Klaus, J.A.; Otsuka, T.; Zhang, Z.; Koehler, R.C.; et al. Poly(ADP-ribose) (PAR) polymer is a death signal. *Proc. Natl. Acad. Sci. USA* **2006**, *103*, 18308–18313. [CrossRef] [PubMed]

357. Yu, S.W.; Wang, H.; Poitras, M.F.; Coombs, C.; Bowers, W.J.; Federoff, H.J.; Poirier, G.G.; Dawson, T.M.; Dawson, V.L. Mediation of poly(ADP-ribose) polymerase-1-dependent cell death by apoptosis-inducing factor. *Science* **2002**, *297*, 259–263. [CrossRef] [PubMed]

358. Yu, S.W.; Andrabi, S.A.; Wang, H.; Kim, N.S.; Poirier, G.G.; Dawson, T.M.; Dawson, V.L. Apoptosis-inducing factor mediates poly(ADP-ribose) (PAR) polymer-induced cell death. *Proc. Natl. Acad. Sci. USA* **2006**, *103*, 18314–18319. [CrossRef]

359. Blenn, C.; Wyrsch, P.; Bader, J.; Bollhalder, M.; Althaus, F.R. Poly(ADP-ribose)glycohydrolase is an upstream regulator of Ca^{2+} fluxes in oxidative cell death. *Cell. Mol. Life Sci. Cmls* **2011**, *68*, 1455–1466. [CrossRef]

360. Yasuda, H.; Shichinohe, H.; Kuroda, S.; Ishikawa, T.; Iwasaki, Y. Neuroprotective effect of a heat shock protein inducer, geranylgeranylacetone in permanent focal cerebral ischemia. *Brain Res.* **2005**, *1032*, 176–182. [CrossRef]

361. Lee, S.H.; Kwon, H.M.; Kim, Y.J.; Lee, K.M.; Kim, M.; Yoon, B.W. Effects of hsp70.1 gene knockout on the mitochondrial apoptotic pathway after focal cerebral ischemia. *Stroke A J. Cereb. Circ.* **2004**, *35*, 2195–2199. [CrossRef]

362. Beere, H.M.; Wolf, B.B.; Cain, K.; Mosser, D.D.; Mahboubi, A.; Kuwana, T.; Tailor, P.; Morimoto, R.I.; Cohen, G.M.; Green, D.R. Heat-shock protein 70 inhibits apoptosis by preventing recruitment of procaspase-9 to the Apaf-1 apoptosome. *Nat. Cell Biol.* **2000**, *2*, 469–475. [CrossRef]

363. Parcellier, A.; Gurbuxani, S.; Schmitt, E.; Solary, E.; Garrido, C. Heat shock proteins, cellular chaperones that modulate mitochondrial cell death pathways. *Biochem. Biophys. Res. Commun.* **2003**, *304*, 505–512. [CrossRef]

364. Gurbuxani, S.; Schmitt, E.; Cande, C.; Parcellier, A.; Hammann, A.; Daugas, E.; Kouranti, I.; Spahr, C.; Pance, A.; Kroemer, G.; et al. Heat shock protein 70 binding inhibits the nuclear import of apoptosis-inducing factor. *Oncogene* **2003**, *22*, 6669–6678. [CrossRef]

365. Matsumori, Y.; Hong, S.M.; Aoyama, K.; Fan, Y.; Kayama, T.; Sheldon, R.A.; Vexler, Z.S.; Ferriero, D.M.; Weinstein, P.R.; Liu, J. Hsp70 overexpression sequesters AIF and reduces neonatal hypoxic/ischemic brain injury. *J. Cereb. Blood Flow Metab.* **2005**, *25*, 899–910. [CrossRef]

366. Eroglu, B.; Kimbler, D.E.; Pang, J.; Choi, J.; Moskophidis, D.; Yanasak, N.; Dhandapani, K.M.; Mivechi, N.F. Therapeutic inducers of the HSP70/HSP110 protect mice against traumatic brain injury. *J. Neurochem.* **2014**, *130*, 626–641. [CrossRef]

367. Proskuryakov, S.Y.; Konoplyannikov, A.G.; Gabai, V.L. Necrosis: A specific form of programmed cell death? *Exp. Cell Res.* **2003**, *283*, 1–16. [CrossRef]

368. Volbracht, C.; Leist, M.; Kolb, S.A.; Nicotera, P. Apoptosis in caspase-inhibited neurons. *Mol. Med. (Camb. Mass.)* **2001**, *7*, 36–48. [CrossRef]

369. Zhu, C.; Wang, X.; Huang, Z.; Qiu, L.; Xu, F.; Vahsen, N.; Nilsson, M.; Eriksson, P.S.; Hagberg, H.; Culmsee, C.; et al. Apoptosis-inducing factor is a major contributor to neuronal loss induced by neonatal cerebral hypoxia-ischemia. *Cell Death Differ.* **2007**, *14*, 775–784. [CrossRef] [PubMed]

370. Kabadi, S.V.; Faden, A.I. Selective CDK inhibitors: Promising candidates for future clinical traumatic brain injury trials. *Neural Regen. Res.* **2014**, *9*, 1578–1580. [CrossRef]

371. Tikka, T.; Fiebich, B.L.; Goldsteins, G.; Keinanen, R.; Koistinaho, J. Minocycline, a tetracycline derivative, is neuroprotective against excitotoxicity by inhibiting activation and proliferation of microglia. *J. Neurosci.* **2001**, *21*, 2580–2588. [CrossRef]

372. Siopi, E.; Llufriu-Daben, G.; Fanucchi, F.; Plotkine, M.; Marchand-Leroux, C.; Jafarian-Tehrani, M. Evaluation of late cognitive impairment and anxiety states following traumatic brain injury in mice: The effect of minocycline. *Neurosci. Lett.* **2012**, *511*, 110–115. [CrossRef]

373. Homsi, S.; Piaggio, T.; Croci, N.; Noble, F.; Plotkine, M.; Marchand-Leroux, C.; Jafarian-Tehrani, M. Blockade of acute microglial activation by minocycline promotes neuroprotection and reduces locomotor hyperactivity after closed head injury in mice: A twelve-week follow-up study. *J. Neurotrauma* **2010**, *27*, 911–921. [CrossRef] [PubMed]

374. Kovesdi, E.; Kamnaksh, A.; Wingo, D.; Ahmed, F.; Grunberg, N.; Long, J.; Kasper, C.; Agoston, D. Acute Minocycline Treatment Mitigates the Symptoms of Mild Blast-Induced Traumatic Brain Injury. *Front. Neurol.* **2012**, *3*. [CrossRef]

375. Wang, X.; Zhu, S.; Drozda, M.; Zhang, W.; Stavrovskaya, I.G.; Cattaneo, E.; Ferrante, R.J.; Kristal, B.S.; Friedlander, R.M. Minocycline inhibits caspase-independent and -dependent mitochondrial cell death pathways in models of Huntington's disease. *Proc. Natl. Acad. Sci. USA* **2003**, *100*, 10483–10487. [CrossRef] [PubMed]

376. Kim, H.S.; Suh, Y.H. Minocycline and neurodegenerative diseases. *Behav. Brain Res.* **2009**, *196*, 168–179. [CrossRef] [PubMed]

377. Casha, S.; Zygun, D.; McGowan, M.D.; Bains, I.; Yong, V.W.; John Hurlbert, R. Results of a phase II placebo-controlled randomized trial of minocycline in acute spinal cord injury. *Brain* **2012**, *135*, 1224. [CrossRef] [PubMed]

378. Wei, J.; Xiao, G.-M. The neuroprotective effects of progesterone on traumatic brain injury: Current status and future prospects. *Acta Pharm. Sin.* **2013**, *34*, 1485–1490. [CrossRef]

379. Guo, Q.; Sayeed, I.; Baronne, L.M.; Hoffman, S.W.; Guennoun, R.; Stein, D.G. Progesterone administration modulates AQP4 expression and edema after traumatic brain injury in male rats. *Exp. Neurol.* **2006**, *198*, 469–478. [CrossRef]

380. Moorthy, K.; Sharma, D.; Basir, S.F.; Baquer, N.Z. Administration of estradiol and progesterone modulate the activities of antioxidant enzyme and aminotransferases in naturally menopausal rats. *Exp. Gerontol.* **2005**, *40*, 295–302. [CrossRef]

381. Robertson, C.L.; Saraswati, M. Progesterone protects mitochondrial function in a rat model of pediatric traumatic brain injury. *J. Bioenerg. Biomembr.* **2015**, *47*, 43–51. [CrossRef]
382. Djebaili, M.; Hoffman, S.W.; Stein, D.G. Allopregnanolone and progesterone decrease cell death and cognitive deficits after a contusion of the rat pre-frontal cortex. *Neuroscience* **2004**, *123*, 349–359. [CrossRef]
383. Skolnick, B.E.; Maas, A.I.; Narayan, R.K.; van der Hoop, R.G.; MacAllister, T.; Ward, J.D.; Nelson, N.R.; Stocchetti, N. A Clinical Trial of Progesterone for Severe Traumatic Brain Injury. *N. Engl. J. Med.* **2014**, *371*, 2467–2476. [CrossRef]
384. Stein, D.G. Embracing failure: What the Phase III progesterone studies can teach about TBI clinical trials. *Brain Inj.* **2015**, *29*, 1259–1272. [CrossRef]
385. Xiong, Y.; Chopp, M.; Lee, C.P. Erythropoietin improves brain mitochondrial function in rats after traumatic brain injury. *Neurol. Res.* **2009**, *31*, 496–502. [CrossRef]
386. Nichol, A.; French, C.; Little, L.; Haddad, S.; Presneill, J.; Arabi, Y.; Bailey, M.; Cooper, D.J.; Duranteau, J.; Huet, O.; et al. Erythropoietin in traumatic brain injury (EPO-TBI): A double-blind randomised controlled trial. *Lancet (Lond. Engl.)* **2015**, *386*, 2499–2506. [CrossRef]
387. Maas, A.I.R.; Menon, D.K.; Lingsma, H.F.; Pineda, J.A.; Sandel, M.E.; Manley, G.T. Re-Orientation of Clinical Research in Traumatic Brain Injury: Report of an International Workshop on Comparative Effectiveness Research. *J. Neurotrauma* **2012**, *29*, 32–46. [CrossRef]
388. High, W.M., Jr.; Briones-Galang, M.; Clark, J.A.; Gilkison, C.; Mossberg, K.A.; Zgaljardic, D.J.; Masel, B.E.; Urban, R.J. Effect of growth hormone replacement therapy on cognition after traumatic brain injury. *J. Neurotrauma* **2010**, *27*, 1565–1575. [CrossRef]
389. Moreau, O.K.; Cortet-Rudelli, C.; Yollin, E.; Merlen, E.; Daveluy, W.; Rousseaux, M. Growth hormone replacement therapy in patients with traumatic brain injury. *J. Neurotrauma* **2013**, *30*, 998–1006. [CrossRef]
390. Powner, D.J.; Boccalandro, C.; Alp, M.S.; Vollmer, D.G. Endocrine failure after traumatic brain injury in adults. *Neurocritical Care* **2006**, *5*, 61–70. [CrossRef]
391. Christopher, A.F.; Kaur, R.P.; Kaur, G.; Kaur, A.; Gupta, V.; Bansal, P. MicroRNA therapeutics: Discovering novel targets and developing specific therapy. *Perspect. Clin. Res.* **2016**, *7*, 68–74. [CrossRef]
392. Sabirzhanov, B.; Stoica, B.A.; Zhao, Z.; Loane, D.J.; Wu, J.; Dorsey, S.G.; Faden, A.I. miR-711 upregulation induces neuronal cell death after traumatic brain injury. *Cell Death Differ.* **2016**, *23*, 654–668. [CrossRef]
393. Ge, X.T.; Lei, P.; Wang, H.C.; Zhang, A.L.; Han, Z.L.; Chen, X.; Li, S.H.; Jiang, R.C.; Kang, C.S.; Zhang, J.N. miR-21 improves the neurological outcome after traumatic brain injury in rats. *Sci. Rep.* **2014**, *4*, 6718. [CrossRef]
394. Han, Z.; Chen, F.; Ge, X.; Tan, J.; Lei, P.; Zhang, J. miR-21 alleviated apoptosis of cortical neurons through promoting PTEN-Akt signaling pathway in vitro after experimental traumatic brain injury. *Brain Res.* **2014**, *1582*, 12–20. [CrossRef]

© 2020 by the authors. Licensee MDPI, Basel, Switzerland. This article is an open access article distributed under the terms and conditions of the Creative Commons Attribution (CC BY) license (http://creativecommons.org/licenses/by/4.0/).

MDPI

Article

Mechanical Stretching-Induced Traumatic Brain Injury Is Mediated by the Formation of GSK-3β-Tau Complex to Impair Insulin Signaling Transduction

Pei-Wen Cheng [1,2,*,†], Yi-Chung Wu [3,4,†], Tzyy-Yue Wong [1,5], Gwo-Ching Sun [6,7] and Ching-Jiunn Tseng [1,2]

1 Department of Education and Research, Kaohsiung Veterans General Hospital, Kaohsiung 813414, Taiwan; wongtzyyyue@gmail.com (T.-Y.W.); cjtseng@vghks.gov.tw (C.-J.T.)
2 Department of Biomedical Science, National Sun Yat-sen University, Kaohsiung 80424, Taiwan
3 Section of Neurology, Zuoying Branch of Kaohsiung Armed Forces General Hospital, Kaohsiung 81342, Taiwan; m870563@yahoo.com
4 School of Medicine, National Defense Medical Center, Neihu, Taipei 11490, Taiwan
5 International Center for Wound Repair and Regeneration, National Cheng Kung University, Tainan 70101, Taiwan
6 Department of Anesthesiology, Kaohsiung Medical University Hospital, Kaohsiung 80756, Taiwan; gcsun39@yahoo.com.tw
7 Faculty of Medicine, College of Medicine, Kaohsiung Medical University, Kaohsiung 80708, Taiwan
* Correspondence: pwcheng@vghks.gov.tw; Tel.: +886-7-3422121 (ext. 71593); Fax: +886-7-3468056
† Pei-Wen Cheng and Yi-Chung Wu contributed equally to this work.

Citation: Cheng, P.-W.; Wu, Y.-C.; Wong, T.-Y.; Sun, G.-C.; Tseng, C.-J. Mechanical Stretching-Induced Traumatic Brain Injury Is Mediated by the Formation of GSK-3β-Tau Complex to Impair Insulin Signaling Transduction. *Biomedicines* **2021**, *9*, 1650. https://doi.org/10.3390/biomedicines9111650

Academic Editors: Kumar Vaibhav, Meenakshi Ahluwalia and Pankaj Gaur

Received: 16 September 2021
Accepted: 6 November 2021
Published: 9 November 2021

Publisher's Note: MDPI stays neutral with regard to jurisdictional claims in published maps and institutional affiliations.

Copyright: © 2021 by the authors. Licensee MDPI, Basel, Switzerland. This article is an open access article distributed under the terms and conditions of the Creative Commons Attribution (CC BY) license (https://creativecommons.org/licenses/by/4.0/).

Abstract: Traumatic brain injury confers a significant and growing public health burden. It is a major environmental risk factor for dementia. Nonetheless, the mechanism by which primary mechanical injury leads to neurodegeneration and an increased risk of dementia-related diseases is unclear. Thus, we aimed to investigate the effect of stretching on SH-SY5Y neuroblastoma cells that proliferate in vitro. These cells retain the dopamine-β-hydroxylase activity, thus being suitable for neuromechanistic studies. SH-SY5Y cells were cultured on stretchable membranes. The culture conditions contained two groups, namely non-stretched (control) and stretched. They were subjected to cyclic stretching (6 and 24 h) and 25% elongation at 1 Hz. Following stretching at 25% and 1 Hz for 6 h, the mechanical injury changed the mitochondrial membrane potential and triggered oxidative DNA damage at 24 h. Stretching decreased the level of brain-derived neurotrophic factors and increased amyloid-β, thus indicating neuronal stress. Moreover, the mechanical injury downregulated the insulin pathway and upregulated glycogen synthase kinase 3β $(GSK-3\beta)^{S9}$/p-Tau protein levels, which caused a neuronal injury. Following 6 and 24 h of stretching, $GSK-3\beta^{S9}$ was directly bound to p-Tau^{S396}. In contrast, the neuronal injury was improved using GSK-3β inhibitor TWS119, which downregulated amyloid-β/p-Taus396 phosphorylation by enhancing ERK1/2T202/Y204 and AktS473 phosphorylation. Our findings imply that the neurons were under stress and that the inactivation of the GSK3β could alleviate this defect.

Keywords: mechanical stimulations; traumatic brain injury (TBI); stretching; neurons; GSK-3β; p-Tau

1. Introduction

Over 60 million new cases of traumatic brain injury (TBI) occur every year worldwide. TBI is the leading cause of death and disability in people aged 1–40 years [1–3]. Despite technological advancements, the treatment for TBI has remained static and is limited to palliative care. The insults leading to TBI include sports injury, crash, falls, and assault [4]. The impact of brain injury can be permanent and can affect an individual in several ways, such as cognition impairment and personality change [4]. The impact is usually negative, which can further affect society. Therefore, it is important to understand the kind of strain that leads to TBI. Moreover, following TBI, even mild physical strain becomes sustained in the neurons. Recent studies have associated TBI with the pathologic accumulation of the neurotoxic proteins Tau, TDP-43, and amyloid-beta, leading to progressive neurodegenerative diseases, including chronic traumatic encephalopathy [5], Alzheimer's disease (AD) [6] and other dementias [7,8]. A moderate to severe TBI is sufficient to increase the risk of developing dementia up to four-fold [6]. Despite these findings, the mechanism by which primary mechanical injury leads to neurodegeneration and increases the risk of dementia-related diseases is unclear. Identifying the interplay between environmental and genetic risk factors for neurodegenerative diseases is critical for the development of therapeutics to mitigate and prevent subsequent pathology. Recently, researchers have attempted the development of in vitro TBI models using a bio-inspired mechanic device. Previous studies have used a high-frequency device to stretch the neurons [9]. This resulted in an injury to the cell bodies and axons. The neurons were found to grow and survive with the rearrangement of cytoskeleton proteins, particularly microtubules, on being subjected to 5% strain. In addition, the aforementioned strain value enhanced the velocity of the synaptic vesicles. On the contrary, an aggressive strain, such as 50% [10], can change the MMP [11], release lactate dehydrogenase, and cause cell death [12,13]. Considering the aforementioned outcomes, mechanical stimulation can alter cell physiology. Therefore, mechanical stimulation can be used for the development of in vitro TBI models based on their impact on neuron growth and survival. Nonetheless, the field of mechanical stimulation on neurons is still unclear.

Reactive oxygen species (ROS) are excessively generated in the injured brain tissue during and after TBI. This can be attributed to changes in the oxygen demand and an abnormal accumulation of cells susceptible to lipid oxidation [14]. Together with the inflammatory response, oxidative stress is supposedly the culprit that leads to neurodegenerative diseases. The aforementioned AD biomarkers can be used to study their expressions and to achieve an in vitro TBI model. Amyloid-β (Aβ) peptides generated from amyloid precursor protein are the hallmark of AD. Aβ peptides accumulate in the brain tissue of patients with AD [15]. The accumulation of Aβ peptides can affect insulin signaling [16], synaptic function, and aberrant gene expressions. In addition, the aforementioned abnormal accumulation plays a role in memory loss. Previous studies contain illustrations representing the diseased neurons with increased Aβ peptides, hyperphosphorylated Tau, alpha-synuclein, and polyglutamine [17]. Furthermore, Tau (or microtubule-associated protein Tau) and α-synuclein proteins are aggregated in AD and PD brains [18–20]. Besides, the change in metabolism following TBI includes an impairment of insulin signaling. Insulin and insulin growth factor (IGF) signaling in the central nervous system regulate cognitive function [21,22]. Insulin/IGF resistance in AD brain leads to decreased phosphoinositol-3-kinase (PI3K)/Akt [23,24] and Wnt/β-catenin signaling [25], as well as increased activation of glycogen synthase kinase 3β (GSK-3β). The impaired insulin/IGF signaling can also disrupt Aβ peptide accumulation [26]. In consequence, the metabolism changes and abnormal protein accumulation promotes oxidative stress in the neurons.

The release of Tau in the cerebrospinal fluid and blood is reportedly a sign of axonal injury. The level of Tau protein increases in Olympic boxers following a mild trauma to the head [27]. Previous studies reported that Tau increase possibly indicates an injury to the central nervous system [28]. The aggregated Tau protein is reportedly toxic, thus was implicated in forming the association between TBI and neurodegenerative diseases. The Tau protein becomes hyperphosphorylated and forms an abnormal aggregate in the cell body of a neuron [21]. It is involved in modulating the binding of microtubule proteins. Thus, the abnormal aggregation of Tau disrupts microtubule formation. This, in turn, disrupts the transmission of neuronal signals and vesicle trafficking in the neurons. We aimed to investigate if mechanical stretching induces oxidative stress and mitochondrial membrane potential (MMP) that causes insulin signaling defects under pathological conditions. Moreover, we intended to identify events critical to the development of Aβ and Tau aggregation. In addition, we clarified if a defect in neuronal insulin signaling triggers the formation of p-Tau^{S396} and pGSK3β^{S9} complex, thus implying neuronal stress. Overall, our results suggest that this neuronal insulin signaling defect is a core mechanism that induces a form of p-Tau^{S396} and pGSK3β^{S9} complex, implying that the neuron cells were under stress and that inactivation of the GSK3β may alleviate this defect.

2. Methods

2.1. Cell Culture

SH-SY5Y neuroblastoma cells were used in this study. The SH-SY5Y cells are frequently used for studying neuron behavior affected by neurotoxic and mechanical injury. The SH-SY5Y cells were maintained in an undifferentiated state. Previously, differentiated SH-SY5Y cells were found resistant to oxidative stress with altered mitochondrial function. The cells were maintained in DMEM, supplemented with 10% FBS, 100 µg/mL penicillin-streptomycin at 37 °C, and 5% CO_2. They were seeded on polydimethylsiloxane (PDMS) at a density of 1.5×10^4/cm^2. The PDMS membrane was stretchable and transparent.

2.2. Stretch Device

The stretch device comprised the following two parts: (1) a primary unit with a strain spindle and (2) a side tray with chambers for cell culture. The cells were cultured in the chamber, filled with a complete medium on the side tray. The side tray was attached to the primary unit with a controlled strain amount. The cells attached to the PDMS membrane were stretched in the biaxial direction. While one end was fixed to the clip, the other end was being stretched.

2.3. Stretched-Injury Model

The cells were stretched following their complete attachment to the PDMS surface. The latter was pre-coated with collagen at 24 h following the seeding. Based on a previous study on moderate and severe stretch-injury model, the neurons were subjected to a severe mechanical stretch comprising a rapid onset strain pulse (25% membrane deformation for 6 and 24 h) at a frequency of 1 Hz. The stretching was performed in a humidified atmosphere at 37 °C and 5% CO_2.

2.4. Cell Alignment Measurement

After stretching, the cells were observed under the microscope (Leica Camera Incorporation, Wetzlar, Germany) with a bright field. Images were taken randomly at magnification 200×. Five fields of view were analyzed, with each field of view having at least 30 cells. The cell orientation or alignment was analyzed using ImageJ software (National Institute of Health, Bethesda, MD, USA) with the angle analysis tool. Cells with angles equal to and greater than $30°$, or equal to and less than $-30°$ were categorized as aligned cells. The cells that were between 30 and $-30°$ were categorized as not aligned.

2.5. Immunoblotting Assay

We resolved the protein (30 µg/sample assessed by the bicinchoninic acid protein assay, Pierce Chemical Co., Rockford, IL, USA) on a 6% polyacrylamide gel and transferred them to a polyvinylidene fluoride membrane (GE Healthcare, Buckinghamshire, UK). The membranes were incubated with appropriate anti-p-TauT231 (ab151559), anti-P-AktS473 (4060, Cell Signaling Technology, Beverly, MA, USA), anti-P-GSK-3βS9 (05-643, EMD Millipore, Billerica, MA, USA), anti-amyloid precursor protein (ab12266), anti-Tau (ab80579), anti-Akt (9272), anti-GSK-3β (07-389), or anti-BDNF antibodies. All the primary antibodies were used at the dilution of 1:1000 in PBST with 5% bovine serum albumin. They were then incubated in an HRP-labelled goat anti-rabbit secondary antibody at 1:10,000. We developed the membranes using an ECL-Plus detection kit (GE Healthcare).

2.6. Measuring BDNF Levels

We measured the BDNF levels using a human BDNF ELISA kit (Life Technologies Corporation Carlsbad, CA, USA). They were detected using a Biochrom Anthos Zenyth 200 rt Microplate Reader (Cambridge, UK).

2.7. Immunofluorescence Assay

The cells were fixed with 4% paraformaldehyde for 20 min, washed with phosphate-buffered saline (PBS), and incubated with Triton X-100 for 10 min (0.5% *v:v*). We performed blocking by incubating the cells with 5% (*w:v*) bovine serum albumin for 30 min. Following washing in PBS, the cells were incubated overnight with primary antibody at 4 °C. We evaluated the DNA damage using anti-8-hydroxy-2-deoxyguanosine (anti-8-OHdG, 1:1000; ab62623 Abcam, Cambridge, UK). Following their binding to primary antibodies, the cells were further incubated with Alexa Fluor-conjugated anti-rabbit, and anti-mouse (Jackson Immunoresearch, 1:1000) for 1 h. They were eventually stained and mounted using the Prolong®DiamondAntifade Mounting Medium, containing 4′,6-diamidino-2-phenylindole (Life Technology). We analyzed them with an Olympus DP71 device (100× and 200× magnification).

2.8. Mitochondrial Membrane Potential

THC-stimulated changes in the mitochondrial membrane potential (MMP) were assessed using the fluorescent reagent tetraethylbenzimidazolylcarbocyanine iodide (JC-1) with the JC-1-Mitochondrial Membrane Potential Assay Kit (Abcam, Cat. no. ab113850) following the manufacturer's protocol. Cells were washed once with 1× dilution buffer and then incubated with 20 µM JC-1 dye in 1× dilution buffer for 10 min at 37 °C, protected from light. JC-1 dye was then removed, cells were washed once with 1× dilution buffer, and 100 µL of fresh 1× dilution buffer was added to each well. The red fluorescence in excitation (535 nm)/emission (590 nm) and green fluorescence excitation/emission (475 nm/530 nm) was measured using a Biochrom Anthos Zenyth 200 rt Microplate Reader (Cambridge, UK). Background fluorescence was subtracted from the fluorescence of treated cells, then the ratio of red (polarized) fluorescence divided by that of green (depolarized) fluorescence was obtained.

2.9. Co-Immunoprecipitation

The Catch and Release Reversible Immunoprecipitation System (Millipore) was used according to the manufacturer's instructions. The proteins were eluted in 70 µL of the elution buffer and subjected to an immunoblotting analysis, using the anti-GSK-3β[S9] and anti-p-Tau[S396] antibodies.

2.10. Statistical Analyses

All measurements were produced at least three times under independent conditions. The results are shown as mean ± standard error of the mean (SEM). Statistics were analyzed with Mann-Whitney U-test. All statistical analyses were carried out on raw data using SPSS, version 20.0 (SPSS Inc, Chicago, IL, USA). * p value < 0.05 indicated a significant result. ** p < 0.01 indicated an extremely significant result.

3. Results

3.1. Mechanical Injury Induces Neuron Injury through Stretching, thus Altering the Mitochondrial Membrane Potential and Inducing Oxidative DNA Damage

Neurons rely on signal transmission for cell-to-cell communication in a neuronal network. The mitochondrial membrane potential plays an important role in maintaining signal transduction in neurons. Herein, we investigated the effect of mechanical stress on neurons to establish an injured neuronal cell model. The neurons aligned perpendicularly to the stretching direction (Figure 1A). Furthermore, the MMP was altered after 6 h of cyclical stretching at 25% and 1 Hz uniaxial deformation, as shown by JC-1 staining. In turn, this alteration substantially increased the ratios of green/red fluorescence compared to those of the control. Thus, stretching triggers MMP collapse by depolarization (Figure 1B). TBI-induced oxidative stress in the brain can lead to neurodegenerative diseases as well. Oxidative stress is a major mediator of the secondary injury that follows TBI [29]. Neuronal death, a hallmark of TBI, where the loss of dopaminergic neurons and dopaminergic dysfunction is observed. Levels of dopamine (DA) were shown to be decreased 24 h after injury in an experimental mouse model of TBI [30]. Therefore, we determined to study if mechanical stress reduces levels of DA and oxidative DNA damage. Following stretching at 25% and 1 Hz for 24 h, the 8-OHdG immunoreactivity had substantially increased compared to the control (Figure 1C), and the levels of DA had substantially decreased compared to the control (Figure 1D). Hence, stretching-mediated mechanical injury altered the mitochondrial membrane potential and increased oxidative DNA damage.

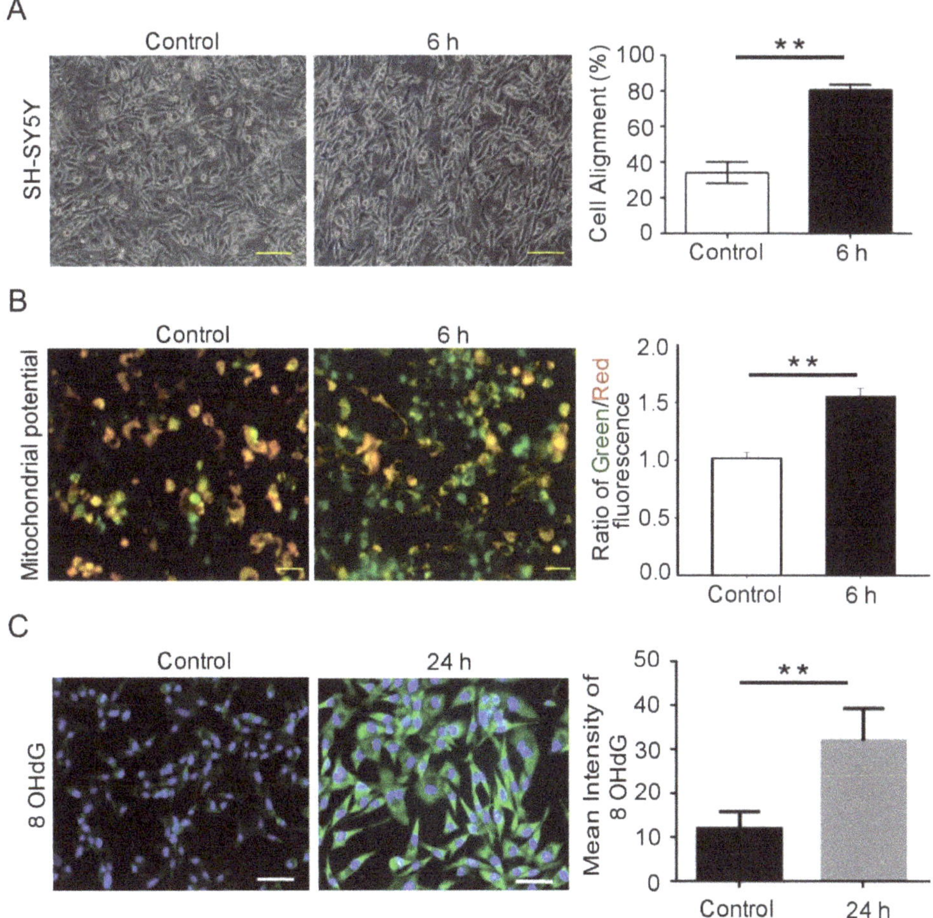

Figure 1. *Cont.*

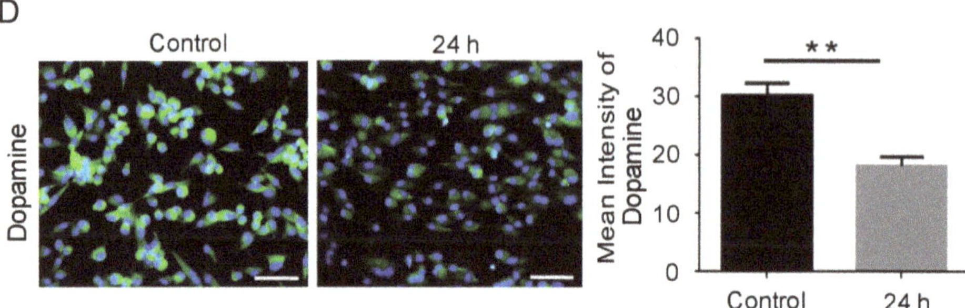

Figure 1. Mechanical injury induces neuron injury through stretching. (**A**) SH-SY5Y neuroblastoma cells are stretched in the uniaxial direction; cell alignment was measured by an angle deviation of 30° in plane, with both ends of each cell. Cell alignment after 6 h was measured using ImageJ. (**B**) The cells were stained with JC-1 probe immediately after 6 h of stretching. The JC-1 probe detects changes in the mitochondrial potential. It was analyzed by flow cytometry for 6 h after stretching. The mechanical stimulation altered the mitochondrial membrane potential. (**C**) A set of representative anti-8-OHdG immunofluorescent stained images (green) depicts an increase in 8-OHdG fluorescence signal at 24 h of stretching, compared to the control. (**D**) A set of representative anti-Dopamine immunofluorescent stained images (green) depicts a decrease in dopamine fluorescence signal at 24 h of stretching, compared to the control. Blue color 4′,6-diamidino-2-phenylindole denotes nuclei. Blue color 4′,6-diamidino-2-phenylindole denotes nuclei. Data are presented as mean ± SD. Scale bar = 50 μm; ** $p < 0.01$. 8-OHdG, anti-8-hydroxy-2-deoxyguanosine.

3.2. BDNF Reduction Is Associated with Increased Amyloid-β/p-Tau

Mechanical stretching triggered neuronal injury by inducing oxidative DNA damage (Figure 1). The BDNF level, amyloid-β and p-Tau^{S396,} was altered following 24 h of cyclical stretching at 25% and 1 Hz uniaxial strain as shown by immunofluorescence and ELISA assay. Cyclical stretching decreased BDNF levels and increased amyloid-β/p-Taus396 in the SH-SY5Y neuroblastoma cells (Figure 2A–C). This BDNF decrease suggests that the neurons were being invoked by stress, whereby BDNF acted as a neuroprotective factor. Moreover, its decrease was accompanied by an increase in aggregated p-Tau^{S396} protein, which is an indication of neuronal injury (Figure 2D). Therefore, we could successfully establish a neuron injury model by mechanical stretching of the neuroblastoma cells.

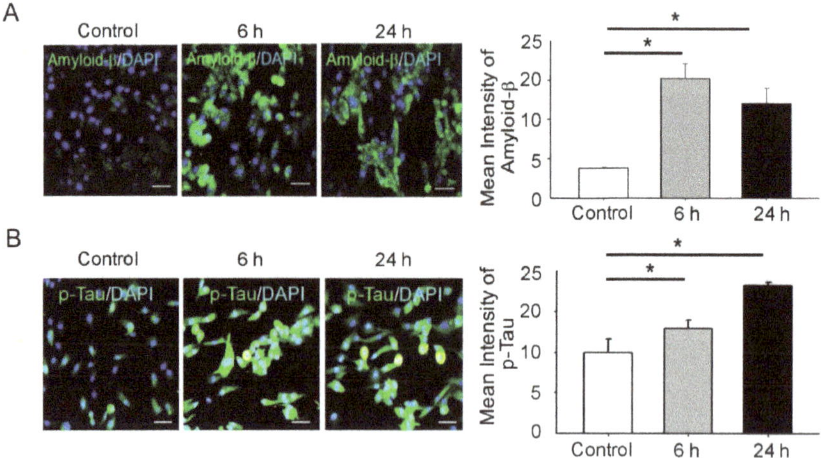

Figure 2. *Cont.*

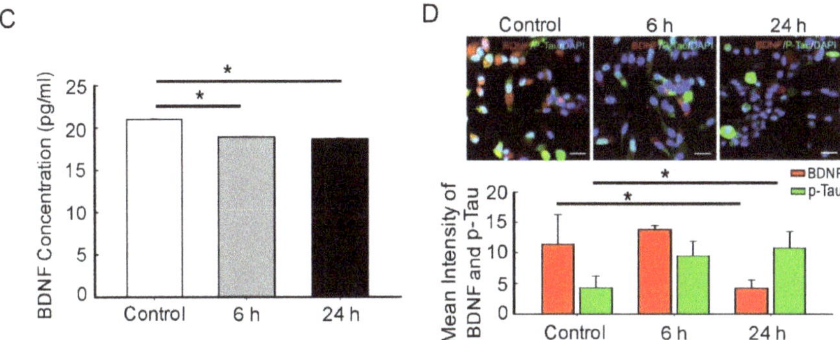

Figure 2. Mechanical injury through stretching upregulates amyloid β and p-Tau levels and decreases BDNF expression. (**A,B**) Representative green fluorescence images of amyloid β and p-Tau^{S396} positive cells in the SH-SY5Y, before and after stretching. The nuclei of the SH-SY5Y cells were counterstained with 4′,6-diamidino-2-phenylindole and exhibit blue fluorescence. Bar graphs representative of the amyloid β and p-Tau^{S396} fluorescence intensity in the SH-SY5Y cells of the indicated groups. The amyloid β and p-Tau^{S396} fluorescence intensity was significantly high in the 25%, 1 Hz group at 6 and 24 h. Scale bar = 200 μm. (**C**) An ELISA assay was performed to analyze BDNF expression. Mechanical injury significantly lowered BDNF levels, compared to the control. (**D**) Immunofluorescence assay was performed to analyze BDNF and p-Tau^{S396} protein expression. Bar graphs representative of the BDNF and p-Tau^{S396} fluorescence intensity in the SH-SY5Y cells of the indicated groups. The p-Tau^{S396} fluorescence intensity was significantly high in the 25%, 1 Hz group at 6 and 24 h. However, BDNF was attenuated at 24 h. The values are presented as means ± SEM (n = 6). Scale bar = 200 μm; * $p < 0.05$. BDNF, brain-derived neurotrophic factor; ELISA, enzyme-linked immunosorbent assay.

3.3. Mechanical Injury through Stretching Upregulated p-GSK3β/p-Tau Protein Levels and Are Associated with a Reduction of the Insulin Pathway

Insulin/IGF resistance in AD brain results in decreased phosphoinositol-3-kinase (PI3K)/Akt [23,24] and increased activation of glycogen synthase kinase 3β (GSK-3β). Furthermore, injured neurons accumulate phosphorylated Tau and GSK3β. Thus, we aimed to determine if AKT/extracellular-signal-regulated kinase (ERK) signaling pathway defects upregulated amyloid-β/p-Tau^{S396}. Both p-Tau^{S396} and p-GSK3β^{S9} protein levels were significantly upregulated, 6 h following the mechanical stretching (Figure 3A). p-AKT/p-ERK, the proteins were significantly downregulated following 6 and 24 h of mechanical stretching (Figure 3B). Thus, mechanical stretching downregulates the p-AKT/p-ERK and upregulates p-GSK3β/p-Tau protein levels, which results in neuronal injury.

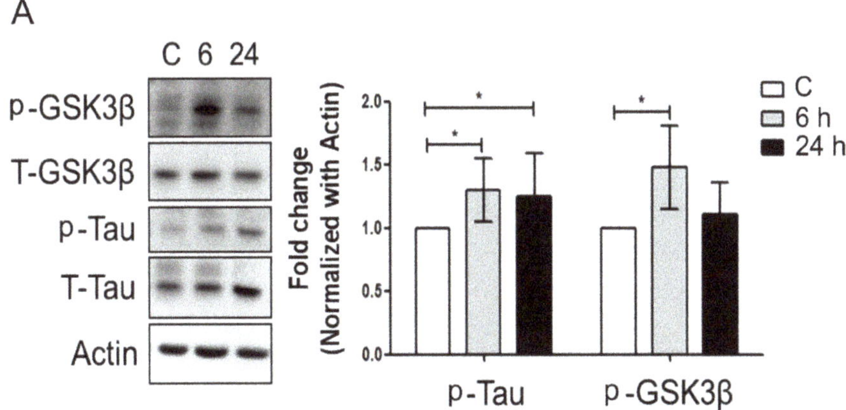

Figure 3. *Cont.*

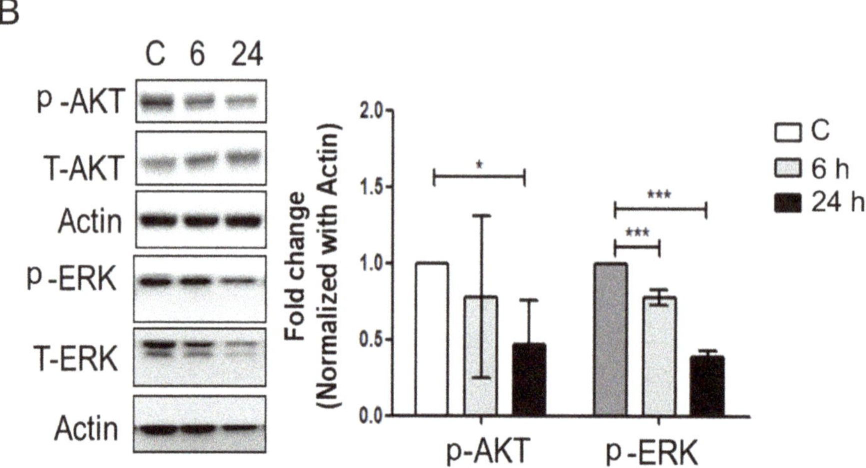

Figure 3. Mechanical injury through stretching upregulates p-Tau and p-GSK3β protein levels and decreases the insulin pathway. (**A**) p-Tau and p-GSK3β; (**B**) p-AKT and p-ERK; proteins expressions. p-GSK3β and p-Tau^{S396} protein levels significantly increased in the 25%, 1 Hz group at 6 and 24 h of stretching. However, the p-AKT and p-ERK protein expressions had the reverse effect. The values are presented means ± SEM (n = 6); *** $p < 0.001$; * $p < 0.05$. GSK3β, glycogen synthase kinase 3; ERK, extracellular-signal-regulated kinase.

3.4. The Interaction between GSK-3β and p-Tau Plays a Crucial Role and Is Associated with the Reduction of the Insulin Pathway

A defect in the insulin pathway is a critical link between Tau and/or Aβ pathologies that define AD [31]. The aforementioned upregulation of pGSK3β/p-Tau protein levels interfered with the insulin pathway. Therefore, we investigated the formation of p-Tau^{S396} and pGSK3β^{S9} protein complex by co-immunoprecipitation. The p-GSK3β^{S9} antibody was able to capture the p-Tau^{S396} protein following 6 and 24 h of mechanical stretching (Figure 4A). In addition, we determined the involvement of GSK-3β in the stretch-induced neuronal injury and insulin pathway. TWS119, the GSK3β phosphorylation inhibitor was administered to the cells at final concentrations of 0, 5, and 10 μM. In combination with 24 h stretching, the p-Tau^{S396} protein level was also downregulated when treated with 10 μM TWS119 (Figure 4B). The administration of TWS119 resulted in an increase in insulin signaling (Figure 4C). The p-GSK3β^{S9} protein level was downregulated when treated with 10 μM TWS119. However, β-catenin has the opposite effect (Figure 4D). Moreover, it leads to a decrease in Tau^{S396} phosphorylation and improves insulin receptor (IR) signaling [32]. Next, we examined the BDNF, Aβ, and Tau^{S396} phosphorylation levels. TWS119 administration resulted in an increase in BDNF levels and a decreased amyloid-β/p-Tau^{S396} in immunofluorescence (Figure 5A–C). Therefore, stretching-mediated mechanical injury downregulated the insulin pathway and upregulated p-GSK3β/p-Tau protein levels, which caused neuronal injury. However, TWS119 attenuates the mechanical stretching-induced activity of Aβ-Tau and improves neuronal injury. Hence, GSK3β promotes the expression of Aβ-Tau, which in turn downregulates IR signaling and BDNF production stretch-induced neuronal injury.

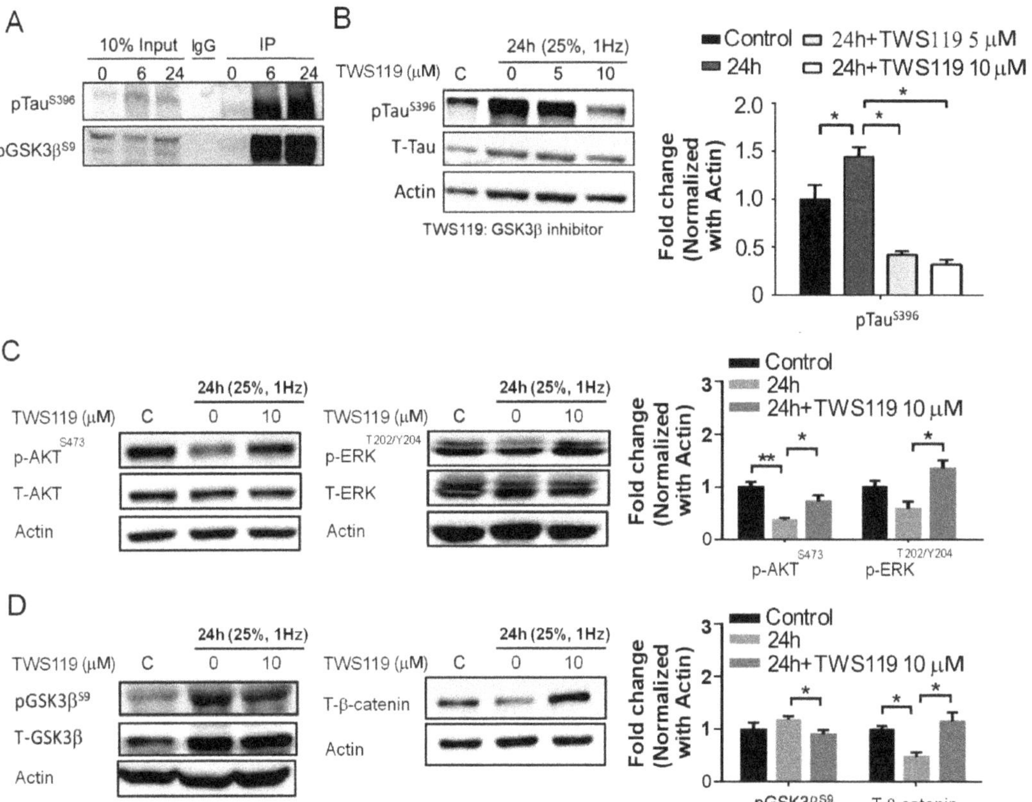

Figure 4. Mechanical injury promotes p-Tau-pGSK3β complex formation and affects the insulin pathway. (**A**) Co-immunoprecipitation was performed to study the p-Tau^{S396} and anti-GSK-3β^{S9} protein complex formation. The input protein comprises 10% of the total lysates, and the remaining 90% protein lysate was incubated with the p-Tau^{S396} or anti-GSK-3β^{S9} primary antibody. Co-immunoprecipitation was performed to study the p-Tau and p-GSK3β protein complex formation. The input protein comprises 10% of the total lysates, and the remaining 90% protein lysate was incubated with the p-GSK3β primary antibody. (**B**) We analyzed the function of p-GSK3β by adding the p-GSK3β inhibitor and TWS119 while the cells were being stretched. p-Tau^{S396} protein levels significantly decreased in the 25%, 1 Hz-TWS119 group. (**C,D**) We conducted Western blot to analyze phosphorylated p-Akt^{S473}, p-ERK$^{T202/Y204}$, p-GSK-3β^{S9}, and T-β-catenin protein in SH-SY5Y cells, before and after the stretching or TWS119 administration. The immunoblot demonstrates greater levels of p-Akt^{S473}, p-ERK$^{T202/Y204}$, p-GSK-3β^{S9}, and T-β-cateninin SH-SY5Y cells, before and after the stretching or TWS119 administration. ** $p < 0.01$; * $p < 0.05$. GSK3β, glycogen synthase kinase 3; ERK, extracellular-signal-regulated kinase.

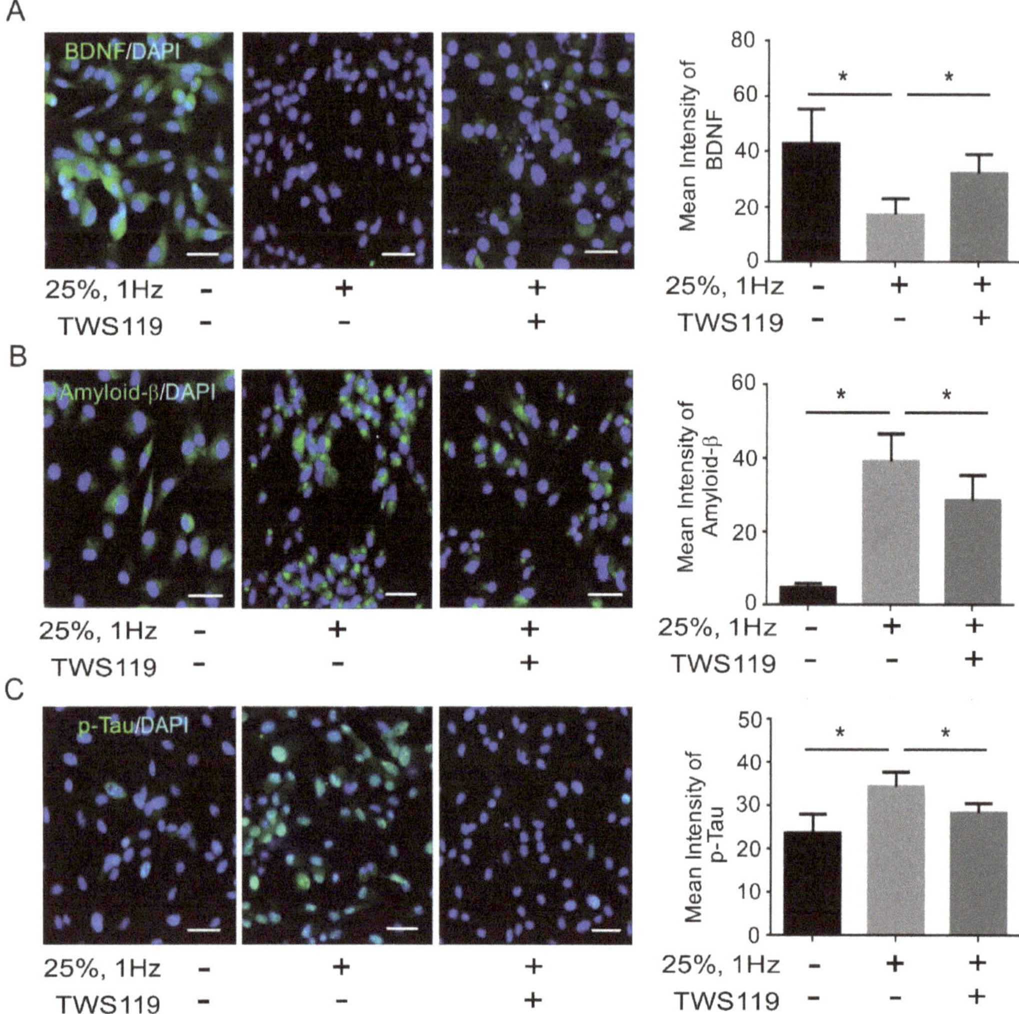

Figure 5. P-GSK3β reduces BDNF levels and upregulates amyloid β/p-Tau expression in the neuron injury model. (**A–C**) Representative green fluorescence images of BDNF, amyloid β, and p-Tau^{S396} positive cells in SH-SY5Y cells, before and after stretching or TWS119 administration. The nuclei of SH-SY5Y cells were counterstained with DAPI and exhibited blue fluorescence. Bar graphs are representative of the BDNF, amyloid β, and p-Tau^{S396} fluorescence intensity in the SH-SY5Y cells of the indicated groups. The BDNF fluorescence intensity increased substantially in the 25%, 1 Hz-TWS119 group, with diminished amyloid β and p-Tau^{S396} intensity. The values are presented as means $\pm$ SEM ($n = 6$); * $p < 0.05$. BDNF, brain-derived neurotrophic factor. Scale bar = 200 μm.

4. Discussion

TBI is defined as damage to the brain that consequently disrupts normal function. Previous in vitro studies on TBI used to stretch and shear forces on either rodent organotypic slices [33] or rudimentary single- or two-cell type cultures [34]. Despite being useful for studying axonal injury, the aforementioned models have limited translational relevance with regards to the extracellular matrix, stiffness, and cell–cell interactions that substantially influence the biophysical properties of mechanical injury in vivo [35]. The target signaling pathway for treating degenerative neuron disease is unclear. However, lost and damaged neurons cannot proliferate in our bodies. Hence, neurodegenerative diseases are usually progressive and are difficult to cure. The neurons generally do not move, are not stretched, or compressed. However, they experience a certain amount of strain during and even after TBI [9]. Considering their less motile behavior, a minor strain on the neurons can result in minor injury. The low motility behavior also implicates the significance of cell–cell communication between neurons. The smallest impact can also affect cell–cell communication by disrupting neuronal transmission.

Recently, researchers have attempted the development of in vitro TBI models using a bio-inspired mechanic device. Previous studies have used a high-frequency device to stretch the neurons [10]. Aggressive stretch can lead to pro-apoptotic neuron death and axonal damage. The neurons usually survive stretching-induced axonal damage, thus enabling a study of the injury to the axons [11]. Despite the role of stretching in promoting neuron growth, axonal growth did not necessarily result in axonal regeneration in previous studies. The axons repaired the damage; however, they were unable to regrow the entire length. Thus, stretch-induced neuronal injury can be used for studying neuron growth, death, and regeneration [12]. The diversity of neuropathology, combined with the heterogeneity in injury distribution, suggests an extensive pathological remodeling of the cellular microenvironment of the brain, the causes of which need to be elucidated to prevent TBI. Considering that mechanical forces are a primary contributor to the etiology of TBI, understanding the biomechanics of injury may elucidate complex injury mechanisms and explain the diverse pathophysiology associated with TBI.

Neuronal death, a hallmark of TBI, is related to the development of neurodegenerative disorders such as Parkinson's disease (PD), where the loss of dopaminergic neurons and dopaminergic dysfunction are observed. Hector et al., to simulate TBI subjected to 0%, 5%, 10%, 15%, 25% and 50% deformation, demonstrated that 24 h after injury, cell viability and apoptosis were determined by lactate dehydrogenase release and DNA fragmentation. Extracellular dopamine (DA) levels increased only at 50%. Levels of DA remained unchanged regardless of treatment. These data support the use of stretch as a model to simulate TBI in vitro in human dopaminergic neurons, replicating the acute effects of TBI in the dopaminergic system [13]. Besides, there is increasing evidence that apoptosis induced by TBI often starts with an accumulation of ROS and induces hypofunction of the striatal dopaminergic system in vitro and in vivo [34,36,37]. In this regard, SH-SY5Y cells retain dopamine-β-hydroxylase activity, thus being suitable for the neuromechanistic study [34,37]. Therefore, we determined to study if mechanical stress reduces levels of DA and oxidative DNA damage. The neurons aligned perpendicularly to the stretching direction (Figure 1A). Furthermore, the MMP was altered following 6 h of cyclical stretching at 25% and 1 Hz uniaxial strain as shown by JC-1 staining. This in turn substantially increased the ratios of green/red fluorescence than the control. Thus, stretching triggers MMP collapse by depolarization (Figure 1B). Following stretching at 25% and 1 Hz for 24 h, the 8-OHdG immunoreactivity had substantially increased, compared to the control (Figure 1C) and the levels of DA had substantially decreased, compared to the control (Figure 1D). Hence, stretching-mediated mechanical injury altered the mitochondrial membrane potential, DA release and increases oxidative DNA damage (Figure 1). Oxidative stress replicates the key transcriptional programs that are hallmarks of neuronal injury, including the upregulation of oxidative stress and aberrant phosphorylation of GSK-3β. Thus, the present founding determined that cyclical stretching increased in DNA damage downregulated BDNF release and increased amyloid-β/p-Tau^{S396} in the SH-SY5Y neuroblastoma cells (Figure 2A–C). This BDNF decrease suggests that the neurons were being invoked by a stress, whereby BDNF acts as a neuroprotective factor. Moreover, its decrease was accompanied with an increase in aggregated p-Tau^{S396} protein, which is an indication of neuronal injury (Figure 2D). In addition, the method accurately recapitulates two key pathological features of neurodegenerative diseases. The aggregation and accumulation of Aβ and Tau are the key pathological markers of TBI, which contribute to the progressive deterioration associated with TBI.

GSK3 is a key kinase contributing to abnormal phosphorylation of the microtubule-binding protein Tau in the process thought to cause neurofibrillary tangles in Alzheimer's disease [38,39]. It is well known that the action of GSK3β is inhibited by phosphorylation of the enzyme on serine 9 following insulin activation [40]. However, insulin resistance causes inhibition of downstream signal transduction and GSK3 remains in its active form, which hyper-phosphorylates several substrates, including Tau protein, in nerve cells. Hyper-phosphorylation of Tau destabilizes the microtubules and induces neurofibrillary tangle formation that leads to AD [41]. Leng et al. demonstrated that GSK3β as a kinase required for high glucose-induced serine332 phosphorylation, ubiquitination, and degradation of IRS1 [42]. Besides, Liberman et al. studies identify Ser332 as the GSK-3 phosphorylation target in IRS-1, indicating its physiological relevance and demonstrating its attenuate insulin signaling [43]. Thus, GSK3β may play a causative role in the regulation of insulin pathway and connecting AD. Previous studies have established a correlation between AD an altered responsiveness to insulin and IGF stimulation in the brain [44]. Furthermore, the aforementioned abnormal accumulation of amyloid-β [45] and Tau aggregates may be linked to insulin/IGF resistance. Therefore, AD is presumably a metabolic disease [46]. Insulin/IGF signaling is responsible for regulating neuronal growth, survival, differentiation, migration, energy metabolism, cytoskeletal assembly, synapse formation, neurotransmitter function, and plasticity [47]. Insulin/IGF resistance in the brain can result in the inhibition of pro-survival and pro-growth signaling pathways. Furthermore, insulin and IGF bind to their own receptors on the cell membrane, leading to the phosphorylation of insulin receptor substrate (IRS) molecules, which eventually activate downstream signals. Accumulating evidence indicates that BDNF and its interaction with ROS may be crucial for neurodegenerative and neuropsychiatric conditions [48]. However, there is no therapeutic approach that appears promising in reducing the progression of TBI to chronic neurodegenerative disease. Increased phosphorylation of Tau was previously correlated to neurodegeneration, in the context of TBI in a rodent model [49] and in numerous clinical studies [5]. Furthermore, we also founding that injured neurons through mechanical stretching accumulate phosphorylated Tau and GSK3β. Both p-TauS396 and p-GSK3βS9 protein levels were significantly upregulated, 6 h following the mechanical stretching (Figure 3A). Besides, p-AKT/p-ERK proteins of survival and cell growth were significantly downregulated following 6 and 24 h of mechanical stretching (Figure 3B). Thus, the present result also demonstrated that stretching-mediated mechanical injury downregulated the insulin pathway and upregulated p-GSK3β/p-Tau protein levels, which caused neuronal injury.

Multiple lines of evidence strongly suggest that the inhibition of GSK-3 is a potential target for the treatment of TBI. GSK-3 constitutively inhibits neuroprotective processes and promotes apoptosis. Following TBI, the receptor tyrosine kinase (RTK) and canonical Wnt signaling pathways inhibit GSK-3 as an innate neuroprotective mechanism against TBI. GSK-3 inhibition via GSK-3 inhibitors and RTK-activating drugs or Wnt signaling is likely to reinforce the innate neuroprotective mechanism. GSK-3 inhibition studies using rodent TBI models have demonstrated that this inhibition produces diverse neuroprotective actions, such as a reduction in the size of the traumatic injury, Tauopathy, Aβ accumulation, and neuronal death, caused by the release and activation of neuroprotective substrates. The above-mentioned effects are correlated with reduced TBI-induced behavioral and cognitive symptoms [50]. We investigated the formation of p-Tau^{S396} and pGSK3β^{S9} protein complex by co-immunoprecipitation. The p-GSK3β^{S9} antibody was able to capture the p-Tau^{S396} protein following 6 and 24 h of mechanical stretching (Figure 4A). In addition, we determined the involvement of GSK-3β in the stretch-induced neuronal injury and insulin pathway. TWS119, the GSK3β phosphorylation inhibitor was administered to the cells at final concentrations of 0, 5, and 10 μM, in combination with 24 h stretching, the p-Tau^{S396} protein level was also downregulated when treated with 10 μM TWS119 (Figure 4B). The administration of TWS119 resulted in an increase in the insulin signaling (Figure 4C). Moreover, it leads to a decrease in Tau^{S396} phosphorylation and improved insulin receptor (IR) signaling. However, TWS119 attenuates the mechanical stretching-induced activity of Aβ-Tau and improves neuronal injury. Hence, GSK3β promotes the expression of Aβ-Tau, which in turn downregulates IR signaling and BDNF production in stretch-induced neuronal injury (Figure 5).

Mechanical stretch injury, which was developed and characterized by Ellis and coworkers, had been used to study the effects of trauma on neurons and astrocytes in vitro [51,52]. Following TBI, neuronal loss is characterized by oxidative stress reaction, mitochondrial dysfunction, neurotoxicity, and neuroinflammation [53,54]. Our findings demonstrate that mechanical stretch injury to SH-SY5Y cells resulted in oxidative stress and the production of MMP that triggered insulin signaling defects and events critical for Aβ and Tau aggregation mimicking pathological conditions. In addition, the neuronal insulin signaling defects promoted the formation of pTau^{S396} and pGSK-3β^{S9} complex, thus further implying the neuronal stress. Thus, inhibition of GSK-3β activity can serve as a potential target for increasing BDNF levels and insulin signaling as well as reducing phosphorylation events, which are critical for Aβ and Tau aggregation. Tws119, an inhibitor of GSK-3β, improves pathological conditions, by positively regulating BDNF and insulin signaling. Overall, the aforementioned neuronal insulin signaling defect is a core mechanism that induces the formation of pTau^{S396} and pGSK-3β^{S9} complex. Hence, the neurons under stress may alleviate this defect upon the inactivation of the GSK-3β (Figure 6). The present findings suggest that the protective role of GSK3β in SH-SY5Y cells that were exposed to mechanical stretch injury was mainly due to its cause insulin signaling defect and Aβ and Tau aggregation. Therefore, our study has some limitations such as SH-SY5Y is a proliferative cell line, which is entirely different from the terminally differentiated neurons. Although SH-SY5Y cells exhibit some features of neurons [37], primary cultured neurons might be more appropriate for investigating endogenous mechanisms after injury.

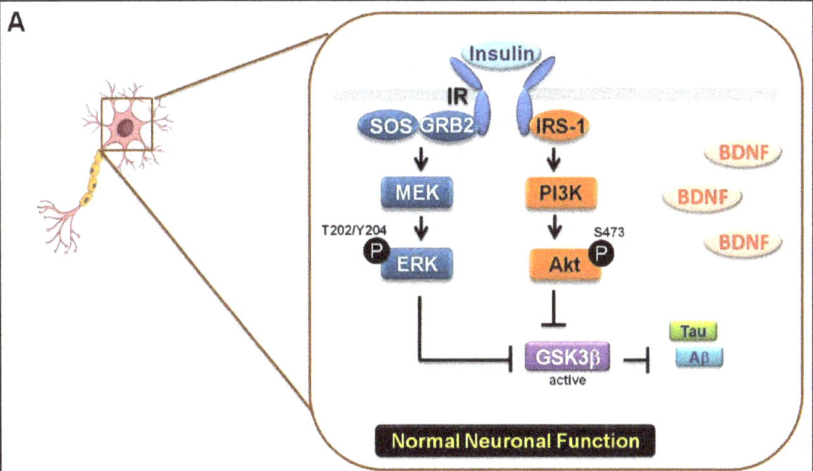

Figure 6. *Cont.*

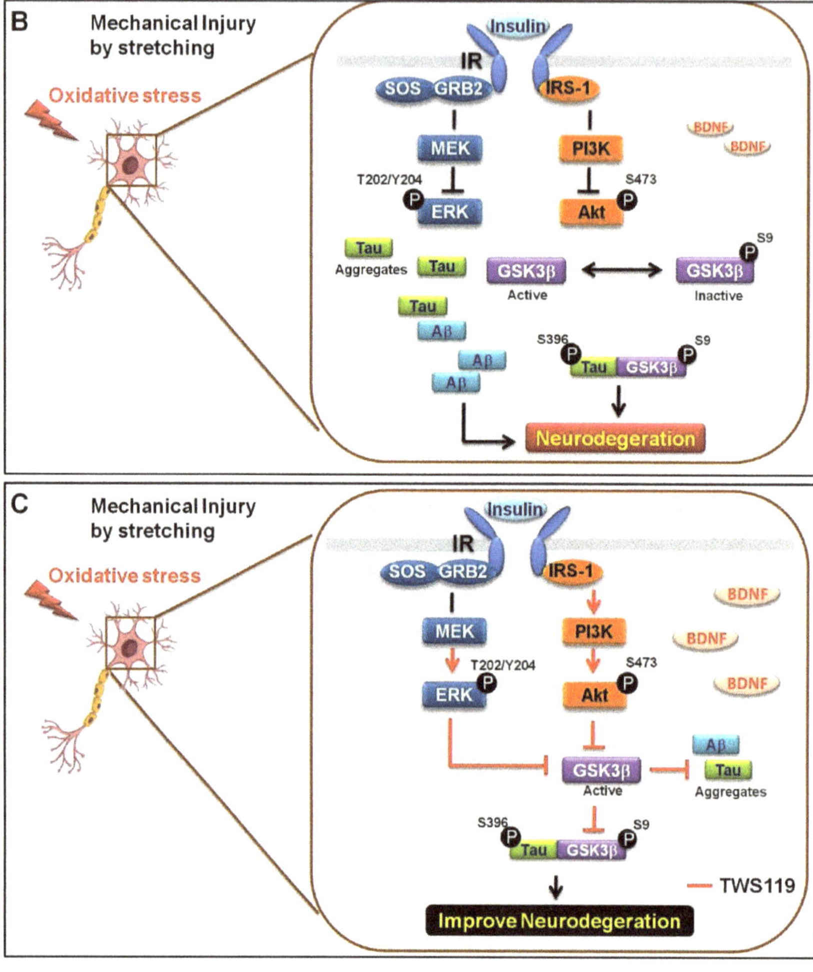

Figure 6. A schematic model suggesting the mechanistic details of the regulation of neurodegenerative diseases by GSK-3β. (**A**) Under normal conditions, the activated PI3K/Akt ERKs signaling pathway leads to the inactivation of GSK-3β, which suppresses the downstream phosphorylation of Tau and Aβ that results in positive regulation of cognitive signaling, such as BDNF and insulin. (**B**) Under pathological conditions, mechanical stretching increases GSK-3β activity, induces oxidative stress, and increases MMP. Upregulation of GSK-3β activity negatively regulates BDNF and insulin signaling and promotes phosphorylation events critical to the development of Aβ and Tau aggregation. (**C**) Tws119, an inhibitor of GSK-3β, improves pathological conditions by positively regulating BDNF and insulin signaling. Moreover, it reduces phosphorylation events critical to the development of Aβ and Tau aggregation. BDNF, brain-derived neurotrophic factor; MMP, mitochondrial membrane potential; and GSK-3, glycogen synthase kinase 3.

5. Conclusions

This is the first study to demonstrate the mechanism by which mechanical stretching stimulates the accumulation of amyloid-β, a neuronal injury biomarker through p-Tau^{S396} and pGSK3β^{S9} protein complex formation. The latter, in turn, downregulates IR signaling and BDNF production. The role of mechanical stretching on neurons requires further investigation to explore the upstream mechanism underlying amyloid-β accumulation. This neuron injury model was developed to mimic neurons under stress. Mechanical stretching does not normally occur on neurons in vivo. Nonetheless, they respond to mechanical cues for intracellular signaling. Therefore, the neuron injury cell model can explain the mechanistic signaling for potential drug discovery in targeting the accumulated amyloid-β, p-Tau^{S396}, and pGSK3β^{S9} protein complex formation.

Author Contributions: The study was conceived and designed by Y.-C.W. P.-W.C. conducted most of the experiments with assistance from T.-Y.W., G.-C.S. and C.-J.T. The paper was written by P.-W.C. All authors have read and agreed to the published version of the manuscript.

Funding: This work was supported by funding from the National Science Council (MOST108-2320-B-075B-001), the Zouying Branch of Kaohsiung Armed Forces General Hospital Kaohsiung (ZBH 108-05, KAFGH-ZY-A-109001, KAFGH-ZY-A-110001), and the Kaohsiung Veterans General Hospital (KSVGH-IGA-110-1).

Institutional Review Board Statement: Not applicable.

Informed Consent Statement: Not applicable.

Data Availability Statement: All data generated or analysed during this study are included in this published article.

Acknowledgments: We gratefully acknowledge Yu-Ju Hsiao for the technical assistance and the accuracy of the data analysis.

Conflicts of Interest: The authors declared that they have no competing interests.

References

1. Dhandapani, S.; Manju, D.; Sharma, B.; Mahapatra, A. Prognostic significance of age in traumatic brain injury. *J. Neurosci. Rural Pract.* **2012**, *3*, 131–135. [CrossRef]
2. Fleminger, S.; Ponsford, J. Long term outcome after traumatic brain injury: More attention needs to be paid to neuropsychiatric functioning. *BMJ Br. Med. J.* **2005**, *331*, 1419–1420. [CrossRef] [PubMed]
3. Dewan, M.C.; Rattani, A.; Gupta, S.; Baticulon, R.E.; Hung, Y.C.; Punchak, M.; Agrawal, A.; Adeleye, A.O.; Shrime, M.G.; Rubiano, A.M.; et al. Estimating the global incidence of traumatic brain injury. *J. Neurosurg.* **2018**, *130*, 1–18. [CrossRef]
4. Blennow, K.; Hardy, J.; Zetterberg, H. The neuropathology and neurobiology of traumatic brain injury. *Neuron* **2012**, *76*, 886–899. [CrossRef] [PubMed]
5. McKee, A.C.; Stern, R.A.; Nowinski, C.J.; Stein, T.D.; Alvarez, V.E.; Daneshvar, D.H.; Lee, H.S.; Wojtowicz, S.M.; Hall, G.; Baugh, C.M.; et al. The spectrum of disease in chronic traumatic encephalopathy. *Brain* **2013**, *136*, 43–64. [CrossRef] [PubMed]
6. Guo, Z.; Cupples, L.A.; Kurz, A.; Auerbach, S.H.; Volicer, L.; Chui, H.; Green, R.C.; Sadovnick, A.D.; Duara, R.; DeCarli, C.; et al. Head injury and the risk of ad in the mirage study. *Neurology* **2000**, *54*, 1316–1323. [CrossRef]
7. Nordstrom, P.; Michaelsson, K.; Gustafson, Y.; Nordstrom, A. Traumatic brain injury and young onset dementia: A nationwide cohort study. *Ann. Neurol.* **2014**, *75*, 374–381. [CrossRef] [PubMed]
8. Gardner, R.C.; Burke, J.F.; Nettiksimmons, J.; Kaup, A.; Barnes, D.E.; Yaffe, K. Dementia risk after traumatic brain injury vs nonbrain trauma: The role of age and severity. *JAMA Neurol.* **2014**, *71*, 1490–1497. [CrossRef]
9. Hall, E.D.; Wang, J.A.; Bosken, J.M.; Singh, I.N. Lipid peroxidation in brain or spinal cord mitochondria after injury. *J. Bioenerg. Biomembr.* **2016**, *48*, 169–174. [CrossRef]
10. Sadigh-Eteghad, S.; Sabermarouf, B.; Majdi, A.; Talebi, M.; Farhoudi, M.; Mahmoudi, J. Amyloid-beta: A crucial factor in alzheimer's disease. *Med. Princ. Pract.* **2015**, *24*, 1–10. [CrossRef]
11. Freund, R.K.; Gibson, E.S.; Potter, H.; Dell'Acqua, M.L. Inhibition of the motor protein eg5/kinesin-5 in amyloid β-mediated impairment of hippocampal long-term potentiation and dendritic spine loss. *Mol. Pharmacol.* **2016**, *89*, 552–559. [CrossRef]
12. Moussaud, S.; Jones, D.R.; Moussaud-Lamodiere, E.L.; Delenclos, M.; Ross, O.A.; McLean, P.J. Alpha-synuclein and tau: Teammates in neurodegeneration? *Mol. Neurodegener.* **2014**, *9*, 43. [CrossRef] [PubMed]
13. Olczak, M.; Niderla-Bielinska, J.; Kwiatkowska, M.; Samojlowicz, D.; Tarka, S.; Wierzba-Bobrowicz, T. Tau protein (mapt) as a possible biochemical marker of traumatic brain injury in postmortem examination. *Forensic Sci. Int.* **2017**, *280*, 1–7. [CrossRef] [PubMed]
14. Waxman, E.A.; Giasson, B.I. Induction of intracellular tau aggregation is promoted by α-synuclein seeds and provides novel insights into the hyperphosphorylation of tau. *J. Neurosci.* **2011**, *31*, 7604–7618. [CrossRef] [PubMed]
15. Koch, J.C.; Bitow, F.; Haack, J.; d'Hedouville, Z.; Zhang, J.N.; Tönges, L.; Michel, U.; Oliveira, L.M.A.; Jovin, T.M.; Liman, J.; et al. Alpha-synuclein affects neurite morphology, autophagy, vesicle transport and axonal degeneration in cns neurons. *Cell Death Dis.* **2015**, *6*, e1811. [CrossRef]
16. de la Monte, S.M. Brain insulin resistance and deficiency as therapeutic targets in alzheimer's disease. *Curr. Alzheimer Res.* **2012**, *9*, 35–66. [CrossRef]

17. Fujisawa, Y.; Sasaki, K.; Akiyama, K. Increased insulin levels after ogtt load in peripheral blood and cerebrospinal fluid of patients with dementia of alzheimer type. *Biol. Psychiatry* **1991**, *30*, 1219–1228. [CrossRef]

18. de la Monte, S.M.; Ganju, N.; Banerjee, K.; Brown, N.V.; Luong, T.; Wands, J.R. Partial rescue of ethanol-induced neuronal apoptosis by growth factor activation of phosphoinositol-3-kinase. *Alcohol. Clin. Exp. Res.* **2000**, *24*, 716–726. [CrossRef]

19. Xu, J.; Yeon, J.E.; Chang, H.; Tison, G.; Chen, G.J.; Wands, J.; de la Monte, S. Ethanol impairs insulin-stimulated neuronal survival in the developing brain: Role of pten phosphatase. *J. Biol. Chem.* **2003**, *278*, 26929–26937. [CrossRef]

20. De Ferrari, G.V.; Inestrosa, N.C. Wnt signaling function in alzheimer's disease. *Brain Res. Brain Res. Rev.* **2000**, *33*, 1–12. [CrossRef]

21. Atwood, C.S.; Obrenovich, M.E.; Liu, T.; Chan, H.; Perry, G.; Smith, M.A.; Martins, R.N. Amyloid-beta: A chameleon walking in two worlds: A review of the trophic and toxic properties of amyloid-beta. *Brain Res. Brain Res. Rev.* **2003**, *43*, 1–16. [CrossRef]

22. Neselius, S.; Zetterberg, H.; Blennow, K.; Randall, J.; Wilson, D.; Marcusson, J.; Brisby, H. Olympic boxing is associated with elevated levels of the neuronal protein tau in plasma. *Brain Inj.* **2013**, *27*, 425–433. [CrossRef]

23. Gill, J.; Merchant-Borna, K.; Jeromin, A.; Livingston, W.; Bazarian, J. Acute plasma tau relates to prolonged return to play after concussion. *Neurology* **2017**, *88*, 595–602. [CrossRef]

24. Ma, M.W.; Wang, J.; Zhang, Q.; Wang, R.; Dhandapani, K.M.; Vadlamudi, R.K.; Brann, D.W. Nadph oxidase in brain injury and neurodegenerative disorders. *Mol. Neurodegener.* **2017**, *12*, 7. [CrossRef]

25. Xu, X.; Cao, S.; Chao, H.; Liu, Y.; Ji, J. Sex-related differences in striatal dopaminergic system after traumatic brain injury. *Brain Res. Bull.* **2016**, *124*, 214–221. [CrossRef]

26. Mullins, R.J.; Diehl, T.C.; Chia, C.W.; Kapogiannis, D. Insulin resistance as a link between amyloid-beta and tau pathologies in alzheimer's disease. *Front. Aging Neurosci.* **2017**, *9*, 118. [CrossRef]

27. Wang, T.; Xie, C.; Yu, P.; Fang, F.; Zhu, J.; Cheng, J.; Gu, A.; Wang, J.; Xiao, H. Involvement of insulin signaling disturbances in bisphenol a-induced alzheimer's disease-like neurotoxicity. *Sci. Rep.* **2017**, *7*, 7497. [CrossRef]

28. Morrison, B., 3rd; Cater, H.L.; Benham, C.D.; Sundstrom, L.E. An in vitro model of traumatic brain injury utilising two-dimensional stretch of organotypic hippocampal slice cultures. *J. Neurosci. Methods* **2006**, *150*, 192–201. [CrossRef] [PubMed]

29. Jing, Y.; Yang, D.; Fu, Y.; Wang, W.; Yang, G.; Yuan, F.; Chen, H.; Ding, J.; Chen, S.; Tian, H. Neuroprotective effects of serpina3k in traumatic brain injury. *Front. Neurol.* **2019**, *10*, 1215. [CrossRef] [PubMed]

30. Hemphill, M.A.; Dauth, S.; Yu, C.J.; Dabiri, B.E.; Parker, K.K. Traumatic brain injury and the neuronal microenvironment: A potential role for neuropathological mechanotransduction. *Neuron* **2015**, *85*, 1177–1192. [CrossRef] [PubMed]

31. McKee, A.C.; Cantu, R.C.; Nowinski, C.J.; Hedley-Whyte, E.T.; Gavett, B.E.; Budson, A.E.; Santini, V.E.; Lee, H.S.; Kubilus, C.A.; Stern, R.A. Chronic traumatic encephalopathy in athletes: Progressive tauopathy following repetitive head injury. *J. Neuropathol. Exp. Neurol.* **2009**, *68*, 709–735. [CrossRef] [PubMed]

32. Morrison, B., 3rd; Elkin, B.S.; Dolle, J.P.; Yarmush, M.L. In vitro models of traumatic brain injury. *Annu. Rev. Biomed. Eng.* **2011**, *13*, 91–126. [CrossRef] [PubMed]

33. Pfister, B.J.; Weihs, T.P.; Betenbaugh, M.; Bao, G. An in vitro uniaxial stretch model for axonal injury. *Ann. Biomed. Eng.* **2003**, *31*, 589–598. [CrossRef]

34. Bar-Kochba, E.; Scimone, M.T.; Estrada, J.B.; Franck, C. Strain and rate-dependent neuronal injury in a 3d in vitro compression model of traumatic brain injury. *Sci. Rep.* **2016**, *6*, 30550. [CrossRef]

35. Rosas-Hernandez, H.; Burks, S.M.; Cuevas, E.; Ali, S.F. Stretch-induced deformation as a model to study dopaminergic dysfunction in traumatic brain injury. *Neurochem. Res.* **2019**, *44*, 2546–2555. [CrossRef]

36. Shin, S.S.; Bray, E.R.; Zhang, C.Q.; Dixon, C.E. Traumatic brain injury reduces striatal tyrosine hydroxylase activity and potassium-evoked dopamine release in rats. *Brain Res.* **2011**, *1369*, 208–215. [CrossRef]

37. Xu, Z.; Liu, Y.; Yang, D.; Yuan, F.; Ding, J.; Chen, H.; Tian, H. Sesamin protects sh-sy5y cells against mechanical stretch injury and promoting cell survival. *BMC Neurosci.* **2017**, *18*, 57. [CrossRef]

38. Hanger, D.P.; Hughes, K.; Woodgett, J.R.; Brion, J.P.; Anderton, B.H. Glycogen synthase kinase-3 induces alzheimer's disease-like phosphorylation of tau: Generation of paired helical filament epitopes and neuronal localisation of the kinase. *Neurosci. Lett.* **1992**, *147*, 58–62. [CrossRef]

39. Medina, M.; Avila, J. Glycogen synthase kinase-3 (gsk-3) inhibitors for the treatment of alzheimer's disease. *Curr. Pharm. Des.* **2010**, *16*, 2790–2798. [CrossRef]

40. Taniguchi, C.M.; Emanuelli, B.; Kahn, C.R. Critical nodes in signalling pathways: Insights into insulin action. *Nat. Rev. Mol. Cell Biol.* **2006**, *7*, 85–96. [CrossRef] [PubMed]

41. Ahmed, F.; Ansari, J.A.; Ansari, Z.E.; Alam, Q.; Gan, S.H.; Kamal, M.A.; Ahmad, E. A molecular bridge: Connecting type 2 diabetes and alzheimer's disease. *CNS Neurol. Disord. Drug Targets* **2014**, *13*, 312–321. [CrossRef]

42. Leng, S.; Zhang, W.; Zheng, Y.; Liberman, Z.; Rhodes, C.J.; Eldar-Finkelman, H.; Sun, X.J. Glycogen synthase kinase 3 beta mediates high glucose-induced ubiquitination and proteasome degradation of insulin receptor substrate 1. *J. Endocrinol.* **2010**, *206*, 171–181. [CrossRef]

43. Liberman, Z.; Eldar-Finkelman, H. Serine 332 phosphorylation of insulin receptor substrate-1 by glycogen synthase kinase-3 attenuates insulin signaling. *J. Biol. Chem.* **2005**, *280*, 4422–4428. [CrossRef] [PubMed]

44. Steen, E.; Terry, B.M.; Rivera, E.J.; Cannon, J.L.; Neely, T.R.; Tavares, R.; Xu, X.J.; Wands, J.R.; de la Monte, S.M. Impaired insulin and insulin-like growth factor expression and signaling mechanisms in alzheimer's disease–is this type 3 diabetes? *J. Alzheimers Dis.* **2005**, *7*, 63–80. [CrossRef] [PubMed]

45. Walsh, D.M.; Klyubin, I.; Fadeeva, J.V.; Cullen, W.K.; Anwyl, R.; Wolfe, M.S.; Rowan, M.J.; Selkoe, D.J. Naturally secreted oligomers of amyloid beta protein potently inhibit hippocampal long-term potentiation in vivo. *Nature* **2002**, *416*, 535–539. [CrossRef] [PubMed]
46. Adolfsson, R.; Bucht, G.; Lithner, F.; Winblad, B. Hypoglycemia in alzheimer's disease. *Acta Med. Scand.* **1980**, *208*, 387–388. [CrossRef]
47. Lewitt, M.S.; Boyd, G.W. The role of insulin-like growth factors and insulin-like growth factor-binding proteins in the nervous system. *Biochem. Insights* **2019**, *12*, 1178626419842176. [CrossRef]
48. Miranda, M.; Morici, J.F.; Zanoni, M.B.; Bekinschtein, P. Brain-derived neurotrophic factor: A key molecule for memory in the healthy and the pathological brain. *Front. Cell Neurosci.* **2019**, *13*, 363. [CrossRef]
49. Kondo, A.; Shahpasand, K.; Mannix, R.; Qiu, J.; Moncaster, J.; Chen, C.H.; Yao, Y.; Lin, Y.M.; Driver, J.A.; Sun, Y.; et al. Antibody against early driver of neurodegeneration cis p-tau blocks brain injury and tauopathy. *Nature* **2015**, *523*, 431–436. [CrossRef]
50. Shim, S.S.; Stutzmann, G.E. Inhibition of glycogen synthase kinase-3: An emerging target in the treatment of traumatic brain injury. *J. Neurotrauma* **2016**, *33*, 2065–2076. [CrossRef]
51. Ahmed, S.M.; Rzigalinski, B.A.; Willoughby, K.A.; Sitterding, H.A.; Ellis, E.F. Stretch-induced injury alters mitochondrial membrane potential and cellular atp in cultured astrocytes and neurons. *J. Neurochem.* **2000**, *74*, 1951–1960. [CrossRef] [PubMed]
52. Arundine, M.; Aarts, M.; Lau, A.; Tymianski, M. Vulnerability of central neurons to secondary insults after in vitro mechanical stretch. *J. Neurosci.* **2004**, *24*, 8106–8123. [CrossRef] [PubMed]
53. Abdul-Muneer, P.M.; Chandra, N.; Haorah, J. Interactions of oxidative stress and neurovascular inflammation in the pathogenesis of traumatic brain injury. *Mol. Neurobiol.* **2015**, *51*, 966–979. [CrossRef] [PubMed]
54. Wang, F.; Franco, R.; Skotak, M.; Hu, G.; Chandra, N. Mechanical stretch exacerbates the cell death in sh-sy5y cells exposed to paraquat: Mitochondrial dysfunction and oxidative stress. *Neurotoxicology* **2014**, *41*, 54–63. [CrossRef]

Article

Transfer RNA-Derived Fragments and isomiRs Are Novel Components of Chronic TBI-Induced Neuropathology

Noora Puhakka *, Shalini Das Gupta, Niina Vuokila and Asla Pitkänen

A.I. Virtanen Institute for Molecular Sciences, University of Eastern Finland, 70211 Kuopio, Finland; shalini.gupta@uef.fi (S.D.G.); niina.vuokila@uef.fi (N.V.); asla.pitkanen@uef.fi (A.P.)
* Correspondence: noora.puhakka@uef.fi

Abstract: Neuroinflammation is a secondary injury mechanism that evolves in the brain for months after traumatic brain injury (TBI). We hypothesized that an altered small non-coding RNA (sncRNA) signature plays a key role in modulating post-TBI secondary injury and neuroinflammation. At 3threemonths post-TBI, messenger RNA sequencing (seq) and small RNAseq were performed on samples from the ipsilateral thalamus and perilesional cortex of selected rats with a chronic inflammatory endophenotype, and sham-operated controls. The small RNAseq identified dysregulation of 2 and 19 miRNAs in the thalamus and cortex, respectively. The two candidates from the thalamus and the top ten from the cortex were selected for validation. In the thalamus, miR-146a-5p and miR-155-5p levels were upregulated, and in the cortex, miR-375-3p and miR-211-5p levels were upregulated. Analysis of isomiRs of differentially expressed miRNAs identified 3′ nucleotide additions that were increased after TBI. Surprisingly, we found fragments originating from 16 and 13 tRNAs in the thalamus and cortex, respectively. We further analyzed two upregulated fragments, 3′tRF-IleAAT and 3′tRF-LysTTT. Increased expression of the full miR-146a profile, and 3′tRF-IleAAT and 3′tRF-LysTTT was associated with a worse behavioral outcome in animals with chronic neuroinflammation. Our results highlight the importance of understanding the regulatory roles of as-yet unknown sncRNAs for developing better strategies to treat TBI and neuroinflammation.

Keywords: TBI; tRNA; miRNA; isomiR; neuroinflammation; RNAseq

Citation: Puhakka, N.; Das Gupta, S.; Vuokila, N.; Pitkänen, A. Transfer RNA-Derived Fragments and isomiRs Are Novel Components of Chronic TBI-Induced Neuropathology. *Biomedicines* **2022**, *10*, 136. https://doi.org/10.3390/biomedicines10010136

Academic Editors: Kumar Vaibhav, Meenakshi Ahluwalia and Pankaj Gaur

Received: 13 December 2021
Accepted: 4 January 2022
Published: 8 January 2022

Publisher's Note: MDPI stays neutral with regard to jurisdictional claims in published maps and institutional affiliations.

Copyright: © 2022 by the authors. Licensee MDPI, Basel, Switzerland. This article is an open access article distributed under the terms and conditions of the Creative Commons Attribution (CC BY) license (https://creativecommons.org/licenses/by/4.0/).

1. Introduction

Traumatic brain injury (TBI) is caused by an external mechanical force to the head [1]. Epidemiologic studies indicate that TBI is most commonly caused by falls and traffic accidents [2]. Importantly, TBI is the most common cause of disability in people under 40 [1]. After the impact, the primary injury triggers a cascade of molecular dysregulation [3,4], leading to chronic neuroinflammation [5,6]. These changes contribute to the development of secondary injuries and chronic post-TBI comorbidities, such as cognitive impairment [7,8] and epilepsy [9]. Studies of the mechanisms of secondary injury and recovery processes are essential to identify molecular targets for therapeutic interventions as there are currently no medications available that alleviate the post-TBI aftermath. Recent technologic advances toward exploring the transcriptome have led to the discovery of several new classes of small non-coding RNAs (sncRNAs) with potentially key regulatory roles in normal brain function and disease [10,11]. Commonly, sncRNAs are defined as transcripts of <200 nucleotides (nts) that lack a protein-coding capacity [12]. Understanding the functional role of sncRNAs might lead to the development of new strategies to combat complex brain disorders like TBI.

Among sncRNAs, microRNAs (miRNAs) and miRNA-regulated mechanisms are most thoroughly studied in brain diseases [13]. MicroRNAs are small (19–22 nts) RNA molecules that regulate the expression of target genes at the post-transcriptional level via sequence-specific (seed) binding to the 3′ untranslated region [14]. After experimental TBI, dysregulated miRNAs reportedly target such processes as cellular functions, transcription, signal transduction, growth, protein modification, and response to stress, which can further impair recovery [15]. While most studies report dysregulation of canonical miRNA, less emphasis has been paid to alterations in the gene expression of miRNA isoforms, i.e., isomiRs [16,17]. Currently, more attention is being focused on isomiRs and researchers are beginning to unravel the complex functions of these molecules and modification events, which have been long underappreciated.

Another class of sncRNAs, transfer RNAs (tRNAs), is best known for its canonical role in protein translation [18]. Interestingly, various stress conditions can induce tRNA cleavage into short tRNA-derived fragments (tRFs) [19]. This process is potentially mediated by endoribonucleases, such as angiogenin and dicer [20]. The biogenesis details of tRFs, however, are not well known [21]. Increasing evidence reveals that tRFs can modulate diverse biologic processes, including cell proliferation and apoptosis [18,19,22–24]. Further, tRFs are reported to display miRNA-like activity, thereby influencing posttranscriptional regulation of their messenger RNA (mRNA) targets [24]. To date, however, few studies have investigated the role of tRFs in brain pathophysiology [10,25–27].

In the present study, we performed genome-wide sequencing of sncRNAs and mRNAs of samples originating from post-TBI animals phenotyped to display chronic neuroinflammation. We hypothesized that an altered sncRNA signature plays a key role in modulating post-TBI neuroinflammation and secondary injury. We analyzed canonical and isomiR profiles, and quantified dysregulated tRFs in the perilesional cortex and ipsilateral thalamus at three months post-TBI. We report an association between increased expression of the full miR-146a profile (canonical miRNA and all isomiRs), 3′tRF-IleAAT and 3′tRF-LysTTT, and a worse behavioral outcome in animals with chronic neuroinflammation after TBI. Our results highlight the importance of understanding the regulatory roles of as-yet unknown sncRNAs for the development of better strategies to treat TBI.

2. Materials and Methods

2.1. Animals

A total of 51 adult male Sprague-Dawley rats (mean body weight at the time of injury, 347–425 g; Harlan Laboratories S.r.l., Udine, Italy) were used in this study. The study design is presented in Figure 1. The rats were individually housed in a controlled environment (temperature $22 \pm 1\ °C$; humidity 50–60%; lights on from 07:00–19:00 h). Pellet food and water were provided ad libitum. All animal procedures were approved by the Animal Ethics Committee of the Provincial Government of Southern Finland and performed in accordance with the guidelines of the European Community Council Directives 2010/63/EU.

During the entire study duration, the overall well-being of the rats, as well as their motor activities, eating and drinking behaviors, and tooth growth were observed daily as previously described in detail [28]. If an animal exhibited signs of pain (e.g., weight loss, abnormal movement or posture, excessive grooming), it was treated with carprofen (Rimadyl®, 5 mg/kg, once per day for 3 days, Zoetis Finland Oy, Helsinki, Finland). Our pre-study-determined humane endpoint was that if the post-injury weight loss exceeded 30%, the rat would be killed by deep anesthesia induced by 4% isoflurane followed by decapitation. None of the animals in this study met this criterion.

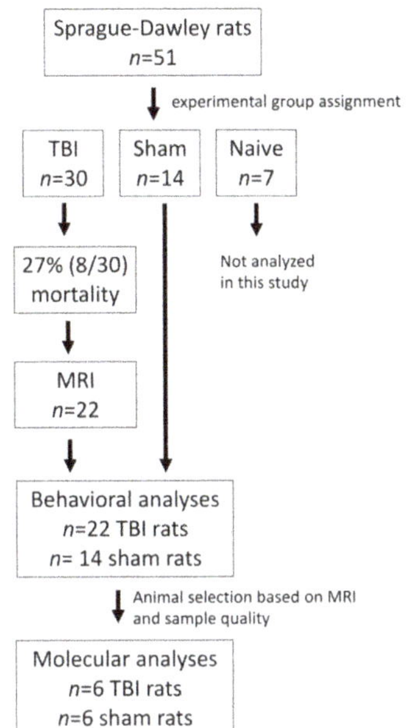

Figure 1. Study design.

2.2. Induction of TBI with Lateral Fluid-Percussion

The procedure for inducing the lateral fluid-percussion injury (FPI) was described previously in detail [29]. Animals (n = 30) were anesthetized with an intraperitoneal (i.p.) injection of a solution containing sodium pentobarbital (58 mg/kg), magnesium sulfate (127.2 mg/kg), propylene glycol (42.8%), and absolute ethanol (11.6%), and placed in a Kopf stereotactic frame (David Kopf Instruments, Tujunga, CA, USA). The skull was exposed with a midline skin incision and the periosteum extracted. The left temporal muscle was gently detached from the lateral ridge. A circular craniectomy (Ø 5 mm) was performed over the left parietal lobe midway between the lambda and bregma, leaving the dura mater intact. The edges of the craniectomy were sealed with a modified Luer-lock cap that was filled with saline while the calvaria was covered with dental acrylate (Selectaplus CN, Dentsply DeTrey GmbH, Dreieich, Germany). Lateral FPI was produced 90 min after the induction of anesthesia by connecting the rat to a fluid-percussion device (AmScien Instruments, Richmond, VA, USA) via a female Luer-lock fitting. The mean severity of the impact was 3.3 ± 0.01 atm. Sham-operated control animals (n = 14) underwent anesthesia and all surgical procedures, but not the impact. Seven naïve animals were included in the cohort but were not used in the present study.

2.3. Composite Neuroscore

To assess somatomotor recovery, the rats (n = 22 TBI animals and n = 14 shams) underwent composite neuroscore testing at baseline (pre-injury), and at 2, 7, 14, 21, and 90 days post-TBI (for details, see [30]). Animals were scored from 0 (severely impaired) to 4 (normal) in the following categories: (1) left and right forelimb contraflexion while suspended by the tail, (2) left and right hindlimb flexion when gently pulled back by the tail, (3) ability to resist left and right lateral pulsion, and (4) ability to stand on an inclined surface. The first 3 categories allow for a maximum of 4 points from both the left and right sides. The last category was scored from 0 to 4, leading to a maximum total score of 28 points.

2.4. Magnetic Resonance Imaging

To distinguish rats with a chronic perilesional inflammatory rim, T2-weighted magnetic resonance imaging (MRI) was performed at 2 months post-TBI (n = 22 TBI animals) using a 7 T Bruker Pharmascan MRI scanner equipped with a volume transmitter coil and surface receiver coil combo [28]. Briefly, the rats were anesthetized with isoflurane (4% for induction and 2% for maintenance) and positioned in a stereotactic holder. Thereafter, T2-weighted images were acquired with fast spin-echo from 25 slices (field of view = 30 mm × 30 mm, matrix slice thickness = 1 mm, TR = 4000 ms, TE = 40 ms).

2.5. Morris Water Maze

The spatial learning and memory performance of the rats (n = 22 TBI animals and n= 14 shams) was tested in a Morris water maze using a 3-day paradigm [31]. At the end of day 3, the platform was removed from the maze apparatus and rats were allowed to swim for 60 s to evaluate their memory of the platform location (probe trial). The time spent in each of the 4 quadrants of the maze was recorded.

2.6. Sampling of Brain Tissue

Rats (n = 22 TBI animals and n = 14 shams) were anesthetized with 5% isoflurane and decapitated, and their brains were quickly removed. Each brain was flushed with 0.9% cold (4 °C) sodium chloride to remove blood and hair and placed onto a slicing matrix on ice (#15007, Rodent Brain Matrix, Ted Pella, Inc., Redding, CA, USA). One 3-mm-thick coronal slice was cut (between −1 and −4 from the bregma), from which the perilesional cortex and ipsilateral thalamus were dissected on top of the glass plate placed on ice. Brain tissue samples were snap-frozen in liquid nitrogen and stored at −70 °C. All tissue preparations were performed in under 12 min (range 9–12 min) after decapitation. The remaining brain tissue was placed in a 10% formalin solution for 3 days, followed by cryoprotection in 20% glycerol in a 0.02 M PB buffer for 2 days. Brain pieces were snap-frozen on dry ice and stored at −70 °C until further processed.

2.7. Histologic Analysis

For histologic analysis, brains pieces were sectioned (1-in-6 series, 30 μm) with a sliding microtome (Leica SM 2000, Leica Microsystems Nussloch GmbH, Nussloch, Germany). The sections were stored in a cryoprotectant solution (30% ethylene glycol, 25% glycerol in 0.05 M sodium phosphate buffer, pH 7.4) until stained.

Nissl staining. To further characterize the lesion pathology (n = 22 TBI animals and n= 14 shams), we performed Nissl staining (see [32]).

Immunohistochemical staining for CD68 and GFAP. Freely floating sections were transferred to 24-well nets for staining. The complete staining protocol for astrocyte (glial fibrillary acidic protein [GFAP]) and microglial/macrophage response (CD68) immunohistochemistry (n = 5 representative post-TBI animals), was previously described in detail by Huusko et al. [32] with slight modifications. As a primary antibody, we used a mouse monoclonal antibody to GFAP (1:2000, #814369, Boehringer, Mannheim, Germany), and a mouse monoclonal antibody to rat CD68 (1:1000, EMD Millipore, Billerica, MD, USA). One representative brain section per rat was selected using the MRI images as a guide for determining the lesion location and extent. For GFAP, sections were incubated in the primary antibody mixture for 2 days. For CD68, the sections were treated for 15 min with 1% sodium borohydride for antigen retrieval prior to incubation in a blocking solution. In addition, 0.1% Triton-X 100 was used in all steps. Similar to GFAP, the sections were incubated in primary antibody for 2 days.

2.8. Isolation of Total RNA from Brain Tissue

RNA was extracted from the perilesional cortex and ipsilateral thalamus (n = 6 TBI animals and n = 6 shams) using a mirVana miRNA isolation kit (#AM1560, Life Technologies (Ambion) Carlsbad, CA, USA), QIAshredder (#79654, Qiagen, Hilden, Germany), and

AllPrep DNA/RNA Mini Kit (#80204, Qiagen) as previously described in detail [33,34]. Briefly, to avoid clogging the spin columns, brain tissue was divided into 2–4 pieces (max ~10 mg each) on dry ice. Each tissue piece was then placed into a 2-mL microcentrifuge tube together with 1 metal ball and 800 µL of Ambion Lysis/binding buffer and homogenized with a TissueLyser (Qiagen) for 3 min (30 Hz). For further homogenization, the lysate was transferred to a QIAshredder spin column and centrifuged (16,000× g) for 2 min at 4 °C. Flow-through lysate was transferred back to the QIAshredder spin column and centrifuged a second time. For DNA separation, tissue lysate was transferred to a Qiagen All Prep DNA spin column and centrifuged (10,000× g) for 1 min at room temperature. Flow-through from the All Prep DNA spin column was used for RNA extraction using a mirVana miRNA isolation kit. Briefly, miRNA homogenate additive (70 µL) was added to the flow-through. The mixture was vigorously vortexed for 30 s and then incubated on ice for 10 min. Acid-phenol:chloroform (700 µL) was then added, mixed, and centrifuged (16,000× g) for 30 s. The aqueous upper phase was transferred to a new microcentrifuge tube. A total of 500 µL of water was added to the lower phase, which was then mixed and centrifuged (16,000× g) for 30 s. The upper aqueous phase was collected into the same tube as the aqueous phase from the previous extraction cycle. Then, 100% ethanol (625 µL) was added to the tube, mixed, and transferred to the mirVana miRNA isolation spin column. Finally, RNA was washed and eluted from the spin column according to instructions provided with the mirVana miRNA isolation kit. Finally, RNA extracted from each brain region was pooled. Thereafter, RNA purification was performed using a miRCURY™ RNA Isolation Kit (#300111, Exiqon, Vedbaek, Denmark) according to the manufacturer's instructions.

The concentrations and 260/280 ratios of the eluted RNA from the perilesional cortex and ipsilateral thalamus were measured using a NanoDrop 1000 spectrophotometer. RNA quality and the RNA integrity number were measured with an Agilent 6000 Nano kit (#5067–1511, Agilent Technologies, Waldbronn, Germany) using an Agilent Bioanalyzer. The RNA integrity numbers (range 8.9–9.2) and 260/280 ratios (range 2.01–2.10) prior to the sequencing service and for samples in the later polymerase chain reaction (PCR) study are shown in Supplementary Table S1.

2.9. Small RNA and RNA Sequencing from Brain Tissue

All sequencing experiments and data analysis (n = 6 TBI animals and n = 6 shams) were conducted at Exiqon Services, Denmark.

2.9.1. Small RNA Sequencing

Library preparation and next-generation sequencing. First, 300 ng of total RNA was converted to microRNA NGS libraries using a NEBNEXT library generation kit (New England Biolabs Inc., Ipswich, MA, USA) according to the manufacturer's instructions. Each individual RNA sample then had adaptors ligated to its 3′ and 5′ ends and converted into cDNA. The cDNA was pre-amplified with specific primers containing sample-specific indexes. After 15 cycles pre-PCR, the libraries were purified on QiaQuick columns, and the insert efficiency was evaluated by a Bioanalyzer 2100 instrument on a high sensitivity DNA chip (Agilent Inc., Santa Clara, CA, USA). The microRNA cDNA libraries were size-fractionated on a LabChip XT (Caliper Inc., Princeton, NJ, USA) and a band representing the adaptors and a 15–40 bp insert was excised according to the manufacturer's instructions. Samples were then quantified by quantitative PCR (qPCR) and concentration standards. Based on the quality of the inserts and the concentration measurements, the libraries were pooled in equimolar concentrations (all concentrations of libraries to be pooled were of the same concentration). The library pool(s) were finally quantified again with qPCR and the optimal concentration of the library pool used to generate the clusters on the surface of a flow cell before sequencing with a v3 sequencing methodology according to the manufacturer' instructions (Illumina Inc., San Diego, CA, USA). Finally, the samples were sequenced on the Illumina NextSeq 500 system. The system uses quality score binning, which enables a more compact storage of raw sequences. The method was tested using only 8 levels (Levels: No call, 6, 15, 22, 27, 33, 37, and 40) of quality and determined to be virtually loss-less (http://res.illumina.com/documents/products/whitepapers/whitepaper_datacompression.pdf; accessed on 16 June 2015).

Data analysis. After the sequencing, intensity correction, base calling, and assigning of Q-scores was performed. Adapters were trimmed off as part of the base calling. Subsequently, the data were quality checked with FastQC [35]. Sequences were mapped against rat reference genome RGSC3.4 and annotated with mirBase 20. Differential expression analyses were performed using the edgeR statistical software package (Bioconductor, http://www.bioconductor.org/; accessed on 10 December 2021). For normalization, the trimmed mean M-value method was used based on log-fold and absolute gene-wise changes in expression levels between samples (TMM normalization).

Identification of isomiRs. IsomiR analysis was performed individually for each sample based on the occurrence of count variants for each detected microRNA. Reads were mapped to known miRNAs according to the annotation in the miRBase release 20 and then investigated for the presence of different isomiRs. These variants were identified by changes in the start or stop position, or the occurrence of mutations within the read. The results for each sample were then merged to generate a single count file with a consistent nomenclature across the samples. Only isomiRs that were present at a level of 5% of the total reads for that miRNA were retained.

Identification of tRNA derived fragments. Reads mapped against small RNA (including tRNA) in the rat reference genome RGSC3.4 were counted and visualized with an Integrative Genomics Visualizer (IGV; [36]). Differential gene expression analysis was performed similarly as for miRNAs to assess overall change in tRFs originating from one particular tRNA. Two candidate tRFs were selected for further experiments as they had a clear cleavage site based on visual analysis with the IGV.

2.9.2. Messenger RNA Sequencing

Library preparation and next-generation sequencing. The library was prepared using a TruSeq® Stranded mRNA Sample preparation kit (Illumina inc.). The starting material (300 ng) of total RNA was mRNA enriched using the oligo (dT) bead system. The isolated mRNA was subsequently fragmented by enzymatic fragmentation. Then, first-strand and second-strand syntheses were performed and the double-stranded cDNA was purified (AMPure XP, Beckman Coulter, Brea, CA, USA). The cDNA was end-repaired and 3′ adenylated, Illumina sequencing adaptors were ligated onto the fragments ends, and the library was purified (AMPure XP). The mRNA stranded libraries were pre-amplified with PCR and purified (AMPure XP). The libraries' size distribution was validated and quality inspected on a Bioanalyzer high sensitivity DNA chip (Agilent Technologies). High quality libraries were quantified by qPCR, the concentration normalized, and the samples pooled according to the project specification (number of reads). The library pool(s) were re-quantified with qPCR and the optimal concentration of the library pool was used to generate clusters on the surface of a flow cell before sequencing on the Nextseq500 instrument using a High Output sequencing kit (75 cycles) according to the manufacturer's instructions (Illumina Inc.).

Data analysis. TopHat (v2.0.11) was used to align the sequencing reads to the reference genome (RGSC3.4) with the sequence aligner Bowtie2 (v.2.2.2). Cufflinks (v2.2.1) was used to take the alignment results from TopHat and assemble the aligned sequences into transcripts, thereby constructing a map or a snapshot of the transcriptome. Reference genome version RGSC3.4 was used. To guide the assembly process, an existing transcript annotation (Rnor_5.0 Ensembl) was used. When comparing groups, Cuffdiff [37] was used to calculate the FPKM (number of fragments per kilobase per million mapped fragments) and test for differential expression and regulation among the assembled transcripts across the submitted samples using the Cufflinks output.

2.10. Pathway Analysis of RNA-Seq Data

Qiagen Ingenuity pathway analysis (IPA®, Qiagen; Content version 31813283) was used to assess regulated molecular pathways and gene networks. IPA analysis was performed using "Experimentally Observed" and "High (predicted)" confidence settings. The cutoff was set to FDR <0.05 (both up/downregulated). After filtering, the dataset from the perilesional cortex included 4653 genes. Similarly, 5256 genes from the ipsilateral thalamus were included in the analysis.

2.11. Validation of the Small RNA-Seq Data with Droplet Digital PCR and Quantitative PCR
2.11.1. microRNA

Reverse transcription (RT) of total RNA from brain tissue with TaqMan chemistry (n = 6 TBI animals and n = 6 shams). Validation of the small RNA-Seq data was performed from the same samples subjected to sequencing (n = 6 TBI and 6 sham-operated controls). All RNA samples were diluted 1:100 with RNAase-free water prior to cDNA synthesis. In the perilesional cortex, validation was performed for the following miRNAs: miR-132-5p (002132), miR-212-5p (463930_mat), miR-375-3p (000564), miR-20a-5p (00580), miR-211-5p (001199), miR-135a-3p (002075), miR-339-5p (002257), miR-34c-3p (000426), miR-128-3p (002216) and miR-17-5p (002308). In the ipsilateral thalamus, validation was performed for miR-146a-5p (002163) and miR-155-5p (002571). For normalization, 4 miRNAs were selected from the small RNA-Seq data as endogenous controls: miR-378a-3p (002243), miR-3594-3p (464929_mat), miR-330-5p (002230), and let-7b-3p (002404). Reverse transcription for all target and control miRNAs was performed with the TaqMan protocol (TaqMan miRNA Reverse Transcriptase Kit #4366596, Applied Biosystems, Foster City, CA, USA https://www.thermofisher.com/document-connect/document-connect.html?url= https%3A%2F%2Fassets.thermofisher.com%2FTFS-Assets%2FLSG%2Fmanuals%2F43640 31_TaqSmallRNA_UG.pdf; accessed on 10 December 2021), according to the manufacturer's instructions. Samples were stored at −20 °C until further processed.

Droplet digital PCR (n = 6 TBI animals and n = 6 shams). Absolute quantification of miRNA copy numbers was performed with droplet digital PCR (ddPCR). For the perilesional cortex, cDNA from miR-378a-3p was diluted 1:6, and miR-128-3p was diluted 1:8 in ddPCR. All other cDNA samples were used undiluted. For the ipsilateral thalamus samples, cDNA from miR-378a-3p and miR-330-5p were diluted 1:4 in ddPCR. Undiluted cDNA was used for all other miRNAs. The cDNA template (1.33 µL) was added to a 20-µL reaction mixture containing 10 µL Bio-Rad 2x ddPCR supermix for probes (#186-3010, Bio-Rad, Hercules, CA USA), 1 µL 20× miRNA PCR primer, and 7.67 µL nuclease-free water. The 20-µL reaction mixtures were loaded in disposable droplet generator cartridges (#186-4008, DG8TM Cartridges for QX100TM/QX200TM Droplet Generator, Bio-Rad, Germany) along with 70 µL of droplet generation oil for probes (#186-3005, Bio-Rad, CA, USA). The cartridges were covered with gaskets (#186-3009, Droplet Generator DG8TM Gaskets, Bio-Rad, Hercules, CA, USA) and placed in the droplet generator (#186-3001, Bio-Rad QX100TM Droplet Generator, Solna, Sweden) for creating the oil-emulsion droplets. The generated droplets were transferred to 96-well PCR plates (Twin. tec PCR plate 96, semi-skirted, colorless, Hamburg, Germany) sealed, thermal-cycled (96-well PTC-200 thermal cycler, MJ Research), and quantified at the end-point in the droplet reader (#186-3001, Bio-Rad QX100TM Droplet Reader, Solna, Sweden) using QuantaSoft software (v1.7, Product code: 186-4011, Bio-Rad, USA). For each sample, the reaction was performed in duplicate, and the average miRNA copies/20 µL well from the 2 replicates was calculated.

To normalize PCR measurements, the total RNA input to the cDNA reaction was first calculated based on the NanoDrop concentrations and the implemented dilution factor (1:100). For each miRNA, the average number of copies/20 µL well was then normalized to the total RNA input, to account for the slight sample-to-sample variations in the RNA input volumes. Normalization was performed as follows:

The miRNA copies normalized to RNA input = (average miRNA copies/20 μL well)/(total RNA input to the 15 μL cDNA reaction)

Next, the geometric mean (GM) of the RNA input-normalized copies from the 4 endogenous controls, miR-378a-3p, miR-3594-3p, miR-330-5p, and let-7b-3p, was calculated. Finally, the target miRNAs were normalized to the geometric mean of the 4 endogenous controls as follows:

Normalized target miRNA copies = (miRNA copies normalized to RNA input)/(GM of RNA input normalized miRNA copies from 4 endogenous controls)

IsomiR analysis of selected candidate miRNAs. For all candidate and control miRNAs tested for validation with ddPCR, the non-normalized isomiR read count list was obtained for each biologic sample from the small RNA-Seq data. For each miRNA, the total read count from all isomiRs, and the number of detected isomiRs were compared among the TBI and sham-operated controls. Next, analysis was focused on the miRNA candidates that were validated with ddPCR: miR-375-3p, miR-211-5p, and miR-146a-5p. We excluded miR-155-5p because it had a low read abundance in all samples, with no detected isomiRs. For miR-375-3p, miR-211-5p, and miR-146a-5p, the top 6 isomiRs (based on read count abundance) from all TBI samples were listed, and among them, the top 2–4 isomiRs that were detected in all the TBI samples were considered. For each of these 2–4 isomiRs, the read count, change involved (if any), percentage of total reads comprising the isomiR, and percentage of canonical sequence reads comprising the isomiR were tabulated and compared between the TBI and sham-operated controls.

Reverse transcription of total RNA from brain tissue with miScript chemistry. Validation of the small RNA-Seq data was performed from the same samples subjected to sequencing (n = 6 TBI animals and 6 sham-operated controls). Synthesis of cDNA was performed with miScript II RT kit (#218161, Qiagen) according to the manufacturer's instructions.

Quantitative PCR analysis of miR-146a with miScript chemistry. Quantitative RT-PCR analysis was performed using miScript chemistry according to the manufacturer's instructions. For this, a miScript SYBR Green PCR Kit (#218073, Qiagen) was used with a rno-miR-146a specific 5′ end primer (MS00000441, Qiagen) and a universal 3′ end primer. For normalization, we used miR-378a-3p (MS00005810, Qiagen) with the formula 2-ΔCt [38]. For each sample, the PCR reactions were performed in triplicate. All samples were run with a LightCycler® 96 Instrument (Roche). A no-template (nuclease-free water) control was run in parallel for all assays.

2.11.2. Transfer RNA Derived Fragments

Reverse transcription of total RNA from brain tissue with miScript chemistry (n = 6 TBI animals and n = 6 shams). Validation of the small RNA-Seq data was performed from the same samples subjected to sequencing (n = 6 TBI and 6 sham-operated controls) as described in the miRNA paragraph for miScript chemistry.

Quantitative PCR (n = 6 TBI animals and n = 6 shams). Relative quantification of tRNA fragments was performed as described for miR-146a with miScript chemistry. Custom designed assays for tRFs derived from the 3′ end of the tRNA-IleAAT (MSC0076231, Qiagen) and tRNA-LysTTT (MSC0076337, Qiagen) were used.

Agarose gel electrophoresis for size separation (n = 3 TBI animals and n = 3 shams). First, 2 μL of RNA (150 ng), 16 μL of formamide, and 2 μL of loading dye were mixed and then heated at 65 °C for 5 min. For size separation of the small RNA, 3% agarose gel (#A9539-500G, Sigma, Tokyo, Japan) was prepared in a 1 × Tris-acetate buffer (TAE) and run 15 min with 80V in 1 × TAE. A GeneRuler low range DNA ladder (#SM1193, Thermo Scientific, Waltham, MA, USA) and Sybr Safe DNA gel stain (#S33102, Invitrogen, Waltham, MA, USA) were used to visualize the RNA size. An appropriate piece of the gel was cut and minced with a microtome blade. Small RNA was eluted from the gel to nuclease-free water (600 μL) at room temperature on a shaker overnight. Complementary cDNA and reverse transcription (RT)-qPCR was performed as described for the miScript chemistry earlier, except data normalization was performed relative to miR-124-3p (MS00005593, Qiagen). Undiluted RNA after elution was directly used for the cDNA reaction.

2.12. In-Silico Prediction and qPCR Validation of mRNA Targets for the Validated miRNAs

In-silico prediction. For the validated miRNA candidates, mRNA target prediction was performed in-silico with TargetScan v7.2. The predicted target list was then compared with the differentially expressed mRNA data (FDR <0.05) obtained from our brain RNA-Seq samples. Only those targets that were predicted in TargetScan and present in our differentially expressed mRNA list with a fold-change (FC) opposite that of the miRNA expression were selected for DAVID gene-Gene Ontology term enrichment analysis. Based on prior literature, evidence of miRNA-mRNA interaction, magnitude of fold-change, and read abundance identified from RNA-Seq; 2 mRNA targets were selected for validation with RT-qPCR, ELAVL2 from the perilesional cortex (predicted target of both miR-375-3p and miR-211-5p) and Syt1 from the ipsilateral thalamus (predicted target of miR-146a-5p).

Complementary DNA (cDNA) synthesis (n = 6 TBI animals and n = 6 shams). The cDNA synthesis was performed using the High Capacity RNA-to-cDNA Kit (Applied Biosystems) according to the manufacturer's instructions (High Capacity RNA-to-cDNA kit Protocol, Part Number 4,387,951 Rev. C). The total amount of RNA was set to 2 µg/20 µL reaction.

Quantitative RT-PCR (n = 6 TBI animals and n = 6 shams). Quantitative RT-PCR was performed in a total volume of 20 µL using 12 ng (RNA equivalents) of cDNA as a template, gene-specific primers and probes (pre-validated TaqMan Gene Expression Assay for Elavl1 and Syt1; Rn01433257_m1 and Rn00436862, Applied Biosystems), and 1× TaqMan Gene Expression Master Mix (#4369016, Applied Biosystems, Waltham, MA, USA). The following program was used in the PCR (StepOne Software v2.1, Applied Biosystems): 1 cycle (95 C, 10 min) and 40 cycles (95 C, 15 s; 60 C, 60 s) in a StepOnePlusTM Real-Time PCR System (Applied Biosystems). The data were normalized to glyceraldehyde 3-phosphate dehydrogenase (Gapdh) mRNA expression (pre-validated Taqman Gene Expression Assay for Gapdh; ID: Rn99999916_s1, Applied Biosystems). Each sample was run in triplicate. A no-template (nuclease-free water) control was run in parallel for all assays.

2.13. In-Silico Prediction and qPCR Vvalidation of mRNA Targets for the Validated tRFs

In-silico prediction. To predict miRNA-like regulation of tRFs, TargetRank [39] was used to find complementary regions from mRNA 3′ untranslated regions (3′UTRs). Seed regions (8-nt long) were formed along the length of the tRF. For gene set enrichment analysis of the predicted targets for the first 8-nt seed region, a gene list of targets for both tRNA-IleAAT and tRNA-LysTTT was created based on a human database (Supplementary Material 2). Thereafter, the acquired gene list was transformed to official rat gene symbols (334 genes). A pre-ranked gene list was created according to the RNA-Seq dataset. The genes were arranged into negative and positive groups according to the fold-change values. Downregulated genes were identified with a minus sign. The genes were then arranged according to their p-value and ranked. Genes with high p-values were ranked closer to zero. The pre-ranked gene list created for the gene enrichment analysis comprised 18893 features (genes). FDR <0.05 was considered statistically significant. After gene set enrichment analysis, all negatively enriched targets (Supplementary Material 2) were included for further Functional Annotation Tool analysis (DAVID Bioinformatics Resources 6.8, NIAID/NIH). A Benjamini–Hochberg-corrected *p*-value <0.05 was used to find the regulated functions. One candidate target (Cplx1) was selected for further proof-of-concept validation.

Complementary DNA (cDNA) synthesis (n = 5 TBI animals and n = 5 shams). The cDNA synthesis was performed as described for the predicted miRNA targets.

Quantitative RT-PCR (n = 5 TBI animals and n = 5 shams). Quantitative RT-PCR was performed as described for the predicted miRNA targets with gene-specific primers and probes (prevalidated TaqMan Gene Expression Assay for Cplx1; Rn02396766_m1, Applied Biosystems).

2.14. qPCR Analysis of tRF Cleaving Enzyme Angiogenin

Complementary DNA (cDNA) synthesis (n = 5 TBI animals and n = 5 shams). The cDNA synthesis was performed as described for the predicted miRNA targets.

Quantitative RT-PCR (n = 5 TBI animals and n = 5 shams). Quantitative RT-PCR was performed as described for the predicted miRNA targets with gene-specific primers and probes (prevalidated TaqMan Gene Expression Assay for angiogenin; Rn03416813_gH, Applied Biosystems).

2.15. Data Analysis and Statistics

No statistical methods were used to predetermine the sample size. Unsupervised hierarchical clustering (R environment) was used to cluster animals and visualize differentially expressed miRNAs and tRFs. The Vienna RNAfold web server (http://rna.tbi.univie.ac.at/cgi-bin/RNAWebSuite/RNAfold.cgi, accessed on 16 June 2015) was used to predict the RNA folding. In RT-qPCR analyses, statistical differences were estimated using the Mann–Whitney U-test (Prism 9). Differences in the neuroscore between 2 groups were assessed using the Mann–Whitney U-test (Prism 9). All correlations were analyzed with the Spearman rank correlation (Prism 9). A $p < 0.05$ was considered to indicate statistical significance. We did not exclude any data points in this study.

3. Results

3.1. Mortality after Lateral FPI

Acute post-TBI mortality was 27% (8/30). The mean post-impact apnea time was 38 ± 3.6 s. Post-impact seizure-like behaviors were observed in 30% (9/30) of the TBI animals.

3.2. Anatomic Analysis and Assessment of Chronic Neuroinflammation

Visual analysis of 1-mm-thick coronal MRI slices revealed substantial variability in the extent of the cortical lesion between the animals (example in the Figure 2A). Moreover, the enhanced T2-weighted signal appeared to be unevenly distributed along the rostrocaudal extent of the cortical lesion. MRI was used to identify animals with a prominent increase in T2-enhancement around the lesion (arrow in Figure 2A), indicating perilesional inflammation (n = 13/22). Nissl-stained sections (Figure 2B) showed detailed cytoarchitectonics of the perilesional area. Next, immunohistochemical staining was used to confirm chronic neuroinflammation suggested by the T2-enhancement. Immunohistochemical staining for CD68 with qualitative analysis showed activated microglia/macrophages in the perilesional cortex (Figure 2C). Further, immunohistochemical staining for GFAP with qualitative analysis showed astrocytic fibers (Figure 2D).

RNA-sequencing was used to assess detailed molecular network alterations after TBI in the perilesional cortex (Figure 3) and ipsilateral thalamus (Figure 4). Basic quality control of RNAseq data suggested experimental success (Supplementary Table S2). Overall, we detected 2773 and 2954 upregulated genes in the perilesional cortex and ipsilateral thalamus, respectively (Supplementary Figure S1). The number of downregulated genes was 2490 in the perilesional cortex and 3044 in the ipsilateral thalamus (Supplementary Figure S1).

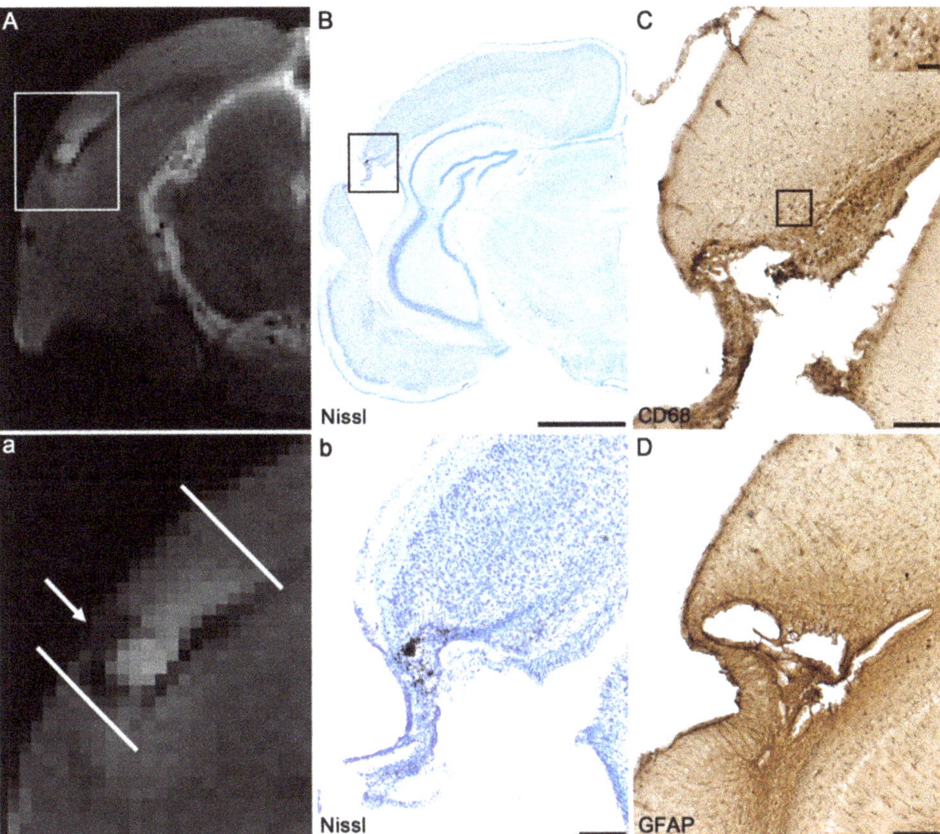

Figure 2. Assessment of chronic neuroinflammation. (**A**) Magnetic resonance imaging (MRI) was used to identify animals with a prominent increase in T2-enhancement around the lesion (arrow in a high magnification representation (**a**)) indicating perilesional inflammation. (**B**) Nissl-stained sections show detailed cytoarchitectonics of the perilesional area. (**C,D**) Immunohistochemical stainings were used to confirm chronic neuroinflammation suggested by T2-enhancement. Immunohistochemical staining for CD68 (macrosialin) showed activated microglia/macrophages in the perilesional cortex (panel (**C**)). Immunohistochemical staining for glial fibrillary acidic protein (GFAP) with qualitative analysis showed astrocytic fibers (panel (**D**)). Scale bars = 2000 µm (**B**), 200 µm (**b,C,D**) and 50 µm (insert in panel (**C**)).

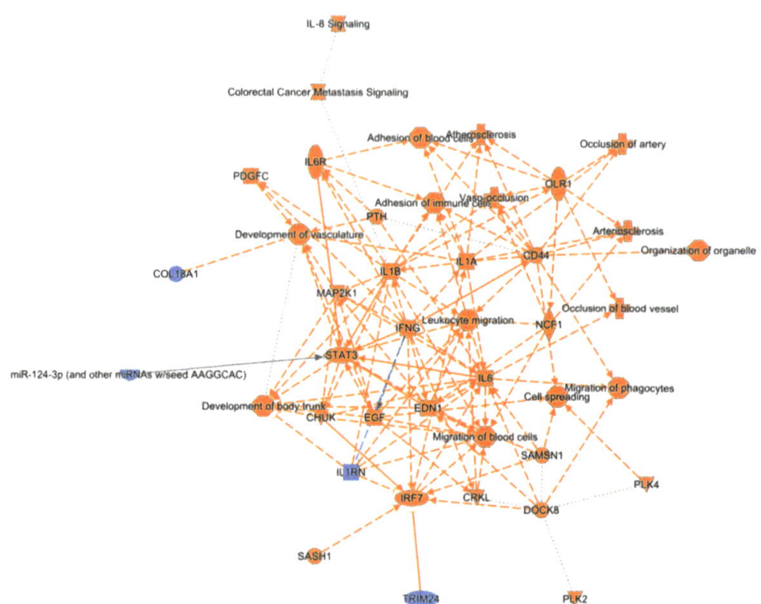

Figure 3. Graphical summary of pathway analysis of differentially expressed genes in the perilesional cortex after traumatic brain injury. Analysis highlighted physiologic functions like organismal survival, tissue morphology, nervous system development and function, and hematologic system development and function.

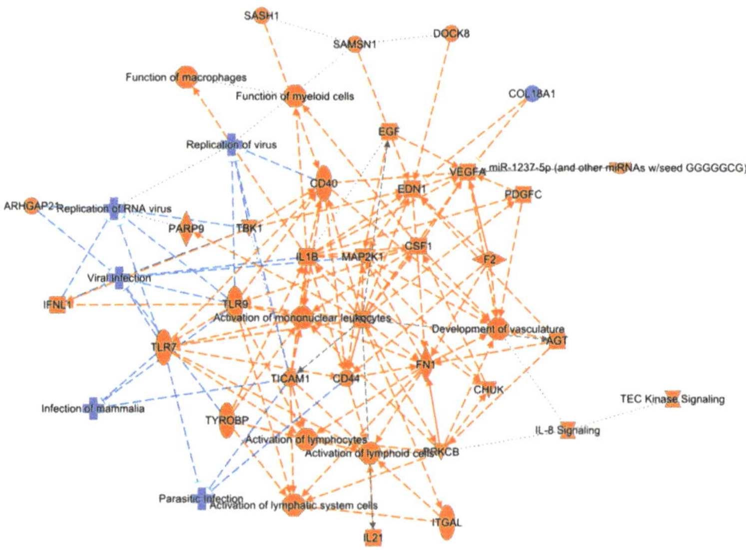

Figure 4. Graphical summary of the pathway analysis of differentially expressed genes in the ipsilateral thalamus after traumatic brain injury. Similar to the perilesional cortex, analysis highlighted physiologic functions, such as organismal survival, tissue morphology, nervous system development and function, and hematologic system development and function. Top regulator effect networks included genes such as Collagen Type XVIII Alpha 1 Chain (COL18A1) and SELPG protein.

Top regulator effect networks included genes like CRK Like Proto-Oncogene Adaptor Protein (CRKL), Polo Like Kinase 2 (PLK2), and Neutrophil Cytosolic Factor 1 (NCF1). In the perilesional cortex, the top five regulated canonical pathways were axonal guidance signaling (38% overlap with dataset, 172/456, $p < 0.001$), colorectal cancer metastasis signaling (42% overlap, 106/252, $p < 0.001$), molecular mechanisms of cancer (37% overlap, 142/382, $p < 0.001$), role of nuclear factor of activated T cells in cardiac hypertrophy (44% overlap, 87/199, $p < 0.001$), and role of macrophages, fibroblasts, and endothelial cells in rheumatoid arthritis (38% overlap, 120/315, $p < 0.001$). Similarly, in the ipsilateral thalamus, the top five regulated canonical pathways were interleukin (IL)-8 signaling (51% 100/197, $p < 0.001$), leucocyte extravasation signaling (49% overlap, 103/210, $p < 0.001$), role of nuclear factor of activated T cells in cardiac hypertrophy (50% overlap, 96/192, $p < 0.001$), G-protein coupled receptor signaling (45% overlap, 123/272, $p < 0.001$), and Huntington's disease signaling (46% overlap, 111/241, $p < 0.001$).

In the perilesional cortex, IPA analysis revealed the top five molecular and cellular functions among differentially expressed genes to be cellular assembly and organization (927 molecules), cellular function and maintenance (1484 molecules), cellular movement (1079 molecules), cellular morphology (1178 molecules), and cellular growth and proliferation (1744 molecules). The top five molecular and cellular functions in the ipsilateral thalamus were cellular assembly and organization (1048 molecules), cellular function and maintenance (1675 molecules), cellular morphology (1321 molecules), molecular transport (1153 molecules), and cell death and survival (1657 molecules).

Based on these data, we also predicted the top five upstream regulators with IPA. In the perilesional cortex, all five of these regulators were activated ($p < 0.001$). These five regulators were TGFB1, tNF, TP53, IFNG, and APP. In the ipsilateral thalamus, of the top five regulators, four (APP, TNF, IFNG, and lipopolysaccharide) were activated and one (L-dopa) was inhibited ($p < 0.001$).

3.3. Differentially Expressed miRNAs Identified from sncRNA-Seq

Overall, quality control analysis of the sncRNAseq suggested good quality (Supplementary Figure S2A–C). In the perilesional cortex, 19 miRNAs were identified to be differentially expressed in TBI animals at 3 months post-injury compared with controls (FDR <0.05, Figure 5A). In the ipsilateral thalamus, only two miRNAs were identified to be differentially expressed with FDR <0.05. From the 19 differentially expressed miRNAs in the perilesional cortex, PCR analysis was performed for the top 10 candidates (ranked based on increasing order of FDR-value; Figure 5B1,B2,C). In the ipsilateral thalamus, the two differentially expressed miRNAs identified from small RNA-Seq were selected for further validation. Among the 10 miRNA candidates tested from the perilesional cortex with ddPCR, upregulation of only miR-375-3p (FC = 3.44, $p < 0.01$, Figure 5B1) and miR-211-5p (FC = 1.45, $p < 0.05$, Figure 5B2) was validated. For the majority (seven of eight) of the remaining candidates, ddPCR showed a similar expression pattern to that observed with small RNA-Seq (Figure 5C). The PCR analysis results, however, were not statistically significant ($p > 0.05$). From the ipsilateral thalamus, both tested candidates, miR-146a-5p (FC = 2.01, $p < 0.05$, Figure 5B3) and miR-155-5p (FC = 2.34, $p < 0.01$, Figure 5B4), were successfully validated with ddPCR. Similar to the perilesional cortex, these two miRNAs were significantly upregulated in the ipsilateral thalamus of the TBI group at 3 months post-injury compared with sham-operated controls.

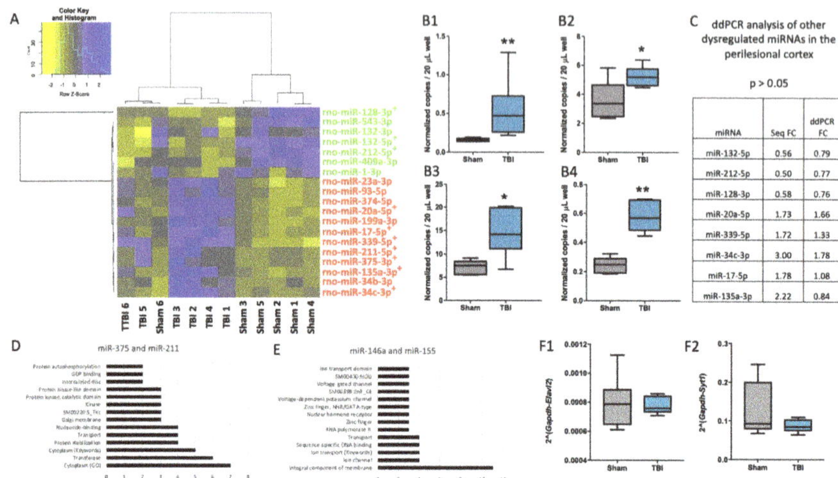

Figure 5. Deregulated microRNAs (miRNAs) after lateral fluid percussion injury. (**A**) Unsupervised hierarchical clustering partly separated sham-operated controls from post-TBI rats according to differentially expressed (**D,E**) miRNAs in the perilesional cortex. A total of 12 upregulated and 7 downregulated miRNAs were found (FDR <0.05). (**B1–B4**) Droplet digital PCR (ddPCR) analysis validated miR-375 (**B1**, $p < 0.01$) and miR-211 (**B2**, $p < 0.05$) upregulation. In the ipsilateral thalamus, only 2 miRNAs were upregulated in small non-coding RNAseq analysis (sncRNAseq, FDR <0.05). Both of these were validated with ddPCR (**B3**, miR-146a, $p < 0.05$; **B4**, miR-155, $p < 0.01$). (**C**) All other sncRNAseq suggested no change in the top 10 DE miRNAs in ddPCR analysis ($p > 0.05$ for all). Most of them (7/8), however, were regulated in the same direction detected in the sncRNAseq analysis. (**D**) Functional pathway analysis of predicted targets of miR-375 and miR-211. (**E**) Functional pathway analysis of predicted targets of miR-146a and miR-155. (**F1,F2**) Quantitative RT-PCR analysis revealed no change in the predicted targets selected for validation (ELAV Like RNA Binding Protein 2 (Elavl2) and Synaptotagmin 1 (Syt1), $p > 0.05$). Statistical significance: *, $p < 0.05$; **, $p < 0.01$.

3.4. mRNA Targets Predicted for the Validated miRNA Candidates

For miR-375-3p, 12/178 (6.7%) of the predicted mRNA targets from TargetScan were also significantly downregulated in our mRNA-Seq data in TBI animals compared with sham-operated controls. For miR-211-5p, 7/98 (7.1%) of the predicted mRNA targets from TargetScan were significantly downregulated in our mRNA-Seq from TBI animals compared with sham-operated controls. The most common functions among these targets were cytoplasmic functions, and transferase and protein stabilization (Figure 5D). Similarly, for miR-146a-5p, 7/137 (5.1%) of the predicted mRNA targets from TargetScan, and for miR-155-5p, 23/288 (7.9%) of the predicted mRNA targets from TargetScan were significantly downregulated in our mRNA-Seq from TBI animals compared with sham-operated controls. The most common functions among these targets were the integral components of membranes, ion channels, and ion transport (Figure 5E). For both miRNAs from the perilesional cortex (miR-375-3p and miR-211-5p), ELAV-like neuron-specific RNA binding protein 2 (ELAVL2) was a common predicted mRNA target. Previous studies revealed a role of miR-375 in regulating dendrite formation and maintenance by affecting the levels of ELAVL4 protein, a paralog of ELAVL2. Thus, ELAVL2 (mRNA-Seq FC = 0.75, FDR <0.05) was selected for validation with RT-qPCR. From the ipsilateral thalamus, synaptotagmin 1 (Syt1) (mRNA-Seq FC = 0.73, FDR <0.05), a predicted target for miR-146a-5p, was selected for validation. Because the read abundance for miR-155-5p in the ipsilateral thalamus was quite low for all samples (TBI: 12.49 (4.61), Sham: 4.56 (0.95), mean (SD)), and there were no commonly predicted targets for miR-146a-5p and miR-155-5p, targets for miR-155-5p were not selected for validation. Quantitative PCR analysis of both targets revealed unaltered gene expression ($p > 0.05$, Figure 5F1,F2).

3.5. IsomiRs Identified for the Validated Differentially Expressed miRNAs

To explain the negative miRNA-target expression findings, we performed a detailed analysis of isomiRs. For miR-375-5p, the total number of reads from all isomiRs, and the number of different isomiR types (canonical (Figure 6A1) and isomiRs (Figure 6A3), $p < 0.05$ for both) identified were significantly higher in the TBI animals compared with the sham-operated controls. For miR-211-5p, the number of detected canonical miRNA species was increased ($p < 0.05$, Figure 6B1). The number of different isomiR types identified was significantly higher in the TBI animals compared with the sham-operated controls (Figure 6B3, $p < 0.05$), but the total number of reads from all isomiRs was not significantly different (Figure 6B4, $p > 0.05$). For miR-146a-5p, the total number of reads from all isomiRs, as well as the number of different isomiR types identified, did not differ significantly between the TBI and sham-operated controls (Figure 6C1–C4, $p > 0.05$). For miR-155-5p, no isomiRs were detected. For the four endogenous controls or non-validated differentially expressed miRNA candidates in both the perilesional cortex and ipsilateral thalamus, the total number of reads from all isomiRs and the number of different isomiR types identified did not differ significantly between the TBI and sham-operated controls ($p > 0.05$). The amount of canonical miRNA and isomiRs related to different miRNAs, however, was highly variable.

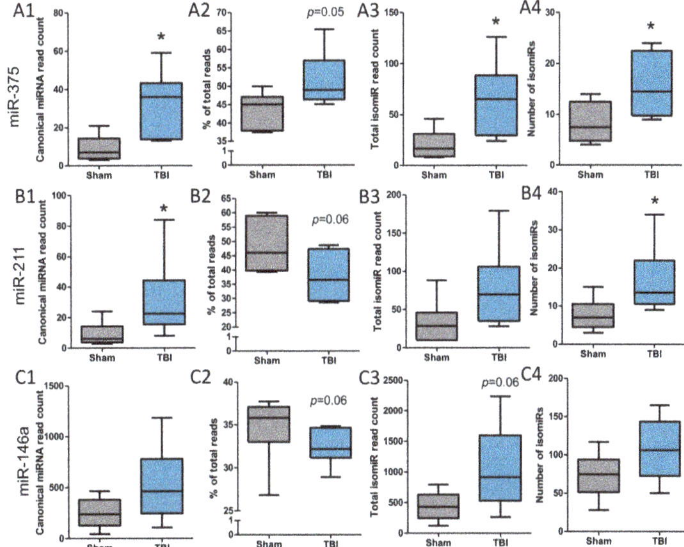

Figure 6. Analysis of canonical (miRbase) miRNA sequence and isomiRs of miR-375, miR-211, and miR-146a. (**A1–A4**) Read count of canonical miR-375 was increased, (**A1**) $p < 0.05$ and there was a trend toward its proportional increase from all miR-375 aligned reads; (**A2**) $p < 0.1$. Additionally, the total isomiR read count (**A3**) $p < 0.05$ and number of isomiRs (**B4**) $p < 0.05$, were increased. ((**B1–4**) Similarly, the read count of canonical miR-211 was increased, (**B1**) $p < 0.05$ and there was a trend toward its proportional increase from all miR-211 aligned reads; (**B2**) $p < 0.1$. Total isomiR read count (**B3**) $p > 0.05$ was not changed, but the number of isomiRs (**B4**) $p < 0.05$ was increased. (**C1–C4**) For miR-146a, we observed no significant changes. The trends for canonical miRNA were opposite for cortical differentially expressed miRNAs (miR-375 and miR-211), however, as the proportion of canonical miR-146a was decreased compared to all sequences aligned to it, $p < 0.1$. Statistical significance: *, $p < 0.05$.

Next, we performed a detailed analysis of different isomiRs observed for validated differentially expressed miRNAs. For miR-375-3p, the canonical miRNA sequence (TTTGTTCG TTCGGCTCGCGTGA) formed on average 32% of the total read count in the TBI animals (32.42 (2.15)), and 35% in the sham animals (34.68 (3.96)) ($p = 0.05$, Figure 6A2). For miR-211-5p, the canonical sequence (TTCCCTTTGTCATCCTTTGCCT) formed on average 52% of the total read count in the TBI animals (51.6 (7.46)), and 44% in the sham animals (43.64 (4.89)) ($p = 0.06$, Figure 6B2). For miR-146a-5p, the canonical sequence (TGAGAACT-GAATTCCATGGGTT) formed on average 38% of the total read count in the TBI animals (37.71 (8.56)), and 48% in the sham animals (48.32 (9.33)) ($p = 0.06$, Figure 6C2).

Details of the top 2–4 isomiRs (among the top six) for miR-375-3p, miR-211-5p, and miR-146a-5p that were detected from all TBI samples are summarized in Figure 6. Most of these were 3′-isomiRs, i.e., additions, deletions, or nucleotide changes to the canonical miRNA sequence that occurred at the 3′ end. The top two isomiRs detected for miR-375-3p and miR-211-5p were not significantly different between the TBI and controls ($p > 0.05$, Figure 7A,B). For miR-146a-5p, among the top four isomiRs, TGAGAACT-GAATTCCATGGGT and TGAGAACTGAATTCCATGGGTTT were significantly more abundant in the TBI animals compared with the controls ($p < 0.05$, Figure 7C). Next, we performed RT-qPCR to detect all 3′ end isomiRs (Figure 7D, FC = 1.7, $p < 0.05$) and were able to detect a difference between the controls and post-TBI animals. Quantitative RT-PCR and primer assay targeting only canonical miR-146a was not able to detect this increase (see Supplementary Figure S3, FC = 1.4, $p > 0.05$). Interestingly, RNA folding assessment predicted that the linear miR-146a form changes to a hairpin-like structure when insertions were detected in the miRNAs 3′ end (Figure 7E).

3.6. Elevation of Transfer RNA-Derived Fragments after TBI

We observed a clear upregulated profile of tRFs detected by small RNA sequencing in both the perilesional cortex (Supplementary Figure S4A) and ipsilateral thalamus (Supplementary Figure S4B) visualized by unsupervised hierarchical clustering. Using IGV to visualize aligned sequences, we selected two candidate tRNAs, tRNA IleAAT and tRNA LysTTT, for further inspection. Upregulation of fragments cleaved from these two tRNAs was evident in both the perilesional cortex (Figure 8A for tRNA IleAAT and Figure 8E for tRNA LysTTT) and the ipsilateral thalamus (Figure 8C for tRNA IleAAT and Figure 8G for tRNA LysTTT). Interestingly, both tRNAs had a clear cleavage site of tRFs in their variable region [40] producing mainly 27–28 nt long fragments from the 3′ end of the tRNA (Figure 8I for tRNA IleAAT and Figure 8J for tRNA LysTTT). Estimation of a possible secondary structure predicted hairpin-like folding of both fragments (Figure 8K1 for 3′tRF-IleAAT and Figure 8K2 for 3′tRF-LysTTT). Quantitative RT-PCR analysis showed a trend toward the upregulation of fragments originating from tRNA IleAAT in the perilesional cortex (FC = 1.5, $p > 0.05$; Figure 8B). Similar to small RNA-Seq analysis, however, RT-qPCR indicated robust upregulation of the fragments from tRNA IleAAT in the ipsilateral thalamus (FC = 5.4, $p < 0.01$; Figure 8D). Instead, RT-qPCR replicated the observed upregulation of fragments cleaved from tRNA LysTTT in both the perilesional cortex (FC = 6.9, $p < 0.01$; Figure 8F) and the ipsilateral thalamus (FC = 5.2, $p < 0.05$; Figure 8H). We used agarose gel electrophoresis to confirm the size of the detected tRFs, as total RNA (including also pre-tRNA and mature tRNA molecules) was used for the RT-qPCR analysis. Indeed, when only 20–50 nt long RNA was selected after the size separation of cortical RNA samples and subsequent RT-qPCR, a similar upregulated profile of 3′tRF-IleAAT (2.1, $p > 0.05$, $n = 3$ per group) and 3′tRF-LysTTT (3.3, $p > 0.05$, $n = 3$ per group) was detected compared with the total RNA samples (Supplementary Figure S5A for 3′tRF-IleAAT and Supplementary Figure S5C for 3′tRF-LysTTT). Analysis indicated a moderate correlation, meaning that the higher the expression in the total RNA sample, the greater the amount of 3′tRF also detected after the size separation (Supplementary Figure S5B for 3′tRF-IleAAT (Spearman r = 0.54, $p > 0.05$) and Supplementary Figure S5D for 3′tRF-LysTTT (Spearman r = 0.75, $p > 0.05$).

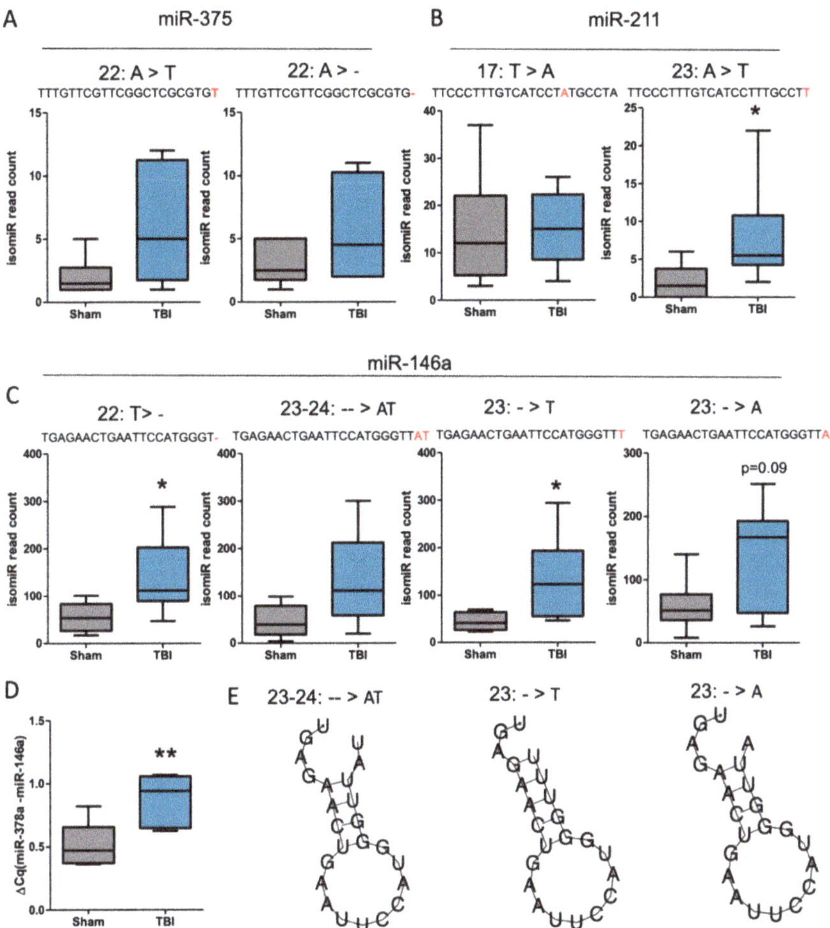

Figure 7. Analysis of isomiR types after traumatic brain injury (TBI). Analysis of cortical differentially expressed miRNAs revealed an increase, especially in 3′ end modifications (**A**,**B**). The miR-211 isomiR 5′–TTCCCTTTGTCATCCTTTGCCTT was significantly increased after TBI (B, $p < 0.05$). For miR-146a, (**C**) several isomiRs were detected. These 2 isomiRs showed increased modifications at the 3′ end (5′– TGAGAACTGAATTCCATGGGT- and 5′– TGAGAACTGAATTCCATGGGTTT; $p < 0.05$). Quantitative RT-PCR chemistry detecting all 3′ end isomiRs (**D**) (FC = 1.7, $p < 0.05$) revealed a difference between controls and post-TBI animals. Primer assay targeting only canonical miR-146a did not detect this increase (see Supplementary Figure S3, FC = 1.4, $p > 0.05$). Interestingly, RNA folding assessment predicted the linear miR-146a form to change to a hairpin-like structure when insertions were detected in the miRNA 3′ end (**E**). Statistical significance: *, $p < 0.05$; **, $p < 0.01$.

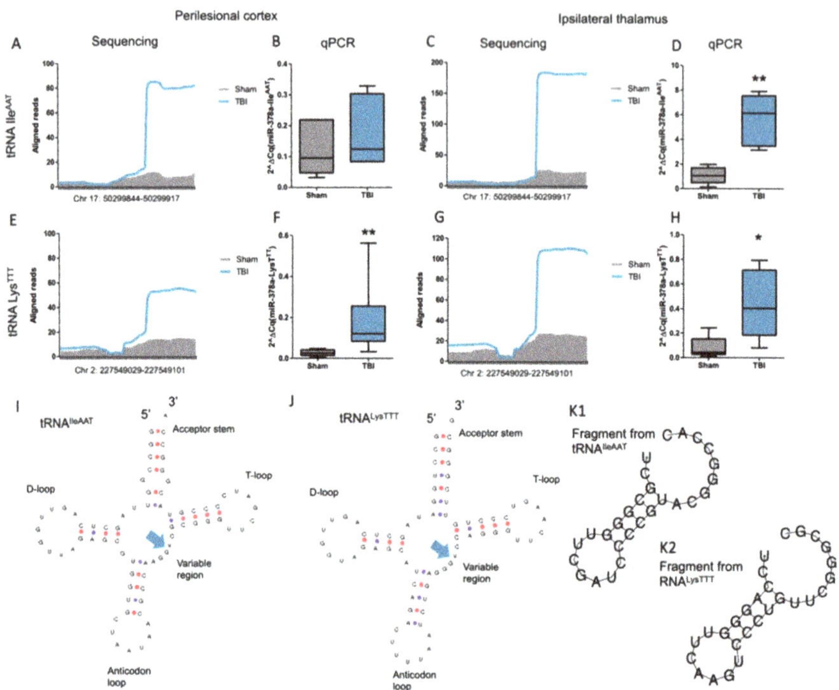

Figure 8. Transfer RNA fragments (tRFs) cleaved from tRNA IleAAT and tRNA LysTTT were upregulated in the perilesional cortex and ipsilateral thalamus after traumatic brain injury (TBI). (**A**) Alignment of the detected reads toward tRNA IleAAT in sham-operated controls (gray shading) and post-TBI rats (blue line) in the perilesional cortex samples. (**B**) Quantitative RT-PCR analysis showed a trend toward upregulation of 3′tRF-IleAAT (FC = 1.5, $p > 0.05$). (**C**) Alignment of the detected reads toward tRNAIleAAT in sham-operated controls (gray shading) and post-TBI rats (blue line) in the ipsilateral thalamus samples. (**D**) Quantitative RT-PCR analysis showed upregulation of 3′tRF-IleAAT (FC = 5.4, $p < 0.01$). (**E**) Alignment of the detected reads toward tRNA LysTTT in sham-operated controls (gray shading) and post-TBI rats (blue line) in the perilesional cortex samples. (**F**) Quantitative RT-PCR analysis showed upregulation of 3′tRF-LysTTT (FC = 6.9, $p > 0.01$). (**G**) Alignment of the detected reads toward tRNALysTTT in sham-operated controls (gray shading) and post-TBI rats (blue line) in the ipsilateral thalamus samples. (**H**) Quantitative RT-PCR analysis showed upregulation of 3′tRF-LysTTT (FC = 5.2, $p < 0.05$). (**I**) Most common cleavage site of fragments from tRNA IleAAT detected based on analysis presented in panels (**A** and **C**). (**J**) Most common cleavage site of fragments from tRNA LysTTT detected based on analysis presented in panels (**E** and **G**). (**K1,K2**) Predicted RNA folding structure for 3′tRFs upregulated after TBI. Statistical significance: *, $p < 0.05$; **, $p < 0.01$.

3.7. mRNA Targets Predicted for the Validated tRF Candidates

We used an 8-nt approach [39] to predict mRNA targets for validated 3′ tRFs (Figure 9A1–A3 for 3′tRF-IleAAT and Figure 9B1–B3 for 3′tRF-LysTTT). Gene set enrichment analysis indicated a negative enrichment of these predicted targets for a combined target list of both tRFs in the ipsilateral thalamus (ES = −0.22, FDR <0.01, Figure 9C). The thalamus was selected for further investigation as both fragments were robustly upregulated in this brain area. We selected one candidate target (Cplx1) of 3′tRF-IleAAT for further validation with RT-qPCR. In the ipsilateral thalamus, Cplx1 was downregulated (FC = 0.78, $p < 0.05$, Figure 9D) and the lower the expression level, the greater the amount of 3′tRF-IleAAT detected in the same sample (Spearman r = 0.43, $p < 0.05$, Figure 9D). Gene expression of Cplx1 showed a trend toward downregulation in the perilesional cortex (FC = 0.4, $p > 0.05$; Spearman r = 0.38, $p > 0.05$; $n = 5$ per group; Supplementary Figure S6) where the upregulation of tRF was also not that high. Further, functional annotation of negatively enriched predicted tRF targets linked them to glycoprotein, protein binding, nucleotide-binding, lipoprotein, and neurogenesis ($p < 0.05$, Figure 9E, Supplementary Spreadsheet 1).

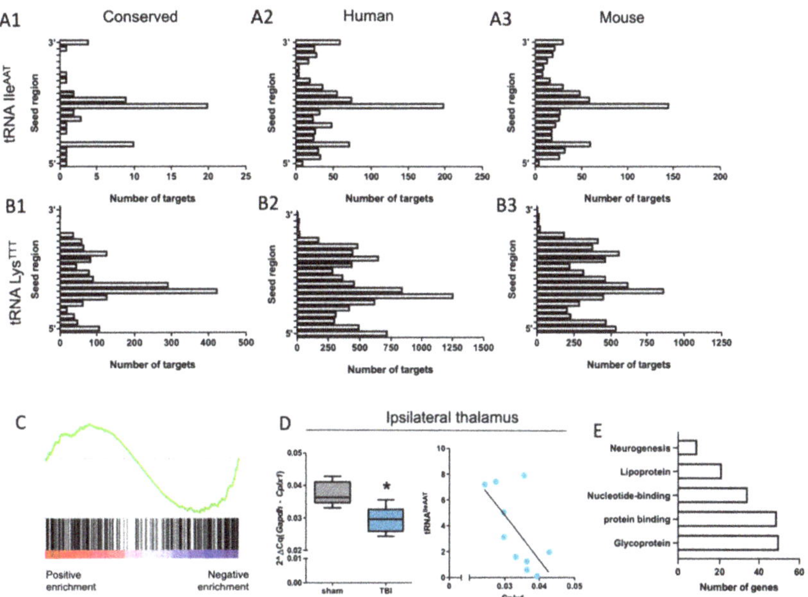

Figure 9. Downregulation of the predicted targets of upregulated tRFs after traumatic brain injury. (**A1–A3**) Number of predicted targets of 3′tRF-IleAAT for conserved, human and mouse 3′-untranslated regions (UTRs) of protein coding genes. (**B1–B3**) Number of predicted targets of 3′tRF-LysTTT for conserved, human, and mouse 3′-UTRs of protein-coding genes. (**C**) Gene set enrichment analysis of predicted targets of both tRFs (combined gene list) in the ipsilateral thalamus revealed negative enrichment (ES = 0.22, FDR <0.01). (**D**) Quantitative RT-PCR analysis validated downregulation of one selected target candidate of 3′tRF-IleAAT, Cplx1 (FC = 0.78, $p < 0.05$). Further, the lower the Cplx1 expression level, the greater the detection of 3′tRF-IleAAT in the same sample (Spearman r = 0.43, $p < 0.05$). (**E**) Functional annotation of negatively enriched predicted tRF targets linked them to glycoprotein, protein binding, nucleotide-binding, lipoprotein, and neurogenesis ($p < 0.05$, DAVID functional annotation tool). Statistical significance: *, $p < 0.05$.

3.8. No Clear Upregulation of tRF-Cleaving Enzyme Angiogenin after TBI

RNAseq experiments revealed that the tRF-cleaving enzyme angiogenin was upregulated in the ipsilateral thalamus (log2FC = 0.47, FDR <0.01, Supplementary Figure S7A). We quantified the expression level of angiogenin with RT-qPCR in both the perilesional cortex (FC = 1.26, $p > 0.05$) and ipsilateral thalamus (FC = 1.62, $p > 0.05$, n = 6 for controls and $n = 5$ for TBI) but did not detect the expected significant upregulation (Supplementary Figure S7B). A moderate correlation, however, was observed between angiogenin expression and the 3′tRF levels (Spearman r = 0.45, $p < 0.05$ for 3′tRF-IleAAT and Spearman r = 0.36, $p < 0.05$ for 3′tRF-LysTTT) (Supplementary Figure S7C for 3′tRF-IleAAT and Supplementary Figure S7D for 3′tRF-LysTTT) in the thalamus.

3.9. Elevated 3′tRF and miR-146a Levels Relate to Worse Behavioral Outcome after TBI

We assessed neuromotor functions with the composite neuroscore at acute and chronic time-points post-TBI and investigated memory deficits in the Morris water maze at 2 months post-TBI (Figure 10A). Our correlation analysis showed that rats exhibiting deficits in the probe trial, represented by less time spent in the correct quadrant, also had the highest expression of the full miR-146a profile (all isomiRs included; Spearman r = 0.43, $p < 0.05$, Figure 10B1), 3′tRF-IleAAT (Spearman r = 0.70, $p < 0.001$; Figure 10B2) and 3′tRF-LysTTT (Spearman r = 0.57, $p < 0.01$; Figure 10B3). As expected, all TBI rats exhibited acute motor deficits compared with the controls at 2 days post-TBI ($p < 0.01$, Figure 10C) with recovery toward subacute at the 7 days ($p < 0.01$, Figure 10C), 14 days ($p < 0.01$, Figure 10C) and 21 days ($p < 0.01$, Figure 10C) post-TBI time-points. When the neuroscore was assessed at 3 months post-TBI, the measured score varied widely between rats. The difference between post-TBI rats and the controls remained significant ($p < 0.01$, Figure 10C). Interestingly, our correlation analysis indicated a relation between the lower neuroscore value at 3 months post-TBI and the increased expression of the full miR-146a profile (all isomiRs included; Spearman r = 0.75, $p < 0.001$; Figure 10(D1)), 3′tRF-IleAAT (Spearman r = 0.78, $p < 0.001$, Figure 10(D2)) and 3′tRF-LysTTT (Spearman r = 0.92, $p < 0.001$; Figure 10D3).

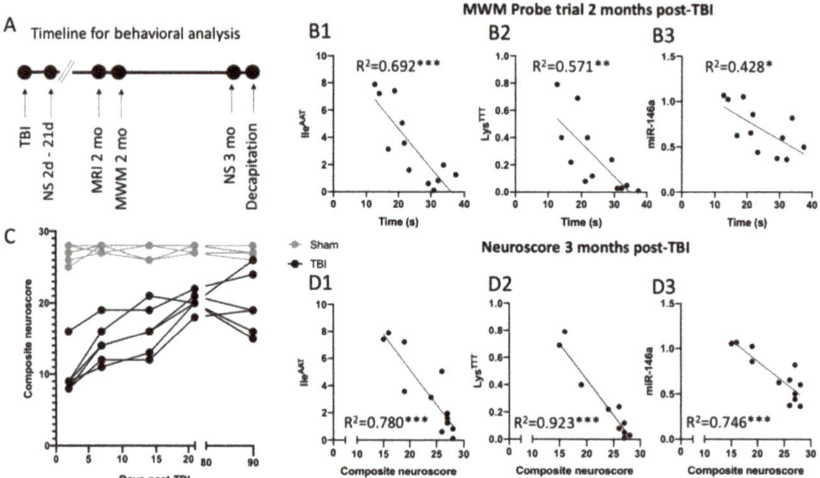

Figure 10. Upregulation of tRFs in the ipsilateral thalamus correlated with a worse behavioral outcome after traumatic brain injury (TBI). (**A**) Neuromotor function was assessed by the composite neuroscore at acute and chronic time-points post-TBI, and memory deficits were investigated with the Morris water maze at 2 months post-TBI. (**B1–B3**) Correlation analysis showed that rats showing deficits in the probe trial, represented by less time spent in the correct quadrant, also had the highest expression of the full miR-146a profile (all isomiRs included; Spearman r = 0.43, $p < 0.05$, panel (**B1**)), 3′tRF-IleAAT (Spearman r = 0.70, $p < 0.001$; panel (**B2**)), and 3′tRF-LysTTT (Spearman r = 0.57, $p < 0.01$; panel (**B3**)). (**C**) TBI rats exhibited acute motor function deficits compared with controls at 2 days post-TBI ($p < 0.01$) with recovery to subacute at 7 days ($p < 0.01$, compared with controls), 14 days ($p < 0.01$, compared with controls) and 21 days ($p < 0.01$, compared with controls) post-TBI time-points. At 3 months post-TBI, high variance in the measured score was observed between rats, but lower scores were still observed compared with controls ($p < 0.01$). (**D1–D3**) Correlation analysis indicated a relation between a lower neuroscore value at 3 months post-TBI and increased expression of the full miR-146a profile (all isomiRs included; Spearman r = 0.75, $p < 0.001$; panel (**D1**)), 3′tRF-IleAAT (Spearman r = 0.78, $p < 0.001$, panel (**D2**)) and 3′tRF-LysTTT (Spearman r = 0.92, $p < 0.001$; panel (**D3**)). Statistical significance: *, $p < 0.05$; **, $p < 0.01$; ***, $p < 0.001$.

4. Discussion

A growing body of evidence demonstrates the involvement of different non-coding RNAs in brain diseases, including TBI [41–44]. MicroRNAs have been identified as key regulators of inflammation after brain injuries [45,46]. In the present study, our objective was to identify chronically altered miRNA and tRF signatures after TBI. Our data indicate that two upregulated miRNAs (miR-155 and miR-146a) in the thalamus at a chronic time post-TBI might have unfavorable effects for recovery after TBI. The inflammatory role of the dysregulated miRNAs in the perilesional cortex, however, is not clear. Further, we revealed a dysregulated profile of isomiRs of quantified differentially expressed miRNAs, and a novel class of small RNAs, the tRFs, involved in post-TBI pathophysiology. For the first time, we report here a relation between upregulated tRFs and an unfavorable behavioral outcome after TBI.

4.1. Chronic Neuroinflammation after TBI

Brain inflammation is a prominent feature after TBI that has long-lasting secondary injury mechanisms and evolves in the brain for months after TBI [47,48]. The control of neuroinflammation may lessen the risk of developing comorbidities, like memory deficits and epilepsy [47,49]. Yet, brain inflammation is sparsely studied at chronic time-points post-TBI. Importantly, many inflammatory mediators remain dysregulated at 3 months post-TBI, as demonstrated by our previous genome-wide transcriptomic analysis [4]. Here we demonstrated a network level dysregulation of protein-coding genes in the perilesional cortex and ipsilateral thalamus. As expected, we observed robust regulation of inflammation-related pathways and regulators in both brain areas. Our analysis also highlighted the regulation of genes involved in organismal injury and abnormalities, and cell and tissue morphology. Perilesional cortex-specific changes were observed in axonal guidance signaling pathways. Further, ipsilateral thalamus-specific changes were observed in functions of molecular transport and cell death, as well as toxic effects of mitochondrial dysfunction. Interestingly, our analysis revealed broad regulation of the pathways and functions connected to hematologic system development and function. Early coagulopathies have been reported in TBI patients and even in patients with isolated severe TBI without severe bleeding [50–52]. Importantly, trauma-induced coagulopathies are independent predictors of poor outcomes and are linked to secondary injuries. How regulation of the hematologic system in the brain influences secondary injuries and relates to chronic neuroinflammation remains to be studied.

4.2. Differentially Expressed miRNAs after TBI

We detected relatively few changes in canonical miRNAs in both brain areas investigated. Based on earlier studies, the neuroinflammatory role of the upregulated miRNAs (miR-375 and miR-211) in the perilesional cortex is not well known. Upregulation of miR-375, however, is considered to protect the brain from ischemic stroke resulting in a reduced infarct volume and decreased cell apoptosis [53]. On the other hand, knockdown of human miR-375 upregulates NF-κB and pro-inflammatory factors, such as tumor necrosis factor-α, IL-1β, IL-6 and IL-8, in the colorectal cancer cell line Caco-2 [54]. These findings suggest the possible anti-inflammatory and neuroprotective roles of miR-375. Further, miR-211 excess regulation is connected to perturbation of learning functions in the Morris water maze [55]; mice with high miR-211 levels lost the capacity to locate the platform quadrant in a probe trial assessing reference memory. Interestingly, in our study, post-TBI rats also showed decreased ability to find the platform in the Morris water maze test and had higher cortical miR-211 levels. On the other hand, higher levels of miR-211 seem to protect the brain from hypersynchronization, and nonconvulsive and convulsive seizures [55]. These findings suggest multileveled, beneficial and harmful, roles of miR-211 in the brain.

MiR-146a is upregulated in the brains of post-TBI patients, animal models of TBI, and similarly in epilepsy patients and animal models of epilepsy [56] and miR-146a is suggested to be involved in the development and progression of seizures, especially through the regulation of inflammation and immune responses [57,58]. A recent study [59] indicated that the silencing of miRNA-146a decreases oxidative stress and the inflammatory response in a rat model of temporal lobe epilepsy (TLE); one of the genes involved in the process was Notch1. This is interesting as we previously reported upregulation of Notch1 in the dentate gyrus at 3 months post-TBI [60]. In this study, Notch1 was upregulated in the perilesional cortex (log2FC = 0.46, FDR < 0.01); however, in the ipsilateral thalamus—the brain area where miR-146a was upregulated—we observed no significant change (Log2FC = 0.26, FDR > 0.05). Notch1 did not come up in our miRNA-target analysis as we were interested in significantly changed targets. It is still possible that miR-146a downregulates the protein level of Notch1 and decreases its mRNA level in the thalamus. Our earlier study showed an increase in miR-155 in the rat and human brain post-TBI [56]. This upregulation was associated with activated glial cells. In vitro studies indicate that human astrocytes acquire a pro-inflammatory phenotype and overexpress miR-155 after stimulation with a pro-inflammatory macrophage-conditioned culture medium. Thus, miR-155-promoted neuroinflammation via astrocyte activation is involved in the secondary injury. Moreover, numerous studies have demonstrated a pro-inflammatory role of miR-155 [46,61].

4.3. Are 3′ isomiRs a Specific Feature of TBI?

General information on the unique mature sequence of each miRNA is cataloged in the miRBase database [62]. These listed sequences are called miRNA canonical forms. The sncRNAseq experiments, however, suggest that miRNA lengths and sequences could be modified in animal and human tissues. These RNA isoforms are called isomiRs [63]. Experiments demonstrate substantial overlap in functional mRNA networks suppressed by both canonical miRNAs and their isomiRs, and that these isomiRs are functionally relevant [63].

Detailed analyses of sncRNAseq data have identified especially high fractions of mature miRNAs containing one or more 3′ nucleotide additions [16,64]. These 3′ nucleotide additions most commonly include single and double nucleotide additions of adenine and/or uridine bases. Studies investigating these modifications have found that only a small fraction of miRNAs derived from a particular miRNA locus are subject to, for example, 3′ adenine addition (0–20%). These additions could therefore provide a mechanism for the cell to regulate the targeting efficiency of the transcripts derived from specific miRNA and this profile has been shown to be cell type-specific [16]. Functional distinction between isomiR production between vertebrates and Drosophila [16] emphasizes that miRNA modifications may have divergent roles across groups of animals and humans. For some evolutionarily conserved miRNA families, however, generally comparable rates of additions have been observed across species [16,65]. Overall, very diverse and contradictory functions are attributed to these addition events [17]. These functions include miRNA targeting efficiency, exosome localization, degradation, stabilization, and influencing associations with argonaute proteins.

In this study we especially observed 3′ nucleotide additions or modifications. Further, our prediction of RNA folding suggested possible alterations in the folding structure of miR-146a. Moreover, we were unable to detect any changes in the predicted targets of the canonical miRNAs analyzed. Whether or not 3′ nucleotide isomiRs are involved in post-TBI pathology and how each detected nucleotide modification/addition affects miRNA functionality in the brain remains to be studied.

4.4. Variable Region Cleaved 3′tRFs Are Upregulated after TBI

Transfer RNAs and tRFs support different cellular processes, from canonical functions in translation to recently unraveled mechanisms as gene expression regulators [24,66]. Interestingly, the recently described TRF-AGO2 association leads to target repression that is similar to miRNA activity, thereby acting on posttranscriptional regulation [66]. In this study, we reported an increase in 3′tRFs. Reports suggest that this particular tRF class regulates gene expression in a miRNA-like manner [67]. Indeed, we found numerous predicted target candidates for these dysregulated tRFs and showed decreased expression of Cplx1. Functional annotation of negatively enriched predicted tRF targets linked them to glycoprotein, protein binding, nucleotide-binding, lipoprotein, and neurogenesis.

As noted, Cplx1 is predicted to be one of the affected targets of 3′tRF-IleAAT. Cplx1 belongs to the complexin/synaphin gene family and positively regulates exocytosis of various cytoplasmic vesicles, such as synaptic vesicles and other secretory vesicles [68]. This protein organizes the soluble NSF attachment protein receptor proteins (SNAREs). Thus, it can prevent SNAREs from releasing neurotransmitters until an action potential arrives at the synapse [68]. Interestingly, a study by Yi et al. [69] reported transient increases in the levels of Cplx1 and Cplx2 proteins in the ipsilateral cortex 6 h following lateral FPI. This increase was followed by a decrease in Cplx1 in the ipsilateral cortex and hippocampus, and a decrease in both complexins in the ipsilateral thalamus at day 3 and day 7 post-injury. The authors suggested that alterations in complexin levels may play an important role in neuronal cell loss following TBI, and thus contribute to the post-TBI pathophysiology. We have now expanded this finding and report chronic downregulation of Cplx1 gene expression in the ipsilateral thalamus still at 3 months post-TBI and its possible regulation through 3′tRF-IleAAT.

Variable region-cleaved tRFs are rarely described in the literature [70] and their biogenesis is not well known; however, tRF cleaving is generally mediated by endoribonucleases, such as angiogenin and dicer [20]. Angiogenin is a component of the acute-phase response that protects the organism from microbial and environmental stress [71]. Knockdown of angiogenin sensitizes cells to stress and promotes stress-induced activation of apoptotic caspases [72]. Treatment of cultured mammalian cells with angiogenin stimulates tRNA cleavage [73]. Research related to amyotrophic lateral sclerosis revealed that angiogenin plays an important role in the endogenous protective pathways of motor neurons exposed to hypoxia [74]. Other studies reported that angiogenin is secreted from motor neurons and taken up by astroglia where it induces tRNA cleavage [75] that targets specific mRNAs together with argonaute proteins modulating cell survival following stress [19]. We found no clear indication of the regulatory role of angiogenin, however, as a single explanatory factor for variable region-cleaved tRFs. We observed no change in the level of dicer (data not shown), another tRF cleaving endonuclease. A recent study suggested that the human RNase T2 generates long and short tRFs in vitro [76]. Interestingly, Rnaset2 was robustly upregulated in our mRNA-Seq analysis in both the perilesional cortex (Log2FC = 0.74, FDR <0.01) and ipsilateral thalamus (Log2FC = 1.39, FDR <0.01) after TBI. Detailed functional in vitro studies are needed to unveil the role of RNase T2 in variable region-cleaved tRF production.

Lastly, our data suggest that increases in the 3′tRFs strongly correlate with miR-146a (Supplementary Figure S8), a miRNA previously linked to reactive astrocytes [77]. Whether 3′tRF-IleAAT and 3′tRF-LysTTT are indeed increased in reactive astrocytes chronically after TBI or whether they co-exist with this pathologic feature remains to be studied. Further, it is not yet known how elevated tRF levels after TBI affect mature tRNA. Whether the overall role of tRFs is pro-inflammatory or anti-inflammatory remains to be explored. Despite a single study demonstrating increased seizure susceptibility and dysfunctional neuronal transmission as a result of mutated tRNA n-Tr20 [26], the link between tRNA and epileptic-associated phenotypes remains unclear. Moreover, Hogg et al., described tRFs that were modulated in an in vitro seizure model and could predict seizure imminence in human epilepsy patients when measured from plasma [25]. Finally, the present study demonstrated a link between a worse behavioral outcome after TBI and increased 3′tRF levels in the perilesional cortex and ipsilateral thalamus.

5. Conclusions

These results suggest a possible functional role of 3′tRF-IleAAT and 3′tRF-LysTTT after TBI and emphasize the importance of future studies to elucidate their mechanisms of action in chronic neuroinflammation and worse recovery outcomes after TBI. Moreover, additional studies are needed to reveal the importance and roles of isomiRs in disease mechanisms and functional outcome.

Supplementary Materials: The following are available online at https://www.mdpi.com/article/10.3390/biomedicines10010136/s1. Methods: Quantitative RT-PCR analysis of canonical miR-146a; Figures S1: Number of upregulated (A) and downregulated (B) genes in the perilesional cortex and ipsilateral thalamus; Figures S2: Mapping of the small non-coding RNA sequencing reads; Figures S3: Quantitative RT-PCR analysis of miR-146a did not reveal significant upregulation (FC = 1.4, $p > 0.05$); Figures S4: Unsupervised hierarchical clustering of differentially expressed transfer RNA fragments; Figures S5: Agarose gel electrophoresis to separate 20–50 nucleotide (nt) long small RNA from the total RNA pool; Figures S6: Gene expression of Cplx1 showed a trend toward downregulation in the perilesional cortex (FC = 0.4, $p > 0.05$); Figures S7: Expression of angiogenin (ANG) after experimental traumatic brain injury; Figures S8: Correlation between miR-146a, miR-155, and deregulated transfer RNA-derived fragments; Table S1: RIN integrity numbers prior the sequencing service and later PCR study; Table S2: Quality control of messenger RNA sequencing; Spreadsheet 1.

Author Contributions: Conceptualization, N.P. and A.P.; methodology, N.P., S.D.G. and N.V.; software, N.P. and S.D.G.; validation, N.P., S.D.G. and N.V.; formal analysis, N.P.; investigation, N.P., S.D.G. and N.V.; resources, A.P.; data curation, N.P., S.D.G. and N.V.; writing—original draft preparation, N.P.; writing—review and editing, S.D.G., N.V. and A.P.; visualization, N.P, S.D.G. and N.V.; supervision, N.P. and A.P.; project administration, N.P. and A.P.; funding acquisition, A.P. All authors have read and agreed to the published version of the manuscript.

Funding: This study was supported by The Medical Research Council of the Academy of Finland (Grants 272249, 273909, 2285733-9), The Sigrid Juselius Foundation, the European Union's Seventh Framework Programme (FP7/2007-2013) under grant agreement n°602102 (EPITARGET).

Institutional Review Board Statement: All animal procedures were approved by the Animal Ethics Committee of the Provincial Government of Southern Finland and performed in accordance with the guidelines of the European Community Council Directives 2010/63/EU.

Data Availability Statement: All RNA-Seq raw data were saved to the NCBI Gene Expression Omnibus (GEO; series accession number GSE192979 (mRNAseq) and GSE192980 (small RNAseq).

Acknowledgments: We thank Jarmo Hartikainen and Merja Lukkari for their excellent technical assistance.

Conflicts of Interest: The authors declare no conflict of interest. The funders had no role in the design of the study; in the collection, analyses, or interpretation of data; in the writing of the manuscript, or in the decision to publish the results.

References

1. Carroll, L.J.; Cassidy, J.D.; Holm, L.; Kraus, J.; Coronado, V.G. Methodological issues and research recommendations for mild traumatic brain injury: The WHO Collaborating Centre Task Force on Mild Traumatic Brain Injury. *J. Rehabil. Med.* **2004**, 113–125. [CrossRef]
2. Maas, A.I.R.; Menon, D.K.; Adelson, P.D.; Andelic, N.; Bell, M.J.; Belli, A.; Bragge, P.; Brazinova, A.; Büki, A.; Chesnut, R.M.; et al. Traumatic brain injury: Integrated approaches to improve prevention, clinical care, and research. *Lancet. Neurol.* **2017**, *16*, 987–1048. [CrossRef]
3. Huang, G.-H.; Cao, X.-Y.; Li, Y.-Y.; Zhou, C.-C.; Li, L.; Wang, K.; Li, H.; Yu, P.; Jin, Y.; Gao, L. Gene expression profile of the hippocampus of rats subjected to traumatic brain injury. *J. Cell. Biochem.* **2019**, *120*, 15776–15789. [CrossRef]
4. Lipponen, A.; Paananen, J.; Puhakka, N.; Pitkänen, A. Analysis of Post-Traumatic Brain Injury Gene Expression Signature Reveals Tubulins, Nfe2l2, Nfkb, Cd44 and S100a4 as Treatment Targets. *Sci. Rep.* **2016**, *6*, 31570. [CrossRef]
5. Johnson, V.E.; Stewart, J.E.; Begbie, F.D.; Trojanowski, J.Q.; Smith, D.H.; Stewart, W. Inflammation and white matter degeneration persist for years after a single traumatic brain injury. *Brain* **2013**, *136*, 28–42. [CrossRef] [PubMed]
6. Thelin, E.P.; Tajsic, T.; Zeiler, F.A.; Menon, D.K.; Hutchinson, P.J.A.; Carpenter, K.L.H.; Morganti-Kossmann, M.C.; Helmy, A. Monitoring the Neuroinflammatory Response Following Acute Brain Injury. *Front. Neurol.* **2017**, *8*, 351. [CrossRef] [PubMed]
7. Fleminger, S.; Oliver, D.L.; Lovestone, S.; Rabe-Hesketh, S.; Giora, A. Head injury as a risk factor for Alzheimer's disease: The evidence 10 years on; a partial replication. *J. Neurol. Neurosurg. Psychiatry* **2003**, *74*, 857–862. [CrossRef] [PubMed]
8. Li, W.; Risacher, S.L.; McAllister, T.W.; Saykin, A.J. Traumatic brain injury and age at onset of cognitive impairment in older adults. *J. Neurol.* **2016**, *263*, 1280–1285. [CrossRef]
9. Walsh, S.; Donnan, J.; Fortin, Y.; Sikora, L.; Morrissey, A.; Collins, K.; MacDonald, D. A systematic review of the risks factors associated with the onset and natural progression of epilepsy. *Neurotoxicology* **2017**, *61*, 64–77. [CrossRef]
10. Li, Q.; Hu, B.; Hu, G.W.; Chen, C.Y.; Niu, X.; Liu, J.; Zhou, S.M.; Zhang, C.Q.; Wang, Y.; Deng, Z.F. TRNA-Derived Small Non-Coding RNAs in Response to Ischemia Inhibit Angiogenesis. *Sci. Rep.* **2016**, *6*, 20850. [CrossRef]

11. Bratkovič, T.; Modic, M.; Camargo Ortega, G.; Drukker, M.; Rogelj, B. Neuronal differentiation induces SNORD115 expression and is accompanied by post-transcriptional changes of serotonin receptor 2c mRNA. *Sci. Rep.* **2018**, *8*, 5101. [CrossRef]

12. Peschansky, V.J.; Wahlestedt, C. Non-coding RNAs as direct and indirect modulators of epigenetic regulation. *Epigenetics* **2014**, *9*, 3–12. [CrossRef] [PubMed]

13. McKiernan, R.C.; Jimenez-Mateos, E.M.; Bray, I.; Engel, T.; Brennan, G.P.; Sano, T.; Michalak, Z.; Moran, C.; Delanty, N.; Farrell, M.; et al. Reduced mature microRNA levels in association with dicer loss in human temporal lobe epilepsy with hippocampal sclerosis. *PLoS ONE* **2012**, *7*, e35921. [CrossRef]

14. Bartel, D.P. MicroRNAs: Genomics, biogenesis, mechanism, and function. *Cell* **2004**, *116*, 281–297. [CrossRef]

15. Redell, J.B.; Liu, Y.; Dash, P.K. Traumatic brain injury alters expression of hippocampal microRNAs: Potential regulators of multiple pathophysiological processes. *J. Neurosci. Res.* **2009**, *87*, 1435–1448. [CrossRef] [PubMed]

16. Burroughs, A.M.; Ando, Y.; de Hoon, M.J.L.; Tomaru, Y.; Nishibu, T.; Ukekawa, R.; Funakoshi, T.; Kurokawa, T.; Suzuki, H.; Hayashizaki, Y.; et al. A comprehensive survey of 3' animal miRNA modification events and a possible role for 3' adenylation in modulating miRNA targeting effectiveness. *Genome Res.* **2010**, *20*, 1398–1410. [CrossRef] [PubMed]

17. Burroughs, A.M.; Ando, Y. Identifying and characterizing functional 3' nucleotide addition in the miRNA pathway. *Methods* **2019**, *152*, 23–30. [CrossRef]

18. Raina, M.; Ibba, M. tRNAs as regulators of biological processes. *Front. Genet.* **2014**, *5*, 171. [CrossRef] [PubMed]

19. Thompson, D.M.; Parker, R. Stressing out over tRNA cleavage. *Cell* **2009**, *138*, 215–219. [CrossRef]

20. Li, S.; Xu, Z.; Sheng, J. tRNA-Derived Small RNA: A Novel Regulatory Small Non-Coding RNA. *Genes* **2018**, *9*, 246. [CrossRef] [PubMed]

21. Oberbauer, V.; Schaefer, M.R. tRNA-Derived Small RNAs: Biogenesis, Modification, Function and Potential Impact on Human Disease Development. *Genes* **2018**, *9*, 607. [CrossRef]

22. Thompson, D.M.; Lu, C.; Green, P.J.; Parker, R. tRNA cleavage is a conserved response to oxidative stress in eukaryotes. *RNA* **2008**, *14*, 2095–2103. [CrossRef] [PubMed]

23. Haussecker, D.; Huang, Y.; Lau, A.; Parameswaran, P.; Fire, A.Z.; Kay, M.A. Human tRNA-derived small RNAs in the global regulation of RNA silencing. *RNA* **2010**, *16*, 673–695. [CrossRef]

24. Krishna, S.; Raghavan, S.; DasGupta, R.; Palakodeti, D. tRNA-derived fragments (tRFs): Establishing their turf in post-transcriptional gene regulation. *Cell. Mol. Life Sci.* **2021**, *78*, 2607–2619. [CrossRef] [PubMed]

25. Hogg, M.C.; Raoof, R.; El Naggar, H.; Monsefi, N.; Delanty, N.; O'Brien, D.F.; Bauer, S.; Rosenow, F.; Henshall, D.C.; Prehn, J.H. Elevation in plasma tRNA fragments precede seizures in human epilepsy. *J. Clin. Investig.* **2019**, *129*, 2946–2951. [CrossRef] [PubMed]

26. Kapur, M.; Ganguly, A.; Nagy, G.; Adamson, S.I.; Chuang, J.H.; Frankel, W.N.; Ackerman, S.L. Expression of the Neuronal tRNA n-Tr20 Regulates Synaptic Transmission and Seizure Susceptibility. *Neuron* **2020**, *108*, 193–208. [CrossRef]

27. Qin, C.; Xu, P.-P.; Zhang, X.; Zhang, C.; Liu, C.-B.; Yang, D.-G.; Gao, F.; Yang, M.-L.; Du, L.-J.; Li, J.-J. Pathological significance of tRNA-derived small RNAs in neurological disorders. *Neural Regen. Res.* **2020**, *15*, 212–221. [CrossRef] [PubMed]

28. Vuokila, N.; Das Gupta, S.; Huusko, R.; Tohka, J.; Puhakka, N.; Pitkänen, A. Elevated Acute Plasma miR-124-3p Level Relates to Evolution of Larger Cortical Lesion Area after Traumatic Brain Injury. *Neuroscience* **2020**, *433*, 21–35. [CrossRef]

29. Kharatishvili, I.; Nissinen, J.P.; McIntosh, T.K.; Pitkänen, A. A model of posttraumatic epilepsy induced by lateral fluid-percussion brain injury in rats. *Neuroscience* **2006**, *140*, 685–697. [CrossRef] [PubMed]

30. Nissinen, J.; Andrade, P.; Natunen, T.; Hiltunen, M.; Malm, T.; Kanninen, K.; Soares, J.I.; Shatillo, O.; Sallinen, J.; Ndode-Ekane, X.E.; et al. Disease-modifying effect of atipamezole in a model of post-traumatic epilepsy. *Epilepsy Res.* **2017**, *136*, 18–34. [CrossRef]

31. Halonen, T.; Nissinen, J.; Jansen, J.A.; Pitkänen, A. Tiagabine prevents seizures, neuronal damage and memory impairment in experimental status epilepticus. *Eur. J. Pharmacol.* **1996**, *299*, 69–81. [CrossRef]

32. Huusko, N.; Römer, C.; Ndode-Ekane, X.E.; Lukasiuk, K.; Pitkänen, A. Loss of hippocampal interneurons and epileptogenesis: A comparison of two animal models of acquired epilepsy. *Brain Struct. Funct.* **2015**, *220*, 153–191. [CrossRef] [PubMed]

33. Peña-Llopis, S.; Brugarolas, J. Simultaneous isolation of high-quality DNA, RNA, miRNA and proteins from tissues for genomic applications. *Nat. Protoc.* **2013**, *8*, 2240–2255. [CrossRef]

34. Lipponen, A.; El-Osta, A.; Kaspi, A.; Ziemann, M.; Khurana, I.; KN, H.; Navarro-Ferrandis, V.; Puhakka, N.; Paananen, J.; Pitkänen, A. Transcription factors Tp73, Cebpd, Pax6, and Spi1 rather than DNA methylation regulate chronic transcriptomics changes after experimental traumatic brain injury. *Acta Neuropathol. Commun.* **2018**, *6*, 17. [CrossRef] [PubMed]

35. Andrews, S. *FastQC, a Quality Control Tool for High throughput Sequence Data*, Version 3; Babraham Bioinformatics: Cambridge, UK, 2010.

36. Thorvaldsdóttir, H.; Robinson, J.T.; Mesirov, J.P. Integrative Genomics Viewer (IGV): High-performance genomics data visualization and exploration. *Brief. Bioinform.* **2013**, *14*, 178–192. [CrossRef]

37. Trapnell, C.; Hendrickson, D.G.; Sauvageau, M.; Goff, L.; Rinn, J.L.; Pachter, L. Differential analysis of gene regulation at transcript resolution with RNA-seq. *Nat. Biotechnol.* **2013**, *31*, 46–53. [CrossRef] [PubMed]

38. Livak, K.J.; Schmittgen, T.D. Analysis of Relative Gene Expression Data Using Real-Time Quantitative PCR and the $2^{-\Delta\Delta CT}$ Method. *Methods* **2001**, *25*, 402–408. [CrossRef]

39. Nielsen, C.B.; Shomron, N.; Sandberg, R.; Hornstein, E.; Kitzman, J.; Burge, C.B. Determinants of targeting by endogenous and exogenous microRNAs and siRNAs. *RNA* **2007**, *13*, 1894–1910. [CrossRef]

40. Chan, P.P.; Lowe, T.M. GtRNAdb 2.0: An expanded database of transfer RNA genes identified in complete and draft genomes. *Nucleic Acids Res.* **2016**, *44*, D184–D189. [CrossRef] [PubMed]
41. Sun, P.; Liu, D.Z.; Jickling, G.C.; Sharp, F.R.; Yin, K.-J. MicroRNA-based therapeutics in central nervous system injuries. *J. Cereb. Blood Flow Metab. Off. J. Int. Soc. Cereb. Blood Flow Metab.* **2018**, *38*, 1125–1148. [CrossRef]
42. Klein, P.; Dingledine, R.; Aronica, E.; Bernard, C.; Blümcke, I.; Boison, D.; Brodie, M.J.; Brooks-Kayal, A.R.; Engel, J.J.; Forcelli, P.A.; et al. Commonalities in epileptogenic processes from different acute brain insults: Do they translate? *Epilepsia* **2018**, *59*, 37–66. [CrossRef]
43. Chen, Z.; Wang, H.; Zhong, J.; Yang, J.; Darwazeh, R.; Tian, X.; Huang, Z.; Jiang, L.; Cheng, C.; Wu, Y.; et al. Significant changes in circular RNA in the mouse cerebral cortex around an injury site after traumatic brain injury. *Exp. Neurol.* **2019**, *313*, 37–48. [CrossRef] [PubMed]
44. Patel, N.A.; Moss, L.D.; Lee, J.-Y.; Tajiri, N.; Acosta, S.; Hudson, C.; Parag, S.; Cooper, D.R.; Borlongan, C.V.; Bickford, P.C. Long noncoding RNA MALAT1 in exosomes drives regenerative function and modulates inflammation-linked networks following traumatic brain injury. *J. Neuroinflammation* **2018**, *15*, 204. [CrossRef]
45. Wu, J.; He, J.; Tian, X.; Luo, Y.; Zhong, J.; Zhang, H.; Li, H.; Cen, B.; Jiang, T.; Sun, X. microRNA-9-5p alleviates blood-brain barrier damage and neuroinflammation after traumatic brain injury. *J. Neurochem.* **2020**, *153*, 710–726. [CrossRef]
46. Henry, R.J.; Doran, S.J.; Barrett, J.P.; Meadows, V.E.; Sabirzhanov, B.; Stoica, B.A.; Loane, D.J.; Faden, A.I. Inhibition of miR-155 Limits Neuroinflammation and Improves Functional Recovery After Experimental Traumatic Brain Injury in Mice. *Neurother. J. Am. Soc. Exp. Neurother.* **2019**, *16*, 216–230. [CrossRef]
47. Corps, K.N.; Roth, T.L.; McGavern, D.B. Inflammation and neuroprotection in traumatic brain injury. *JAMA Neurol.* **2015**, *72*, 355–362. [CrossRef]
48. Ramlackhansingh, A.F.; Brooks, D.J.; Greenwood, R.J.; Bose, S.K.; Turkheimer, F.E.; Kinnunen, K.M.; Gentleman, S.; Heckemann, R.A.; Gunanayagam, K.; Gelosa, G.; et al. Inflammation after trauma: Microglial activation and traumatic brain injury. *Ann. Neurol.* **2011**, *70*, 374–383. [CrossRef]
49. Webster, K.M.; Sun, M.; Crack, P.; O'Brien, T.J.; Shultz, S.R.; Semple, B.D. Inflammation in epileptogenesis after traumatic brain injury. *J. Neuroinflammation* **2017**, *14*, 10. [CrossRef]
50. Albert, V.; Arulselvi, S.; Agrawal, D.; Pati, H.P.; Pandey, R.M. Early posttraumatic changes in coagulation and fibrinolysis systems in isolated severe traumatic brain injury patients and its influence on immediate outcome. *Hematol. Oncol. Stem Cell Ther.* **2019**, *12*, 32–43. [CrossRef] [PubMed]
51. Zhang, J.; Jiang, R.; Liu, L.; Watkins, T.; Zhang, F.; Dong, J. Traumatic brain injury-associated coagulopathy. *J. Neurotrauma* **2012**, *29*, 2597–2605. [CrossRef] [PubMed]
52. Stein, S.C.; Smith, D.H. Coagulopathy in traumatic brain injury. *Neurocrit. Care* **2004**, *1*, 479–488. [CrossRef]
53. Ou, J.; Kou, L.; Liang, L.; Tang, C. MiR-375 attenuates injury of cerebral ischemia/reperfusion via targeting Ctgf. *Biosci. Rep.* **2017**, *37*. [CrossRef] [PubMed]
54. Wu, C.-P.; Bi, Y.-J.; Liu, D.-M.; Wang, L.-Y. Hsa-miR-375 promotes the progression of inflammatory bowel disease by upregulating TLR4. *Eur. Rev. Med. Pharmacol. Sci.* **2019**, *23*, 7543–7549. [CrossRef]
55. Bekenstein, U.; Mishra, N.; Milikovsky, D.Z.; Hanin, G.; Zelig, D.; Sheintuch, L.; Berson, A.; Greenberg, D.S.; Friedman, A.; Soreq, H. Dynamic changes in murine forebrain miR-211 expression associate with cholinergic imbalances and epileptiform activity. *Proc. Natl. Acad. Sci. USA* **2017**, *114*, E4996–E5005. [CrossRef] [PubMed]
56. Korotkov, A.; Puhakka, N.; Das Gupta, S.; Vuokila, N.; Broekaart, D.W.M.; Anink, J.J.; Heiskanen, M.; Karttunen, J.; van Scheppingen, J.; Huitinga, I.; et al. Increased expression of miR142 and miR155 in glial and immune cells after traumatic brain injury may contribute to neuroinflammation via astrocyte activation. *Brain Pathol.* **2020**, *30*, 897–912. [CrossRef]
57. van Scheppingen, J.; Iyer, A.M.; Prabowo, A.S.; Mühlebner, A.; Anink, J.J.; Scholl, T.; Feucht, M.; Jansen, F.E.; Spliet, W.G.; Krsek, P.; et al. Expression of microRNAs miR21, miR146a, and miR155 in tuberous sclerosis complex cortical tubers and their regulation in human astrocytes and SEGA-derived cell cultures. *Glia* **2016**, *64*, 1066–1082. [CrossRef]
58. Wang, X.; Yin, F.; Li, L.; Kong, H.; You, B.; Zhang, W.; Chen, S.; Peng, J. Intracerebroventricular injection of miR-146a relieves seizures in an immature rat model of lithium-pilocarpine induced status epilepticus. *Epilepsy Res.* **2018**, *139*, 14–19. [CrossRef] [PubMed]
59. Huang, H.; Cui, G.; Tang, H.; Kong, L.; Wang, X.; Cui, C.; Xiao, Q.; Ji, H. Silencing of microRNA-146a alleviates the neural damage in temporal lobe epilepsy by down-regulating Notch-1. *Mol. Brain* **2019**, *12*, 102. [CrossRef] [PubMed]
60. Puhakka, N.; Bot, A.M.; Vuokila, N.; Debski, K.J.; Lukasiuk, K.; Pitkänen, A. Chronically dysregulated NOTCH1 interactome in the dentate gyrus after traumatic brain injury. *PLoS ONE* **2017**, *12*, e0172521. [CrossRef] [PubMed]
61. Zhang, L.; Liu, C.; Huang, C.; Xu, X.; Teng, J. miR-155 Knockdown Protects against Cerebral Ischemia and Reperfusion Injury by Targeting MafB. *Biomed Res. Int.* **2020**, *2020*, 6458204. [CrossRef]
62. Griffiths-Jones, S.; Grocock, R.J.; van Dongen, S.; Bateman, A.; Enright, A.J. miRBase: MicroRNA sequences, targets and gene nomenclature. *Nucleic Acids Res.* **2006**, *34*, D140–D144. [CrossRef] [PubMed]
63. Cloonan, N.; Wani, S.; Xu, Q.; Gu, J.; Lea, K.; Heater, S.; Barbacioru, C.; Steptoe, A.L.; Martin, H.C.; Nourbakhsh, E.; et al. MicroRNAs and their isomiRs function cooperatively to target common biological pathways. *Genome Biol.* **2011**, *12*, R126. [CrossRef] [PubMed]

64. Landgraf, P.; Rusu, M.; Sheridan, R.; Sewer, A.; Iovino, N.; Aravin, A.; Pfeffer, S.; Rice, A.; Kamphorst, A.O.; Landthaler, M.; et al. A mammalian microRNA expression atlas based on small RNA library sequencing. *Cell* **2007**, *129*, 1401–1414. [CrossRef] [PubMed]

65. Muljo, S.A.; Kanellopoulou, C.; Aravind, L. MicroRNA targeting in mammalian genomes: Genes and mechanisms. *Wiley Interdiscip. Rev. Syst. Biol. Med.* **2010**, *2*, 148–161. [CrossRef]

66. Avcilar-Kucukgoze, I.; Kashina, A. Hijacking tRNAs From Translation: Regulatory Functions of tRNAs in Mammalian Cell Physiology. *Front. Mol. Biosci.* **2020**, *7*, 610617. [CrossRef]

67. Kuscu, C.; Kumar, P.; Kiran, M.; Su, Z.; Malik, A.; Dutta, A. tRNA fragments (tRFs) guide Ago to regulate gene expression post-transcriptionally in a Dicer-independent manner. *RNA* **2018**, *24*, 1093–1105. [CrossRef]

68. Kümmel, D.; Krishnakumar, S.S.; Radoff, D.T.; Li, F.; Giraudo, C.G.; Pincet, F.; Rothman, J.E.; Reinisch, K.M. Complexin cross-links prefusion SNAREs into a zigzag array. *Nat. Struct. Mol. Biol.* **2011**, *18*, 927–933. [CrossRef] [PubMed]

69. Yi, J.-H.; Hoover, R.; McIntosh, T.K.; Hazell, A.S. Early, transient increase in complexin I and complexin II in the cerebral cortex following traumatic brain injury is attenuated by N-acetylcysteine. *J. Neurotrauma* **2006**, *23*, 86–96. [CrossRef]

70. Xie, Y.; Yao, L.; Yu, X.; Ruan, Y.; Li, Z.; Guo, J. Action mechanisms and research methods of tRNA-derived small RNAs. *Signal Transduct. Target. Ther.* **2020**, *5*, 109. [CrossRef]

71. Olson, K.A.; Verselis, S.J.; Fett, J.W. Angiogenin is regulated in vivo as an acute phase protein. *Biochem. Biophys. Res. Commun.* **1998**, *242*, 480–483. [CrossRef] [PubMed]

72. Ivanov, P.; O'Day, E.; Emara, M.M.; Wagner, G.; Lieberman, J.; Anderson, P. G-quadruplex structures contribute to the neuroprotective effects of angiogenin-induced tRNA fragments. *Proc. Natl. Acad. Sci. USA* **2014**, *111*, 18201–18206. [CrossRef]

73. Yamasaki, S.; Ivanov, P.; Hu, G.-F.; Anderson, P. Angiogenin cleaves tRNA and promotes stress-induced translational repression. *J. Cell Biol.* **2009**, *185*, 35–42. [CrossRef] [PubMed]

74. Sebastià, J.; Kieran, D.; Breen, B.; King, M.A.; Netteland, D.F.; Joyce, D.; Fitzpatrick, S.F.; Taylor, C.T.; Prehn, J.H.M. Angiogenin protects motoneurons against hypoxic injury. *Cell Death Differ.* **2009**, *16*, 1238–1247. [CrossRef]

75. Skorupa, A.; King, M.A.; Aparicio, I.M.; Dussmann, H.; Coughlan, K.; Breen, B.; Kieran, D.; Concannon, C.G.; Marin, P.; Prehn, J.H.M. Motoneurons secrete angiogenin to induce RNA cleavage in astroglia. *J. Neurosci.* **2012**, *32*, 5024–5038. [CrossRef] [PubMed]

76. Megel, C.; Hummel, G.; Lalande, S.; Ubrig, E.; Cognat, V.; Morelle, G.; Salinas-Giegé, T.; Duchêne, A.-M.; Maréchal-Drouard, L. Plant RNases T2, but not Dicer-like proteins, are major players of tRNA-derived fragments biogenesis. *Nucleic Acids Res.* **2019**, *47*, 941–952. [CrossRef] [PubMed]

77. Iyer, A.; Zurolo, E.; Prabowo, A.; Fluiter, K.; Spliet, W.G.M.; van Rijen, P.C.; Gorter, J.A.; Aronica, E. MicroRNA-146a: A key regulator of astrocyte-mediated inflammatory response. *PLoS ONE* **2012**, *7*, e44789. [CrossRef]

 *biomedicines*

Article

Exosomal microRNA Differential Expression in Plasma of Young Adults with Chronic Mild Traumatic Brain Injury and Healthy Control

Rany Vorn [1], Maiko Suarez [2], Jacob C. White [3], Carina A. Martin [1], Hyung-Suk Kim [1], Chen Lai [1], Si-Jung Yun [4], Jessica M. Gill [5,6] and Hyunhwa Lee [7,*]

1   National Institute of Nursing Research, National Institutes of Health, Bethesda, MD 20814, USA; rany.vorn@nih.gov (R.V.); carina.martin@nih.gov (C.A.M.); kimhy@mail.nih.gov (H.-S.K.); laichi@mail.nih.gov (C.L.)
2   School of Medicine, University of Nevada, Las Vegas, NV 89102, USA; suarem2@unlv.nevada.edu
3   College of Liberal Arts, University of Nevada, Las Vegas, NV 89154, USA; whitej31@unlv.nevada.edu
4   Yotta Biomed, LLC, Bethesda, MD 20817, USA; sijungyun@yottabiomed.com
5   School of Nursing and Medicine, Johns Hopkins University, Baltimore, MD 21205, USA; JessicaGill@jhu.edu
6   Center for Neuroscience and Regenerative Medicine, Uniformed Services University of the Health Science, Bethesda, MD 20814, USA
7   School of Nursing, University of Nevada, Las Vegas, NV 89154, USA
*   Correspondence: hyunhwa.lee@unlv.edu

Citation: Vorn, R.; Suarez, M.; White, J.C.; Martin, C.A.; Kim, H.-S.; Lai, C.; Yun, S.-J.; Gill, J.M.; Lee, H. Exosomal microRNA Differential Expression in Plasma of Young Adults with Chronic Mild Traumatic Brain Injury and Healthy Control. *Biomedicines* **2022**, *10*, 36. https://doi.org/10.3390/biomedicines10010036

Academic Editors: Kumar Vaibhav, Meenakshi Ahluwalia and Pankaj Gaur

Received: 12 October 2021
Accepted: 21 December 2021
Published: 24 December 2021

Publisher's Note: MDPI stays neutral with regard to jurisdictional claims in published maps and institutional affiliations.

Copyright: © 2021 by the authors. Licensee MDPI, Basel, Switzerland. This article is an open access article distributed under the terms and conditions of the Creative Commons Attribution (CC BY) license (https://creativecommons.org/licenses/by/4.0/).

Abstract: Chronic mild traumatic brain injury (mTBI) has long-term consequences, such as neurological disability, but its pathophysiological mechanism is unknown. Exosomal microRNAs (exomiRNAs) may be important mediators of molecular and cellular changes involved in persistent symptoms after mTBI. We profiled exosomal microRNAs (exomiRNAs) in plasma from young adults with or without a chronic mTBI to decipher the underlying mechanisms of its long-lasting symptoms after mTBI. We identified 25 significantly dysregulated exomiRNAs in the chronic mTBI group ($n = 29$, with 4.48 mean years since the last injury) compared to controls ($n = 11$). These miRNAs are associated with pathways of neurological disease, organismal injury and abnormalities, and psychological disease. Dysregulation of these plasma exomiRNAs in chronic mTBI may indicate that neuronal inflammation can last long after the injury and result in enduring and persistent post-injury symptoms. These findings are useful for diagnosing and treating chronic mTBIs.

Keywords: mild traumatic brain injury; microRNA; exomiRNA; exosome

1. Introduction

More than 5.5 million mild traumatic brain injuries (mTBI) are reported annually in the United States [1]. In most individuals with mTBI, the symptoms can resolve within days to weeks; yet for 10–15% of mTBI patients, symptoms last longer than three months and can result in disability [2,3]. Repetitive mTBIs, which usually contribute to chronic mTBI with unresolved and persistent symptoms, are common in athletes and military personnel, and these individuals have a higher risk of chronic neurologic impairment [4,5]. Over 25% of individuals with long-term mTBI consequences are not able to return to work 1-year post-injury [1]. Chronic mTBI becomes a major health concern due to life-long disabilities and long-term consequences that severely compromise the affected individuals' quality of life [6–8]. mTBI costs each patient $36,000 for rehabilitation [9] and the entire nation nearly $17 billion each year [10]. However, molecular mechanisms critical to chronic mTBI symptoms are currently unknown.

MicroRNAs (miRNAs) are small, single-stranded non-coding RNAs that regulate gene expression at the post-transcriptional level of target messenger RNA [11,12]. Circulating miRNAs are found in biofluids such as serum, plasma, saliva, and cerebrospinal fluid. They are stable and resistant to RNase digestion because they are encapsulated into extracellular vesicles called exosomes [13]. Exosomal miRNAs (exomiRNAs) in biofluids are potential biomarker candidates for diagnosis and prognosis of neurodegenerative disorders [14] such as Alzheimer's disease (AD), Parkinson's disease, and TBI due to their stability, and have the ability to regulate hundreds of target genes [13,15]. Dysregulated miRNAs may also reflect changes at the molecular level after the TBI [16].

We performed this comparative research to investigate differential expression of plasma exomiRNAs in the participants exhibiting chronic mTBIs such as repetitive head injuries compared with healthy controls. Identified plasma exomiRNAs were analyzed using bioinformatic knowledge-based Ingenuity Pathways Analysis (IPA) to understand the molecular pathway associated with plasma exomiRNAs after injury.

2. Materials and Methods

2.1. Study Protocol

Participants were recruited from the protocols granted by the Biomedical Institutional Review Board Committee at the University of Nevada, Las Vegas (UNLV) (Protocols No. 1048342 & 975928). Voluntary participants of this study were recruited via flyers and email advertisements addressing members of the UNLV campus, including the Military and Veteran Services Center and the Las Vegas community. They were voluntary participants aged 18 years or older with or without a self-reported history of mTBI. Those with mTBI histories were exposed to closed head trauma with loss of consciousness for less than 30 min and post-traumatic amnesia for less than 24 h, following the mTBI diagnosis guidelines defined by the American Congress of Rehabilitation Medicine [17]. The Neurobehavioral Symptom Inventory (NSI) and the Rivermead Post-Concussion Symptoms Questionnaire (RPQ) were used to assess the participant's symptom experiences following mTBI. The NSI consists of 22 symptoms on a Likert scale of 0 to 4 (0 = none, 1 = mild, 2 = moderate, 3 = severe, and 4 = very severe), with a total sum score range of 0 to 88, and can be divided into three subscales affective, cognitive, and somatic/somatosensory [18]. The NSI is known to be reliable and valid for measuring post-concussive symptoms in TBI patients [19]. The RPQ is a 16-item questionnaire with a Likert scale of 0 to 4 (0 = none, 1 = mild, 2 = moderate, 3 = severe, and 4 = very severe) with a total sum score range of 0 to 64. The RPQ exhibits excellent internal consistency in TBI patients at all levels of severity—mild, moderate, and severe TBI [20]. Higher total scores of NSI and RPQ indicate more severe symptoms. Subjects with a previous or current diagnosis of neuropsychiatric disease (e.g., multiple sclerosis, attention deficit hyperactivity disorder, mood disorders, substance use disorders, etc.) or who were currently on any prescription drugs were excluded.

2.2. Plasma Collection

A phlebotomist drew blood in 4 mL blood collection tubes via venipuncture from each participant in the designated lab. Each 4 mL participant blood sample was centrifuged at $1900 \times g$ at 4 °C for 10 min. Once centrifuged, the plasma was carefully aspirated into DNase- and RNase-free Eppendorf tubes without disturbing the underlying buffy coat layer where most of the white blood cells were collected. The plasma samples were centrifuged in the Eppendorf tubes at $3000 \times g$ at 4 °C for an additional 15 min to remove cellular debris. The cleared supernatant was transferred to polypropylene cryovial tubes and stored at −70 °C in the UNLV Applied Biomedical Research Lab until purification.

2.3. ExomiRNA Purification

Plasma samples were removed from −70 °C storage, thawed, and centrifuged at $3000 \times g$ for 5 min to remove cryoprecipitates. The supernatant was transferred to a new tube to begin total RNA purification. Plasma exosomal total RNA purification was performed using the exoRNeasy Serum Plasma kit (Cat. # 77064, Qiagen, Hilden, Germany) according to the manufacturer's protocol. In short, up to 1 mL or 4 mL of plasma was used, depending on the use of the Midi or Maxi kit, respectively. Both kits yielded similar output volumes. Plasma samples were washed with various buffer solutions provided by the manufacturer. Once washed, 700 μL of QIAzol lysis solution was added and centrifuged at $5000 \times g$ for 5 min to create a lysate solution. Then chloroform was combined with the lysate and centrifuged at $12{,}000 \times g$ for 15 min, after which the aqueous phase containing RNA was isolated. Total RNA purification was thereafter carried out using the RNeasy MinElute spin column, additional buffer solutions, and RNase-free water. Purified RNA samples were shipped overnight on dry ice to the Intramural Research Program laboratory at the National Institute of Nursing Research (NINR), in the National Institutes of Health (NIH) for additional processing.

2.4. ExomiRNA Profiling

Plasma exomiRNA analysis was performed in a nCounter MAX/FLEX Analysis System using nCounter® Human v3 miRNA Expression Panels (NanoString Technologies, Seattle, WA, USA) that contained 798 unique miRNA probes. Probes for housekeeping genes such as ribosomal protein L10, beta-actin, beta-2-microglobulin, glyceraldehyde 3-phosphate dehydrogenase, and ribosomal protein L19, as well as endogenous miRNAs that were incorporated in the code sets, were used for analysis in addition to positive and negative controls. The complete RNA samples were prepared and run according to the manufacturer's protocol. Counts of the reporter probes were obtained using the nCounter Digital Analyzer. Raw data were analyzed by nSolver Software version 4.0 (NanoString Technologies), and code count normalization was carried out by calculating the geometric mean of the ligation factors. Multiple testing correction with Benjamini-Hochberg's False Discovery Rate (FDR; $p < 0.100$) was used as a cutoff for the differential dysregulation with statistical significance.

2.5. Network Analysis

The biological pathways associated with dysregulated plasma exomiRNAs were determined using Ingenuity Pathways Analysis (IPA) software (Ingenuity Systems Inc., Redwood City, CA, USA). All differentially expressed plasma exomiRNAs with FDR correction of less than 0.100 were uploaded into IPA for core pathway analysis. MicroRNA Target Filters were then used to identify the plasma exomiRNA-regulated mRNAs and enriched pathways. The Ingenuity Expert Findings, Ingenuity ExpertAssist Findings, miRecords, TarBase, and TargetScan Human source were used to assess gene target predictions, and miRWalk web tools were used for confirmation.

2.6. Statistical Analysis

Statistical analysis was conducted using SPSS version 28.0.0.0 (IBM Corp., Armonk, NY, USA). Demographic and clinical characteristics were compared between groups using Chi-square (χ^2) and an independent sample t-test. The significance level was set at 0.05 in all tests.

3. Results

3.1. Demographics of the Study Population

A total of 40 participants, including 29 individuals with a history of mTBI and 11 without, participated in the study. The study participants were aged 19 to 36 years (24.8 ± 5.219). More than half of the study participants were female (52.5%) or White (57.5%). There were no statistical differences based on demographic characteristics, including age, gender, race, and BMI, between the two groups. In the chronic mTBI group ($n = 29$), 65.5% reported having more than 1 injury (41.4% with 2 or 3 injuries and 27.6% with 4 or more injuries), and the average number was 2.55 injuries (SD = 1.325). The average number of years since the last injury was 4.48 years (SD = 5.000). Only 1 participant in the mTBI group reported that they experienced both a brief loss of consciousness and post-traumatic amnesia resulting from the injury (3.4%). Two additional participants with a history of mTBI reported having post-traumatic amnesia (6.9%). The most common causes of injuries were sports-related activities (48.3%) (e.g., boxing, mixed martial arts [MMA] training, skating, football, etc.), followed by head hitting hard objects (e.g., sharp edge, metal materials, etc.) (20.7%), and high-level falls (17.2%). A few cases were related to military service activities (10.3%) or car accidents (3.4%). The demographics and clinical characteristics of the participants are presented in Table 1.

Table 1. Demographic and Clinical Characteristics of Chronic mTBI and Healthy Control Participants (n = 40).

Characteristic	Overall (n = 40)	Chronic mTBI (n = 29)	Control (n = 11)	χ^2 or t	p
Demographic					
Age, Mean (SD)	24.80 (5.22)	25.59 (5.36)	22.73 (4.41)	1.576	0.123
Gender, n (%)					
Males	19 (47.5)	15 (51.7)	4 (36.4)	0.755	0.488
Females	21 (52.5)	14 (48.3)	7 (63.6)		
Weight (kg), Mean (SD)	69.83 (14.64)	70.18 (13.29)	68.91 (18.45)	0.243	0.810
Height (cm), Mean (SD)	167.52 (10.94)	168.26 (10.48)	165.56 (12.38)	0.691	0.494
BMI, Mean (SD)	24.74 (3.77)	24.65 (3.40)	24.98 (4.81)	−0.247	0.807
Ethnicity/Race, n (%)					
Hispanic	6 (15.0)	4 (13.8)	2 (18.2)	1.038	0.904
White	23 (57.5)	17 (58.6)	6 (54.5)		
Black	1 (2.5)	1 (3.4)	0 (0.0)		
Asian	8 (20.0)	6 (20.7)	2 (18.2)		
Other	2 (5.0)	1 (3.4)	1 (9.1)		
Handedness, n (%)					
Right	37 (92.5)	27 (93.1)	10 (90.9)	0.055	1.000
Left	3 (7.5)	2 (6.9)	1 (9.1)		
Education, n (%)					
In college	31 (77.5)	22 (75.9)	9 (81.8)	0.162	1.000
In graduate school	9 (22.5)	7 (24.1)	2 (18.2)		
Marital Status, n (%)					
Single	36 (90.0)	26 (89.7)	10 (90.9)	0.412	0.814
Married	4 (10.0)	3 (10.3)	1 (9.1)		
Employment Status, n (%)					
Yes	30 (75.0)	23 (79.3)	7 (63.6)	1.045	0.418
No	10 (25.0)	6 (20.7)	4 (36.4)		
Clinical					
RPQ Total, Mean (SD)	12.58 (12.42)	16.76 (12.14)	1.55 (2.07)	6.505	<0.001
NSI Total, Mean (SD)	15.43 (14.06)	19.86 (13.91)	3.73 (4.65)	5.489	<0.001
Somatic/Sensory, Mean (SD)	5.58 (5.81)	7.34 (5.83)	0.91 (1.81)	5.304	<0.001
Cognitive, Mean (SD)	2.75 (2.88)	3.55 (2.89)	0.64 (1.50)	4.156	<0.001
Affective, Mean (SD)	7.10 (6.42)	8.97 (6.56)	2.18 (1.94)	5.023	<0.001
Injury Characteristics					
Number of Injuries, Mean (SD)		2.55 (1.33)	N/A		
Single Injury		9 (31.0)	N/A		
Multiple Injuries		20 (69.0)	N/A		
Time since the last Injury (years), Mean (SD)		4.48 (5.00)	N/A		
Mechanism of Injury, n (%)					
Sports-related		14 (48.3)	N/A		
Head hit		6 (20.7)	N/A		
High-level Falls		5 (17.2)	N/A		
Military-related		3 (10.3)	N/A		
Car accident		1 (3.4)	N/A		

mTBI, mild traumatic brain injury; BMI, body mass index; RPQ, Rivermead Post-Concussion Symptoms Questionnaire; NSI, Neurobehavioral Symptom Inventory.

3.2. Differential Expression of Plasma ExomiRNAs

We assessed the expression levels of plasma exomiRNAs in the chronic mTBI group compared to control subjects. After normalization with ligation factors, we identified 25 plasma exomiRNAs differentially expressed in chronic mTBI compared to healthy control with an adjusted p-value < 0.100. Among them, 4 plasma exomiRNAs were upregulated, and 21 plasma exomiRNAs were downregulated in the chronic mTBI group compared with the control (Table 2).

Table 2. Dysregulated ExomiRNAs following Chronic mTBI.

Probe Name	Target Sequence	Log2FC	Adjusted p-Value
Upregulated			
hsa-miR-520e	AAAGUGCUUCCUUUUUGAGGG	0.98	0.03
hsa-miR-499b-3p	AACAUCACUGCAAGUCUUAACA	0.83	0.01
hsa-miR-520b	AAAGUGCUUCCUUUUAGAGGG	0.43	0.04
hsa-miR-4488	AGGGGGCGGGCUCCGGCG	0.42	0.03
Downregulated			
hsa-miR-625-5p	AGGGGGAAAGUUCUAUAGUCC	−0.96	0.08
hsa-miR-421	AUCAACAGACAUUAAUUGGGCGC	−1.39	0.08
hsa-miR-664a-3p	UAUUCAUUUAUCCCCAGCCUACA	−1.43	0.08
hsa-miR-28-3p	CACUAGAUUGUGAGCUCCUGGA	−1.49	0.04
hsa-miR-125a-5p	UCCCUGAGACCCUUUAACCUGUGA	−2.10	0.04
hsa-miR-222-3p	AGCUACAUCUGGCUACUGGGU	−2.14	0.09
hsa-miR-140-5p	CAGUGGUUUUACCCUAUGGUAG	−2.17	0.07
hsa-miR-98-5p	UGAGGUAGUAAGUUGUAUUGUU	−2.32	0.09
hsa-miR-148a-3p	UCAGUGCACUACAGAACUUUGU	−2.63	0.06
hsa-miR-423-5p	UGAGGGGCAGAGAGCGAGACUUU	−2.65	0.09
hsa-miR-107	AGCAGCAUUGUACAGGGCUAUCA	−2.75	0.07
hsa-miR-181a-5p	AACAUUCAACGCUGUCGGUGAGU	−2.81	0.09
hsa-miR-374a-5p	UUAUAAUACAACCUGAUAAGUG	−2.86	0.09
hsa-miR-340-5p	UUAUAAAGCAAUGAGACUGAUU	−2.87	0.07
hsa-miR-29b-3p	UAGCACCAUUUGAAAUCAGUGUU	−2.95	0.05
hsa-miR-191-5p	CAACGGAAUCCCAAAAGCAGCUG	−3.03	0.08
hsa-miR-199a-3p	ACAGUAGUCUGCACAUUGGUUA	−3.13	0.05
hsa-miR-126-3p	UCGUACCGUGAGUAAUAAUGCG	−3.13	0.09
hsa-miR-23a-3p	AUCACAUUGCCAGGGAUUUCC	−3.36	0.04
hsa-miR-142-3p	UGUAGUGUUUCCUACUUUAUGGA	−3.39	0.07
hsa-miR-223-3p	UGUCAGUUUGUCAAAUACCCCA	−3.62	0.04

ExomiRNA, exosomal microRNAs.

3.3. Pathway Analysis

Pathway analysis using IPA revealed that dysregulated plasma exomiRNAs are related to neurological disease, organismal injury and abnormalities, and psychological disorders (Table 3). Top network analysis showed that these plasma exomiRNAs are associated with connective tissue disorders, inflammatory disease, organismal injury, and network abnormalities. The specific plasma exomiRNAs associated in this network (Figure 1) were hsa-miR-103-3p (synonym to hsa-miR-107), hsa-miR-126a-5p, hsa-miR-140-5p, hsa-miR-142-3p, hsa-miR-199a-3p, hsa-miR-221-3p (synonym to hsa-miR-222-3p), hsa-miR-23a-3p, hsa-miR-291a-3p (synonym to hsa-miR-520e), hsa-miR-374b-5p, hsa-miR-423-5p, hsa-miR-625-5p, hsa-miR-664-3p. These plasma exomiRNAs are directly or indirectly associated with acyl-CoA synthetase long-chain family member 6 (*ACSL6*), EPH receptor B6 (*EPHB6*), growth arrest specific 5 (*GAS5*), GNAS antisense RNA 1 (*Gnasas1*), hepatocellular carcinoma upregulated EZH2-associated long non-coding RNA (*HEIH*), homeobox A11 (*HOXA11*), phosphatase and tensin homolog (*PTEN*), resolvin D1, ribosomal protein S 15 (*RPS15*), tumor necrosis factor (*TNF*), vasohibin 1 (*VASH1*), and vascular endothelial growth factor A (*VEGFA*). Dysregulated exomiRNAs targeting mRNA associated with neuroinflammation pathways are presented in Figure 2 and Supplementary Table S1.

Table 3. Top IPA Biological Functions and Disease Pathway.

Diseases and Disorders	*p*-Value Range	Number of Molecules
Neurological disease	4.58×10^{-2}–4.85×10^{-14}	14
Organismal injury and abnormality	4.95×10^{-2}–4.85×10^{-14}	23
Psychological disease	4.58×10^{-2}–4.85×10^{-14}	13
Cancer	4.95×10^{-2}–7.96×10^{-13}	21
Reproductive system disease	4.85×10^{-2}–1.49×10^{-12}	20

Molecular and Cellular Functions	*p*-Value Range	Number of Molecules
Cell cycle	4.02×10^{-2}–2.46×10^{-6}	4
Cellular movement	4.88×10^{-2}–2.46×10^{-6}	12
Cellular development	4.47×10^{-2}–5.25×10^{-6}	12
Cellular growth and proliferation	4.47×10^{-2}–5.25×10^{-6}	12
Cell death and survival	3.99×10^{-2}–8.71×10^{-5}	11

Physiological System Development and Function	*p*-Value Range	Number of Molecules
Organismal development	3.74×10^{-2}–6.45×10^{-9}	11
Organismal functions	8.69×10^{-4}–7.19×10^{-4}	2
Tissue morphology	8.90×10^{-5}–1.72×10^{-3}	3
Hematological system development and functions	4.47×10^{-2}–1.98×10^{-3}	6
Immune cell trafficking	2.74×10^{-2}–1.98×10^{-3}	2

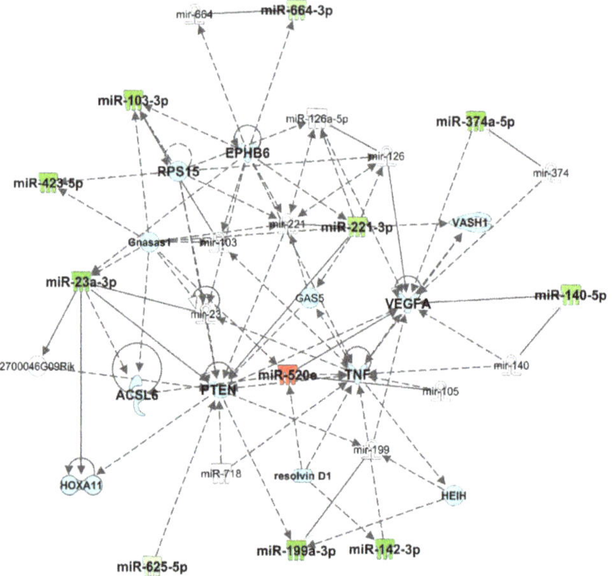

Figure 1. Top network identified by Ingenuity Pathway Analysis (IPA) for chronic mild traumatic brain injury (mTBI) versus control: Connective Tissue Disorders, Inflammatory Disease, and Organismal Injury and Abnormalities. Green indicates genes that are downregulated, and red indicates genes that are upregulated, where increased color saturation represents more extreme down- or upregulated in the dataset. Solid lines represent direct interactions, nontargeting interactions, or correlations between chemicals, proteins, or RNA. Dotted lines represented indirect interaction. Arrowed lines represent activation, causation, expression, localization, membership, modification, molecular cleavage, phosphorylation, protein–DNA interactions, protein–TNA interaction, binding regulation, and transcription. Shapes represent molecule type (double circle = complex/group; square = cytokine; diamond = enzyme; inverted triangle = kinase; triangle = phosphatase; oval = transcription regulator; trapezoid = transporter; circle = other). Reprinted from Ingenuity Pathway Analysis under a CC BY 4.0 license, with permission from QIAGEN Silicon Valley, original copyright 2000–2021.

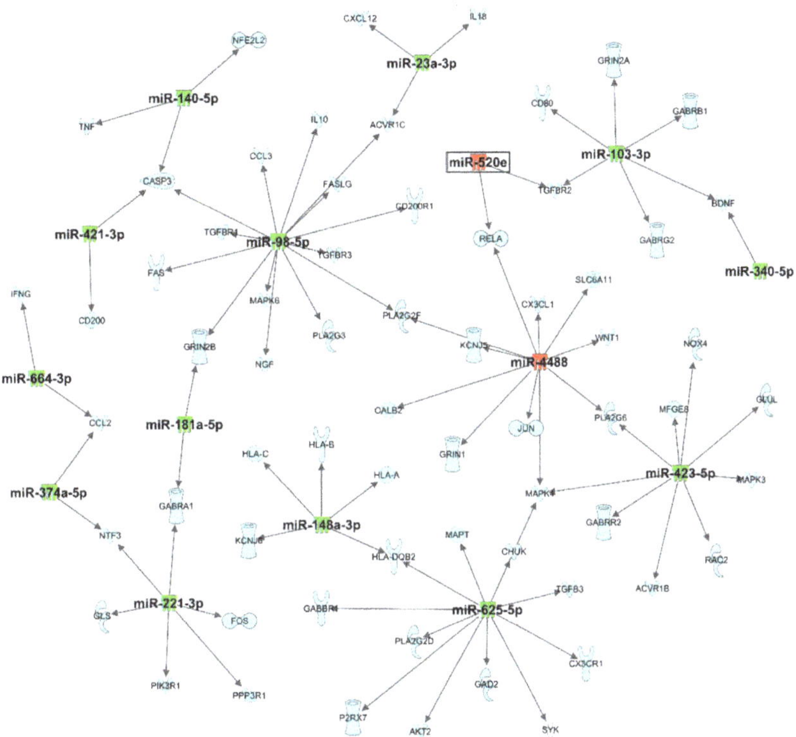

Figure 2. Target filter analysis of dysregulated exomiRNA targeting mRNA associated with neuroin-flammation pathways. Green indicates genes that are downregulated, and red indicates genes that are upregulated, where increased color saturation represents more extreme down- or upregulated in the dataset. Reprinted from Ingenuity Pathway Analysis under a CC BY 4.0 license, with permission from QIAGEN Silicon Valley, original copyright 2000–2021.

4. Discussion

In this study, we reported on several dysregulated plasma exomiRNAs in chronic mTBIs compared to healthy controls. We used multiplexed nCounter miRNA assays developed by NanoString technology to quantify low expression levels of exomiRNAs in plasma. The nCounter system is more sensitive than microarray for quantification of gene expression, with more accurate and reliable detection of very low expression levels [21]. The current study identified 25 dysregulated exomiRNAs that were associated with chronic mTBI. Pathway analysis showed that 14 plasma exomiRNAs were related to neurological disease, 23 plasma exomiRNAs were related to organismal injury and abnormalities, and 13 plasma exomiRNAs were related to psychological disease. Our results offered insights into the molecular mechanisms underlying injuries to brain function after mTBI.

In recent years, the role of miRNAs is indicated in the early diagnosis and progression of diseases including cancer, AD, and Parkinson's disease [13–16]. Because they are short, miRNAs move freely across the blood-brain barrier (BBB) into the peripheral circulation, which can reflect changes in brain function due to a TBI [11,12,22]. Disruption of the BBB after injury promotes systemic inflammatory factors, neurotoxins, and pathogens into the brain and leads to neuronal damage [23]. Pathway analysis revealed that these plasma exomiRNAs are closely associated with *VEGFA*. VEGF is predominantly expressed in endothelial cells and plays an essential role in vascular development and neuroprotection [24,25]. Increased peripheral circulating levels of VEGF have been linked to alterations in BBB permeability following the infiltration of immune cells after a TBI [23,26]. Elevated concentrations of plasma VEGF and TNF-α protein were reported previously in mTBI patients [27].

Microglia and astrocytes are key mediators of neuroinflammation in the central nervous system (CNS) because of the release of inflammatory cytokines after brain injury. Ongoing neuroinflammation was observed when microglia became activated, even after 1 to 18 years of a single moderate-to-severe TBI [28–30]. Inflammatory cytokines were elevated in the plasma of mTBI patients within 24 h of injury [31], even up to a year later [32]. Many of the plasma exomiRNAs identified in our study were consistently found in the exosome and associated with neuroinflammation [5]. Previous studies reported that downregulation of hsa-miR-223-3p was associated with sporadic amyotrophic lateral sclerosis patients [33,34]. Downregulation of mitochondria-associated miR-142-3p was reported in a preclinical model of severe TBI [35]. As confirmed by target miRNAs and pathway analyses, these plasma exomiRNAs were found in brain cells and highly expressed in microglia [35], which are involved in neuroinflammation signaling and the glutamate signaling pathway through targeting the *SLC1A3* gene [36,37].

Our targeted analysis showed that downregulated hsa-miR-223-3p, has-miR-29b-3p, and has-miR-107, present in the chronic mTBI group, targeted the *NFIA* gene. *NFIA* is a member of the transcription factor nuclear factor I (NFI) family (*NFIB*, *NFIC*, and *NFIX*) and plays an essential role in neural development in the CNS [38]. NFIA is expressed in mature astrocytes and plays an essential role in retrieving forms of memory. A recent preclinical study showed that loss of *NFIA* inhibited neurotransmission and memory loss [39]. The major neurotransmitter in the CNS includes gamma-aminobutyric acid (GABA) and glutamate, which are involved in synapse plasticity, learning, and memory formation [40]. The targeted analysis showed that dysregulated has-miR-107 targeted the GABA receptor subunit gamma1 (*GABRG1*) and GABA A receptor subunit beta1 (*GABRB1*) genes. Mutation in the $GABA_A$ receptor was associated with a neurological disorder and dysregulated $GABA_A$ subunit gene expression as observed in the preclinical TBI model [41,42].

Downregulation of hsa-miR-107 (synonym to hsa-miR-103-3p) in our current study is consistent with previous findings [5]. miR-107 was reported as a synaptic region marker involved in the synaptogenesis signaling pathway targeting the synuclein gamma (*SNCG*), synaptotagmin 2 (*SYT2*), *SYT6*, and brain-derived neurotrophic factor (*BDNF*) genes (Supplementary Figure S1) [43]. *SNCG* genes are predominantly expressed in neuronal tissue and play an important role in synaptic plasticity and dopamine regulation [44]. Overexpressed SNCG was associated with neurodegenerative disease pathogenesis in clinical and preclinical studies [45–47]. In addition to neurodegeneration, overexpressed SNCG was linked to poor prognosis in cancer and a preclinical TBI model [48,49]. A previous study showed that miR-107 was downregulated in AD and enhanced disease progression by regulating the beta-site amyloid precursor cleaving enzyme1 (*BACE1*) gene [50]. BACE1 is predominantly expressed in neurons and is responsible for the generation of amyloid-beta [51]. Overall, our findings suggest that chronic mTBI is associated with exomiRNA dysregulation, which may impact neurodegenerative disorders.

A major strength of this study is that the cases and controls were well matched. Although this study provides biological molecular insights into chronic mTBI, it was constrained by a small sample size. The majority of our sample population consisted of White individuals, which limits the genetic diversity and generalizability to all chronic mTBI populations. We were also unable to differentiate plasma exomiRNAs expression in single or multiple injuries due to our small sample size. Additional studies in a larger cohort should investigate the effect of single and repetitive injury on exomiRNAs expression level. Repetitive injury is a major risk for developing chronic neurological symptoms or impaired behaviors [52]. Future studies with larger cohorts will validate our observations that plasma exomiRNAs are associated with symptom deficits following brain injury.

In summary, chronic mTBI is associated with dysregulated exomiRNAs in plasma. These exomiRNAs were associated with inflammation processes that potentially link to neurological disability. Our data may aid in understanding the pathophysiological mechanism underlying the long-term impacts of mTBI.

Supplementary Materials: The following are available online at https://www.mdpi.com/article/10.3390/biomedicines10010036/s1, Figure S1: Target filter analysis of dysregulated exomiRNA targeting mRNA associated with neuroinflammation and synaptogenesis pathways. Table S1: microRNA Target Filter (IPA).

Author Contributions: H.L. conceptualized and supervised the overall project. H.L. and R.V. contributed to designing the experiment, interpreting the data, and writing the manuscript. H.L., M.S. and J.C.W. administered the project, recruited human subjects, and collected data. M.S. and J.C.W. processed and purified plasma exomiRNAs. R.V. conducted exomiRNA profiling. R.V. and H.L. analyzed and interpreted the data. M.S., J.C.W., C.A.M., H.-S.K., S.-J.Y., C.L. and J.M.G. reviewed and edited the manuscript. H.L. and J.M.G. acquired funding for the project. All authors have read and agreed to the published version of the manuscript.

Funding: This research was funded by the School of Nursing of the University of Nevada, Las Vegas (UNLV; 2221-257-554G), the American Psychiatric Nurses Association (APNA; GR009253), and the National Institute of Nursing Research (NINR) Intramural Program. The research was also supported by the UNLV 2017-2020 Top Tier Doctoral Graduate Research Assistantship and the UNLV School of Medicine Medical Student Research Program.

Institutional Review Board Statement: The study was conducted according to the guidelines of the Declaration of Helsinki and approved by the Institutional Review Board Committee of the University of Nevada, Las Vegas (Protocols No. 1048342 & 975928, approved on 25 May 2017, and 28 June 2017).

Informed Consent Statement: Informed consent was obtained from all subjects involved in the study.

Data Availability Statement: Data that support the findings of this study are available upon reasonable request from any qualified investigator to the corresponding author.

Acknowledgments: This study was supported by the UNLV School of Nursing, 2017–2020 UNLV Faculty Top Tier Doctoral Graduate Research Assistantship Program, and the 2019 UNLV School of Medicine Research Program for Medical Students. Blood samples were collected via venipuncture by 2 Registered Nurses, Hannah Greilish and Florence Chipondaminga. The publication fees for this article were supported by the UNLV University Libraries Open Article Fund. Figures 1 and 2 (Top networks) were reprinted from Ingenuity Pathway Analysis under a CC BY 4.0 license, with permission from QIAGEN Silicon Valley, original copyright 2000–2021.

Conflicts of Interest: The authors declare no conflict of interest. The funders had no role in designing the study, in the collection, analyses, or interpretation of data, in the writing of the manuscript or the decision to publish the results.

References

1. Langlois, J.A.; Rutland-Brown, W.; Wald, M.M. The epidemiology and impact of traumatic brain injury: A brief overview. *J. Head Trauma Rehabil.* **2006**, *21*, 375–378. [CrossRef]
2. Dikmen, S.; Machamer, J.; Temkin, N. Mild traumatic brain injury: Longitudinal study of cognition, functional status, and post-traumatic symptoms. *J. Neurotrauma* **2017**, *34*, 1524–1530. [CrossRef]
3. Frieden, T.R.; Houry, D.; Baldwin, G. Traumatic Brain Injury in the United States: Epidemiology and Rehabilitation. *CDC NIH Rep. Congr.* **2015**, 1–74. Available online: https://www.cdc.gov/traumaticbraininjury/pdf/tbi_report_to_congress_epi_and_rehab-a.pdf (accessed on 7 August 2021).
4. Lehman, E.J.; Hein, M.J.; Baron, S.L.; Gersic, C.M. Neurodegenerative causes of death among retired national football league players. *Neurology* **2012**, *79*, 1970–1974. [CrossRef] [PubMed]
5. Devoto, C.; Lai, C.; Qu, B.X.; Guedes, V.A.; Leete, J.; Wilde, E.; Walker, W.C.; Diaz-Arrastia, R.; Kenney, K.; Gill, J. Exosomal MicroRNAs in Military Personnel with Mild Traumatic Brain Injury: Preliminary Results from the Chronic Effects of Neurotrauma Consortium Biomarker Discovery Project. *J. Neurotrauma* **2020**, *37*, 2482–2492. [CrossRef] [PubMed]
6. Seabury, S.A.; Gaudette, É.; Goldman, D.P.; Markowitz, A.J.; Brooks, J.; McCrea, M.A.; Okonkwo, D.O.; Manley, G.T.; Adeoye, O.; Badjatia, N.; et al. Assessment of Follow-up Care After Emergency Department Presentation for Mild Traumatic Brain Injury and Concussion: Results From the TRACK-TBI Study. *JAMA Netw. Open* **2018**, *1*, e180210. [CrossRef] [PubMed]

7. Smith, D.H.; Johnson, V.E.; Stewart, W. Chronic neuropathologies of single and repetitive TBI: Substrates of dementia? *Nat. Rev. Neurol.* **2013**, *9*, 211–221. [CrossRef] [PubMed]

8. Centers for Disease Control and Prevention. Surveillance Report of Traumatic Brain Injury-Related Emergency Department Visits, Hospitalizations, and Deaths—United States, 2014. 2019. Available online: https://www.cdc.gov/traumaticbraininjury/pdf/TBI-Surveillance-Report-FINAL_508.pdf (accessed on 7 August 2021).

9. McGregor, K.; Pentland, B. Head injury rehabilitation in the U.K.: An economic perspective. *Soc. Sci. Med.* **1997**, *45*, 295–303. [CrossRef]

10. Centers for Disease Control and Prevention (CDC). *Report to Congress on Mild Traumatic Brain Injury in the United States: Steps to Prevent a Serious Public Health Problem*; CDC: Atlanta, GA, USA, 2003.

11. Friedman, R.C.; Farh, K.K.H.; Burge, C.B.; Bartel, D.P. Most mammalian mRNAs are conserved targets of microRNAs. *Genome Res.* **2009**, *19*, 92–105. [CrossRef] [PubMed]

12. Peng, Y.; Croce, C.M. The role of MicroRNAs in human cancer. *Signal Transduct. Target. Ther.* **2016**, *1*, 15004. [CrossRef]

13. Mitchell, P.S.; Parkin, R.K.; Kroh, E.M.; Fritz, B.R.; Wyman, S.K.; Pogosova-Agadjanyan, E.L.; Peterson, A.; Noteboom, J.; O'Briant, K.C.; Allen, A.; et al. Circulating microRNAs as stable blood-based markers for cancer detection. *Proc. Natl. Acad. Sci. USA* **2008**, *105*, 10513–10518. [CrossRef] [PubMed]

14. Watson, C.N.; Belli, A.; Di Pietro, V. Small non-coding RNAs: New class of biomarkers and potential therapeutic targets in neurodegenerative disease. *Front. Genet.* **2019**, *10*, 364. [CrossRef]

15. Xi, Y.; Nakajima, G.; Gavin, E.; Morris, C.G.; Kudo, K.; Hayashi, K.; Ju, J. Systematic analysis of microRNA expression of RNA extracted from fresh frozen and formalin-fixed paraffin-embedded samples. *RNA* **2007**, *13*, 1668–1674. [CrossRef]

16. Atif, H.; Hicks, S.D. A Review of MicroRNA Biomarkers in Traumatic Brain Injury. *J. Exp. Neurosci.* **2019**, *13*, 1–12. [CrossRef] [PubMed]

17. Kay, T.; Harrington, D.; Adams, R. Definition of mild traumatic brain injury: American congress of rehabilitation medicine. *J. Head Trauma Rehabil.* **1993**, *8*, 74–85. [CrossRef]

18. Caplan, L.J.; Ivins, B.; Poole, J.H.; Vanderploeg, R.D.; Jaffee, M.S.; Schwab, K. The structure of postconcussive symptoms in 3 us military samples. *J. Head Trauma Rehabil.* **2010**, *25*, 447–458. [CrossRef]

19. King, P.R.; Donnelly, K.T.; Donnelly, J.P.; Dunnam, M.; Warner, G.; Kittleson, C.J.; Bradshaw, C.B.; Alt, M.; Meier, S.T. Psychometric study of the Neurobehavioral Symptom Inventory. *J. Rehabil. Res. Dev.* **2012**, *49*, 879–888. [CrossRef] [PubMed]

20. Medvedev, O.N.; Theadom, A.; Barker-Collo, S.; Feigin, V. Distinguishing between enduring and dynamic concussion symptoms: Applying Generalisability theory to the Rivermead Post Concussion Symptoms Questionnaire (RPQ). *PeerJ* **2018**, *6*, e5676. [CrossRef] [PubMed]

21. Geiss, G.K.; Bumgarner, R.E.; Birditt, B.; Dahl, T.; Dowidar, N.; Dunaway, D.L.; Fell, H.P.; Ferree, S.; George, R.D.; Grogan, T.; et al. Direct multiplexed measurement of gene expression with color-coded probe pairs. *Nat. Biotechnol.* **2008**, *26*, 317–325. [CrossRef]

22. Wienholds, E.; Plasterk, R.H.A. MicroRNA function in animal development. *FEBS Lett.* **2005**, *579*, 5911–5922. [CrossRef] [PubMed]

23. Sulhan, S.; Lyon, K.A.; Shapiro, L.A.; Huang, J.H. Neuroinflammation and blood–brain barrier disruption following traumatic brain injury: Pathophysiology and potential therapeutic targets. *J. Neurosci. Res.* **2020**, *98*, 19–28. [CrossRef]

24. Zacchigna, S.; Lambrechts, D.; Carmeliet, P. Neurovascular signalling defects in neurodegeneration. *Nat. Rev. Neurosci.* **2008**, *9*, 169–181. [CrossRef] [PubMed]

25. Argandona, E.G.; Bengoetxea, H.; Ortuzar, N.; Bulnes, S.; Rico-Barrio, I.; Vicente Lafuente, J. Vascular Endothelial Growth Factor: Adaptive Changes in the Neuroglialvascular Unit. *Curr. Neurovascular Res.* **2012**, *9*, 72–81. [CrossRef] [PubMed]

26. Nasr, I.W.; Chun, Y.; Kannan, S. Neuroimmune responses in the developing brain following traumatic brain injury. *Exp. Neurol.* **2019**, *320*, 112957. [CrossRef]

27. Edwards, K.A.; Pattinson, C.L.; Guedes, V.A.; Peyer, J.; Moore, C.; Davis, T.; Devoto, C.; Turtzo, L.C.; Latour, L.; Gill, J.M. Inflammatory Cytokines Associate With Neuroimaging After Acute Mild Traumatic Brain Injury. *Front. Neurol.* **2020**, *11*, 348. [CrossRef]

28. Johnson, V.E.; Stewart, J.E.; Begbie, F.D.; Trojanowski, J.Q.; Smith, D.H.; Stewart, W. Inflammation and white matter degeneration persist for years after a single traumatic brain injury. *Brain* **2013**, *136*, 28–42. [CrossRef] [PubMed]

29. Loane, D.J.; Kumar, A.; Stoica, B.A.; Cabatbat, R.; Faden, A.I. Progressive neurodegeneration after experimental brain trauma: Association with chronic microglial activation. *J. Neuropathol. Exp. Neurol.* **2014**, *73*, 14–29. [CrossRef] [PubMed]

30. Scott, G.; Zetterberg, H.; Jolly, A.; Cole, J.H.; De Simoni, S.; Jenkins, P.O.; Feeney, C.; Owen, D.R.; Lingford-Hughes, A.; Howes, O.; et al. Minocycline reduces chronic microglial activation after brain trauma but increases neurodegeneration. *Brain* **2018**, *141*, 459–471. [CrossRef] [PubMed]

31. Kim, J.; Krichevsky, A.; Grad, Y.; Hayes, G.D.; Kosik, K.S.; Church, G.M.; Ruvkun, G. Identification of many microRNAs that copurify with polyribosomes in mammalian neurons. *Proc. Natl. Acad. Sci. USA* **2004**, *101*, 360–365. [CrossRef] [PubMed]

32. Chaban, V.; Clarke, G.J.B.; Skandsen, T.; Islam, R.; Einarsen, C.E.; Vik, A.; Damås, J.K.; Mollnes, T.E.; Håberg, A.K.; Pischke, S.E. Systemic Inflammation Persists the First Year after Mild Traumatic Brain Injury: Results from the Prospective Trondheim Mild Traumatic Brain Injury Study. *J. Neurotrauma* **2020**, *37*, 2120–2130. [CrossRef]

33. Benigni, M.; Ricci, C.; Jones, A.R.; Giannini, F.; Al-Chalabi, A.; Battistini, S. Identification of miRNAs as Potential Biomarkers in Cerebrospinal Fluid from Amyotrophic Lateral Sclerosis Patients. *Neuromol. Med.* **2016**, *18*, 551–560. [CrossRef] [PubMed]

34. Joilin, G.; Leigh, P.N.; Newbury, S.F.; Hafezparast, M. An Overview of MicroRNAs as Biomarkers of ALS. *Front. Neurol.* **2019**, *10*, 186. [CrossRef]
35. Wang, W.X.; Visavadiya, N.P.; Pandya, J.D.; Nelson, P.T.; Sullivan, P.G.; Springer, J.E. Mitochondria-associated microRNAs in rat hippocampus following traumatic brain injury. *Exp. Neurol.* **2015**, *265*, 84–93. [CrossRef] [PubMed]
36. Landgraf, P.; Rusu, M.; Sheridan, R.; Sewer, A.; Iovino, N.; Aravin, A.; Pfeffer, S.; Rice, A.; Kamphorst, A.O.; Landthaler, M.; et al. A Mammalian microRNA Expression Atlas Based on Small RNA Library Sequencing. *Cell* **2007**, *129*, 1401–1414. [CrossRef] [PubMed]
37. Sonkoly, E.; Ståhle, M.; Pivarcsi, A. MicroRNAs and immunity: Novel players in the regulation of normal immune function and inflammation. *Semin. Cancer Biol.* **2008**, *18*, 131–140. [CrossRef] [PubMed]
38. Deneen, B.; Ho, R.; Lukaszewicz, A.; Hochstim, C.J.; Gronostajski, R.M.; Anderson, D.J. The Transcription Factor NFIA Controls the Onset of Gliogenesis in the Developing Spinal Cord. *Neuron* **2006**, *52*, 953–968. [CrossRef] [PubMed]
39. Huang, A.Y.S.; Woo, J.; Sardar, D.; Lozzi, B.; Bosquez Huerta, N.A.; Lin, C.C.J.; Felice, D.; Jain, A.; Paulucci-Holthauzen, A.; Deneen, B. Region-Specific Transcriptional Control of Astrocyte Function Oversees Local Circuit Activities. *Neuron* **2020**, *106*, 992–1008. [CrossRef] [PubMed]
40. Guerriero, R.M.; Giza, C.C.; Rotenberg, A. Glutamate and GABA Imbalance Following Traumatic Brain Injury. *Curr. Neurol. Neurosci. Rep.* **2015**, *15*, 27. [CrossRef]
41. Drexel, M.; Puhakka, N.; Kirchmair, E.; Hörtnagl, H.; Pitkänen, A.; Sperk, G. Expression of GABA receptor subunits in the hippocampus and thalamus after experimental traumatic brain injury. *Neuropharmacology* **2015**, *88*, 122–133. [CrossRef] [PubMed]
42. Yuan, H.; Low, C.M.; Moody, O.A.; Jenkins, A.; Traynelis, S.F. Ionotropic GABA and glutamate receptor mutations and human neurologic diseases. *Mol. Pharmacol.* **2015**, *88*, 203–217. [CrossRef]
43. Sheinerman, K.S.; Toledo, J.B.; Tsivinsky, V.G.; Irwin, D.; Grossman, M.; Weintraub, D.; Hurtig, H.I.; Chen-Plotkin, A.; Wolk, D.A.; McCluskey, L.F.; et al. Circulating brain-enriched microRNAs as novel biomarkers for detection and differentiation of neurodegenerative diseases. *Alzheimer's Res. Ther.* **2017**, *9*, 89. [CrossRef] [PubMed]
44. Rokad, D.; Ghaisas, S.; Harischandra, D.S.; Jin, H.; Anantharam, V.; Kanthasamy, A.; Kanthasamy, A.G. Role of neurotoxicants and traumatic brain injury in α-synuclein protein misfolding and aggregation. *Brain Res. Bull.* **2017**, *133*, 60–70. [CrossRef] [PubMed]
45. Goedert, M. Alpha-synuclein and neurodegenerative diseases. *Nat. Rev. Neurosci.* **2001**, *2*, 492–501. [CrossRef] [PubMed]
46. Acosta, S.A.; Tajiri, N.; de la Pena, I.; Bastawrous, M.; Sanberg, P.R.; Kaneko, Y.; Borlongan, C.V. Alpha-Synuclein as a pathological link between chronic traumatic brain injury and parkinson's disease. *J. Cell. Physiol.* **2015**, *230*, 1024–1032. [CrossRef] [PubMed]
47. Ninkina, N.; Peters, O.; Millership, S.; Salem, H.; van der Putten, H.; Buchman, V.L. γ-Synucleinopathy: Neurodegeneration associated with overexpression of the mouse protein. *Hum. Mol. Genet.* **2009**, *18*, 1779–1794. [CrossRef]
48. Surgucheva, I.; He, S.; Rich, M.C.; Sharma, R.; Ninkina, N.N.; Stahel, P.F.; Surguchov, A. Role of synucleins in traumatic brain injury—An experimental in vitro and in vivo study in mice. *Mol. Cell. Neurosci.* **2014**, *63*, 114–123. [CrossRef] [PubMed]
49. Winder, A.D.; Maniar, K.P.; Wei, J.J.; Liu, D.; Scholtens, D.M.; Lurain, J.R.; Schink, J.C.; Buttin, B.M.; Filiaci, V.L.; Lankes, H.A.; et al. Synuclein-γ in uterine serous carcinoma impacts survival: An NRG Oncology/Gynecologic Oncology Group study. *Cancer* **2017**, *123*, 1144–1155. [CrossRef]
50. Wang, W.X.; Rajeev, B.W.; Stromberg, A.J.; Ren, N.; Tang, G.; Huang, Q.; Rigoutsos, I.; Nelson, P.T. The expression of microRNA miR-107 decreases early in Alzheimer's disease and may accelerate disease progression through regulation of β-site amyloid precursor protein-cleaving enzyme 1. *J. Neurosci.* **2008**, *28*, 1213–1223. [CrossRef]
51. Zhao, J.; Fu, Y.; Yasvoina, M.; Shao, P.; Hitt, B.; O'Connor, T.; Logan, S.; Maus, E.; Citron, M.; Berry, R.; et al. β-site amyloid precursor protein cleaving enzyme 1 levels become elevated in neurons around amyloid plaques: Implications for Alzheimer's disease pathogenesis. *J. Neurosci.* **2007**, *27*, 3639–3649. [CrossRef]
52. Guedes, V.A.; Kenney, K.; Shahim, P.; Qu, B.X.; Lai, C.; Devoto, C.; Walker, W.C.; Nolen, T.; Diaz-Arrastia, R.; Gill, J.M. Exosomal neurofilament light: A prognostic biomarker for remote symptoms after mild traumatic brain injury? *Neurology* **2020**, *94*, e2412–e2423. [CrossRef]

 *biomedicines*

Article

Prophylactic Activation of Shh Signaling Attenuates TBI-Induced Seizures in Zebrafish by Modulating Glutamate Excitotoxicity through Eaat2a

James Hentig [1,2,3], Leah J. Campbell [1,2,3], Kaylee Cloghessy [1,2,3], Mijoon Lee [4], William Boggess [4] and David R. Hyde [1,2,3,*]

[1] Department of Biological Sciences, Galvin Life Science Building, University of Notre Dame, Notre Dame, IN 46556, USA; jhentig@nd.edu (J.H.); lcampbe4@nd.edu (L.J.C.); kcloghes@nd.edu (K.C.)
[2] Center for Zebrafish Research, Galvin Life Science Building, University of Notre Dame, Notre Dame, IN 46556, USA
[3] Center for Stem Cells and Regenerative Medicine, Galvin Life Science Building, University of Notre Dame, Notre Dame, IN 46556, USA
[4] Department of Chemistry and Biochemistry, University of Notre Dame, Notre Dame, IN 46556, USA; mlee12@nd.edu (M.L.); wboggess@nd.edu (W.B.)
* Correspondence: dhyde@nd.edu; Tel.: +1-574-631-8054

Citation: Hentig, J.; Campbell, L.J.; Cloghessy, K.; Lee, M.; Boggess, W.; Hyde, D.R. Prophylactic Activation of Shh Signaling Attenuates TBI-Induced Seizures in Zebrafish by Modulating Glutamate Excitotoxicity through Eaat2a. *Biomedicines* **2022**, *10*, 32. https://doi.org/10.3390/biomedicines10010032

Academic Editor: Bruno Meloni

Received: 3 November 2021
Accepted: 22 December 2021
Published: 24 December 2021

Publisher's Note: MDPI stays neutral with regard to jurisdictional claims in published maps and institutional affiliations.

Copyright: © 2021 by the authors. Licensee MDPI, Basel, Switzerland. This article is an open access article distributed under the terms and conditions of the Creative Commons Attribution (CC BY) license (https://creativecommons.org/licenses/by/4.0/).

Abstract: Approximately 2 million individuals experience a traumatic brain injury (TBI) every year in the United States. Secondary injury begins within minutes after TBI, with alterations in cellular function and chemical signaling that contribute to excitotoxicity. Post-traumatic seizures (PTS) are experienced in an increasing number of TBI individuals that also display resistance to traditional anti-seizure medications (ASMs). Sonic hedgehog (Shh) is a signaling pathway that is upregulated following central nervous system damage in zebrafish and aids injury-induced regeneration. Using a modified Marmarou weight drop on adult zebrafish, we examined PTS following TBI and Shh modulation. We found that inhibiting Shh signaling by cyclopamine significantly increased PTS in TBI fish, prolonged the timeframe PTS was observed, and decreased survival across all TBI severities. Shh-inhibited TBI fish failed to respond to traditional ASMs, but were attenuated when treated with CNQX, which blocks ionotropic glutamate receptors. We found that the Smoothened agonist, purmorphamine, increased Eaat2a expression in undamaged brains compared to untreated controls, and purmorphamine treatment reduced glutamate excitotoxicity following TBI. Similarly, purmorphamine reduced PTS, edema, and cognitive deficits in TBI fish, while these pathologies were increased and/or prolonged in cyclopamine-treated TBI fish. However, the increased severity of TBI phenotypes with cyclopamine was reduced by cotreating fish with ceftriaxone, which induces Eaat2a expression. Collectively, these data suggest that Shh signaling induces Eaat2a expression and plays a role in regulating TBI-induced glutamate excitotoxicity and TBI sequelae.

Keywords: traumatic brain injury; blunt-force TBI; post-traumatic seizures; zebrafish; sonic hedgehog signaling; purmorphamine

1. Introduction

Traumatic brain injuries (TBIs) have tremendous and lasting impacts because they can result in various sequelae and are one of the leading causes of disease burden in the United States [1–3]. Post-traumatic seizures (PTS) are a common consequence following TBI [3,4]. Development of PTS is influenced by age at time of injury, type of TBI (blunt vs. penetrating), and injury severity [4–6]. Although the incidence rate varies greatly, PTS has been reported as high as 30% in military veterans with TBI, while civilians displayed a 20–50% increased risk of TBI-induced PTS compared to the development of non-acquired epilepsy [7,8]. Furthermore, PTS can result in a cyclic injury pattern in which TBI results in PTS, and during seizure event, another TBI occurs from fall-like events.

Several anti-seizure medications (ASMs) exist that rely largely on γ-aminobutyric acid (GABA)- regulated mechanisms to combat uncontrolled excitatory-driven seizures [9]. However, a large portion of individuals that experience TBI-induced PTS are resistant to many ASMs [10,11]. This could be due to TBI resulting in a secondary injury that includes glutamic excitotoxicity [12].

Although rodents are the traditional neurotrauma model due to their recapitulation of a variety of human neurobehavioral deficits following injury, alternative models such as zebrafish have recently emerged [13–17]. Zebrafish possess many features that are advantageous for studying neurotrauma such as an extensively characterized behavioral repertoire [18], recapitulate a variety of human pathologies following a TBI [17], and several orthologs of excitatory amino acid transporters (Eaat), which are thought to play key roles in TBI excitotoxicity and epileptogenesis [19,20]. It was recently demonstrated using a zebrafish *eaat2a* mutant that loss of the glutamate transporter exhibited defects in neuronal function and epileptic seizures originating in the hindbrain of developing zebrafish [21], although it remains unknown if upregulating Eaat2 expression can suppress PTS.

We recently described a scalable blunt-force zebrafish TBI model that produced brain edema, neuroinflammation, cognitive deficits, and PTS, followed by neuronal regeneration [17]. Following injury, components of the Shh signaling pathway were rapidly and highly upregulated, and Shh was found to be important in the injury-induced proliferation. The use of purmorphamine, a Smoothened agonist and Shh signaling activator, was previously shown to ameliorate a variety of disorders, improve behavioral and cognitive outcomes, reduce neuroinflammation, and minimize edema following CNS trauma [22,23]. However, few studies have examined the effects of modulating Shh signaling in relation to PTS and regulating excitotoxicity.

Here, we examined the mechanistic interface between Shh and Eaat2 and demonstrated that prophylactic activation of Shh via purmorphamine attenuated TBI-induced PTS by reducing extracellular glutamate. In contrast, cyclopamine-induced inhibition of the innate Shh response following TBI increased PTS events and prolonged injury-induced edema and cognitive deficits. These data suggest that prophylactic activation of Shh upregulates Eaat2 and combats excitotoxicity, resulting in reduced injury pathologies.

2. Materials and Methods

2.1. Fish Lines and Maintenance

Adult wild-type *albino*^b4 zebrafish [24] (Danio rerio) of both sexes were maintained in the Center for Zebrafish Research at the University of Notre Dame Freimann Life Sciences Center. This study used approximately equal numbers of male and female adult zebrafish, 6–18 months old, and 3 to 5 cm in length. All experimental protocols in this study were approved by the University of Notre Dame Animal Care and Use Committee protocol # 18-03-4558.

2.2. TBI Induction via Modified Marmarou Weight Drop

Zebrafish were anesthetized in 1:1000 2-phenoxyethanol (2-PE, Sigma-Aldrich, St. Louis, MO, USA) until unresponsive to tail pinch as described in approved IACUC protocol. 2-PE was used because it is approved for anesthesia in zebrafish and has no effect on neuronal function even at higher concentrations [25]. A modified Marmarou weight drop protocol for zebrafish [17,26] was used to administer a blunt-force injury. Following anesthetic, fish were secured onto a clay mold that stabilized the body and exposed the zebrafish head. Undamaged fish were anesthetized, secured to the damage rig, and then returned to an aerated recovery tank. For TBI, fish were anesthetized and either a 1.5 g or 3.3 g ball bearing weight was dropped down a shaft of either 7.6 or 12.7 cm length to produce the desired force. All TBIs were induced between 4:00 and 8:00 p.m., and following the TBI, fish were placed into an aerated tank to recover, and then placed in standard living conditions. At the desired end points, fish were euthanized in 2-PE (1:500 dilution) until either unresponsive to tail pinch or no operculum movement for 30 min, whichever came last as described in an approved IACUC protocol.

2.3. Survival and Seizure

Following blunt-force injury and recovery from anesthesia, zebrafish were assessed for post-traumatic seizures within the first 1 h after injury and for 30 min every 12 hpi up to 5 dpi. Fish were observed and scored for displaying the following clonic-seizure metrics defined in the zebrafish behavior catalog [18]: ataxia (ZBC 1.9), bending (ZBC 1.16), circling (ZBC 1.32), and corkscrew swimming (ZBC 1.37). To assess survival and latent seizure activity, untreated and cyclopamine/purmorphamine-treated fish were observed for 30 min every 12 hpi and observed for the same metrics: mortality was determined as no operculum movement over the 30 min observation period, and latent post-traumatic seizures as defined by the same ZBC metrics. Shh-modulated survival and seizure activity was calculated as an average of each group (n = 100 fish per control/experimental group).

2.4. Fluid Content Measurement

Edema was measured using a modified Hoshi protocol [26,27]. Whole brains (undamaged, sTBI, or sTBI treated fish n = 9 fish control/experimental group) were isolated from control and injured fish at 1 or 3 dpi, weighed, dried at 60 °C for 12 h, and weighed again, with the percent fluid calculated using the following formula:

$$\%\text{Fluid} = \frac{\text{wet weight} - \text{dry weight}}{\text{wet weight}} \times 100\%$$

2.5. Learning

Learning was assessed as previously described [28]. Treated and untreated naïve undamaged control or blunt-force-damaged fish were individually placed into the shuttle box and examined at 1, 3, 5, 7, or 14 dpi, and assessed for learning with no fish being tested twice. A positive trial was recorded when the fish swam to the other stall in response to the red-light visual stimulus However, a failed trial was recorded if the fish failed to swim to the other stall within the 15 s of the red-light exposure, followed by a simultaneous pulsating electric current (20 V, 1 mA) for up to 15 s (10 electrical shocks/15 s). This was repeated until the fish successfully completed 5 consecutive trials. The number of trials that each fish required to learn were determined for each treatment group and averaged for each experiment, (n = 9 fish control/experimental group). A two-way ANOVA and Tukey's post hoc test were performed to statistically compare the undamaged control fish and the different damage groups.

2.6. Memory

Memory was assessed as previously described [28]. Purmorphamine-treated and untreated naïve undamaged fish were individually placed into the shuttle box testing apparatus. Fish were acclimated in a dark and quiet room for 15 min. A red-light visual and a pulsating electric current (20 V, 1 mA) were simultaneously applied until the fish swam halfway across the testing tank. Presentation of the visual stimulus and electrical current was repeated 25 times (training) with 30 s of rest between intervals. Fish were then tested 15 min later by exposing the fish only to the red-light stimulus (initial testing period) 25 times. The number of successful trials, in which the fish did not require the electric shock to cross the tank, was counted to generate the initial testing baseline. Once all fish were tested, they were randomly selected for either the undamaged control group or were administered a severe blunt-force injury. To assess immediate recall, undamaged and experimental groups were then subjected to a post-sham/injury testing period consisting of 25 iterations 4 h/hpi. To test delayed recall, four days after the initial training and testing period, fish were randomly selected and given a severe blunt-force injury and allowed to recover for 4 h/hpi. Both undamaged and experimental groups were then subjected to a post-sham/injury testing period of 25 iterations. For both recall examinations, the percent difference in the number of successful post-sham/injury trials relative to the number of initial successful trials was calculated and averaged for each group.

2.7. Pharmacological Agents

Sonic hedgehog and EAAT2 modulation was performed as illustrated in Figure 1.

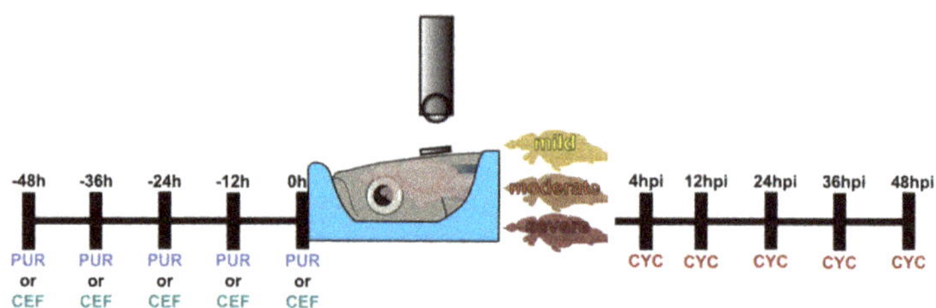

Figure 1. Experimental timeline of Shh modulation. Graphical representation of the dosing schedule of purmorphamine, ceftriaxone, and cyclopamine in relation to the timing of the TBI.

Purmorphamine: Undamaged fish were intraperitoneally (IP) injected with ~40 μL of 10 μM purmorphamine (Sigma-Aldrich, St. Louis, MO, USA) using a 30 gauge needle every 12 h for 48 h (0, 12, 24, 36, and 48 h).

Cyclopamine: Fish were exposed to sTBI and IP injected with ~40 μL of 2 mM cyclopamine (Sigma-Aldrich, St. Louis, MO, USA) using a 30 gauge needle at 4, 12, 24, 36, and 48 hpi.

Ceftriaxone: Undamaged fish were IP injected with ~40 μL of 10 mM ceftriaxone (Sigma-Aldrich, St. Louis, MO, USA) using a 30 gauge needle ever 12 h for 48 h (0, 12, 24, 36, and 48 h).

6-Cyano-7-nitroquinoxaline-2,3-dione (CNQX): Fish were exposed to sTBI and IP injected with ~40 μL of 2 mM CNQX (Sigma-Aldrich, St. Louis, MO, USA) and 2 mM cyclopamine using a 30 gauge needle at 4, 12, 24, 36, and 48 hpi.

Valproic acid (VPA): Fish were exposed to sTBI and IP injected with ~40 μL of 2 mM VPA (Sigma-Aldrich, St. Louis, MO, USA) and 2 mM cyclopamine using a 30 gauge needle at 4, 12, 24, 36, and 48 hpi.

Gabapentin (GABAP): Fish were exposed to sTBI and IP injected with ~40 μL of 3 mM GABAP (Sigma-Aldrich, St. Louis, MO, USA) and 2 mM cyclopamine using a 30 gauge needle at 4, 12, 24, 36, and 48 hpi.

2.8. Glutamate Excitotoxicity

Undamaged fish treated with either purmorphamine, cyclopamine, ceftriaxone, or a combination of the three were placed in a tank containing 5 mM glutamate (Sigma-Aldrich, St. Louis, MO, USA) in the tank water for 30 min. The time of the first seizure event per fish was recorded (once a fish experienced a seizure event, they were removed and placed into a second tank with 5 mM glutamate for the remainder of the 30 min as to avoid counting a second seizure as another unique event), and the percent of fish in each cohort that experienced at least one seizure was quantified. Following the 30 min glutamate exposure, all surviving fish were transferred to a tank containing only standard tank water and assessed for an additional 30 min for survival, and the percent of fish in each cohort that survived was quantified ($n = 90$ fish control/experimental group). Statistical analysis was performed with a one-way ANOVA followed by Tukey's post hoc test.

2.9. Quantitative Real-Time PCR (qRT-PCR)

Total RNA was isolated and purified from whole telencephalons and cerebellums from ten adult untreated and purmorphamine-treated undamaged fish using Trizol extraction. The RNA was converted to cDNA from 1 ug of RNA using qScript cDNA SuperMix (VWR International, Radnor, PA, USA) as previously described [29]. Taqman probes were used according to the manufacturer's instructions with 10 ng of cDNA. Taqman probes (Thermo Fisher) *gli1* (Dr03093665_m1), *eaat1a* (Dr03120588_m1), *eaat2a* (Dr03119707_m1), *eaat3* (Dr03108401_m1), and *il1β* (Dr03114367_g1) were used for quantitative real-time PCR (qRT-PCR) and the data were normalized to 18 s rRNA (Hs03003631_g1) in each well. Data were acquired using the ABI StepOnePlus Real-Time PCR System (Applied Biosystems). Cycling conditions were as follows: 2 min at 50 °C, 10 min at 95 °C, and 40 cycles of 15 s at 95 °C and 1 min at 60 °C with data collection occurring following the 60 °C extension step. The $\Delta\Delta$CT values were calculated and used to determine the log2-fold changes [30] of *gli1*, *eaat1a*, *eaat2a*, and *eaat3*. Expression levels were examined in technical and biological triplicate.

2.10. Microdialysis and Glutamate Quantification

Post-injury extracellular glutamate concentrations were determined using a modified Puppala protocol [31]. Undamaged, sTBI, and sTBI fish treated with either purmorphamine, cyclopamine, or ceftriaxone were collected at 30 min post-injury, 12 hpi, 36 hpi, or 5 dpi and were euthanized with 1:500 2-PE. Each fish was immobilized by modeling clay, and the skull removed. A microdialysis probe (CMA7, 2 mm membrane length; CMA/Microdialysis, N. Chelmsford, MA, USA), which was rinsed and equilibrated for 30 min, was slowly inserted under a dissection microscope into the cerebellum along the rostral-caudal axis. The corpus cerebelli was dialyzed with sterile Ringer's solution (147 mM Na^+, 2.3 mM Ca^{2+}, 4 mM K^+, and 155.6 mM Cl^-; pH = 6.0) at a flow rate of 5 μL/min. Dialysate fractions were collected for 30 min and immediately stored at -20 °C. Following the manufacturer's instructions (Glutamate-Glo™ Assay, Promega, Madison, WI, USA), samples were thawed and 50 μL was incubated for 1 h with 50 μL mastermix of Luciferin detection solution, reductase, reductase substrate, glutamate dehydrogenase, and NAD. Samples were analyzed on a SpectraMax M5 plate reader (Molecular Devices, San Jose, CA, USA), and sample RLU values were compared to a glutamate serial dilution standard curve to quantify glutamate concentration. Comparisons were made using a one-way ANOVA followed by either Tukey's or Dunnett's multiple comparison post hoc test.

2.11. Immunoblot

Total protein collection and immunoblotting techniques were performed as previously described [32], with minor modifications. Five cerebellums were pooled for each experimental replicate. Both chicken anti-Eaat2a (1:1000, gift of Neuhauss lab) [33] and rabbit anti-Gapdh (1:2500; cat #: ab210113, Abcam, Cambridge, UK) polyclonal anti-sera were incubated on membranes (0.45 μm PVDF, VWR, Radnor, PA, USA) that were blocked overnight at 4 °C in 1 × TBS/5% nonfat dry milk/0.1% Tween 20. Membranes were washed 4 × 15 min in 1 × TBS/0.1% Tween 20. The membranes were incubated with anti-chicken or anti-rabbit peroxidase secondary antibody (1:10,000, chicken, cat #: A9046, Sigma-Aldrich, St. Louis, MO, USA; rabbit, NA934, VWR International, Radnor, PA, USA), washed, and detected as described previously [32]. The blot was incubated with the ECL-Prime detection system (Fischer Scientific, Waltham, MA, USA) and exposed to X-ray film.

2.12. Statistical Analysis

With regard to all data within this study, except for microdialysis and immunoblot, each experiment was obtained from three independent trials of at least 3 fish per trial. The data are expressed as the mean $\pm$ SE of the mean. Data sets were analyzed in Prism 8 (GraphPad) with either Student's *t*-test for single pairwise comparisons with control, one-way or two-way ANOVA followed by either Tukey's or Dunnett's post hoc test for multiple comparisons, or the Friedman test for repeated-measures data. The statistical test used is stated in each figure legend.

3. Results

3.1. Shh Inhibition Increases TBI-Induced Post-Traumatic Seizures

Shh was identified as a key pathway that is activated following damage to the central nervous system (CNS) in the adult zebrafish and is required for injury-induced regeneration [17,34,35]. It was shown that following TBI in zebrafish and rodents, genes encoding Shh signaling components were upregulated as early as 6 h post-injury (hpi) [17,36]. Additionally, Shh activation attenuated sequelae following various forms of CNS trauma [23,37–39]. We examined undamaged and TBI fish that were either untreated or treated with the Smo antagonist cyclopamine (a Shh signaling inhibitor) at 4, 12, 24, 36, and 48 hpi. Undamaged fish, whether untreated or cyclopamine-treated, displayed no seizure behavior, while untreated TBI fish displayed modest, but significant, PTS in a severity-dependent manner (Figure 2A–C). No PTS was observed within 1 hpi following a mild TBI (miTBI, Figure 2A), while moderate TBI (moTBI) induced PTS in a significant percentage of fish immediately following injury (10.66% $\pm$ 1.37%, Figure 2B) and this was further elevated following severe TBI (sTBI, 17.44% $\pm$ 1.54%, Figure 2C). Untreated TBI fish exhibited a rapid decrease in seizure behavior, with no clonic-tonic seizures observed in moTBI by 1 day post-injury (dpi) and in sTBI by 1.5 dpi (Figure 2B,C). In contrast, cyclopamine-treated TBI fish (TBI/CYC) displayed a significantly increased percentage of fish with PTS compared to their matched counterparts. A significantly increased percentage of miTBI/CYC fish experienced PTS at 12 hpi (29.66% $\pm$ 6.10%, $p < 0.01$) and the percentage peaked at 1.5 dpi (40% $\pm$ 5.76%, $p < 0.01$, Figure 2A). Similarly, and in a severity-dependent manner, increased percentages of moTBI/CYC (39.4% $\pm$ 4.43%, $p < 0.01$) and sTBI/CYC (63.33% $\pm$ 7.99%, $p < 0.01$) fish were observed at 12 hpi and peaked at 1.5 dpi (moTBI/CYC: 52.57% $\pm$ 5.70%, $p < 0.01$, sTBI/CYC: 72.33% 3.73%, $p < 0.01$, Figure 2B,C). Although cyclopamine treatment ceased at 2 dpi, all severities continued to experience heightened seizure activity for a period of time. The percentage of fish exhibiting PTS noticeably decreased in miTBI and moTBI by 3.5 and 4 dpi, respectively (Figure 2A,B). However, sTBI/CYC fish continued to experience significant seizure activity out to 5 dpi ($p < 0.01$, Figure 2C).

We also observed stark differences in survival between TBI fish and cyclopamine-treated TBI fish (Figure 2D–F). Over 5 days following injury, all miTBI fish survived (Figure 2D), and although moTBI and sTBI experienced some mortality, they largely survived (moTBI: 96.66%, sTBI: 85.71%, Figure 2E,F). In contrast, only 56.66% of miTBI/CYC fish survived 5 dpi (Figure 2D), while survival continued to decrease in both moTBI/CYC (46.66%, Figure 2E) and sTBI/CYC fish (28.12%, Figure 2F). For all severities, the greatest mortality occurred at 12 hpi, shortly after initiating Shh pathway inhibition. For miTBI/CYC and moTBI/CYC, all death occurred during cyclopamine treatment with no mortality following the cessation of cyclopamine. In contrast, sTBI/CYC fish continued to experience mortality following the end of cyclopamine treatment, although slowed, out to 5 dpi (Figure 2F). Collectively, these data suggest that inhibiting Shh signaling plays a role in seizure activity and overall survival following TBI.

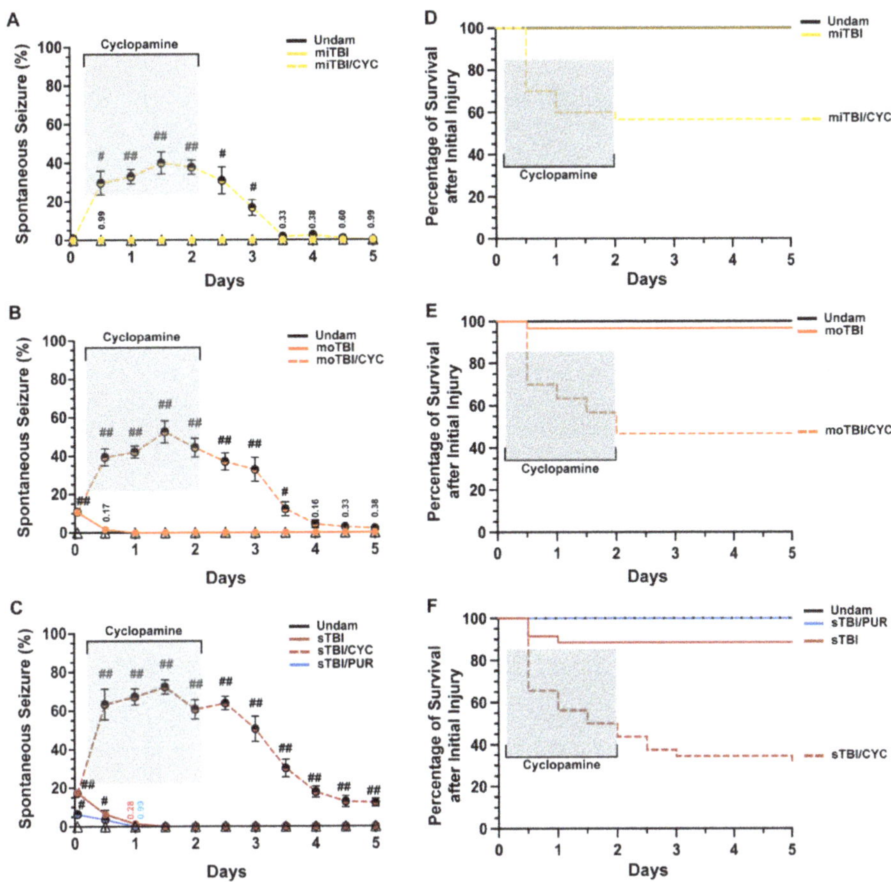

Figure 2. Modulation of Shh alters frequency of TBI-induced PTS events. (**A–C**) Percentage of fish that displayed spontaneous seizure events spanning the timeframe from within 1 hpi to 5 dpi with and without Shh modulation. (**D–F**) Percentage of survival across 5 days of fish who initially survived the primary injury across mi-, mo-, and sTBI with and without Shh modulation. Statistical analyses of the repeated-measures data were performed with the Friedman test, $n = 100$ fish per control/experimental group, grey box denotes period of cyclopamine administration, # $p < 0.05$, ## $p < 0.01$.

3.2. Blocking Ionotropic Glu Receptors Inhibits TBI-Induced PTS

Neuronal signaling is finely tuned and small deviations that go uncontrolled can manifest in various aberrant neurobehaviors, including seizures [40,41]. Developmental and acquired events can induce seizures, usually resulting from misregulation of neurotransmitters glutamate or GABA [42,43]. We examined if TBI-induced seizures that were exacerbated following inhibition of Shh signaling were predominately glutamate or GABA driven. We coinjected sTBI fish that were treated with cyclopamine at 4, 12, 24, 36, 48 hpi with either the ionotropic glutamate receptor antagonist 6-cyano-7-nitroquinoxaline-2,3-dione (CNQX), valproic acid (VPA) a common first-line ASM [44], or the synthetic GABA analog gabapentin (GABAP). Undamaged fish treated with cyclopamine displayed no seizures, while sTBI/CYC fish experienced rapid and significantly increased percentage of fish with PTS seizure at 12 hpi (58.33% ± 9.36%, $p < 0.01$, Figure 3A), peak percentage at 1.5 dpi (70.16% ± 5.29%, $p < 0.01$) and remained significantly elevated after ceasing cyclopamine (5 dpi: 10.83% ± 2.71%, $p < 0.01$, Figure 3A). Cyclopamine-treated sTBI that were coinjected with either VPA or GABAP (VPA/CYC and GABAP/CYC, respectively) showed similar percentages of PTS-experiencing sTBI fish that were only treated with cyclopamine. At 12 hpi, both VPA/CYC- and GABAP/CYC-treated fish experienced significantly increased percentages of fish with PTS (VPA/CYC: 57.5% ± 3.35%, $p < 0.01$, GABAP/CYC: 62.5% ± 5.59%, $p < 0.01$, Figure 3A). In contrast, the percentage of sTBI cyclopamine-treated fish that were coinjected with CNQX (CNQX/CYC) with PTS was significantly decreased at 12 hpi compared to sTBI/CYC fish ($p < 0.05$), although still increased compared to undamaged cyclopamine-injected fish (13% ± 1.34%, $p < 0.01$, Figure 3A). By 1 dpi, VPA and GABAP treatments resulted in a noticeably decreased percentage of PTS fish, while CNQX treatment significantly reduced the percentage of PTS fish greater than either VPA or GABAP treatment (CNQX/CYC vs. VPA/CYC, $p < 0.01$, CNQX/CYC vs. GABAP/CYC, $p < 0.01$, VPA/CYC vs. GABAP/CYC, $p = 0.26$, Figure 3A). Additionally, CNQX reduced the percentage of PTS to a level similar to Undam/CYC by 1.5 dpi compared to 3 dpi for VPA/CYC- and GABAP/CYC-treated fish (Figure 3A). Similarly, cyclopamine treatment (sTBI/CYC) negatively impacted survival (5 dpi: 27.5%), as did sTBI/CYC fish cotreated with GABAP (5 dpi: 33.33%, Figure 3B). sTBI fish cotreated with cyclopamine and VPA displayed a slight improvement in survival (55.6%). In contrast, sTBI/CYC fish that were cotreated with CNQX experienced dramatically improved survival (88.89%, Figure 3B), to a level that was similar to undamaged fish. Collectively, these data suggest that traditional GABA-driven ASMs had minimal influence on TBI-induced PTS and survival, while blocking glutamate receptors with CNQX attenuated PTS and improved survival, suggesting that TBI-induced PTS may be due to glutamic excitotoxicity.

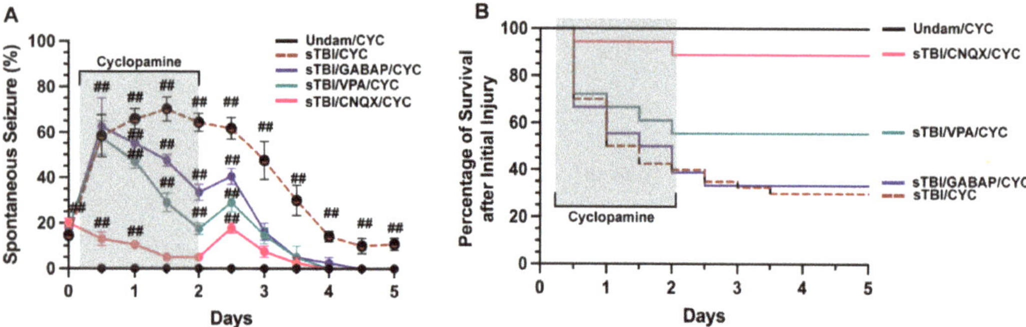

Figure 3. CNQX attenuates PTS in Shh-inhibited fish. (**A**) Percentage of undamaged and sTBI that were Shh inhibited (cyclopamine-treated) and cotreated with either valproic acid (VPA), gabapentin (GABAP), or CNQX that displayed spontaneous seizure events across 5 dpi. (**B**) Quantification of survival across 5 days of undamaged or sTBI Shh-inhibited fish cotreated with VPA, GABAP, or CNQX. Statistical analysis of the repeated-measures data was performed with the Friedman test, $n = 100$ fish per control/experimental group, grey box denotes period of cyclopamine/(VPA or GABAP or CNQX) administration, ## $p < 0.01$.

3.3. Shh Activation Combats Excitotoxicity and Upregulates Eaat2a

TBIs result in a wide breadth of pathologies, with excitotoxicity contributing to many TBI-induced sequelae [43]. Previously, Shh signaling was reported to reduce excitotoxicity following middle cerebral artery occlusion in rodents [45,46], though it was never examined following TBI in adult zebrafish. To test if Shh signaling could affect glutamate-induced excitotoxicity, undamaged fish were either left untreated, treated with the Smo agonist purmorphamine, which is a Shh signaling activator, or treated with both purmorphamine and cyclopamine. Fish were then challenged with 5 mM glutamate in their tank water for 30 min and the time of the first seizure, percentage of the cohort to seize, and the overall survival of the cohort was quantified. Untreated fish experienced rapid seizure onset with the first seizure recorded an average of 9.26 min (±0.86 min), the majority of the fish (93.3% ± 3.3%) experienced a seizure, and survival was low (10.6% ± 5.8%, Figure 4A–C).

In contrast, purmorphamine-treated fish significantly warded off seizure activity, with the average first recorded seizure event at 18.43 min ($\pm$1.07 min, $p < 0.01$), significantly experienced fewer fish with a seizure during the glutamate challenge (21.6% $\pm$ 6.0%, $p < 0.01$), and survival was significantly improved (85% $\pm$ 7.63%, $p < 0.01$, Figure 4A–C). The attenuation observed in purmorphamine-treated fish was lost when co-treated with cyclopamine, as time of first seizure was reduced to 7.57 min ($\pm$0.82 min, $p = 0.61$), 90% ($\pm$5.77%, $p = 0.95$) of the fish displayed seizure behavior, and the cohort experienced low survival (8.66% $\pm$ 5.92%, $p = 0.99$, Figure 4A–C).

The findings from the glutamate challenge suggested that Shh signaling activation may upregulate one or more excitatory amino acid transporters (Eaat), which would reduce excessive extracellular glutamate following TBI. Both humans and zebrafish have multiple Eaat, though Eaat1 (Glast) and Eaat2 (Glt-1) are the most prominent. To identify which, if any, Eaat may be influenced by Shh activation, we performed qRT-PCR using RNA collected from telencephalons and cerebellums of untreated and purmorphamine-treated undamaged fish. As an internal control, we confirmed Shh activation by observing increased *gli1* expression (downstream targe of Shh signaling) in undamaged purmorphamine-treated brains compared to untreated brains (Figure 4D,E). We then examined *eaat1a*, *eaat2a*, and *eaat3* expression (paralogs *eaat1b* and *eaat2b* are expressed at extremely low levels in the zebrafish brain) [47] and observed increased expression of only *eaat2a* in both the telencephalon and the cerebellum of purmorphamine-treated undamaged brains (Figure 4D,E).

Ceftriaxone was reported to upregulate Eaat2 expression in rodents and zebrafish [48,49]. We treated undamaged fish with 10 mM ceftriaxone every 12 h for 60 h and then collected telencephalons and cerebellums for qRT-PCR analysis. We did not see upregulation of *gli1*, *eaat1a, or eaat3*, but did observe increased *eaat2a* expression following ceftriaxone treatment in both the telencephalon and the cerebellum (Figure 4F,G). We also observed increased protein expression of Eaat2a in both purmorphamine and ceftriaxone-treated undamaged brains relative to untreated controls (Figure 4H). In the glutamate challenge, the severe negative effects observed in the purmorphamine/cyclopamine double-treated fish were suppressed by the ceftriaxone treatment and upregulating *eaat2a* expression. Both triple-treated (purmorphamine/cyclopamine/ceftriaxone) and ceftriaxone-only-treated fish exhibited delayed seizure onset with average time of first seizures observed at 19.54 min ($\pm$1.11 min) and 21.25 min ($\pm$1.2 min), respectively (Figure 4A). Both the triple-treated and ceftriaxone-only-treated groups experienced significantly lower percentage of seizures compared to untreated fish (triple treated: 13.3% $\pm$ 3.3%, $p < 0.01$, ceftriaxone only: 14.33% $\pm$ 2.96%, $p < 0.01$, Figure 4B), and both groups exhibited significantly improved survival (triple treated: 90% $\pm$ 5.77%, $p < 0.01$, ceftriaxone only: 96.6% $\pm$ 1.6%, $p < 0.01$, Figure 4C).

We next examined the temporal expression of *eaat2a* relative to injury. We performed qRT-PCR of *eaat2a* in sTBI fish at 12 hpi, 1, 2, 3, 7, and 14 dpi with Shh modulation (Figure 4I). Following injury, untreated sTBI fish exhibited upregulated *eaat2a* expression at 12 hpi, which then gradually decreased to near undamaged levels by 7 and 14 dpi (Figure 4I). In sTBI/CYC fish the initial expression was noticeably repressed during cyclopamine treatment through 2 dpi, followed by a robust increase in *eaat2a* expression between 3 and 7 dpi that returned near basal levels at 14 dpi (Figure 4I). In contrast, fish that were treated with both ceftriaxone and cyclopamine exhibited a rapid increase in *eaat2a* expression that peaked at 2 dpi, before decreasing through 14 dpi, though it was still elevated relative to untreated fish (Figure 4I). Similarly, sTBI/PUR fish displayed elevated *eaat2a* expression from 12 hpi through 2 dpi, before declining and reaching untreated undamaged levels by 14 dpi (Figure 4I).

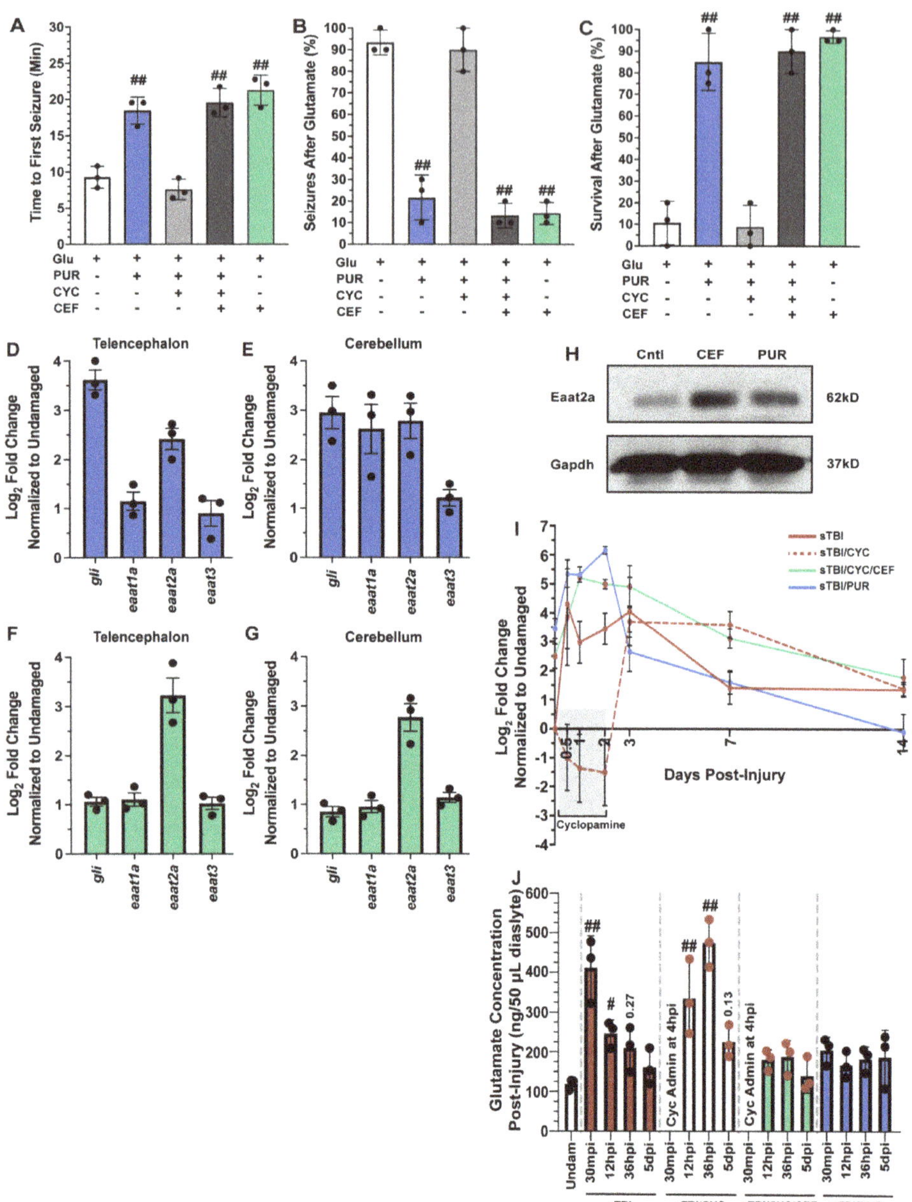

Figure 4. Shh activation combats excitotoxicity and upregulates Eaat2a. Undamaged fish with and without Shh modulation and ceftriaxone treatment were exposed to 5 mM glutamate in fish water (*n* = 90 fish per control/experimental group). (**A**) Quantification of time to first seizure, (**B**) percent of the group to display at least 1 seizure event, and (**C**) percent of the group that survived for 1 h. (**D,E**) Expression of Shh component, *gli,* and excitatory amino acid transporter (*eaat*) genes by qRT-PCR revealed that purmorphamine increased *gli* and *eaat2a* mRNAs in both undamaged telencephalons and cerebellums (*n* = 3 per control/experimental group, with 5 pooled telencephalons or cerebellums/trial). (**F,G**) Administration of ceftriaxone increased *eaat2a* mRNA expression levels without Shh activation in undamaged telencephalons and cerebellums (*n* = 3 per control/experimental group, with 5 pooled telencephalons or cerebellums/trial). (**H**) Expression of Eaat2a in the undamaged

cerebellum (5 pooled cerebellums per control/experimental group) following either purmorphamine or ceftriaxone treatment was assessed by immunoblot, using GAPDH as a loading control (H, top and lower bands, respectively). Eaat2a protein expression was increased following either purmorphamine or ceftriaxone treatment compared to controls. (**I**) Expression of *eaat2a* by qRT-PCR in sTBI fish with either Shh signaling activated, inhibited, or inhibited and cotreated with ceftriaxone at 12 hpi to 14 dpi (n = 3 per control/experimental group, with 5 pooled cerebellums/group). (**J**) Microdialysis was performed to quantify extracellular glutamate levels in undamaged and sTBI fish with and without Shh modulation and ceftriaxone treatment at 30 min post-injury, 12 and 36 hpi, and 5 dpi (n = 3 per control/experimental group). In panels A-C, the plus sign (+) denotes application of the reagent and the minus sign (-) denotes the absence of the reagent. Grey box denotes period of cyclopamine administration. Statistical analyses were performed with a one-way ANOVA followed by Dunnett's multiple comparison post hoc test, # $p < 0.05$, ## $p < 0.01$.

We next examined the effect of manipulating Shh activation on TBI-induced glutamate excitotoxicity. Due to the heightened and peak seizure activity observed following cyclopamine treatment at 1.5 days (36 hpi), we examined the concentration of extracellular glutamate at multiple time-points (30 min, 12 h, 36 h, and 5 days post-injury) from sTBI or sTBI fish treated with either cyclopamine, ceftriaxone/cyclopamine, or purmorphamine, and compared to uninjured controls. Glutamate levels in sTBI fish 30 min post-injury (mpi) and 12 hpi were significantly higher compared to undamaged fish (sTBI 30 mpi: 411.66 ng $\pm$ 46.36 ng, $p < 0.01$, sTBI 12 hpi: 246.3 ng $\pm$ 19.53 ng, $p < 0.05$, Undam: 118.66 ng $\pm$ 7.83 ng, Figure 4J), and although still elevated, decreased in sTBI fish at 36 hpi (sTBI 36 hpi: 209 ng $\pm$ 32.57, $p = 0.27$, Figure 4J). First cyclopamine administration occurred at 4 hpi as it has been shown that injury-induced upregulation of Shh signaling components occurred as early as 6 hpi [17]. Therefore, assessment of glutamate levels in cyclopamine-treated fish at 30 mpi was not possible. Cyclopamine-treated sTBI fish at 12 hpi possessed significantly elevated glutamate levels compared to undamaged fish (sTBI/CYC 12 hpi: 334.6 ng $\pm$ 54.29 ng, $p < 0.01$), which was further elevated at 36 hpi (sTBI/CYC 36 hpi: 474.3 ng $\pm$ 34.36 ng) compared to both undamaged ($p < 0.01$) and sTBI 36 hpi fish ($p < 0.01$, Figure 4J). Cyclopamine treatment ended at 60 hpi, and while glutamate levels were still elevated at 5 dpi, they were statistically not different than undamaged levels (sTBI/CYC 5 dpi: 225 ng $\pm$ 23.47 ng, $p = 0.13$, Figure 4J). The sTBI/CYC/CEF fish exhibited glutamate levels following injury that were not statistically different from undamaged fish at any timepoint (Figure 4J). A similar effect was seen in purmorphamine-treated fish, where extracellular TBI-induced glutamate levels were slightly elevated, but not statistically different, relative to undamaged fish at any time point (Figure 4J). Collectively, these data suggest that purmorphamine-mediated Shh attenuates excitotoxicity by regulating extracellular glutamate via Eaat2a.

3.4. Shh Activation Reduces Edema, Neuroinflammation, and PTS following TBI

Increased brain edema and neuroinflammation are highly correlated with increased PTS following TBI [50], while glutamic excitotoxicity increases edema and neuroinflammation [51]. Based on our data, we hypothesized that reducing glutamic excitotoxicity and extracellular glutamate levels post-TBI would also decrease brain edema. We quantified edema, measured as percentage of brain fluid, in undamaged and sTBI fish at 1, 3, 5, 7, and 14 dpi with and without Shh modulation. Following injury, sTBI fish displayed significantly increased brain edema at 1 and 3 dpi compared to undamaged brains (undam: 73.4% $\pm$ 0.55%, sTBI 1 dpi: 85.3% $\pm$ 1.13%, $p < 0.01$, 3 dpi: 79.3% $\pm$ 0.84%, $p < 0.05$, Figure 5A). By 5 dpi, brain edema in sTBI fish is reduced near undamaged levels and remains through 14 dpi. In contrast, sTBI with cyclopamine displayed prolonged and significantly elevated edema 1–5 dpi compared to undamaged brains (sTBI/CYC 1 dpi: 84% $\pm$ 1.43%, $p < 0.01$, 3 dpi: 82.88% $\pm$ 1.51%, $p < 0.01$, 5 dpi: 78.1% $\pm$ 1.08%, $p < 0.05$) before decreasing to undamaged levels by 7 dpi (Figure 5A). However, when ceftriaxone is administered to upregulate *eaat2a* expression in sTBI/CYC fish no significant brain edema was observed at any time point (Figure 5A). Similarly, when sTBI fish were treated with purmorphamine (sTBI/PUR), we again observed no significant brain edema following TBI at any time point (Figure 5A).

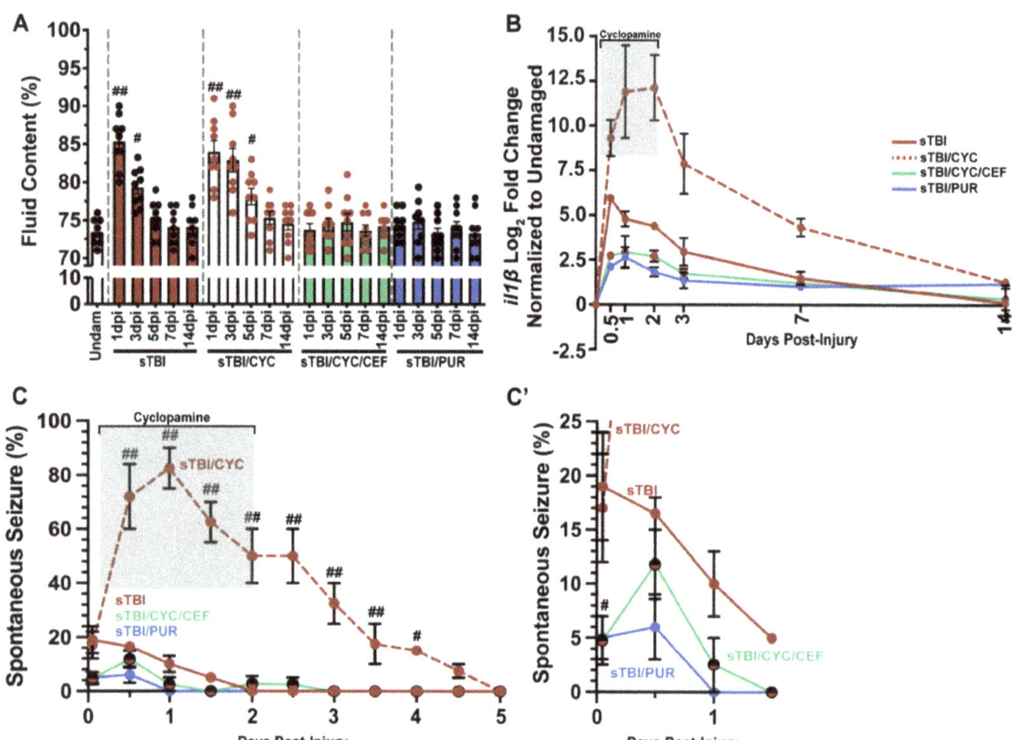

Figure 5. Prophylactic Shh activation attenuates TBI-induced edema, neuroinflammation, and PTS. (**A**) Histogram quantifying brain edema in undamaged and sTBI fish with and without Shh modulation and ceftriaxone treatment across 1 to 14 dpi (*n* = 9 fish per control/experimental group). (**B**) Expression of *il1β* by qRT-PCR in the cerebellum of sTBI fish with and without Shh modulation and ceftriaxone treatment compared undamaged fish (*n* = 3 per control/experimental group, with 5 pooled cerebellums/trial). (**C**) Percentage of fish that displayed spontaneous seizure events spanning the timeframe of from within 1 hpi to 5 dpi with and without Shh modulation and ceftriaxone treatment (*n* = 100 fish per control/experimental group). (**C′**) Expanded view of early post-traumatic seizure percentage. Grey box denotes period of cyclopamine administration. Statistical analyses were performed with either a one-way ANOVA followed by Tukey's multiple comparison post hoc test (**A**), a two-way ANOVA followed by Dunnett's multiple comparison post hoc test (**B**), or the Friedman test of the repeated-measures data (**C**), # *p* < 0.05, ## *p* < 0.01.

In relation to edema, Shh was also reported to affect neuroinflammation through the regulation il1β expression [39,52], which is often used as a critical biomarker in human TBI [53] and known to have a role in zebrafish injury-induced regeneration [54]. We performed qRT-PCR on cerebellar RNA following damage and Shh modulation to assess the expression of il1β. sTBI fish displayed moderate increase in il1β that gradually declined returning near undamaged levels of expression by 7 and 14 dpi (Figure 5B). In contrast, sTBI fish treated with cyclopamine displayed greater expression of il1β, which persisted through the entire cyclopamine treatment followed by a decrease in expression, though it was still elevated at 14 dpi (Figure 5B). However, when sTBI fish were treated with cyclopamine and ceftriaxone, a large reduction in il1β expression was observed relative to sTBI and sTBI/CYC fish (Figure 5B). Similarly, purmorphamine treatment exhibited a reduced il1β expression similar to sTBI/CYC/CEF fish (Figure 5B).

The influence of Shh modulation on PTS (Shh activation decreasing PTS, and inhibition increasing PTS, Figure 2C), and glutamic excitotoxicity suggested that PTS might be suppressed by CEF-mediated increase in *eaat2a* expression. As observed previously (Figure 2), cyclopamine-treated sTBI fish displayed increased seizure activity that persisted significantly to 4 dpi compared to untreated sTBI fish (sTBI/CYC: 15% ± 2.48%, $p < 0.05$, Figure 5C). In contrast, when sTBI fish were treated with cyclopamine and ceftriaxone to upregulate *eaat2a*, the percentage of fish exhibiting PTS was significantly reduced (sTBI/CEF/CYC: 4.75% ± 0.92%, $p < 0.05$, Figure 5C,C'). Similarly, purmorphamine-treated sTBI fish displayed significantly reduced percentage of PTS fish compared to untreated sTBI fish (sTBI/PUR: 5% ± 0.81%, sTBI: 19% ± 2.08%, $p < 0.05$), with no seizure behavior observed in purmorphamine-treated fish after 12 hpi (Figure 5C,C'). Collectively, these data suggest that the attenuation seen in purmorphamine-treated fish may be due to the reduction in brain edema, neuroinflammation, and glutamate excitotoxicity, areas known to be influenced by Shh modulation.

3.5. Prophylactic Shh Activation Attenuates TBI-Induced Cognitive Deficits

Purmorphamine treatment has also been shown to improve the neurobehavioral and cognitive recovery following CNS trauma [23,37–39]. Zebrafish was previously shown to have a significant learning and memory impairment following sTBI, which rapidly recovered in 4–7 dpi [17,28]. Therefore, we examined the potential of purmorphamine preconditioning to impact TBI-induced cognitive deficits in zebrafish. We assessed associative learning with the shuttle-box assay [17,28], which uses a visual stimulus and an electrical shock as a negative behavioral reinforcement. Undamaged control fish required an average of 20 ± 2.22 trials (Figure 6A) to master the assay (completing 5 consecutive positive trials without the negative reinforcement). Untreated sTBI fish exhibited significant deficits at 1, 3, and 5 dpi (sTBI 1 dpi: 75 ± 4.6 trials, $p < 0.01$, 3 dpi: 45 ± 3.22 trials, $p < 0.01$, 5 dpi: 36 ± 3.6 trials, $p < 0.05$) relative to undamaged control fish, which gradually decreased over time to return statistically to undamaged levels at 7 dpi (22 ± 1.38 trials, Figure 6A). Significant cognitive deficits persisted in Shh-inhibited fish out to 7 dpi relative to undamaged fish (sTBI/CYC 1 dpi: 82 ± trials, $p < 0.01$, 3 dpi: 80 ± trials, $p < 0.01$, 5 dpi: 67 ± trials, $p < 0.01$, 7 dpi: 59 ± trials, $p < 0.01$) before returning to near undamaged levels at 14 dpi (sTBI/CYC 14 dpi: 16 ± trials, $p = 0.98$). Additionally, the sTBI/CYC fish exhibited significant deficits at 3–7 dpi compared to untreated sTBI fish ($p < 0.01$, Figure 6A). However, these deficits in the sTBI/CYC fish were rescued by co-treating fish with ceftriaxone, which did not exhibit any learning deficit at any time point post-injury (sTBI/CEF/CYC 1 dpi: 17 ± 2.42 trials, $p = 0.99$, 14 dpi: 14 ± 1.59 trials, $p = 0.99$, Figure 6A). Similarly, preconditioning sTBI fish with purmorphamine blocked any significant learning deficit at any time (1–14 dpi) following injury relative to undamaged controls, (sTBI/PUR 1 dpi: 19 ± 1.56 trials, $p = 0.99$, 14 dpi: 16 ± 1.7 trials, $p = 0.98$, Figure 6A).

We next assessed the potential of purmorphamine preconditioning to ameliorate the deficits seen in memory following TBI [17,28]. To assess immediate and delayed recall, we again used the shuttle box assay [28]. Untreated sTBI fish displayed significant immediate memory deficit, with a decrease of −48.77% ± 3.36% ($p < 0.01$) in successful trials when retested 4 hpi, while undamaged fish exhibited a slight increase of 5% ± 1.62% in successful trials when retested 4 h following testing period 1 (Figure 6B). sTBI fish that were purmorphamine preconditioned also displayed significant immediate memory deficits (−14.66% ± 2.99%, $p < 0.01$); however, they significantly outperformed untreated sTBI fish ($p < 0.01$, Figure 6B). Neither group displayed significant delayed memory deficits suggesting that prophylactic Shh activation may dampen the cognitive learning and immediate memory deficits observed following sTBI.

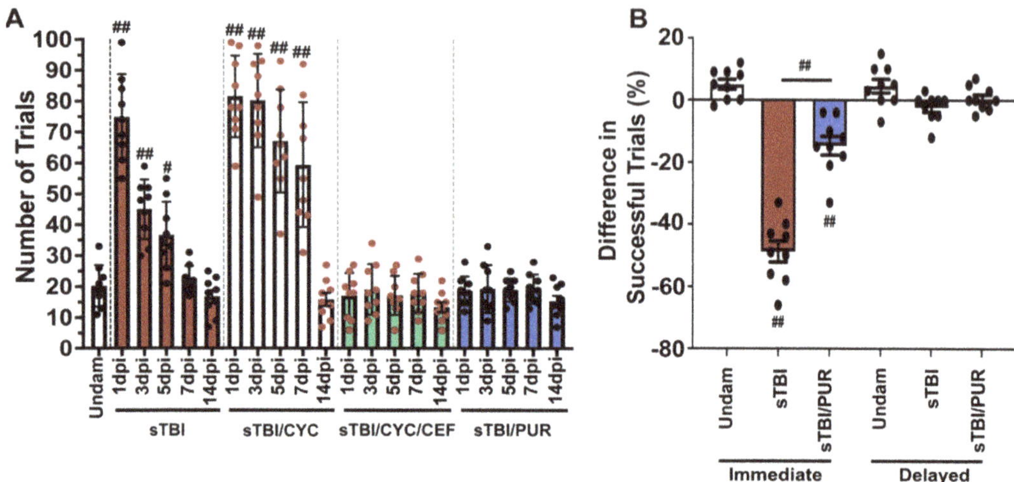

Figure 6. Cognitive impairments are attenuated by prophylactic Shh activation. (**A**) Histogram quantifying the number of trials required for associative learning assay of undamaged and either untreated or Shh-modulated sTBI fish at 1, 3, 5, 7 and 14 dpi (*n* = 9 fish per control/experimental group). (**B**) Quantification of immediate and delayed recall of undamaged, sTBI, and sTBI/PUR fish (*n* = 9 fish per control/experimental group). Statistical analyses were performed with a one-way ANOVA followed by Tukey's multiple comparison post hoc test, # *p* < 0.05, ## *p* < 0.01.

4. Discussion

Smo agonists have been demonstrated to play intricate roles in regenerative therapies [37,55] and purmorphamine has been shown to provide neuroprotective effects and greatly influence recovery of cognitive performance in rodents following neurological insult and aged postoperative cognitive dysfunction [22,37,56]. We examined the regulation of TBI-induced excitotoxicity through the mechanistic link between Shh signaling and Eaat2a. We demonstrated that inhibiting the Shh response following TBI exacerbated PTS, while preconditioning with purmorphamine significantly reduced multiple facets of injury-induced pathologies, rapidly increased cognitive recovery, and in some instances attenuated the deficit altogether. To our knowledge, the use of purmorphamine in zebrafish following any mechanism of TBI has not been examined. Although several studies have demonstrated the inhibition of cell proliferation in the zebrafish brain following cyclopamine [57–59], we could not find studies illuminating the persistence of injury-induced pathologies or cognitive deficits associated with the application of cyclopamine.

The need to further examine TBI-induced PTS and identify potential therapeutics has become more and more imperative. Over the past decade, the number of ASMs developed has increased, however these therapeutics largely rely on a similar mechanism of action and drug-resistant epilepsy has increased among patients with PTS [10,11]. Traditional first line ASMs are GABA agonists, and rodent models revealed refractory PTS to these ASMs [60,61]. Cho et al. [62] recently examined Valproate (valproic acid, VPA) and other first-line ASMs (Carbamazepine, and Phenytoin) in their adult zebrafish TBI-PTS model [15] and observed no significant reduction in seizure activity and persistent cognitive deficits. We recognized the potential difference in mechanism in inducing TBI and tested both first and second-line ASMs in our TBI model. We similarly observed no improvement in TBI-induced seizures, edema, and cognitive deficits (data not shown). In contrast, our findings that blocking glutamate receptors or increasing expression of the glutamate transporter Eaat2a reduced TBI-induced PTS supported similar observations of EAAT2 reducing PTS in other models [63–65].

Modulation of Shh signaling has been considered a potential therapeutic for various ailments, though much of this work focused on Shh inhibition and use of Smo antagonists as potential cancer targets [66,67]. ERIVEDGE®, the only current FDA-approved Smo antagonist [68], made it through clinical evaluation and is approved for advanced basal cell carcinoma. The mechanistic consequences of Shh activation are less known and long-term Shh activation may stimulate tumor growth [69]. Furthermore, the results of Shh signaling activators have been inconsistent. For example, in epileptic rodents, Feng et al. [70] reported that increased Shh expression exacerbated seizure activity and negatively regulated extracellular glutamate, while others reported data suggesting that Shh activity upregulated EAAT2 expression, reduced astrocyte reactivity, and combated excitotoxicity [45,46,71,72]. Other studies reported that Shh activation is neuroprotective [23,37–39] and is effective in regenerative therapies [35,73,74]. However, much work needs to be done to provide insights into the mechanistic actions contributing these positive neurological outcomes. Although our study is far from exhaustive, it suggests that low and short doses of prophylactic purmorphamine may provide a neuroprotective effect against TBI-induced PTS. This therapeutic protocol may be amendable to targeted populations at higher risk of TBI and may provide protection beyond current safety measures, such as helmets.

Author Contributions: Conceptualization, D.R.H. and J.H.; Methodology, D.R.H., J.H., M.L. and W.B.; Validation, D.R.H. and J.H.; Formal Analysis, J.H.; Investigation, J.H. and M.L.; Writing—Original Draft Preparation, D.R.H. and J.H.; Writing—Review and Editing, D.R.H., J.H., L.J.C., K.C., M.L. and W.B.; Visualization, D.R.H. and J.H.; Supervision, D.R.H. and W.B.; Project Administration, D.R.H.; Funding Acquisition, D.R.H. and J.H. All authors have read and agreed to the published version of the manuscript.

Funding: This project was funded, in part, with support from the Indiana Spinal Cord & Brain In-jury Research Fund from the Indiana State Department of Health (#EPAR1723) to D.R.H. Its contents are solely the responsibility of the providers and do not necessarily represent the official views of the Indiana State Department of Health. This work was also supported by grants from the National Science Foundation Graduate Research Fellowship Program #DGE-1841556 (JTH), LTC Neil Hyland Fellowship of Notre Dame (JTH), Sentinels of Freedom Fellowship (JTH), Pat Tillman Scholarship (JTH), the Center for Zebrafish Research at the University of Notre Dame, and the Center for Stem Cells and Regenerative Medicine at the University of Notre Dame.

Institutional Review Board Statement: This study was approved and conducted according to the guidelines of the University of Notre Dame Institutional Animal Care and Use Committee (IACUC Protocol # 21-02-6445, most recent approval 22 March 2021).

Informed Consent Statement: Not applicable.

Data Availability Statement: Data are contained within this article.

Acknowledgments: The authors would like to thank the Hyde lab members for their thoughtful discussions, the Neuhauss lab for their generous gift of Eaat2a antibody, and the Freimann Life Sciences Center technicians for zebrafish care and husbandry. Figure 1 was made with BioRender.com (accessed on 2 November 2021).

Conflicts of Interest: The authors declare no conflict of interest.

References

1. Kushner, D. Mild traumatic brain injury: Toward understanding manifestations and treatment. *Arch. Intern. Med.* **1998**, *158*, 1617–1624. [CrossRef]
2. Rabinowitz, A.R.; Levin, H.S. Cognitive sequelae of traumatic brain injury. *Psychiatr. Clin. N. Am.* **2014**, *37*, 1–11. [CrossRef] [PubMed]
3. Campbell, J.N.; Gandhi, A.; Singh, B.; Churn, S.B. Traumatic brain injury causes a tacrolimus-sensitive increase in non-convulsive seizures in a rat model of post-traumatic epilepsy. *Int. J. Neurol. Brain Disord.* **2014**, *1*, 1–11. [PubMed]
4. Rumalla, K.; Smith, K.A.; Letchuman, V.; Gandham, M.; Kombathula, R.; Arnold, P.M. Nationwide incidence and risk factors for posttraumatic seizures in children with traumatic brain injury. *J. Neurosurg. Pediatr.* **2018**, *22*, 684–693. [CrossRef] [PubMed]
5. Annegers, J.F.; Hauser, W.A.; Coan, S.P.; Rocca, W.A. A population-based study of seizures after traumatic brain injuries. *N. Engl. J. Med.* **1998**, *1338*, 20–24. [CrossRef] [PubMed]

6.   Frey, L.C. Epidemiology of posttraumatic epilepsy: A critical review. *Epilepsia.* **2003**, *44* (Suppl. 10), 11–17. [CrossRef]
7.   Lowenstein, D.H. Epilepsy after head injury: An overview. *Epilepsia.* **2009**, *50* (Suppl. 2), 4–9. [CrossRef] [PubMed]
8.   Lucke-Wold, B.P.; Nguyen, L.; Turner, R.C.; Logsdon, A.F.; Chen, Y.W.; Smith, K.E.; Huber, J.D.; Matsumoto, R.; Rosen, ⌐
      Tucker, E.S.; et al. Traumatic brain injury and epilepsy: Underlying mechanisms leading to seizure. *Seizure* **2015**, *33*, 1
      [CrossRef]
9.   Farber, N.B.; Jiang, X.P.; Heinkel, C.; Nemmers, B. Antiepileptic drugs and agents that inhibit voltage-gated sodium chanr
      prevent NMDA antagonist neurotoxicity. *Mol. Psychiatry* **2002**, *7*, 726–733. [CrossRef]
10.   Larkin, M.; Meyer, R.M.; Szuflita, N.S.; Severson, M.A.; Levine, Z.T. Post-traumatic, drug-resistant epilepsy and review of seizu
      control outcomes from blinded, randomized controlled trials of brain stimulation treatments for drug-resistant epilepsy. *Cure.*
      2016, *8*, e744.
11.   Sheng, J.; Liu, S.; Qin, H.; Li, B.; Zhang, X. Drug-resistant epilepsy and surgery. *Curr. Neuropharmacol.* **2018**, *16*, 17–28. [CrossRe
      [PubMed]
12.   Tehse, J.; Taghibiglou, C. The overlooked aspect of excitotoxicity: Glutamate-independent excitotoxicity in traumatic brai
      injuries. *Eur. J. Neurosci.* **2019**, *49*, 1157–1170. [CrossRef]
13.   Skaggs, K.; Goldman, D.; Parent, J.M. Excitotoxic brain injury in adult zebrafish stimulates neurogenesis and long-distance
      neuronal integration. *Glia* **2014**, *62*, 2061–2079. [CrossRef] [PubMed]
14.   McCutcheon, V.; Park, E.; Liu, E.; Sobhebidari, P.; Tavakkoli, J.; Wen, X.Y.; Baker, A.J. A novel model of traumatic brain injury ir
      adult zebrafish demonstrates response to injury and treatment comparable with mammalian models. *J. Neurotrauma* **2017**, *3*
      1382–1393. [CrossRef]
15.   Cho, S.J.; Park, E.; Telliyan, T.; Baker, A.; Reid, A.Y. Zebrafish model of posttraumatic epilepsy. *Epilepsia* **2020**, *61*, 1774–178'
      [CrossRef]
16.   Alyenbaawi, H.; Kanyo, R.; Locskai, L.F.; Kamali-Jamil, R.; DuVal, M.G.; Bai, Q.; Wille, H.; Burton, E.A.; Allison, W.T. Seizures ar
      a druggable mechanistic link between TBI and subsequent tauopathy. *Elife* **2021**, *10*, e58744. [CrossRef]
17.   Hentig, J.; Cloghessy, K.; Lahne, M.; Jung, Y.J.; Petersen, R.A.; Morris, A.C.; Hyde, D.R. Zebrafish blunt-force TBI induces
      heterogenous injury pathologies that mimic human TBI and responds with sonic hedgehog-dependent cell proliferation across
      the neuroaxis. *Biomedicines* **2021**, *9*, 861. [CrossRef]
18.   Kalueff, A.V.; Gebhardt, M.; Stewart, A.M.; Cachat, J.M.; Brimmer, M.; Chawla, J.S.; Craddock, C.; Kyzar, E.J.; Roth, A.; Landsman,
      S.; et al. Towards a comprehensive catalog of zebrafish behavior 1.0 and beyond. *Zebrafish* **2013**, *10*, 70–86. [CrossRef]
19.   Hu, S.; Sheng, W.S.; Ehrlich, L.C.; Peterson, P.K.; Chao, C.C. Cytokine effects on glutamate uptake by human astrocytes.
      Neuroimmunomodulation **2000**, *7*, 153–159. [CrossRef]
20.   Fogal, B.; Hewett, S.J. Interleukin-1beta: A bridge between inflammation and excitotoxicity? *J. Neurochem.* **2008**, *106*, 1–23.
      [CrossRef] [PubMed]
21.   Hotz, A.L.; Jamali, A.; Rieser, N.N.; Niklaus, S.; Aydin, E.; Myren-Svelstad, S.; Lalla, L.; Jurisch-Yaksi, N.; Yaksi, E.; Neuhauss, S.C.F.
      Loss of glutamate transporter *eaat2a* leads to aberrant neuronal excitability, recurrent epileptic seizures, and basal hypoactivity.
      Glia **2021**, *70*, 196–214. [CrossRef] [PubMed]
22.   Yin, S.; Bai, X.; Xin, D.; Li, T.; Chu, X.; Ke, H.; Han, M.; Chen, W.; Li, X.; Wang, Z. Neuroprotective effects of the sonic hedgehog
      signaling pathway in ischemic injury through promotion of synaptic and neuronal health. *Neural Plast.* **2020**, *2020*, 8815195.
      [CrossRef] [PubMed]
23.   Rahi, S.; Gupta, R.; Sharma, A.; Mehan, S. Smo-Shh signaling activator purmorphamine ameliorates neurobehavioral, molecular,
      and morphological alterations in an intracerebroventricular propionic acid-induced experimental model of autism. *Hum. Exp.*
      Toxicol. **2021**, *40*, 1880–1898. [CrossRef]
24.   Tsetskhladze, Z.R.; Canfield, V.A.; Ang, K.C.; Wentzel, S.M.; Reid, K.P.; Berg, A.S.; Johnson, S.L.; Kawakami, K.; Cheng, K.C.
      Functional assessment of human coding mutations affecting skin pigmentation using zebrafish. *PLoS ONE* **2012**, *7*, e47398.
      [CrossRef]
25.   Machnik, P.; Schirmer, E.; Glück, L.; Schuster, S. Recordings in an integrating central neuron provide a quick way for identifying
      appropriate anesthetic use in fish. *Sci. Rep.* **2018**, *8*, 17541. [CrossRef]
26.   Hentig, J.; Cloghessy, K.; Dunseath, C.; Hyde, D.R. A scalable model to study the effects of blunt-force injury in adult zebrafish.
      J. Vis. Exp. **2021**, *31*, 171. [CrossRef]
27.   Hoshi, Y.; Okabe, K.; Shibasaki, K.; Funatsu, T.; Matsuki, N.; Ikegaya, Y.; Koyama, R. Ischemic brain injury leads to brain edema
      via hyperthermia-induced TRPV4 activation. *J. Neurosci.* **2018**, *38*, 5700–5709. [CrossRef] [PubMed]
28.   Hentig, J.; Cloghessy, K.; Hyde, D.R. Shuttle box assay as an associative learning tool for cognitive assessment in learning and
      memory studies using adult zebrafish. *J. Vis. Exp.* **2021**, *173*, e62745. [CrossRef]
29.   Campbell, L.J.; Hobgood, J.S.; Jia, M.; Boyd, P.; Hipp, R.I.; Hyde, D.R. Notch3 and DeltaB maintain Müller glia quiescence and act
      as negative regulators of regeneration in the light-damaged zebrafish retina. *Glia* **2021**, *69*, 546–566. [CrossRef]
30.   Kassen, S.C.; Ramanan, V.; Montgomery, J.E.; TBurket, C.; Liu, C.G.; Vihtelic, T.S.; Hyde, D.R. Time course analysis of gene
      expression during light-induced photoreceptor cell death and regeneration in albino zebrafish. *Dev. Neurobiol.* **2007**, *67*, 1009–1031.
      [CrossRef]
31.   Puppala, D.; Maaswinkel, H.; Mason, B.; Legan, S.J.; Li, L. An in vivo microdialysis study of light/dark-modulation of vitreal
      dopamine release in zebrafish. *J. Neurocytol.* **2004**, *33*, 193–201. [CrossRef] [PubMed]

32. Nelson, C.M.; Gorsuch, R.A.; Bailey, T.J.; Ackerman, K.M.; Kassen, S.C.; Hyde, D.R. Stat3 defines three populations of Müller glia and is required for initiating maximal müller glia proliferation in the regenerating zebrafish retina. *J. Comp. Neurol.* **2012**, *520*, 4294–4311. [CrossRef]

33. Niklaus, S.; Cadetti, L.; Vom Berg-Maurer, C.M.; Lehnherr, A.; Hotz, A.L.; Forster, I.C.; Gesemann, M.; Neuhauss, S.C.F. Shaping of signal transmission at the photoreceptor synapse by EAAT2 gluta EAAT2 Glutamate Transporters. *eNeuro* **2017**, *4*, ENEURO.0339-16.2017. [CrossRef]

34. Sirko, S.; Behrendt, G.; Johansson, P.A.; Tripathi, P.; Costa, M.; Bek, S.; Heinrich, C.; Tiedt, S.; Colak, D.; Dichgans, M.; et al. Reactive glia in the injured brain acquire stem cell properties in response to sonic hedgehog. *Cell Stem Cell.* **2013**, *12*, 426–439. [CrossRef] [PubMed]

35. Thomas, J.L.; Morgan, G.W.; Dolinski, K.M.; Thummel, R. Characterization of the pleiotropic roles of Sonic Hedgehog during retinal regeneration in adult zebrafish. *Exp. Eye Res.* **2018**, *166*, 106–115. [CrossRef] [PubMed]

36. Amankulor, N.M.; Hambardzumyan, D.; Pyonteck, S.M.; Becher, O.J.; Joyce, J.A.; Holland, E.C. Sonic hedgehog pathway activation is induced by acute brain injury and regulated by injury-related inflammation. *J. Neurosci.* **2009**, *29*, 10299–10308. [CrossRef] [PubMed]

37. Chechneva, O.V.; Mayrhofer, F.; Daugherty, D.J.; Krishnamurty, R.G.; Bannerman, P.; Pleasure, D.E.; Deng, W. A Smoothened receptor agonist is neuroprotective and promotes regeneration after ischemic brain injury. *Cell Death Dis.* **2014**, *5*, e1481. [CrossRef]

38. Hu, Q.; Li, T.; Wang, L.; Xie, Y.; Liu, S.; Bai, X.; Zhang, T.; Bo, S.; Xin, D.; Xue, H.; et al. Neuroprotective effects of a Smoothened receptor agonist against early brain injury after experimental subarachnoid hemorrhage in rats. *Front. Cell Neurosci.* **2017**, *10*, 306. [CrossRef]

39. Liu, D.; Bai, X.; Ma, W.; Xin, D.; Chu, X.; Yuan, H.; Qiu, J.; Ke, H.; Yin, S.; Chen, W.; et al. Purmorphamine attenuates neuro-inflammation and synaptic impairments after hypoxic-ischemic injury in neonatal mice via Shh signaling. *Front. Pharmacol.* **2020**, *11*, 204. [CrossRef]

40. Scharfman, H.E. The neurobiology of epilepsy. *Curr. Neurol. Neurosci. Rep.* **2007**, *7*, 348–354. [CrossRef]

41. Briggs, S.W.; Galanopoulou, A.S. Altered GABA signaling in early life epilepsies. *Neural Plast.* **2011**, *2011*, 527605. [CrossRef]

42. Barker-Haliski, M.; White, H.S. Glutamatergic mechanisms associated with seizures and epilepsy. *Cold Spring Harb. Perspect. Med.* **2015**, *5*, a022863. [CrossRef]

43. Guerriero, R.M.; Giza, C.C.; Rotenberg, A. Glutamate and GABA imbalance following traumatic brain injury. *Curr. Neurol. Neurosci. Rep.* **2015**, *15*, 27. [CrossRef]

44. Löscher, W.; Nau, H. Valproic acid: Metabolite concentrations in plasma and brain, anticonvulsant activity, and effects on GABA metabolism during subacute treatment in mice. *Arch. Int. Pharmacodyn. Ther.* **1982**, *257*, 20–31.

45. Zhang, D.; Wang, H.; Liu, H.; Tao, T.; Wang, N.; Shen, A. nNOS translocates into the nucleus and interacts with Sox2 to protect neurons against early excitotoxicity via promotion of Shh transcription. *Mol. Neurobiol.* **2016**, *53*, 6444–6458. [CrossRef] [PubMed]

46. Yang, C.; Qi, Y.; Sun, Z. The role of sonic hedgehog pathway in the development of the central nervous system and aging-related neurodegenerative diseases. *Front. Mol. Biosci.* **2021**, *8*, 711710. [CrossRef] [PubMed]

47. Gesemann, M.; Lesslauer, A.; Maurer, C.M.; Schönthaler, H.B.; Neuhauss, S.C. Phylogenetic analysis of the vertebrate excita-tory/neutral amino acid transporter (SLC1/EAAT) family reveals lineage specific subfamilies. *BMC Evol. Biol.* **2010**, *10*, 117. [CrossRef] [PubMed]

48. Lee, S.G.; Su, Z.Z.; Emdad, L.; Gupta, P.; Sarkar, D.; Borjabad, A.; Volsky, D.J.; Fisher, P.B. Mechanism of ceftriaxone induction of excitatory amino acid transporter-2 expression and glutamate uptake in primary human astrocytes. *J. Biol. Chem.* **2008**, *283*, 13116–13123. [CrossRef]

49. Agostini, J.F.; Costa, N.L.F.; Bernardo, H.T.; Baldin, S.L.; Mendes, N.V.; de Pieri Pickler, K.; Manenti, M.C.; Rico, E.P. Ceftriaxone attenuated anxiety-like behavior and enhanced brain glutamate transport in zebrafish subjected to alcohol withdrawal. *Neurochem. Res.* **2020**, *45*, 1526–1535. [CrossRef]

50. Sharma, R.; Leung, W.L.; Zamani, A.; O'Brien, T.J.; Casillas Espinosa, P.M.; Semple, B.D. Neuroinflammation in post-traumatic epilepsy: Pathophysiology and tractable therapeutic targets. *Brain Sci.* **2019**, *9*, 318. [CrossRef]

51. Belov Kirdajova, D.; Kriska, J.; Tureckova, J.; Anderova, M. Ischemia-triggered glutamate excitotoxicity from the perspective of glial cells. *Front. Cell. Neurosci.* **2020**, *14*, 51. [CrossRef] [PubMed]

52. Wang, Y.; Jin, S.; Sonobe, Y.; Cheng, Y.; Horiuchi, H.; Parajuli, B.; Kawanokuchi, J.; Mizuno, T.; Takeuchi, H.; Suzumura, A. Interleukin-1β induces blood-brain barrier disruption by downregulating Sonic hedgehog in astrocytes. *PLoS ONE* **2014**, *9*, e110024. [CrossRef] [PubMed]

53. Murray, K.N.; Parry-Jones, A.R.; Allan, S.M. Interleukin-1 and acute brain injury. *Front. Cell. Neurosci.* **2015**, *9*, 18. [CrossRef]

54. Hasegawa, T.; Hall, C.J.; Crosier, P.S.; Abe, G.; Kawakami, K.; Kudo, A.; Kawakami, A. Transient inflammatory response mediated by interleukin-1β is required for proper regeneration in zebrafish fin fold. *Elife* **2017**, *6*, e22716. [CrossRef]

55. Wang, J.; Lu, J.; Bond, M.C.; Chen, M.; Ren, X.R.; Lyerly, H.K.; Barak, L.S.; Chen, W. Identification of select glucocorticoids as Smoothened agonists: Potential utility for regenerative medicine. *Proc. Natl. Acad. Sci. USA* **2010**, *107*, 9323–9328. [CrossRef] [PubMed]

56. Shao, S.; Wang, G.L.; Raymond, C.; Deng, X.H.; Zhu, X.L.; Wang, D.; Hong, L.P. Activation of sonic hedgehog signal by purmorphamine, in a mouse model of Parkinson's disease, protects dopaminergic neurons and attenuates inflammatory response by mediating PI3K/AKt signaling pathway. *Mol. Med. Rep.* **2017**, *16*, 1269–1277. [CrossRef] [PubMed]

57. Reimer, M.M.; Kuscha, V.; Wyatt, C.; Sörensen, I.; Frank, R.E.; Knüwer, M.; Becker, T.; Becker, C.G. Sonic hedgehog is a polarized signal for motor neuron regeneration in adult zebrafish. *J. Neurosci.* **2009**, *29*, 15073–15082. [CrossRef] [PubMed]

58. Wullimann, M.F.; Umeasalugo, K.E. Sonic hedgehog expression in zebrafish forebrain identifies the teleostean pallidal signaling center and shows preglomerular complex and posterior tubercular dopamine cells to arise from shh cells. *J. Comp. Neurol.* **2020**, *528*, 1321–1348. [CrossRef] [PubMed]

59. Male, I.; Ozacar, A.T.; Fagan, R.R.; Loring, M.D.; Shen, M.C.; Pace, V.A.; Devine, C.A.; Lawson, G.E.; Lutservitz, A.; Karlstrom, R.O. Hedgehog signaling regulates neurogenesis in the larval and adult zebrafish hypothalamus. *eNeuro* **2020**, *7*, ENEURO.0226-20.2020. [CrossRef]

60. Riban, V.; Bouilleret, V.; Pham-Lê, B.T.; Fritschy, J.M.; Marescaux, C.; Depaulis, A. Evolution of hippocampal epileptic activity during the development of hippocampal sclerosis in a mouse model of temporal lobe epilepsy. *Neuroscience* **2002**, *112*, 101–111. [CrossRef]

61. Maroso, M.; Balosso, S.; Ravizza, T.; Iori, V.; Wright, C.I.; French, J.; Vezzani, A. Interleukin-1β biosynthesis inhibition reduces acute seizures and drug resistant chronic epileptic activity in mice. *Neurotherapeutics* **2011**, *8*, 304–315. [CrossRef]

62. Cho, S.J.; Park, E.; Baker, A.; Reid, A.Y. Post-traumatic epilepsy in zebrafish is drug-resistant and impairs cognitive function. *J. Neurotrauma* **2021**, *38*, 3174–3183. [CrossRef] [PubMed]

63. Kong, Q.; Chang, L.C.; Takahashi, K.; Liu, Q.; Schulte, D.A.; Lai, L.; Ibabao, B.; Lin, Y.; Stouffer, N.; Das Mukhopadhyay, C.; et al. Small-molecule activator of glutamate transporter EAAT2 translation provides neuroprotection. *J. Clin. Investig.* **2014**, *124*, 1255–1267. [CrossRef] [PubMed]

64. Young, D.; Fong, D.M.; Lawlor, P.A.; Wu, A.; Mouravlev, A.; McRae, M.; Glass, M.; Dragunow, M.; During, M.J. Adenosine kinase, glutamine synthetase and EAAT2 as gene therapy targets for temporal lobe epilepsy. *Gene Ther.* **2014**, *21*, 1029–1040. [CrossRef] [PubMed]

65. Yang, Z.; Wang, J.; Yu, C.; Xu, P.; Zhang, J.; Peng, Y.; Luo, Z.; Huang, H.; Zeng, J.; Xu, Z. Inhibition of p38 MAPK signaling regulates the expression of EAAT2 in the brains of epileptic rats. *Front. Neurol.* **2018**, *9*, 925. [CrossRef]

66. Berman, D.M.; Karhadkar, S.S.; Hallahan, A.R.; Pritchard, J.I.; Eberhart, C.G.; Watkins, D.N.; Chen, J.K.; Cooper, M.K.; Taipale, J.; Olson, J.M.; et al. Medulloblastoma growth inhibition by hedgehog pathway blockade. *Science* **2002**, *297*, 1559–1561. [CrossRef]

67. Watkins, D.N.; Berman, D.M.; Burkholder, S.G.; Wang, B.; Beachy, P.A.; Baylin, S.B. Hedgehog signalling within airway epithelial progenitors and in small-cell lung cancer. *Nature* **2003**, *422*, 313–317. [CrossRef]

68. Axelson, M.; Liu, K.; Jiang, X.; He, K.; Wang, J.; Zhao, H.; Kufrin, D.; Palmby, T.; Dong, Z.; Russell, A.M.; et al. Food and Drug Administration approval: Vismodegib for recurrent, locally advanced, or metastatic basal cell carcinoma. *Clin. Cancer Res.* **2013**, *19*, 2289–2293. [CrossRef]

69. Montagnani, V.; Stecca, B. Role of protein kinases in hedgehog pathway control and implications for cancer therapy. *Cancers* **2019**, *11*, 449. [CrossRef]

70. Feng, S.; Ma, S.; Jia, C.; Su, Y.; Yang, S.; Zhou, K.; Liu, Y.; Cheng, J.; Lu, D.; Fan, L.; et al. Sonic hedgehog is a regulator of extracellular glutamate levels and epilepsy. *EMBO Rep.* **2016**, *17*, 682–694. [CrossRef]

71. Ugbode, C.I.; Smith, I.; Whalley, B.J.; Hirst, W.D.; Rattray, M. Sonic hedgehog signaling mediates astrocyte crosstalk with neurons to confer neuroprotection. *J. Neurochem.* **2017**, *142*, 429–443. [CrossRef] [PubMed]

72. Gonzalez-Reyes, L.E.; Chiang, C.C.; Zhang, M.; Johnson, J.; Arrillaga-Tamez, M.; Couturier, N.H.; Reddy, N.; Starikov, L.; Capadona, J.R.; Kottmann, A.H.; et al. Sonic Hedgehog is expressed by hilar mossy cells and regulates cellular survival and neurogenesis in the adult hippocampus. *Sci. Rep.* **2019**, *9*, 17402. [CrossRef] [PubMed]

73. Jin, Y.; Raviv, N.; Barnett, A.; Bambakidis, N.C.; Filichia, E.; Luo, Y. The shh signaling pathway is upregulated in multiple cell types in cortical ischemia and influences the outcome of stroke in an animal model. *PLoS ONE* **2015**, *10*, e0124657. [CrossRef] [PubMed]

74. Yamada, Y.; Ohazama, A.; Maeda, T.; Seo, K. The Sonic Hedgehog signaling pathway regulates inferior alveolar nerve regeneration. *Neurosci. Lett.* **2018**, *671*, 114–119. [CrossRef]

 *biomedicines*

Review

Pharmacological and Therapeutic Approaches in the Treatment of Epilepsy

Shampa Ghosh [1], Jitendra Kumar Sinha [2,*], Tarab Khan [2], Kuramkote Shivanna Devaraju [3], Prabhakar Singh [4], Kumar Vaibhav [5] and Pankaj Gaur [6,*]

[1] ICMR-National Institute of Nutrition (NIN), Tarnaka, Hyderabad 500007, India; g.shampa17@gmail.com
[2] Amity Institute of Neuropsychology and Neurosciences (AINN), Amity University UP, Noida 201303, India; tarabkhan55@gmail.com
[3] Department of Biochemistry, Karnatak University, Dharwad 580003, India; ksdevaraju@kud.ac.in
[4] Department of Anatomy, All India Institute of Medical Sciences (AIIMS), Ansari Nagar, New Delhi 110029, India; prabhakar.singh@aiims.edu
[5] Department of Neurosurgery, Medical College of Georgia, Augusta University, Augusta, GA 30912, USA; kvaibhav@augusta.edu
[6] Lombardi Comprehensive Cancer Center, Georgetown University Medical Center, Washington, DC 20007, USA
* Correspondence: jksinha@amity.edu (J.K.S.); pg684@georgetown.edu (P.G.); Tel.: +91-120-4392971 (J.K.S.); +1-(706)-589-8628 (P.G.)

Citation: Ghosh, S.; Sinha, J.K.; Khan, T.; Devaraju, K.S.; Singh, P.; Vaibhav, K.; Gaur, P. Pharmacological and Therapeutic Approaches in the Treatment of Epilepsy. *Biomedicines* **2021**, *9*, 470. https://doi.org/10.3390/biomedicines9050470

Academic Editor: Jacopo Meldolesi

Received: 4 April 2021
Accepted: 21 April 2021
Published: 25 April 2021

Publisher's Note: MDPI stays neutral with regard to jurisdictional claims in published maps and institutional affiliations.

Copyright: © 2021 by the authors. Licensee MDPI, Basel, Switzerland. This article is an open access article distributed under the terms and conditions of the Creative Commons Attribution (CC BY) license (https://creativecommons.org/licenses/by/4.0/).

Abstract: Epilepsy affects around 50 million people across the globe and is the third most common chronic brain disorder. It is a non-communicable disease of the brain that affects people of all ages. It is accompanied by depression, anxiety, and substantially increased morbidity and mortality. A large number of third-generation anti-epileptic drugs are available, but they have multiple side-effects causing a decline in the quality of life. The inheritance and etiology of epilepsy are complex with multiple underlying genetic and epigenetic mechanisms. Different neurotransmitters play intricate functions to maintain the normal physiology of various neurons. If there is any dysregulation of neurotransmission due to aberrant transmitter levels or their receptor biology, it can result in seizures. In this review, we have discussed the roles played by various neurotransmitters and their receptors in the pathophysiology of epilepsy. Drug-resistant epilepsy (DRE) has remained one of the forefront areas of epilepsy research for a long time. Understanding the mechanisms underlying DRE is of utmost importance because of its high incidence rate among epilepsy patients and increased risks of psychosocial problems and premature death. Here we have enumerated various hypotheses of DRE. Further, we have discussed different non-conventional therapeutic strategies, including combination therapy and non-drug treatment. The recent studies supporting the modern approaches for the treatment of epilepsy have been deliberated with particular reference to the mTOR pathway, breakdown of the blood-brain barrier, and inflammatory pathways.

Keywords: anti-convulsants; anti-epileptic drugs; drug targets; epileptogenesis; non-communicable disease; seizures; transcriptional modifications; pseudo-resistance

1. Introduction

Epilepsy is a group of chronic non-communicable neurological disorders categorized by spontaneous recurrent seizures [1,2]. These seizures result from episodes of abnormal electrical activity in the brain. The process by which epilepsy develops in an otherwise normal brain is called epileptogenesis. Epilepsy may result from a head injury, brain tumors, brain infections like meningitis or encephalitis, stroke, birth defects, and sometimes even altered levels of entities like blood sugar or sodium [3]. It is the third most common chronic neurological disorder. It is a life-shortening brain disorder that affects around 50 million people or 1% of the world population. Although it is found across the globe, 80% of epileptic patients live in low- and middle-income countries. Epilepsy is characterized by

15 different types of seizures and 30 types of epilepsy syndromes and is accompanied by substantial comorbidity, depression, increased mortality, and anxiety [4,5].

During the last 30 years, there has been a huge advancement in the treatment of many types of seizures due to the introduction of over 15 third-generation anti-epileptic drugs (AEDs) [6]. Around 70–80% of patients enter remission with present AEDs who have new onset of epilepsy. Among 20–30% of patients, these medications are not able to control seizures. Besides, there is no AED that can prevent the development of the disease before the occurrence of the first seizure. Unfortunately, there is also drug-resistant epilepsy that is not controlled by or responds to AEDs. This shows the urgent need to develop appropriate therapeutic strategies to tackle the complex situation of epilepsy. Devising better treatment paradigms for epilepsy using various pharmaceutical and therapeutic approaches would need a better understanding of the different clinical and experimental strategies for the development and discovery of more efficient treatment methods that can help prevent and control the diseases [7–9].

2. Role of Genes, Genetics and Inheritance

The newly emerging genetic technology has played a significant role in the discovery of a variety of genes that are associated with epilepsy. A list of the few genes related to epilepsy is summarized in Table 1. The advancement of genomic techniques and gene sequencing has substantially enhanced the knowledge about the genetic variations taking place in the human genome. Studies estimate that there is an underlying genetic cause in about half of all cases of epilepsy [10]. Currently, with the emerging research of epigenetic biomarkers, MicroRNAs (miRNAs) have been assumed to play a significant role. MiRNA molecules 19-25 nucleotides long regulate gene expression as post-transcriptional modifiers [11]. Differential expression of more than 100 miRNAs has been reported in epilepsy. Among them, because of their predominant role in biological processes related to epilepsy, such as neurodegeneration, neuronal growth, and neuroinflammation, miR-132, miR-155, and miR-146a have been highlighted primarily [11]. Some cases of genetic mutations result in the core symptom of epilepsy, while changes in a few of the genes are responsible for malformations in the gross development of the brain that cause seizures. There are around 84 genes classified as epilepsy genes based on the OMIM database results. Mutation in these classified genes leads to epilepsy as a core symptom. There are approximately 73 genes that are categorized as neurodevelopment-associated epilepsy genes [10]. Twenty-four genetic variants have been identified by two large genome-wide association studies that are linked with epilepsy [12]. The defect in the *FMR1* gene is responsible for causing Fragile X Syndrome. Fragile X syndrome is one of the abnormalities associated with epilepsy and is characterized by intellectual dysfunction and abnormal behavior. According to the OMIM database (https://www.ncbi.nlm.nih.gov/omim, accessed on 24 April 2021), there are about 536 genes responsible for causing associated diseases of epilepsy [10,13].

Table 1. List of genes associated with the pathophysiology of epilepsy. The pattern of inheritance, age of onset, and functional categories has been given in concurrence with the associated genes.

S.No.	Epilepsy Genes	Functional Category	Pattern of Inheritance	Type Of Syndrome	S.No.
1.	*ALDH7A1*	Enzyme	Autosomal recessive	Pyridoxine dependent epilepsy.	Neonatal period
2.	*KCNQ2*	Potassium channel	Autosomal dominant	Benign familial neonatal seizures.	
3.	*GABRA1*	Receptor of GABA A	Autosomal dominant	Early infantile epileptic encephalopathy.	
4.	*SCN8A, SCN2A*	Sodium channel	Autosomal dominant	Benign familial neonatal seizures. Early infantile epileptic encephalopathy.	
5.	*CHD2*	Enzyme	Autosomal dominant	Childhood onset epileptic encephalopathy.	
6.	*STX1B*	Transport across membrane	Autosomal dominant	Generalized epilepsy.	
7.	*TBC1D24*	Modulator of enzyme	Autosomal recessive	Familial infantile myoclinic epilepsy. Early infantile epileptic encephalopathy.	
8.	*NECAP1*	Not classified	Autosomal recessive	Early infantile epileptic encephalopathy.	
9.	*UBA5, GNAO1*	Enzyme	Not known	Early infantile epileptic encephalopathy.	
	HCN1	HCN channel			
10.	*GPR98*	Receptor	Autosomal dominant	Familial febrile seizures.	Infancy and early childhood
11.	*KCNMA1*	Potassium channel	Autosomal dominant	Generalized epilepsy with paroxysmal dyskinesia.	
12.	*STRGAL3, WWOX*	Enzyme	Autosomal recessive	Early infantile epileptic encephalopathy.	
13.	*PRRT2*	Not classified	Autosomal dominant	Benign familial infantile seizures.	
14.	*SLC6A1*	Transporter	Autosomal dominant	Myoclinic atonic epilepsy.	
15.	*ARHGEF9*	Modulator of enzyme	X linked recessive	Early infantile epileptic encephalopathy.	
16.	*SCN9A*	Sodium channel	Autosomal dominant	Dravet Syndrome. Familial febrile seizures.	
17.	*CDKL5*	Enzyme	X linked dominant	Early infantile epileptic encephalopathy.	
18.	*GRIN2A*	NMDA receptor	Autosomal dominant	Focal epilepsy with speech disorder.	
19.	*STRGAL5*	Enzyme	Autosomal recessive	Amish infantile epilepsy.	
20.	*CACNA1H*	Calcium channel	Not known	Childhood absence epilepsy.Idiopathic generalized epilepsy.	
21.	*ALG13*	Enzyme	X linked	Early infantile epileptic encephalopathy.	

Table 1. *Cont.*

S.No.	Epilepsy Genes	Functional Category	Pattern of Inheritance	Type Of Syndrome	S.No.
22.	*CPA6*	Enzyme	Autosomal dominant	Familial temporal lobe epilepsy.	Juvenile phase and later
	LGI1	Not classified			
23.	*CACNB4*	Calcium channel	Autosomal dominant	Juvenile myoclinic epilepsy. Idiopathic generalized epilepsy.	
24.	*EFHC1*	Signalling molecule	Autosomal dominant	Juvenile absence epilepsy. Juvenile myoclinic epilepsy.	
25.	*CLCN2*	Chloride channel	Autosomal dominant	Juvenile generalized epilepsy. Juvenile absence epilepsy. Juvenile myoclinic epilepsy.	
26.	*ADRA2B*	Receptor	Autosomal dominant	Familial adult myoclinic epilepsy.	
27.	*GABRD*	Receptor of GABA A	Autosomal dominant	Generalized epilepsy with febrile seizures. Juvenile myoclinic epilepsy.	
28.	*CASR*	Receptor	Not known	Idiopathic generalized epilepsy.	
29.	*DEPDC5*	Not classified	Autosomal dominant	Familial focal epilepsy.	Unspecified
30.	*CHRNB2*	Acetylcholine receptor	Unknown	Nocturnal frontal lobe epilepsy.	
31.	*KCNC1*	Potassium channel	Autosomal dominant	Progressive myoclinic epilepsy.	
32.	*GOSR2*	Transport across membrane	Autosomal recessive	Progressive myoclinic epilepsy.	
	CERS1	Enzyme			
	LMNB2	Protein for cytoskeleton			
	KCTD7	Not classified			

One crucial feature of epilepsy is cognitive dysfunction. Deregulation of cognitive dysfunction is linked to activity-dependent transcription, which is essential for the transition of neuronal plasticity to long-term from short-term [9]. Therefore, it has been put forward that long-term memory formation, which necessitates activation of transcriptional programs for various processes such as learning, depends on the chromatin modifications of activated neurons [14]. Early infantile epileptic encephalopathy (EIEE) is marked by the presence of intellect deficits and seizures. The X-linked diseases are responsible for exhibiting many severe phenotypes in males as compared to females. However, epilepsy caused by mutations in protocadherin 19 (PCDH19) causes epilepsy in females that are heterozygous but not in males that are hemizygous [15]. Protocadherins (PCDHs) play an essential role in various neurological processes such as synaptogenesis, axon guidance, etc. They are considered as the most prominent family of cell-cell adhesion molecules [16]. The genes that encode for protocadherins are present throughout the genome and are responsible for demonstrating an overlapping and exclusive pattern of expression in the mature brain and also during the development of CNS in several populations of neurons. The mechanism that underlies this X-linked pattern of inheritance of epilepsy is not known [15]. Researchers are working to decipher the basic mechanisms as well as to find appropriate therapeutic interventions.

3. Epigenetics Involved in Epilepsy

Recent studies show epigenetics playing essential roles in temporal lobe epilepsy [17]. Therefore, studying the role of epigenetic changes in the development of the disease has become an emerging topic in the area of research. The knowledge of epigenetic mechanisms helps in providing the putative conceptual framework in the development of therapies that can help in the prevention of the disease. Epileptogenesis should be considered as a target point for developing therapy when there is an increment in the severity and frequency of impulsive recurrent seizures. Various processes that take place along with epileptogenesis are mossy fiber sprouting, dysfunction of adenosine together with gliosis, aberrant connectivity, neuronal cell loss, and neuroinflammation [18,19]. TCF4, MECP2, UBE3A, and CHD2 are some of the regulatory genes associated with epilepsy. Among these, the CHD2 gene is responsible for encoding a protein that remodels chromatin, and deregulation of CHD2 might have a downstream effect on other genes [20].

Epigenetic modifications are also responsible for many of the pathological changes that take place during epileptogenesis. Many changes have been shown to occur in the central nervous system cells that alter the gene expression due to DNA methylation and histone acetylation and methylation [12,21]. Moreover, these changes happen quickly and regularly. DNA methylation-reliant alterations in the process of transcription of genes get induced by even a single change in neural synchronization, which ultimately induces a cascade of transcription factors leading to long-term alterations [21]. When DNA methylation was deliberated globally with the help of antibody capture and in the BRD2 gene promoter, variations in the lymphoblastoid cell lines were observed in epileptic patients. Furthermore, in the hippocampus of epileptic patients, alterations in DNA methylation were observed [12].

Various processes like histone modifications that involve either adding up or eliminating the acetyl or methyl groups are suggested to be associated with epileptogenesis. According to the hypothesis of DNA methylation implicated in epileptogenesis, seizures can induce epigenetic modifications and can exaggerate the process of epileptogenesis. DNA hypermethylation, along with the amplified activity of DNA methylating enzymes has been implicated in the development of experimental and human epilepsy [22]. During brain development, DNA methyltransferases (DNMTs) have been shown to perform a significant role and expressed more predominantly in neurons than glial cells. Moreover, in the adult brain, studies have exhibited the functional significance of DNMTs in processes such as memory, learning, behavior, and synaptic plasticity. Thus, DNA methylation is speculated in mediating the process of epileptogenesis by leaving an imprint on physiological processes and gene expression, which might lead to the development of temporal lobe epilepsy later in life [23,24].

Histone modification is one of the epigenetic mechanisms that have the considerable potential to alter the neuronal expression of genes by exerting their additive effects in a correlated manner [17]. The principal role of histone proteins is to support the tertiary and quaternary structure of the DNA. Disruption in these vital epigenetic machinery leads to various disorders such as epilepsy, autism, Rett syndrome, etc. Epigenetic alterations that occur at the histone tail are responsible for influencing the structure of the chromatin that ultimately modifies the approachability of transcriptional regulators. One of the significant histone modifications that have been linked to epilepsy is lysine acetylation. In several animal models, the process of histone deacetylation by restraining histone deacetylase enzymes proved to be beneficial in preventing the symptoms of epilepsy [25].

DNA methylation is dependent on several biochemical enzymatic reactions. One such reaction pathway is the S-adenosylmethionine-dependent transmethylation pathway, which is controlled by glycine and adenosine under the regulation of adenosine kinase (ADK). In chronic epilepsy, it is observed that there is an increase in the ADK and a resulting decline in adenosine, which leads to elevated DNA methylation in the brain. Thus, interference with methylation of DNA gives the new conceptual prospect to control and prevent epilepsy. Glycine modifying therapies can also be regarded as an alternative opportunity that affects the process of DNA methylation and, ultimately, the process of epileptogenesis. Therefore, understanding the epigenetics of epileptogenesis might help in the discovery and development of therapeutic interventions [18,26]. Ketone bodies are found to play an important role of signaling molecules as well in addition to be a source of energy. Class 1 histone deacetylases (HDACs) are inhibited by β–hydroxybutyrate (BHB), which is considered as a ketone body with signaling functions. HDACs are responsible for the subtraction of acetyl groups from histone tails. These epigenetic modifications by ketone bodies are believed to be regulating gene expression and other epigenome modifications, which can implicate the treatment of several human diseases like epilepsy [24,27,28].

4. Neurotransmitter Release Machinery in Epilepsy

4.1. Glutamate Receptors

Glutamate is an excitatory neurotransmitter responsible for stimulating an increase in calcium and sodium conduction through ligand-gated ion channels (Figure 1). A wide spectrum of anti-convulsant properties is displayed by AMPA antagonists and NMDA antagonists in animal models with acute and chronic epilepsy [29]. Once seizures begin, the activity-dependent plasticity of the glutamate receptors becomes a vital feature of the epileptic brain. Epileptogenesis has been linked to the receptor pore properties through the mechanism of mutations of the calcium-impermeable and AMPA-sensitive glutamate receptors. AMPA receptors in the central nervous system contain several subunits like GluR1, GluR2, GluR3, and GluR4. Mutation of the GluR2 (also known as GluA2 or GRIA2) subunit leads to the formation of heteromeric AMPA receptors, which lack the essential pore lining arginine site, which is responsible for conferring both calcium impermeability as well as single-channel conductance of three-fold attenuation. Mice bearing this mutation gather AMPA receptors with increased calcium permeability and a neurological phenotype of cell death and various severe seizures [30,31]. Additionally, glutamate, which is released at the synapse that acts on the metabotropic and ionotropic receptors, is responsible for the stimulation and escalation of the seizure activity [29]. The Na^+-dependent transporters remove the glutamate from the extracellular space and synaptic cleft. Reduced expression of the glutamate transporters can induce seizures. Genetic manipulations related to the functioning of the glutamate receptor proteins in rodent models can enhance the threshold of seizures [32].

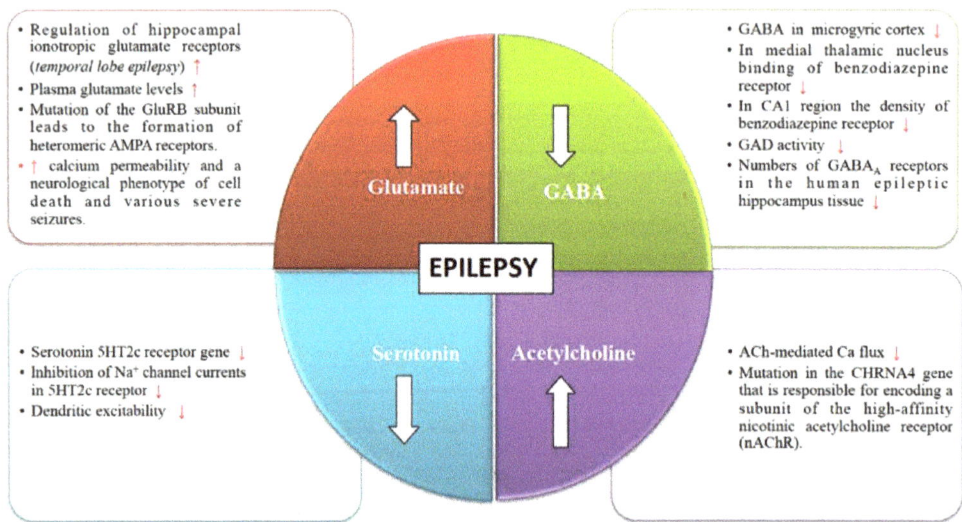

Figure 1. Neurotransmitter mediated changes in epilepsy. Different changes have been observed in the brain due to the increase in the levels of glutamate and acetylcholine and a decrease in the levels of serotonin and GABA. Such modulations due to various physiological conditions or comorbid situations have been reported, which should be taken into account while designing any pharmacological interventions in the epileptic patients.

4.2. GABA Receptors

Various pieces of evidence from several clinical and experimental data underline the role of GABA in the mechanism and management of epilepsy (Figure 1). The synaptic inhibition by GABA plays an essential role in regulating neuronal excitability, which has been linked to epilepsy. GABAergic neurotransmission that is controlled by Cl⁻ permeable GABA$_A$ receptors can exhibit both seizure-repressing and -stimulating activity [33]. GABAergic functions have shown abnormalities in the acquired and genetic animal models of epilepsy. Several studies have shown changes in the GABA receptor densities and concentrations of GABA in the human epileptic tissue. It has been found that there are reduced numbers of GABA$_A$ receptors in the human epileptic hippocampus tissue [34]. Many researchers have shown that seizures are suppressed by GABA agonists and are produced by GABA antagonists. GABA agonists are anticonvulsants whereas GABA antagonists are proconvulsant. Seizures are caused by drugs that inhibit the synthesis of GABA. Some of the GABA synthesis inhibitors include L-allyglycine, isoniazid, thiosemicarbazide, and 4-deoxypyridoxine. Therefore, inhibition of the synthesis of GABA is known to be epileptogenic [34]. Effective anti-convulsants like barbiturates and benzodiazepines (BZDs) function by increasing the GABA mediated inhibition. Barbiturates are responsible for extending the opening time of the chloride channels. Benzodiazapenes also amplify the chloride channel opening rate by enhancing the binding of the GABA to its receptors [35]. Drugs like vigabatrin and tiagabine are effective anti-convulsants that enhance synaptic GABA by decreasing GABA catabolism or reuptake, ultimately responsible for inhibiting seizure activity. Drugs that play a significant role in increasing synaptic GABA have been shown to be effective AEDs [36].

4.3. Cholinergic Receptors

The structural and functional diversity of the neuronal nicotinic acetylcholine receptors (nAChRs) perform modulatory functions throughout the mammalian brain (Figure 1). Nicotinic receptors are involved in various developmental mechanisms such as memory, attention, and learning. Disruptions in cholinergic mechanisms can lead to several disorders such as epilepsy, Parkinson's disease, dementia, schizophrenia, autism, Alzheimer's disease, etc. [37]. Functional nAChRs are broadly dispersed throughout the central nervous system and are situated on dendrites, axon terminals, and cell bodies and mediate synaptic neurotransmission [38]. According to genetic studies conducted in several animal models and epileptic patients, it has been illustrated that the activity of nAChRs is altered in certain types of epilepsy such as juvenile myoclonic epilepsy (JME) and autosomal dominant nocturnal frontal lobe epilepsy (ADNFLE) [39]. Cholinergic receptors are broadly dispersed on both inhibitory and excitatory interneurons present in and outside of the frontal lobe. Pentamers, which are assembled in the subunit pattern of (2α-3β), form the widely held nicotinic receptors of the brain. Five mutations in these two subunits have been related to autosomal dominant frontal lobe epilepsy [40]. Acetylcholinergic neurons play an essential role in the mental development in the pedunculopontine tegmental nucleus. Disturbance in the nAChRs can give rise to epilepsy [41].

4.4. Serotonin Receptors

Serotonergic neurotransmission has been shown to have a potential role in epilepsy [42]. Serotonin receptors (5-HTRs) have been therefore considered as promising candidate targets for the development of new AEDs (Figure 1). Studies show 5-HT regulates a wide variety of focal and generalized seizures. Agents like 5-hydroxytryptophan and 5-HT reuptake blockers are known to increase the extracellular serotonin levels and hence inhibit focal as well as generalized seizures. On the other hand, depletion of serotonin levels in the brain is known to subside the threshold for different evoked convulsions [43]. Selective serotonin reuptake inhibitors (SSRIs) can enhance the levels of serotonin at synapse but reduces the synthesis of the serotonin in the brain. Persistent administration of SSRIs are known to be one of the cause for the decreased synthesis of serotonin and increases seizure susceptibility [44]. Spontaneous seizures and a decreased threshold for audiogenic seizures have been linked to the targeted subtraction of the serotonin 5HT2c receptor gene. The 5HT2c receptor plays a significant role in modulating a persistent Na^+ current. It is shown that in cortical neurons, 5-HT2a/c receptor activation is reduced via rapidly inactivating Na+ currents by reducing the maximal current amplitude and shifting fast inactivation voltage dependence. 5-HT2a/c receptor stimulation also reduced the amplitude of persistent Na+ current without altering its activation voltage dependence. This modulation is mediated by a PKC-dependent mechanism and further reduces the dendritic excitability. Seizure susceptibility may be enhanced by the loss of inhibition of Na+ channel currents in 5HT2c receptor [45,46].

5. Drug Resistant Epilepsy

Drug-resistant epilepsy (DRE) is also known as refractory epilepsy or pharmaco-resistant epilepsy. It can be defined as a failure of two or more sufficient trials of tolerated, chosen, and appropriately used AEDs regimens, which can be administered as monotherapies or in combination to get relief from seizures. Around one out of four patients with seizures develop DRE [47]. DRE patients have increased risks of injuries, psychosocial problems, and premature death [48–50]. Some of the hypothesized mechanisms underlying the cause of DRE are discussed below.

5.1. Alterations in the Drug Targets

This hypothesis states that sensitivity to the treatment is decreased due to the alterations in the cellular targets of the drugs. The α_2 subunit of the neuronal Na^+ channel, which is encoded by the SCN2A gene, has been found to be related to anti-epileptic drug resistance [47]. However, this hypothesis cannot explain the contributory role of the alterations in drug targets in causing epilepsy in patients resistant to several drugs with different modes of action [51,52].

5.2. The Inability of the Drugs to Reach Their Targets

This transporter hypothesis postulates that at the epileptic target, drug resistance may apply to the overexpression of the multidrug efflux transporters. P-glycoprotein is the most widely researched efflux transporter whose primary function is to maintain the integrity of the blood–brain barrier by decreasing the cerebral build-up of the substrate drugs [47]. Up-regulation of efflux transporters such as P-glycoprotein in capillaries and abnormal expression in neuronal and glial cells has been described in various studies on patients with DRE [53].

5.3. Real Targets Missed by the Drugs

Presently, the primary usage of the AEDs is only to prevent seizures rather than focusing on the pathogenic processes that are causing the disease. Autoantibodies to the ion channels that are associated with neuronal inhibition and excitation, including voltage-gated ion channels and NMDA and GABA receptors, have been identified in patients with seizures. Such cases have been seen predominantly in multiple circumstances of encephalitis and occult cancer. However, these patients usually are unable to respond to conventional anti-epileptic drugs [47,54,55].

6. Non-Conventional Therapeutic Strategies

The available AEDs control the generating tendency of seizures and are effective in about 60% of individuals. Additionally, most of the existing medications have adverse drug reactions and hypersensitivity issues. It further aggravates by the increased incidence of mental health problems, depression, anxiety, and suicidal tendencies [56]. In such intricate situations, one of the important ways to cope up is the management of the disorder through patient care and devising different therapeutic strategies.

6.1. Ruling Out Pseudo-Resistance

It is the phenomenon in which the seizures persevere because the fundamental disorder has not been adequately treated. So, it is imperative to rule out or correct the underlying disorder before the drug treatment can be considered to have failed. There can be several situations in which misdiagnosis of epilepsy can take place. Cardiac arrhythmia, vasovagal syncope, and other neurological disorders such as migraine and transient ischemic attacks are some of the conditions that imitate epileptic seizures [47,57]. The patient's behavior and lifestyle can also be the possible cause of pseudo-resistance. Alcohol addiction, stress, sleep deprivation, and drug abuse are also the common factors that can cause seizures [58]. Therefore, it is essential to first rule out the chances of misdiagnosis before starting the medication paradigm or changing it once started.

6.2. Combination Therapy

According to many studies and research data, the combination of drugs helps in controlling the disease. The same drugs at differential dosages have been seen to suit different patients. This may be due to the fact that at any instance, separate pathophysiological mechanisms can occur in the same individual. Such a situation would warrant the usage of different therapeutic strategies at different levels for better results. Various natural or herbal-based drugs like Cicuta virosa and Nux vomica have been shown to be effective in reducing seizure activity and other physiological parameters in animal models of epilepsy [9,59]. Several animal models have supported evidence that a combination of lamotrigine and sodium valproate aids in the management of partial-onset and generalized seizures. Other usually recommended combinations include valproate with ethosuximide for controlling absence seizures and lamotrigine with topiramate for controlling a wide range of seizures [47,60].

6.3. Non-Drug Treatment

Patients who are suffering from DRE have surgery as an alternative method for treatment, mainly if they have a surgically remedial disease like unilateral hippocampal sclerosis or other curable lesions. Therefore, depending on the indication, a number of surgical procedures can be performed to treat and control epilepsy after the deliberation of further trials of anti-epileptic drugs [61]. In children with DRE, the ketogenic diet is used as a management strategy. These are also available as formula-based ketogenic diets for infants [47]. A ketogenic diet seems to be helpful in controlling many types of seizures. Through the epigenetic mechanism, a ketogenic diet is known to be responsible for regulating gene expression. It has been found that dietary intake of donors of methyl groups like choline can have an intense effect on the process of DNA methylation. Ketogenic diets with high-fat and low-carbohydrate content are observed to be highly useful in the treatment of refractory seizures in kids [24]. Additionally, various studies on the micronutrient deficiencies like vitamin B12 show possible involvement in the pathway and hence may prove to be beneficial in future therapeutic strategies [62,63]. In adults and adolescents, a device called the vagus nerve stimulator has been approved for use as an adjunctive therapy with partial-onset seizures that are resistant to anti-epileptic drugs [47,64]. A vagus nerve stimulator is used to generate a pulse and is implanted in the upper chest of the patient, which then provides electrical current to the vagus nerve of the neck [65].

7. Modern Approaches for Treatment

Approved AEDs for the treatment of epilepsy work by various mechanisms that mainly include the modulation of voltage-dependent ion channels, activation of GABA, and inhibition of glutamate receptors. Several capable pathways including shared ones with neurodegenerative disorders [66] and potential drug targets have been identified by many researchers and some of them are discussed here.

7.1. MTOR Pathway

This signaling pathway is the mammalian target of rapamycin, which is responsible for regulating cell growth, cell proliferation, cell differentiation, and cell metabolism in the brain. Many studies have shown that dysregulation of mTOR is responsible for the pathogenesis of acquired forms of epilepsy, such as temporal lobe epilepsy [6]. Therefore, therapeutic intervention in this pathway can lead to the discovery of more tolerable anti-epileptic and anti-epileptogenic drugs [67,68].

7.2. Inflammatory Pathways

Various studies and researches have shown shreds of evidence, which states that inflammatory mediators that are released by cells of the brain and peripheral immune cells are implicated in the foundation of seizures and the process of epileptogenesis [6]. Data and facts have emerged that changes in the inflammatory and immune pathways might be the consequences as well as the origin of the different types of epilepsy [69].

7.3. Breakdown of Blood-Brain Barrier

Irrespective of their etiology, dysfunction of the blood-brain barrier is a characteristic of epileptogenic injuries of the brain. Any harm to the blood-brain barrier microvasculature during the brain injury leads to the leakage of the serum albumin into the microenvironment of the cerebral cortex, which induces a signaling cascade in astrocytes, which results in local inflammation by activating the transforming growth factor β receptor (TGFβR) [70]. Dysfunction of the astrocytes leads to the impairment in the homeostasis of the extracellular brain environment, which further results in the increased excitability of the neurons. TGFβR is a remarkable novel target that interferes with the process of epileptogenesis. This is because of the obstruction of TGFβ signaling in the albumin is responsible for reversing the inflammation and transcriptional modifications linked with activated glia and thus prevents the progression of epileptogenesis [6,71]. Various approaches that are being employed in the therapeutics associated with epilepsy have been summarized in Table 2.

8. Conclusions

Several applicable models are available to study and understand the process of epileptogenesis [72]. However, there are still many challenges in dealing with epileptogenesis. There are so many emerging approaches that are under consideration for the treatment of DRE. Various researchers are working on pharmacotherapy and many other therapeutic approaches that can aid in the treatment of epilepsy. Complementary alternative medicines like Cicuta virosa [59] and Nux vomica [9] could be used in combination with lower doses of effective drugs for better results. Such newer combinations of drugs need more research to be done to find their efficacy. Many capable pathways including epigenomic maintenance through dietary intervention [73] and potential drug targets have been identified by many researchers that provide the approach for the treatment. The discovery of a wide variety of anti-epileptic drugs has provided great advancement in treating different types of seizures [72]. Further, understanding the role of different neurotransmitters and their effect on epileptogenesis can help in the emergence of novel treatment strategies for epilepsy in near future.

Table 2. Different approaches used for the treatment of epilepsy.

S. No.	Treatment Approaches	Interventions Used	Action Mechanism	Main Uses	References
A.	PHARMACEUTICAL APPROACHES	Gabapentin	Ca^{2+} blockage	Used for generalised and focal seizures.	[74,75]
	(Anti-epileptic drugs)	Carbamazepine	Na^+ channel blockage	Decrease nerve impulses that are responsible for causing seizures.	[74]
		Lamotrigine	Na^+ channel blockage	Used as a first-line drug for generalized and focal seizures.	[60,75]
		Tiagabine	GABA potentiation	Used for partial seizures in adjunctive therapy.	[74]
		Zonisamide	Na^+ channel blockage	Used for generalized and focal seizures.	[74,75]
		Vigabatrin	GABA potentiation	Used for infantile spasms and for focal onset of seizures.	[74]
		Perampanel	Glutamate (AMPA) antagonist	Used for partial seizures with focal onset.	[74,75]
B.	THERAPEUTIC APPROACHES	Progressive muscle relaxation	Tense a group of muscles while breathing in and relaxes them while breathing out.	Improves sleep and overall well-being. Enhances control over epilepsy by the patients.	[76]
		Yoga	Release tension in key joints through combination of body postures.	Decrease in automatic dysfunction during onset of seizures.	[76,77]
		Cognitive behavioural therapy	Restructuring of maladaptive thought patterns.	Improvement in anxiety and depression and enhanced psychosocial functioning.	[76,78]
		Vagus nerve stimulation	Used to generate impulse through electric current in vagus nerve.	Used as an adjunctive therapy for partial onset of seizures.	[65]
C.	NATURAL APPROACHES	Ketogenic diet	Neurotransmitter modulation in brain by ketone bodies.	Successful in reducing seizures and enhancing motor function.	[27,64,79]
		Vitamin D3	Increase Ca^{2+} uptake and decrease neuronal excitability.	Produces anti-convulsant effect and prevent seizures.	[80,81]
		Herbal treatments	Found to be involved in potentiation of GABAergic activity in brain.	Herbal medications control epileptic seizures and reduce side effects and increase cognitive effects of AEDs.	[82,83]

Author Contributions: Conceptualization: S.G., J.K.S., and P.G.; writing—original draft preparation: T.K. and S.G.; writing—review and editing: S.G., J.K.S., T.K., K.S.D., P.S., K.V., and P.G. All authors have read and agreed to the published version of the manuscript.

Funding: This research received no external funding.

Institutional Review Board Statement: Not applicable.

Informed Consent Statement: Not applicable.

Data Availability Statement: Not applicable.

Conflicts of Interest: The authors declare no conflict of interest. The funders had no role in the design of the study; in the collection, analyses, or interpretation of data; in the writing of the manuscript, or in the decision to publish the results.

References

1. Scharfman, H.E. The neurobiology of epilepsy. *Curr. Neurol. Neurosci. Rep.* **2007**, *7*, 348–354. [CrossRef]
2. Kobylarek, D.; Iwanowski, P.; Lewandowska, Z.; Limphaibool, N.; Szafranek, S.; Labrzycka, A.; Kozubski, W. Advances in the Potential Biomarkers of Epilepsy. *Front. Neurol.* **2019**, *10*, 685. [CrossRef] [PubMed]
3. Fisher, R.S.; Acevedo, C.; Arzimanoglou, A.; Bogacz, A.; Cross, J.H.; Elger, C.E.; Engel, J.; Forsgren, L.; French, J.A.; Glynn, M.; et al. ILAE Official Report: A practical clinical definition of epilepsy. *Epilepsia* **2014**, *55*, 475–482. [CrossRef] [PubMed]
4. Schmidt, D.; Sillanpää, M. Evidence-based review on the natural history of the epilepsies. *Curr. Opin. Neurol.* **2012**, *25*, 159–163. [CrossRef] [PubMed]
5. Berg, A.T.; Berkovic, S.F.; Brodie, M.J.; Buchhalter, J.; Cross, J.H.; Boas, W.V.E.; Engel, J.; French, J.; Glauser, T.A.; Mathern, G.W.; et al. Revised terminology and concepts for organization of seizures and epilepsies: Report of the ILAE Commission on Classification and Terminology, 2005–2009. *Epilepsia* **2010**, *51*, 676–685. [CrossRef]
6. Löscher, W.; Klitgaard, H.; Twyman, R.E.; Schmidt, D. New avenues for anti-epileptic drug discovery and development. *Nat. Rev. Drug Discov.* **2013**, *12*, 757–776. [CrossRef] [PubMed]

7. Schmidt, D. Is antiepileptogenesis a realistic goal in clinical trials? Concerns and new horizons. *Epileptic Disord* **2012**, *14*, 105–113. [CrossRef]

8. Brodie, M.J.; Barry, S.J.E.; Bamagous, G.A.; Norrie, J.D.; Kwan, P. Patterns of treatment response in newly diagnosed epilepsy. *Neurology* **2012**, *78*, 1548–1554. [CrossRef]

9. Mishra, P.; Mittal, A.K.; Rajput, S.K.; Sinha, J.K. Cognition and memory impairment attenuation via reduction of oxidative stress in acute and chronic mice models of epilepsy using antiepileptogenic Nux vomica. *J. Ethnopharmacol.* **2021**, *267*, 113509. [CrossRef]

10. Wang, J.; Lin, Z.-J.; Liu, L.; Xu, H.-Q.; Shi, Y.-W.; Yi, Y.-H.; He, N.; Liao, W.-P. Epilepsy-associated genes. *Seizure* **2017**, *44*, 11–20. [CrossRef]

11. Martins-Ferreira, R.; Chaves, J.; Carvalho, C.; Bettencourt, A.; Chorão, R.; Freitas, J.; Samões, R.; Boleixa, D.; Lopes, J.; Ramalheira, J.; et al. Circulating microRNAs as potential biomarkers for genetic generalized epilepsies: A three microRNA panel. *Eur. J. Neurol.* **2020**, *27*, 660–666. [CrossRef] [PubMed]

12. Caramaschi, D.; Hatcher, C.; Mulder, R.H.; Felix, J.F.; Cecil, C.A.M.; Relton, C.L.; Walton, E. Epigenome-wide association study of seizures in childhood and adolescence. *Clin. Epigenet.* **2020**, *12*, 1–13. [CrossRef] [PubMed]

13. Grønskov, K.; Brøndum-Nielsen, K.; Dedic, A.; Hjalgrim, H. A nonsense mutation in FMR1 causing fragile X syndrome. *Eur. J. Hum. Genet.* **2011**, *19*, 489–491. [CrossRef] [PubMed]

14. Fernandez-Albert, J.; Lipinski, M.; Lopez-Cascales, M.T.; Rowley, M.J.; Martin-Gonzalez, A.M.; Del Blanco, B.; Corces, V.G.; Barco, A. Immediate and deferred epigenomic signatures of in vivo neuronal activation in mouse hippocampus. *Nat. Neurosci.* **2019**, *22*, 1718–1730. [CrossRef] [PubMed]

15. Pederick, D.T.; Richards, K.L.; Piltz, S.G.; Kumar, R.; Mincheva-Tasheva, S.; Mandelstam, S.A.; Dale, R.C.; Scheffer, I.E.; Gecz, J.; Petrou, S.; et al. Abnormal Cell Sorting Underlies the Unique X-Linked Inheritance of PCDH19 Epilepsy. *Neuron* **2018**, *97*, 59–66.e5. [CrossRef] [PubMed]

16. Lefebvre, J.L.; Kostadinov, D.; Chen, W.V.; Maniatis, T.; Sanes, J.R. Protocadherins mediate dendritic self-avoidance in the mammalian nervous system. *Nat. Cell Biol.* **2012**, *488*, 517–521. [CrossRef]

17. Toth, M. Epigenetic Neuropharmacology: Drugs Affecting the Epigenome in the Brain. *Annu. Rev. Pharmacol. Toxicol.* **2021**, *61*, 181–201. [CrossRef]

18. Eboison, D. The Biochemistry and Epigenetics of Epilepsy: Focus on Adenosine and Glycine. *Front. Mol. Neurosci.* **2016**, *9*, 26. [CrossRef]

19. Kobow, K.; Blümcke, I. The emerging role of DNA methylation in epileptogenesis. *Epilepsia* **2012**, *53*, 11–20. [CrossRef]

20. Symonds, J.D. CHD2 epilepsy: Epigenetics and the quest for precision medicine. *Dev. Med. Child Neurol.* **2019**, *62*, 549–550. [CrossRef]

21. Feng, J.; Zhou, Y.; Campbell, S.L.; Le, T.; Li, E.; Sweatt, J.D.; Silva, A.J.; Fan, G. Dnmt1 and Dnmt3a maintain DNA methylation and regulate synaptic function in adult forebrain neurons. *Nat. Neurosci.* **2010**, *13*, 423–430. [CrossRef] [PubMed]

22. Henshall, D.C.; Kobow, K. Epigenetics and Epilepsy. *Cold Spring Harb. Perspect. Med.* **2015**, *5*, a022731. [CrossRef] [PubMed]

23. De Nijs, L.; Choe, K.; Steinbusch, H.; Schijns, O.E.M.G.; Dings, J.; Hove, D.L.A.V.D.; Rutten, B.P.F.; Hoogland, G. DNA methyltransferase isoforms expression in the temporal lobe of epilepsy patients with a history of febrile seizures. *Clin. Epigenetics* **2019**, *11*, 118. [CrossRef] [PubMed]

24. Boison, D.; Rho, J.M. Epigenetics and epilepsy prevention: The therapeutic potential of adenosine and metabolic therapies. *Neuropharmacol.* **2020**, *167*, 107741. [CrossRef] [PubMed]

25. Younus, I.; Reddy, D.S. Epigenetic interventions for epileptogenesis: A new frontier for curing epilepsy. *Pharmacol. Ther.* **2017**, *177*, 108–122. [CrossRef]

26. Boison, D. Adenosine Kinase: Exploitation for Therapeutic Gain. *Pharmacol. Rev.* **2013**, *65*, 906–943. [CrossRef]

27. Longo, R.; Peri, C.; Cricrì, D.; Coppi, L.; Caruso, D.; Mitro, N.; De Fabiani, E.; Crestani, M. Ketogenic Diet: A New Light Shining on Old but Gold Biochemistry. *Nutrients* **2019**, *11*, 2497. [CrossRef]

28. Jago, S.C.S.; Lubin, F.D. Epigenetic Therapeutic Intervention for a Rare Epilepsy Disorder. *Epilepsy Curr.* **2020**, *20*, 111–112. [CrossRef]

29. Chapman, A.G. Glutamate and Epilepsy. *J. Nutr.* **2000**, *130*, 1043S–1045S. [CrossRef]

30. Salpietro, V.; SYNAPS Study Group; Dixon, C.L.; Guo, H.; Bello, O.D.; Vandrovcova, J.; Efthymiou, S.; Maroofian, R.; Heimer, G.; Burglen, L.; et al. AMPA receptor GluA2 subunit defects are a cause of neurodevelopmental disorders. *Nat. Commun.* **2019**, *10*, 1–16. [CrossRef]

31. Kumar, S.S.; Bacci, A.; Kharazia, V.; Huguenard, J.R. A Developmental Switch of AMPA Receptor Subunits in Neocortical Pyramidal Neurons. *J. Neurosci.* **2002**, *22*, 3005–3015. [CrossRef] [PubMed]

32. Meldrum, B.S.; Akbar, M.T.; Chapman, A.G. Glutamate receptors and transporters in genetic and acquired models of epilepsy. *Epilepsy Res.* **1999**, *36*, 189–204. [CrossRef]

33. Wang, Y.; Wang, Y.; Chen, Z. Double-edged GABAergic synaptic transmission in seizures: The importance of chloride plasticity. *Brain Res.* **2018**, *1701*, 126–136. [CrossRef]

34. Treiman, D.M. GABAergic Mechanisms in Epilepsy. *Epilepsia* **2001**, *42*, 8–12. [CrossRef] [PubMed]

35. Johnson, E.W.; De Lanerolle, N.C.; Kim, J.H.; Sundaresan, S.; Spencer, D.D.; Mattson, R.H.; Zoghbi, S.S.; Baldwin, R.M.; Hoffer, P.B.; Seibyl, J.P.; et al. "Central" and "peripheral" benzodiazepine receptors: Opposite changes in human epileptogenic tissue. *Neurology* **1992**, *42*, 811. [CrossRef]

36. Grunewald, R.A.; Thompson, P.J.; Corcoran, R.; Corden, Z.; Jackson, G.D.; Duncan, J.S. Effects of vigabatrin on partial seizures and cognitive function. *J. Neurol. Neurosurg. Psychiatry* **1994**, *57*, 1057–1063. [CrossRef]

37. Dani, J.A.; Bertrand, D. Nicotinic Acetylcholine Receptors and Nicotinic Cholinergic Mechanisms of the Central Nervous System. *Annu. Rev. Pharmacol. Toxicol.* **2007**, *47*, 699–729. [CrossRef]

38. Raggenbass, M.; Bertrand, D. Nicotinic receptors in circuit excitability and epilepsy. *J. Neurobiol.* **2002**, *53*, 580–589. [CrossRef]

39. Ghasemi, M.; Hadipour-Niktarash, A. Pathologic role of neuronal nicotinic acetylcholine receptors in epileptic disorders: Implication for pharmacological interventions. *Rev. Neurosci.* **2015**, *26*, 199–223. [CrossRef]

40. Bertrand, D.; Picard, F.; Le Hellard, S.; Weiland, S.; Favre, I.; Phillips, H.; Berkovic, S.F.; Malafosse, A.; Mulley, J. How Mutations in the nAChRs Can Cause ADNFLE Epilepsy. *Epilepsia* **2002**, *43*, 112–122. [CrossRef]

41. Hayashi, M.; Nakajima, K.; Miyata, R.; Tanuma, N.; Kodama, T. Lesions of Acetylcholine Neurons in Refractory Epilepsy. *ISRN Neurol.* **2012**, *2012*, 1–6. [CrossRef] [PubMed]

42. Bourin, M. Mechanisms of Action of Anxiolytics. In *Psychiatry and Neuroscience Update*; Springer Science and Business Media LLC: New York, NY, USA, 2021; pp. 195–211.

43. Maia, G.H.; Brazete, C.S.; Soares, J.I.; Luz, L.L.; Lukoyanov, N.V. Serotonin depletion increases seizure susceptibility and worsens neuropathological outcomes in kainate model of epilepsy. *Brain Res. Bull.* **2017**, *134*, 109–120. [CrossRef] [PubMed]

44. Singh, T.; Goel, R.K. Managing epilepsy-associated depression: Serotonin enhancers or serotonin producers? *Epilepsy Behav.* **2017**, *66*, 93–99. [CrossRef]

45. Krishnakumar, A.; Abraham, P.M.; Paul, J.; Paulose, C. Down-regulation of cerebellar 5-HT2C receptors in pilocarpine-induced epilepsy in rats: Therapeutic role of Bacopa monnieri extract. *J. Neurol. Sci.* **2009**, *284*, 124–128. [CrossRef] [PubMed]

46. Bagdy, G.; Kecskeméti, V.; Riba, P.; Jakus, R. Serotonin and epilepsy. *J. Neurochem.* **2007**, *100*, 857–873. [CrossRef] [PubMed]

47. Kwan, P.; Schachter, S.C.; Brodie, M.J. Drug-Resistant Epilepsy. *N. Engl. J. Med.* **2011**, *365*, 919–926. [CrossRef]

48. Mohanraj, R.; Norrie, J.; Stephen, L.J.; Kelly, K.; Hitiris, N.; Brodie, M.J. Mortality in adults with newly diagnosed and chronic epilepsy: A retrospective comparative study. *Lancet Neurol.* **2006**, *5*, 481–487. [CrossRef]

49. Lawn, N.D.; Bamlet, W.R.; Radhakrishnan, K.; O'Brien, P.C.; So, E.L. Injuries due to seizures in persons with epilepsy: A population-based study. *Neurology* **2004**, *63*, 1565–1570. [CrossRef]

50. McCagh, J.; Fisk, J.E.; Baker, G.A. Epilepsy, psychosocial and cognitive functioning. *Epilepsy Res.* **2009**, *86*, 1–14. [CrossRef]

51. Loup, F.; Picard, F.; Yonekawa, Y.; Wieser, H.-G.; Fritschy, J.-M. Selective changes in GABAA receptor subtypes in white matter neurons of patients with focal epilepsy. *Brain* **2009**, *132*, 2449–2463. [CrossRef]

52. Remy, S.; Beck, H. Molecular and cellular mechanisms of pharmacoresistance in epilepsy. *Brain* **2005**, *129*, 18–35. [CrossRef]

53. Tishler, D.M.; Weinberg, K.I.; Hinton, D.R.; Barbaro, N.; Annett, G.M.; Raffel, C. MDR1 Gene Expression in Brain of Patients with Medically Intractable Epilepsy. *Epilepsia* **1995**, *36*, 1–6. [CrossRef] [PubMed]

54. Vincent, A.; Irani, S.R.; Lang, B. The growing recognition of immunotherapy-responsive seizure disorders with autoantibodies to specific neuronal proteins. *Curr. Opin. Neurol.* **2010**, *23*, 144–150. [CrossRef]

55. McKnight, K.; Jiang, Y.; Hart, Y.; Cavey, A.; Wroe, S.; Blank, M.; Shoenfeld, Y.; Vincent, A.; Palace, J.; Lang, B. Serum antibodies in epilepsy and seizure-associated disorders. *Neurology* **2005**, *65*, 1730–1736. [CrossRef]

56. Elger, C.E.; Schmidt, D. Modern management of epilepsy: A practical approach. *Epilepsy Behav.* **2008**, *12*, 501–539. [CrossRef]

57. Smith, D.; DeFalla, B.; Chadwick, D. The misdiagnosis of epilepsy and the management of refractory epilepsy in a specialist clinic. *Qjm: Int. J. Med.* **1999**, *92*, 15–23. [CrossRef]

58. Faught, E.; Duh, M.S.; Weiner, J.R.; Guerin, A.; Cunnington, M.C. Nonadherence to anti-epileptic drugs and increased mortality: Findings from the RANSOM Study. *Neurology* **2008**, *71*, 1572–1578. [CrossRef]

59. Mishra, P.; Sinha, J.K.; Rajput, S.K. Efficacy of Cicuta virosa medicinal preparations against pentylenetetrazole-induced seizures. *Epilepsy Behav.* **2021**, *115*, 107653. [CrossRef] [PubMed]

60. Brodie, M.; Yuen, A. Lamotrigine substitution study: Evidence for synergism with sodium valproate? *Epilepsy Res.* **1997**, *26*, 423–432. [CrossRef]

61. Spencer, S.; Huh, L. Outcomes of epilepsy surgery in adults and children. *Lancet Neurol.* **2008**, *7*, 525–537. [CrossRef]

62. Ghosh, S.; Sinha, J.K.; Muralikrishna, B.; Putcha, U.K.; Raghunath, M. Chronic transgenerational vitamin B12 deficiency of severe and moderate magnitudes modulates adiposity-probable underlying mechanisms. *BioFactors* **2017**, *43*, 400–414. [CrossRef]

63. Ghosh, S.; Sinha, J.K.; Khandelwal, N.; Chakravarty, S.; Kumar, A.; Raghunath, M. Increased stress and altered expression of histone modifying enzymes in brain are associated with aberrant behaviour in vitamin B12 deficient female mice. *Nutr. Neurosci.* **2020**, *23*, 714–723. [CrossRef] [PubMed]

64. Neal, E.G.; Chaffe, H.; Schwartz, R.H.; Lawson, M.S.; Edwards, N.; Fitzsimmons, G.; Whitney, A.; Cross, J.H. The ketogenic diet for the treatment of childhood epilepsy: A randomised controlled trial. *Lancet Neurol.* **2008**, *7*, 500–506. [CrossRef]

65. Milby, A.H.; Halpern, C.H.; Baltuch, G.H. Vagus nerve stimulation in the treatment of refractory epilepsy. *J. Neurosurg.* **2009**, *6*, 228–237. [CrossRef]

66. Ghosh, S.; Durgvanshi, S.; Agarwal, S.; Raghunath, M.; Sinha, J.K. Current Status of Drug Targets and Emerging Therapeutic Strategies in the Management of Alzheimer's Disease. *Curr. Neuropharmacol.* **2020**, *18*, 883–903. [CrossRef] [PubMed]

67. Vezzani, A. Before Epilepsy Unfolds: Finding the epileptogenesis switch. *Nat. Med.* **2012**, *18*, 1626–1627. [CrossRef] [PubMed]
68. Ryther, R.C.C.; Wong, M. Mammalian Target of Rapamycin (mTOR) Inhibition: Potential for Antiseizure, Antiepileptogenic, and Epileptostatic Therapy. *Curr. Neurol. Neurosci. Rep.* **2012**, *12*, 410–418. [CrossRef]
69. Vezzani, A.; Friedman, A.; Dingledine, R.J. The role of inflammation in epileptogenesis. *Neuropharmacol.* **2013**, *69*, 16–24. [CrossRef]
70. Frigerio, F.; Frasca, A.; Weissberg, I.; Parrella, S.; Friedman, A.; Vezzani, A.; Noe', F.M. Long-lasting pro-ictogenic effects induced in vivo by rat brain exposure to serum albumin in the absence of concomitant pathology. *Epilepsia* **2012**, *53*, 1887–1897. [CrossRef]
71. Cacheaux, L.P.; Ivens, S.; David, Y.; Lakhter, A.J.; Bar-Klein, G.; Shapira, M.Y.; Heinemann, U.; Friedman, A.; Kaufer, D. Transcriptome Profiling Reveals TGF- Signaling Involvement in Epileptogenesis. *J. Neurosci.* **2009**, *29*, 8927–8935. [CrossRef]
72. Grone, B.P.; Baraban, S.C. Animal models in epilepsy research: Legacies and new directions. *Nat. Neurosci.* **2015**, *18*, 339–343. [CrossRef] [PubMed]
73. Ghosh, S.; Sinha, J.K.; Raghunath, M. Epigenomic maintenance through dietary intervention can facilitate DNA repair process to slow down the progress of premature aging. *IUBMB Life* **2016**, *68*, 717–721. [CrossRef]
74. Ventola, C.L. Epilepsy Management: Newer Agents, Unmet Needs, and Future Treatment Strategies. *P T Peer-Rev. J. Formul. Manag.* **2014**, *39*, 776–792.
75. Schmidt, D.; Schachter, S.C. Drug treatment of epilepsy in adults. *BMJ* **2014**, *348*, g254. [CrossRef] [PubMed]
76. Haut, S.R.; Gursky, J.M.; Privitera, M. Behavioral interventions in epilepsy. *Curr. Opin. Neurol.* **2019**, *32*, 227–236. [CrossRef]
77. Panebianco, M.; Sridharan, K.; Ramaratnam, S. Yoga for epilepsy. *Cochrane Database Syst. Rev.* **2017**, *2017*, CD001524. [CrossRef]
78. Ramaratnam, S.; A Baker, G.; Goldstein, L.H. Psychological treatments for epilepsy. *Cochrane Database Syst. Rev.* **2016**, *2*, CD002029. [CrossRef]
79. Çubukçu, D.; Güzel, O.; Arslan, N. Effect of Ketogenic Diet on Motor Functions and Daily Living Activities of Children With Multidrug-Resistant Epilepsy: A Prospective Study. *J. Child Neurol.* **2018**, *33*, 718–723. [CrossRef] [PubMed]
80. Pendo, K.; DeGiorgio, C.M. Vitamin D3 for the Treatment of Epilepsy: Basic Mechanisms, Animal Models, and Clinical Trials. *Front. Neurol.* **2016**, *7*, 218. [CrossRef]
81. Holló, A.; Clemens, Z.; Kamondi, A.; Lakatos, P.; Szűcs, A. Correction of vitamin D deficiency improves seizure control in epilepsy: A pilot study. *Epilepsy Behav.* **2012**, *24*, 131–133. [CrossRef] [PubMed]
82. Liu, W.; Ge, T.; Pan, Z.; Leng, Y.; Lv, J.; Li, B. The effects of herbal medicine on epilepsy. *Oncotarget* **2017**, *8*, 48385–48397. [CrossRef] [PubMed]
83. Sucher, N.J.; Carles, M.C. A pharmacological basis of herbal medicines for epilepsy. *Epilepsy Behav.* **2015**, *52*, 308–318. [CrossRef] [PubMed]

biomedicines

Article

Chronic Alcoholism and HHV-6 Infection Synergistically Promote Neuroinflammatory Microglial Phenotypes in the *Substantia Nigra* of the Adult Human Brain

Nityanand Jain [1,*], Marks Smirnovs [1], Samanta Strojeva [2], Modra Murovska [2] and Sandra Skuja [1,*]

[1] Joint Laboratory of Electron Microscopy, Institute of Anatomy and Anthropology, Rīga Stradiņš University, LV-1010 Riga, Latvia; marks.smirnovs@rsu.lv

[2] Institute of Microbiology and Virology, Rīga Stradiņš University, LV-1067 Riga, Latvia; samanta.strojeva@rsu.lv (S.S.); modra.murovska@rsu.lv (M.M.)

* Correspondence: nityapkl@gmail.com (N.J.); sandra.skuja@rsu.lv (S.S.); Tel.: +371-673-204-21 (N.J. & S.S.)

Abstract: Both chronic alcoholism and human herpesvirus-6 (HHV-6) infection have been identified as promoters of neuroinflammation and known to cause movement-related disorders. *Substantia Nigra* (SN), the dopaminergic neuron-rich region of the basal ganglia, is involved in regulating motor function and the reward system. Hence, we hypothesize the presence of possible synergism between alcoholism and HHV-6 infection in the SN region and report a comprehensive quantification and characterization of microglial functions and morphology in postmortem brain tissue from 44 healthy, age-matched alcoholics and chronic alcoholics. A decrease in the perivascular CD68+ microglia in alcoholics was noted in both the gray and white matter. Additionally, the CD68+/Iba1− microglial subpopulation was found to be the dominant type in the controls. Conversely, in alcoholics, dystrophic changes in microglia were seen with a significant increase in Iba1 expression and perivascular to diffuse migration. An increase in CD11b expression was noted in alcoholics, with the Iba1+/CD11b− subtype promoting inflammation. All the controls were found to be negative for HHV-6 whilst the alcoholics demonstrated HHV-6 positivity in both gray and white matter. Amongst HHV-6 positive alcoholics, all the above-mentioned changes were found to be heightened when compared with HHV-6 negative alcoholics, thereby highlighting the compounding relationship between alcoholism and HHV-6 infection that promotes microglia-mediated neuroinflammation.

Keywords: microglia; neuroinflammation; chronic alcoholism; HHV-6; *Substantia Nigra*

Citation: Jain, N.; Smirnovs, M.; Strojeva, S.; Murovska, M.; Skuja, S. Chronic Alcoholism and HHV-6 Infection Synergistically Promote Neuroinflammatory Microglial Phenotypes in the *Substantia Nigra* of the Adult Human Brain. *Biomedicines* **2021**, 9, 1216. https://doi.org/10.3390/biomedicines9091216

Academic Editors: Kumar Vaibhav, Meenakshi Ahluwalia and Pankaj Gaur

Received: 26 August 2021
Accepted: 11 September 2021
Published: 14 September 2021

Publisher's Note: MDPI stays neutral with regard to jurisdictional claims in published maps and institutional affiliations.

Copyright: © 2021 by the authors. Licensee MDPI, Basel, Switzerland. This article is an open access article distributed under the terms and conditions of the Creative Commons Attribution (CC BY) license (https://creativecommons.org/licenses/by/4.0/).

1. Introduction

A linear double-stranded DNA virus, human herpesvirus-6 (HHV-6) is a ubiquitous β-herpesvirus, first isolated from the peripheral blood mononuclear cells of patients with lymphoproliferative disorders [1,2]. It has been postulated that the salivary glands act as the major and persistent reservoir of the virus in humans, especially given its frequent detection in saliva and the suggested role of saliva in viral transmission [3–6]. A small proportion (1–2%) of the cases has been reported to be associated with vertical transmission of the virus during pregnancy [7,8]. A vast majority of children by the age of 2 years become seropositive as the seroconversion begins as soon as the protection from passive maternal antibodies starts to wear off [1]. The virus is known to exist in two close variants: HHV-6A and HHV-6B. Whilst HHV-6A has not yet been etiologically linked to any disease [1,9,10], HHV-6B is known to cause Roseola infantum (sixth's disease; exanthema subitum), which occasionally presents with neurological complications like febrile seizures or encephalitis [9].

The neurotropic nature of the virus, coupled with its ability to infect a broad range of cells/tissues in vivo including the endothelium [11] and brain [12,13], leads to a wide spectrum of clinical complications. Additionally, it has been speculated that it plays a role in central nervous system (CNS) -related diseases including meningoencephalitis [14,15], Alzheimer's disease [16,17], multiple sclerosis [18,19], and mood disorders [20,21]. Often, active infection in immunocompromised patients has been described as a fulminant multifocal demyelinating disease [22–24]. Although active HHV-6 infection predominately affects the limbic system [25], other brain regions like the olfactory cortex have also been reported to be altered [26–28]. Reports of patients presenting chorea-like involuntary movements and other movement disorders in HHV-6 infection [25,29,30] leads us to speculate the possible involvement of basal ganglion, a region associated with the "motor circuit" and related movement disorders [31].

Substantia Nigra (SN), also known as the black substance, is the dopaminergic neuron-rich region that modulates motor movement and reward functions as part of the basal ganglia [32]. It is divided anatomically into two subregions: *Pars Compacta* (SNpc) and *Pars Reticulata* (SNpr). SNpc is composed of densely packed neurons with high concentrations of neuromelanin (a dark polymer pigment formed by the oxidative polymerization of dopamine or noradrenaline) [32,33], while SNpr mainly consists of GABAergic (gamma-aminobutyric acid) inhibitory neurons. SNpc exerts its effects on the motor cortex either via direct or indirect pathways. In the direct pathway, projections from SNpc reach D1 (dopamine) receptors in the striatum, causing inhibition of the GPi (globus pallidus interna) and SNpr, leading to disinhibition of the thalamic nuclei and allowing the required movement to occur [33]. In the indirect pathway, projection synapses with D2 striatal receptors cause the GPe (globus pallidus externa) to be relatively excited, thereby inhibiting subthalamic nuclei and activating GPi and SNpr, which finally leads to inhibition of motor function [34].

Alcohol (ethanol) has long been an established modifier of brain activity. At low-to-moderate doses, it has been recognized as a dose-dependent stimulant of motor activity, while at high doses, it has been shown to cause ataxic effects (uncontrolled, abnormal movements) [35–37]. Further, the development of alcohol tolerance and addiction represent another modulating effect of alcohol on the dopamine-mediated reward system of the brain. Chronic alcohol use dysregulates the immune system, causing increased susceptibility to infections and inflammatory reactions in the brain and peripheral organs [38]. Studies in animal models have shown that β-carbolines and their derivatives (found in abundance in alcoholic beverages), upon in vivo metabolization, form compounds resembling 1-methyl-4-phenylpyridinium ions (MPP$^+$), a neurotoxicant involved in the pathogenesis of idiopathic Parkinson's disease [39]. Moreover, these derivatives have been shown to induce early-onset neurodegenerative changes, glial activation in SNpc, and a significant long-term decrease in spontaneous motor activity [39].

Both HHV-6 and alcohol have been identified as promoters of neuroinflammation. Neuroinflammation, a broad term that does not correlate to the classical characteristics seen in peripheral inflammation, represents a chronic, CNS-specific, and glial-mediated inflammation-like response [40]. Microglial cells, the immune sentinels in the brain parenchyma that are thought to orchestrate a potent neuroinflammatory response [41], account for about 5–10% of the total cell population in the brain. Together with nonparenchymal brain macrophages (perivascular, meningeal, and choroid plexus macrophages), each of these cell types occupies a specific niche, thereby covering the entire CNS [42]. Under physiological conditions, mature microglial cells (so-called "resting/surveillance microglia") are usually highly ramified with fine, long processes and small somata; however, upon activation in response to pathogens or DAMPs (damage-associated molecular patterns) [42], a change in surface molecule expression along with transformation from ramified to ameboid (so-called "activated microglia") shape via a multistep activation cascade is seen [43,44]. While the initial microglial response may provide neuroprotection by eliminating the distress source and restoring tissue homeostasis, in cases of persistent stimuli like HHV-6, microglial cells can become chronically active, resulting in upregulation of pro-inflammatory cytokines which jeopardizes neuronal survival [45–47].

Evidence from previous studies led us to investigate the potential role and synergistic effects of these partners-in-crime, i.e., chronic alcoholism and HHV-6 infection, in the disruption of brain homeostasis and increased neuroinflammatory response in the SN region of the brain. Additionally, we explored the various subpopulations of microglial cells, providing valuable insights into the stages of transformation which these cells undergo, their function, and quantification in both normal and alcoholism-related conditions, thereby advancing our knowledge of brain homeostasis in the hope of achieving better diagnostic and treatment modalities for patients.

2. Materials and Methods

2.1. Autopsy Brain Tissue Collection and Characteristics

Brain autopsy specimens from 44 individuals were provided by the Latvian State Centre for Forensic Medical Examination. Postmortems were performed between 7 and 30 h after death. A complete medical history including evidence of alcoholism (confirmed by ethanol levels in the blood coupled with tissue changes seen in the liver, pancreas, and heart) was provided by a certified pathologist. The autopsy brain samples were collected between 2007–2012 and preserved in paraffin blocks following the appropriate protocols. Brain tissues for the SN region were obtained using the human brain map atlas [48]. Patients with infections, diabetes, or respiratory system pathologies were excluded from the study. The inclusion criteria included previous history of alcoholism in accordance with the criteria established by Harper et al. [49].

The autopsies of alcoholic subjects evidenced alcohol abuse and liver cirrhosis. Additionally, ascites, pancreatitis, and cardiomyopathy were found [50]. In the laboratory, patient specimens were grouped based on age and exposure to alcohol, and assigned internal codes as shown in Table 1. Group A included 13 control individuals with a median age of 31 ± 6.79 years and no history of alcohol consumption. Control group individuals presented no neuropathological abnormalities upon postmortem examination and had no history of major psychiatric illness [50]. Group B included 13 age-matched alcoholic individuals with a median age of 31 ± 4.85 years. Group C consisted of 18 non-age-matched alcoholic individuals with a median age of 49.5 ± 8.66 years.

Table 1. Grouping of individuals along with their age and gender.

Group A (Control Group)			Group B (Age-Matched Alcoholics)			Group C (Non-Age-Matched Alcoholics)		
Individual Code	Age (Years)	Gender (Male/ Female)	Individual Code	Age (Years)	Gender (Male/ Female)	Individual Code	Age (Years)	Gender (Male/ Female)
A1	34	M	B1	36	F	C1	48	M
A2	31	M	B2	23	M	C2	55	M
A3	27	M	B3	31	M	C3	60	M
A4	23	M	B4	26	M	C4	50	F
A5	32	M	B5	33	M	C5	63	M
A6	37	M	B6	34	F	C6	45	M
A7	33	M	B7	25	F	C7	45	M
A8	22	M	B8	30	M	C8	55	M
A9	36	M	B9	35	M	C9	45	F
A10	17	M	B10	22	M	C10	60	F
A11	37	M	B11	35	M	C11	66	M
A12	20	M	B12	34	M	C12	63	M
A13	26	M	B13	29	M	C13	44	F
						C14	49	M
						C15	45	F
						C16	60	M
						C17	38	F
						C18	40	M
Group median age	31 ± 6.79		Group median age	31 ± 4.85		Group median age	49.5 ± 8.66	

The protocol for the present study was approved by the Ethics Committee of Rīga Stradiņš University (Decision No. 6-1/12/9) dated 26 November 2020, as per the provisions of the Declaration of Helsinki. Informed consent was provided by the next of kin for the autopsy.

2.2. Immunohistochemistry (IHC), Double Immunohistochemistry, and Immunofluorescence (IF)

Routine hematoxylin and eosin (H&E) staining was done using formalin-fixed paraffin-embedded (FFPE) tissue sections which were examined using a light microscope to verify that the slides contained the regions of interest for the present study (SNpc and SNpr). Upon verification, the sections were prepared for standard immunohistochemistry (IHC) and immunofluorescence (IF) reactions [51]. Following the manufacturer's guidelines, tissue sections were incubated overnight with primary antibodies, as described in Table 2.

Table 2. Description and characteristics of the primary antibodies.

Primary Antibody *	Antibody Characteristics **	Clone	Dilution	Manufacturer
CD68	Monoclonal mouse AB against human AG	Kp-1	1:200	Cell Marque (USA)
CD11b	Monoclonal rabbit AB against human AG	EP45	1:100	Epitomics (USA)
Iba1	Monoclonal rabbit AB against human AG	EP289	1:150	Epitomics (USA)
HHV-6 (20)	Monoclonal mouse AB against viral lysate	-	1:200	Santa Cruz (USA)

* CD68—cluster of differentiation 68; CD11b—cluster of differentiation 11b; Iba1—ionizing calcium-binding adaptor molecule 1; and HHV-6—human herpesvirus-6. ** AB, antibody; AG, antigen.

CD68 (Cluster of differentiation 68), also known as LAMP-4 (Lysosomal associated membrane protein 4), is a transmembrane glycoprotein that is associated with the cellular, endosomal, and lysosomal compartments [52,53]. A scavenger receptor that binds oxidized low-density lipoproteins (oxLDL), CD68 is highly expressed in cells of macrophage lineage with low expression in lymphocytes, fibroblasts, and endothelial cells [54]. Whilst CD68 can be expressed by resting microglia, it is commonly considered as a marker of activated microglia due to its role in phagocytotic activities [55,56].

CD11b (Cluster of differentiation 11b), a commonly used peptide marker for activated and resting microglia, is the α-subunit of CR3 (Complement receptor 3), an integrin that is involved in adhesion processes [57,58]. Iba1 (Ionized calcium-binding adapter molecule1) is a peptide encoded by the gene AIF1 (Allograft inflammatory factor 1). As a cytoplasmic protein, Iba-1 is primarily and constitutively expressed by both activated and resting microglial cells in the brain tissue. Like CD68, it has also been implicated in phagocytic processes [59,60].

HHV-6 (20) is an antibody that is raised against viral lysate and can be used to detect the HHV-6A and HHV-6B subtypes [61,62]. The presence of the respective antigens was determined either by using the HiDef DetectionTM HRP Polymer system (Cell Marque, Rocklin, CA, USA) and 3,3′ diaminobenzidine (DAB) tetrahydrochloride kit (DAB + Chromogen and DAB + Substrate buffer, Cell Marque, Rocklin, CA, USA) or the goat antimouse IgG (H + L) antibody, Alexa Fluor® 488 conjugate (Thermo Fisher Scientific, Invitrogen, UK, 1:300). For nuclear visualization, counterstaining with Mayer's hematoxylin (Sigma Aldrich, St. Louis, MO, USA) or 4′,6-diamidino-2-phenylindole (DAPI) (Thermo Fisher Scientific, Invitrogen, UK, 1:3000) was done, respectively.

In order to detect the two different antigens (double IHC staining), both the HiDef DetectionTM HRP Polymer system and the HiDef DetectionTM Alk Phos Polymer system were used successively, followed by counterstaining with a DAB substrate kit (brown color) and Permanent Red Chromogen Kit (red color), respectively (Cell Marque, Rocklin, CA, USA). Tissue rinsing, dehydration, clearing, and mounting in Roti® Histokitt (Carl Roth, Karlsruhe, Germany) or Prolong Gold with DAPI (Thermo Fisher Scientific, Invitrogen, UK) was done as per the manufacturer's protocols. Treatment with Sudan Black B solution (Sigma Aldrich, St. Louis, MO, USA) was done to reduce tissue autofluorescence. Positive and negative controls were prepared for each antibody reaction (as per the manufacturer's guidelines). Negative controls were prepared using tris(hydroxymethyl)amino -methane (TRIS) solution in place of the primary antibody.

After conventional immunostaining, the expression of marker proteins was manually evaluated using a Leica light microscope (LEICA, LEITZ DMRB, Wetzlar, Germany) and Glissando Slide Scanner (Objective Imaging Ltd., Cambridge, UK) in 10 randomly selected visual fields per sample per region. IHC markers were considered positive if brown-stained cell or cell clusters were observed. A quantitative scoring system was used for positively stained cells or cell groups by two independent observers. Immunofluorescence (IF) was used to confirm the localization of the viral proteins using a Nikon Eclipse Ti-E confocal microscope (Nikon, Brighton, MI, USA).

2.3. DNA Extraction and HHV-6 Detection Using Nested Polymerase Chain Reaction (nPCR)

Fine tissue sections of 0.5–1 mm^2 were obtained from the formalin-fixed, paraffin-embedded (FFPE) brain tissue blocks to isolate DNA using the commercial blackPREP FFPE DNA kit (Analytik Jena AG, Germany) following the manufacturer's protocol. The concentration of the extracted DNA was determined using a NanoDrop ND-1000 Spectrophotometer (Thermo Fisher Scientific, Waltham, MA, USA). The measurements were taken according to the manufacturer's protocol. The quality of genomic DNA (gDNA) was determined by detecting the β-globin gene sequence with the polymerase chain reaction (PCR) method using appropriate primers [63]. DNA obtained from *Substantia Nigra* (SN) tissue was considered as qualitative if 200 bp products were acquired by PCR. We found that all samples were β-globin positive.

Detection of viral genomic sequences in the isolated DNA from SN tissue was done using nested polymerase chain reaction (nPCR). HHV-6 was detected by a two-step PCR reaction with primers targeting the viral U3 gene that encodes the main capsid proteins for both variants HHV-6A and HHV-6B [64]. Each experiment was performed along with positive, negative, and water controls. For positive controls, HHV-6A and HHV-6B genomic DNA (Advanced Biotechnologies Inc., Columbia, MD, USA) was used. For negative controls, DNA samples obtained from practically healthy HHV-6 negative blood donors were used.

2.4. HHV-6A and HHV-6B Variant Detection Using HindIII Restriction Endonuclease

HHV-6A and HHV-6B were differentiated according to the methodology described by Lyall and Cubie [65]. A two-step PCR reaction was performed again to differentiate between the two subtypes of the virus with primers targeting the HHV-6 large tegument protein (LTP) gene following HindIII restriction analysis. The primers used for the PCR reaction are summarized in Table 3. The obtained nPCR amplicons were digested with HindIII restriction endonuclease (Thermo Scientific, Waltham, MA, USA) according to the manufacture's protocol. HindIII cleaves the HHV-6B positive sample into two fragments of 66 bp and 97 bp (base pair), but does not cleave HHV-6A at all. All PCR results were visualized using 1.7% agarose electrophoreses gel.

Table 3. Primers used for two-step PCR reaction for detection of HHV-6 variants.

Primer	Primer Sequence
Outer primer O1	5′-AGTCATCACGATCGGCGTGCTATC-3′
Outer primer O2	5′-TATCTAGCGCAATCGCTATGTCG-3′
Inner primer I3	5′-TCGACTCTCACCCTACTGAACGAG-3′
Inner primer I4	5′-TGACTAGAGAGCGACAAATTGGAG-3′

2.5. Viral Load Determination Using Real-Time PCR (RT-PCR)

Positive DNA samples from FFPE whole SN region (SNpc and SNpr) blocks were used for HHV-6 load detection using the HHV-6 Real-TM Quant (Sacace Biotechnologies, Como, Italy) commercial kit in accordance with the manufacturer's instructions. The β-globin gene was used as the internal control which serves as an amplification control for each processed specimen and can aid in the identification of possible reaction inhibitions. HHV-6 viral load > 10 copies/10^6 cells were considered to be significantly increased.

2.6. Statistical Analysis

The data collected from each visual field (quantitative counting of the immune-positive structures in IHC) were stored in spreadsheets using MS Excel (Microsoft Office 365) Distribution of the dataset was checked using the Shapiro-Wilk test for normality ($p < 0.05$ indicates a violation of normality). Since all our datasets violated the conditions of normality, nonparametric tests were used for analysis. Kruskal-Wallis ANOVA (analysis of variance) was used for intergroup analysis with post hoc tests and Bonferroni correction. The intragroup analysis was done using the related-samples Wilcoxon signed rank test. $p < 0.05$ was considered statistically significant. The statistical analysis was done using SPSS (IBM Corp. Released 2020; IBM SPSS Statistics for Windows, Version 27.0; Armonk, NY, USA: IBM Corp), while graphs were prepared using R studio and MS Excel.

3. Results

3.1. Alcoholics Showed a Significant Decrease in CD68+ Cells Than Controls in Both Gray and White Matter

White matter showed more CD68 positive (CD68+; significant only in SNpr) cells in *Substantia Nigra* (Table 4), in line with previous studies which indicated that the basal level of phagocytosis was higher in white matter than gray matter [66–68]. Overall, a decrease in the number of CD68+ cells in both gray and white matter was noted in alcoholics when compared with controls (Table 4). Although the decrease was not significant in SNpc, a significant decrease in CD68+ cells was seen in the gray matter of SNpr between Group A and Group C ($p = 0.025$). Similarly, in the white matter of SNpr, a significant decrease was seen between controls and alcoholics (Groups A–C $p = 0.000$; Groups B–C $p = 0.021$).

Table 4. Distribution of CD68+ cells per visual field in different regions of *Substantia Nigra* (SN).

Region	Group A	Group B	Group C	p Value [†]
Pars Compacta (SNpc)				
Gray matter	6.50 ± 0.27	6.32 ± 0.20	6.22 ± 0.15	0.908
White matter	6.55 ± 0.25	6.07 ± 0.18	5.97 ± 0.20	0.360
p value [‡]	0.749	0.145	0.519	-
Pars Reticulata (SNpr)				
Gray matter	6.50 ± 0.25	6.34 ± 0.22	5.61 ± 0.25	0.031 **
White matter	7.40 ± 0.29	6.75 ± 0.25	5.86 ± 0.19	<0.001 **
p value [‡]	0.001 **	0.002 **	0.020 **	-

[†] p value was calculated for Kruskal-Wallis ANOVA (intergroup analysis). [‡] p value was calculated for related-samples Wilcoxon Signed Rank Test (intragroup analysis). The numbers represent the average number of CD68+ cells per visual field ± S.E. (standard error). ** indicates a significant difference between groups ($p < 0.05$ is considered significant with Bonferroni correction for Kruskal-Wallis ANOVA and without correction for related-samples Wilcoxon Signed Rank Test).

In *Pars Compacta* (SNpc), alcoholics showed less diffuse CD68+ cells than controls; however, the difference was not found to be statistically significant in both gray and white matter (Figure 1; Kruskal Wallis $p = 0.784$ and 0.683, respectively). Perivascularly, however, there were significant differences observed in the white matter (Figure 1b). Alcoholics showed significantly less CD68+ cells than controls (Groups A–B and A–C; $p = 0.003$ and 0.027, respectively). No significant differences were noted in the gray matter perivascularly ($p = 0.374$).

In *Pars Reticulata* (SNpr), perivascularly, both alcoholic groups (Groups B and C) showed a decrease in the number of CD68+ cells when compared with controls (Group A) in both gray and white matter (Figure 2). A significant decrease in nonmatched alcoholics was noted in both gray and white matter when compared with controls ($p < 0.001$). Further, a significant decrease was noted in the gray matter between age-matched alcoholics and controls ($p < 0.001$). Conversely, alcoholics showed an increase in the number of diffuse CD68+ cells when compared with controls in both the gray and white matter. While significant differences were observed in the gray matter (Groups A–C $p < 0.001$; Groups B–C $p = 0.024$), the differences were not statistically significant in the white matter ($p = 0.187$).

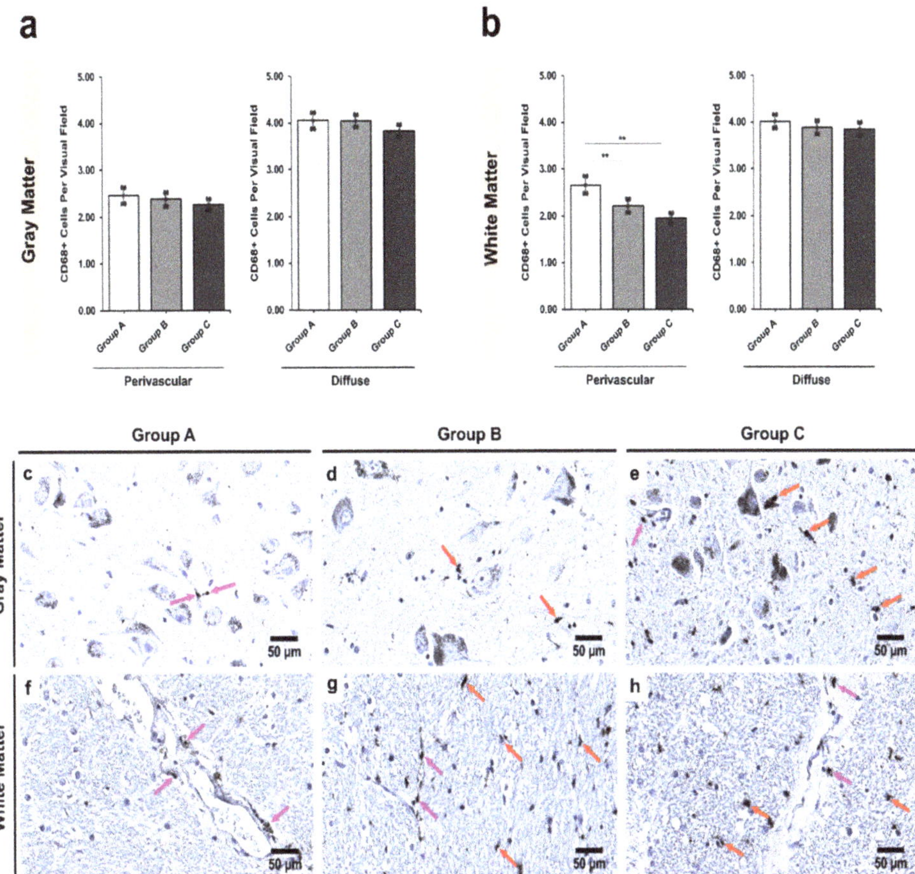

Figure 1. Intergroup analysis of CD68+ cells per visual field in *Pars Compacta* (SNpc) for the three studied groups in both (**a**) gray and (**b**) white matter. The analysis was done for both perivascular and diffuse locations. The bar plots indicate the average number of CD68+ cells ± S.E. (Standard Error) seen per visual field. ** indicates a significant difference between the groups ($p < 0.05$ with Bonferroni correction is considered significant). Distribution of CD68+ cells per visual field in (**c**,**f**) Group A (controls); (**d**,**g**) Group B (age-matched alcoholics) and (**e**,**h**) Group C (non-age-matched alcoholics). Pink arrows indicate perivascular CD68+ cells while red arrows indicate diffuse CD68+ cells. Original magnification, 400×. Scale bars, 50 μm.

3.2. Alcoholics Showed a Significant Increase in Iba1+ Cells Compared to Controls in Both Gray and White Matter

The distribution of Iba1+ cells was found to be significantly more abundant in the white matter in the SNpc (Groups A and B) and SNpr (Group A; Table 5) when compared with gray matter. In contrast to the average number of CD68+ cells, alcoholics showed significantly more Iba1+ cells when compared with controls (Table 5) in both SNpc and SNpr. In both gray and white matter of SNpc, controls had significantly less Iba1+ cells in comparison to both age-matched alcoholics ($p < 0.001$) and non-age-matched alcoholics ($p < 0.001$). In SNpr, however, there were no significant differences between controls and age-matched alcoholics. Rather, significant differences in both gray and white matter were noted between Groups A–C and Groups B–C ($p < 0.001$).

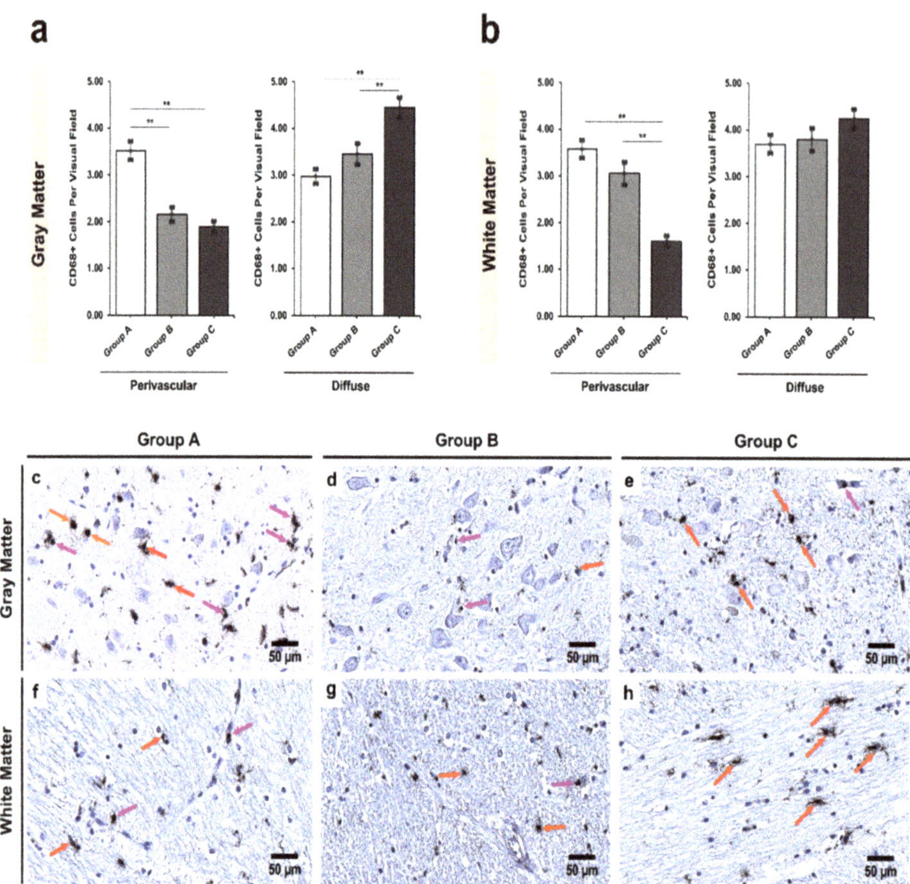

Figure 2. Intergroup analysis of CD68+ cells per visual field in *Pars Reticulata* (SNpr) for the three studied groups in both (**a**) gray and (**b**) white matter. The analysis was done for both perivascular and diffuse locations. The bar plots indicate the average number of CD68+ cells ± S.E. (Standard Error) seen per visual field. ** indicates a significant difference between the groups ($p < 0.05$ with Bonferroni correction is considered significant). Distribution of CD68+ cells per visual field in (**c,f**) Group A (controls); (**d,g**) Group B (age-matched alcoholics) and (**e,h**) Group C (non-age-matched alcoholics). Pink arrows indicate perivascular CD68+ cells; red arrows indicate diffuse CD68+ cells while orange arrows indicate monocytes (seen in the lumen of micro-vessel). Original magnification, 400×. Scale bars, 50 μm.

Table 5. Distribution of Iba1+ cells per visual field in different regions of *Substantia Nigra* (SN).

Region	Group A	Group B	Group C	*p* Value [†]
Pars Compacta (SNpc)				
Gray matter	2.97 ± 0.15	4.41 ± 0.16	4.72 ± 0.17	<0.001 **
White matter	3.50 ± 0.17	4.57 ± 0.14	4.80 ± 0.17	<0.001 **
p value [‡]	<0.001 **	<0.001 **	0.232	-
Pars Reticulata (SNpr)				
Gray matter	1.97 ± 0.13	2.20 ± 0.15	3.27 ± 0.14	<0.001 **
White matter	2.30 ± 0.16	2.45 ± 0.16	3.31 ± 0.13	<0.001 **
p value [‡]	<0.001 **	0.342	0.752	-

[†] *p* value was calculated for Kruskal-Wallis ANOVA (intergroup analysis). [‡] *p* value was calculated for related-samples Wilcoxon Signed Rank Test (intragroup analysis). The numbers represent the average number of Iba1+ cells per visual field ± S.E. (standard error). ** indicates a significant difference between groups ($p < 0.05$ is considered significant with Bonferroni correction for Kruskal-Wallis ANOVA and without correction for related-samples Wilcoxon Signed Rank Test).

In the SNpc, both age-matched and chronic alcoholics showed a significant increase in perivascular Iba1+ cells in both the gray and white matter compared with the controls ($p < 0.001$; Figure 3). Similar results were observed in the diffuse locations in both gray and white matter ($p < 0.001$; Figure 3). The effect of age-related changes could be ignored here due to significant differences between the controls and age-matched alcoholics ($p = 0.001$—perivascular and $p = 0.001$—diffuse).

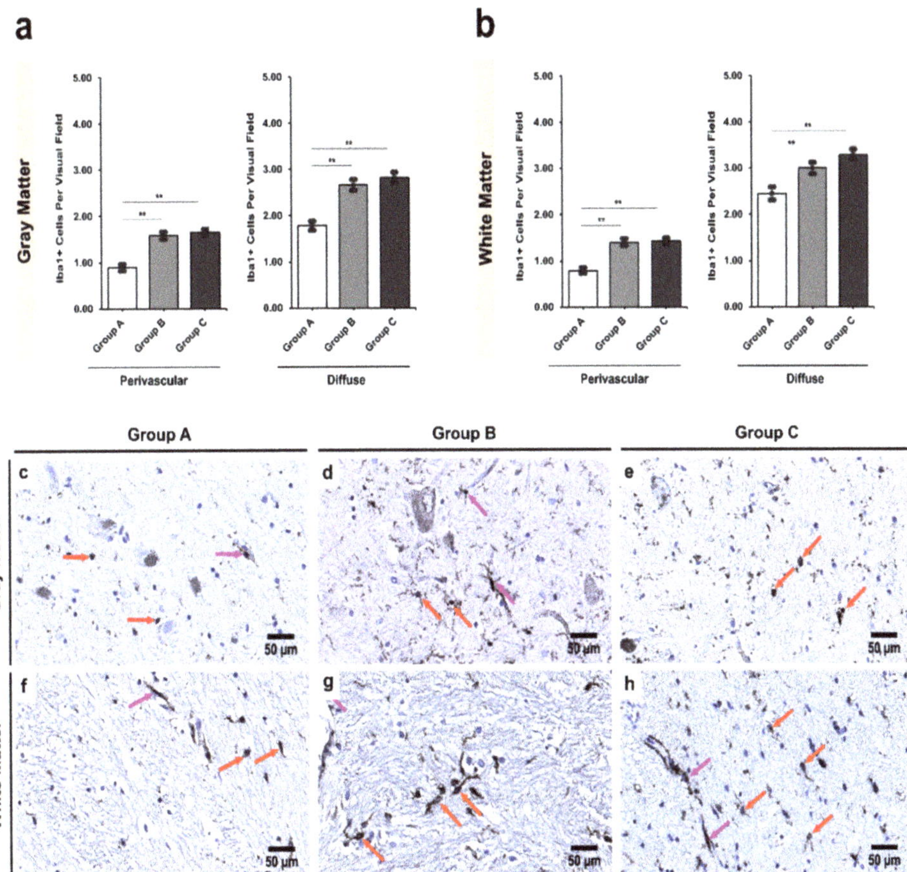

Figure 3. Intergroup analysis of Iba1+ cells per visual field in *Pars Compacta* (SNpc) for the three studied groups in both (**a**) gray and (**b**) white matter. The analysis was done for both perivascular and diffuse locations. The bar plots indicate the average number of Iba1+ cells ± S.E. (Standard Error) seen per visual field. ** indicates a significant difference between the groups ($p < 0.05$ with Bonferroni correction is considered significant). Distribution of Iba1+ cells per visual field in (**c,f**) Group A (controls); (**d,g**) Group B (age-matched alcoholics) and (**e,h**) Group C (non-age-matched alcoholics). Pink arrows indicate perivascular Iba1+ cells while red arrows indicate diffuse Iba1+ cells. Original magnification, 400×. Scale bars, 50 μm.

In the SNpr, however, no statistically significant differences were observed between controls and age-matched alcoholics (Figure 4), though age-matched alcoholics had more Iba1+ cells (Table 5). Furthermore, the chronic alcoholics showed a significant increase in Iba1+ cells compared to controls in both gray and white matter ($p < 0.001$), and in both perivascular ($p = 0.006$) and diffuse locations ($p = 0.009$). Similar results were obtained upon comparison of age-matched and chronic alcoholics ($p = 0.002$ and 0.001, respectively; Figure 4).

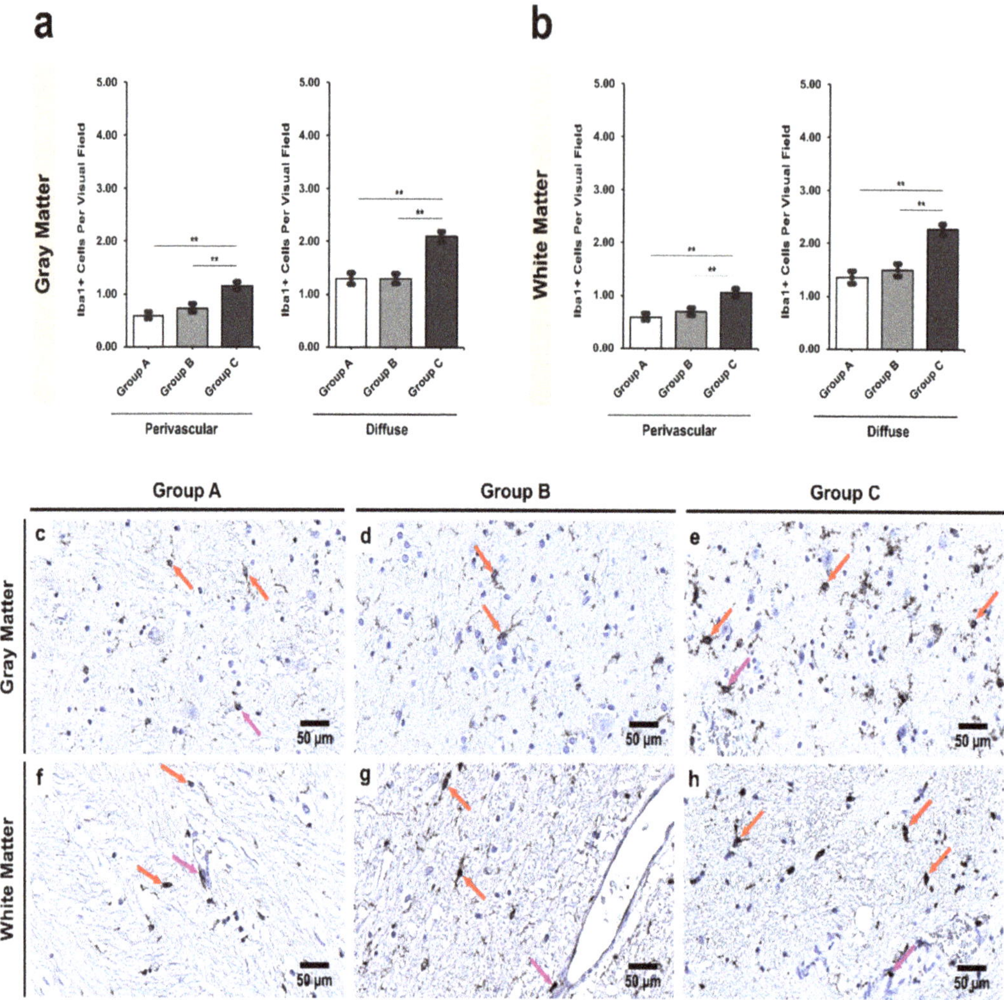

Figure 4. Intergroup analysis of Iba1+ cells per visual field in *Pars Reticulata* (SNpr) for the three studied groups in both (**a**) gray and (**b**) white matter. The analysis was done for both perivascular and diffuse locations. The bar plots indicate the average number of Iba1+ cells ± S.E. (Standard Error) seen per visual field. ** indicates a significant difference between the groups ($p < 0.05$ with Bonferroni correction is considered significant). Distribution of Iba1+ cells per visual field in (**c,f**) Group A (controls); (**d,g**) Group B (age-matched alcoholics) and (**e,h**) Group C (non-age-matched alcoholics). Pink arrows indicate perivascular Iba1+ cells while red arrows indicate diffuse Iba1+ cells. Original magnification, 400×. Scale bars, 50 μm.

3.3. Alcoholics Showed Significantly More CD11b+ Cells Than Controls in Both Gray and White Matter

In SNpc, there was a significant increase in CD11b+ population in non-age-matched alcoholics compared with controls (Table 6). There were also significant differences between age-matched alcoholics and controls in both gray and white matter ($p = 0.000$ and 0.034, respectively). In SNpr, although there were more CD11b+ cells noted in alcoholics in both gray and white matter, the difference was not statistically significant. Like Iba1 and CD68, CD11b was also found to be more abundant in white matter than gray matter.

Table 6. Distribution of CD11b+ cells per visual field in different regions of *Substantia Nigra* (SN).

Region	Group A	Group B	Group C	p Value [†]
Pars Compacta (SNpc)				
Gray matter	1.02 ± 0.07	1.47 ± 0.07	1.58 ± 0.09	<0.001 **
White matter	1.34 ± 0.09	1.60 ± 0.07	1.73 ± 0.09	0.005 **
p value [‡]	0.002 **	0.191	0.165	-
Pars Reticulata (SNpr)				
Gray matter	0.72 ± 0.08	0.83 ± 0.08	0.88 ± 0.07	0.143
White matter	0.79 ± 0.09	0.83 ± 0.07	0.95 ± 0.09	0.724
p value [‡]	0.815	0.270	0.603	-

[†] p value was calculated for Kruskal-Wallis ANOVA (intergroup analysis). [‡] p value was calculated for related-samples Wilcoxon Signed Rank Test (intragroup analysis). The numbers represent the average number of CD11b+ cells per visual field ± S.E. (standard error). ** indicates a significant difference between groups ($p < 0.05$ is considered significant with Bonferroni correction for Kruskal-Wallis ANOVA and without correction for related-samples Wilcoxon Signed Rank Test).

In the SNpc, the alcoholics showed a mild increase in CD11b+ cells perivascularly in both the gray and white matter, though the results were not statistically significant ($p = 0.330$ and 0.368, respectively). However, in the diffuse locations in both the gray and white matter, there was a significant increase between controls and age-matched alcoholics ($p = 0.001$ and 0.009, respectively; Figure 5). A similar significant increase was noted between controls and chronic alcoholics in both the gray and white matter ($p = 0.002$ and 0.048, respectively; Figure 5).

In the SNpr, however, no significant differences in CD11b+ cell numbers were noted in either gray or white matter in both perivascular ($p = 0.218$ and 0.614, respectively) and diffuse ($p = 0.528$ and 0.811, respectively) locations (Figure 6).

3.4. Exposure to Alcohol Induces Stronger Iba1 Expression Than CD11b in Microglial Cells

Overall, we observed four distinct microglial subpopulations based on the antibodies used in the present study (CD68, Iba1, and CD11b). These subpopulations were CD68+/Iba1+; CD68+/Iba1−; Iba1+/CD11b+ and Iba1+/CD11b− (Figure 7). In controls, the vast majority of microglial cells expressed CD68 but did not express Iba1 (CD68+/Iba1−) in either gray or white matter in SNpc and SNpr. However, a sharp decrease of this subpopulation was noted in age-matched alcoholics, which was followed by continued decline in non-age-matched alcoholics. Conversely, prolonged exposure to alcohol led to an increase in expression of Iba1+ microglial subpopulation (an increase by factor of about $1.7\times$ to $2.4\times$).

Within the Iba1+ microglial population, both cells with and without CD11b expression increased with prolonged exposure to alcohol (non-age-matched alcoholics). In both the regions of *Substantia Nigra* (SNpc and SNpr), we found that there were discrepancies in the distribution of CD11b+ and CD11b− subpopulations. In all regions, CD11b− microglia was the dominant subpopulation (even in controls; Figure 8). In fact, while the increase of CD11b+ subpopulation in alcoholics was around $1.4\times$ to $1.6\times$ in comparison with controls, the increase of CD11b− subpopulation was around $1.6\times$ to $2.2\times$ compared with controls.

3.5. Morphological Characterization of Microglial Subpopulations

The findings from the previous section suggest a rather sequential and preferential expression of certain immunohistochemical markers as microglial cells undergo the different stages of morphological transformations in response to prolonged alcohol exposure. The CD68+/Iba1− subpopulation, which was dominant in the controls (Figure 9a,b), showed different morphologic characteristics in SNpc and SNpr. Whilst in SNpc, microglia were sparsely branched, in the SNpr, microglia showed complexity in cell processes. On the other hand, in alcoholics, the CD68+/Iba1+ subpopulation was dominant, which was less ramified and possessed fewer branches and/or beaded processes (Figure 9c,d). Furthermore, such dystrophic changes were more prominently noticed in the SNpr.

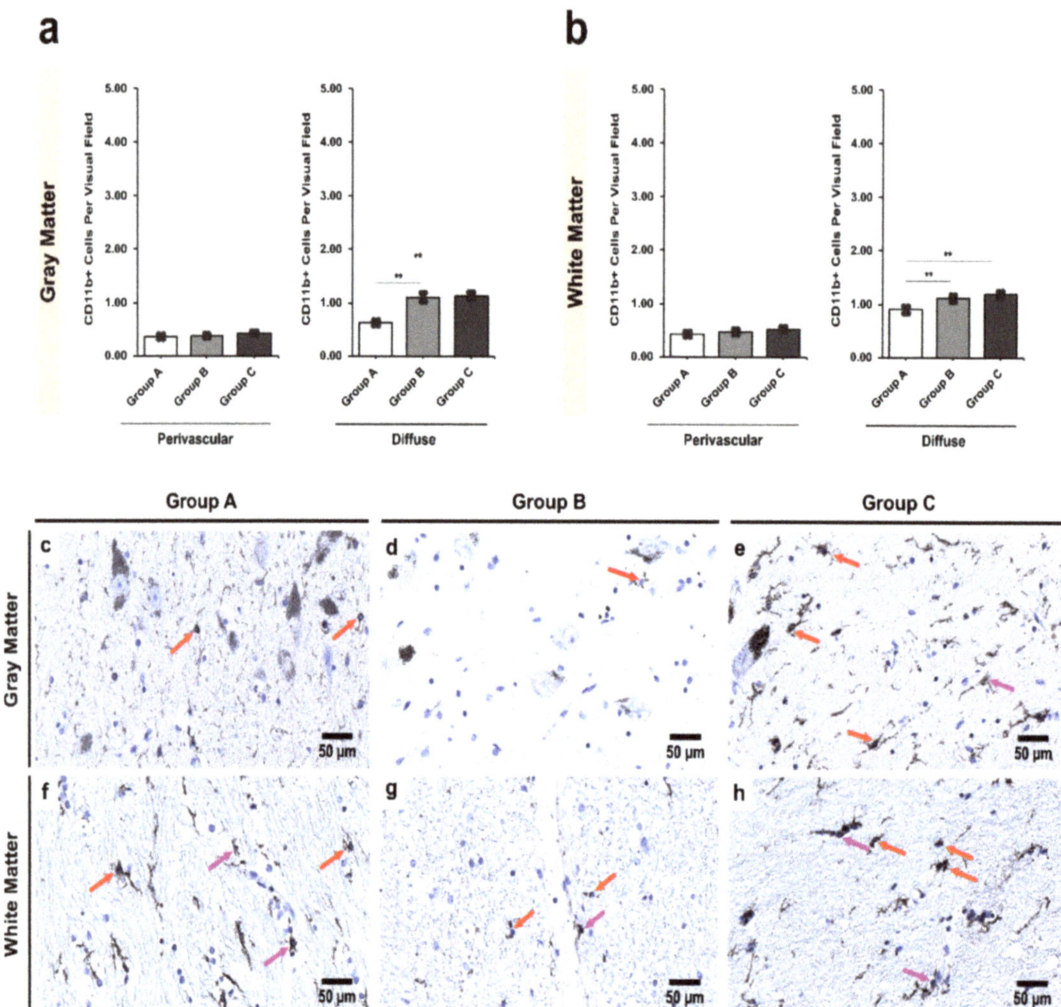

Figure 5. Intergroup analysis of CD11b+ cells per visual field in *Pars Compacta* (SNpc) for the three studied groups in both (**a**) gray and (**b**) white matter. The analysis was done for both perivascular and diffuse locations. The bar plots indicate the average number of CD11b+ cells ± S.E. (Standard Error) seen per visual field. ** indicates a significant difference between the groups ($p < 0.05$ with Bonferroni correction is considered significant). Distribution of CD11b+ cells per visual field in (**c,f**) Group A (controls); (**d,g**) Group B (age-matched alcoholics) and (**e,h**) Group C (non-age-matched alcoholics). Pink arrows indicate perivascular CD11b+ cells while red arrows indicate diffuse CD11b+ cells. Original magnification, $400\times$. Scale bars, 50 µm.

3.6. Alcoholics Showed Detectable HHV-6 Positivity and Viral Load

As expected, none of the individuals in the control group (Group A) showed positive HHV-6 immunostaining (Figure 10a). This was also confirmed with a negative result for HHV-6 genomic sequences (using nPCR). Amongst the alcoholics (Groups B and C), 25% (8/31) of the individuals were found to be positive for HHV-6 genomic sequences, with viral detection seen more commonly in the gray matter (88% of positive individuals) than the white matter (63% of positive individuals) in the SN region (Figure 10b,c). In all positive individuals, only the HHV-6B variant was detected, with an average viral load of 101,207.97 copies/10^6 cells.

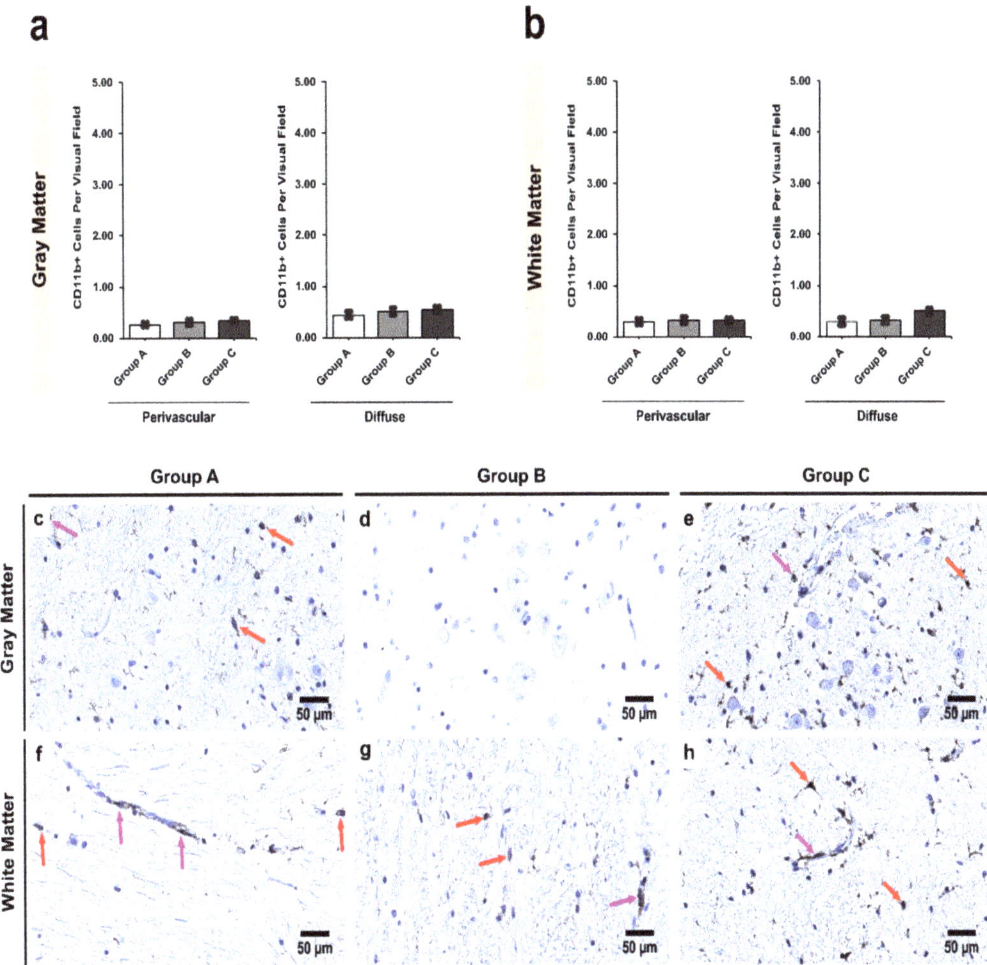

Figure 6. Intergroup analysis of CD11b+ cells per visual field in *Pars Reticulata* (SNpr) for the three studied groups in both (**a**) gray and (**b**) white matter. The analysis was done for both perivascular and diffuse locations. The bar plots indicate the average number of CD11b+ cells ± S.E. (Standard Error) seen per visual field. Distribution of CD11b+ cells per visual field in (**c**,**f**) Group A (controls); (**d**,**g**) Group B (age-matched alcoholics) and (**e**,**h**) Group C (non-age-matched alcoholics). Pink arrows indicate perivascular CD11b+ cells while red arrows indicate diffuse CD11b+ cells. Original magnification, 400×. Scale bars, 50 μm.

3.7. Alcohol and HHV-6 Infection in a Synergistic and Potentiating Relationship Cause Disruption of Homeostasis in the SN Region

To evaluate the synergistic relationship between alcoholism and HHV-6, we next evaluated their combined effects on microglia (CD68, CD11b and Iba1). This was achieved by comparing the HHV-6 positive alcoholics (from both Groups B and C) to the HHV-6 negative alcoholics (from both Groups B and C) and the controls (Figure 11). HHV-6 positive alcoholics showed greater decrease in the number of CD68+ cells per visual field than HHV-6 negative alcoholics, though the difference was not statistically significant ($p = 0.287$).

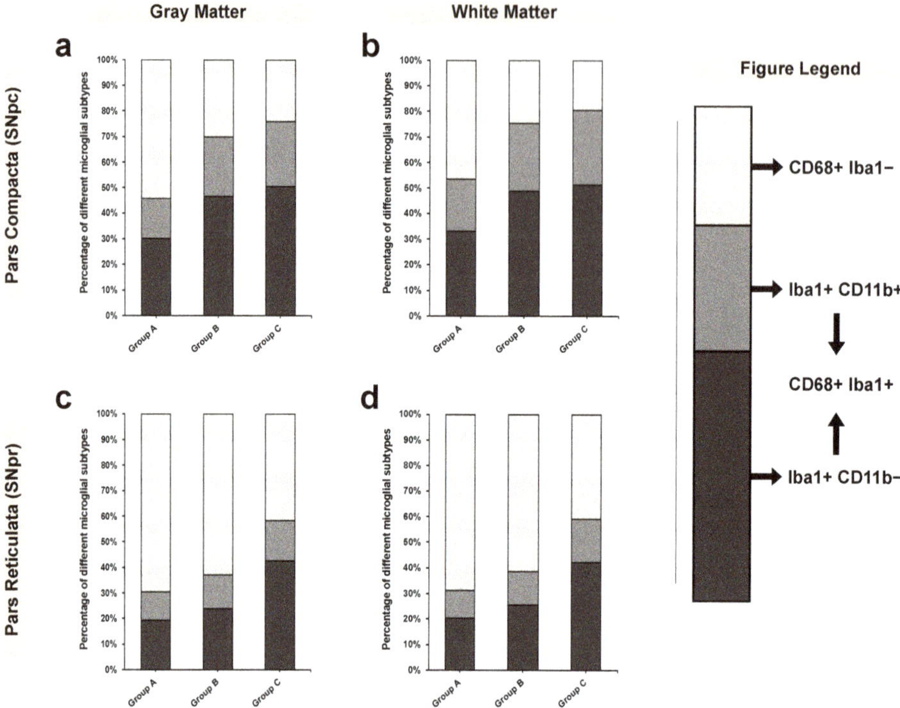

Figure 7. Distribution of different microglial subpopulations (in %) in both SNpc and SNpc for all three groups. The data has been shown for (**a,c**) gray and (**b,d**) white matter. The figure legend describes the various subpopulations identified including CD68+/Iba1− and CD68+/Iba1+. Within CD68+/Iba1+, two subtypes were identified namely, Iba1+/Cd11b+ and Iba1+/CD11b−.

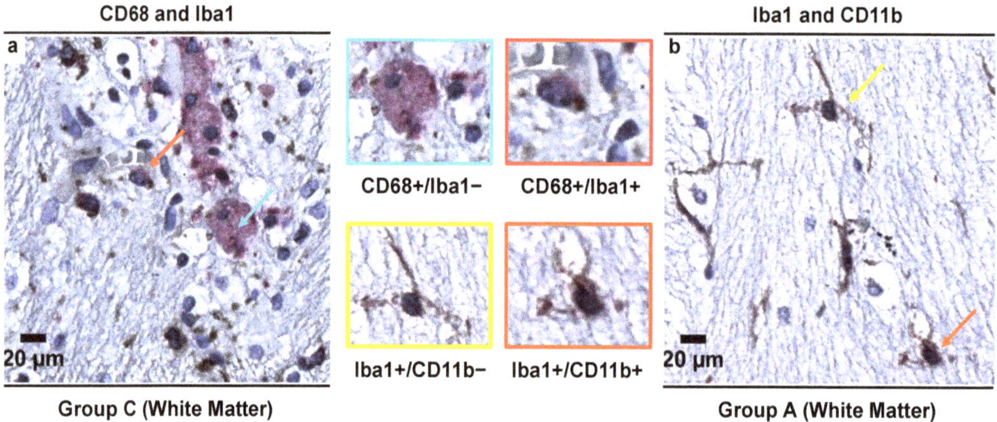

Figure 8. Representative images from double immunohistochemical staining (using two IHC markers in a single slide) showing different microglial subpopulations in the SN region. (**a**) Double CD68/Iba1 IHC staining showing two distinct subpopulations of microglia—CD68+/Iba1− (blue arrow) and CD68+/Iba1+ (red arrow) in the white matter of the alcoholics (Group C); (**b**) Double Iba1/CD11b IHC staining showing two distinct subpopulations of microglia—Iba1+/CD11b− (yellow arrow) and Iba1+/CD11b+ (orange arrow) in the white matter of the controls (Group A). In both images, brown color (DAB) indicates Iba1+ structures while red color (Permanent red) indicates CD68+ or CD11b+ structures. Original magnification, 400×. Scale bars, 20 μm.

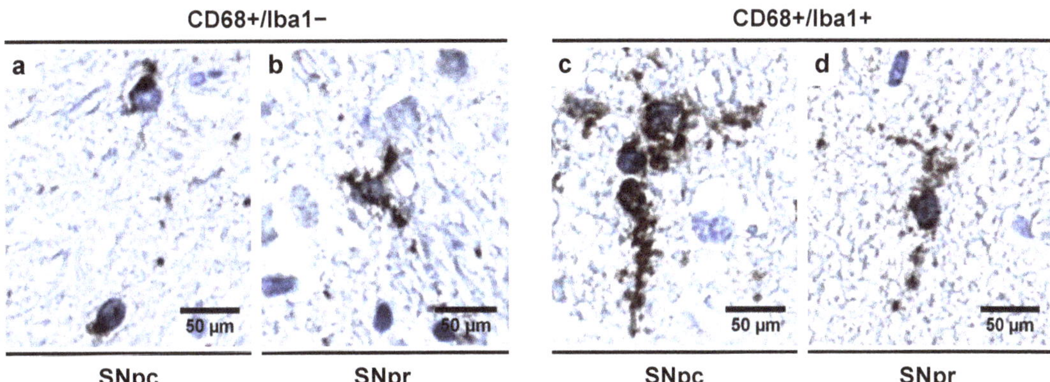

Figure 9. Morphology of different microglial subpopulations as seen in (**a**,**c**) SNpc and (**b**,**d**) SNpr. In CD68+/Iba1− subpopulation (dominant in controls in SN region), (**a**) microglia is sparsely branched in SNpc while (**b**) show complexity in cell processes in SNpr. This subpopulation represents microglia in *"controlled phagocytotic activated state"*. In CD68+/Iba1+ subpopulation (dominant in alcoholics in SN region), (**c**,**d**) microglia shows dystrophic changes seen as by appearance of beaded processes in SNpc and SNpr. Original magnification, 400×. Scale bars, 50 µm.

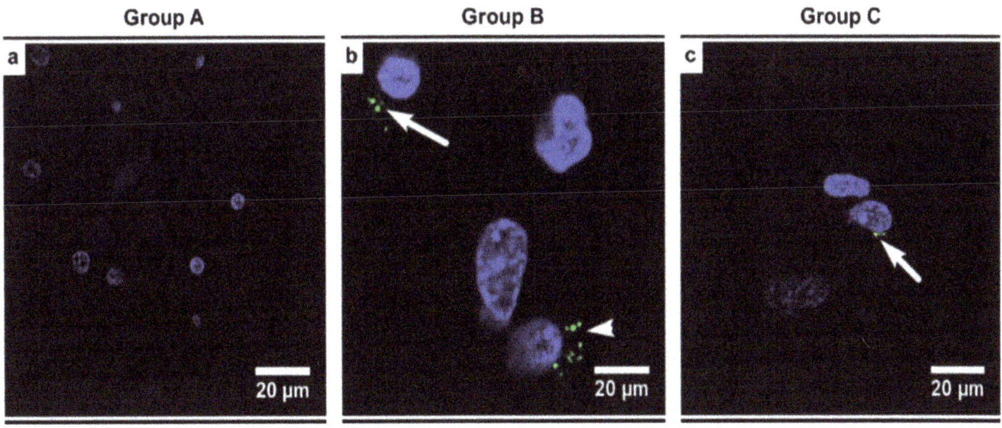

Figure 10. Representative images illustrating HHV-6 viral proteins (seen as green spots) using immunofluorescence (IF) in the gray matter of the alcoholics (confocal microscopy; original magnification × 1000). White arrows indicate the clusters of HHV-6 antigens—(**a**) controls (Group A) showed no positivity whilst (**b**) age-matched alcoholics (Group B) and (**c**) chronic alcoholics (Group C) showed HHV-6 immunopositivity in the cell cytoplasm.

For Iba1+ cells, there was a significant increase in HHV-6 positive alcoholics from both controls (2× increase; $p = 0.002$) and HHV-6 negative alcoholics (1.34× increase; $p = 0.028$). Further, although the difference between CD11b+ cells remained nonsignificant ($p = 0.265$), HHV-6 positive alcoholics showed an increase of about 1.4× from controls and 1.1 × from HHV-6 negative alcoholics (Figure 11c). This discrepancy in the expression induction of Iba1+ and CD11b+ in microglial cells demonstrates the synergistic effects of HHV-6 infection in terms of potentiating a stronger alcoholic induction of Iba1 and a relatively milder induction of CD11b (Figure 7), leading to the emergence of Iba1+/CD11b− subpopulation (Figure 8).

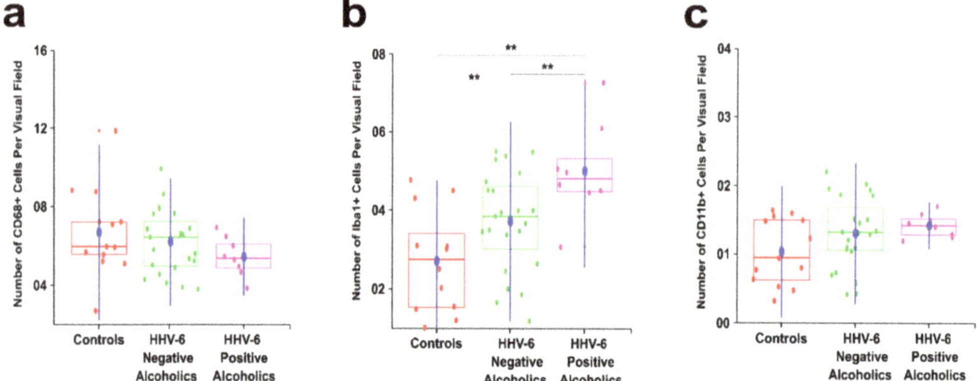

Figure 11. Intergroup analysis between HHV-6 positive alcoholics, HHV-6 negative alcoholics, and controls. The blue dot indicates the mean of the respective group. Red color indicates controls, green color indicates HHV-6 negative alcoholics and pink color indicates HHV-6 positive alcoholics. ** indicates a significant difference between the groups (Kruskal Wallis ANOVA $p < 0.05$ with Bonferroni correction is considered significant). (**a**) Boxplot showing the average number of CD68+ cells per visual field in the SN region; (**b**) Boxplot showing the average number of Iba1+ cells per visual field in the SN region; (**c**) Boxplot showing the average number of CD11b+ cells per visual field in the SN region.

4. Discussion

The potential for alcoholism and viral infection to disrupt brain homeostasis has long been a topic of interest and research. Yet, a lot of questions remain unanswered. In the present study, we report a comprehensive characterization of microglial functions, morphology, and quantification in individuals in the SN region, along with the specific changes that microglial cells undergo due to co-exposure to alcohol and HHV-6 infection. Since neuroinflammatory changes are also seen in the normal aging process, the inclusion of age-matched alcoholics (Group B) gives us the power to specifically exclude aging as an underlying factor in the study. Additionally, with the inclusion of postmortem human brain tissue from relatively young adults, the insights gained reflect the closest possible representation to changes in the adult human brain.

4.1. Abundance of CD68+/Iba1− Microglial Subpopulation in Controls Indicates a Special Physiological "Controlled Phagocytotic Activated State"

Iba1 has been reported to be a universal marker for all microglial subpopulations [56,69,70]. However, emerging evidence suggests otherwise [71]. Recent experiments on the murine microglia BV2 cell line showed that silencing Iba1 protein in microglial cells leads to a significant decrease in cell migration, proliferation, and cell adhesion capabilities [72]. Further, the authors demonstrated that such silencing leads to an increase in phagocytic activity along with upregulation of P2 × 7 (ATP-activated P2 purinergic receptors) functioning [72]. This is of immense interest since, in our study, we found that in controls, 45–55% of the entire microglial population in SNpc and 65–70% of the entire microglial population in SNpr did not demonstrate Iba1 expression (Figures 7 and 8). Such findings indicate that in the SN region, microglial cells show reduced motility but a high level of phagocytic activity. This specific state of activated microglia is different from the one induced by disease-specific activating mechanisms [72].

The above-mentioned findings compelled us to further explore the reasons behind the prevalence of such specific microglial phenotype in controls. Firstly, the SN region has one of the highest densities of microglial cells (around 12% of the total brain microglia population), with SNpr being denser than SNpc (Table 4) [73,74]. Secondly, we speculate that such a high proportion of microglia in a relatively smaller region (compared with cortex) gives microglial cells the ability to monitor the entire region sufficiently without the need for excessive migration-related activities. Thirdly, in a recent study by Ayata et al., the authors demonstrated that the microglial clearance activity rate is different in different regions of the brain, with microglial epigenetic regulators restricting clearance activity in certain regions (striatum, cortex) while promoting it in others (cerebellum) [75]. The authors further linked this epigenetic regulation to the rate of neuronal attrition [75].

Finally, De Biase et al. found that in a functional state (i.e., in the absence of pathology), microglia in SNpr exhibited structural complexity in the cell processes with a significant percentage of cell volume being occupied by lysosomes (10% of the cell volume), while in SNpc, microglial cells were sparsely branched with 6% cell volume being occupied by lysosomes [73], in line with our results (Figure 9a,b). In both SNpc and SNpr, the authors found that the lysosomal content (by % volume occupied) in microglia is more than that of other regions of the basal ganglia [73], leading us to correlate this with the increased baseline phagocytotic activity due to non-expression of Iba1 [72]. Our results, combined with those presented in the literature, lead us to postulate that the microglia in the SN region, under normal physiological conditions, are in what we are calling a *"controlled phagocytic activated state"* which needs further investigation (especially in the context of Parkinson's and other neurodegenerative diseases).

4.2. Decrease in the Number of CD68+ Microglia with Increase in Iba1+ Expression Shows Microglial Dystrophy Which Leads to Compensatory Mobility from Perivascular to Diffuse Locations

Alcohol has been shown in vitro (BV2 microglial cell line) to accelerate the production of reactive oxygen species (ROS) in microglial cells, which subsequently leads to activation of PARP (poly(ADP-ribose) polymerase) and oxidative-stress sensitive TRPM2 (transient receptor potential melastatin-related 2) channels, ultimately causing microglial death [76]. Additionally, alcohol exposure in rats leads to an increase in dystrophic microglia [77]. Although the classical hypothesis posits that hyperactivated and hypertrophic microglia are responsible for responding to stress or injury and are responsible for neurodegenerative changes, emerging evidence from animal studies suggests that alcohol-induced suppression of normal microglial function (dystrophic changes) also leads to neuronal cell death and subsequent neurodegenerative changes [77–79].

Long-term and constant microglial activation is thought to cause immune exhaustion and microglial burn-out, leading to dystrophic changes which are seen as cells being less ramified and possessing fewer branches and/or beaded microglial processes (Figure 9c,d), which ultimately causes cytorrhexis (accidental cell death) [80,81]. Additionally, previous studies have indicated that both microglial dystrophy and microglial activation are simultaneous processes [82]. Such dystrophic changes alter the distribution and quantity of CD68+ microglial cells in alcoholics (Table 4), which then require compensatory changes.

We hypothesize two major compensatory changes that the microglial cells demonstrate. The first is an increase in motility-related activities. This is supported by the hypothesis that if silencing of Iba1 leads to decreased cellular motility and migration [72], the increased expression would have the opposite effects. Moreover, it has been shown that such migrations are needed to remove cellular debris and provide support to salvageable cells in regions of damage [41,77,83]. It appears that in the SN region, such migratory movements are rather directed towards diffuse locations, away from perivascular locations (Figures 1 and 2). The second change is the decrease in phagocytic activities. Our results also support this notion, as we saw a decrease in the number of CD68+ cells (since CD68 is related to phagocytosis; Table 4) and the fact that Iba1 expression probably leads to a slowdown in phagocytosis [72]. It has been demonstrated that alcohol may be cytotoxic to microglia since there is a complete lack of phagocytic microglia after an acute alcohol binge [78]. Although we saw an increase in Iba1 expression, the authors reported otherwise [77,78]. We speculate that such differences are due to the region of the brain investigated, which further bolsters the findings that microglial regulation is not uniform across the different regions of the brain [73,75,84].

4.3. Alcohol and HHV-6 Infection Co-Induce and Accelerate Microglial Dystrophy

In a study by Bortolotti et al., spheroid 3D models of peripheral blood monocyte-derived microglia from healthy donors were infected with HHV-6A [45]. The authors reported a significant uptick in Iba1 and substance P (associated with neuroinflammation) expression and concluded that microglial cells were permissive to HHV-6A infection, which ultimately resulted in increased Aβ1-42 (beta-amyloid; implicated in Alzheimer's disease) expression [45]. Further, they demonstrated that HHV-6A infection leads to microglial cell migration to the site of infection (under paracrine effect), bringing their results in-line with our findings. However, in contrast to our findings, the authors reported that the ramified microglial morphology was predominant in response to HHV-6A infection [45]. Such morphological differences can be attributed to the effects of alcohol, as discussed above. Like HHV-6A, its cousin HHV-6B has also been shown to infect and dysregulate microglial function [85,86].

Leibovitch et al. demonstrated increased Iba1 expression in HHV-6B infected marmosets (primates) suffering from experimental autoimmune encephalomyelitis [87]. According to the fertile field hypothesis, in a heightened immune state (which can be induced by pathogens or self-antigens), there is a lower threshold for autoreactivity due to the expansion of autoreactive cells against the backdrop of immune response disbalance [88]. Whilst based on multiple clinical studies, the role of alcohol and HHV-6 individually remains controversial in the pathogenesis of Alzheimer's disease (AD) and multiple sclerosis (MS), we hypothesize that joint exposure to alcohol and HHV-6 is the key trigger required for clinical manifestations and progression of AD and MS (this hypothesis requires in vitro and in vivo validation). In such a case, we hypothesize, that HHV-6A and HHV-6B exposure (or reactivation, immune suppression, alcohol-mediated activation) creates a fertile field which would probably lead to AD- and MS-like symptoms, while alcohol exploits the lowered autoreactivity threshold to potentiate the neuroinflammatory and neurodegenerative changes.

4.4. Dominance of Iba1+/CD11b− Microglial Subpopulation in Alcoholics Leads to Chronic Inflammation, Hyperalgesia, and Allodynia

Our results indicate that the expression of CD11b was weaker than Iba1 (Figures 7 and 8), which led to the subsequent dominance of Iba1+/CD11b− microglial cells in alcoholics in both gray and white matter in SNpr and SNpc (up to 40–50% of all microglial/macrophage population). Previous studies in mice-brain-derived microglial cells have demonstrated that CD11b silences TLR4 (toll-like receptor 4) -induced inflammatory responses [89]. Such physiological silencing in controls (wild-type mice) leads to increased microglial production of IL-10 and TGF-β (anti-inflammatory cytokines), coupled with decreased production of IL-6 and TNF-α (pro-inflammatory cytokines) [89], thereby indicating that silencing of CD11b promotes a pro-inflammatory microenvironment. Further, deficiency of CD11b has been shown to promote lipopolysaccharide-induced reactive oxygen species production, leading to the mice being more susceptible to endotoxin shock [86,90].

Recent studies have shown the involvement of SNpc and SNpr in the nociceptive pathways. SNpc and the ventral tegmental area receives nociceptive-related afferents from the parabrachial nucleus (PBN) and transfers these signals to subthalamic nucleus (STN) [91]. From STN, efferent signals reach SNpr which, in turn, sends signals to superior colliculus and PBN [92,93], thereby creating an anatomic nociceptive signal processing circuit. Alcohol has been implicated in modulating the reward mechanisms in STN (lesions in STN have been shown to decrease the motivation for alcohol intake) [94,95]. Such elevation of activity in STN by alcohol could lead to chronic pain and associated symptoms, as seen in Parkinson's disease [91,96] and other neurodegenerative disease such as AD, motor neuron disease, Huntington's disease, spinocerebellar ataxia, and spinal muscular atrophy [96,97]. Apart from the direct effects of alcohol on STN to increase pain sensitivity, we speculate that the dominance of CD11b− microglia in SN region also plays an amplifying role. CD11b−microglia-deficient mice models have been shown to be more susceptible to thermal and mechanical allodynia, which, in turn, have been attributed to increased microglial inflammatory response [89]. Altogether, our results indicate that attenuation of microglial CD11b in SN due to alcohol exposure leads to chronic and sustained inflammation, which leads to paradoxical withdrawal hyperalgesia and allodynia (via STN circuit), complaints commonly reported as part of alcohol withdrawal process [98,99].

4.5. Limitations of the Present Study

Nonetheless, the results obtained in the present study raise more questions than answers and are constrained by some limitations. Firstly, the number of investigated individuals per group was relatively low, and investigations in larger cohorts are needed. Secondly, to obtain a comprehensive view of the neuroinflammatory and degenerative changes seen in SN, a wider spectrum of immune markers is needed. Thirdly, we only investigated the role of microglia in alcohol and HHV-6 infection co-exposure-mediated changes in SN, and since other regulatory cells like astrocytes, macrophages, monocytes, neutrophils, etc. also take part in mediating these changes, future studies are needed to elucidate the specific roles of these cells. Finally, since HHV-6 infection precedes alcohol exposure (99% of us contract HHV-6 by the age of 2–3 years), it is difficult to elucidate whether HHV-6 infection amplifies alcohol-mediated damage or the other way around. Studies in animal models and cell cultures may shed some light on this matter in the future.

5. Conclusions

In the present study, we showed that the neuroinflammation-associated microglia changes in chronic alcoholics synergizes with the neuroinflammatory changes associated with HHV-6 infection, which results in a new, enhanced neuroinflammatory phenotype when both conditions are present. Further, the largely coherent agreement between the data from non-age matched alcoholics and age matched alcoholics enabled us to exclude ageing as an underlying covariate in the analysis. Finally, the following three conclusions are noteworthy.

Firstly, CD68+/Iba1− is the predominant microglial subpopulation in physiological conditions in the SN region and represents a *"controlled phagocytotic activated state"* with less migration-related and more phagocytotic activity. Alcoholics showed a significant decline of this subpopulation in both SNpc and SNpr.

Secondly, alcohol- and virus-induced immune exhaustion leads to progressive dystrophic degeneration of microglia via upregulation of Iba1 expression, leading to the emergence of CD68+/Iba1+ microglia (seen with less branching and beaded processes), which may be associated with neuroinflammatory changes. Microglia then compensate for these changes by increasing their migratory activities. Such a phenomenon is universal in all the regions of SN (SNpc and SNpr).

Finally, an indiscriminate increase in Iba1 expression when compared to CD11b in microglial cells in response to HHV-6 and alcohol exposure leads to the emergence of various subpopulations with Iba1+/CD11b− microglia being associated with the sustaining and promotion of neuroinflammatory changes in the SN region.

Author Contributions: N.J. and S.S. (Sandra Skuja) conceptualized the study while S.S. (Sandra Skuja), N.J., M.S., M.M. and S.S. (Samanta Strojeva) were responsible for methodology. Data and statistical analysis were done by N.J. while visualizations were done by N.J and S.S. (Sandra Skuja). Validation of the study protocol, project supervision, and funding acquisition was done by S.S. (Sandra Skuja). Original draft was prepared by N.J., while revisions were done by S.S. (Sandra Skuja), N.J., M.S., M.M. and S.S. (Samanta Strojeva). All authors have read and approved the final version of the manuscript for publication.

Funding: The present study was funded by Fundamental & Applied Research Projects (FLPP), Latvian Council of Science wide no. lzp-2020/2-0069 (The role of human herpesvirus-6 infection and alcohol abuse in the development of neuroinflammation).

Institutional Review Board Statement: The study was conducted according to the guidelines of the Declaration of Helsinki and approved by the Ethics Committee of Rīga Stradiņš University (Decision No. 6-1/12/9).

Informed Consent Statement: In this study, informed patient consent was waived due to use of postmortem specimens taken during the conventional autopsies. Protocols for obtaining postmortem brain tissue complied with all institutional guidelines especially in respect to identity confidentiality.

Data Availability Statement: All the data used in this study are available from the corresponding author upon request.

Conflicts of Interest: The authors declare no conflict of interest.

References

1. Louten, J. Herpesviruses. In *Essential Human Virology*; Elsevier (Academic Press): Cambridge, MA, USA, 2016; pp. 235–256, ISBN 978-0-12-800947-5.
2. Salahuddin, S.; Ablashi, D.; Markham, P.; Josephs, S.; Sturzenegger, S.; Kaplan, M.; Halligan, G.; Biberfeld, P.; Wong-Staal, F.; Kramarsky, B.; et al. Isolation of a New Virus, HBLV, in Patients with Lymphoproliferative Disorders. *Science* **1986**, *234*, 596–601. [CrossRef] [PubMed]
3. Collot, S.; Petit, B.; Bordessoule, D.; Alain, S.; Touati, M.; Denis, F.; Ranger-Rogez, S. Real-Time PCR for Quantification of Human Herpesvirus 6 DNA from Lymph Nodes and Saliva. *J. Clin. Microbiol.* **2002**, *40*, 2445–2451. [CrossRef]
4. Di Luca, D.; Mirandola, P.; Ravaioli, T.; Frigatti, A.; Bovenzi, P.; Monini, P.; Cassai, E.; Sighinolfi, L.; Dolcetti, R. Human Herpesviruses 6 and 7 in Salivary Glands and Shedding in Saliva of Healthy and Human Immunodeficiency Virus Positive Individuals. *J. Med. Virol.* **1995**, *45*, 462–468. [CrossRef] [PubMed]
5. Mukai, T.; Yamamoto, T.; Kondo, T.; Kondo, K.; Okuno, T.; Kosuge, H.; Yamanishi, K. Molecular Epidemiological Studies of Human Herpesvirus 6 in Families. *J. Med. Virol.* **1994**, *42*, 224–227. [CrossRef]
6. Tanaka-Taya, K.; Kondo, T.; Mukai, T.; Miyoshi, H.; Yamamoto, Y.; Okada, S.; Yamanishi, K. Seroepidemiological Study of Human Herpesvirus-6 and -7 in Children of Different Ages and Detection of These Two Viruses in Throat Swabs by Polymerase Chain Reaction. *J. Med. Virol.* **1996**, *48*, 88–94. [CrossRef]
7. Boutolleau, D.; Cointe, D.; Dejean, A.G.; Mace, M.; Agut, H.; Keros, L.G.; Ingrand, D. No Evidence for a Major Risk of Roseolovirus Vertical Transmission during Pregnancy. *Clin. Infect. Dis.* **2003**, *36*, 1634–1635. [CrossRef] [PubMed]
8. Dahl, H.; Fjaertoft, G.; Norsted, T.; Wang, F.; Mousavi-Jazi, M.; Linde, A. Reactivation of Human Herpesvirus 6 during Pregnancy. *J. Infect. Dis.* **1999**, *180*, 2035–2038. [CrossRef]
9. De Bolle, L.; Naesens, L.; De Clercq, E. Update on Human Herpesvirus 6 Biology, Clinical Features, and Therapy. *Clin. Microbiol. Rev.* **2005**, *18*, 217–245. [CrossRef]
10. Ablashi, D.; Agut, H.; Alvarez-Lafuente, R.; Clark, D.A.; Dewhurst, S.; DiLuca, D.; Flamand, L.; Frenkel, N.; Gallo, R.; Gompels, U.A.; et al. Classification of HHV-6A and HHV-6B as Distinct Viruses. *Arch. Virol.* **2014**, *159*, 863–870. [CrossRef]
11. Caruso, A.; Rotola, A.; Comar, M.; Favilli, F.; Galvan, M.; Tosetti, M.; Campello, C.; Caselli, E.; Alessandri, G.; Grassi, M.; et al. HHV-6 Infects Human Aortic and Heart Microvascular Endothelial Cells, Increasing Their Ability to Secrete Proinflammatory Chemokines. *J. Med. Virol.* **2002**, *67*, 528–533. [CrossRef]
12. Chan, P.K.S.; Ng, H.-K.; Hui, M.; Cheng, A.F. Prevalence and Distribution of Human Herpesvirus 6 Variants A and B in Adult Human Brain. *J. Med. Virol.* **2001**, *64*, 42–46. [CrossRef]
13. Donati, D.; Akhyani, N.; Fogdell-Hahn, A.; Cermelli, C.; Cassiani-Ingoni, R.; Vortmeyer, A.; Heiss, J.D.; Cogen, P.; Gaillard, W.D.; Sato, S.; et al. Detection of Human Herpesvirus-6 in Mesial Temporal Lobe Epilepsy Surgical Brain Resections. *Neurology* **2003**, *61*, 1405–1411. [CrossRef]
14. Ishiguro, N.; Yamada, S.; Takahashi, T.; Takahashi, Y.; Togashi, T.; Okuno, T.; Yamanishi, K. Meningo-Encephalitis Associated with HHV-6 Related Exanthem Subitum. *Acta Paediatr.* **1990**, *79*, 987–989. [CrossRef]
15. Suga, S.; Yoshikawa, T.; Asano, Y.; Kozawa, T.; Nakashima, T.; Kobayashi, I.; Yazaki, T.; Yamamoto, H.; Kajita, Y.; Ozaki, T.; et al. Clinical and Virological Analyses of 21 Infants with Exanthem Subitum (Roseola Infantum) and Central Nervous System Complications. *Ann. Neurol.* **1993**, *33*, 597–603. [CrossRef]
16. Eimer, W.A.; Vijaya Kumar, D.K.; Navalpur Shanmugam, N.K.; Rodriguez, A.S.; Mitchell, T.; Washicosky, K.J.; György, B.; Breakefield, X.O.; Tanzi, R.E.; Moir, R.D. Alzheimer's Disease-Associated β-Amyloid Is Rapidly Seeded by Herpesviridae to Protect against Brain Infection. *Neuron* **2018**, *99*, 56–63.e3. [CrossRef] [PubMed]
17. Readhead, B.; Haure-Mirande, J.-V.; Funk, C.C.; Richards, M.A.; Shannon, P.; Haroutunian, V.; Sano, M.; Liang, W.S.; Beckmann, N.D.; Price, N.D.; et al. Multiscale Analysis of Independent Alzheimer's Cohorts Finds Disruption of Molecular, Genetic, and Clinical Networks by Human Herpesvirus. *Neuron* **2018**, *99*, 64–82.e7. [CrossRef] [PubMed]
18. Dunn, N.; Kharlamova, N.; Fogdell-Hahn, A. The Role of Herpesvirus 6A and 6B in Multiple Sclerosis and Epilepsy. *Scand. J. Immunol.* **2020**, *92*. [CrossRef] [PubMed]
19. Opsahl, M.L. Early and Late HHV-6 Gene Transcripts in Multiple Sclerosis Lesions and Normal Appearing White Matter. *Brain* **2005**, *128*, 516–527. [CrossRef] [PubMed]

20. Prusty, B.K.; Gulve, N.; Govind, S.; Krueger, G.R.F.; Feichtinger, J.; Larcombe, L.; Aspinall, R.; Ablashi, D.V.; Toro, C.T. Active HHV-6 Infection of Cerebellar Purkinje Cells in Mood Disorders. *Front. Microbiol.* **2018**, *9*, 1955. [CrossRef]

21. Kobayashi, N.; Oka, N.; Takahashi, M.; Shimada, K.; Ishii, A.; Tatebayashi, Y.; Shigeta, M.; Yanagisawa, H.; Kondo, K. Human Herpesvirus 6B Greatly Increases Risk of Depression by Activating Hypothalamic-Pituitary -Adrenal Axis during Latent Phase of Infection. *iScience* **2020**, *23*, 101187. [CrossRef]

22. Beović, B.; Pecaric-Meglic, N.; Marin, J.; Bedernjak, J.; Muzlovic, I.; Cizman, M. Fatal Human Herpesvirus 6-Associated Multifocal Meningoencephalitis in an Adult Female Patient. *Scand. J. Infect. Dis.* **2001**, *33*, 942–944. [CrossRef]

23. Drobyski, W.R.; Knox, K.K.; Majewski, D.; Carrigan, D.R. Fatal Encephalitis Due to Variant B Human Herpesvirus-6 Infection in a Bone Marrow-Transplant Recipient. *N. Engl. J. Med.* **1994**, *330*, 1356–1360. [CrossRef] [PubMed]

24. Novoa, L.J.; Nagra, R.M.; Nakawatase, T.; Edwards-Lee, T.; Tourtellotte, W.W.; Cornford, M.E. Fulminant Demyelinating Encephalomyelitis Associated with Productive HHV-6 Infection in an Immunocompetent Adult. *J. Med. Virol.* **1997**, *52*, 301–308. [CrossRef]

25. Seeley, W.W.; Marty, F.M.; Holmes, T.M.; Upchurch, K.; Soiffer, R.J.; Antin, J.H.; Baden, L.R.; Bromfield, E.B. Post-Transplant Acute Limbic Encephalitis: Clinical Features and Relationship to HHV6. *Neurology* **2007**, *69*, 156–165. [CrossRef]

26. Provenzale, J.M.; van Landingham, K.E.; Lewis, D.V.; Mukundan, S.; White, L.E. Extrahippocampal Involvement in Human Herpesvirus 6 Encephalitis Depicted at MR Imaging. *Radiology* **2008**, *249*, 955–963. [CrossRef] [PubMed]

27. Webb, D.W.; Bjornson, B.H.; Sargent, M.A.; Hukin, J.; Thomas, E.E. Basal Ganglia Infarction Associated with HHV-6 Infection. *Arch. Dis. Child.* **1997**, *76*, 362–364. [CrossRef]

28. Wainwright, M.S.; Martin, P.L.; Morse, R.P.; Lacaze, M.; Provenzale, J.M.; Coleman, R.E.; Morgan, M.A.; Hulette, C.; Kurtzberg, J.; Bushnell, C.; et al. Human Herpesvirus 6 Limbic Encephalitis after Stem Cell Transplantation. *Ann. Neurol.* **2001**, *50*, 612–619. [CrossRef]

29. Singh, N.; Paterson, D.L. Encephalitis caused by human herpesvirus-6 in transplant recipients: Relevance of a Novel Neurotropic Virus. *Transplantation* **2000**, *69*, 2474–2479. [CrossRef]

30. Pulickal, A.S.; Ramachandran, S.; Rizek, P.; Narula, P.; Schubert, R. Chorea and Developmental Regression Associated with Human Herpes Virus-6 Encephalitis. *Pediatric Neurol.* **2013**, *48*, 249–251. [CrossRef]

31. DeLong, M.; Wichmann, T. Changing Views of Basal Ganglia Circuits and Circuit Disorders. *Clin. EEG Neurosci.* **2010**, *41*, 61–67. [CrossRef]

32. Sonne, J.; Reddy, V.; Beato, M.R. *Neuroanatomy, Substantia Nigra*; StatPearls Publishing: Treasure Island, FL, USA, 2021. Available online: https://www.ncbi.nlm.nih.gov/books/NBK536995/ (accessed on 25 June 2021).

33. Fedorow, H.; Tribl, F.; Halliday, G.; Gerlach, M.; Riederer, P.; Double, K. Neuromelanin in Human Dopamine Neurons: Comparison with Peripheral Melanins and Relevance to Parkinson's Disease. *Prog. Neurobiol.* **2005**, *75*, 109–124. [CrossRef]

34. Javed, N.; Cascella, M. *Neuroanatomy, Globus Pallidus*; StatPearls Publishing: Treasure Island, FL, USA, 2021. Available online: https://www.ncbi.nlm.nih.gov/books/NBK557755/ (accessed on 25 June 2021).

35. Phillips, T.J.; Shen, E.H.; McKinnon, C.S.; Burkhart-Kasch, S.; Lessov, C.N.; Palmer, A.A. Forward, Relaxed, and Reverse Selection for Reduced and Enhanced Sensitivity to Ethanol's Locomotor Stimulant Effects in Mice. *Alcohol. Clin. Exp. Res.* **2002**, *26*, 593–602. [CrossRef] [PubMed]

36. Dudek, B.C.; Phillips, T.J. Distinctions among Sedative, Disinhibitory, and Ataxic Properties of Ethanol in Inbred and Selectively Bred Mice. *Psychopharmacology* **1990**, *101*, 93–99. [CrossRef] [PubMed]

37. Arizzi-LaFrance, M.N.; Correa, M.; Aragon, C.M.G.; Salamone, J.D. Motor Stimulant Effects of Ethanol Injected into the *Substantia Nigra Pars Reticulata*: Importance of Catalase-Mediated Metabolism and the Role of Acetaldehyde. *Neuropsychopharmacology* **2006**, *31*, 997–1008. [CrossRef] [PubMed]

38. Leclercq, S.; de Timary, P.; Delzenne, N.M.; Stärkel, P. The Link between Inflammation, Bugs, the Intestine and the Brain in Alcohol Dependence. *Transl. Psychiatry* **2017**, *7*, e1048. [CrossRef]

39. Ostergren, A.; Fredriksson, A.; Brittebo, E.B. Norharman-induced motoric impairment in mice: Neurodegeneration and glial activation in *Substantia Nigra*. *J. Neural. Transm.* **2006**, *113*, 313–329. [CrossRef] [PubMed]

40. Streit, W.J.; Mrak, R.E.; Griffin, W.S.T. Microglia and Neuroinflammation: A Pathological Perspective. *J. Neuroinflamm.* **2004**, *1*, 1–14. [CrossRef] [PubMed]

41. Bachiller, S.; Jiménez-Ferrer, I.; Paulus, A.; Yang, Y.; Swanberg, M.; Deierborg, T.; Boza-Serrano, A. Microglia in Neurological Diseases: A Road Map to Brain-Disease Dependent-Inflammatory Response. *Front. Cell Neurosci.* **2018**, *12*, 488. [CrossRef]

42. Li, Q.; Barres, B.A. Microglia and Macrophages in Brain Homeostasis and Disease. *Nat. Rev. Immunol.* **2018**, *18*, 225–242. [CrossRef] [PubMed]

43. Heindl, S.; Gesierich, B.; Benakis, C.; Llovera, G.; Duering, M.; Liesz, A. Automated Morphological Analysis of Microglia after Stroke. *Front. Cell. Neurosci.* **2018**, *12*, 106. [CrossRef]

44. Wolf, Y.; Yona, S.; Kim, K.-W.; Jung, S. Microglia, Seen from the CX3CR1 Angle. *Front. Cell. Neurosci.* **2013**, *7*, 26. [CrossRef] [PubMed]

45. Bortolotti, D.; Gentili, V.; Rotola, A.; Caselli, E.; Rizzo, R. HHV-6A Infection Induces Amyloid-Beta Expression and Activation of Microglial Cells. *Alzheimer's Res. Ther.* **2019**, *11*, 1–11. [CrossRef] [PubMed]

46. Tang, Y.; Le, W. Differential Roles of M1 and M2 Microglia in Neurodegenerative Diseases. *Mol. Neurobiol.* **2016**, *53*, 1181–1194. [CrossRef] [PubMed]

47. Jarrahi, A.; Braun, M.; Ahluwalia, M.; Gupta, R.V.; Wilson, M.; Munie, S.; Ahluwalia, P.; Vender, J.R.; Vale, F.L.; Dhandapani, K.M.; et al. Revisiting Traumatic Brain Injury: From Molecular Mechanisms to Therapeutic Interventions. *Biomedicines* **2020**, *8*, 389. [CrossRef] [PubMed]

48. Dauber, W.; Feneis, H.; Feneis, H. *Pocket Atlas of Human Anatomy: Founded by Heinz Feneis*, 5th rev. ed.; Thieme Stuttgart: New York, NY, USA, 2007; ISBN 978-3-13-511205-3.

49. Harper, C.; Dixon, G.; Sheedy, D.; Garrick, T. Neuropathological alterations in alcoholic brains. Studies arising from the New South Wales Tissue Resource Centre. *Prog. Neuro-Psychopharmacol. Biol. Psychiatry* **2003**, *27*, 951–961. [CrossRef]

50. Skuja, S.; Groma, V.; Ravina, K.; Tarasovs, M.; Cauce, V.; Teteris, O. Protective reactivity and alteration of the brain tissue in alcoholics evidenced by SOD1, MMP9 immunohistochemistry, and electron microscopy. *Ultrastruct. Pathol.* **2013**, *37*, 346–355. [CrossRef]

51. Skuja, S.; Svirskis, S.; Murovska, M. Human Herpesvirus-6 and -7 in the Brain Microenvironment of Persons with Neurological Pathology and Healthy People. *Int. J. Mol. Sci.* **2021**, *22*, 2364. [CrossRef]

52. Chistiakov, D.A.; Killingsworth, M.C.; Myasoedova, V.A.; Orekhov, A.N.; Bobryshev, Y.V. CD68/Macrosialin: Not Just a Histochemical Marker. *Lab. Investig.* **2017**, *97*, 4–13. [CrossRef]

53. Holness, C.; Simmons, D. Molecular Cloning of CD68, a Human Macrophage Marker Related to Lysosomal Glycoproteins. *Blood* **1993**, *81*, 1607–1613. [CrossRef]

54. Gottfried, E.; Kunz-Schughart, L.A.; Weber, A.; Rehli, M.; Peuker, A.; Müller, A.; Kastenberger, M.; Brockhoff, G.; Andreesen, R.; Kreutz, M. Expression of CD68 in Non-Myeloid Cell Types. *Scand. J. Immunol.* **2008**, *67*, 453–463. [CrossRef]

55. Hopperton, K.E.; Mohammad, D.; Trépanier, M.O.; Giuliano, V.; Bazinet, R.P. Markers of Microglia in Post-Mortem Brain Samples from Patients with Alzheimer's Disease: A Systematic Review. *Mol. Psychiatry* **2018**, *23*, 177–198. [CrossRef]

56. Walker, D.G.; Lue, L.-F. Immune Phenotypes of Microglia in Human Neurodegenerative Disease: Challenges to Detecting Microglial Polarization in Human Brains. *Alzheimer Res. Ther.* **2015**, *7*, 56. [CrossRef]

57. Perego, C.; Fumagalli, S.; De Simoni, M.-G. Temporal Pattern of Expression and Colocalization of Microglia/Macrophage Phenotype Markers Following Brain Ischemic Injury in Mice. *J. Neuroinflamm.* **2011**, *8*, 174. [CrossRef]

58. Fischer, H.-G.; Reichmann, G. Brain Dendritic Cells and Macrophages/Microglia in Central Nervous System Inflammation. *J. Immunol.* **2001**, *166*, 2717–2726. [CrossRef]

59. Bachstetter, A.D.; Van Eldik, L.J.; Schmitt, F.A.; Neltner, J.H.; Ighodaro, E.T.; Webster, S.J.; Patel, E.; Abner, E.L.; Kryscio, R.J.; Nelson, P.T. Disease-Related Microglia Heterogeneity in the Hippocampus of Alzheimer's Disease, Dementia with Lewy Bodies, and Hippocampal Sclerosis of Aging. *Acta Neuropathol. Commun.* **2015**, *3*, 32. [CrossRef]

60. Streit, W.J.; Braak, H.; Xue, Q.-S.; Bechmann, I. Dystrophic (Senescent) Rather than Activated Microglial Cells Are Associated with Tau Pathology and Likely Precede Neurodegeneration in Alzheimer's Disease. *Acta Neuropathol.* **2009**, *118*, 475–485. [CrossRef] [PubMed]

61. Santpere, G.; Telford, M.; Andrés-Benito, P.; Navarro, A.; Ferrer, I. The Presence of Human Herpesvirus 6 in the Brain in Health and Disease. *Biomolecules* **2020**, *10*, 1520. [CrossRef]

62. Siddon, A.; Lozovatsky, L.; Mohamed, A.; Hudnall, S.D. Human Herpesvirus 6 Positive Reed-Sternberg Cells in Nodular Sclerosis Hodgkin Lymphoma. *Br. J. Haematol.* **2012**, *158*, 635–643. [CrossRef] [PubMed]

63. Vandamme, A.-M.; Fransen, K.; Debaisieux, L.; Marissens, D.; Sprecher, S.; Vaira, D.; Vandenbroucke, A.T.; Verhofstede, C.; Van Dooren, S.; Goubau, P.; et al. Standardisation of Primers and an Algorithm for HIV-1 Diagnostic PCR Evaluated in Patients Harbouring Strains of Diverse Geographical Origin. *J. Virol. Methods* **1995**, *51*, 305–316. [CrossRef]

64. Sultanova, A.; Cistjakovs, M.; Gravelsina, S.; Chapenko, S.; Roga, S.; Cunskis, E.; Nora-Krukle, Z.; Groma, V.; Ventina, I.; Murovska, M. Association of Active Human Herpesvirus-6 (HHV-6) Infection with Autoimmune Thyroid Gland Diseases. *Clin. Microbiol. Infect.* **2017**, *23*, e1–e5. [CrossRef] [PubMed]

65. Lyall, E.G.H.; Cubie, H.A. Human Herpesvirus-6 DNA in the Saliva of Paediatric Oncology Patients and Controls. *J. Med. Virol.* **1995**, *47*, 317–322. [CrossRef]

66. Carlson, S.L.; Parrish, M.E.; Springer, J.E.; Doty, K.; Dossett, L. Acute Inflammatory Response in Spinal Cord Following Impact Injury. *Exp. Neurol.* **1998**, *151*, 77–88. [CrossRef] [PubMed]

67. Zrzavy, T.; Machado-Santos, J.; Christine, S.; Baumgartner, C.; Weiner, H.L.; Butovsky, O.; Lassmann, H. Dominant Role of Microglial and Macrophage Innate Immune Responses in Human Ischemic Infarcts: Inflammation in Ischemic Lesions. *Brain Pathol.* **2018**, *28*, 791–805. [CrossRef] [PubMed]

68. Lee, J.; Hamanaka, G.; Lo, E.H.; Arai, K. Heterogeneity of Microglia and Their Differential Roles in White Matter Pathology. *CNS Neurosci. Ther.* **2019**, *25*, 1290–1298. [CrossRef] [PubMed]

69. Ohsawa, K.; Imai, Y.; Kanazawa, H.; Sasaki, Y.; Kohsaka, S. Involvement of Iba1 in Membrane Ruffling and Phagocytosis of Macrophages/Microglia. *J. Cell Sci.* **2000**, *113*, 3073–3084. [CrossRef] [PubMed]

70. Ito, D.; Imai, Y.; Ohsawa, K.; Nakajima, K.; Fukuuchi, Y.; Kohsaka, S. Microglia-Specific Localisation of a Novel Calcium Binding Protein, Iba1. *Mol. Brain Res.* **1998**, *57*, 1–9. [CrossRef]

71. Waller, R.; Baxter, L.; Fillingham, D.J.; Coelho, S.; Pozo, J.M.; Mozumder, M.; Frangi, A.F.; Ince, P.G.; Simpson, J.E.; Highley, J.R. Iba1−/CD68+ Microglia are a Prominent Feature of Age-Associated Deep Subcortical White Matter Lesions. *PLoS ONE* **2019**, *14*, e0210888. [CrossRef] [PubMed]

72. Gheorghe, R.-O.; Deftu, A.; Filippi, A.; Grosu, A.; Bica-Popi, M.; Chiritoiu, M.; Chiritoiu, G.; Munteanu, C.; Silvestro, L.; Ristoiu, V. Silencing the Cytoskeleton Protein Iba1 (Ionized Calcium Binding Adapter Protein 1) Interferes with BV2 Microglia Functioning. *Cell. Mol. Neurobiol.* **2020**, *40*, 1011–1027. [CrossRef] [PubMed]

73. De Biase, L.M.; Schuebel, K.E.; Fusfeld, Z.H.; Jair, K.; Hawes, I.A.; Cimbro, R.; Zhang, H.-Y.; Liu, Q.-R.; Shen, H.; Xi, Z.-X.; et al. Local Cues Establish and Maintain Region-Specific Phenotypes of Basal Ganglia Microglia. *Neuron* **2017**, *95*, 341–356.e6. [CrossRef]

74. Lawson, L.J.; Perry, V.H.; Dri, P.; Gordon, S. Heterogeneity in the Distribution and Morphology of Microglia in the Normal Adult Mouse Brain. *Neuroscience* **1990**, *39*, 151–170. [CrossRef]

75. Ayata, P.; Badimon, A.; Strasburger, H.J.; Duff, M.K.; Montgomery, S.E.; Loh, Y.-H.E.; Ebert, A.; Pimenova, A.A.; Ramirez, B.R.; Chan, A.T.; et al. Epigenetic Regulation of Brain Region-Specific Microglia Clearance Activity. *Nat. Neurosci.* **2018**, *21*, 1049–1060. [CrossRef] [PubMed]

76. Sha'fie, M.S.A.; Rathakrishnan, S.; Hazanol, I.N.; Dali, M.H.I.; Khayat, M.E.; Ahmad, S.; Hussin, Y.; Alitheen, N.B.; Jiang, L.-H.; Syed Mortadza, S.A. Ethanol Induces Microglial Cell Death via the NOX/ROS/PARP/TRPM2 Signalling Pathway. *Antioxidants* **2020**, *9*, 1253. [CrossRef]

77. Marshall, S.A.; McClain, J.A.; Wooden, J.I.; Nixon, K. Microglia Dystrophy Following Binge-Like Alcohol Exposure in Adolescent and Adult Male Rats. *Front. Neuroanat.* **2020**, *14*. [CrossRef] [PubMed]

78. Marshall, S.; Geil, C.; Nixon, K. Prior Binge Ethanol Exposure Potentiates the Microglial Response in a Model of Alcohol-Induced Neurodegeneration. *Brain Sci.* **2016**, *6*, 16. [CrossRef]

79. Peng, H.; Geil Nickell, C.R.; Chen, K.Y.; McClain, J.A.; Nixon, K. Increased Expression of M1 and M2 Phenotypic Markers in Isolated Microglia after Four-Day Binge Alcohol Exposure in Male Rats. *Alcohol* **2017**, *62*, 29–40. [CrossRef]

80. Streit, W.J.; Xue, Q.-S.; Tischer, J.; Bechmann, I. Microglial Pathology. *Acta neuropathol. Commun.* **2014**, *2*, 142. [CrossRef]

81. Baron, R.; Babcock, A.A.; Nemirovsky, A.; Finsen, B.; Monsonego, A. Accelerated Microglial Pathology Is Associated with A β Plaques in Mouse Models of A Lzheimer's Disease. *Aging Cell* **2014**, *13*, 584–595. [CrossRef]

82. Johnson, E.A.; Dao, T.L.; Guignet, M.A.; Geddes, C.E.; Koemeter-Cox, A.I.; Kan, R.K. Increased Expression of the Chemokines CXCL1 and MIP-1α by Resident Brain Cells Precedes Neutrophil Infiltration in the Brain Following Prolonged Soman-Induced Status Epilepticus in Rats. *J. Neuroinflamm.* **2011**, *8*, 41. [CrossRef]

83. Tremblay, M.-E.; Stevens, B.; Sierra, A.; Wake, H.; Bessis, A.; Nimmerjahn, A. The Role of Microglia in the Healthy Brain. *J. Neurosci.* **2011**, *31*, 16064–16069. [CrossRef] [PubMed]

84. He, J.; Crews, F.T. Increased MCP-1 and Microglia in Various Regions of the Human Alcoholic Brain. *Exp. Neurol.* **2008**, *210*, 349–358. [CrossRef] [PubMed]

85. Patt, S.; Gertz, H.J.; Gerhard, L.; Cervós-Navarro, J. Pathological Changes in Dendrites of *Substantia Nigra* Neurons in Parkinson's Disease: A Golgi Study. *Histol. Histopathol.* **1991**, *6*, 373–380. [PubMed]

86. Albright, A.V.; Lavi, E.; Black, J.B.; Goldberg, S.; O'Connor, M.J.; Gonzalez-Scarano, F. The Effect of Human Herpesvirus-6 (HHV-6) on Cultured Human Neural Cells: Oligodendrocytes and Microglia. *J. Neurovirol.* **1998**, *4*, 486–494. [CrossRef] [PubMed]

87. Leibovitch, E.C.; Caruso, B.; Ha, S.K.; Schindler, M.K.; Lee, N.J.; Luciano, N.J.; Billioux, B.J.; Guy, J.R.; Yen, C.; Sati, P.; et al. Herpesvirus Trigger Accelerates Neuroinflammation in a Nonhuman Primate Model of Multiple Sclerosis. *Proc. Natl. Acad. Sci. USA* **2018**, *115*, 11292–11297. [CrossRef] [PubMed]

88. von Herrath, M.G.; Fujinami, R.S.; Whitton, J.L. Microorganisms and Autoimmunity: Making the Barren Field Fertile? *Nat. Rev. Microbiol.* **2003**, *1*, 151–157. [CrossRef] [PubMed]

89. Yang, M.; Xu, W.; Wang, Y.; Jiang, X.; Li, Y.; Yang, Y.; Yuan, H. CD11b−Activated Src Signal Attenuates Neuroinflammatory Pain by Orchestrating Inflammatory and Anti-Inflammatory Cytokines in Microglia. *Mol. Pain* **2018**, *14*, 174480691880815. [CrossRef]

90. Han, C.; Jin, J.; Xu, S.; Liu, H.; Li, N.; Cao, X. Integrin CD11b Negatively Regulates TLR-Triggered Inflammatory Responses by Activating Syk and Promoting Degradation of MyD88 and TRIF via Cbl-B. *Nat. Immunol.* **2010**, *11*, 734–742. [CrossRef]

91. Coizet, V.; Dommett, E.J.; Klop, E.M.; Redgrave, P.; Overton, P.G. The Parabrachial Nucleus Is a Critical Link in the Transmission of Short Latency Nociceptive Information to Midbrain Dopaminergic Neurons. *Neuroscience* **2010**, *168*, 263–272. [CrossRef]

92. Alexander, G.E.; DeLong, M.R.; Strick, P.L. Parallel Organization of Functionally Segregated Circuits Linking Basal Ganglia and Cortex. *Annu. Rev. Neurosci.* **1986**, *9*, 357–381. [CrossRef]

93. Gurney, K.; Prescott, T.J.; Redgrave, P. A Computational Model of Action Selection in the Basal Ganglia. I. A New Functional Anatomy. *Biol. Cybern.* **2001**, *84*, 401–410. [CrossRef]

94. Lardeux, S.; Baunez, C. Alcohol Preference Influences the Subthalamic Nucleus Control on Motivation for Alcohol in Rats. *Neuropsychopharmacology* **2008**, *33*, 634–642. [CrossRef]

95. Pelloux, Y.; Baunez, C. Targeting the Subthalamic Nucleus in a Preclinical Model of Alcohol Use Disorder. *Psychopharmacology* **2017**, *234*, 2127–2137. [CrossRef] [PubMed]

96. Pautrat, A.; Rolland, M.; Barthelemy, M.; Baunez, C.; Sinniger, V.; Piallat, B.; Savasta, M.; Overton, P.G.; David, O.; Coizet, V. Revealing a Novel Nociceptive Network That Links the Subthalamic Nucleus to Pain Processing. *eLife* **2018**, *7*. [CrossRef] [PubMed]

97. de Tommaso, M.; Arendt-Nielsen, L.; Defrin, R.; Kunz, M.; Pickering, G.; Valeriani, M. Pain in Neurodegenerative Disease: Current Knowledge and Future Perspectives. *Behav. Neurol.* **2016**, *2016*, 7576292. [CrossRef] [PubMed]

98. Angst, M.S.; Clark, J.D. Opioid-Induced Hyperalgesia. *Anesthesiology* **2006**, *104*, 570–587. [CrossRef]
99. Avegno, E.M.; Lobell, T.D.; Itoga, C.A.; Baynes, B.B.; Whitaker, A.M.; Weera, M.M.; Edwards, S.; Middleton, J.W.; Gilpin, N.W. Central Amygdala Circuits Mediate Hyperalgesia in Alcohol-Dependent Rats. *J. Neurosci.* **2018**, *38*, 7761–7773. [CrossRef]

Review

The Childhood-Onset Neurodegeneration with Cerebellar Atrophy (CONDCA) Disease Caused by *AGTPBP1* Gene Mutations: The Purkinje Cell Degeneration Mouse as an Animal Model for the Study of this Human Disease

Fernando C. Baltanás [1,*], María T. Berciano [2], Eugenio Santos [1] and Miguel Lafarga [3]

[1] Lab.1, CIC-IBMCC, University of Salamanca-CSIC and CIBERONC, 37007 Salamanca, Spain; esantos@usal.es

[2] Department of Molecular Biology and Centro de Investigación Biomédica en Red sobre Enfermedades Neurodegenerativas (CIBERNED), University of Cantabria-IDIVAL, 39011 Santander, Spain; berciant@unican.es

[3] Department of Anatomy and Cell Biology and Centro de Investigación Biomédica en Red sobre Enfermedades Neurodegenerativas (CIBERNED), University of Cantabria-IDIVAL, 39011 Santander, Spain; lafargam@unican.es

* Correspondence: baltanas@usal.es; Tel.: +34-923294801

Abstract: Recent reports have identified rare, biallelic damaging variants of the *AGTPBP1* gene that cause a novel and documented human disease known as childhood-onset neurodegeneration with cerebellar atrophy (CONDCA), linking loss of function of the AGTPBP1 protein to human neurodegenerative diseases. CONDCA patients exhibit progressive cognitive decline, ataxia, hypotonia or muscle weakness among other clinical features that may be fatal. Loss of AGTPBP1 in humans recapitulates the neurodegenerative course reported in a well-characterised murine animal model harbouring loss-of-function mutations in the *AGTPBP1* gene. In particular, in the Purkinje cell degeneration (*pcd*) mouse model, mutations in *AGTPBP1* lead to early cerebellar ataxia, which correlates with the massive loss of cerebellar Purkinje cells. In addition, neurodegeneration in the olfactory bulb, retina, thalamus and spinal cord were also reported. In addition to neurodegeneration, *pcd* mice show behavioural deficits such as cognitive decline. Here, we provide an overview of what is currently known about the structure and functional role of AGTPBP1 and discuss the various alterations in AGTPBP1 that cause neurodegeneration in the *pcd* mutant mouse and humans with CONDCA. The sequence of neuropathological events that occur in *pcd* mice and the mechanisms governing these neurodegenerative processes are also reported. Finally, we describe the therapeutic strategies that were applied in *pcd* mice and focus on the potential usefulness of *pcd* mice as a promising model for the development of new therapeutic strategies for clinical trials in humans, which may offer potential beneficial options for patients with *AGTPBP1* mutation-related CONDCA.

Keywords: AGTPBP1; CCP1; CONDCA; neurodegeneration; NNA1; pcd

Citation: Baltanás, F.C.; Berciano, M.T.; Santos, E.; Lafarga, M. The Childhood-Onset Neurodegeneration with Cerebellar Atrophy (CONDCA) Disease Caused by *AGTPBP1* Gene Mutations: The Purkinje Cell Degeneration Mouse as an Animal Model for the Study of this Human Disease. *Biomedicines* **2021**, *9*, 1157. https://doi.org/10.3390/biomedicines9091157

Academic Editor: Kumar Vaibhav

Received: 16 August 2021
Accepted: 2 September 2021
Published: 4 September 2021

Publisher's Note: MDPI stays neutral with regard to jurisdictional claims in published maps and institutional affiliations.

Copyright: © 2021 by the authors. Licensee MDPI, Basel, Switzerland. This article is an open access article distributed under the terms and conditions of the Creative Commons Attribution (CC BY) license (https://creativecommons.org/licenses/by/4.0/).

1. Introduction

Childhood-onset neurodegeneration with cerebellar atrophy (CONDCA; OMIM 618276) is a recently identified, rare and severe autosomal recessive disease that affects the central and peripheral nervous systems. Individuals present an early global developmental delay resulting in cognitive decline and motor performance alterations, among other clinical features [1–3]. The severity of the disease is variable, and CONDCA can even result in death during childhood. Whole exome sequencing studies on CONDCA patients have identified different damaging biallelic variants of the *AGTPBP1* gene [1–4], linking *AGTPBP1* loss of function to human neurodegenerative diseases. Nevertheless, the deleterious effects of AGTPBP1 protein loss of function in animal models, especially in mouse models, have been known for some time [5–11].

Two decades ago, studies characterising genes involved in axonal regeneration in mice led to the identification of a 4-kb-long transcript hotspot. Due to its proposed role and nuclear localization, the gene encoding this transcript was named Nna1 (Nervous system Nuclear protein induced by Axotomy) [7]. Although the precise role of the protein encoded by the *AGTPBP1* gene remained elusive for several years, structural analysis of the AGTPBP1 protein revealed that it belonged to a new subfamily (M14D) of the M14 metallocarboxypeptidase family [12–14]. Subsequent functional analyses revealed that AGTPBP1, which acts as an enzyme, participates in post-translational modifications (PTMs) of tubulin. In particular, AGTPBP1 acts as a deglutamylase, catalysing the removal of polyglutamates at the C-terminal region of tubulin [15,16]. Other functions not directly related to tubulin processing, including maintenance of chromosomal stability and regulation of mitochondrial energy metabolism, were proposed for AGTPBP1 [17–19].

More than 40 years ago, the Purkinje cell degeneration (*pcd*) mouse model, which harbours a mutation that is autosomal recessive and displays distinct neurological deficits, causing profound ataxic behaviour, was established [5]. After extensive studies, the mutant gene responsible for the pcd mutation was mapped to mouse chromosome 13 and identified as the *AGTPBP1* gene [10].

The pcd mutation causes selective postnatal degeneration of certain neuronal populations, including the mitral cells (MCs) in the olfactory bulb (OB) [20], photoreceptors in the retina [6], certain subpopulations of thalamic neurons [21] and the Purkinje cells (PCs) in the cerebellum [10], the latter being responsible for the cerebellar ataxia of *pcd* mice. Moreover, a recent analysis has revealed that these animals also undergo peripheral nerve and spinal motor neuron degeneration [1]. Similarly, recent works have shown that excessive tubulin polyglutamylation in neurons, which results from AGTPBP1 dysfunction, alters the axonal transport of vesicles and appears to be the main mechanism of neurodegeneration [22–24]. In addition to ataxia, *pcd* mice also exhibit progressive cognitive impairments [9,25].

Interestingly, the well-characterised neurological deficits in *pcd* mice closely mimic the pathophysiology and clinical manifestations reported in CONDCA patients. Thus, the *pcd* mouse is an ideal animal model for investigating other probable but not yet characterised clinical alterations in patients with CONDCA. Moreover, potential therapeutic options for preventing, or at least attenuating, the neurodegenerative course in CONDCA patients could be assessed using this animal model.

Here, we summarise the current knowledge about the structure and function of the *AGTPBP1* gene in different cell types and tissues. We also review the variety of alterations in m*AGTPBP1* in *pcd* mice and their relationships with the pathological variants of the h*AGTPBP1* gene reported in CONDCA patients. Finally, we focus on the potential usefulness of the *pcd* mouse model as a suitable model for the clinical assessment of new pharmacological strategies and therapies that may offer possible treatment options for patients with *AGTPBP1* mutation-induced CONDCA.

2. The *AGTPBP1* Gene

2.1. Genomic Structure and Organisation

NNA1/AGTPBP1 contains a putative Walker A-box ATP/GTP binding motif (GXXGKS), which is highly conserved throughout evolution. According to the function and location of the protein encoded by this gene, it has also been called CCP1 (Cytosolic CarboxyPeptidase 1). Henceforth, we will use AGTPBP1 as the preferred term when referring to the gene or the protein.

The genomic regions occupied by the *AGTPBP1* locus vary widely between different organisms, but the intron/exon distribution of this gene is highly conserved. For example, the h*AGTPBP1* gene is located on chromosome 9 (Chr 9q21.33 position: 85,546,539–85,742,029; 26 exons), whereas the m*AGTPBP1* gene is located on chromosome 13 (Chr13 position: 59,445,742–59,585,227; 26 exons) (Figure 1A).

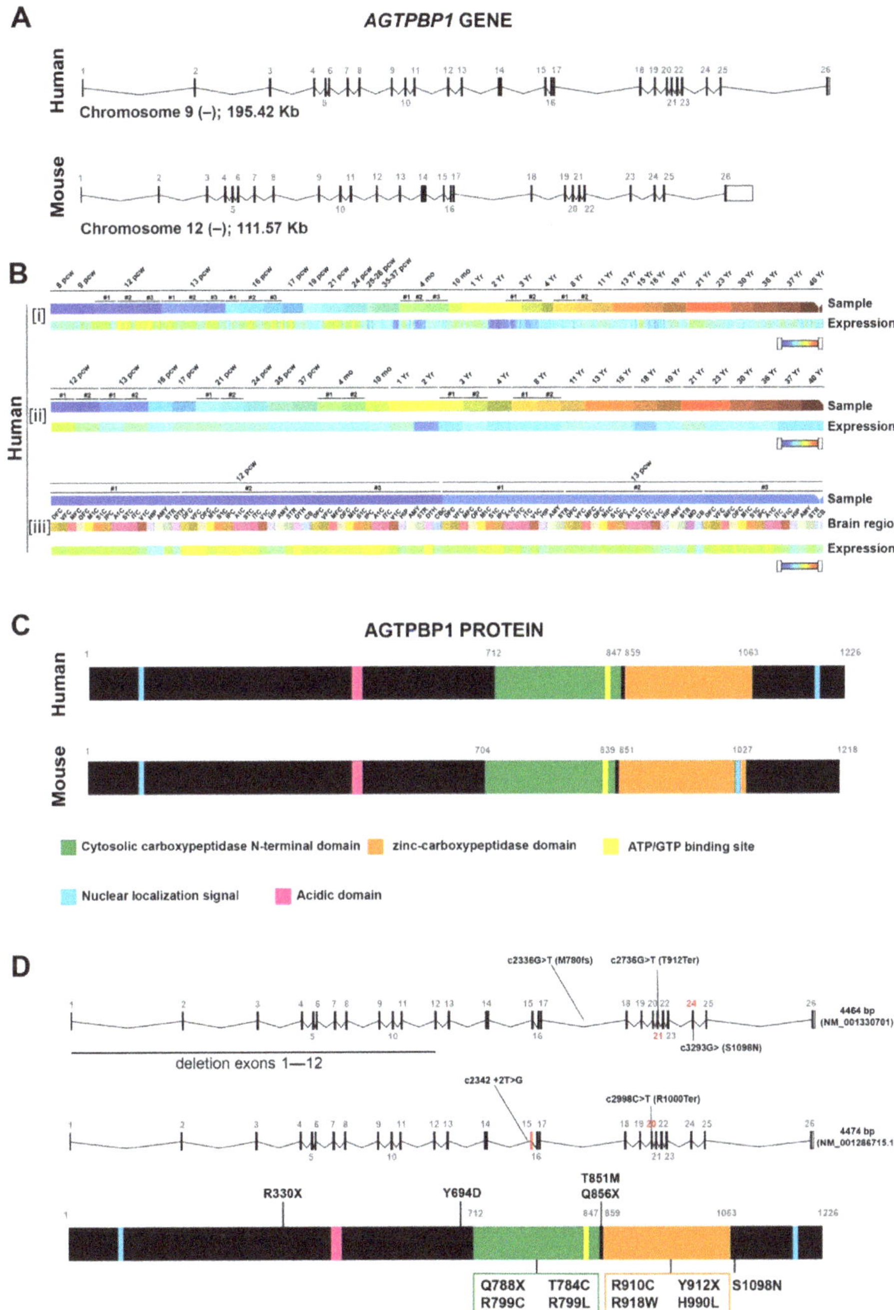

Figure 1. Genomic structure, organisation, expression and intron/exon distribution of AGTPBP1 in humans and in mice. (**A**) Genomic organisation of human *AGTPBP1* and mouse *AGTPBP1* loci. Figure assembled using current data from the Ensembl genome browser database (http://www.ensembl.org/index.html) (accessed on 10 July 2021). Chromosomal location, chromosome strand (+ or −) used for transcription, and the size (Kb) of the genomic stretches containing the *AGTPBP1* locus are also indicated for each species. Exons (solid vertical boxes) are numbered. The open box indicates the alternatively spliced region of the exon. (**B**) Schematic representation of [i] overall expression of the h*AGTPBP1* gene throughout

development in all studied brain structures, [ii] the specific expression of the h*AGTPBP1* gene in the cerebellum throughout development, and [iii] region-specific hAGTPBP1 gene expression at 12–13 pcw. (**C**) Schematic representation of the primary structure of the human and mouse AGTPBP1 proteins. The structural domains and their relative positions are indicated. (**D**) Schematic representation of the human *AGTPBP1* gene structure and the encoded protein and the locations of variations found in patients with AGTPBP1 mutations. DFC: dorsolateral prefrontal cortex; VFC: ventrolateral prefrontal cortex; MFC: anterior (rostral) cingulate (medial prefrontal) cortex; OFC: orbital frontal cortex; M1C: primary motor cortex; S1C: primary somatosensory cortex; IPC: inferior parietal cortex; A1C: primary auditory cortex; STC: superior temporal cortex; ITC: inferolateral temporal cortex; V1C: primary visual cortex; HIP: hippocampus; AMY: amygdaloid complex; STR: striatum; DTH: dorsal thalamus; CB: cerebellum; CBC: cerebellar cortex. PWC: postconception weeks; MO: Months; Yr: Year.

2.2. Expression Pattern

h*AGTPBP1* mRNA or protein is detectable in practically all human cells, tissues and organs tested. Of note, the mRNA and protein expression levels of h*AGTPBP1* differ significantly depending on the specific organ and tissue.

According to the consensus dataset of the Human Protein Atlas database obtained based on transcriptomic analyses of human tissue and organs, the mRNA expression of h*AGTPBP1* is the highest in bone marrow (https://www.proteinatlas.org/ENSG00000 135049-AGTPBP1/tissue) (accessed on 10 July 2021) followed by many regions of the central nervous system, including the spinal cord, pons and medulla, corpus callosum, cerebellar cortex, hippocampal formation and olfactory region, among others. h*AGTPBP1* mRNA expression is generally higher in the brain than in all other non-brain tissues. For example, low mRNA expression levels are detected in the ovary, liver, stomach, small intestine, lung, adrenal gland, spleen, and thymus, among other tissues. Similar distribution patterns of m*AGTPBP1* mRNA were found in mouse tissues [7,12]. In brain regions, m*AGTPBP1* is expressed preferably in differentiating neurons rather than in proliferating precursors/progenitors. [7].

Using the BrainSpan Developmental Transcriptome database, which contains data related to human gene expression in 16 specific brain structures obtained using RNA sequencing and exon microarray analysis of 42 brain samples spanning pre- and post-natal development, both the temporal and regional specificity of h*AGTPBP1* gene expression in the brain can be examined (https://www.brainspan.org/rnaseq/searches?exact_match= false&search_term=%22NNA1%22&search_type=gene) (accessed on 10 July 2021). Regarding the temporal pattern of h*AGTPBP1* expression, overall higher h*AGTPBP1* gene expression is detected during embryonic development than during childhood and adulthood (Figure 1B, panel i), with the expression of the gene peaking at approximately 12–13 postconception weeks (pcw). In contrast, the lowest h*AGTPBP1* gene transcription is found in brain samples from children (2–8 years old) (Figure 1B, panel i). A very similar expression pattern is found in the cerebellum and cerebellar cortex throughout development (Figure 1B, panel ii). With regards to regional-specific expression, in the specific time windows in which the h*AGTPBP1* gene shows the highest transcription levels (~12–13 pcw), the highest h*AGTPBP1* gene expression levels are found in the prefrontal and frontal cortices and the dorsal thalamus, whereas the lowest levels are detected in the hippocampus, cerebellum and striatum (Figure 1B, panel iii).

3. The AGTPBP1 Protein

3.1. Modular Domain Structure

The h*AGTPBP1* gene encodes a 1226-amino acids (aa) protein (https://www.uniprot. org/uniprot/Q9UPW5) (accessed on 10 July 2021) containing a P-loop ATP/GTP-binding motif and a nucleotide-binding site. The primary structure of the hAGTPBP1 protein is a sequential, linearly organised modular configuration featuring conserved distribution of two well-defined domains: the cytosolic carboxypeptidase N-terminal domain (aa 712–847), which is highly conserved among M14D subfamily members [12,26], and the catalytic zinc-carboxypeptidase domain (aa 859–1063; Figure 1C). The ATP/GTP binding site is at aa 820–825, and an active catalytic site is found at position 970. Using cNLS mapper software (http://nls-mapper.iab.keio.ac.jp/cgi-bin/NLS_Mapper_form.cgi) (accessed on 10 July 2021), two bipartite nuclear localisation signals (NLSs) were predicted: a signal at the N-terminal region (aa 120–149) and a signal situated at the C-terminus (aa 1139–1165). In addition, other key residues directly related to the function of the protein were identified, including zinc-binding sites in the hAGTPBP1 protein (residues 920, 923 and 1017) [12]. Importantly, the hAGTPBP1 protein is highly evolutionarily conserved, sharing ~72.5% identity with the AGTPBP1 protein in D. melanogaster and D. rerio and 87.2% identity with the AGTPBP1 protein in mice, supporting the notion that this gene, especially the catalytic domains, is highly conserved in metazoans [26–28]. The m*AGTPBP1* gene encodes a 1218-aa protein (https://www.uniprot.org/uniprot/Q641K1) (accessed on 10 July 2021) that also contains a cytosolic carboxypeptidase N-terminal domain (aa 704–839; Figure 1C) and a zinc-carboxypeptidase domain (aa 851–1027; Figure 1C). The ATP/GTP binding site is at aa 810–817, and the active site is located at residue 962. Whereas mutation of the ATP/GTP binding site has no effect in vivo, preservation of a functional zinc-binding domain is essential for neuronal survival [14,29,30]. Zinc-binding sites in the mAGTPBP1 protein are found at residues 912, 915 and 1009. The NLSs are positioned at both the N-terminal domain (aa 144–151) and C-terminal domain (aa 996–1016) [9].

3.2. Expression Pattern

hAGTPBP1 protein expression level data have revealed that the protein is expressed at the highest level in the testis (https://www.proteinatlas.org/ENSG00000135049-AGTPBP1/tissue) (accessed on 10 July 2021). Moderate protein expression levels are also observed in human brain regions, such as the cerebral cortex and cerebellum, and other organs, including the lung, stomach, kidney, pancreas, and muscle.

At the cellular level, the AGTPBP1 protein is detected both in the cytoplasm, mainly in vesicles and mitochondria, and in the nucleus, particularly in the nucleolus (https://www.proteinatlas.org/ENSG00000135049-AGTPBP1/cell) (accessed on 10 July 2021), as reported for mAGTPBP1 [7,18,26].

3.3. Role of the AGTPBP1 Protein

Microtubules are major components of the cytoskeleton composed of α- and β-tubulin heterodimer subunits that polymerise to form tubular and polar structures [31]. Both tubulin subunits are subject to certain PTMs, including the tyrosination/detyrosination and polyglutamylation of α-tubulin [32]. The tyrosination/detyrosination cycle involves the reversible removal and re-addition of a tyrosine residue at the C-terminus of α-tubulin. Two enzymes participate in this cycle: tubulin tyrosine ligase (TTL) and tubulin carboxypeptidase (TubCP). The latter cleaves the C-terminal tyrosine residue of α-tubulin, resulting in Glu-tubulin. The tyrosine residue can be re-added by TTL, forming Tyr-tubulin again. Another variant of non-tyrosinable α-tubulin (Δ2-tubulin) that lacks the Glu C-terminal residue was defined.

Several years ago, the enzymes catalysing tubulin PTMs were fully characterised [33]. Cytosolic carboxypeptidase (CCP) subfamily comprises six members (CCP1-CCP6), [12,13]. CCP enzymes process amino acid residues from the C-terminus [15,34] and are also involved in reversing polyglutamylation catalysed by polyglutamylases from the tubulin-tyrosine ligase-like family and in removing glutamate residues from the C-terminus, thus transforming α-tubulin into non-tyrosinable Δ2-tubulin [15] (Figure 2A,B). In particular, CCP1, encoded by the *AGTPBP1* gene, acts as a tubulin deglutamylase, generating Δ2-tubulin from Glu-tubulin, and also counteracts the tubulin–tyrosine ligase-like mediated reaction and then shortens the long glutamate chains generated by polyglutamylases [15,16,35]. In addition, myosin light chain kinase and telokin proteins were also identified as substrates for CCP1 [15] (Figure 2C).

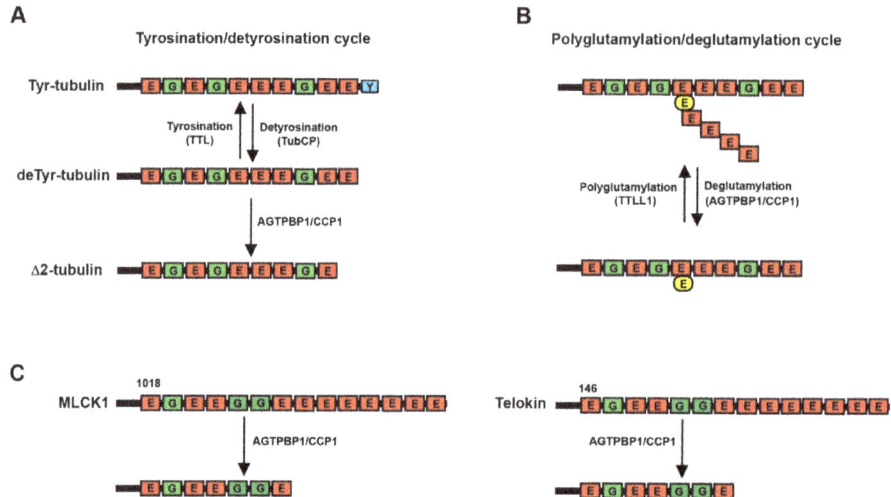

Figure 2. AGTPBP1/CCP1-mediated tubulin posttranslational modifications. (**A,B**) Schematic representation of the two AGTPBP1/CCP1-mediated tubulin posttranslational modifications. (**C**) Schematic representation of the deglutamylation of MLCK1 (left) and telokin (right) catalysed by AGTPBP1/CCP1. The numbers indicate the amino acid residues. MLCK1: Myosin light chain kinase 1. TTL: Tubulin Tyrosine Ligase; TTLL1: Tubulin Tyrosine Ligase-Like 1; TubCP: Tubulin Carboxypeptidase. Scheme is a modification of [15].

Although AGTPBP1 contains two NLSs [9] and is expressed in the nucleus [7], its contribution to nuclear physiology remains uncertain; however, it was suggested that this protein may be involved in chromatin remodelling [7].

4. *AGTPBP1* Mutation-Related Childhood-Onset Neurodegeneration with Cerebellar Atrophy (CONDCA)

Shashi and colleagues (2018) identified biallelic variants in the *AGTPBP1* gene in patients suffering from an infantile-onset neurodevelopmental disorder known as childhood-onset neurodegeneration with cerebellar atrophy (CONDCA) [1]. At the time this review was written, a total of 18 individuals from 15 different families were reported to carry damaging CONDCA-associated, biallelic, loss-of-function variants of the *AGTPBP1* gene (Table 1). The affected individuals, most of them unrelated, were born from both consanguineously related and unrelated parents, and the disease appeared to equally affect both males and females (Table 2).

Complete exome sequencing analysis of CONDCA patient samples revealed rare allelic variants of the *AGTPBP1* gene, including (i) missense mutations predicted to result in single aa substitutions; (ii) a canonical splice-site change causing disturbance of the open reading frame; (iii) de novo heterozygous stop-gain variants predicted to generate a premature stop codon; (iv) an intronic splice-site change predicted to lead to the in-frame absence of 29 highly conserved aa; and (v) a 12-exon-long genomic deletion (Table 1; Figure 1D) [1–4]. The majority of these variants affect the highly conserved N-terminal regions and zinc-carboxypeptidase protein motifs (Figure 1D). As a consequence, the function of the AGTPBP1 protein is abrogated due to a reduction in mRNA transcription, nonsense-mediated mRNA decay, early protein truncation and protein misfolding into a structure that is highly susceptible to proteasome-dependent proteolysis [1].

Table 1. Identified biallelic variants in the *AGTPBP1* gene. M: Male F: Female. * means substitution.

Patient (Age; Sex; Consanguinity)	Allelic Variant	Consequence	Region Affected	Reference
2-year-old; F; NO	NM_001330701 c.2336-1G>T	Transversion in intron 17. Results in a splice site aberration, a frameshift and premature termination (M780fs)	Cytosolic carboxypeptidase N-terminal domain	[1]
	NM_001330701 c.2736delC	Deletion in exon 21. Results in a frameshift and premature termination (T912Ter)	Zinc-carboxypeptidase domain	[1]
12-month-old; M; YES	NM_001330701 c.2752C>T	Transition in exon 21. Results in R918W substitution	Zinc-carboxypeptidase domain	[1]
7-month-old; M; YES	-	Deletion of exons 1 to 12	Non-defined	[1]
Not available	NM_015239.2 c.2632C>T	Cerebellar hypoplasia and lower motor neuron degeneration. Results in R878W substitution	Zinc-carboxypeptidase domain	[4]
4-year-old; M; NO	NM_001286715 c.2351A>G	Results in a T784C substitution	Non-defined	[2]
	NM_001286715 c.2998C>T	Results in a frameshift and premature termination (R1000Ter)	Zinc-carboxypeptidase domain	
15-month-old; M; YES	NM_001286715 c.2342C>T+2T>G	Skips exon 15 (loss of 29 highly conserved aa)	Non-defined modular domain	[2]
5-year-old; F; NO	NM_001330701 c.2752C>T	Transition in exon 21. Results in R918W substitution	Zinc-carboxypeptidase domain	[1]
	NM_001330701 c.2080T>G	Transversion in exon 15. Results in a Y694D substitution	Non-defined	
16-month-old; F; YES	NM_001330701 c.2566C>T	Homozygous transition in exon 19. Results in a Q856 *	Non-defined	[1]
8-year-old; M; YES	NM_001330701 c.2395C>T	Results in a R799C substitution	Cytosolic carboxypeptidase N-terminal domain	[1]
8-year-old; M; YES	NM_001330701 c.2566C>T	Results in a P799C substitution	Cytosolic carboxypeptidase N-terminal domain	[1]
7-month-old; M; YES	NM_001330701 c.2396G>T	Results in a P799L substitution	Cytosolic carboxypeptidase N-terminal domain	[3]
2-year-old; M; YES	NM_001330701 c.2396G>T	Results in a P799L substitution	Cytosolic carboxypeptidase N-terminal domain	[3]
20-month-old; F; NO	NM_001330701 c.988C>T	Results in a R330 *	Non-defined	[1]
		Results in a Y912X substitution	Zinc-carboxypeptidase domain	
8-year-old; M; YES and 5-year-old; F; YES	NM_001330701 c.2728C>T	Results in a R910C substitution	Zinc-carboxypeptidase domain	[1]
3 infant sibs; YES	NM_001330701 c.2362C>T	Transition in exon 18. Results in a Q788 *	Cytosolic carboxypeptidase N-terminal domain	[1]
14-year-old; M; NO	NM_001330701 c.2552C>T	Transition in exon 19, resulting in a T851M substitution	Non-defined	[1]
	NM_001330701 c.2969A>T	transversion in exon 22, resulting in a H990L substitution	Zinc-carboxypeptidase domain	
21-month-old; M; YES	NM_001330701 c.3293G>A	Mutation in exon 24, resulting in a S1098N substitution	3′ end domain	[36]
17-year-old; F; YES	NM_001330701 c.3293G>A	Mutation in exon 24, resulting in an S1098N substitution	3′ end domain	[36]

Table 2. Clinical findings reported for patients harboring *AGTPBP1*-mutated gene.

Feature	Data from [1–4,36]
Onset	Birth to 20 months
Gender	10F, 9M Not available (1)
Consanguinity	14/19 Not available (1)
Progressive degenerative course	19/20
Microcephaly	11/20 Not available (1)
Motor delay	20/20
Hypotonia	19/20 Not available (1)
Muscle weakness	16/18 Not available (2)
Muscle weakness pattern	Tetraparesis/plegia (8) Lower limb (2) Neck (3) Diaphragm/intercostal (4) Not specified (8)
Muscle atrophy	9/18
Tongue fasciculations	7/20 Not available (13/20)
Tendon reflexes	Low or absent (10/20) Normal (3/20) Increased (6/29) Not available (1)
Ataxia	Yes (6) Not available (12)
Dystonia	5/20 Not available (1)
Spasticity	7/20 Not available (3/18)
Respiratory distress	9/20 Not available (1)
Feeding difficulties	13/20 Not available (1)
Eye movement abnormalities	Detected (12/20) Not detected (6/20) Not available (2/20)
Hearing	Impaired (1/20) Normal (5/20) Not available (14/20)
Cognitive delay	17/20 Not available (3/20)
Brain MRI	Cerebellar atrophy (18/20) Dysplastic corpus callosum (6/20) Small pons (1/20) Enlarged CSF spaces (1/20)
Nerve conduction studies	Motor neuropathy (2/20) Axonal motor neuropathy (5/20) Normal (1/18) Not available (12/20)
Electromyography	Denervation (5/20) Neurogenic (2/20) Normal (1/20) Not available (12/20)

The clinical findings of patients with AGTPBP1 mutations are summarised in Table 2. Overall, the patients display early-onset developmental delays (between birth and 20 months of age) with a progressive degenerative course, mainly characterised by hypotonia and generalised muscle weakness, frequently causing tetraparesis. Brain MRI revealed detected cerebellar atrophy with respect to non-affected, control individuals (Figure 3A,B). Other brain alterations, such as microcephaly or dysplastic corpus callosum, were also frequently detected. Alterations in tendon reflexes were observed in almost all individuals. Muscle atrophy was found in half of the patients (Figure 3C,D), and other regularly detected clinical manifestations included feeding problems, eye movement abnormalities, respiratory insufficiency, spasticity, tongue fasciculations and dystonia. Other clinical features, such as bilateral hearing loss and hand tremors, were sporadically detected.

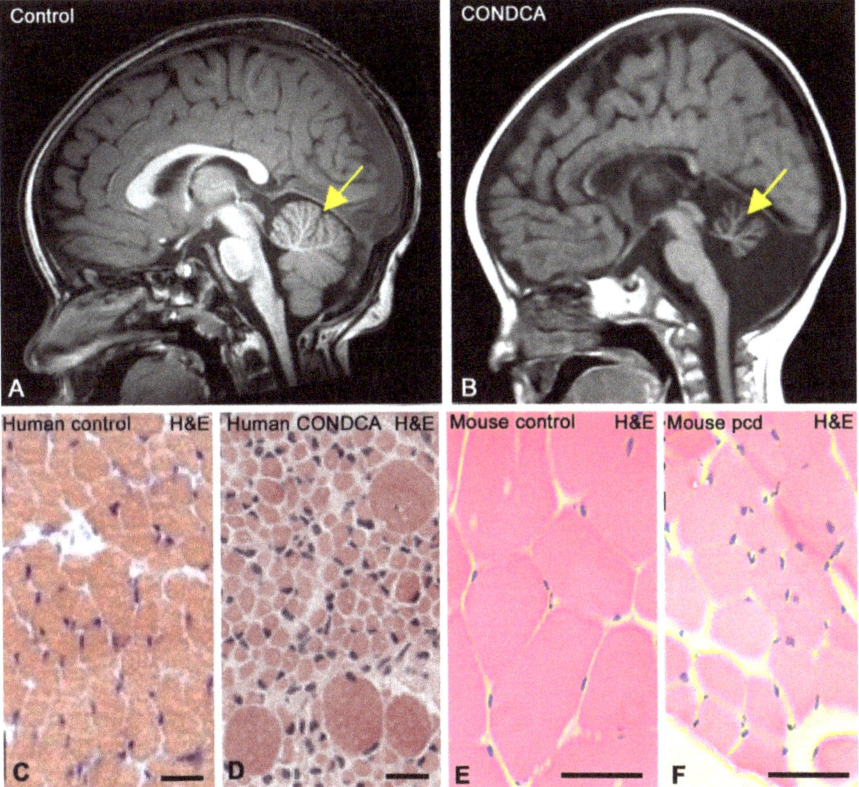

Figure 3. Cranial magnetic resonance imaging (MRI) of a control (**A**) and a CONDCA patient (**B**). Whereas the 20-month-old female healthy control patient (**A**) shows a typical well-developed cerebellum (yellow arrow), severe cerebellar atrophy (yellow arrow) is observed in the 24-month-old female CONDCA patient (**B**). (**A**) Courtesy of Dr. Ana Canga, "Hospital Universitario Marqués de Valdecilla", Santander (Spain). (**B**) Adapted from [3]. Copyright 2019 American Journal of Medical Genetics. (**C,D**) Haematoxylin-eosin (H&E)-stained skeletal muscle tissue biopsies at 7 months of age from healthy control (**C**) and CONDCA patients (**D**). Note the fibre atrophy with a few interspersed hypertrophic fibres in the muscle tissue of the patient. Scale bars: 50 μm. Adapted with permission from [1]. Copyright 2018 The EMBO Journal. (**E,F**) Haematoxylin-eosin (**E,H**)-stained skeletal muscle tissue cross-sections from control (**E**) and *pcd* (**F**) mice. Note the muscle atrophy and the notable reduction in muscle fibre size in the *pcd* mouse. Scale bars: 30 μm. Adapted with permission from [37]. Copyright 2018 Journal of Tissue Engineering and Regenerative Medicine.

Electrophysiological recordings revealed motor neuropathy affecting the lower limbs and arms. In particular, electromyography studies have detected signs of denervation causing muscle atrophy in proximal and distal muscles, including the tibialis anterior and posterior deltoid, suggesting degeneration of both peripheral nerve motor fibres and spinal cord α-motor neurons. In contrast, sensory nerve action potentials seemed to be unaffected, suggesting that the neuropathy was mainly of motor origin. Progressive advancement of the neurological disorders resulted in the death of 7 patients out of the 18 individuals carrying damaging variants of the *AGTPBP1* gene.

It is relevant to mention that in a very recently published paper, two members of a consanguineous family harbouring a novel homozygous variant (c.3293G>A) at the 3′ end of the *AGTPBP1* gene (Figure 1D) showed no signs of cerebellar atrophy [36]. Further analysis is needed to examine whether only AGTPBP1 mutations affecting the catalytic domains of the protein are directly associated with cerebellar atrophy.

At the cellular level, deleterious accumulation of polyglutamylated tubulin was detected in biopsies taken from the quadriceps muscles of CONDCA patients with AGTPBP1 mutations [1]. Additionally, this hyperglutamylation is directly related to neurodegeneration in mice and humans, most likely due to deficiency in microtubule-based axonal transport [38].

Interestingly, a recent study has identified the *AGTPBP1* gene as being the most significant gene coexpressed with the amyotrophic lateral sclerosis (ALS)-linked gene C90RF72 and revealed a positive correlation between the expression of their respective mRNAs [39]. These findings suggest that AGTPBP1 is an interacting partner of C90rf72 that contributes to the regulation of important neuronal functions [39]. This raises questions regarding the potential role of AGTPBP1 in other human neurological disorders.

5. The *pcd* Mouse as an Animal Model for Studying *AGTPBP1* Mutation-Related CONDCA

More than 40 years ago, a spontaneous recessive mutation that causes early cerebellar ataxia and is associated with rapid degeneration of cerebellar PCs was identified [5]. This mutation was subsequently called pcd [5]. To date, damaging mutations in AGTPBP1, including spontaneous mutant variants, chemically induced variants, transgenic alleles and conditional knockout, were reported (Table 3). The most severe pathological alleles are pcd^{1J}, pcd^{3J}, pcd^{5J} and pcd^{JWG} [9]. The pcd^{2J} allele is a less severe hypomorphic variant that causes a mild phenotype and the development of ataxia much later than other pcd mutants but not thalamic degeneration (Table 3). A recently generated AGTPBP1 KO mouse model in which exons 21–22 are deleted [40], exhibits similar pathological features as those reported for pcd1J, pcd^{3J} or pcd^{5J} (Table 3).

Loss of function of the *AGTPBP1* gene was subsequently identified as being responsible for the *pcd* phenotype [7,10]. *pcd* mice have a smaller body and lower body weight than their wild-type counterparts [35,40]. In addition to PC degeneration, the pcd mutation leads to postnatal degeneration of other distinct neuronal populations, including MCs in the OB [20], photoreceptors in the retina [6], a certain subpopulation of thalamic neurons [21] and peripheral nerve and spinal motor neurons [1].

Interestingly, the time course of degeneration of affected neuronal populations in *pcd* mice markedly differs among them, making this animal model highly appropriate for studying different neuronal degenerative processes caused by the same mutation. Degeneration of photoreceptors is slow and takes approximately one year [6,41] whereas the death of thalamic neurons and MCs takes up to 6 and 4 months, respectively [20,21]. In contrast, degeneration of PCs occurs extremely quickly, is severe and occurs between 3 and 4 weeks of age. In fact, almost all PCs are eliminated at a well-defined time point, approximately by 1 month after birth. To date, no clear findings supporting marked variations in these neurodegenerative processes exist. Since most *pcd* mice survive, secondary neuronal death and remodelling of neuronal networks, as consequences of primary neurodegeneration, were reported in the affected regions.

Interestingly, cerebellar atrophy is one of the earliest signs of the disease manifested in CONDCA patients, which resembles the early PC degeneration and the cerebellar atrophy showed by *pcd* mice. Nevertheless, although the correspondence of other clinical findings between CONDCA patients and *pcd* mice is established (Table 4) it still remains unknown if the developmental course of the pathologies is comparable in both groups. With the currently available data, the assumption of similar developmental phases of the disease between CONDCA patients and *pcd* mice cannot be yet established.

Table 3. Summary of AGTPBP1 reported mutant alleles. ENU: N-ethyl-N-nitrosourea.

Allele Name Mutation	Mutation	Clinical Features	Genetic Mutation in *AGTPBP1*
Agtpbp1^{Pcd-1J}	Spontaneous	Reduced body size; Ataxia; cerebellar atrophy; postnatal degeneration of thalamic neurons, PCs, MCs and retinal photoreceptors; male infertility; female partial fertility.	Unknown (possibly in regulatory region)
Agtpbp1^{Pcd-2J}	Spontaneous	Hylomorphic allele with reduced	Insertion (~7.8Kb) between exons 14–15
Agtpbp1^{Pcd-3J}	Spontaneous	Reduced body size; Ataxia; Cerebellar atrophy; postnatal degeneration of thalamic neurons, PCs, MCs and photoreceptors; male infertility; female partial fertility; Reduced number of antral follicles.	Deletion (~12.2 Kb) between intron 5 and exon 8
Agtpbp1^{Pcd-4J}	ENU-induced mutagenesis	Ataxia; degeneration of PCs	Unknown
Agtpbp1^{Pcd-5J}	Spontaneous	Ataxia; Degeneration of PCs and MCs	Insertion of an aspartic acid residue (D775) in exon 18
Agtpbp1^{Pcd-6J}	ENU-induced mutagenesis	Ataxia; cerebellar and testicular atrophy; postnatal degeneration of PCs, MCs and photoreceptors; decreased skeletal muscle fiber size; male infertility.	Unknown
Agtpbp1^{Pcd-7J}	Spontaneous	Ataxia; postnatal degeneration of PCs; enlarged hippocampus; abnormal hearing	Unknown
Agtpbp1^{Pcd-8J}	Spontaneous	Affectation of behavior; low size body; Alteration of nervous system development, reproductive, and vision.	Unknown
Agtpbp1^{Pcd-9J}	Spontaneous	Ataxia, but has a slightly later onset than that caused by the original *pcd* allele.	Unknown
Agtpbp1$^{Pcd-Tg(Dhfr)1jwg}$	Transgene insertion	Ataxia; degeneration of PCs, MCs and photoreceptor cells; some male infertility, female partial fertility; degeneration of sperm	Random gene disruption
Agtpbp1^{Drunk}	Mutagenesis	Degeneration of Purkinje cells and photoreceptor cells; Male infertility	Unknown
Agtpbp1^{Rio}	Mutagenesis	Tremor and abnormal sperm	Unknown
Agtpbp1^{babe}	ENU-induced mutagenesis	Ataxia; paraparesis	P804 arginine to a termination codon
Agtpbp1$^{Pcd-Btlr}$	ENU-induced mutagenesis	Ataxia; degeneration of PCs, MCs and photoreceptor cells; male infertility, oligozoospermia and teratozoospermia	a T-to-A transversion in the donor splice site of intron 11
Agtpbp1$^{Pcd-m2Btlr}$	ENU-induced mutagenesis	Tremors; decreased body size; reduced activated sperm motility	an A to G transition; destroys the acceptor splice site of intron 7 of the gene
Agtpbp1$^{Pcd-Sid}$	Spontaneous	Reduced body size; Ataxia; Cerebellar atrophy.	Deletion of exon 7
Agtpbp1$^{Gt(IST13517F11)Tigm}$	Gene trapped allele	one ES cell; unclassified	Chr13:59477801-59478055 bp (-);Chr13:59477801-59477979 bp (-)
Agtpbp1$^{Gt(OST186151)Lex}$	Gene trapped allele	Lex-1 (ES Cell)	Chr13:59531904-59544452 bp (-)
Agtpbp1$^{Gt(OST188387)Lex}$	Gene trapped allele	Lex-1 (ES Cell)	Chr13:59531902-59533237 bp (-)
Agtpbp1$^{Gt(OST252171)Lex}$	Gene trapped allele	Lex-1 (ES Cell)	Chr13:59531904-59544452 bp (-)
Agtpbp1$^{Gt(OST300426)Lex}$	Gene trapped allele	Lex-1 (ES Cell)	Chr13:59531904-59544452 bp (-)
Agtpbp1$^{Gt(OST300428)Lex}$	Gene trapped allele	Lex-1 (ES Cell)	Chr13:59536248-59536374 bp (-)
Agtpbp1$^{Gt(OST301743)Lex}$	Gene trapped allele	Lex-1 (ES Cell)	Chr13:59531913-59536374 bp (-)
pcd^{KO}	Knock-out	Ataxia; cerebellar atrophy, postnatal degeneration of PCs and photoreceptors.	Deletion of exons 21 and 22

In summary, the *pcd* mouse is a suitable animal model for studying ataxia and cerebellar atrophy with genetic, clinical and histopathological characteristics similar to those of human CONDCA patients (Table 4). The following sections describe in detail the neuropathological signs and mechanisms of neurological dysfunction in *pcd* mice.

5.1. Degeneration in the Cerebellum

One of the main phenotypic hallmarks of the pcd mutation in mice is early-onset cerebellar atrophy, which is mainly associated with drastic and premature primary degeneration of PCs [19,40], starting in the vermis and progressively advancing to the cerebellar hemispheres [42]. As degeneration proceeds, other cerebellar neuronal populations, such as granule cells and neurons of the inferior olivary complex and deep cerebellar nuclei (DCN), subsequently degenerate. However, this secondary degeneration process is much slower, taking approximately one year, probably as a consequence of primary PC death [9]. Consequently, there is an additional thinning of the molecular and granule cell layers with the subsequent worsening of the cerebellar atrophy [43,44].

Table 4. Comparison of pathological findings between CONDCA patients and the pcd mutant mouse. N.E: Not examined.

Physiopathological Feature	CONDCA Patients	*pcd* Mice
Early-onset	YES	YES
Progressive degenerative course	YES	YES
Microcephaly	YES	YES
Motor delay	YES	YES
Hypotonia	YES	N.E
Muscle weakness	YES	YES
Muscle atrophy	YES	YES
Tongue fasciculations	Frequent	N.E
Alteration of tendon reflexes	Frequent	N.E
Ataxia	Frequent	YES
Dystonia	Frequent	N.E
Spasticity	Frequent	N.E
Respiratory distress	Frequent	N.E
Feeding difficulties	Frequent	YES
Eye movement abnormalities	Frequent	N.E
Defective hearing	Occasional	YES
Cognitive delay	YES	YES
Motor and axonal motor neuropathy	Frequent	YES
Denervation	Frequent	YES
Olfactory dysfunction	N.E	YES
Visual deficiency	N.E	YES
Defective sperm	N.E	YES

At the molecular level, analysis of the transcriptional signature in the cerebella of *pcd* mice has shown that the vast majority of genes with altered transcriptional levels are related to functional categories such as cell death, developmental disorders, survival and glial responses [45,46].

5.1.1. Degeneration of Purkinje Cells

One of the phenotypic hallmarks of the pcd mutation in mice is early-onset cerebellar atrophy, which is mainly associated with drastic primary degeneration of PCs. Between two and four weeks of age, PCs rapidly degenerate, with only a few PCs remaining in lobule X of the cerebellar vermis, which is preserved for a few additional weeks (Figure 4A,B) [10,17,19,47]. The degeneration of a massive number of PCs in *pcd* mice sequentially involves an initial "preneurodegenerative" stage, from postnatal day (P) 15 to P20, during which both cytoplasmic and nuclear alterations occur [17,25,48,49], followed by a degenerative stage (P25–45), in which all cerebellar PCs degenerate (Figure 4C,D) [17,48], leading to an alteration in cerebellar-related motor performance. Interestingly, heterozygous *pcd* mice show a significant reduction in the number of PCs at P300, an observation that supports the idea that heterozygosity of the AGTPBP1 mutation may influence the ageing process, causing moderate PC degeneration [50].

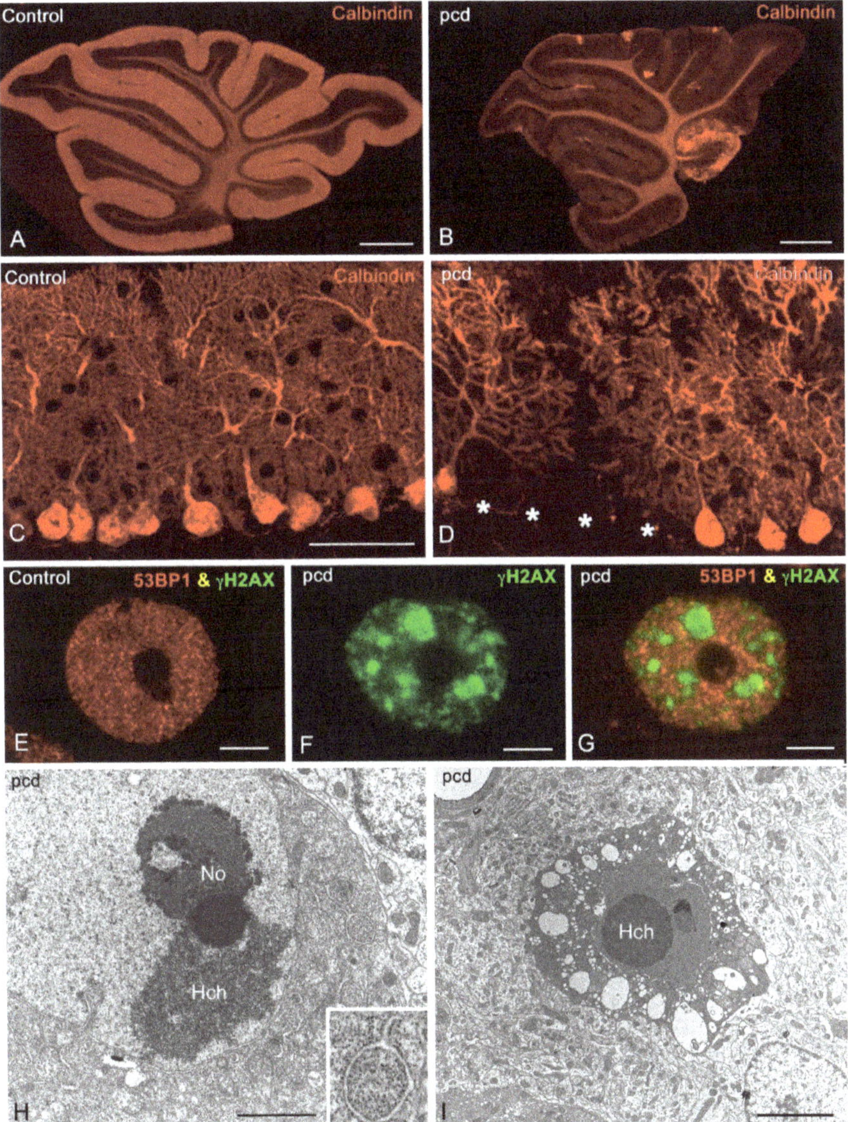

Figure 4. Degenerative features of the Purkinje cells in the *pcd* mice. (**A**,**B**) Representative confocal microscopy images of sagittal sections of the vermis of P30 control (**A**) and P30 (**B**) *pcd* mutant mice immunolabelled for calbindin D-28k. Note that in the *pcd* mouse there was a dramatic reduction in calbindin immunostaining in the molecular and PC layers resulting from the massive loss of PCs at P30 and that only PCs located in lobule X remained (arrow in B). Scale bars: 1 mm. (**C**,**D**) High magnification of calbindin D-28K immunolabelling of PC perikarya and their dendritic trees in control (**C**) and *pcd* mice (**D**) at P20. Note the loss of PCs (white asterisks) in the *pcd* mouse. Scale bar: 100 μm. (**E**–**G**) Confocal microscopy images of PC nuclei from control (**E**) and *pcd* mice (**F**,**G**) at P20 double immunolabelled for the modified histone γH2AX (green), a marker of DNA double-strand breaks at sites of DNA damage, and p53-binding protein 1 (P53BP1, red), a key DNA repair factor. (**E**) Note the absence of γH2AX labelling and the typical diffuse nucleoplasmic distribution of 53BP1, in the control PC nucleus, excluding the nucleolus. (**F**,**G**) In contrast, the nucleus of the *pcd* mouse shows prominent nuclear foci of DNA damage immunostained for γH2AX (**F**). Although the DNA repair factor 53BP1 was expressed in the nucleoplasm, it was not concentrated in γH2AX-positive nuclear foci of DNA lesions (**G**), indicating defective DNA repair. Scale bars: 5 μm. (**H**) Electron

microscopy image of PCs from *pcd* mice at P20. Free polyribosomes were replaced by densely packed monoribosomes. Cytoplasmic portions containing monoribosomes appear sequestered in autophagic vacuoles bound by isolated RE cisternae (insert). Scale bar: 1 μm. (**I**) Electron microscopy image of mutant apoptotic PC Scale bars: 5 μm. (**A,B**) Adapted with permission from [17]. Copyright 2011 The Journal of Biological Chemistry. (**C,D,H**) Adapted with permission from [48]. Copyright 2011 Brain Pathology. (**E–G**) Adapted with permission from [49]. Copyright 2019 Neurobiology of Disease.

At the early stages of PC degeneration, alterations in the cytoarchitecture of the PCs, including disrupted dendrites, soma and axons, are already noticeable [40,46]. Ultrastructural analysis of axonal torpedoes has revealed organelle accumulation and cytoplasmic densification but the preservation of the myelin sheath [46]. This finding suggests that degeneration of axons is the primary defect and rules out the possibility that PCs degenerate following demyelination, as observed in the cerebral cortex in AGTPBP1-deficient mice [22]. AGTPBP1 loss of function affects dendritic tree development and architecture; however, these alterations do not seem to be directly involved in PC death [19,25,40].

Another pre-degenerative characteristic of PCs is a reduction in perikaryal size [25]. Intriguingly, one of the earliest cytoplasmic morphological features of pre-degenerative PCs is the accumulation of free polyribosomes [19,47,51]. Polyribosome accumulation correlates with endoplasmic reticulum (ER) stress in the PCs of *pcd* mice. In the initial stages of PC degeneration, the stacks of ER cisterns tend to disappear, and PCs show a prominent mass of densely packed free polyribosomes at the basal pole of PC somas. As degeneration proceeds, polyribosomes are disassembled into free monoribosomes [48]. A fraction of these free monoribosomes is sequestered in autophagic vacuoles for lysosomal degradation through a process termed ribophagy (Figure 4H) [48].

At the molecular level, upregulation of the expression of ER stress-related substrates and the unfolded protein response accompanied by downregulation of the expression of initiation factors for translation were detected in the mutant PCs [40,51,52]. Accumulation of hyperglutamylated tubulin in AGTPBP1-deficient PCs directly correlates with ER stress [51]. Other cytoplasmic alterations, such as the presence of abnormal mitochondria and reduced mitochondrial complex 1 activity, were also observed in mutant PCs [18,53].

The AGTPBP1 gene contains an NLS (Figure 1B) and encodes nuclear and cytoplasmic proteins [7]. However, little is known about its potential role in the neuronal nucleus. We performed an extensive analysis of the effects of AGTPBP1-deficiency in the PC's nuclear compartments involved in RNA transcription and processing and DNA damage repair, evaluating the impact of their dysfunction on neuronal homeostasis and survival [17,48,49]. During the pre-degenerative stage, in PCs, there is a progressive large-scale reorganisation of chromatin into large, transcriptionally silent, heterochromatin domains associated with the accumulation of DNA damage, which is one of the first signs of the PC pre-neurodegenerative stage of in *pcd* mice [17].

To avoid the detrimental effects of DSBs, neurons exhibit a strong DNA repair response. However, growing evidence has indicated that defective DNA repair is the basis for brain ageing and several degenerative disorders [54]. In this context, a progressive accumulation of unrepaired DNA damage was detected in the PCs of *pcd* mice (Figure 4E–G) [17,48]. Accordingly, defective DNA repair was implicated in the pathogenesis of several ataxias with a PC degeneration phenotype [55–60].

In addition to DNA damage and epigenetic changes in chromatin conformation, other nuclear compartments are affected in the mutant PCs. In particular, the disassembly of the Cajal bodies [61] and nucleolar disruption are directly correlated with the activation of nucleolar stress and defective ribosome biogenesis [48,49]. Nucleolar stress has been associated with several neurodegenerative human disorders, including Alzheimer's and Parkinson's diseases, ALS and spinal muscular atrophy, among others [62–68]. Moreover, a reduction in both nucleolar size and ribosome biogenesis occurs during ageing and is a key risk factor related to the onset of neurodegenerative disorders [54,69].

Consistent with all defined cytoplasmic and nuclear alterations, dysfunctional PCs in *pcd* mice ultimately activate the caspase-mediated apoptotic pathway (Figure 4I), including by upregulating the expression of neuronal apoptosis facilitators such as Bim3 and Bcl2l11 [9,17–19,48,51,52,70]. In summary, some types of nuclear rearrangement and pathological alterations observed in PCs harbouring the pcd mutation resemble the cellular alterations described in certain human neurodegenerative diseases [66,71,72].

5.1.2. Alterations in Other Neuronal Types in the Cerebella of *pcd* Mice

In the cerebella of *pcd* mice, late secondary death of granule cells and neurons in the DCN and the inferior olivary complex, which is likely due to PC degeneration, occurs [9,43,73]. Remarkably, granule cells express the AGTPBP1 gene [7,10]. However, a significant reduction in granule cell number is only detected at 6 months of age, which progresses throughout the lifespan of the animal [9].

The loss of presynaptic afferents from PCs to the DCN results in decreased neuron survival (~30%) at advanced stages of neurodegeneration [44,74]. As a consequence of the abrogation of its major cortical target, the inferior olivary complex becomes atrophied subsequent to partial denervation in *pcd* mice [40,43,75]. In particular, the number of neurons in the inferior olivary complex is reduced by half at 10 months of age, which then causes a reduction in the number of climbing fibres that reach the cerebellar cortex [75].

Finally, the disappearance of PCs in *pcd* mice stimulates severe gliosis [40,46] and strikingly, oligodendrocytes and their precursors are markedly affected in the cerebellar cortex in *pcd* mice [46].

5.1.3. Reorganisation of Cerebellar Circuitry in *pcd* Mice after Purkinje Cell Loss

The compensatory mechanisms in the cerebellar circuitry following PC loss in *pcd* mice were examined [76]. Electrophysiological studies have revealed that despite the absence of PC-mediated tonic inhibition in the vestibular nucleus (VN) in *pcd* mice, spontaneous activity is not greater in AGTPBP1-deficient neurons [76]. In addition, abrogation of PC input did not underlie disinhibition in neurons from the VN in *pcd* mice [76]. The influence of PC input loss in VN neurons in regulating muscle contraction through the vestibulospinal pathway was also assessed. The response phase is slightly modified in the *pcd* mice in comparison with control littermates. This effect may be partially involved in motor impairment [76].

The impact of AGTPBP1 deficiency on cerebellar neurotransmission was also evaluated. Abrogation of inputs originating from PCs reduces GABAergic inhibitory innervation and decreases the density of GABAA receptors in the DCN and VN [74,76]. In contrast, the glycinergic system was promoted in the DCN in *pcd* mutant mice [74]. On the other hand, an increase in the number of glutamatergic synapses in both the DCN and the VN was detected, most likely due to enhanced mossy fibre innervation of the DCN and a secondary effect of reduced GABAergic-mediated inhibition in DCN neurons [74].

PCs and granule cells are specific targets for serotonergic neurons projecting from raphe nuclei and other brain areas. The serotonergic centrifugal system in the cerebella of *pcd* mice becomes dysfunctional once PCs are completely depleted [77]. In particular, *pcd* mice showed a higher 5-HT-IR fibre density [78] and a reduction in the 5-HIAA/5-HT ratio [77]. Enhanced synthesis of 5-HT transporters and receptors was also detected in both the cerebellar cortex and DCN [77,79]. Together, these data point to a reduction in serotonergic modulation, indicating a decrease in serotonergic turnover in the *pcd* cerebella [77].

Noradrenergic axon terminals from the locus coeruleus that reach the cerebellar cortex are preserved in *pcd* mice despite the absence of PCs, the targets of noradrenergic projections [80,81]. In addition, an increase in the density of norepinephrine fibres, most likely due to a reduction in cerebellar mass in *pcd* mice, was also detected [81]. Likewise, a moderate increase in the levels of noradrenergic transporters and adrenergic receptors upon PC loss was observed [82]. Regarding the dopaminergic system, PC loss induces an increase in the levels of dopamine transporters in the DCN but a significant reduction in these levels in the molecular layer of the cerebellar cortex [83].

5.1.4. Alterations in Cerebellar-Dependent Tasks

A large battery of motor tests were used to evaluate the degree of cerebellar atrophy and the consequent impairment of locomotor coordination, as well as progressive weakness of musculature and cognitive decline in AGTPBP1-depleted mice.

Analysis of ataxic gait has revealed irregularly spaced and shorter steps in 4-week-old mutants, with this deficit becoming worse as cerebellar degeneration proceeds [14,40]. Severe early impairments in motor performance, in the rotarod test, which progressively deteriorates in parallel with PC death, were observed in *pcd* mice [14,17,19,25,40,84,85]. It was demonstrated that in the treadmill motor assessment, *pcd* mice have decreased body movement coordination [84]. Similarly, front-hind interlimb and whole-body coordination deficits were characterised using LocoMouse [86].

The grip strength and wire hang tests have revealed that *pcd* mice experience muscle weakness [35,40]. In addition, the balance beam test has revealed that 4-week-old AGTPBP1 KO mice fall significantly more frequently than their wild-type counterparts [40].

Delayed eye-blinking conditioning appears to be severely affected and altered cerebellum-dependent learning is altered in adult *pcd* mice [87], whereas trace eyeblink conditioning is unimpaired, suggesting that the cerebellum plays an indispensable role in the neuronal circuitry regulating this response [88]. The effects of cerebellar dysfunction on spatial learning in adult *pcd* mice were determined using the Morris water maze test [89–91]. The novel object recognition test has revealed that long-term memory in *pcd* mice is affected in the late stages of PC degeneration, and the results of the social preference test have suggested that PC loss in *pcd* mice affects social interaction [25]. Likewise, the results of the forced swimming test have suggested that *pcd* mice exhibit depressive-like behaviour [91].

5.2. Degeneration in the Olfactory Bulb

5.2.1. Degeneration of Mitral Cells

The OB is considerably smaller in *pcd* mice than in control animals, mainly due to the loss of MCs, the principal relay neurons in the olfactory pathway [5]. The loss of MCs is accompanied by reductions in the size of glomeruli and the thickness of the external plexiform layer. Other bulbar layers or neural elements are apparently unaffected by the loss of AGTPBP1 [8].

Degeneration of MCs occurs later and more slowly than that of PCs, taking place from P60 to P90 days [8,20]. While there is an extensive amount of data on the mechanisms underlying PC death, there is limited information regarding the mechanisms involved in MC degeneration. Similar to that of mutant PCs, degeneration of MCs is associated with ER stress, transcriptional repression, DNA damage and disruption of nucleoli and Cajal bodies, which ultimately cause apoptosis [20]. As in the cerebellum, tubulin hyperglutamylation in the OB was suggested to be a determinant of MC death in *pcd* mice [15,16].

MC degeneration induces reactive glial activation of astrocytes and microglia in the OB. However, this response is milder than that detected in the cerebellum [46]. Curiously, bulbar oligodendrocytes are not affected in *pcd* mice [46]. Differential glial responses observed in the cerebellum and the OB seem to correlate with the degree of neurodegeneration in each brain region and physiological AGTPBP1 expression levels [46].

5.2.2. Reorganisation of Synaptic Circuitry after Mitral Cell Loss

MCs establish reciprocal dendrodendritic synapses with bulbar granule cells. Although the pcd mutation does not compromise the viability of granule cells, MC degeneration prevents afferent inputs from contacting granule cells. Some granule cells establish new reciprocal dendrodendritic synapses with unaffected tufted cells [92]. However, it should be noted that mutant granule cells have an effect on the dendritic tree, including shortening dendrites and reducing the number of spines [93]. In contrast, afferent inputs reaching the OB from olfactory receptor cells are slightly affected by the loss of MCs [8].

MCs send axonal efferent inputs to the lateral olfactory tract. A general decrease in the thickness of the olfactory tract was found in *pcd* mice [92], supporting the notion that the number of synapses declines upon MC degeneration. In addition, the diameter of terminal boutons increases, as does the number of multiple synaptic contacts, in *pcd* mutants, suggesting further compensatory mechanisms for the loss of MC presynaptic terminals [92].

Centrifugal afferences from secondary olfactory structures to the OB upon MC loss were also examined. Strengthening of the centrifugal input to the OB from the anterior olfactory nucleus after MC loss was detected in *pcd* mice and is accompanied by complete loss of bilaterality in olfactory connections due to degeneration of the anterior commissure [94]. These results point to a dramatic reorganisation of this essential olfactory circuit between the anterior olfactory nucleus and the OB upon MC degeneration.

Regarding the dopaminergic system in the OB, autoradiography studies have shown that dopamine receptor and transporter levels are not affected by AGTPBP1 loss of function [83]. Accordingly, tyrosine hydroxylase activity and immunoreactivity in OB juxtaglomerular neurons are more preserved in *pcd* mutants after MC degeneration than in heterozygous littermates [95].

The serotonergic system undergoes adaptive changes after, but not before, MC loss [96]. Degeneration of MCs causes a decrease in serotonergic input received by the OB, whereas the number of serotonergic cells in the raphe nuclei remains constant. In this regard, the neurotrophin BDNF and its main receptor TrkB exhibit altered expression in the OBs of *pcd* animals even before the loss of MCs [96].

Although the expression of noradrenaline transporters is not affected by MC degeneration, variations in adrenergic receptors in some olfactory regions were defined, suggesting a local regulation of the NA system in regions influenced by MC loss [82]. The *pcd* mice also show reorganisation of zincergic centrifugal projections from the anterior olfactory nucleus to the OB, indicating that plasticity occurs in response to MC loss [97].

5.2.3. Neural Plasticity in the Olfactory Bulb after Mitral Cell Loss

Neural progenitor cells from the rostral migratory stream differentiate into bulbar interneurons that modulate MC activity. Interestingly, changes in the proliferation rate, tangential and radial migration patterns and survival of newly generated neurons in *pcd* mice were reported. Consequently, the absence of MCs in these mutants elicits differences in the final destination of the newly generated interneurons. Moreover, the depletion of MCs also alters the survival of the newly generated interneurons, in accordance with the decrease in the number of synaptic targets available [98].

5.2.4. Alterations in Olfactory Task Performance after Mitral Cell Loss

Despite the importance of the olfactory system in learning and affective behaviour in mice [99], little information about the potentially deleterious consequences of MC degeneration on olfactory-related task performance in *pcd* mice is available. Using precision olfactometry, Diaz and colleagues showed that after MC death, *pcd* mutants exhibit poor odourant detection ability and limited odour discrimination ability [100]. In particular, *pcd* mice are able to detect elevated, but not low, concentrations of odourants and discriminate them in a crude manner, suggesting the involvement of MCs in fine odour transmission and processing [100,101].

5.3. Degeneration in the Thalamus

Discrete populations of thalamic neurons degenerate in *pcd* mice between P50 and P60 and are nearly absent at P90 [21,102]. Thus, massive neuronal degeneration is observed in the central division of the mediodorsal nucleus, the ventral medial geniculate, posterior ventromedial and submedial nuclei, as well as portions of the ventrolateral and posteromedial nuclei that immediately surround the medial division of the ventrobasal complex. Degenerating thalamic neurons in the ventral medial geniculate nucleus, the main auditory thalamic area, show degenerative cellular hallmarks that resemble those reported for mutant PCs and MCs [102,103].

The electrophysiological and molecular changes in the ventral medial geniculate nucleus in *pcd* mice were also examined [103]. Likewise, a progressive decrease in auditory evoked potentials and NMDA receptor-dependent fast oscillations in the auditory cortex were detected in *pcd* mice [103].

Changes in the regional thalamic distribution of noradrenaline uptake sites, as well as in the expression of adrenergic receptors, were described following thalamic neuron loss in *pcd* mice [82]. In addition, increased levels of dopamine receptors were found in the centromedian thalamic nucleus in *pcd* mice [83].

5.4. Degeneration in the Retina

The onset of retinal degeneration of photoreceptors in *pcd* mice occurs between 3 and 5 weeks of age, when approximately 50% of receptors are quickly lost [104,105]. Afterward, degeneration progresses quite slowly, with approximately 10% of the photoreceptors remaining by one year of age [104,105], and rods degenerating faster than cones [105]. The main photoreceptor alterations include abnormal accumulation of "bead-like" vesicles and ribosomes, disruption of the Golgi apparatus, and a significant reduction in the number of connecting cilia [106], which ultimately lead to the death of photoreceptors by apoptosis [6,40,107]. In addition, the pcd mutation in photoreceptors increases their vulnerability to the cellular stress produced by constant light exposure [108]. Progressive accumulation of polyglutamylated tubulin was detected in parallel with the degeneration of *pcd* mutant photoreceptors [106]. Progressive loss of dendrites and disorganisation of axon terminals in retinal bipolar cells were also reported in parallel with degeneration of photoreceptors [41].

Consistent with cellular alterations, electroretinography of the *pcd* mutant retina has revealed a progressive reduction in the amplitude of electrical signals in both rods and cones at advanced stages of degeneration in comparison with that in the control retina [41].

5.5. Degeneration of Other Neuronal Types

Quantitative estimation of the number of α-motor neurons in the ventral horn of the lumbar spinal cord has revealed an approximately 50% reduction in the number of these cells in *pcd* mice compared with control mice, which is accompanied by dysregulation of tubulin polyglutamylation [1]. Moreover, peripheral nerve degeneration with reduced motor nerve caliber, significant loss of myelinated axons and altered axon morphology were reported in *pcd* mice [1].

As mentioned, one of the main pathological hallmarks in CONDCA patients is muscle weakness. Additionally, AGTPBP1 mRNA was found to be expressed in mouse skeletal muscle [12]. Muscle tissue organisation in *pcd* mice was found to be hardly affected, with no obvious accumulation of collagen or fibrosis. However, the diameter of skeletal myofibres is reduced compared with that of control myofibres, most likely due to ataxia-derived atrophy (Figure 3E,F) [37] resulting from the decrease in the number of α-motor neuron axons innervating the skeletal muscle [1]. As in other tissues affected by AGTPBP1 loss of function, whole-protein extracts from *pcd* skeletal muscle exhibit higher levels of tubulin polyglutamylation than those from control skeletal muscle [15].

5.6. Therapeutic Strategies

Due to AGTPBP1 mutation-mediated primary neuronal death and the occurrence of secondary neurodegenerative processes in the *pcd* mouse brain, this mouse model may serve as an attractive model for investigating new neuroprotective strategies to prevent, or at least attenuate, neurodegeneration. Most of the experimental therapies assessed in the *pcd* mice presented here aimed to reverse cerebellar degeneration in *pcd* mutants. Stem cell-based neuroregeneration or the use of molecules with neuroprotective potential are the main experimental approaches that were assessed in *pcd* mice.

5.6.1. Stem Cell-Based Transplantation

Under neurodegenerative conditions, grafted cells from healthy donors may provide neurotransmitters with neuroprotective potential, replace degenerated neurons and provide trophic support to surviving neurons. Based on this notion, embryonic cerebellar grafts appear to be a potential therapeutic strategy not only to replace PCs but also to prevent secondary neuronal death. Cells from solid embryonic cerebellar grafts from healthy donors implanted into 3-month-old *pcd* mutants were able to migrate, settle and establish functional synapses in the host cerebellar cortex [109–113]. Similarly, suspended normal embryonic cerebellar cells transplanted into the *pcd* mouse cerebellum were shown to survive and integrate with the degenerative harmful host environment, develop the characteristic PC cytoarchitecture and re-establish host-to-graft afferent innervation, while also ameliorating motor deficits [114–119].

Bone marrow-derived stem cell (BMDSC) transplantation also appears to be a therapeutic option for ameliorating neurodegeneration in *pcd* mice. Initial studies have indicated that grafted BMDSCs in *pcd* mice are able to migrate and reach the degenerating cerebellum and OB, although most of them differentiate into glial cells [120]. Posterior bone marrow transplantation notably improves skeletal muscle tissue organisation rather than attenuating neurodegeneration, which correlates with a partial but significant restoration of locomotor performance [37]. Thus, recovery of muscular dysfunction appears to be the basis of this locomotor improvement. BMDSC transplantation in *pcd* mice also results in the attenuation of MC degeneration and an associated improvement in odour detection [100]. Limitations of this approach include that the delivery of healthy BMDSCs to the damaged site is not fast enough to stop neuronal loss over time. Thus, optimisation of this technique by ensuring a regular supply of healthy stem cells through continuous, daily transplants would increase the population of pluripotent cells that reach the target tissue and potentially fuse with unaffected mutant PCs, increasing their survival [121].

An additional limitation is the physical barrier of the granule cell layer, which impedes healthy grafted cells from reaching the PC layer. In addition, the complexity of cerebellar circuitry is too great for it to be finely reconstructed. Moreover, the purposed fate of grafted cells is strictly regulated by a large variety of factors that may vary according to each degenerative environment. Therefore, neurotransplantation and stem cell-based therapy in patients with cerebellar degeneration are still far from being practical [122].

5.6.2. Preservation of Degenerating Neurons in *pcd* Mice

Other experimental approaches have aimed to protect and preserve mutant PCs through the administration of neuroprotective molecules, the exogenous administration of functional AGTPBP1 and the directed modulation of specific signalling pathways involved in the degeneration of PCs in *pcd* mice.

The neuroprotective role of insulin-like growth factor (IGF-I) was shown, and IGF-I has been extensively used to treat several neurodegenerative disorders. Interestingly, reduced levels of IGF-I were found in patients with cerebellar dysfunction and ataxia [123]. Consistently, IGF-1 administration in *pcd* mice was shown to significantly increase body weight and survival and improve motor performance [124]. Interestingly, administration of IGF-1 in patients with autosomal dominant cerebellar ataxia delays the progression of the disease and appears to be a potentially promising therapeutic option for CONDCA patients [125]. In addition, oleoylethanolamide, an endocannabinoid compound, was proposed to prevent neuronal damage, delay PC death and ameliorate cognitive decline in *pcd* mice [126].

Affectation in the auditory cortex in *pcd* mice following thalamic neuron degeneration is closely related to a marked upregulation of NMDA expression. Accordingly, the administration of an NMDA antagonist restores the electrophysiological response evoked in the auditory cortex in *pcd* mice [103].

Another important target of neuroprotective agents is the modulation of the glial response. Accordingly, a harmful glial response in the cerebella of *pcd* mice could be directly related to the rapid degeneration of PCs [46]. In this regard, attenuation of glial activation following minocycline administration delays the death of PCs in *pcd* mice and mildly improves their locomotor performance [52]. Thus, administration of glial activation inhibitors or genetic modulation of the glial response may be considered potential therapeutic approaches to ameliorate neurodegenerative disorders [127].

5.6.3. Genetically Mediated Therapeutic Approaches

Genetically mediated restoration of functional AGTPBP1 expression in both the cerebellum and the retina is sufficient to rescue PC and photoreceptor degeneration in *pcd* mice [14,30]. As mentioned above, AGTPBP1 loss of function leads to excessive tubulin polyglutamylation, which seems to be the main cause of neurodegeneration in *pcd* mice. Based on this notion, the abnormal accumulation of polyglutamylation in the mutant cerebellum may be rescued by inactivation of the polyglutamylase tubulin-tyrosine ligase-like 1 (TTL1), resulting in almost complete preservation of PCs in the *pcd* mouse cerebellum [22]. This raises the possibility that pharmacologically mediated regulation of enzymes counteracting AGTPBP1-mediated reactions, such as TTL1, could be considered a new therapeutic approach for the treatment of AGTPBP1-related diseases.

6. Conclusions

Here, we summarise the most recent experimental findings that support *AGTPBP1* as the gene responsible for the development of CONDCA in humans. Additionally, the pathogenic events that occur in *pcd* mice, which harbours loss-of-function mutations in the *AGTPBP1* gene, are described in-depth and summarised in the graphical abstract. The fact that the pathological characteristics of the pcd mutation share clear similarities with those of CONDCA patients indicates that AGTPBP1 function is essential to the development of neurological disorders not only in mice but also in humans. In this regard, it is evident that the *pcd* mouse appears as a promising model for the development of new therapeutic strategies for clinical trials in humans. Polyglutamylation inhibition was recently described as a promising therapeutic option for CONDCA patients. This line of investigation together with others involving the use of *pcd* mice and those included in this review may be considered in the future as therapeutic options for CONDCA treatment.

Author Contributions: F.C.B. wrote the review. M.T.B., E.S. and M.L. participated in revising and editing the final version of the manuscript and the figures. All authors have read and agreed to the published version of the manuscript.

Funding: This work was supported by funding received by the Institute of Health Carlos III (CIBERONC, CB16/12/00352 and CIBERNED, CB/05/0037), the Valdecilla Research Institute (IDI-VAL, Santander Spain), the Regional Government of Castile and Leon, funds received by the European Regional Development Fund (SA264P18-UIC076; University of Salamanca, Spain) and the Samuel-Solórzano Barruso Foundation (FS/32-2020, University of Salamanca, Spain).

Institutional Review Board Statement: Not applicable.

Informed Consent Statement: Not applicable.

Acknowledgments: We thank Ana Canga from the "Hospital Universitario Marqués de Valdecilla", Santander (Spain) for gently provide MRI images from a control individual.

Conflicts of Interest: The authors declare no conflict of interest.

References

1. Shashi, V.; Magiera, M.M.; Klein, D.; Zaki, M.; Schoch, K.; Rudnik-Schöneborn, S.; Norman, A.; Lopes Abath Neto, O.; Dusl, M.; Yuan, X.; et al. Loss of tubulin deglutamylase CCP1 causes infantile-onset neurodegeneration. *EMBO J.* **2018**, *37*, e100540. [CrossRef] [PubMed]
2. Sheffer, R.; Gur, M.; Brooks, R.; Salah, S.; Daana, M.; Fraenkel, N.; Eisenstein, E.; Rabie, M.; Nevo, Y.; Jalas, C.; et al. Biallelic variants in AGTPBP1, involved in tubulin deglutamylation, are associated with cerebellar degeneration and motor neuropathy. *Eur. J. Hum. Genet.* **2019**, *27*, 1419–1426. [CrossRef] [PubMed]

3. Karakaya, M.; Paketci, C.; Altmueller, J.; Thiele, H.; Hoelker, I.; Yis, U.; Wirth, B. Biallelic variant in AGTPBP1 causes infantile lower motor neuron degeneration and cerebellar atrophy. *Am. J. Med. Genet. A* **2019**, *179*, 1580–1584. [CrossRef] [PubMed]

4. Maddirevula, S.; Alzahrani, F.; Al-Owain, M.; Al Muhaizea, M.A.; Kayyali, H.R.; AlHashem, A.; Rahbeeni, Z.; Al-Otaibi, M.; Alzaidan, H.I.; Balobaid, A.; et al. Autozygome and high throughput confirmation of disease genes candidacy. *Genet. Med.* **2019**, *21*, 736–742. [CrossRef]

5. Mullen, R.J.; Eicher, E.M.; Sidman, R.L. Purkinje cell degeneration, a new neurological mutation in the mouse. *Proc. Natl. Acad. Sci. USA* **1976**, *73*, 208–212. [CrossRef]

6. Blanks, J.C.; Mullen, R.J.; Lavail, M.M. Retinal degeneration in thepcd cerebellar mutant mouse. II. Electron microscopic analysis. *J. Comp. Neurol.* **1982**, *212*, 231–246. [CrossRef]

7. Harris, A.; Morgan, J.I.; Pecot, M.; Soumare, A.; Osborne, A.; Soares, H.D. Regenerating motor neurons express Nna1, a novel ATP/GTP-binding protein related to zinc carboxypeptidases. *Mol. Cell. Neurosci.* **2000**, *16*, 578–596. [CrossRef]

8. Greer, C.A.; Shepherd, G.M. Mitral cell degeneration and sensory function in the neurological mutant mouse Purkinje cell degeneration (PCD). *Brain Res.* **1982**, *235*, 156–161. [CrossRef]

9. Wang, T.; Morgan, J.I. The Purkinje cell degeneration (pcd) mouse: An unexpected molecular link between neuronal degeneration and regeneration. *Brain Res.* **2007**, *1140*, 26–40. [CrossRef]

10. Fernandez-Gonzalez, A.; La Spada, A.R.; Treadaway, J.; Higdon, J.C.; Harris, B.S.; Sidman, R.L.; Morgan, J.I.; Zuo, J. Purkinje cell degeneration (pcd) phenotypes caused by mutations in the axotomy-induced gene, Nna1. *Science* **2002**, *295*, 1904–1906. [CrossRef]

11. Zhao, X.; Onteru, S.K.; Dittmer, K.E.; Parton, K.; Blair, H.T.; Rothschild, M.F.; Garrick, D.J. A missense mutation in AGTPBP1 was identified in sheep with a lower motor neuron disease. *Heredity* **2012**, *109*, 156–162. [CrossRef] [PubMed]

12. Kalinina, E.; Biswas, R.; Berezniuk, I.; Hermoso, A.; Aviles, F.X.; Fricker, L.D. A novel subfamily of mouse cytosolic carboxypeptidases. *FASEB J.* **2007**, *21*, 836–850. [CrossRef] [PubMed]

13. Vega, M.R.; Sevilla, R.G.; Hermoso, A.; Lorenzo, J.; Tanco, S.; Diez, A.; Fricker, L.D.; Bautista, J.M.; Aviles, F.X. Nna1-like proteins are active metallocarboxypeptidases of a new and diverse M14 subfamily. *FASEB J.* **2007**, *21*, 851–865. [CrossRef] [PubMed]

14. Wang, T.; Parris, J.; Li, L.; Morgan, J.I. The carboxypeptidase-like substrate-binding site in Nna1 is essential for the rescue of the Purkinje cell degeneration (pcd) phenotype. *Mol. Cell. Neurosci.* **2006**, *33*, 200–213. [CrossRef]

15. Rogowski, K.; van Dijk, J.; Magiera, M.M.; Bosc, C.; Deloulme, J.-C.; Bosson, A.; Peris, L.; Gold, N.D.; Lacroix, B.; Bosch Grau, M.; et al. A family of protein-deglutamylating enzymes associated with neurodegeneration. *Cell* **2010**, *143*, 564–578. [CrossRef]

16. Berezniuk, I.; Vu, H.T.; Lyons, P.J.; Sironi, J.J.; Xiao, H.; Burd, B.; Setou, M.; Angeletti, R.H.; Ikegami, K.; Fricker, L.D. Cytosolic carboxypeptidase 1 is involved in processing α- and β-tubulin. *J. Biol. Chem.* **2012**, *287*, 6503–6517. [CrossRef]

17. Baltanás, F.C.; Casafont, I.; Lafarga, V.; Weruaga, E.; Alonso, J.R.; Berciano, M.T.; Lafarga, M. Purkinje Cell Degeneration in pcd Mice Reveals Large Scale Chromatin Reorganization and Gene Silencing Linked to Defective DNA Repair. *J. Biol. Chem.* **2011**, *286*, 28287–28302. [CrossRef]

18. Chakrabarti, L.; Zahra, R.; Jackson, S.M.; Kazemi-Esfarjani, P.; Sopher, B.L.; Mason, A.G.; Toneff, T.; Ryu, S.; Shaffer, S.; Kansy, J.W.; et al. Mitochondrial Dysfunction in NnaD Mutant Flies and Purkinje Cell Degeneration Mice Reveals a Role for Nna Proteins in Neuronal Bioenergetics. *Neuron* **2010**, *66*, 835–847. [CrossRef]

19. Li, J.; Gu, X.; Ma, Y.; Calicchio, M.L.; Kong, D.; Teng, Y.D.; Yu, L.; Crain, A.M.; Vartanian, T.K.; Pasqualini, R.; et al. Nna1 mediates purkinje cell dendritic development via lysyl oxidase propeptide and NF-κB signaling. *Neuron* **2010**, *68*, 45–60. [CrossRef]

20. Valero, J.; Berciano, M.T.; Weruaga, E.; Lafarga, M.; Alonso, J.R. Pre-neurodegeneration of mitral cells in the pcd mutant mouse is associated with DNA damage, transcriptional repression, and reorganization of nuclear speckles and Cajal bodies. *Mol. Cell. Neurosci.* **2006**, *33*, 283–295. [CrossRef]

21. O'Gorman, S.; Sidman, R.L. Degeneration of thalamic neurons in? Purkinje cell degeneration? mutant mice. I. Distribution of neuron loss. *J. Comp. Neurol.* **1985**, *234*, 277–297. [CrossRef] [PubMed]

22. Magiera, M.M.; Bodakuntla, S.; Žiak, J.; Lacomme, S.; Marques Sousa, P.; Leboucher, S.; Hausrat, T.J.; Bosc, C.; Andrieux, A.; Kneussel, M.; et al. Excessive tubulin polyglutamylation causes neurodegeneration and perturbs neuronal transport. *EMBO J.* **2018**, *37*, e100440. [CrossRef]

23. Strzyz, P. Neurodegenerative polyglutamylation. *Nat. Rev. Mol. Cell Biol.* **2019**, *20*, 1. [CrossRef]

24. Bodakuntla, S.; Schnitzler, A.; Villablanca, C.; Gonzalez-Billault, C.; Bieche, I.; Janke, C.; Magiera, M.M. Tubulin polyglutamylation is a general traffic control mechanism in hippocampal neurons. *J. Cell Sci.* **2020**, *133*, jcs241802. [CrossRef] [PubMed]

25. Muñoz-Castañeda, R.; Díaz, D.; Peris, L.; Andrieux, A.; Bosc, C.; Muñoz-Castañeda, J.M.; Janke, C.; Alonso, J.R.; Moutin, M.-J.; Weruaga, E. Cytoskeleton stability is essential for the integrity of the cerebellum and its motor- and affective-related behaviors. *Sci. Rep.* **2018**, *8*, 3072. [CrossRef]

26. De La Vega Otazo, M.R.; Lorenzo, J.; Tort, O.; Avilés, F.X.; Bautista, J.M. Functional segregation and emerging role of cilia-related cytosolic carboxypeptidases (CCPs). *FASEB J.* **2013**, *27*, 424–431. [CrossRef]

27. Lyons, P.J.; Sapio, M.R.; Fricker, L.D. Zebrafish cytosolic carboxypeptidases 1 and 5 are essential for embryonic development. *J. Biol. Chem.* **2013**, *288*, 30454–30462. [CrossRef]

28. Kimura, Y.; Kurabe, N.; Ikegami, K.; Tsutsumi, K.; Konishi, Y.; Kaplan, O.I.; Kunitomo, H.; Iino, Y.; Blacque, O.E.; Setou, M. Identification of tubulin deglutamylase among *Caenorhabditis elegans* and mammalian cytosolic carboxypeptidases (CCPs). *J. Biol. Chem.* **2010**, *285*, 22936–22941. [CrossRef] [PubMed]

29. Wu, H.Y.; Wang, T.; Li, L.; Correia, K.; Morgan, J.I. A structural and functional analysis of Nna1 in Purkinje cell degeneration (pcd) mice. *FASEB J.* **2012**, *26*, 4468–4480. [CrossRef]

30. Chakrabarti, L.; Eng, J.; Martinez, R.A.; Jackson, S.; Huang, J.; Possin, D.E.; Sopher, B.L.; La Spada, A.R. The zinc-binding domain of Nna1 is required to prevent retinal photoreceptor loss and cerebellar ataxia in Purkinje cell degeneration (pcd) mice. *Vis. Res.* **2008**, *48*, 1999–2005. [CrossRef]

31. Conde, C.; Cáceres, A. Microtubule assembly, organization and dynamics in axons and dendrites. *Nat. Rev. Neurosci.* **2009**, *10*, 319–332. [CrossRef]

32. Bulinski, J.C. Microtubules and Neurodegeneration: The Tubulin Code Sets the Rules of the Road. *Curr. Biol.* **2019**, *29*, R28–R30. [CrossRef] [PubMed]

33. Janke, C. The tubulin code: Molecular components, readout mechanisms, functions. *J. Cell Biol.* **2014**, *206*, 461–472. [CrossRef] [PubMed]

34. Tanco, S.; Tort, O.; Demol, H.; Aviles, F.X.; Gevaert, K.; Van Damme, P.; Lorenzo, J. C-terminomics Screen for Natural Substrates of Cytosolic Carboxypeptidase 1 Reveals Processing of Acidic Protein C termini. *Mol. Cell. Proteom.* **2015**, *14*, 177–190. [CrossRef] [PubMed]

35. Gilmore-Hall, S.; Kuo, J.; Ward, J.M.; Zahra, R.; Morrison, R.S.; Perkins, G.; La Spada, A.R. CCP1 promotes mitochondrial fusion and motility to prevent Purkinje cell neuron loss in pcd mice. *J. Cell Biol.* **2018**, *218*, 206–219. [CrossRef]

36. Türay, S.; Eröz, R.; Başak, A.N. A novel pathogenic variant in the 3′ end of the AGTPBP1 gene gives rise to neurodegeneration without cerebellar atrophy: An expansion of the disease phenotype? *Neurogenetics* **2021**, *22*, 127–132. [CrossRef] [PubMed]

37. Díaz, D.; Piquer-Gil, M.; Recio, J.S.; Martínez-Losa, M.M.; Alonso, J.R.; Weruaga, E.; Álvarez-Dolado, M. Bone marrow transplantation improves motor activity in a mouse model of ataxia. *J. Tissue Eng. Regen. Med.* **2018**, *12*, e1950–e1961. [CrossRef]

38. Akhmanova, A.; Hoogenraad, C.C. More is not always better: Hyperglutamylation leads to neurodegeneration. *EMBO J.* **2018**, *37*, e101023. [CrossRef]

39. Kitano, S.; Kino, Y.; Yamamoto, Y.; Takitani, M.; Miyoshi, J.; Ishida, T.; Saito, Y.; Arima, K.; Satoh, J.-I. Bioinformatics Data Mining Approach Suggests Coexpression of AGTPBP1 with an ALS-linked Gene C9orf72. *J. Cent. Nerv. Syst. Dis.* **2015**, *7*, 15–26. [CrossRef] [PubMed]

40. Zhou, L.; Hossain, M.I.; Yamazaki, M.; Abe, M.; Natsume, R.; Konno, K.; Kageyama, S.; Komatsu, M.; Watanabe, M.; Sakimura, K.; et al. Deletion of exons encoding carboxypeptidase domain of Nna1 results in Purkinje cell degeneration (pcd) phenotype. *J. Neurochem.* **2018**, *147*, 557–572. [CrossRef] [PubMed]

41. Marchena, M.; Lara, J.; Aijón, J.; Germain, F.; de la Villa, P.; Velasco, A. The retina of the PCD/PCD mouse as a model of photoreceptor degeneration. A structural and functional study. *Exp. Eye Res.* **2011**, *93*, 607–617. [CrossRef] [PubMed]

42. Zhang, W.; Ghetti, B.; Lee, W.H. Decreased IGF-I gene expression during the apoptosis of Purkinje cells in pcd mice. *Dev. Brain Res.* **1997**, *98*, 164–176. [CrossRef]

43. Ghetti, B.; Norton, J.; Triarhou, L.C. Nerve cell atrophy and loss in the inferior olivary complex of "Purkinje cell degeneration" mutant mice. *J. Comp. Neurol.* **1987**, *260*, 409–422. [CrossRef] [PubMed]

44. Triarhou, L.C.; Norton, J.; Ghetti, B. Anterograde transsynaptic degeneration in the deep cerebellar nuclei of Purkinje cell degeneration (pcd) mutant mice. *Exp. Brain Res.* **1987**, *66*, 577–588. [CrossRef] [PubMed]

45. Ford, G.D.; Ford, B.D.; Steele, E.C.; Gates, A.; Hood, D.; Matthews, M.A.B.; Mirza, S.; MacLeish, P.R. Analysis of transcriptional profiles and functional clustering of global cerebellar gene expression in PCD3J mice. *Biochem. Biophys. Res. Commun.* **2008**, *377*, 556–561. [CrossRef]

46. Baltanás, F.C.; Berciano, M.T.; Valero, J.; Gómez, C.; Díaz, D.; Alonso, J.R.; Lafarga, M.; Weruaga, E. Differential glial activation during the degeneration of Purkinje cells and mitral cells in the PCD mutant mice. *Glia* **2013**, *61*, 254–272. [CrossRef] [PubMed]

47. Landis, S.C.; Mullen, R.J. The development and degeneration of Purkinje cells in pcd mutant mice. *J. Comp. Neurol.* **1978**, *177*, 125–143. [CrossRef]

48. Baltanás, F.C.; Casafont, I.; Weruaga, E.; Alonso, J.R.; Berciano, M.T.; Lafarga, M. Nucleolar Disruption and Cajal Body Disassembly are Nuclear Hallmarks of DNA Damage-Induced Neurodegeneration in Purkinje Cells. *Brain Pathol.* **2011**, *21*, 374–388. [CrossRef]

49. Baltanás, F.C.; Berciano, M.T.; Tapia, O.; Narcis, J.O.; Lafarga, V.; Díaz, D.; Weruaga, E.; Santos, E.; Lafarga, M. Nucleolin reorganization and nucleolar stress in Purkinje cells of mutant PCD mice. *Neurobiol. Dis.* **2019**, *127*, 312–322. [CrossRef] [PubMed]

50. Díaz, D.; Recio, J.S.; Weruaga, E.; Alonso, J.R. Mild cerebellar neurodegeneration of aged heterozygous PCD mice increases cell fusion of Purkinje and bone marrow-derived cells. *Cell Transplant.* **2012**, *21*, 1595–1602. [CrossRef]

51. Li, J.; Snyder, E.Y.; Tang, F.H.F.; Pasqualini, R.; Arap, W.; Sidman, R.L. Nna1 gene deficiency triggers Purkinje neuron death by tubulin hyperglutamylation and ER dysfunction. *JCI Insight* **2020**, *5*, e136078. [CrossRef]

52. Kyuhou, S.I.; Kato, N.; Gemba, H. Emergence of endoplasmic reticulum stress and activated microglia in Purkinje cell degeneration mice. *Neurosci. Lett.* **2006**, *396*, 91–96. [CrossRef] [PubMed]

53. Pollard, A.K.; Craig, E.L.; Chakrabarti, L. Mitochondrial complex 1 activity measured by spectrophotometry is reduced across all brain regions in ageing and more specifically in neurodegeneration. *PLoS ONE* **2016**, *11*, e0157405. [CrossRef] [PubMed]

54. Chow, H.; Herrup, K. Genomic integrity and the ageing brain. *Nat. Rev. Neurosci.* **2015**, *16*, 672–684. [CrossRef]

55. Date, H.; Onodera, O.; Tanaka, H.; Iwabuchi, K.; Uekawa, K.; Igarashi, S.; Koike, R.; Hiroi, T.; Yuasa, T.; Awaya, Y.; et al. Early-onset ataxia with ocular motor apraxia and hypoalbuminemia is caused by mutations in a new HIT superfamily gene. *Nat. Genet.* **2001**, *29*, 184–188. [CrossRef] [PubMed]

56. Enokido, Y.; Tamura, T.; Ito, H.; Arumughan, A.; Komuro, A.; Shiwaku, H.; Sone, M.; Foulle, R.; Sawada, H.; Ishiguro, H.; et al. Mutant huntingtin impairs Ku70-mediated DNA repair. *J. Cell Biol.* **2010**, *189*, 425–443. [CrossRef]
57. Suraweera, A.; Becherel, O.J.; Chen, P.; Rundle, N.; Woods, R.; Nakamura, J.; Gatei, M.; Criscuolo, C.; Filla, A.; Chessa, L.; et al. Senataxin, defective in ataxia oculomotor apraxia type 2, is involved in the defense against oxidative DNA damage. *J. Cell Biol.* **2007**, *177*, 969–979. [CrossRef] [PubMed]
58. Takashima, H.; Boerkoel, C.F.; John, J.; Saifi, G.M.; Salih, M.A.M.; Armstrong, D.; Mao, Y.; Quiocho, F.A.; Roa, B.B.; Nakagawa, M.; et al. Mutation of TDP1, encoding a topoisomerase I-dependent DNA damage repair enzyme, in spinocerebellar ataxia with axonal neuropathy. *Nat. Genet.* **2002**, *32*, 267–272. [CrossRef]
59. Rass, U.; Ahel, I.; West, S.C. Defective DNA Repair and Neurodegenerative Disease. *Cell* **2007**, *130*, 991–1004. [CrossRef]
60. Shackelford, D.A. DNA end joining activity is reduced in Alzheimer's disease. *Neurobiol. Aging* **2006**, *27*, 596–605. [CrossRef] [PubMed]
61. Lafarga, M.; Casafont, I.; Bengoechea, R.; Tapia, O.; Berciano, M.T. Cajal's contribution to the knowledge of the neuronal cell nucleus. *Chromosoma* **2009**, *118*, 437–443. [CrossRef] [PubMed]
62. Rieker, C.; Engblom, D.; Kreiner, G.; Domanskyi, A.; Schober, A.; Stotz, S.; Neumann, M.; Yuan, X.; Grummt, I.; Schütz, G.; et al. Nucleolar disruption in dopaminergic neurons leads to oxidative damage and parkinsonism through repression of mammalian target of rapamycin signaling. *J. Neurosci.* **2011**, *31*, 453–460. [CrossRef] [PubMed]
63. Hetman, M.; Pietrzak, M. Emerging roles of the neuronal nucleolus. *Trends Neurosci.* **2012**, *35*, 305–314. [CrossRef] [PubMed]
64. Parlato, R.; Kreiner, G. Nucleolar activity in neurodegenerative diseases: A missing piece of the puzzle? *J. Mol. Med.* **2013**, *91*, 541–547. [CrossRef]
65. Garcia-Esparcia, P.; Sideris-Lampretsas, G.; Hernandez-Ortega, K.; Grau-Rivera, O.; Sklaviadis, T.; Gelpi, E.; Ferrer, I. Altered mechanisms of protein synthesis in frontal cortex in Alzheimer disease and a mouse model. *Am. J. Neurodegener. Dis.* **2017**, *6*, 15–25.
66. Hernández-Ortega, K.; Garcia-Esparcia, P.; Gil, L.; Lucas, J.J.; Ferrer, I. Altered Machinery of Protein Synthesis in Alzheimer's: From the Nucleolus to the Ribosome. *Brain Pathol.* **2016**, *26*, 593–605. [CrossRef] [PubMed]
67. Tapia, O.; Narcís, J.O.; Riancho, J.; Tarabal, O.; Piedrafita, L.; Calderó, J.; Berciano, M.T.; Lafarga, M. Cellular bases of the RNA metabolism dysfunction in motor neurons of a murine model of spinal muscular atrophy: Role of Cajal bodies and the nucleolus. *Neurobiol. Dis.* **2017**, *108*, 83–99. [CrossRef] [PubMed]
68. Haeusler, A.R.; Donnelly, C.J.; Periz, G.; Simko, E.A.J.; Shaw, P.G.; Kim, M.-S.; Maragakis, N.J.; Troncoso, J.C.; Pandey, A.; Sattler, R.; et al. C9orf72 nucleotide repeat structures initiate molecular cascades of disease. *Nature* **2014**, *507*, 195–200. [CrossRef]
69. Mattson, M.P.; Magnus, T. Ageing and neuronal vulnerability. *Nat. Rev. Neurosci.* **2006**, *7*, 278–294. [CrossRef]
70. Berezniuk, I.; Fricker, L.D. A defect in cytosolic carboxypeptidase 1 (Nna1) causes autophagy in Purkinje cell degeneration mouse brain. *Autophagy* **2010**, *6*, 558–559. [CrossRef]
71. El-Bazzal, L.; Rihan, K.; Bernard-Marissal, N.; Castro, C.; Chouery-Khoury, E.; Desvignes, J.P.; Atkinson, A.; Bertaux, K.; Koussa, S.; Lévy, N.; et al. Loss of Cajal bodies in motor neurons from patients with novel mutations in VRK1. *Hum. Mol. Genet.* **2019**, *28*, 2378–2394. [CrossRef]
72. Pessina, F.; Gioia, U.; Brandi, O.; Farina, S.; Ceccon, M.; Francia, S.; d'Adda di Fagagna, F. DNA Damage Triggers a New Phase in Neurodegeneration. *Trends Genet.* **2020**, 337–354.
73. Zhou, L.; Araki, A.; Nakano, A.; Sezer, C.; Harada, T. Different types of neural cell death in the cerebellum of the ataxia and male sterility (AMS) mutant mouse. *Pathol. Int.* **2006**, *56*, 173–180. [CrossRef] [PubMed]
74. Blosa, M.; Bursch, C.; Weigel, S.; Holzer, M.; Jäger, C.; Janke, C.; Matthews, R.T.; Arendt, T.; Morawski, M. Reorganization of synaptic connections and perineuronal nets in the deep cerebellar nuclei of purkinje cell degeneration mutant mice. *Neural Plast.* **2016**, *2016*, 2828536. [CrossRef] [PubMed]
75. Triarhou, L.C.; Ghetti, B. Stabilisation of neurone number in the inferior olivary complex of aged "Purkinje cell degeneration" mutant mice. *Acta Neuropathol.* **1991**, *81*, 597–602. [CrossRef]
76. Grüsser-Cornehls, U.; Bäurle, J. Mutant mice as a model for cerebellar ataxia. *Prog. Neurobiol.* **2001**, *63*, 489–540. [CrossRef]
77. Ghetti, B.; Perry, K.W.; Fuller, R.W. Serotonin concentration and turnover in cerebelum and other brain regions of pcd mutant mice. *Brain Res.* **1988**, *458*, 367–371. [CrossRef]
78. Triarhou, L.C.; Ghetti, B. Serotonin-immunoreactivity in the cerebellum of two neurological mutant mice and the corresponding wild-type genetic stocks. *J. Chem. Neuroanat.* **1991**, *4*, 421–428. [CrossRef]
79. Le Marec, N.; Hébert, C.; Amdiss, F.; Botez, M.I.; Reader, T.A. Regional distribution of 5-HT transporters in the brain of wild type and "Purkinje cell degeneration" mutant mice: A quantitative autoradiographic study with [3H]citalopram. *J. Chem. Neuroanat.* **1998**, *15*, 155–171. [CrossRef]
80. Felten, D.L.; Felten, S.Y.; Perry, K.W.; Fuller, R.W.; Nurnberger, J.I.; Ghetti, B. Noradrenergic innervation of the cerebellar cortex in normal and in Purkinje cell degeneration mutant mice: Evidence for long term survival following loss of the two major cerebellar cortical neuronal populations. *Neuroscience* **1986**, *18*, 783–793. [CrossRef]
81. Roffler-Tarlov, S.; Landis, S.C.; Zigmond, M.J. Effects of Purkinje cell degeneration on the noradrenergic projection to mouse cerebellar cortex. *Brain Res.* **1984**, *298*, 303–311. [CrossRef]
82. Strazielle, C.; Lalonde, R.; Hébert, C.; Reader, T.A. Regional brain distribution of noradrenaline uptake sites, and of $\alpha1$-, $\alpha2$- and β-adrenergic receptors in PCD mutant mice: A quantitative autoradiographic study. *Neuroscience* **1999**, *94*, 287–304. [CrossRef]

83. Delis, F.; Mitsacos, A.; Giompres, P. Dopamine receptor and transporter levels are altered in the brain of Purkinje Cell Degeneration mutant mice. *Neuroscience* **2004**, *125*, 255–268. [CrossRef]

84. Le Marec, N.; Lalonde, R. Sensorimotor learning and retention during equilibrium tests in Purkinje cell degeneration mutant mice. *Brain Res.* **1997**, *768*, 310–316. [CrossRef]

85. Wu, H.Y.; Rong, Y.; Correia, K.; Min, J.; Morgan, J.I. Comparison of the Enzymatic and Functional Properties of Three Cytosolic Carboxypeptidase Family Members. *J. Biol. Chem.* **2015**, *290*, 1222–1232. [CrossRef] [PubMed]

86. Machado, A.S.; Darmohray, D.M.; Fayad, J.; Marques, H.G.; Carey, M.R. A quantitative framework for whole-body coordination reveals specific deficits in freely walking ataxic mice. *Elife* **2015**, *4*, e07892. [CrossRef]

87. Chen, L.; Bao, S.; Lockard, J.M.; Kim, J.J.; Thompson, R.F. Impaired classical eyeblink conditioning in cerebellar-lesioned and Purkinje cell degeneration (pcd) mutant mice. *J. Neurosci.* **1996**, *16*, 2829–2838. [CrossRef]

88. Brown, K.L.; Agelan, A.; Woodruff-Pak, D.S. Unimpaired trace classical eyeblink conditioning in Purkinje cell degeneration (pcd) mutant mice. *Neurobiol. Learn. Mem.* **2010**, *93*, 303–311. [CrossRef]

89. Goodlett, C.R.; Hamre, K.M.; West, J.R. Dissociation of spatial navigation and visual guidance performance in Purkinje cell degeneration (pcd) mutant mice. *Behav. Brain Res.* **1992**, *47*, 129–141. [CrossRef]

90. Lalonde, R.; Strazielle, C. The effects of cerebellar damage on maze learning in animals. *Cerebellum* **2003**, *2*, 300–309.

91. Tuma, J.; Kolinko, Y.; Vozeh, F.; Oendelint, J. Mutation-related differences in exploratory, spatial, and depressive-like behavior in pcd and Lurcher cerebellar mutant mice. *Front. Behav. Neurosci.* **2015**, *9*, 116. [CrossRef]

92. Bartolomei, J.C.; Greer, C.A. The organization of piriform cortex and the lateral olfactory tract following the loss of mitral cells in PCD mice. *Exp. Neurol.* **1998**, *154*, 537–550. [CrossRef]

93. Greer, C.A. Golgi analyses of dendritic organization among denervated olfactory bulb granule cells. *J. Comp. Neurol.* **1987**, *257*, 442–452. [CrossRef]

94. Recio, J.S.; Weruaga, E.; Gomez, C.; Valero, J.; Briñón, J.G.; Alonso, J.R. Changes in the connections of the main olfactory bulb after mitral cell selective neurodegeneration. *J. Neurosci. Res.* **2007**, *85*, 2407–2421. [CrossRef]

95. Baker, H.; Greer, C.A. Region-specific consequences of PCD gene expression in the olfactory system. *J. Comp. Neurol.* **1990**, *293*, 125–133. [CrossRef] [PubMed]

96. Gómez, C.; Curto, G.G.; Baltanás, F.C.; Valero, J.; O'Shea, E.; Colado, M.I.; Díaz, D.; Weruaga, E.; Alonso, J.R. Changes in the serotonergic system and in brain-derived neurotrophic factor distribution in the main olfactory bulb of pcd mice before and after mitral cell loss. *Neuroscience* **2012**, *201*, 20–33. [CrossRef] [PubMed]

97. Airado, C.; Gómez, C.; Recio, J.S.; Baltanás, F.C.; Weruaga, E.; Alonso, J.R. Zincergic innervation from the anterior olfactory nucleus to the olfactory bulb displays plastic responses after mitral cell loss. *J. Chem. Neuroanat.* **2008**, *36*, 197–208. [CrossRef] [PubMed]

98. Valero, J.; Weruaga, E.; Murias, A.R.; Recio, J.S.; Curto, G.G.; Gómez, C.; Alonso, J.R. Changes in cell migration and survival in the olfactory bulb of the pcd/pcd mouse. *Dev. Neurobiol.* **2007**, *67*, 839–859. [CrossRef]

99. Chu, M.W.; Li, W.L.; Komiyama, T. Balancing the Robustness and Efficiency of Odor Representations during Learning. *Neuron* **2016**, *92*, 174–186. [CrossRef]

100. Díaz, D.; Lepousez, G.; Gheusi, G.; Alonso, J.R.; Lledo, P.M.; Weruaga, E. Bone marrow cell transplantation restores olfaction in the degenerated olfactory bulb. *J. Neurosci.* **2012**, *32*, 9053–9058. [CrossRef]

101. Díaz, D.; Gómez, C.; Muñoz-Castañeda, R.; Baltanás, F.; Alonso, J.R.; Weruaga, E. The olfactory system as a puzzle: Playing with its pieces. *Anat. Rec.* **2013**, *296*, 1383–1400. [CrossRef]

102. O'Gorman, S. Degeneration of thalamic neurons in "Purkinje cell degeneration" mutant mice. II. Cytology of neuron loss. *J. Comp. Neurol.* **1985**, *234*, 298–316. [CrossRef] [PubMed]

103. Kyuhou, S.-i.; Gemba, H. Fast cortical oscillation after thalamic degeneration: Pivotal role of NMDA receptor. *Biochem. Biophys. Res. Commun.* **2007**, *356*, 187–192. [CrossRef] [PubMed]

104. Chakrabarti, L.; Neal, J.T.; Miles, M.; Martinez, R.A.; Smith, A.C.; Sopher, B.L.; La Spada, A.R. The Purkinje cell degeneration 5J mutation is a single amino acid insertion that destabilizes Nna1 protein. *Mamm. Genome* **2006**, *17*, 103–110. [CrossRef] [PubMed]

105. LaVail, M.M.; Blanks, J.C.; Mullen, R.J. Retinal degeneration in the pcd cerebellar mutant mouse. I. Light microscopic and autoradiographio analysis. *J. Comp. Neurol.* **1982**, *212*, 217–230. [CrossRef] [PubMed]

106. Grau, M.B.; Masson, C.; Gadadhar, S.; Rocha, C.; Tort, O.; Sousa, P.M.; Vacher, S.; Bieche, I.; Janke, C. Alterations in the balance of tubulin glycylation and glutamylation in photoreceptors leads to retinal degeneration. *J. Cell Sci.* **2017**, *130*, 938–949. [CrossRef]

107. Okubo, A.; Sameshima, M.; Unoki, K.; Uehara, F. The ultrastructural study of ribosomes in photoreceptor inner segments of the pcd cerebellar mutant mouse. *Jpn. J. Ophthalmol.* **1995**, *39*, 152–161.

108. LaVail, M.M.; Gorrin, G.M.; Yasumura, D.; Matthes, M.T. Increased susceptibility to constant light in nr and pcd mice with inherited retinal degenerations. *Investig. Ophthalmol. Vis. Sci.* **1999**, *40*, 1020–1024.

109. Gardette, R.; Alvarado-Mallart, R.M.; Crepel, F.; Sotelo, C. Electrophysiological demonstration of a synaptic integration of transplanted purkinje cells into the cerebellum of the adult purkinje cell degeneration mutant mouse. *Neuroscience* **1988**, *24*, 777–789. [CrossRef]

110. Sotelo, C.; Alvarado-Mallart, R.M. Embryonic and adult neurons interact to allow Purkinje cell replacement in mutant cerebellum. *Nature* **1987**, *327*, 421–423. [CrossRef]

111. Sotelo, C.; Alvarado-Mallart, R.M. Reconstruction of the defective cerebellar circuitry in adult purkinje cell degeneration mutant mice by Purkinje cell replacement through transplantation of solid embryonic implants. *Neuroscience* **1987**, *20*, 1–22. [CrossRef]
112. Sotelo, C.; Alvarado-Mallart, R.M.; Gardette, R.; Crepel, F. Fate of grafted embryonic purkinje cells in the cerebellum of the adult "purkinje cell degeneration" mutant mouse. I. Development of reciprocal graft-host interactions. *J. Comp. Neurol.* **1990**, *295*, 165–187. [CrossRef] [PubMed]
113. Gardette, R.; Crepel, F.; Alvarado-Mallart, R.M.; Sotelo, C. Fate of grafted embryonic purkinje cells in the cerebellum of the adult "purkinje cell degeneration" mutant mouse. II. Development of synaptic responses: An in vitro study. *J. Comp. Neurol.* **1990**, *295*, 188–196. [CrossRef] [PubMed]
114. Triarhou, L.C.; Low, W.C.; Ghetti, B. Serotonin fiber innervation of cerebellar cell suspensions intraparenchymally grafted to the cerebellum of pcd mutant mice. *Neurochem. Res.* **1992**, *17*, 475–482. [CrossRef]
115. Sotelo, C.; Alvarado-Mallart, R.M. Growth and differentiation of cerebellar suspensions transplanted into the adult cerebellum of mice with heredodegenerative ataxia. *Proc. Natl. Acad. Sci. USA* **1986**, *83*, 1135–1139. [CrossRef]
116. Chang, A.C.; Ghetti, B. Embryonic cerebellar graft development during acute phase of gliosis in the cerebellum of pcd mutant mice. *Chin. J. Physiol.* **1993**, *36*, 141–149.
117. Triarhou, L.C.; Zhang, W.; Lee, W.H. Graft-induced restoration of function in hereditary cerebellar ataxia. *Neuroreport* **1995**, *6*, 1827–1932. [CrossRef]
118. Triarhou, L.C.; Zhang, W.; Lee, W.H. Amelioration of the behavioral phenotype in genetically ataxic mice through bilateral intracerebellar grafting of fetal Purkinje cells. *Cell Transplant.* **1996**, *5*, 269–277. [CrossRef]
119. Zhang, W.; Lee, W.H.; Triarhou, L.C. Grafted cerebellar cells in a mouse model of hereditary ataxia express IGF-I system genes and partially restore behavioral function. *Nat. Med.* **1996**, *2*, 65–71. [CrossRef] [PubMed]
120. Recio, J.S.; Álvarez-Dolado, M.; Díaz, D.; Baltanás, F.C.; Piquer-Gil, M.; Alonso, J.R.; Weruaga, E. Bone marrow contributes simultaneously to different neural types in the central nervous system through different mechanisms of plasticity. *Cell Transplant.* **2011**, *20*, 1179–1192. [CrossRef] [PubMed]
121. Díaz, D.; del Pilar, C.; Carretero, J.; Alonso, J.R.; Weruaga, E. Daily bone marrow cell transplantations for the management of fast neurodegenerative processes. *J. Tissue Eng. Regen. Med.* **2019**, *13*, 1702–1711. [CrossRef] [PubMed]
122. Cendelin, J. Transplantation and Stem Cell Therapy for Cerebellar Degenerations. *Cerebellum* **2016**, *15*, 48–50. [CrossRef] [PubMed]
123. Torres-Aleman, I.; Barrios, V.; Lledo, A.; Berciano, J. The insulin-like growth factor I system in cerebellar degeneration. *Ann. Neurol.* **1996**, *39*, 335–342. [CrossRef] [PubMed]
124. Carrascosa, C.; Torres-Aleman, I.; Lopez-Lopez, C.; Carro, E.; Espejo, L.; Torrado, S.; Torrado, J.J. Microspheres containing insulin-like growth factor I for treatment of chronic neurodegeneration. *Biomaterials* **2004**, *25*, 707–714. [CrossRef]
125. Sanz-Gallego, I.; Rodriguez-de-Rivera, F.J.; Pulido, I.; Torres-Aleman, I.; Arpa, J. IGF-1 in autosomal dominant cerebellar ataxia—Open-label trial. *Cerebellum Ataxias* **2014**, *1*, 13. [CrossRef] [PubMed]
126. Pérez-Martín, E.; Muñoz-Castañeda, R.; Moutin, M.J.; Ávila-Zarza, C.A.; Muñoz-Castañeda, J.M.; Del Pilar, C.; Alonso, J.R.; Andrieux, A.; Díaz, D.; Weruaga, E. Oleoylethanolamide Delays the Dysfunction and Death of Purkinje Cells and Ameliorates Behavioral Defects in a Mouse Model of Cerebellar Neurodegeneration. *Neurotherapeutics* **2021**. [CrossRef]
127. Henry, R.J.; Ritzel, R.M.; Barrett, J.P.; Doran, S.J.; Jiao, Y.; Leach, J.B.; Szeto, G.L.; Wu, J.; Stoica, B.A.; Faden, A.I.; et al. Microglial depletion with CSF1R inhibitor during chronic phase of experimental traumatic brain injury reduces neurodegeneration and neurological deficits. *J. Neurosci.* **2020**, *40*, 2960–2974. [CrossRef]

Review

Alzheimer's Disease: A Molecular View of β-Amyloid Induced Morbific Events

Rajmohamed Mohamed Asik [1,2,3], Natarajan Suganthy [4], Mohamed Asik Aarifa [3], Arvind Kumar [5], Krisztián Szigeti [6,7], Domokos Mathe [6,7,8], Balázs Gulyás [1,2,9], Govindaraju Archunan [3,10,*] and Parasuraman Padmanabhan [1,2,*]

1   Lee Kong Chian School of Medicine, Nanyang Technological University, Singapore 636921, Singapore; n1906466a@e.ntu.edu.sg (R.M.A.); balazs.gulyas@ntu.edu.sg (B.G.)
2   Cognitive Neuroimaging Centre, 59 Nanyang Drive, Nanyang Technological University, Singapore 636921, Singapore
3   Department of Animal Science, Bharathidasan University, Tiruchirappalli 620024, Tamil Nadu, India; aarifam91@gmail.com
4   Department of Nanoscience and Technology, Alagappa University, Karaikudi 630003, Tamil Nadu, India; suganthy.n@gmail.com
5   Centre for Cellular and Molecular Biology, Hyderabad 500007, Telangana, India; akumar@ccmb.res.in
6   Department of Biophysics and Radiation Biology, Semmelweis University, 1094 Budapest, Hungary; krisztian.szigeti@gmail.com (K.S.); domokos.mathe@hcemm.eu (D.M.)
7   CROmed Translational Research Centers, 1094 Budapest, Hungary
8   In Vivo Imaging Advanced Core Facility, Hungarian Center of Excellence for Molecular Medicine (HCEMM), 1094 Budapest, Hungary
9   Department of Clinical Neuroscience, Karolinska Institute, 17176 Stockholm, Sweden
10 Marudupandiyar College, Thanjavur 613403, Tamil Nadu, India
*   Correspondence: garchu56@yahoo.co.in (G.A.); ppadmanabhan@ntu.edu.sg (P.P.)

Citation: Mohamed Asik, R.; Suganthy, N.; Aarifa, M.A.; Kumar, A.; Szigeti, K.; Mathe, D.; Gulyás, B.; Archunan, G.; Padmanabhan, P. Alzheimer's Disease: A Molecular View of β-Amyloid Induced Morbific Events. *Biomedicines* **2021**, *9*, 1126. https://doi.org/10.3390/biomedicines9091126

Academic Editors: Kumar Vaibhav, Meenakshi Ahluwalia and Pankaj Gaur

Received: 29 June 2021
Accepted: 27 August 2021
Published: 1 September 2021

Publisher's Note: MDPI stays neutral with regard to jurisdictional claims in published maps and institutional affiliations.

Copyright: © 2021 by the authors. Licensee MDPI, Basel, Switzerland. This article is an open access article distributed under the terms and conditions of the Creative Commons Attribution (CC BY) license (https://creativecommons.org/licenses/by/4.0/).

Abstract: Amyloid-β (Aβ) is a dynamic peptide of Alzheimer's disease (AD) which accelerates the disease progression. At the cell membrane and cell compartments, the amyloid precursor protein (APP) undergoes amyloidogenic cleavage by β- and γ-secretases and engenders the Aβ. In addition, externally produced Aβ gets inside the cells by receptors mediated internalization. An elevated amount of Aβ yields spontaneous aggregation which causes organelles impairment. Aβ stimulates the hyperphosphorylation of tau protein via acceleration by several kinases. Aβ travels to the mitochondria and interacts with its functional complexes, which impairs the mitochondrial function leading to the activation of apoptotic signaling cascade. Aβ disrupts the Ca^{2+} and protein homeostasis of the endoplasmic reticulum (ER) and Golgi complex (GC) that promotes the organelle stress and inhibits its stress recovery machinery such as unfolded protein response (UPR) and ER-associated degradation (ERAD). At lysosome, Aβ precedes autophagy dysfunction upon interacting with autophagy molecules. Interestingly, Aβ act as a transcription regulator as well as inhibits telomerase activity. Both Aβ and p-tau interaction with neuronal and glial receptors elevate the inflammatory molecules and persuade inflammation. Here, we have expounded the Aβ mediated events in the cells and its cosmopolitan role on neurodegeneration, and the current clinical status of anti-amyloid therapy.

Keywords: amyloid beta; Alzheimer's disease; inflammation; gene regulation; organelle dysfunction

1. Introduction

AD is a chronic debilitating neurological illness constituting 80% of dementia, primarily affecting the aging population above 65 years old. This devastating disorder is the one of the sixth leading causative of fatality worldwide, which has turned into a scourge of 21st century creating huge socioeconomic havoc globally and providing a burden to the caretakers [1]. AD is characterized by gradual neuronal degeneration that affects the cognitive function with severe memory impairment, lack of thinking, behavioral and social skills worsening activities of daily life. Epidemiological survey reveals that the global prevalence of AD increased in aging population where the incidence rate is higher in developed countries while in developing countries it is less than 1%. The World Health Organization (WHO) reported that globally around 47.5 million individuals are affected by AD as of 2018 with 7.7 million newcases every year, which is estimated to increase to 75.6 million by 2030 and 135.5 million by 2050, with expected cost of care at $1 trillion per annum [2,3]. Progressive changes in the brain of AD from asymptomatic to symptomatic transformation such as memory and behavioral disruptions are termed as AD continuum. AD is categorized as mild, moderate, and severe depending on the degree of symptoms that affects the day today activities [4–6]. In general, AD is classified into early onset AD (EOAD) and late onset AD (LOAD) based on the age prevalence. EOAD constitutes 5% of the total AD cases and is prevalent in people below the age of 65 years. EOAD is also termed as familial type AD (FAD) caused due to autosomal dominant mutation of genes coding for the APP, Presenilin 1 and 2 (PS1 and 2) located in Chromosomes 21, 14 and 1, respectively. People with Down's syndrome (21st trisomy) have a higher risk for EOAD [7,8]. LOAD also termed as sporadic AD which frequently occurs in the elderly population above the age of 65 years, accounting 95% of total AD cases. Major genetic risk factor for LOAD is gene encoding for Apolipoprotein E (APOE) involved in cholesterol metabolism (Chromosome 19), which exists in three forms of APOE2, APOE3, APOE4. Among the alleles ApoE4 exhibited threefold increased risk for AD development [9]. In addition, other factors such as aging, diet, lifestyle, environmental factors, and chronic metabolic disorders such as type II diabetes mellitus, hypertension and vascular disorders intensify the pathogenesis of LOAD [10]. Pathological trademarks of AD involve classical positive lesion comprising of amyloid plaques composed of Aβ peptides in the synaptic terminal of brain parenchyma and in the cerebral blood vessels leading to congophilic angiopathy/cerebral amyloid angiopathy (CAA); NFTs composed of paired helical filaments with hyperphosphorylated tau in the axonal region, neuropil threads, and dystrophic neurites accompanied by microgliosis and astrogliosis [11]. In addition, pathological features such as neuronal loss and synaptic dysfunction representing the core negative features of AD were also observed along with plaques in the cortical mantle and tangles in limbic and association cortices. Neuropathology of AD is associated with neuronal loss and atrophy in the temporofrontal cortex inducing further deposition of amyloid plaques and tangled bundles of fibers leading to enhanced migration of monocytes and macrophages in cerebral cortex provoking neuroinflammation [12,13]. Aβ peptide is the key factor in AD pathogenesis and fibrogenesis of Aβ peptide, triggers a cascade of events such as hyperphosphorylation of tau and NFT formation, ER stress, disruption in Ca^{2+} homeostasis, mitochondrial dysfunction, microgliosis and astrogliosis inducing neuroinflammation eventually leading to neuronal death [14]. Therapeutic intervention approved by FDA for AD includes cholinesterase inhibitors (Donepezil, Galantamine, Rivastigmine and Tacrine) and N-methyl-D-aspartate receptor (NMDAR) agonist (memantine) which are effective only for mild to moderate dementia and antipsychotic drugs for treatment of behavioral disturbances [15,16]. These therapeutic interventions possess severe side effects and have limited therapeutic efficacy on cognitive function, as these drugs only relieve the symptoms with no effect on progression of disease. Currently researchers are focusing on development of disease modifying drugs, which can slow or reverse the progression of disease [17]. However, most of these drug molecules failed in the clinical trials due to the mystery in understanding AD pathogenesis. Hence, the present review focuses on unravelling the molecular mechanism and biochemical pathways leading to pathogenesis of AD, which might help researchers in reassessing AD pathogenesis in different perception, thereby providing novel ideas for identification of therapeutic strategies to combat AD.

2. APP Processing

2.1. Post-Translational Modification of APP Alters Aβ Production

APP, a type I integral membrane protein is found in both mammalian and non-mammalian cells. Three members of APP are present in mammals which are APP, the APP-like protein-1, and 2 (APLP1 and APLP2) [18]. In humans, APP is encoded in the chromosome arm 21 (21q21.3) with approximately 240 kb size which has 18 exons [19]. The promotor sequence of APP does not have TATA or CAAT boxes but has a sus sequence where the transcription factors such as SP-1, AP-1, and AP-4 binds to the promotor site and commences the gene expression [20]. SP1 is a zinc finger protein that binds to the GC rich region of APP promoter, which facilitates the binding of RNA polymerase II. Recent reports reveal that binding of lead (Pb^{2+}) on the Zn^{2+} site of zinc finger 3 of SP1 (SP1-f3) protein promotes the APP overexpression and it could cause a high chance of amyloidogenesis [21]. The alternative splicing of APP mRNA engenders several isoforms, which vary from 365 to 770 amino acid residues. The Aβ1-42 is embedded in the proteins such as APP695, APP751, and APP770. The APP mRNA is localized in multiple tissues of the body. However, the APP695 is predominantly present in neuronal cells and the other two isoforms are expressed in other tissues [22,23]. Pre-mature APP protein undergoes several post-translational modifications (PTM) including N- and O- glycosylation, sumoylation, phosphorylation, ubiquitination, sulfation, and palmitoylation (Figure 1) [24–26]. N- and O- glycosylation occurs inside the ER, the oligosaccharyltransferases catalyses the N- glycosylation in Asn467 and Asn496 [27]. O- glycosylation is identified in several sites of APP770 including Thr291, Thr292, Thr576, and Thr353 [28], and Ser597, Ser606, Ser611, Thr616, Thr634, Thr635, Ser662 and Ser680 [24]. Furthermore, the single β-N-acetylglucosamine (GlcNAc) residue is added in serine or threonine residue that leads to the formation of O-GalNAcylation, which plays an important role in non-amyloidogenic processing of APP [29]. In GC, APP undergoes phosphorylation in 10 residues of both ecto- and cytoplasmic domains, which are Ser198, Ser206, Tyr653, Tyr682, Tyr687, Ser655, Ser675, Thr654, Thr668 and Thr686 [25,30]. The cytoplasmic Ser655 and Thr654 are phosphorylated by Ca^{2+}/calmodulin-dependent protein kinase II (CaMKII) [31]. A decreased level of Ser655 phosphorylation and enhanced level of Thr668 phosphorylation stimulates Aβ generation [25,32]. Tyr682 and Tyr687 phosphorylation have been found in the AD brain but not in a healthy brain, as well as in APP overexpressed cells [33]. About 10% APP undergoes palmitoylation process in the ER by the DHHC-7 and DHHC-21 (palmitoyl acyltransferases) at the site of Cys186 and Cys187. Palmitoylated APP enriched in the lipid rafts where the Beta-Site APP Cleaving Enzyme 1 (BACE-1) level is also higher facilitating the Aβ production. Thus, suppressing the APP palmitoylation has therapeutic competence against Aβ through targeting both α- and β-secretase cleavage [26]. The Lys649, Lys 650, Lys651 and Lys688 are the ubiquitination sites of APP where the attachment of ubiquitin associates in the protein degradation, interaction, and the trafficking process. In addition, the sumoylation of APP at Lys587 and Lys595 is catalysed by the enzyme small ubiquitin-like modifier 1 and 2 (SUMO-1 and SUMO-2). Both the ubiquitination and sumoylation process decreases the Aβ level for example high-level of APP sumoylation reduce the Aβ production [34–36]. The sulfation occurs at Tyr217 and Tyr262 residues of APP, however the exact sulfation sites and functions are not completely discovered [37]. On the other hand, the PTMs analysis of AD molecules revealed citrullination (Arg301), phosphorylation (Ser366, Ser441), and methylation (Lys624, Lys699) of APP protein [38]. In addition, citrullination (Arg5→Cit) and deamidation (Asn27→Asp) of Aβ fragments affects the fibrillation rate of Aβ [39]. The physiological function of APP is still unclear however the overexpression of wild-type APP promotes the cell proliferation, neurotoxic and neurotrophic protective effects [40]. Overall, the PTM of APP venues, its bi-directional therapeutics opportunity via enhancing ubiquitination, O-GalNAcylation, simulation and diminishing phosphorylation and palmitoylation.

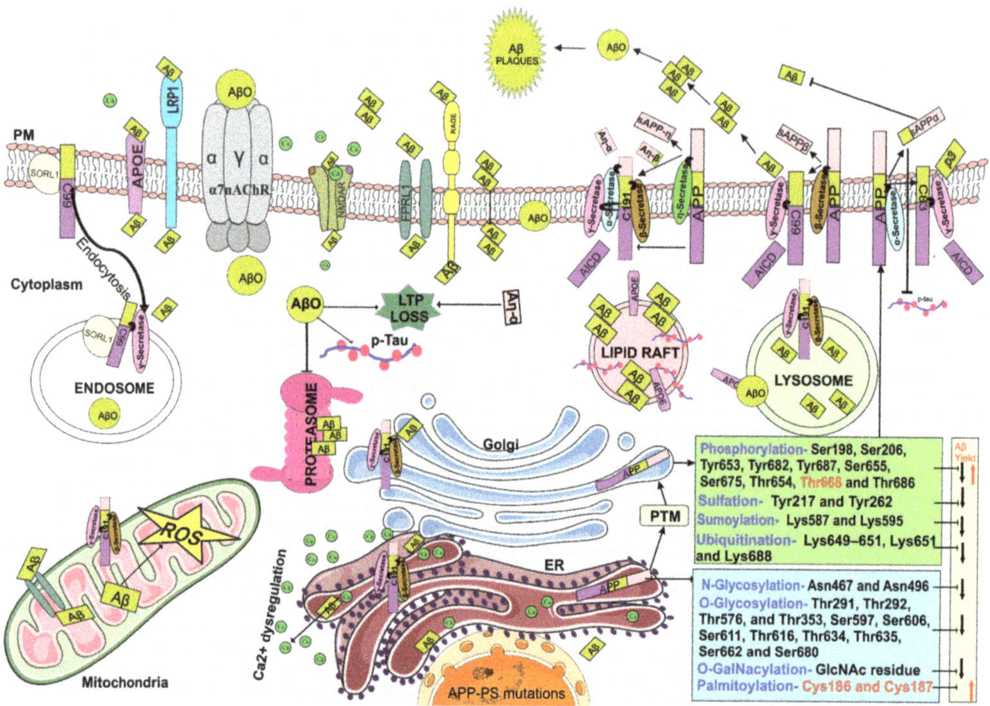

Figure 1. APP processing and Aβ generation: The PTM of APP alters Aβ production. The APP processing by the secretases (α, γ, β and η) at plasma membrane (PM), mitochondria, ER, lysosome, GC and lipid rafts generate its metabolites notably Aβ. Aβ undergoes oligomerization and plague formation at extra cellular matrix (ECM) and the ECM Aβ enters the cytoplasm through direct PM passing and receptor mediated internalization. Intra cellular Aβ accumulation in the cell organelle impairs its physiological functions.

2.2. APP-Secretases Processing and Aβ Generation

APP is processed by membrane proteases such as α-, β-, γ- and η-secretases. Cleavage of α- or β-secretases, followed by γ- secretase leads to generation of non-pathological (by α-secretase) or pathological (by β-secretase) fragments (Figure 1) [41]. α-secretase is a type of metalloprotease and disintegrin (ADAM) family member. Further, ADAM 9, 10 and 17 has α-secretase activity. ADAM 10 is the primary secretase that cleaves the APP in neurons. α-secretase resides within the Aβ domain of APP which cleaves and produces the extracellular soluble APPα (sAPPα) and C-terminal fragment (CTF)-α-83 or C83 [42]. BACE-1 is a type-I transmembrane aspartyl protease which cleaves APP at Asp1 or Glu11 sites generating the soluble APPβ (sAPPβ), the CTFβ-99 (C99) or CTFβ-89 (C89) [43]. A673V mutation in APP exhibits shifts BACE-1 cleavage from Glu11 to Asp1 site, increasing the level of C99 and C99/C89 ratio [44]. θ-secretase or BACE-2 is homologous to the BACE-1, which cleaves APP at Phe19 of the Aβ domain and produces CTF-80. However, BACE-2 transgenic mice did not show Aβ overproduction and cognitive deficits [45,46]. γ- secretase is a complex protein comprising of PS1 and PS2, nicastrin, anterior pharynx-defective 1 (APH-1) and presenilin enhancer (PEN-2) subunits. PS1 and PS2 are the catalytic unit and nicastrin, APH-1 and PEN-2 are the regulatory units that mainly functions in substrate recognition [47,48]. Several reports reveal that the γ- secretase activating protein (GSAP) regulates the γ- secretase specificity, which induce conformational change of PS1 and substrate recognition. GSAP significantly and selectively elevates the Aβ production without altering γ- secretase normal functions such as notch cleavage [49–51]. γ- secretase cleaves the C83 and C99 resulting in the production of fragments such as P3, varied length of Aβ(38-49), and APP intracellular domain (AICD) [42]. C89 processing by γ- secretase reveals generation of truncated Aβ11-40/42 (Aβ') peptides [52]. In addition, the Aβ from the AD brain exhibits prominent modifications in N- terminal site of Aβ including truncation (Aβ'), glutamate conversion to pyroglutamate and isomerization of L-Asp to D-Asp, which results in the loss of N-terminal charge that turns the peptide more hydrophobic [53]. γ- secretase assay with recombinant APP-CTFs in human induced pluripotent stem cell (iPSC) or cell-free system showed elevated level of 42:40 ratio higher in Aβ' > Aβ > P3 [54]. η-secretase is a membrane-bound matrix-metalloproteinase such as MT5-MMP, which cleaves APP at 504/505 residue and generates truncated sAPP-η and CTF-η (C191). ADAM-10, BACE-1 and γ- secretase involved in CTF-η cleavage produces Aη-α and Aη-β peptides of ECM, ACID into the cytoplasm. Hippocampal long-term potentiation (LTP) is lowered upon the treatment of synthetic Aη-α in mice hippocampal slices [55]. About 440 mutations in APP, PS1, and PS2 were identified to be linked to FAD (https://www.alzforum.org/mutations searched on 28 August 2021), and several pathogenic research models were developed thereafter. PS mutations decreased γ-secretase sensitivity, which increases the number of cuts in a single substrate. γ- secretase cleavage in multiple sites of C99 (γ, ς and ε-sites) generates intermediate products 43, 45, 46, 48, 49 and 51 amino acids, which were further cleaved producing final product Aβ40/Aβ42. FAD APP mutations become partially resistant to the γ- secretase cleavage. The N-terminal APP mutations show subtle effect on γ- secretase cleavage efficiency and Aβ40/42 specificity. However, the C-terminal APP mutations show skewed cleavage and aggregation prone Aβ42 specificity [56]. Several Aβ degrading enzymes (ADEs) were reported which catalysts the proteolytic degradation [57]. Currently, researchers are focusing on the disease-modifying anti-amyloid therapy targets to alter Aβ production/aggregation which includes ADEs, BACE-1, PS1, and GSAP (detailed [49,57]).

2.3. Intra-Cellular Aβ

Aβ is observed as an extracellular product of APP, but later the incidences of intracellular Aβ production were reported. APP is found in several compartments such as ER, trans-Golgi network (TGN), mitochondrial membranes, endosomes, and lysosomes (Figure 1) [58–60]. In particular, the exosomes play a significant role in transporting the APP and APP-CTF where the Aβ release could occur wherever the β- and γ- secretases are co-localized within the exosomes. AD patient's extracellular vehicles (EVs) or exosomes has increased levels of cytotoxic Aβ and prion protein (PrP) which get transferred to neighbor neurons that could serve as a diagnostic and in therapeutic applications [61–63]. The intracellular Aβ liberation is found in cells with APP$_{Swe}$ but not in the cells with wild type APP [64]. Sortilin-related receptor 1 (SORL1) plays a critical role in late-onset AD, where it recovers the uncut APP from the PM through internalization into endosomes [65]. Endosomes are acidic in nature where the BACE-1 actively interacts with APP and generates Aβ42 [66]. The decreasing intracellular Aβ lead to extracellular plaque formation, and this is evident in Down syndrome patients [67]. In addition, the reuptake of the extracellular Aβ is found in the cells through the Aβ interaction with several biomolecules including lipids, proteins, and proteoglycans. Aβ can instantly aggregate to form lower molecular weight dimers to fibrils that lead to amyloid plaques in AD brain. In Aβ aggregation, the intermolecular hydrogen bonds of β-strand of Aβ peptides form the cross-β structural pattern, and the hydrogen-bonded, parallel β-sheeted Aβ peptides which induce fibril formation in the presence of tissue transglutaminase (tTg). These Aβ aggregates exhibit toxic effects mostly from dimers itself, which interrupts learning and memory and LTP [68,69]. It was anticipated that the elevated amount of Aβ oligomers (AβOs) may be non-specifically conducted through the direct interaction with negatively charged phospholipid bilayers [70]. Receptor-mediated Aβ internalization was found in α7 nicotinic acetylcholine receptor (α7nAChR) [71], APOE [72], formyl peptide receptor-like 1 (FPRL1) [73], NMDA [74] and receptor for advanced glycation end products (RAGE) [75] receptors. Aβ oligomerization is found intracellularly in multiple sites of the cell such as plasma, endosomal and lysosomal membranes, and lipid rafts [76,77]. The Tg2576 AD model exhibit Aβ dimers which predominantly accumulate in the lipid rafts along with APOE and p-tau indicating the fact that lipid rafts are the important interaction site for the above proteins [78]. Recent reports reveal that AβO directly inhibits the ubiquitin-proteasome system, which is shown in both animals and cell lines. The proteasome inhibition upon Aβ interaction leads to accumulation of tau protein. On the other hand, Aβ internalization is also detected in the mitochondria, which interferes in its function by diminishing the ETC III and IV and reduced oxygen consumption [79]. Christian and his colleagues reported that Aβ42 treatment in mouse neuronal primary cells and hippocampal slices induced tau phosphorylation (p-tau), while sAPPα treatment decreased the levels of Aβ42 and p-tau proteins [80]. In addition, the BACE-1 and GSK3β activities were also observed to be dropped in cell culture and APP-PS animal models, associated with decline in tau hyperphosphorylation upon treatment with sAPPα [81]. sAPPα activates mitogen-activated protein kinase (MAPK)/extracellular signal-regulated kinase (ERK) which results in neurotrophic and neuroprotective activity [82]. The LTP of transgenic mice is recovered by sAPPα through microglial invasion on the sites of Aβ deposits, which upregulates insulin-degrading enzymes (IDE) leading to Aβ clearance and restoration of spatial learning and synaptic plasticity [83].

3. Tau Pathology

Major neuropathological hall mark of AD in addition to senile plaque is the presence of intra neuronal neurofibrillary tangles composed of hyperphosphorylated tau protein. Tau protein is a microtubule-associated phosphoprotein (MAP) abundant in axons of the neurons engaged in boosting the tubulin assembly into microtubules and its stabilization in the brain [84]. Microtubule associated protein tau gene is encoded by Chromosome 17. Tau exists in six isoforms (352–445 amino acids) formed by alternate splicing of exon 2, 3 and 10, which vary by the presence or absence of 29 or 58 amino acids insert in the N-terminal part and by three or four microtubule binding repeats (3R or 4R) in the C-terminal end. Exon 10 on alternative splicing forms two isoforms, four repeats isoform tau (4R-tau) or three repeats isoform tau (3R-tau), which play a vital role in microtubule binding [85]. In addition, examination of healthy adult brain regions shows N-terminal fragments of tau ranging from 40 kDa and 45 kDa and C-terminal fragments of tau ranging from 17 kDa to 25 kDa in the age group between 18 and 108 years where the truncations of tau at the Asp421 and Glu391 residues are believed to be an aggregation promoting factor of AD [86]. Tau protein consists of hydrophilic acidic residue rich N-terminal domain, flanked by basic proline rich domain, microtubule binding domain (MBD) and downstream is the C-terminal domain. In the adult brain, the tau protein consists of two to three moles of phosphate per mole of the protein, with a rich source of proline-glycine motifs facilitating the folding of tau protein [87]. Post translational modification of tau primarily serine/threonine directed phosphorylation and glycosylation modulates the affinity of tau for microtubules. Phosphorylation of tau, which starts during brain embryonic development of the fetal brain is comparatively higher when compared to adult brain [88]. In addition, deamidation (Asn279→Asp279) of 4R-tau involving degeneration of several brain regions acts as a biomarker for AD where the deamidated sequence specific antibodies helps to detect the marker protein [89]. A major role of the tau protein is to facilitate assembly of the microtubules and maintain its structure for normal axoplasmic flow. Assembly of microtubules is dependent on the extent of phosphorylation, as hyperphosphorylation affects the assembly and stability of microtubules. In addition to its interaction with tubulin, tau also binds with the SH3 domain of Src family tyrosine kinases revealing its role in cell signalling [90]. Tau protein plays a putative role in maintaining the stability of chromosomes [91].

3.1. Aβ Peptide Promotes Hyperphosphorylation of Tau Protein

Tau pathology in AD is mainly contributed by the oxidative stress and inflammatory response induced by Aβ toxicity. Several kinases are engaged in the phosphorylation of tau protein (p-tau) which includes proline dependent kinases such as glycogen synthase kinase-

3 (GSK3), dual specificity tyrosine-phosphorylation-regulated kinase 1A/B (Dyrk1A/B), cyclin-dependent protein kinase-5 (CDK5), and mitogen activated protein kinases (MAPK) (p38, Erk1/2, and JNK1/2/3); proline independent kinase such as tau-tubulin kinase 1/2 (casein kinase 1α/1δ/1ε/2), microtubule affinity regulating kinases, phosphorylase kinase, cAMP-dependent protein kinase A, C and N, CaMKII and tyrosine protein kinases such as Src family kinase (SFK) members (Src, Lck, Syk, and Fyn) and Abelson family kinase members, ABL1 and ABL2 (ARG) which were considered as target sites for AD therapeutics (Figure 2) [92]. Among the tau specific kinases, GSK3β is the key enzyme in hyperphosphorylation of tau protein, which phosphorylates at least 15 residues of Tau protein. In AD, soluble and fibrillar Aβ peptides tend to be more toxic rather than amyloid plaques. Several pieces of evidence revealed the inter-relationship between Aβ toxicity and tau pathology, which remains still unclear. One hypothesis proposed that soluble Aβ binding to α7nAChR irreversibly activates tau specific kinases promoting persistent p-tau protein at three proline directed sites that enhances tau protein dislocation from axon to dendrites and synaptic junction, disrupting axon/dendritic transport followed by neurofibrillary lesion, dendritic breakdown leading to neurofibrillary tangles formation [93]. Evidence also revealed that enhanced intracellular Ca^{2+} due to Aβ induced mitochondrial dysfunction and ER stress activates caplapin dependent cleavage of P35 and P39 generating P25 and P29 fragments, which in turn activates Cdk5 followed by activation of GSK3β leading to hyperphosphorylation of tau protein forming NFT in AD brain as observed in transgenic animals [94,95]. The enhanced tau phosphorylation is associated with difficulty in spatial learning in transgenic mice expressing increased GSK3β [96]. King et al., stated that oxidative stress induces nitration of tyrosine residue in tau protein and disulphide bridge formation between tau protein facilitates abnormal folding and detachment of tau from microtubules thereby, promoting aggregation [97]. Transient activation of PKA in AD induces continuous hyperphosphorylation of tau in both PKA and non-PKA sites making tau protein more susceptible to successive phosphorylation by GSK3β, revealing the fact that stimulation of PKA plays a crucial part in the commencement of the AD [98]. Another enzyme, MAP affinity regulating kinase (MARK) also phosphorylates the KxGs motif present in MBD of tau protein. Although upstream regulation of MARK is not clear, a recent study revealed that GSK3β activates MARK2 inducing phosphorylation of Ser-262 of tau [99]. In addition, ERK2 activation in neurons also promotes p-tau reducing its ability to stabilize microtubules [100]. The rate of p-tau depends on the initial phosphorylation site Ser 396 and Ser 235, following the phosphorylation of predecessor Ser 400 [101], and prime phosphorylation of Thr 231 [102], which has significant role in modulating the function of tau protein, i.e., stabilizing the microtubules. In addition to tau phosphorylation, these kinases are also involved in APP processing also, e.g., GSK3, especially GSK3α, promotes Aβ formation from APP offering new approach, that inhibition of GSK3α might attenuate amyloid plaques and NFT formation [103]. CDK5-p25 regulates APP processing leading to Aβ production by phosphorylating Thr 668 residue of APP. Other factors such as stress in the ER and life stress also influence p-tau [104]. A recent study reported that increased expression of P75 neurotrophin receptor (p75NTR) the pan-receptor for Aβ peptide promoted Aβ neurotoxicity by inducing the production of Aβ via endocytosis of APP and BACE-1, phosphorylation of tau protein via calpain/CDK5 and AKT/GSK3β pathways. Calpain promotes truncation and activation of GSK3β. The CDK5 activator protein p25 preferentially binds with and activates GSK3β [105]. In vitro studies revealed a cross talk between the calpain/CDK5 and AKT/GSK3β pathways downstream of Aβ/p75NTR signalling in the regulation of p-tau levels in AD (Figure 2) [106].

Three major phosphatases PP1, PP2A, PP2B and PP2C play a major role in dephosphorylation of tau protein among, which PP2A has predominant role whose activity is reduced by 30% in the AD brain [107]. As an inhibitor of PP2A, Aβ deposition and estrogen deficiency causes phosphorylation of Y307 subunit of PP2A inactivating the protein ability to dephosphorylate hyperphosphorylated tau protein, which in turn leads to the NFT formation [108]. Another report revealed that abnormal increase in mitochondrial

ROS level in Aβ inhibits PP2A and PP5 activating JNK and Erk1/2 pathways leading to apoptosis of neuronal cells [109]. Bolmont et al. reported that when tau mutant transgenic mice were intracerebroventricularly administrated with APP transgenic mice brain extract NFT formation was observed, depicting the fact that Aβ acts as upstream factor of tau pathology [110]. Another report revealed that Aβ induces upregulation of gene coding for tyrosine kinase 1A (DYRK1A) promoting hyper-phosphorylation of tau protein, causing microtubule disassembly. Similarly, the cytotoxicity of Aβ oligomers is promoted by the tau protein, revealing the fact that Aβ triggers tau pathology, while tau protein intercede toxicity of the Aβ protein and both acts synergistically, enhancing AD pathology [111,112]. Despite these studies, the connection between the GSK3β and PP2A with Aβ induced oxidative stress is still elusive. The p-tau and aggregation are also influenced by PTM of tau including glycosylation, nitration, truncation, acetylation, sumoylation, ubiquitination, and polyamination.

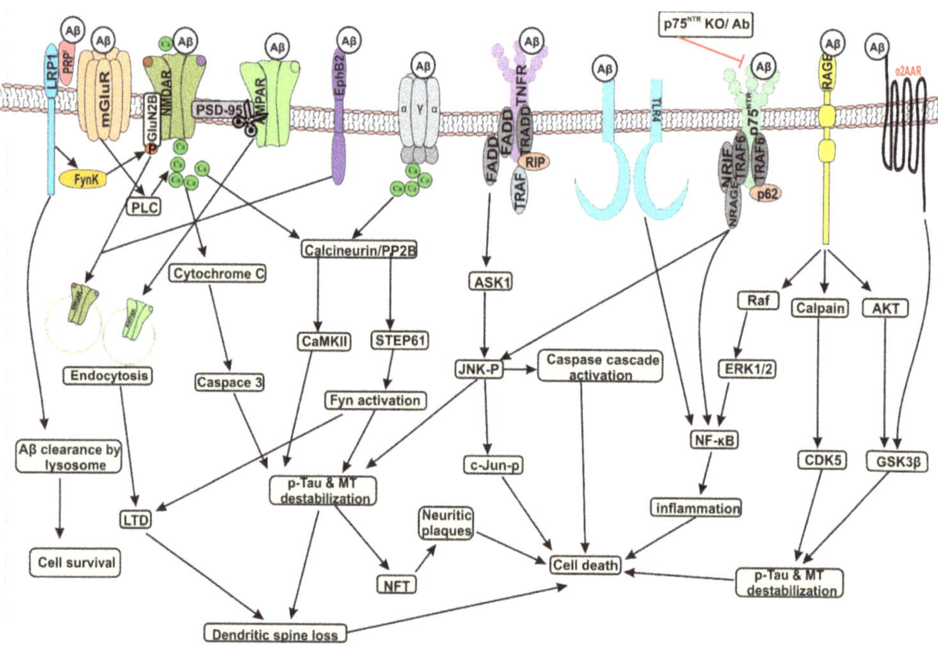

Figure 2. Signaling of Aβ in memory impairment, tau hyperphosphorylation and cell death in AD: LRP1 receptor facilitates the Aβ clearances as well activates the FynK. Aβ binding with receptors such as mGluR, NMDAR, AMPAR, EphB2 triggers the LTD via NMDAR and AMPAR surface removal. In other hand, prolonged activation of NMDAR elevates the intracellular Ca^{2+} level which increases caspase 3. The inflammation is induced by the activation of NF-κB upon Aβ mediated activation of TLR-4, p75NTR and RAGE receptors. The tau hyperphosphorylation is induced by several kinases including GSK3β, CDK5, JNK, FynK upon Aβ binding with most of the receptors which destabilizes microtubules, forms the NFT and neurotic plaques. The LTD, NFT, inflammation, increased caspase cascade leads to dendritic spine loss and cell death.

3.2. Molecular Mechanism of p-Tau Mediated Neurodegeneration

In AD brain cytosol, 40% of abnormally hyperphosphorylated tau exists in oligomeric and nonfilamentous form. Hyperphosphorylated tau forms of PHF in the soma and thread-like lesion termed as neurophil neuritis found in the grey or white matter and senile plaque are associated with dystrophic neuritis. The concentration of total tau and p-tau in CSF determines the stage of AD [113]. Hyperphosphorylated tau not only bind with normal tau protein promoting self-assembly into PHF, but it also binds with microtubule associated proteins (MAP1 and MAP2) and actin disrupting the self-assembled microtubules leading to zeiosis of cell membrane ultimately causing neurofibrillary degeneration (Figure 2) [114]. Hyperphosphorylated tau is released within the vesicles formed by pinching of destabilized membrane, which are taken up by the neighboring cells by a process called endocytosis, thereby sequestering the healthy tau protein to hyperphosphorylated form via its prion like nature spreading from one neuronal cell to other disrupting its cytoskeletal structure, organelle destabilization, interrupting protein synthesis ultimately leading to induction of zeiosis [115]. Extracellular soluble tau is taken up by the healthy neurons through muscarinic receptor, which play significant role in the neuronal signal transduction [116]. Based on these facts, Morozova et al., reported that uptake of hyperphosphorylated tau via muscarinic receptor, in addition to endocytosis, also translocates into the nucleus, alters the protein expression, move to synapse, and impairs the mitochondrial function [117]. As the level of phosphorylated tau increases in neurons with disease progression, it disrupts the microtubule stability and damages the cytoskeletal components triggering neurodegeneration [118]. Pathways involved in removal of damaged, misfolded, and aggregated protein are the ubiquitin-proteasome and autophagy lysosomal pathways. The brains of AD patients were characterized by accumulation of polyubiquitylated tau proteins with reduced proteasome activity. The proteasome activity reflects the amount of PHF in AD brain and its activity is inhibited by the hyperphosphorylation of tau protein illustrating the fact that hyperphosphorylation and proteasome inhibition are inter-related [119].

4. Aβ in Mitochondria Dysfunction

The central nervous system has high metabolic demand which is met by the abundant mitochondria present in the neurons. Mitochondria regulate the life and death cycle through various cellular regulatory process such as ATP production, maintaining intracellular Ca^{2+} homeostasis, reactive oxygen species production, detoxification and apoptosis [120]. Mitochondria generates ROS such as superoxide ($O_2{}^-$), hydroxyl radical (OH), and hydrogen peroxide (H_2O_2) which are counterbalanced by the antioxidants such as super oxide dismutase, catalase, glutathione peroxidase (GPX), glutathione reductase (GR) and sirtuins (mammalian class III histone deacetylases) [121,122]. Imbalance in the proxidant and antioxidant level leads to protein oxidation, lipid peroxidation and DNA damage altering the mitochondrial membrane potential, disturbing Ca^{2+} homeostasis, enhancing cytochrome C ultimately leading to apoptosis. Sirtuins (Sirt) in the mitochondria of neurons regulates the transcription and antioxidant enzymes (SOD2 and catalases) activities. Sirt1 regulates the peroxisome proliferator-activated receptor-gamma coactivator1 alpha (PGC-1α), APP metabolism and Aβ level [123]. Sirt3 play an important role in 8-oxoguanidine DNA glycosylase 1 (OGG1) stabilization and deacetylation, promoting mtDNA repair via nuclear enzyme. Aβ-induced reduction in Sirt1 and Sirt3, upregulates the tau level and acetylation [124,125]. The dynamin-related protein 1 (Drp1), mitofusins1 and 2 (Mfn1 and Mfn2) and optical atrophy 1 (Opa1) are the key fission/fusion proteins that regulates the mitochondrial dynamics via maintaining the assembly and stability of ECT super-complex structure, altering mitochondrial structure and distribution of mitochondria throughout the neurons [126].

Mitochondrial Dysfunction by Aβ

Mitochondrial dysfunction was found in the post-mortem brain samples of AD patients and their platelets, AD transgenic mice and in-vitro over expressed mutant APP or external Aβ treatment [127–130]. Post-mortem brain samples of the AD patients exhibit an elevated level of oxidative damage, increased mtDNA and cytochrome oxidase in neuronal cytoplasm, which indicates accumulation of mitochondrial products upon degradation [128]. The glucose metabolism of the AD patient's brain is down-regulated, and the late-onset AD brain showed reduced mitochondrial membrane potential through metabolic shift from mitochondrial oxidative system to glycolysis [129,130]. EOAD patient's mRNA investigation revealed down-regulation of complex I and upregulation of complexes III and IV of ETC when compared to normal subjects indicating the highest demand in energy production [131]. AD patient platelets isolated and fused with human neuroblastoma (SH-SY5Y) and teratocarcinoma (NT2) cells showed reduced endogenous mtDNA. These cells showed elevated Aβ production along with mitochondrial dysfunction with lower cytochrome oxidase activity, higher free radical generation, reduced mitochondrial membrane potential and altered calcium homeostasis [127]. The transgenic mice models such as APP$_{swe/Lon}$ and double transgenic APP/PS1 showed reduction in the mitochondrial chaperon hsp70, reduced glucose metabolism, impaired Cu/Zn SOD activity, decreased mitochondrial membrane potential and ATP level, increased mitochondrial permeability transition, a decline in respiratory function, and increased mitochondrial oxidative stress [132–134]. Overexpression of APP$_{Swe}$ on PC12 cells depicts increased level of oxidative stress and mitochondrial dysfunction mediated by various caspases and the stress-activated protein kinase pathway, elevated level of ROS, reduced ATP generation [135,136]. In addition, APP$_{Swe}$ MC17 cells exhibited elevated level of Aβ production, which causes the imbalance in the mitochondrial fission/fusion protein where the dynamin-like protein 1 (DLP1) and OPA1 levels are decreased and the Fis1 level is enhanced. These changes results in mitochondrial fragmentation which promotes mitochondrial and neuronal dysfunction [137]. Expression of transcription genes (anti-oxidative and mitochondria-related proteins) in AβO treated BV2 (glial) and SH-SY5Y (neuronal) cells, transgenic mice (Tg-AD), and human AD brain revealed enhanced expression of Sod2, Dnm1l, Bcl2 genes and reduction of Gpx4, Sirt1, Sirt3, mt-Nd1, Sdha and Mfn2 genes. In addition, reduction in cell viability was observed with enhanced ROS production and impaired MMP Tg-AD mice showed significant down-regulation of Sirt1, Mfn1 and mt-Nd1 and upregulation of Dnm1l. In addition, the human AD brain showed alteration in microRNA pattern, which is responsible for the reduced Sirt1 expression [138]. SIRT1 activates the transcription of PGC-1α, which improves the antioxidant capacity inducing the expression of SOD and GPX in cells. Reduction in PGC-1α causes impairment in mitochondrial biogenesis, the one-month aged AD mice showed no significant AβO deposition with elevated PGC-1α, whereas the six months old AD mice exhibited the high level of AβO deposition with significant reduction in PGC-1α [139].

APP has a binding motif for TOM40 that impairs the routine function of mitochondria, the APP binding with TOM40 impedes the COX (IV and Vb) transportation, which results in diminished COX activity and increased ROS production [140]. Mitochondria-associated ER membrane is a communication point between the ER and mitochondria.

Mitochondria-associated membrane (MAM) play a crucial role in calcium transport, phospholipids synthesis and mitochondrial fission–fusion dynamics. Remarkably, MAM has enhanced the level of C99 and γ- secretase where the Aβ is generated via amyloidogenic pathway [58,141]. In addition, the C99 accumulation at MAM activates sphingomyelinase that generates ceramides, which leads to inhibition of mitochondrial respiration and apoptosis [142,143]. Further, the Aβ co-localization is found with the complex II of the mitochondrial ETC, which indicates that the Aβ pass the outer and inner membranes of the mitochondria [144]. Hansson Petersen et al. studied the import mechanism of Aβ into the mitochondria using immunohistochemistry, immunoblotting, immunoelectron microscopy and flow cytometry techniques [145]. Decline in Aβ transportation was observed in mitochondria exposed to antibodies towards mitochondrial receptors Tom20 or Tom70, or the general mitochondrial import pore of the outer membrane Tom40, indicating the fact that Aβ gets into mitochondria via the TOM40 complex, which translocate Aβ into the inner membrane via TIM22 complex. Aβ interacts with different mitochondrial proteins upon entry into the matrix including amyloid binding alcohol dehydrogenase (ABAD), Complex V and cyclophilin D. Complex V of ETC produces ATP, when Aβ interacts with Complex V α-subunit disturbing its energy production (Figure 3). Alcohol dehydrogenase catalyses the reduction of nicotinamide adenine dinucleotide (NAD) to NADH. ABAD has a direct link with the Aβ to mitochondrial toxicity, the crystal structure of Aβ bound ABAD shows the deformation of active site that affects the NAD binding. ABAD over expressed mice in Aβ enriched environment reveal elevated oxidative stress and impaired memory [146]. Complex V is regulated by the addition of O-linked N-acetylglucosamine (O-GlcNAcylation) which is catalysed by O-GlcNAc transferase, Aβ interrupt the binding between ATP5A and O-GlcNAc transferase [147]. The mitochondrial permeability transition pore (mPTP) regulates the apoptotic pathway via Ca^{2+} and apoptotic signalling molecules from the matrix [148]. The Ca^{2+} enters the mitochondria through either mitochondrial membrane Ca^{2+} uniporter (MCU) or voltage-dependent anion channel (VDAC) which results in unlocking the mPTP and generates the ROS [149]. The mPTP is regulated by the Cyclophilin D, the interaction of Aβ with Cyclophilin D inhibits Cyclophilin D leaving the mPTP pore open [150]. Further, the specific loss of oligomycin sensitivity conferring protein (OSCP), a subunit of Complex V upon interaction with Aβ results in mPTP activation, decreased ATP production, and increased oxidative stress (Figure 3) [151]. The oligomeric Aβ treatment showed an entry of extracellular calcium into the mitochondria, which causes the mitochondrial mediated apoptosis via opening the mPTP channels and releasing the cytochrome C [152]. Cym1/PreP proteasome degrades the pre-sequence of the protein and Aβ peptides. In oxidizing condition, the Cys527 and Cys90 can form a disulphide bridge that results in PreP inactivation elevating the amount of Aβ in mitochondria. Aβ inhibits the degradation of pre-sequence peptides, which results in the dysfunction of preprotein processing leading to imbalanced organellar proteome and multiple mitochondrial defects including reduced membrane potential, oxygen conception and increased ROS [153,154]. Aβ exposed brain vascular endothelial cells show increased level of inter mitochondrial calcium (Ca^{2+}) and reduced cytosolic Ca^{2+} resulting in enhanced amount of oxygen conception, higher ATP production and increased ROS generation, which in turn contribute to cerebrovascular dysfunction [155].

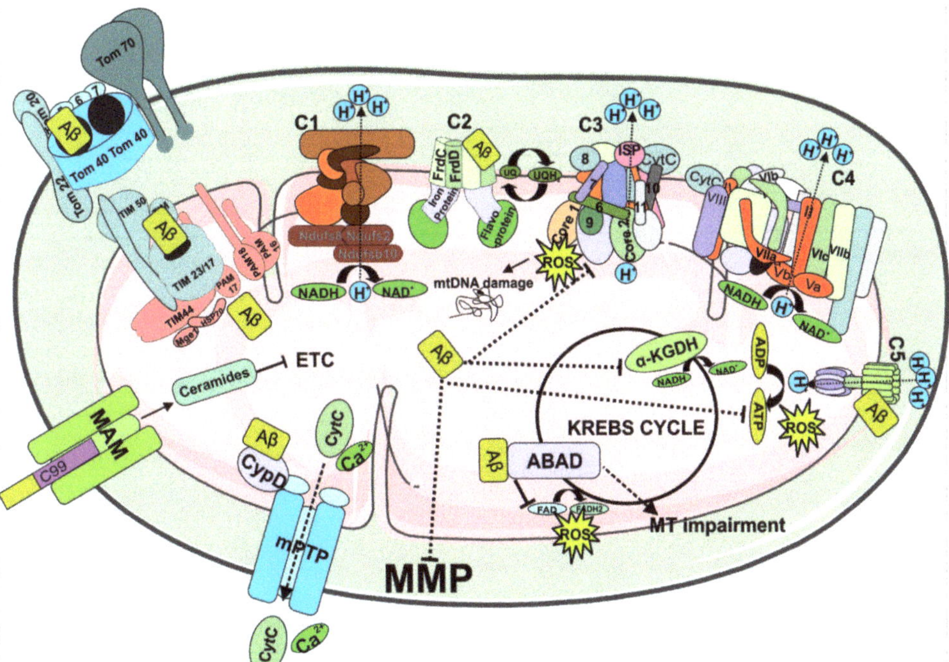

Figure 3. Mechanism of Aβ induced mitochondrial dysfunction: Aβ enters the mitochondria (MT) through the TOM40 and TIM 23 complex. Aβ interacts with the Aβ-binding alcohol dehydrogenase (ABAD), Cyclophilin D and electron transport complexes (ETC) (C2, C3 and C5) which enhance the ROS production, Cytochrome C liberation, ROS mediated MT-DNA damage, MT impairment and mPTP opening where CytC exported to the cytoplasm. In addition, MAM bound APP fragment (C99) activates ceramides that inhibits the ETC.

5. Aβ in ER Stress

The ER is a functional organelle that coordinates the proteostasis of the eukaryotic cells including protein biosynthesis, folding, assembly, trafficking, and ruining. Accurate protein folding according to intracellular or extracellular signal is crucial for normal cell survival and physiological functions [156]. The ER maintains the Ca^{2+} homeostasis, where the lumen of the ER has the utmost level of Ca^{2+} ions within the cell via the active transportation of Ca^{2+} through Ca^{2+} ATPase channel [157]. Ca^{2+} plays a vital in the processing and folding of new-born proteins, since these mechanisms are strictly Ca^{2+} dependent so high Ca^{2+} level is required for the proper functioning [158]. Any disruption in ER function such as dysregulation of Ca^{2+} homeostasis, inhibition of PTM and hypoxia heaps the accumulation of unfolded or misfolded proteins, which leads to the ER stress due to the huge unwanted protein load and the long-term accumulation of these proteins induce the cell damage [159]. To decrease the unfolded protein level, the cells activate several cellular systems including UPR and ERAD, which protects the cells against the toxic proteins that augments the ER capacity and quality control [160].

ER Dysfunction by Aβ

Several lines of evidence including cultured cells, animals and human brain slices reveal the relationship between the ER stress and Aβ. The most widely proposed connection is Ca^{2+} where the Aβ induced Ca^{2+} dysregulation triggers the ER stress mediated cell death. Analysis of AD brain revealed proteostasis dysfunction, Ca^{2+} dysregulation, and elevated level of molecular chaperones such as heat shock protein-27 (HSP27) and 78-kDa glucose-regulated protein or binding immunoglobulin protein (GRP78/BiP), the characteristic features of ER stress [161,162]. Interestingly, the FAD mutations in PS1 and PS2, and Aβ triggers inositol 1,4,5-trisphosphate receptor (IP3R) and ryanodine receptor (RyR) promoting release of Ca^{2+} ions from the ER, the early pathogenic sign in AD [163,164]. The detrimental role of Aβ is broadly studied in neuronal cells where the AβO increases the Ca^{2+} influx into the cells via stimulation of NMDA receptor (Figure 4) [165]. However, in cortical neurons, Aβ treatment prompted Ca^{2+} discharge from ER into the cytoplasm via IP3R and RyR, provoking ROS production, which in turn disrupts MMP enhancing the release of cytochrome C eventually leading to caspase-mediated apoptosis [166]. The crosstalk between the ER and mitochondria via MAMs play a significant role in the induction of apoptosis. The MAM, Mfn2 dimers restrains the tight junction between both the organelles. Sigma-1 receptor (Sig-1R) recognizes the ER-IP3R released Ca^{2+} concentrations facilitating its diffusion into the mitochondria [167]. During ER stress, the adaptive response from ER stimulates the increased mitochondrial metabolism and energy production. However, the maladaptive response from the ER impairs the mitochondria progression and triggers the cell death signalling which is directly proportional to the level of Ca^{2+} exchange between the two organelles [168,169]. The excessive ER Ca^{2+} discharge promotes the rapid accumulation of the toxic proteins including unfolded and misfolded proteins that leads to activation of UPR signalling. PERK activation and its signalling are engaged in the cognitive impairment of AD mice, where the increased p-eIF2α reduces the global protein synthesis inducing synaptic dysfunction and neurodegeneration. However, several reports suggest that Aβ treatment promotes ER stress. The APP/PS1 transgenic mice and Aβ treated SH-SY5Y cells showed enhanced level of GRP78, p-eIF2α, p-PERK, CHOP, and ATF-6 [170,171]. The E693Δ mutation in APP expresses high level of AβO, not fibrillation provoking ER stress and TGC dysfunction in cultured cells [172]. In the drosophila AD model, expression of XBP1 decreases the Aβ neurotoxicity where it inhibits the Aβ mediated overloading of the Ca^{2+} in the cytoplasm [173]. Multiple pieces of evidence reveal IRE1 mediated neurodegeneration in AD. IRE1 interaction with PS1 activates the proapoptotic pathway via JNK. A high level of JNK3 and phosphorylated JNK is observed in the post-mortem brain samples and cerebrospinal fluid (CSF) of AD patients, which is correlated with the Aβ42 levels [174]. A mechanism postulated that the IRE1 activates proapoptotic signaling via forming a complex between tumor necrosis factor receptor-associated factor 2 (TRAF2) and apoptosis signal-regulating kinase 1 (ASK1) that activates the various proteins and signalling pathways including NF-κB, JNK, caspase-12 and p38MAPK mediated CHOP [175–177]. Analysis of AD mice brain (Tg2576) samples reveal a substantial increase in the level XBP1 mRNA splicing in comparison with the age matched controls. The elevated level of XBP1 induces activation of CHOP, caspase cascade including caspase-3, 4 and 12 which pave the way to cell death [178]. The pathogenic role of IRE1-XBP1 is validated in various samples including the AD human tissue, IRE1 knockout (IRE1cKO) 5xFAD mice and XBP1 silenced Neuro2A cells. The human AD brains samples showed enhanced levels of p-IRE1 and spiced XBP1, whereas the healthy control brain samples reveal lower or undetectable levels of p-IRE1 and spiced XBP1. The IRE1cKO-5xFAD mice exhibits the characteristic of at least 50% reductions in the levels of the APP, Aβ monomers and Aβ plaques with improved learning and memory, and restoration of LTP in comparison with 5xFAD control. The level of APP is dramatically declined in the IRE1cKO-5xFAD mice, but not APP mRNA, which suggests that UPR controls the APP at post-translational level. Similarly, the XBP1 over expressed cells show increased APP expression, which is attenuated in the XBP1 silenced cells. Overall, XBP1 further elevates the expression of APP and production Aβ which facilitates the neurodegeneration by the maladaptive UPR (Figure 4) [179].

Bundles of literature evidence reveal relationship between the Aβ and ubiquitin-proteasome system (UPS). The E3 enzyme performs an important role in the ubiquitination process, which connects the Ub substrate protein and proteasome. The enzymes Parkin, HRD1 and UCHL-1 regulates the ERAD mechanisms including ubiquitination, ER membrane translocation, and proteasomal degradation [180]. The hippocampal and cortex regions of the AD patient showed co-residents of Aβ and parkin, which suggest that the parkin ubiquitinates the Aβ. However, the parkin over expressed AD mice showed Aβ-parkin ubiquitination, but the parkins expression was downregulated, and increased extra cellular plaque formation was observed, which indicates that the Ub-Aβ suppresses the parkins activity [181]. In the AD patient brain, the mRNA and protein expression studies revealed that the level of HRD1 is drastically decreased in comparison with healthy brain. Moreover, the biochemical analysis of Aβ is significantly increased in AD brain, which correlates with the HRD1 expression. On the other hand, the HRD1 over expressed SHSY5Y cells exhibits decreased expression of APP and Aβ production inside the ER. The immunocytochemistry analysis expresses that colocalization of HRD1 with APP in proline-rich sites promotes ubiquitination and degradation. The mutant and wild type HRD1 expressed HEK293 cells alters the APP ubiquitination, the wildtype HRD1 promotes the APP degradation whereas the mutant HRD1 elevates the APP expression and Aβ generation [35]. The UCHL-1 is a deubiquitinating enzyme that maintains the cellular ubiquitin process. In AD transgenic mice, downregulation of UCHL-1 was observed which, impaired the expression of BDNF/TrkB. Similarly, the Aβ treated cortical or hippocampal neurons showed downregulation of BDNF trafficking and signaling (including ERK5 activation and CREB-dependent gene regulation) due to the decreased level of UCHL-1, which was reversed by overexpression of UCHL-1 [182]. A report reveals that Aβ diffuses into the cytoplasm from the lumen of the ER where the Aβ is ruined by the proteasome and IDE [183]. In contrast, the proteasome inhibition by lactacystin decreases the Aβ degradation in the primary cortical and astrocyte cells. In addition, alteration in the ubiquitin-proteasome leads to the unusual accumulation of Aβ (Figure 4) [184]. The AD brain shows increased accumulation of Ub-protein complex, which results in proteasomal dysfunction induced by Aβ. It is hypothesised that the Aβ get into the active site of 20S proteasome subunits and inhibits its chymotrypsin-like activity. Further, Aβ aggregates acts as a competitive substrate of the chymotrypsin-like activity of 20S proteasome where the proteasomal functions get impaired elevating the level of AβO in the AD patients (Figure 4) [185]. On the other hand, the membralin is an ER protein with predicted transmembrane loops and does not have any domains. A recent genome wide association approach revealed that a 500bp single nucleotide polymorphism is linked with LOAD. Interactome network analysis discovers that the membralin is an ERAD component that retains homeostasis by degradation of membrane and pathological substrates including nicastrin [186]. However, the mRNA and protein expression of membralin is reduced in AD brain, where significant enhancement of γ-secretase activity was observed. The membralin shRNA treated mice brain and N2a cells showed downregulation of membralin in dentate gyrus region and elevated level of CHOP and XBP1. In addition, a significant rise in the amount of nicastrin and Aβ plaque load was noted [187].

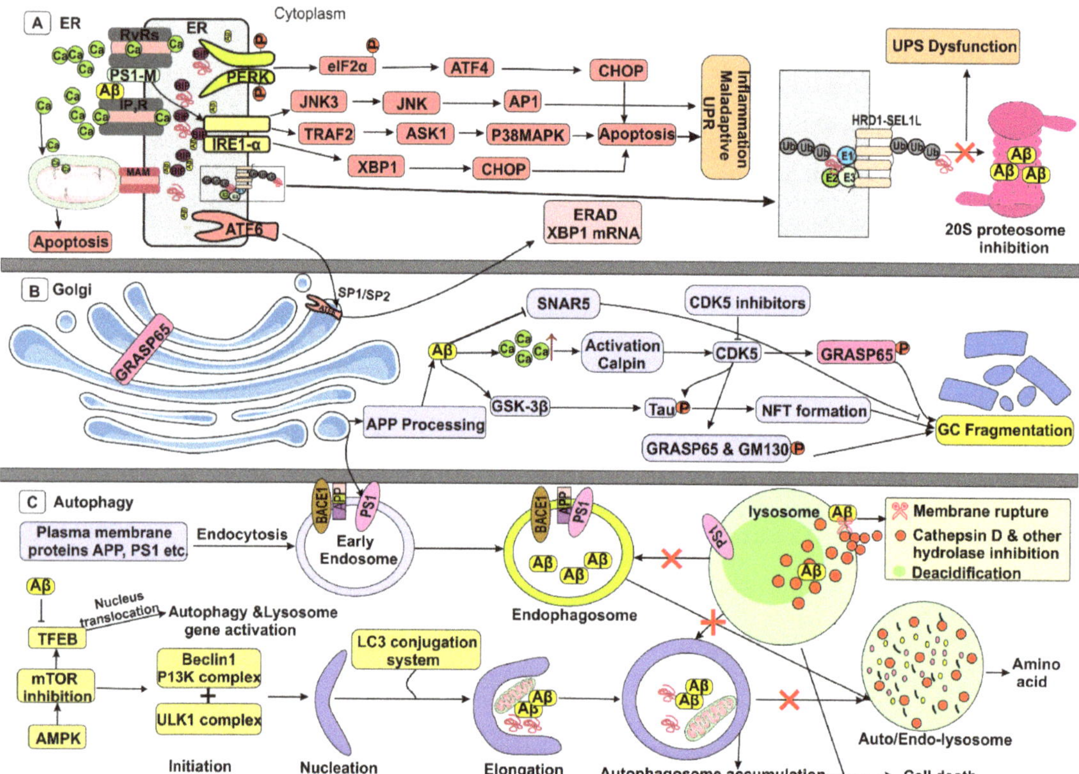

Figure 4. The molecular mechanisms of Aβ prompted ER, GC and Lysosome dysfunction: Aβ induces the ER stress via increasing Ca^{2+} outflux from ER through IP_3R and RyR which accelerates the apoptosis via mitochondrial Cyt C release. The maladaptive UPR signalling in ER is facilitated via PERR, IRE1 receptors where the cells undergo apoptosis and inflammation. Aβ interacts with 20S proteosome and inhibits its functions such as degradation of the unfolded proteins. ATF6 is transported and processed in GC which acts as a transcription factor, transcribes the genes related to ERAD and XBP1. GC fragmentation is induced by the activation of several kinases which phosphorylates the GRASP65 and GM130. Autophagy and lysosome gene transcription is impeded via TFEB inhibition. On the other hand, the autophagy elongation is perturbed, as well as inhibition of autophagosome/endophagosome fusion with lysosome enhances its accumulation which drives cell death. In addition, Aβ ruptures the lysosome membrane, deacidify the niche and inhibits its hydrolases. All together the stress recovery mechanism is completely affected by the Aβ.

6. Autophagy/Lysosomal Dysfunction

6.1. Autophagy Physiology

The cells undergo clearance of damaged organelles and unwanted protein aggregates via lysosome mediated degradation termed as autophagy. Autophagy supports the organelles to regulate its homeostasis and maintains the cellular nutrient level. Based on consignment transport the autophagy is broadly divided into three distinct groups such as micro-autophagy, chaperone-mediated autophagy (CMA), and macro-autophagy. Micro-autophagy involves direct invagination of the cytoplasmic portion by the lysosomes for degradation. CMA deals with the specific KFERQ sequence of the cytoplasmic proteins targeted by the Hsc 70 and its co-chaperones where this complex is transported to the lumen of the lysosome via membrane receptor lysosome associated membrane protein type 2A (LAMP-2A) for degradation. The p38-MAPK regulates the CMA process via phosphorylation and activation of LAMP-2A where inhibition of MAPK specifically inhibits the CMA [188,189]. Macro-autophagy (here referred as autophagy) is regulated in three steps, nucleation, elongation, and lysosomal degradation. In brief, autophagy nucleation is required for the Atg1/Unc-51-like kinase (ULK) complex to recruit the crucial proteins for autophagosome formation. Autophagy initiation is controlled by phosphorylation of ULK1 by the mammalian target of rapamycin complex 1 (mTORC1). Under nutrients rich condition the mTORC1 is recruited to lysosome membrane by Rag complex where the mTORC is activated by GTP-bound Rheb. The activated mTORC1 phosphorylates the ULK1 at Ser757, which suppresses the autophagy initiation as well as the transcription factor EB (TFEB). Under stress, mTORC1 becomes inactivated that results in activation of TFEB and AMPK mediated by galectins. The galectins interaction with the Rag complex inactivates the mTORC1 upon dissociation where the activated TFEB translocate into nucleus and activates the autophagy and lysosomal genes. In addition, AMPK activation phosphorylates the ULK1, which further phosphorylates the proteins FAK family kinase interacting protein of 200 kDa (FIP200) and the autophagy-related proteins ATG13 and ATG101. Further, ULK1 activates beclin1-vacuolar protein sorting 34 (VSP34) complex, which works as a Class III phosphatidylinositol 3-kinase (PI3KCIII) to generate phosphatidylinositol 3-phosphate (PI3P) that hires its binding proteins for the phagophore nucleation. The phagophore elongation depends on either ATG8 (LC3 and GABARAP) or ATG12 (ATG12 and ATG5) ubiquitin-like conjugation systems. The E1 (ATG7) and E2 (ATG10) ubiquitin ligases conjugates the ATG12 and ATG5, further ATG12-ATG5 binds to the ATG16L1 which primes the recruitment of microtubule associated proteins LC3. The ATG5-ATG12- ATG16L1 complex produces LC3-II that helps the extension and closure of the phagophore to become mature autophagosome. At last, the lysosome fusion with the autophagosome takes place where the lysosome hydrolases digest the contents of the autophagosome [190,191].

6.2. Autophagy Impairment by Aβ

Several lines of evidence prove that autophagy impairment plays a vital role in the neurodegeneration. In the AD brain, the abundant accumulation of autophagic vacuoles (AVs) in dystrophic dendrites illustrates impaired AVs maturation (Figure 4). Subsequently, the AVs act as a reservoir for Aβ, the purified AVs are enriched with β-CTF (C99) together with the components of γ- secretase. Further, the genesis of AD specific EVs containing MHC class-type markers where the disruption of autophagic proteins such as GABARAP and LAMP1 are the symptomatic appearance of AD patients which aids the AD clinical diagnostics and treatment [192]. In addition, the rapamycin-induced mTOR inhibition activates the autophagy where the γ- secretase complex translocates predominantly from the endosomes to AV [193,194]. Fedeli et al. uncovered that PS2 mutations impair autophagy through clogging the autophagosome-lysosome fusion process, which due to the lessened recruitment of GTPase RAB7 to autophagosome and altered Ca^{2+} homeostasis [195]. Beclin-1- an autophagy regulating protein is diminished in AD patients. Baclin-1 deleted mice exhibited reduction in autophagy and impaired lysosome, thereby accelerating intra-neuronal Aβ accumulation [196]. Beclin-1 stimulates degradation of the PM-APP and its metabolites via endosomes and endo-lysosome which is negatively regulated by AKT [197]. Autophagic proteins such as ATG5, ATG 12, and LC3 were discovered in association with the Aβ plaque in AD cells [198]. In the drosophila model, the ATG1, ATG8a, and ATG18 were downregulated depending on the age, subsequently the autophagy induction is decreased with increase in Aβ production [199]. Accretion of mutant APP and Aβ in the hippocampus cells of APP mice explicates the decreased level of autophagy protein (ATG5, LC3BL1, and LC3BII) [200]. The LC3-II is ambiguously increased along with accumulated Aβ, which indicates the induction of autophagy, whereas LC3-II lysosomal degradation is hindered. A prenylated protein Rab7 is vital for autophagy progression, which colocalizes with LC3-II, though the Aβ treatment has dropped its colocalization with LC3-II (Figure 4) [201]. Further, Aβ disturbs the functions of the dynein and kinesin which are essential for the axonal anterograde and retrograde transportation. Aβ within the AVs competitively impedes the pairing of dynein and snapin, and its complex, which is necessary for AVs cargo transportation towards the perinuclear space for lysosomal degradation. Aβ precisely interacts with the dynein, axonemal, intermediate chain (DNAIC) which clogs the organization of dynein-snapin motor-adaptor complexes [202,203]. Likewise, the dynein subunit dynactin-P50 expression is downregulated in AD brain with APOE mutations and the dynactin-P50 colocalized with Aβ plaques [204]. AβO inhibits the bidirectional axonal transport via the endogenous activation of casein kinase 2 (CK2) which is reversed by CK2 inhibitors. Both AβO and CK2 enhance the phosphorylation of kinesin-1 light chains (LLCs) which causes the kinesin-1 liberation from vesicular loads that perturb the fast-axonal transport [202]. Autophagy is transcriptionally regulated by the NRF2 (nuclear factor, erythroid 2 like 2) where it activates autophagic genes. NRF2 deficient mice along with APPV717 and MAPTP301L mutations exhibit elevated levels of APPV717 and MAPTP301L proteins and the expression P62, NDP52, ULK1, ATG5 and GABARAPL1 is reduced [205]. The reverse translational studies (man to mice) revealed that the NRF2 deficiency increases the p-tau and AβO, which renders significant rise in oxidative and inflammatory stress [206]. Furthermore, the increased BACE1 and mRNA stabilizing antisense (BACE-1-AS) is silenced by NRF2 by linking with antioxidant response elements, whereas the NRF2 deficiency upregulates the BACE-1 and BACE-1-AS expression and Aβ generation thereby enhancing the cognitive impairment [207]. Overall, the data suggest that the activation of NRF2 may act as a potential therapeutic target to reduce AD pathogenesis.

Lysosome plays a key role in maintaining the protein homeostasis of a cell where proteins are degraded in the heterogeneous compartments, the autolysososme or endolysosome. Several lines of evidence prove the lysosomal proteolytic failure could potentially cause the accumulation of intermediate autophagy compartments autophagososmes, autolysosome and AVs in the neurons [193]. Mutations in PS1 or deletion of PS1 aggravate the autophagy pathology. Apart from participating in γ-secretase complex, the PS1 is involved in lysosome acidification and accelerates the autophagosome-lysosome fusion [208]. The reports explore that PS1 acts as a chaperon for v-ATPase (vacuolar ATPase) where it mediates N-glycosylation of V0a1 subunit which facilitates the lysosomal acidification through pumping the protons whereas the PS1$^{-/-}$ impairs the lysosomal acidification [209]. Coen et al., postulated an alternative hypothesis that the PS1$^{-/-}$ cells exhibit the lysosomal Ca^{2+} efflux which may impair the lysosomal fusion capacity [210]. Immunogold electron microscopy studies reveal the coresidents of mature nicastrin, PS-1, and APP with lysosomal associated membrane protein-1 (cAMP-1) in lysosomal membrane where the Aβ production occurs within the lysosome [57]. The intracellular Aβ is unaffected by proteases which last at least 48 h in the cultured neuron. Aβ exposure in neuronal cells evokes the oxidative stress which disrupts the membrane proton gradient via damaging the lysosomal membrane which is blocked by treatment with either methylamine or n-propyl gallate that prevents the lysosomal leakage [211]. Several reports showed that the function of lysosomal enzymes was diminished in the AD patients impairing the toxic Aβ clearance and autophagic functions (Figure 4) [212]. The acidic environment accelerates the lysosomal enzyme activity, failure of which leads to protein clearance deficiency. A master protein TFEB regulates the functions of autophagy and lysosomes via manifesting the lysosomal enzymes and producing the membrane proteins upon translocating to the nucleus. Further, osteopetrosis-associated transmembrane protein 1 (OSTM1) in collaboration with chloride channel 7 (CLCN7) controls the lysosomal pH and Aβ clearance, which is closely regulated by the nucleus translocation of the TFEB. Aβ treated microglial cells showed dose dependent nuclear diminution of TFEB, with parallel increase in cytoplasm. Similarly, OSTM1 expression was significantly downregulated along with poor lysosome acidification, which suggest that inhibition of the TFEB nuclear translocation mediates the lysosomal dysfunction [213]. A lysosomal aspartic protease, Cathepsin D is responsible for the degradation of aged and toxic proteins including Aβ. Cathepsin D zymogen is activated in an acidic compartment by cleaving the pro-peptide. Cathepsin D is highly expressed in the AD brain as an initial event. APP-C99 impairs the lysosomal functions by increasing the lysosomal pH and inactivating the cathepsin D and other hydrolases, while silencing APP or inhibiting BACE-1 rescues the cathepsin functions [214]. Aβ treated neuroblastoma cells exhibit cellular alterations as observed in AD such as external administration of Aβ was initially detected in the clathrin-positive organelles, and later in lysosomes. Further, cellular Aβ uptake facilitates the formation of autophagosome and destruction of lysosomal membranes, leaking its contents into the cytoplasm. Notably, the cells showed enhanced autophagosomes invagination in the nuclear envelopes, which showed the connection between autophagosomes accumulation and cell death [215]. Gowrishanker et al. demonstrated that Aβ accumulated vesicles is dwelled within the swollen axons of the neurons, where the LAMP1 staining explores the recruitment of LAMP1 by autophagosome fused late endosome for maturation to lysosomes. These lysosomes have Aβ plaques within its compartment where it contains very lower concentrations of cathepsins B, D, and L (as well as AEP) than the lysosomes present in the soma of a cell. Further, the enhanced BACE1 expression within the swollen axons indicates that the defective axonal lysosome transport and maturation might boost up the Aβ production [216].

7. Aβ-Accelerated Golgi Fragmentation

7.1. APP Processing in GC

GC is a warehouse of cells where the lipids and proteins are processed and sorted as well as transported to various destinations. In addition, GC is playing a vital role in ion homeostasis, apoptosis, and stress sensing in mammals. GC is assembled with closely connected parallel cisterna known as Golgi stacks which are laterally associated by tubules forming a continuous ribbon that helps for the protein processing and transporting to its targets. The Golgi stacks has two distinct sides, cis- and trans-, where the cis-Golgi network located near ER for the entry and processing of substances and TGN is inhabited near the PM to deliver the products to its destination. This transportation amongst these organelles is mediated by the coat vesicles. GC plays an important role in processing the synthesized APP and BACE-1 and transportation through the secretory pathway. Tan et al. discovered the transportation route of BACE-1 which is different from APP transportation. In primary neurons and HeLa cells, BACE-1 transportation is facilitated by the AP-1 and Arf1/4 dependent manner, as well as BACE-1, which is recycled through the endosomal pathway. Inhibition of BACE-1 transportation increases the amyloidogenic cleavage of APP and Aβ production [217]. On other hand, the APP transportation is facilitated via recruitment of Arl5b-AP4 but not AP-1 to the GC. The perturbation in either Arl5b/AP4 raises the APP accumulation but not BACE-1 which indicates the diverted transportation of APP and BACE-1 [218]. Despite the literature proving that APP processing occurs in the endosomes, Choy et al. investigated the post-endocytic trafficking events in Aβ through the RNAi technique. HRS and TSG101 reduction present the APP at early endosomes and reduces the Aβ production. In opposition, diminution of CHMP6 and VPS4 rerouted the APP from endosome to GC for APP processing; where VSP35 mediated retrograde transport is needed for Aβ production. It has been suggested that the GC may be one of the intracellular sites for Aβ production [59].

7.2. Golgi Fragmentation by Aβ

The Golgi stack and ribbon organization are maintained by a complex molecular system such as Golgi matrix proteins (GRASP55, GASP65, GM130, Golgin-45, Golgin-84,

Golgin-160), tethering proteins (p115/SNARE protein), microtubule related motor proteins, signaling proteins and proteins related to pH and Ca^{2+} homeostasis [219]. Under physiological conditions, the Golgi fragmentation is crucial to begin the mitosis, which is a highly organized process. Several kinases such as Cdc2, GSK3β, RAF/MEK1/ERK1c, Plk1, and Plk3 phosphorylates number of Golgi proteins, which facilitate the Golgi dispersion for mitosis. After cytokinesis, the fragmented Golgi reunites and serves its normal functions. However, under the pathological conditions such as apoptosis, the Golgi fragmentation is irreversible due to the caspase-mediated proteolytic cleavage of several Golgi proteins including golgin, t-SNARE syntaxin 5 and GRASP-65 [220]. Notably, many neurodegenerative diseases such as Alzheimer's and Parkinson's exhibit the Golgi fragmentation as a common hallmark. However, the Golgi fragmentation in neurodegeneration is an earlier irreversible process, which triggers the apoptosis. Several reports from both AD patients and animal models depict the Golgi fragmentation as a consequence of ER stress and oxidative/nitrosative insults or excitotoxins. In the AD brain, Aβ deposition alters the morphology of GC by dropping the mitochondrial membrane potential and release of cytochrome c in the cytoplasm. The immunogold-electron microscopic studies on AD mice (APP-PS1) model and Aβ-treated BV-2 cells showed Golgi cisternae fragmentation mediated by the COPI depletion which affects the intra-Golgi transport through Aβ deposition [221]. The Golgi morphological defects were observed in both AD animal and cells models such as Golgi fragmentation. The GC of AD mice ($APP_{Swe}/PS1_{\Delta E9}$) hippocampal and cortical tissues were fragmented with swollen cisternae, but in wild type mice normal ribbon-like organization was observed [222]. Similarly, $APP_{Swe}/PS1_{\Delta E9}$ overexpressed CHO cells reveal Golgi fragmentation, which decreases the APP trafficking and increases the Aβ production [222].

The molecular view of Aβ mediated Golgi fragmentation is reported by Joshi et al. where the AD mice and $APP_{swe}/PS1_{\Delta E9}$ transfected cells exhibit an elevated amount of Aβ. Aβ induced CDK-5 activation facilitates the Golgi fragmentation via phosphorylation of Golgi proteins such as GRASP65 which is rescued by CDK-5 inhibition as well as expression of non-phosphorylatable GRASP65 mutants that also reduces Aβ production [223]. However, Golgi fragmentation in the Aβ treated cells is reversible upon removal of Aβ treated cell culture media. The electron microscope analysis portraits that the GC ribbons are disconnected with shorter and low numbers of cisternae per stack, with numerous vesicles adjacent to each stack. The relationship between the CDK5 and Golgi fragmentation in AD is explored upon the Aβ and glutamate treatment in the PC12, SHSY5Y cells where the cells undergo Golgi fragmentation via phosphorylation of GM130 which leads to cell death. GM130 acts as a substrate for CDK5 which impedes the binding of GM130 and vesicle docking protein p115 [224]. Furthermore, the CDK5 induces the p-tau and formation intracellular NFT. Reports indicates that siRNA mediated Golgin-84 diminution in HEK293 cells induce Golgi fragmentation and p-tau is mediated by the CDK5 and ERK kinases which indicates that depletion of Golgin-84 activates the CDK5 and ERK kinases [225]. On the other hand, GSK3β is activated by the Aβ which indicates that the GSK3β activation could develop a feedforward loop that promotes further APP amyloidogenic processing via activation of BACE-1 [222]. Further, GSK3β facilitates p-tau, which disturbs the microtubule network inducing Golgi fragmentation and neuronal malfunction [226]. In addition, many reports indicate that the JNK activity is highly linked with AD progression through higher Aβ generation and NFT formation. JNK2 plays a vital role in the separation of Golgi stacks via phosphorylation of GRASP65, whereas RNAi, or JNK inhibitors mediated JNK2 inhibition can restore the Golgi ribbon [227]. On other hand, the Golgi fragmentation in progressive motor neuronopathy mice lack TBCE and TBCE-depleted motor neurons with defective Golgi engaged microtubules and decreased COPI vesicles diminishing the recruitment of p115/GM130 proteins and SNARE mediated vesicle fusion. siRNA mediated SNARE Syx5 inhibition facilitates the Golgi fragmentation, similar to the reports of the Golgi fragmentation in the AD brain cells (Figure 4). In contrast, overexpression of Syx5 shows accretion of APP in the ER, which restrains the APP processing towards

Aβ [228,229]. Recently, Suga et al. explored that Syx5 works as a stress rescuing component that involves in the neuronal cell survival. In detail, the apoptosis inducers decrease the expression of Syx5, whereas the ER stress inducers increased the levels of Syx5 and Bet1 protein expression. Syx5 deletion during apoptosis or ER stress causes the cell highly vulnerable. Further, Golgi stress increased the expression of Syx5 and concurrently reduced the Aβ production. These data suggest that the Syx5 is a common stress reducing agent for both ER and Golgi, where the Syx5 reduction in AD is due to the Aβ mediated inhibition of stress response [230]. Hence, protecting the Golgi structure and function denotes a new comprehension to reduce the Aβ generation and its toxicity.

8. Aβ in Gene Regulation

8.1. Aβ as Transcription Factor

Nucleus is a global control center of a cell where the genetic information is replicated and transcribed, which regulates the cellular behaviours. Bundles of reports confirm that as observed in other organelles, Aβ translocate into the nucleus, interacts with several nuclear proteins, and alters the gene expression. The soluble Aβ translocation into the nucleus is confirmed multiple techniques such as chemical testing of nuclear fragments, biotin labelled Aβ confocal imaging and transmission electron microscopic analysis of cultured cells. Possibly, the Aβ is passed directly into the nucleus through the channel-like pores. Remarkably, this study also explores the involvement of Aβ in nuclear signaling, the ChIP assay shows the specific interaction of Aβ with the LRP1 and KAI1 promotors, which potentially decreases the mRNA expression of the candidate genes [231]. Both LRP1 and KAI1 protect the neuronal cells against Aβ neurotoxicity. Aβ acts as a putative transcription factor for AD linked genes such as APOE, APP and BACE1. The electrophoretic mobility assay reveals that Aβ is precisely docked with the Aβ interacting domain (AβID) of the nucleus with the consensus of "KGGRKTGGGG" where any mutation in it, the peptide-DNA interaction is neglected [232]. Hence, Aβ itself act as a transcription factor and can control the transcription of candidate genes. The Aβ-chromatin interaction is discovered in the polymorphic APP-promotor CAT fusion clones transfected PC12 cells using ChIP assay. This transfected cells when supplemented with Aβ elucidates the DNA sequence specific response where it regulates its own amyloidogenic proteins such as APP and BACE-1 that induces more Aβ production [233]. Similarly, the transcriptional regulation of Aβ is validated upon studying several genes including (I) the amyloidogenic genes such as ADAM10, BACE1, PS1, PS2, Nicastrin and APP, (II) AD risk genes APOE and TREM2, (III) learning and memory factors genes such as NMDAR and PKC zeta, (IV) kinases which contribute for p-tau including GSK3α, GSK3β and Cdk5 and (V) enzyme 1α-hydroxylase (1αOHase). The qRT-PCR analysis explores the upregulation of amyloidogenic and p-tau related genes, which generate toxic Aβ and p-tau, and the downregulation of NMDARs, ApoE, Trem2, and 1αOHase genes [234]. DNA microarray analysis of Aβ treated neuroblastoma cells explore the upregulation of the insulin-like growth factor binding proteins 3 and 5 (IGFBP3/5). The qRT-PCR results confirm the above finding that the expression level of IGFBP3/5 is two-fold increased. Literatures indicate that IGFBP3/5 contributes to p-Tau. Further, the immunohistochemistry studies support these data illustrating higher expression of IGFBPswere observed in the hippocampal and cortical neurons. Further, the proteomic studies CSF of human AD, illustrates the appearance of elevated amount of IGFBP. These data suggest that both transcriptional and translational regulation of IGFBP by Aβ could be an early biomarker for AD [235]. In addition to Aβ, several reports implicate that the secretase cleaved fragments undergo nuclear translocation and controls the transcription regulation. In an AD patient's brain, the γ-secretase cleaved ~6 kDa CTF-APP-like protein 2 translocate to nucleus and interacts with CP2 transcription factor where it upregulates the expression of GSK3β which contribute variety of pathological events for neurodegeneration [236]. Similarly, the γ-secretase cleaved APP-CTF in cytoplasm binds with an adapter protein Fe65, the confocal and FRET analysis discloses colocalization of GFP-APP-CTP and myc-Fe65 and translocation to the nucleus. Taken together, the APP-CT-Fe65 complex can potentially modify the transcription of the cells [65].

8.2. Telomerase, Spliceosome Inhibition and DNA Methylation by Aβ

Several reports discovered the relationship between the telomerase and the AD pathology where the shortened telomerase involved in AD progression whereas the increased telomerase activity protects the neurons from the neurotoxic aggregates [237–240]. The cellular senescence also known as DNA damage response (DDR) is a prime factor of age-linked diseases. In the sight of DNA damage, the DDR coordinate the DNA damage through cell cycle arrest until the DNA damage is recovered, also DDR facilitates the perpetual growth arrest if the cells are failed to repair the damage [241]. Increasing evidence shows that the telomeres are engaged in development of neurodegeneration including AD [238,239]. In tissues and peripheral blood cells preferentially aged patient's leucocyte, the telomere length is highly associated with AD risk and cognitive deficits. Few studies showed contrariety data in the telomerase length; however, numerous reports confirm that the leucocyte in reference to the age showed the excessive telomerase loss leading to AD development [242]. Recently, Wang et al. discovered that the AβO potentially inhibited the telomerase activity through interacting with DNA-RNA templates and RNA templates of telomerase and blocking the telomeric DNA elongation [240]. In addition, the Aβ colocalized telomere is also observed as a cause of telomerase inhibition. In contrast, overexpression of catalytic subunit of the telomerase decreases the Aβ induced cell apoptosis, which might be capable of defending the age-related neurodegeneration [237]. On the other hand, the epigenetic regulation including DNA methylation and histone modification controls the gene expression. DNA methylation obstructs the transcription factors binding with the DNA via transferring the methyl group to the cytosine CpG dinucleotides, which is catalysed by specific DNA methyltransferases. DNA methylation plays an important role in gene silencing/inactivation. The AD brain shows loss of DNA methylation through the increased concentrations of S-adenosylhomocysteine, a potential inhibitor of methyltransferase. Induction of DNA hypomethylation in the promotors of the AD associated genes including APP, PS1 and BACE1 accelerates abnormal expression of these genes, which leads to increased production and accumulation of Aβ [243]. Aβ-treated samples are digested with a methylation-sensitive (HpaII) or a methylation-insensitive (MspI) restriction endonuclease for the DNA microarray analysis. The results showed significant methylation changes in the genomic loci with highly enriched cell-fate genes, which control the apoptosis and neuronal differentiation involved in potentially inducing the brain contraction and memory deficits in AD [244]. The AD brain cortex and AD patient's lymphocytes are analysed for epigenetic alterations at the promotor regions of several genes including PS1, and APOE. The PS1 is usually hypomethylated which triggers unusual Aβ generation, whereas the APOE has a bimodal structure where at most it is found in a hypermethylated state. Hence, the data suggest that the concurrent manifestation of both hyper- and hypo-methylation could potentially contribute for AD progression [245]. The Aβ mediated epigenetic regulation including DNA methylation/demethylation of a specific promotor is widely studied. The Aβ induced oxidative stress is a prime causative of DNA hypermethylation in an aging brain. Chen et al. discovered that the Aβ suppresses the neprilysin (NEP) promotors via DNA methylation [246]. The HPLC and methylation specific PCR studies of the Aβ treated endothelial cells showed increased NEP methylation which suppresses the NEP mRNA and protein expression. The NEP is a zinc metalloproteinase, which facilitates Aβ clearance in AD mice. Thus, the results suggest that the DNA methylation of NEP is a consequence of Aβ accumulation. In addition, the modification in the alternative splicing is highly linked with the AD progression. The LOAD is characterized with U1 small nuclear ribonucleoprotein (snRNP) tangle-like deposition due to the aberrant genetic mutations in PS1and APP, which results in unusual APP processing leading to formation of snRNP aggregates [247]. Aβ treated neuroblastoma cells were subjected to the proteomic analysis to expound the early events of AD pathogenesis. Remarkably, the bioinformatics results imply the downregulation of ribosomal biogenesis and splicing process. Further, Western blot analysis explicated the downregulation of each splicing steps facilitating the downregulation of every subunit of the spliceosome. These results suggest that the spliceosome dysfunction is a consequence of Aβ deposition. Overall, Aβ is acting as gene regulating factor upon interacting with telomerase/telomeres, epigenetic and transcriptional regulation, and aberrant spliceosome function which contributes to high risk of AD progression.

9. Signalling Mechanism of Aβ Leading to Memory Impairment and Cell Death
9.1. Receptor Mediated Long Term Potentiation Inhibition

It is well known that the extracellular Aβ interacts with the surface of the brain cells where the Aβ activates several signalling mechanisms unusually that triggers the cells either to survive or die. As a good sign, Aβ interaction with the receptors such as low-density lipoprotein receptor-related protein 1 (LRP1), low-density lipoprotein receptor (LDLR), scavenger receptors A1 and A2 (SCARA1 and SCARA2) facilitates the Aβ uptake and clearance [248–250]. In addition, Aβ binding to the macrophage receptor with collagenous structure (MARCO) activates the extracellular signal regulated kinase 1/2 (ERK1/2) signalling pathway, which reduces inflammation [251]. In contrast, various receptors on synapse show toxic effect, causing the synaptic dysfunction and neurodegeneration. The receptors of Aβ transduce the specific intracellular changes via activating the extracellular factors either directly or association with other molecules. The AMPARs and NMDAR are the ligand gated ionotropic glutamate receptors, and the mGluRs regulates the learning and memory via LTP and long-term depression (LTD) at excitatory synapses. The lower synaptic signal activates AMPARs, and the stronger synaptic signals unblock the NMDARs, which results in increased number of AMPARs on the post synaptic membrane [252]. The increased LTP was observed when there is over-expression of AMPAR at postsynaptic membrane; in contrast, some reports reveal that elimination of AMPARs increases the LTD [253,254]. The role of AMPARs in AD is still unclear, however, the results suggest that the AMPARs are downregulated during the preliminary stage of AD. Aβ25-35 treated rat embryonic hippocampal cells showed elevated level of caspase activity, which leads to enzymatic degradation of the AMPAR not NMDAR [255]. Apart from the enzymatic cleavage, Aβ directly interacts with AMPAR and modulates it functions. Iontophoretically exposed aggregated Aβ1–42 on the hippocampal CA1 neurons reduces the AMPA-induced neuronal firing, but NMDA-evoked neuronal firing was enhanced, which suggest that the LTP disruption and attenuation of field excitatory postsynaptic potential (fEPSP) [256]. On the other hand, lower synaptic stimuli trigger either NMDARs to generate NMDA-mediated LTD or mGluRs to make mGluR-dependent LTD, which prompt the removal of postsynaptic AMPAR [257]. Patient-specific human iPSCs derived neurons produced Aβ exhibits synaptotoxic mediated cell death showing impaired axonal vesicle clusters, postsynaptic loss of AMPAR and rise in Aβ mediated tau phosphorylation [258]. Numerous protein kinases and phosphatases play a vital role in generation of LTP and LTD. AβO interrupts the postsynaptic Ca^{2+} signalling through increasing the accessibility of glutamate molecules to the NMDAR. The enhanced activation of NMDAR causes abnormal redox reactions as well as increased Ca^{2+} influx into neurons that activate the Ca^{2+}-dependent protein phosphatase calcineurin/PP2B and protein phosphatase 2A (PP2A) [259]. Activation of calcineurin further activates or deactivate the target proteins via dephosphorylation. The synaptic dysfunction is mediated by the surface removal and endocytosis of AMPAR. Aβ-stimulated AMPAR endocytosis is reliant on the activation of calcineurin/PP2B, which is mediated by downregulation of CaMKII [260,261]. Like AMPARs, Aβ also can prompt NMDARs internalization, which is mediated by dephosphorylation of GluN2B (NMDAR subunit) of p-Tyr1472 in the striatal-enriched protein tyrosine phosphatase (STEP) [262]. AβO downregulates the glutamate transporters EAAT1 and EAAT2 of glial cells, which disrupts the glutamate uptake causing the glutamate overflow at synaptic cleft that over-activates the GluN2B [263].

NMDAR and AMPAR interactions with post-synaptic density scaffolding protein (PSD-95) at post-synaptic membrane regulate the protein assembly and neural plasticity [264]. Aβ exposed cortical neurons exhibits CDK5 and NMDAR mediated reduction on PSD-95 that leads to synaptic dysfunction and surface AMPAR removal [265]. Co-immunoprecipitation studies on human post-mortem AD brain and AβO treated murine neurons shows that the Aβ directly interacts with PSD-95 at post synaptic membrane causing synaptic loss [266]. Cellular PrPC has high affinity with AβO, such molecular associations of AβO-PrPC found only in the AD brains, but not control brains [267]. AβO-PrPC inhibits the hippocampal LTP that manifests the memory deficit in an AD mouse model [268]. AβO bound PrPC influences the activation of Fyn kinase that in switches the GluN2B phosphorylation, resulting in NMDARs surface removal. The AβO-PrPC complex demands both mGluR5 and LRP1 co-receptors to activate the Fyn [269,270]. In addition, the Fyn activation steers on tau phosphorylation [271]. PrPC resides at cholesterol- and sphingolipid-abundant, detergent-resistant lipid rafts. The saturated acyl chains of glycosylphosphatidylinositol trigger the N-terminal signal interaction with the heparan sulfate proteoglycan, glypican-1 [272]. PrPC knockout or anti-PrPC antibodies rescues the AβO-stimulated synaptic dysfunction and spatial memory, which indicates that the PrPC play a crucial role on AD pathogenesis [267]. In addition, the AβO activates α7-nAChR which increases the presynaptic Ca^{2+} level and disrupts the rafts by cholesterol depletion [273]. AβO activated α7-nAChR causes the elevated cytosolic Ca^{2+}, calcineurin activation and dephosphorylation and activation of STEP61. The enhanced STEP61 inactivates Fyn and lowers the NMDAR exocytosis because of the GluN2B dephosphorylation mediated NMDAR internalization [261]. Likewise, AβO binds to the receptor tyrosine kinase EphB2 resulting in its degradation that causes reduction in NMDA receptor subunits such as GluN2B, which leads to impairment in NMDAR-mediated synaptic activity and cognitive function. In opposite, the EphB2 overexpressed AD Tg mice reverses the deficits of NMDAR-dependent LTP and cognitive impairments [274]. Further, Ephrin A4 (EphA4) was discovered as a putative Aβ receptor. Aβ mediated EphA4 activation leads to repression of LTP and spine loss in AD transgenic mice where the EphA4 shRNA or EphA4 inhibitors/antagonists inhibits these deficits [275,276]. Overall, the Aβ interaction with various receptors stimulates the neurotoxicity via NMDAR (Figure 2).

9.2. Receptor Mediated Cells Death Induced by Aβ

Ligand binding cell surface death receptors (DR) are the tumour necrosis factor (TNF) gene superfamily receptors that confer caspase mediated death pathway. DR contains cysteine rich extracellular domain and intra cellular death domain. The receptors including TNF receptor 1 (TNFR1), Fas receptor (FasR), TRAIL receptor 1 and 2 (TRAIL-R1 and R2), p75NTR and lymphoid cell specific receptors CD30, CD40 and CD27 facilitating the death signaling [277]. Interestingly, the Aβ has high affinity with these receptors and activates the apoptotic pathway. Ivins et al. hypothesised that Aβ may activate the Fas/TNFR mediated apoptosis signalling. The Aβ treated hippocampal neurons exhibit recruitment of caspase-8 and FADD proteins during the apoptotic event, which was prevented by the pre-treatment of caspase-8 specific inhibitor IETD-fmk and viral mediated dominant negative FADD gene delivery. Ivins and colleagues concluded that both the caspase-8 and FADD requirement in apoptosis support that the cell death might be initiated through Fas/TNFR family receptors upon interaction with Aβ (Figure 2) [278]. The vascular Aβ mediated extrinsic apoptotic signalling mechanism discovered using the human brain microvascular endothelial cells treated Aβ40 or its vasculotropic variants E22Q or L34V. The apoptotic cell death facilitated via binding of AβO with the (TRAIL) death receptors DR4 and DR5 followed by the activation of caspase-8 and caspase-9. Further, the caspase-8 inhibitor FLICE-like inhibitory protein (cFLIP) downregulated the mitochondrial path associated with the BH3-interacting domain death agonist (BID) cleavage. DR4 and DR5 up-regulation and co-localization with AβO indicate the receptor specific interaction, which was attenuated upon RNA silencing of both DR4 and DR5 [279].

Numerous reports discovered that the p75NTR, a nerve growth factor (NGF) receptor mediated cell death. The p75NTR is structurally similar to the p55 TNF and Fas receptors. Alteration in the p75NTR expression promotes cell death or survival, a decreased p75NTR increases cell survival whereas increased p75NTR induces apoptosis by silencing Trk-mediated survival signals. On the other hand, Aβ mediated cell death via p75NTR is altered by the NGF, the NGF binding instead of Aβ inhibits the p75NTR death signalling, however, the Aβ-p75NTR induced death is found in PC12 cells [280], NIH 3T3 cells [281], human neuroblastoma cells [282] and hippocampal neurons [283]. Aβ exposed mutant (p75NTR$^{-/-}$) mice reveals the least cell death on the hippocampus compared with wild-type mice [284]. Aβ facilitates the apoptosis in cultured neurons through the activation of JNK–c-Jun–Fas ligand–Fas pathway [284]. Knowles et al. investigated the AβO interaction on the surface of p75NTR using fluorescence resonance energy transfer (FRET)-imaging technique. The role of p75NTR in Aβ-induced neuronal death and c-Jun expression is validated using p75NTR$^{-/-}$ mutant mice derived neuronal cultures and p75NTR$^{-/-}$ AD mice model. The results reveal that neurodegeneration through p75NTR requires AβO interaction on the surface domain of p75NTR [285]. A high level of Ca^{2+} inflow is reported as an important factor for AD, which is mediated via upregulated L-type Ca^{2+} channel in AD mice not in wild type mice. The overexpression of p75NTR prevented the Ca^{2+} channel current, but Aβ1-42 treatment significantly increased the Ca^{2+} channel current, due to the blockage or decreasing expression of p75NTR. The Aβ1-42 induced Ca^{2+} channel current activation is removed when the p75NTR expression is dropped (Figure 2) [286]. The high mobility group box 1 (HMGB1) acts as a proinflammatory mediator and it activates the inflammatory response via docking RAGE and Toll-like receptor 4 (TLR-4). The RAGE entails significantly in neurodegeneration by the action of several signalling moieties such as CaMK-β-AMPK, the RAGE/(ERK1/2), GSK-3β, and NF-κB, which directs the Aβ and p-tau pathology [287–289] while the TLR-4 acts as an immune receptor elucidating the immune response.

10. Inflammation a Central Mechanism in AD

Extensive research on pathogenesis leading to AD revealed that gap exists in between core pathologies, Aβ plaques and NFT in understanding AD pathogenesis. Paramount evidence indicated the existence of inflammatory markers in the brain of AD patients and preclinical AD model system, other than the neuropathological hallmarks senile plaques and PHF, indicating that inflammation acts as interlink between early lesion senile plaques and the later lesion NFT in AD pathogenesis [290]. Inflammation acts as double-edged sword, in a healthy brain, as acute inflammation plays the role of defense mechanism against various infection, toxin and injury clearing the invading pathogen or injurious agent. On the other hand, imbalance between pro and anti-inflammatory mediators due to Aβ accumulation in AD leads to chronic inflammation, characterized by the activation of microglial cells, which was observed to be accumulated around Aβ plaques in AD brain and transgenic animal models [291]. Microglial cells and astrocytes in its activated state release mediators of inflammation-like cytokines, chemokines, complements, monocyte chemoattractant, ROS and prostaglandins, etc., disrupting the balance between the normal neurophysiologic conditions associated with cognition, learning and memory. Inflammatory mediators activate more glial cells and astrocytes to release cytokines, which promotes the migration of monocytes and lymphocytes across the BBB towards Aβ accumulated site in brain of AD individuals triggering inflammatory response (Figure 5) [292,293]. Initially, constant inflammatory response was considered as causative for neuronal loss in AD patients, later substantial evidence revealed that persistent immune response facilitates and exacerbate both Aβ and NFT pathologies. Hence, inflammation is considered as a driving force, which induces or accelerates the pathogenesis of AD. Epidemiological studies revealed the linkage between the polymorphisms in the immune molecule involved in AD and the role of nonsteroidal anti-inflammatory drugs in attenuating the incidence of AD. The degree of inflammatory response depends on the level of Aβ, tau and ubiquitin and

APOE ε4 as observed in AD sub types. McGeers et al. observed enhanced expression of HLA-DR (human leukocyte antigen, antigen D related) a Class II major histocompatibility complex (MHC) in microglial cells around the senile plaques [294]. Further, Giometto et al. revealed the presence of high level of complement and acute phase proteins in AD blood sample depicting the fact that both immune and inflammatory response synergistically triggers AD pathogenesis [295]. Scientific evidence on the use of anti-inflammatory drugs for the treatment of rheumatoid arthritis, in transgenic mice and human showed convincing results on reduction in AD pathogenesis revealing the fact that inflammation play key role in AD pathogenesis [296]. Microglia on chronic activation produces several proinflammatory mediators such as ROS, RNS and chemokines. Increase in the level of interleukin 1 (IL-1) enhancing the level of cerebral Aβ deposit was observed in deceased patient affected by head trauma, illustrating the facts that IL-1 promotes amyloidogenic processing of APP enhancing the level of Aβ peptide.

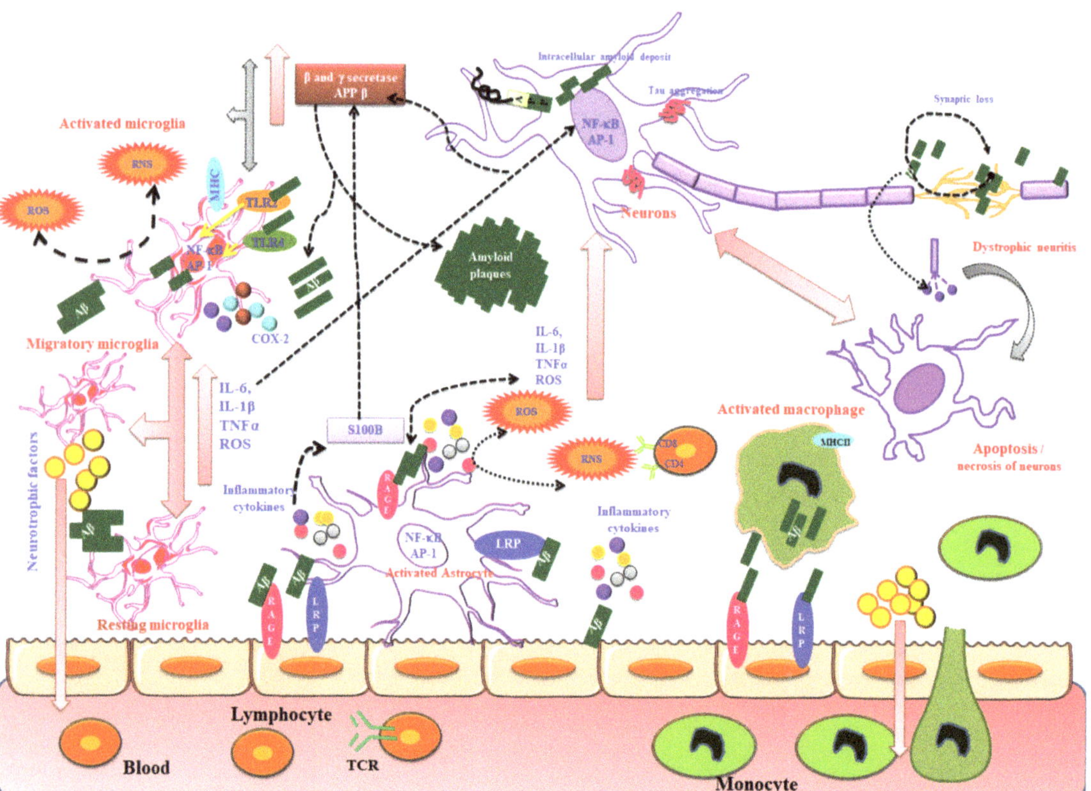

Figure 5. Neuroinflammation in the pathogenesis of Alzheimer's disease: Aβ activates microglial cells via TLR and RAGE receptors which stimulates the NF-κB and AP-1 transcriptional factors leading to release of inflammatory cytokines (IL-1, IL-6, TNFα), ROS and RNS inducing oxidative/nitrosative stress mediated neuronal damage. Inflammatory cytokines stimulate astrocytes leading to amplification of inflammatory signals inducing neurotoxic effect. Chemokines also attracts peripheral immune cells towards amyloid plaque exacerbating inflammatory response.

IL-1β enhances the release of IL-6 which activates CdK5 inducing hyperphosphorylation of tau [297]. These reports reveal that incidence of neuroinflammation occurs before the neuropathological hall marks aggravating Aβ load and hyperphosphorylation of tau protein, which in turn further activates the inflammatory pathway revealing the inter-relationship between these apparently contrasting core pathologies leading to AD (Figure 5).

10.1. Cellular Mediators Involved in Neuroinflammation

10.1.1. Microglial Cells

Microglial cells are the specialized macrophages found in CNS which plays a vital role in restoring the brain homeostasis via inflammatory response, phagocytosis of Aβ plaques and NFT. Microglial cells in the resting state exist in ramified morphology portrayed with small cell body, which interacts with neurons and other glial cells via signaling mechanism through numerous receptors for neurotransmitters, cytokines maintaining the neurons healthy [298]. Microglia interacts with Aβ peptide through several receptors such as scavenger receptors (SR-SCARA-1, MARCO, SCARB-1, CD36 and RAGE); G protein-coupled receptors (GPCRs- formyl peptide receptor 2 (FPR2) and chemokine-like receptor 1 (CMKLR1)), and toll-like receptors (TLRs- TLR2, TLR4, and the co-receptor CD14). Receptors SCARA-1, SCARB-1, MARCO and CMKLR1 interact with Aβ promoting its cellular uptake, during which RAGE activates microglial cells to release proinflammatory molecules, while other receptors such as TLR, CD36 and FPR2 exhibit dual functions (Table 1) [299]. In early AD pathogenesis, Aβ peptide acts as primary driver triggering microglial cells towards plaques and provokes phagocytosis of Aβ peptide, but on prolonged activation results in exacerbation of AD pathology. Overall, neuroinflammation in AD is caused by microglial priming on interaction of Aβ peptide with receptors (SR1, GPCR, TLRs). Aβ fibrils recognizes the complex CD36-α6β1-CD47 leading to generation of ROS, providing signal for heterodimerization of TLR4-TLR6 transmitting signal for activation of NLRP3 a component of inflammasomes in microglial cells. Inflammasomes is an intracellular multiprotein complex composed of NLR family pyrin domain containing three (NLRP3), apoptosis-associated speck-like protein containing a caspase-recruitment domain (ASC) and procaspase- 1, which acts as first line of defense. Inflammasomes acts as platform for activation of caspase-1, which activates cytokines IL-1β and IL-18 the key mediators of inflammation [300]. CD36 promotes entry of Aβ into the lysosome inducing destabilization, dysfunction of lysosomes with consequent release of cathepsin B into cytosol. Cathepsin B induces NLRP3 dependent caspase-1 activation of IL-1β and IL-18, which in turn triggers the release of several chemokines (IL-1, IL-18 and TNF-α), chemotactic mediators stimulating nuclear factor-kappa-B (NFκB) dependent pathway [301,302]. Cytokines and chemokines exacerbate Aβ accumulation leading to activation of microglial cells enhancing the production of proinflammatory mediators provoking neurodegeneration in cyclic manner (Figure 5) [303]. Aβ peptide promoted activation of microglial cells, releasing proinflammatory mediators which in turn induced microgliosis and astrogliosis decreasing the efficiency of microglial cells to phagocytize Aβ peptide, reduction in Aβ degrading enzyme affecting the clearance of Aβ peptide leading to the deposition of amyloid plaques [304]. Although the Aβ clearance is compromised, persistent immune response induces production of proinflammatory mediators by microglial cells recruiting additional microglial cells towards plaque creating halo of activated microglial cells surrounding plaques. In addition, peripheral macrophages are also attracted towards the Aβ plaque deposition to clear Aβ peptide exacerbating neuroinflammation contributing to neurodegeneration. In AD patients, microglial cells exhibit a mixture of classical and alternate activation pathways causing irreparable damage resulting in continuous neurodegeneration.

10.1.2. Astrocytes

Activated astrocytes are distributed near the vicinity of amyloid deposits in cortical pyramidal neurons in early stages of AD, which contribute to clearance of plaques by degradation and phagocytosis of accumulated Aβ in parenchyma. Like microglial cells, TLRs and RAGE pathways activate astrocytes promoting local inflammation intensifying neuronal death. Astrocyte activation causes disruption of normal activities essential for normal neuronal function leading to local neuron depolarization ultimately leading to neuronal damage. Retro-splenial cortex administration of oligomeric Aβ forms in rats revealed the presence of activated astrocyte associated with activated NF-κB, signaling molecules leading to inflammation (COX-2, TNFα and IL-1β) and expression of cell surface receptors such as SRs, proteoglycans and lipoprotein receptors, via which it binds to Aβ peptides [305,306]. Astrocytes on activation express inflammation associated factor S100β leading to dystrophic neuritis in AD patients. Activation of NF-κB regulates the secretion of chemokine and cell adhesion molecules enhancing infiltration of peripheral lymphocyte enhancing neuroinflammation ultimately leading to neurodegeneration. Astrocytes, in its activated stage, protects the brain, however, on extreme activation, it aggravates damage to the neurons hastening the progression of AD.

10.1.3. Oligodendrocytes

Abnormalities in the white matter and myelin sheath have been observed in asymptomatic FAD preferentially under PS1 mutation [307]. Mutation in PS1 and Aβ accumulation alters the function and differentiation of oligodendrocyte inducing abnormal patterns in myelin basic protein (MBP), affecting homeostasis of oligodendrocytes [308]. As a result, the trophic supports provided by these cells to neurons are lost and neurons become vulnerable to oxidative stress and inflammation provoking neurodegeneration.

10.1.4. Neurons

Neurons are also involved in inflammatory response which is evident by the presence of proinflammatory mediators such as COX-2-derived prostanoids, cytokines such as IL-1β and IL-18, complement and macrophage colony-stimulating factor. In addition, iNOS the inflammation induced enzyme expression in degenerating neurons is also observed in brain of AD individuals substantiating the involvement of neurons in inflammation [309]. Neurons generally produce TRM2, CD22, CD200, CD59 and fractalkine to suppress inflammation and these molecules were observed to be deficient in AD [310]. Studies on the expression of mediators of inflammation in neuronal cells are not yet completely investigated and it remains still elusive.

10.2. Inflammatory Mediators in AD

Aβ deposition activates the microglial cells and astrocytes to acute immune response provoking the release of mediators of inflammation-like complement factors, cytokines and chemokines (IL-1, IL-6 and TNF-α) and transforming growth factor β (TGF-β) which exhibits cascade of events with both beneficial and harmful effects.

10.2.1. Complement System

The complement system regulates T-helper cell differentiation and response in adaptive immune response. In the brain, the complements are produced locally, which are dysregulated during brain trauma and neurodegenerative AD [311]. In the brain of AD patients, an enhanced level of complement components of classical pathways (C1q, C3b, C4d, C5b-9, and MAC) were observed in the vicinity of senile plaques along with microglial cells illustrating the relationship between Aβ peptide aggregation, activation of classical complement pathway and inflammatory response [312]. A recent report showed that the interaction of C1q/C3b with Aβ aggregates, and NFT activates the classical pathway, which on subsequent interaction with the C1q receptor of microglial cells, activates it leading to the clearance of Aβ and tau aggregates by phagocytosis, together with unwanted inflammation causing neurotoxicity [313]. Complement receptor 1 (CR1), widely found on the surface of phagocytic cells with binding affinity for C3b and C4b play a vital role in phagocytosis. Chibnik et al. [314] studied the genome-wide association screening (GWAS) in AD patients, where they observed an inter-relationship between CR1 gene variants with impairment in cognitive function associated with enhanced formation of amyloid plaque. Complement fragment C5a in neuronal excitotoxicity induced apoptosis, promotes chemotaxis and glial cell activation leading to the development of neurodegenerative disease. Although complement activation in AD play beneficial role in Aβ clearance, its activation becomes deregulated, promoting unwanted inflammation leading to neurotoxicity which remains unclear and need to be studied.

10.2.2. Chemokines

Chemokines in the CNS were synthesized by astrocytes and microglial and its receptors are highly localized in the neurons which might be responsible for inflammation mediated neurodegeneration [315]. Expression of chemokines and their receptors promotes communication between microglia and neuronal cells leading to commencement of local inflammatory response promoting phagocytosis of Aβ peptide in early AD. This inflammation also contributes to Tau pathology accelerating progression of the disease [316]. Enhanced chemokine level recruits the phagocytic and microglial cells which co-localizes near the senile plaques during the process of chronic inflammation in AD. Enhanced level of CCL2 in CSF correlates with cognitive decline and IL-8 production by neurons is related to formation of brain-derived neurotrophic factor (BDNF) [317]. Overall, chemokines in the CNS promotes migration of local and peripheral immune cells to establish an immune response, which on chronic production, leads to inflammation mediated neurodegeneration in AD.

10.2.3. Cytokines

Cytokines are non-structural soluble proteins produced by immune cells such as microglial cells and astrocytes in the CNS which play a significant role in the development of the brain during embryonic stages. An elevated level of pro-inflammatory cytokines (IL-1β, IL-6, IL-10, TNF-α, and TGF-β) was found in the CSF and brain of AD patients, illustrating the role of cytokines in aggravating AD pathology. The transgenic animal model expressing mutant hAPP protein revealed the link between the cytokines level and Aβ aggregates [318]. Aβ aggregates activate the microglial cells to produce pro-inflammatory cytokines, which in turn, activate microglial cells leading to microgliosis and astrogliosis, amplifying the cytokine level leading to neurodegeneration, amyloidosis due to upregulation of β and γ secretase affecting learning and spatial memory [319].

10.3. Proinflammatory Mediators

10.3.1. Interleukin 1 (IL-1)

IL-1 up-regulated in early AD induces endothelial APP-β at mRNA expression which can be inter-related to enhanced Aβ, level in AD patients [320]. Microglial cells surrounding neuritic plaques (NPs) in AD patients produce IL-1, which promotes S100β synthesis in reactive astrocytes leading to dystrophic neurite formation ultimately leading to neuronal death [321]. In addition, IL-1 also promotes p38-MAP kinase activity enhancing tau hyperphosphorylation and enhances the level of neurotrophin-3 and neurogenin-1 promoting neurogenesis through promoting outgrowth. We also found that IL-1β increased mRNA and protein levels of Wnt5a, promotes neurogenesis through the Wnt5a/RhoA/ROCK/JNK pathway [322]. IL-1β is the master regulator in brain inflammatory cascade regulating the level of TNFα and IL-6. An elevated level of IL-1β was observed in the cerebral cortex and the hippocampal region of brain tissue from AD patients. IL-1β interacts with receptor (IL-1R) widely found in the dentate gyrus and pyramidal cells of hippocampal region of brain, which are highly susceptible to early AD pathogenesis early development of AD pathology [323]. IL-1β regulates the synthesis and secretion of APP in glial cells, upregulates β-secretase activity leading to increased amyloidogenic processing of APP enhancing Aβ burden, creating a vicious cycle where enhanced Aβ load, in turn, activates microglial cells leading to IL-1β production [324].

10.3.2. IL-6

IL-6 is a multifaceted cytokine exhibiting dual role as anti-inflammatory myokine and proinflammatory cytokine depending on the condition, thereby, maintaining homeostasis of neuronal tissue. Serum and CSF of LOAD patients showed enhanced level of IL-6. Aβ aggregates induces the glial cells to produce IL-6 which enhances APP transcription, promotes tau hyperphosphorylation through activation of Cdk5 via cdk5/p35 pathway contributing to NFT formation serving as a bridge between AD core pathologies [325].

10.3.3. Tumor Necrotic Factor Alpha (TNF-α)

TNF-α exhibits its biological activity via TNFR1 and TNFR2 receptors which was observed to be overexpressed in the hippocampal tissue and CSF of MCI and AD patients. Aβ aggregates stimulates microglial cells to produce TNF α through activation of NFκB pathway, which induces the pro-inflammatory factors involved in neuronal survival such as calbindin, Mn-SOD enzyme, BCl-2 protein and on contrary also activates glutaminase in microglial cells leading to glutamate induced excitotoxicity promoting neurodegenerative disorders [326]. TNF–α increases Aβ load by upregulating β and γ secretase activity enhancing amyloidogenic processing of APP protein. TNF–α enhances the cell adhesion molecule in vascular endothelial cells facilitating the migration of phagocytic cells and lymphocytes towards stress induced areas promoting inflammatory response [327].

10.3.4. NF-κB

Transcription factor NF-κB acts as primary regulator of inflammation, which is activated in response to proinflammatory signals, TNF-α or IL-1. Aβ activates NF-κB via RAGE widely found in the glial cells and neurons in the vicinity of senile plaques enhancing the release of inflammatory markers which activates microglial cells and astrocytes provoking release of proinflammatory mediators, intensifying inflammation resulting in neurodegeneration. In addition, NF-κB also promotes TNG-α induced β secretase transcription increasing Aβ burden. Several studies revealed that utilization of NF-κB inhibitors and NSAIDs reduce NF-κB activity which lowered the Aβ1-42 level [328,329].

10.4. Anti-Inflammatory Mediators

10.4.1. TGF-β (Tumour Growth Factor- β)

In AD patients, the level TGF-β is enhanced in the CSF, serum and brain microvascular endothelial cells which induces secretion of pro-inflammatory cytokines (IL-1β and TGF-α) [330]. TGF- β1 is the most abundant isoform of TGFβ secreted by astrocytes and its receptors widely distributed in neurons, astrocytes and microglial cells. TGF β1 primarily involves in neuroprotection by inhibiting Aβ production and deposition, regulating neuroinflammation, inhibiting GSK3β thereby attenuating tau hyperphosphorylation and enhancing the expression of antiapoptotic protein Bcl-2 and Bcl-xl [331]. TGF-β1 level is found to be decreased in plasma of AD patients. Deficiency in TGF-β1 induces impairment in TGF-β1 mediated Smad signaling pathway leading to Smad2/3 phosphorylation present in the hippocampal neurons accumulated with NFT and Aβ plaques [332].

10.4.2. IL-10

Interleukin 10 (IL-10) secreted by microglial cells and astrocytes in healthy neurons in response to proinflammatory mediator inhibits cytokines such as IL-1α, IL-1β, TNF-α, IL-6 and MCP-1 to restore the brain homeostasis. IL10 level is increased in AD patients which serves as biomarker for diagnosis and progression of AD. Scientific evidence revealed that in some population, IL-10 polymorphism enhances the risk of AD [333,334].

10.5. Inflammatory Mediators

10.5.1. Cyclooxygenases (COX)

In AD the microglial cells surrounding neuritic plaques (NPs) exhibited increased expression of COX-1 suggesting inflammation, while COX-2 expression in hippocampal CA3 region causes neurotoxicity depending on the level of NFT and Aβ and the observed cognitive impairment [335]. COX-2 increases the γ-secretase activity promoting amyloidogenic processing of APP enhancing the formation of amyloid plaque in parenchyma and prostaglandin E2 synthesis [336]. Increase in caspase-3 immunoreactivity and phosphorylation of retinoblastoma protein causing cell growth suppression was observed in triple transgenic mice model (hAPP/PS1/hCOX-2) due to increased expression of COX-2. The above fact was further substantiated by the primary cultures of cortical and hippocampal neurons derived from transgenic mice exhibited apoptotic mediated cell death [337]. Early stages of AD, Aβ aggregates promote IL-1β to enhance the COX-2 expression leading to synthesis of prostaglandin [338].

10.5.2. Nitric Oxide (NO)

In a healthy state, the expression of inducible NOS (iNOS) is less, however under inflammatory conditions it is enhanced in microglia and astrocytes leading to an increased level of NO provoking oxidative stress mediated neuronal damage, synaptic dysfunction, and apoptosis of neurons [339]. Examination of neuronal tissue of AD patients illustrated that other than iNOS Aβ induced IL-1β and TNF-α also promotes NO and peroxynitrate release, which induces oxidative stress mediated mitochondrial damage and enhanced γ-secretase activity promoting Aβ formation [329,340]. Further, the NO leads to the formation NFT and thereby accelerates the pathogenesis in AD [341].

Overall, the scientific evidence reveals that inflammation exhibits significant role in the initiation and progression of AD pathogenesis. Several hypotheses revealed that production and accumulation of Aβ and inflammation converge and synergize the progression of this neurodegenerative disease. Inflammation starts very early in AD even before the formation of amyloid plaques as innate immune response to clear the Aβ fibrils, which becomes intense during the progression of disease and ends up in cell mediated immunity. In the late inflammatory response, the Aβ plaques induces microglial priming followed by recruitment of glial cells around amyloid plaques with predominant phagocytic activity for the removal of toxic Aβ fragments. Activated astroglial cells, together with peripheral monocytes, invade CNS forming secondary cellular corolla surrounding amyloid plaques. These immune cells produce bulk cytokines and chemokines which along with immune-related molecules such as antibodies, complement, complement-related proteins, MHC proteins, and inflammasome protein complexes activates late inflammatory response of late AD pathology. TREM2-dependent activation of microglia with disease-mitigating properties supports the fact that late AD inflammation represents a tissue-resolution stage. In the end stages of AD, CNS inflammation becomes less relevant as it declines with senescence. Despite much evidence, there remains a knowledge gap on the cells associated with AD and the pathway which link Aβ accumulation and on-going inflammation. Unravelling these mechanisms will help in identifying new therapeutic molecules in combating AD.

Table 1. Receptors associated with inflammatory response in AD.

Model System	CNS Cells Expressing Receptor	Ligands	Role in AD Pathogenesis	References
Complement receptors (CR1, CR3) CD88	Neurons, microglial cells, astrocytes, and oligodendrocyte	C3b, C4b C3 C5a	Neuroinflammation, uptake and clearance of Aβ	[342–344]
FPRL1 and FPRL2	Macrophages, glial cells, astrocytes	Aβ	Proinflammation, Aβ_{42} internalization, formation of fibrillar aggregates, phagocyte chemotaxis and oxidant stress	[251,298,345,346]
Scavenger receptor (SR-A) CD36 (SR-B) LDLR, RAGE, LRP1	Microglia, human monocytes astrocytes Macrophages neurons	Aβ, β-sheet fibrils, HMGB1	Aβ clearance, synaptic dysfunction, neuroinflammation, production of chemokines, and neurotoxic ROS, NLRP3 activation	[248–250,287,288,302,347–351]
Toll-like receptors (TLR2 TLR4, TLR2, TLR9)	Microglia, astrocytes	LPS, Aβ	LTP deficit and neuronal death, neuroinflammation, Aβ uptake and clearance.	[288,289,352–354]
CX3CR1	Microglia, neurons, astrocytes	CX3CL	Neuroprotection against AD. AD patients showed reduced level of CX3CR1 which led to enhanced activation of microglial cells with enhanced tau phosphorylation	[355]
TREM2	Microglia and neurons	Aβ	Microglial depolarization, apoptosis activation of Wnt/β-catenin leads to inflammation.	[356,357]
CD33	Microglia	Aβ	Increased expression of CD33 attenuates Aβ uptake leading facilitating plaque formation	[358]
NALP3/NLRP3	Microglia, macrophage cells	Aβ and other mediators	Enhanced caspase-1 activity leading to IL-1β and IL-18 mediated neuroinflammation	[300–302]

11. Conclusions

To start with review elaborated in detail, the genomics and proteomics modifications of APP and secretases mediated amyloidogenic processing of APP in both cell membrane and cellular organelles leading to the release of insoluble Aβ peptides which tends to aggregate as oligomers and plaques in synaptic junction causing organelle dysfunction and disease progression. Despite of multiple etiological factors, mounting evidence hypothesised that Aβ is the key triggering factor inducing AD pathogenesis via hyperphosphorylation of tau protein, ER stress, Golgi stress/fragmentation, mitochondrial dysfunction, lysosome dysfunction, inflammation, obstruction of the synaptic communication and genomic dysregulation. Current symptomatic treatment approved by FDA for the AD therapeutics involves AChE inhibitors and NMDA receptor antagonist which can only slow down the progression of disease. As AD is a complex disorder involving several biochemical pathways, drugs with multipotent targeting ability are needed for the AD therapy. Several treatment strategies have been proposed which showed positive results in preclinical trials but suffered limitation under human trial due to blood brain barrier reducing the bioavailability of the drug to brain. Hence, an effective treatment strategy for the prevention and cure of AD is still at the developmental stage.

Recent evidence suggests that nutritional supplementation rich in antioxidants, vitamin B12 and folic acid attenuates the fibrillation of Aβ α-synuclein and p-tau, consequently, inhibits Aβ mediated toxicity and attenuates the neuronal inflammation [359–361]. Hence combinatorial therapy of a nutritional diet along with less toxic natural drugs inhibiting Aβ production will be effective for AD therapy. Drugs screened for AD therapy should abide the following properties (i) targeting the amyloidogenic pathway proteins either in genomic/proteomic level, (ii) potentially activating the enzymes, or directly modifying the Aβ hydrophobic properties upon binding, and (iii) developing epitope-specific monoclonal antibodies. However, most of the pharmaceutical industries approaching in the above aspects, observed that the drugs screened were toxic in nature and less efficient due to poor bioavailability. Few drugs have entered successfully into Phase3 clinical trials [362]. Complete understanding of disease pathogenesis, pharmacokinetics and bioavailability of drug is necessary to solve the puzzle in AD therapy. As this review unwinds all the plausible mechanisms leading to AD pathogenesis, it provides new insightsinto identifying the key targets for the treatment of AD.

Author Contributions: R.M.A., N.S., and M.A.A. led the writing of the manuscript, prepared all the figures, and edited the manuscript. A.K., K.S., and D.M. provided the new information upon data collection and contributed to editing the manuscript. B.G., P.P., and G.A. supervised the writing and co-edited the manuscript. All authors have read and agreed to the published version of the manuscript.

Funding: RMA thanks to Science and Engineering Research Board and Indian Council of Medical Research, New Delhi, India for providing financial support through SERB—Overseas Visiting Doctoral Fellowship (ODF/2018/001044/dt. 27 May 2019).

Institutional Review Board Statement: Not applicable.

Informed Consent Statement: Not applicable.

Data Availability Statement: Not applicable.

Acknowledgments: R.M.A.thanks to Science and Engineering Research Board and Indian Council of Medical Research, New Delhi, India for providing financial support through SERB—Overseas Visiting Doctoral Fellowship (ODF/2018/001044/dt. 27 May 2019) and Senior Research Fellow (No. 2017-2853/SCRT-BMS). NS thanks the University Grants Commission, New Delhi, India, for the financial support through UGC-Start-Up Grant [Ref. No. F. 30-381/2017 (BSR), dated 6 July 2017] and RUSA—Phase 2.0 grant (No. F. 24-51/2014-U, Policy (TNMulti-Gen), Dept. of Edn. Govt. of India dated 9 October 2018). DM thanks HCEMM funded by the European Union's Horizon 2020 Research and Innovation Programme under Grant Agreement Number 739593 KSz and DM's neuroinflammation imaging research is financed by the Thematic Excellence Programme (TKP) of the Ministry of Innovation and Technology of Hungary, within the framework of the BIOImaging Excellence Programme at Semmelweis University. Part of this work was supported by EU COST Action CA16122 and the National Research, Development and Innovation Office of Hungary (NKFIA; NVKP-16-1-2016-0017 National Heart Program. BG and PP thanks to Data Science and AI Research (DSAIR) Centre of NTU (Project Number ADH-11/2017-DSAIR), Imaging Probe Development Platform (IPDP) and the support from the Cognitive Neuro Imaging Centre (CONIC) at NTU. GA acknowledges with thanks to University Grants Commission, New Delhi, India for the award of UGC-BSR Faculty Fellow (No. F.18-1/2011(BSR) dt. 4 January 2017). PP and BG acknowledge the support from Lee Kong Chian School of Medicine and Nanyang Technological University, Singapore.

Conflicts of Interest: The authors declare no conflict of interest.

References

1. Tiwari, S.; Atluri, V.; Kaushik, A.; Yndart, A.; Nair, M. Alzheimer's disease: Pathogenesis, diagnostics, and therapeutics. *Int. J. Nanomed.* **2019**, *14*, 5541. [CrossRef]
2. Nichols, E.; Szoeke, C.E.; Vollset, S.E.; Abbasi, N.; Abd-Allah, F.; Abdela, J.; Aichour, M.T.E.; Akinyemi, R.O.; Alahdab, F.; Asgedom, S.W. Global, regional, and national burden of Alzheimer's disease and other dementias, 1990–2016: A systematic analysis for the Global Burden of Disease Study 2016. *Lancet Neurol.* **2019**, *18*, 88–106. [CrossRef]
3. Alzheimer's Association report. Alzheimer's Disease Facts and Figures. *Alzheimers Dement.* **2020**, *16*, 391–460.
4. Albert, M.S.; DeKosky, S.T.; Dickson, D.; Dubois, B.; Feldman, H.H.; Fox, N.C.; Gamst, A.; Holtzman, D.M.; Jagust, W.J.; Petersen, R.C. The diagnosis of mild cognitive impairment due to Alzheimer's disease: Recommendations from the National Institute on Aging-Alzheimer's Association workgroups on diagnostic guidelines for Alzheimer's disease. *Alzheimers Dement.* **2011**, *7*, 270–279. [CrossRef]
5. McKhann, G.M.; Knopman, D.S.; Chertkow, H.; Hyman, B.T.; Jack, C.R., Jr.; Kawas, C.H.; Klunk, W.E.; Koroshetz, W.J.; Manly, J.J.; Mayeux, R. The diagnosis of dementia due to Alzheimer's disease: Recommendations from the National Institute on Aging-Alzheimer's Association workgroups on diagnostic guidelines for Alzheimer's disease. *Alzheimers Dement.* **2011**, *7*, 263–269. [CrossRef] [PubMed]
6. Jack, C.R., Jr.; Albert, M.S.; Knopman, D.S.; McKhann, G.M.; Sperling, R.A.; Carrillo, M.C.; Thies, B.; Phelps, C.H. Introduction to the recommendations from the National Institute on Aging-Alzheimer's Association workgroups on diagnostic guidelines for Alzheimer's disease. *Alzheimers Dement.* **2011**, *7*, 257–262. [CrossRef] [PubMed]

7. Ballard, C.; Khan, Z.; Clack, H.; Corbett, A. Nonpharmacological treatment of Alzheimer disease. *Can. J.Psychiatry* **2011**, *56*, 589–595. [CrossRef]

8. Giri, M.; Zhang, M.; Lü, Y. Genes associated with Alzheimer's disease: An overview and current status. *Clin. Interv. Aging* **2016**, *11*, 665. [CrossRef]

9. Mahley, R.W. Apolipoprotein E: From cardiovascular disease to neurodegenerative disorders. *J. Mol. Med.* **2016**, *94*, 739–746. [CrossRef] [PubMed]

10. Mayeux, R.; Stern, Y. Epidemiology of Alzheimer disease. *Cold Spring Harb. Perspect. Med.* **2012**, *2*, a006239. [CrossRef] [PubMed]

11. Crews, L.; Masliah, E. Molecular mechanisms of neurodegeneration in Alzheimer's disease. *Hum. Mol. Genet.* **2010**, *19*, R12–R20. [CrossRef]

12. Thakur, A.; Kamboj, P.; Goswami, K. Pathophysiology and management of Alzheimer's disease: An overview. *J. Anal. Pharm. Res.* **2018**, *9*, 226–235. [CrossRef]

13. DeTure, M.A.; Dickson, D.W. The neuropathological diagnosis of Alzheimer's disease. *Mol. Neurodegener.* **2019**, *14*, 32. [CrossRef] [PubMed]

14. Kumar, A.; Singh, A. A review on Alzheimer's disease pathophysiology and its management: An update. *Pharmacol. Rep.* **2015**, *67*, 195–203. [CrossRef]

15. Folch, J.; Petrov, D.; Ettcheto, M.; Abad, S.; Sánchez-López, E.; García, M.L.; Olloquequi, J.; Beas-Zarate, C.; Auladell, C.; Camins, A. Current research therapeutic strategies for Alzheimer's disease treatment. *Neural Plast.* **2016**, *2016*, 8501693. [CrossRef]

16. Cao, J.; Hou, J.; Ping, J.; Cai, D. Advances in developing novel therapeutic strategies for Alzheimer's disease. *Mol. Neurodegener.* **2018**, *13*, 64. [CrossRef]

17. Kurz, A.; Perneczky, R. Novel insights for the treatment of Alzheimer's disease. *Prog. Neuro-Psychopharmacol. Biol. Psychiatry* **2011**, *35*, 373–379. [CrossRef]

18. Wasco, W.; Gurubhagavatula, S.; Paradis, M.D.; Romano, D.M.; Sisodia, S.S.; Hyman, B.T.; Neve, R.L.; Tanzi, R.E. Isolation and characterization of APLP2 encoding a homologue of the Alzheimer's associated amyloid β protein precursor. *Nat. Genet.* **1993**, *5*, 95–100. [CrossRef]

19. Pearson, P.; Price, D.; Gearhart, J. Introduction and expression of the 400 kilobase amyloid precursor protein gene in transgenic mice (corrected). *Nat. Genet.* **1993**, *5*, 22–30.

20. Izumi, R.; Yamada, T.; Yoshikai, S.-I.; Sasaki, H.; Hattori, M.; Sakaki, Y. Positive and negative regulatory elements for the expression of the Alzheimer's disease amyloid precursor-encoding gene in mouse. *Gene* **1992**, *112*, 189–195. [CrossRef]

21. Gao, Q.; Dai, Z.; Zhang, S.; Fang, Y.; Yung, K.K.L.; Lo, P.K.; Lai, K.W.C. Interaction of Sp1 and APP promoter elucidates a mechanism for Pb^{2+} caused neurodegeneration. *Arch. Biochem. Biophys.* **2020**, *681*, 108265. [CrossRef] [PubMed]

22. Liu, X.; Yu, X.; Zack, D.J.; Zhu, H.; Qian, J. TiGER: A database for tissue-specific gene expression and regulation. *BMC Bioinform.* **2008**, *9*, 271. [CrossRef]

23. Sisodia, S.S.; Koo, E.H.; Hoffman, P.N.; Perry, G.; Price, D.L. Identification and transport of full-length amyloid precursor proteins in rat peripheral nervous system. *J. Neurosci.* **1993**, *13*, 3136–3142. [CrossRef] [PubMed]

24. Halim, A.; Brinkmalm, G.; Rüetschi, U.; Westman-Brinkmalm, A.; Portelius, E.; Zetterberg, H.; Blennow, K.; Larson, G.; Nilsson, J. Site-specific characterization of threonine, serine, and tyrosine glycosylations of amyloid precursor protein/amyloid β-peptides in human cerebrospinal fluid. *Proc. Natl. Acad. Sci. USA* **2011**, *108*, 11848–11853. [CrossRef]

25. Lee, M.-S.; Kao, S.-C.; Lemere, C.A.; Xia, W.; Tseng, H.-C.; Zhou, Y.; Neve, R.; Ahlijanian, M.K.; Tsai, L.-H. APP processing is regulated by cytoplasmic phosphorylation. *J. Cell Biol.* **2003**, *163*, 83–95. [CrossRef] [PubMed]

26. Bhattacharyya, R.; Barren, C.; Kovacs, D.M. Palmitoylation of amyloid precursor protein regulates amyloidogenic processing in lipid rafts. *J. Neurosci.* **2013**, *33*, 11169–11183. [CrossRef] [PubMed]

27. Yazaki, M.; Tagawa, K.; Maruyama, K.; Sorimachi, H.; Tsuchiya, T.; Ishiura, S.; Suzuki, K. Mutation of potential N-linked glycosylation sites in the Alzheimer's disease amyloid precursor protein (APP). *Neurosci. Lett.* **1996**, *221*, 57–60. [CrossRef]

28. Perdivara, I.; Petrovich, R.; Allinquant, B.; Deterding, L.J.; Tomer, K.B.; Przybylski, M. Elucidation of O-glycosylation structures of the β-amyloid precursor protein by liquid chromatography-mass spectrometry using electron transfer dissociation and collision induced dissociation. *J. Proteome Res.* **2009**, *8*, 631–642. [CrossRef]

29. Jacobsen, K.T.; Iverfeldt, K. O-GlcNAcylation increases non-amyloidogenic processing of the amyloid-β precursor protein (APP). *Biochem. Biophys. Res. Commun.* **2011**, *404*, 882–886. [CrossRef]

30. Walter, J.; Capell, A.; Hung, A.Y.; Langen, H.; Schnölzer, M.; Thinakaran, G.; Sisodia, S.S.; Selkoe, D.J.; Haass, C. Ectodomain phosphorylation of β-amyloid precursor protein at two distinct cellular locations. *J. Biol. Chem.* **1997**, *272*, 1896–1903. [CrossRef]

31. Gandy, S.; Czernik, A.J.; Greengard, P. Phosphorylation of Alzheimer disease amyloid precursor peptide by protein kinase C and Ca^{2+}/calmodulin-dependent protein kinase II. *Proc. Natl. Acad. Sci. USA* **1988**, *85*, 6218–6221. [CrossRef]

32. Buxbaum, J.D.; Gandy, S.E.; Cicchetti, P.; Ehrlich, M.E.; Czernik, A.J.; Fracasso, R.P.; Ramabhadran, T.V.; Unterbeck, A.J.; Greengard, P. Processing of Alzheimer beta/A4 amyloid precursor protein: Modulation by agents that regulate protein phosphorylation. *Proc. Natl. Acad. Sci. USA* **1990**, *87*, 6003–6006. [CrossRef]

33. Rebelo, S.; Vieira, S.I.; Esselmann, H.; Wiltfang, J.; da Cruz e Silva, E.F.; da Cruz e Silva, O.A.B. Tyrosine 687 phosphorylated Alzheimer's amyloid precursor protein is retained intracellularly and exhibits a decreased turnover rate. *Neurodegener. Dis.* **2007**, *4*, 78–87. [CrossRef] [PubMed]

34. Zhang, Y.-Q.; Sarge, K.D. Sumoylation of amyloid precursor protein negatively regulates Aβ aggregate levels. *Biochem. Biophys. Res. Commun.* **2008**, *374*, 673–678. [CrossRef] [PubMed]

35. Kaneko, M.; Koike, H.; Saito, R.; Kitamura, Y.; Okuma, Y.; Nomura, Y. Loss of HRD1-mediated protein degradation causes amyloid precursor protein accumulation and amyloid-β generation. *J. Neurosci.* **2010**, *30*, 3924–3932. [CrossRef] [PubMed]

36. Morel, E.; Chamoun, Z.; Lasiecka, Z.M.; Chan, R.B.; Williamson, R.L.; Vetanovetz, C.; Dall'Armi, C.; Simoes, S.; Du Jour, K.S.P.; McCabe, B.D. Phosphatidylinositol-3-phosphate regulates sorting and processing of amyloid precursor protein through the endosomal system. *Nat. Commun.* **2013**, *4*, 2250. [CrossRef] [PubMed]

37. Weidemann, A.; König, G.; Bunke, D.; Fischer, P.; Salbaum, J.M.; Masters, C.L.; Beyreuther, K. Identification, biogenesis, and localization of precursors of Alzheimer's disease A4 amyloid protein. *Cell* **1989**, *57*, 115–126. [CrossRef]

38. Deolankar, S.C.; Patil, A.H.; Koyangana, S.G.; Subbannayya, Y.; Prasad, T.S.K.; Modi, P.K. Dissecting Alzheimer's disease molecular substrates by proteomics and discovery of novel post-translational modifications. *Omics J. Integr. Biol.* **2019**, *23*, 350–361. [CrossRef]

39. Osaki, D.; Hiramatsu, H. Itrullination and deamidation affect aggregation properties of amyloid β-proteins. *Amyloid* **2016**, *23*, 234–241. [CrossRef] [PubMed]

40. Zheng, H.; Koo, E.H. Biology and pathophysiology of the amyloid precursor protein. *Mol. Neurodegener.* **2011**, *6*, 27. [CrossRef]

41. Müller, U.C.; Deller, T.; Korte, M. Not just amyloid: Physiological functions of the amyloid precursor protein family. *Nat. Rev. Neurosci.* **2017**, *18*, 281–298. [CrossRef]

42. Yuan, X.-Z.; Sun, S.; Tan, C.-C.; Yu, J.-T.; Tan, L. The role of ADAM10 in Alzheimer's disease. *J. Alzheimers Dis.* **2017**, *58*, 303–322. [CrossRef]

43. Zhang, X.; Song, W. The role of APP and BACE1 trafficking in APP processing and amyloid-β generation. *Alzheimers Res. Ther.* **2013**, *5*, 46. [CrossRef]

44. Zhang, S.; Wang, Z.; Cai, F.; Zhang, M.; Wu, Y.; Zhang, J.; Song, W. BACE1 cleavage site selection critical for amyloidogenesis and Alzheimer's pathogenesis. *J. Neurosci.* **2017**, *37*, 6915–6925. [CrossRef] [PubMed]

45. Sun, X.; He, G.; Song, W. BACE2, as a novel APP θ-secretase, is not responsible for the pathogenesis of Alzheimer's disease in Down syndrome. *FASEB J.* **2006**, *20*, 1369–1376. [CrossRef] [PubMed]

46. Azkona, G.; Levannon, D.; Groner, Y.; Dierssen, M. In vivo effects of APP are not exacerbated by BACE2 co-overexpression: Behavioural characterization of a double transgenic mouse model. *Amino Acids* **2010**, *39*, 1571–1580. [CrossRef]

47. Zhang, Z.; Nadeau, P.; Song, W.; Donoviel, D.; Yuan, M.; Bernstein, A.; Yankner, B.A. Presenilins are required for γ-secretase cleavage of β-APP and transmembrane cleavage of Notch-1. *Nat. Cell Biol.* **2000**, *2*, 463–465. [CrossRef] [PubMed]

48. Zhang, S.; Zhang, M.; Cai, F.; Song, W. Biological function of Presenilin and its role in AD pathogenesis. *Transl. Neurodegener.* **2013**, *2*, 15. [CrossRef]

49. He, G.; Luo, W.; Li, P.; Remmers, C.; Netzer, W.J.; Hendrick, J.; Bettayeb, K.; Flajolet, M.; Gorelick, F.; Wennogle, L.P. Gamma-secretase activating protein is a therapeutic target for Alzheimer's disease. *Nature* **2010**, *467*, 95–98. [CrossRef] [PubMed]

50. Wong, E.; Liao, G.P.; Chang, J.C.; Xu, P.; Li, Y.-M.; Greengard, P. GSAP modulates γ-secretase specificity by inducing conformational change in PS1. *Proc. Natl. Acad. Sci. USA* **2019**, *116*, 6385–6390. [CrossRef] [PubMed]

51. Angira, D.; Chikhale, R.; Mehta, K.; Bryce, R.A.; Thiruvenkatam, V. Tracing the GSAP–APP C-99 Interaction Site in the β-Amyloid Pathway Leading to Alzheimer's Disease. *ACS Chem. Neurosci.* **2019**, *10*, 3868–3879. [CrossRef] [PubMed]

52. Liu, K.; Doms, R.W.; Lee, V.M.-Y. Glu11 site cleavage and N-terminally truncated Aβ production upon BACE overexpression. *Biochemistry* **2002**, *41*, 3128–3136. [CrossRef] [PubMed]

53. Miravalle, L.; Calero, M.; Takao, M.; Roher, A.E.; Ghetti, B.; Vidal, R. Amino-terminally truncated Aβ peptide species are the main component of cotton wool plaques. *Biochemistry* **2005**, *44*, 10810–10821. [CrossRef]

54. Siegel, G.; Gerber, H.; Koch, P.; Bruestle, O.; Fraering, P.C.; Rajendran, L. The Alzheimer's disease γ-secretase generates higher 42: 40 ratios for β-amyloid than for p3 peptides. *Cell Rep.* **2017**, *19*, 1967–1976. [CrossRef]

55. Willem, M.; Tahirovic, S.; Busche, M.A.; Ovsepian, S.V.; Chafai, M.; Kootar, S.; Hornburg, D.; Evans, L.D.; Moore, S.; Daria, A. η-Secretase processing of APP inhibits neuronal activity in the hippocampus. *Nature* **2015**, *526*, 443–447. [CrossRef]

56. Xu, T.-H.; Yan, Y.; Kang, Y.; Jiang, Y.; Melcher, K.; Xu, H.E. Alzheimer's disease-associated mutations increase amyloid precursor protein resistance to γ-secretase cleavage and the Aβ42/Aβ40 ratio. *Cell Discov.* **2016**, *2*, 16026. [CrossRef]

57. Sikanyika, N.L.; Parkington, H.C.; Smith, A.I.; Kuruppu, S. Powering amyloid beta degrading enzymes: A possible therapy for Alzheimer's disease. *Neurochem. Res.* **2019**, *44*, 1289–1296. [CrossRef]

58. Pasternak, S.H.; Bagshaw, R.D.; Guiral, M.; Zhang, S.; Ackerley, C.A.; Pak, B.J.; Callahan, J.W.; Mahuran, D.J. Presenilin-1, nicastrin, amyloid precursor protein, and γ-secretase activity are co-localized in the lysosomal membrane. *J. Biol. Chem.* **2003**, *278*, 26687–26694. [CrossRef] [PubMed]

59. Schreiner, B.; Hedskog, L.; Wiehager, B.; Ankarcrona, M. Amyloid-β peptides are generated in mitochondria-associated endoplasmic reticulum membranes. *J. Alzheimers Dis.* **2015**, *43*, 369–374. [CrossRef] [PubMed]

60. Choy, R.W.Y.; Cheng, Z.; Schekman, R. Amyloid precursor protein (APP) traffics from the cell surface via endosomes for amyloid β (Aβ) production in the trans-Golgi network. *Proc. Natl. Acad. Sci. USA* **2012**, *109*, E2077–E2082. [CrossRef]

61. Fiandaca, M.S.; Kapogiannis, D.; Mapstone, M.; Boxer, A.; Eitan, E.; Schwartz, J.B.; Abner, E.L.; Petersen, R.C.; Federoff, H.J.; Miller, B.L.; et al. Identification of preclinical Alzheimer's disease by a profile of pathogenic proteins in neurally derived blood exosomes: A case-control study. *Alzheimers Dement.* **2015**, *11*, 600–607. [CrossRef] [PubMed]

62. Perez-Gonzalez, R.; Gauthier, S.A.; Kumar, A.; Levy, E. The exosome secretory pathway transports amyloid precursor protein carboxyl-terminal fragments from the cell into the brain extracellular space. *J. Biol. Chem.* **2012**, *287*, 43108–43115. [CrossRef] [PubMed]

63. Sinha, M.S.; Ansell-Schultz, A.; Civitelli, L.; Hildesjö, C.; Larsson, M.; Lannfelt, L.; Ingelsson, M.; Hallbeck, M. Alzheimer's disease pathology propagation by exosomes containing toxic amyloid-beta oligomers. *Acta Neuropathol.* **2018**, *136*, 41–56. [CrossRef]

64. Martin, B.L.; Schrader-Fischer, G.; Busciglio, J.; Duke, M.; Paganetti, P.; Yankner, B.A. Intracellular accumulation of β-amyloid in cells expressing the Swedish mutant amyloid precursor protein. *J. Biol. Chem.* **1995**, *270*, 26727–26730. [CrossRef]

65. Rogaeva, E.; Meng, Y.; Lee, J.H.; Gu, Y.; Kawarai, T.; Zou, F.; Katayama, T.; Baldwin, C.T.; Cheng, R.; Hasegawa, H. The neuronal sortilin-related receptor SORL1 is genetically associated with Alzheimer disease. *Nat. Genet.* **2007**, *39*, 168–177. [CrossRef] [PubMed]

66. Kinoshita, A.; Whelan, C.; Smith, C.; Berezovska, O.; Hyman, B. Direct visualization of the gamma secretase-generated carboxyl-terminal domain of the amyloid precursor protein: Association with Fe65 and translocation to the nucleus. *J. Neurochem.* **2002**, *82*, 839–847. [CrossRef]

67. Gyure, K.A.; Durham, R.; Stewart, W.F.; Smialek, J.E.; Troncoso, J.C. Intraneuronal Aβ-amyloid precedes development of amyloid plaques in Down syndrome. *Arch. Pathol. Lab. Med.* **2001**, *125*, 489–492. [CrossRef]

68. Walsh, D.M.; Klyubin, I.; Fadeeva, J.V.; Cullen, W.K.; Anwyl, R.; Wolfe, M.S.; Rowan, M.J.; Selkoe, D.J. Naturally secreted oligomers of amyloid β protein potently inhibit hippocampal long-term potentiation in vivo. *Nature* **2002**, *416*, 535–539. [CrossRef]

69. Cleary, J.P.; Walsh, D.M.; Hofmeister, J.J.; Shankar, G.M.; Kuskowski, M.A.; Selkoe, D.J.; Ashe, K.H. Natural oligomers of the amyloid-β protein specifically disrupt cognitive function. *Nat. Neurosci.* **2005**, *8*, 79–84. [CrossRef]

70. Wilcox, K.C.; Marunde, M.R.; Das, A.; Velasco, P.T.; Kuhns, B.D.; Marty, M.T.; Jiang, H.; Luan, C.-H.; Sligar, S.G.; Klein, W.L. Nanoscale synaptic membrane mimetic allows unbiased high throughput screen that targets binding sites for Alzheimer's-associated Aβ oligomers. *PLoS ONE* **2015**, *10*, e0125263. [CrossRef]

71. Yang, W.; Ma, K.; Chen, X.; Shi, L.; Bu, G.; Hu, X.; Han, H.; Liu, Y.; Qian, Y. Mitogen-activated protein kinase signaling pathways are involved in regulating α7 nicotinic acetylcholine receptor-mediated amyloid-β uptake in SH-SY5Y cells. *Neuroscience* **2014**, *278*, 276–290. [CrossRef]

72. Bu, G.; Cam, J.; Zerbinatti, C. LRP in Amyloid-β Production and Metabolism. *Ann. N. Y. Acad. Sci.* **2006**, *1086*, 35–53. [CrossRef]

73. Iribarren, P.; Zhou, Y.; Hu, J.; Le, Y.; Wang, J.M. Role of formyl peptide receptor-like 1 (FPRL1/FPR2) in mononuclear phagocyte responses in Alzheimer disease. *Immunol. Res.* **2005**, *31*, 165–176. [CrossRef]

74. Snyder, E.M.; Nong, Y.; Almeida, C.G.; Paul, S.; Moran, T.; Choi, E.Y.; Nairn, A.C.; Salter, M.W.; Lombroso, P.J.; Gouras, G.K. Regulation of NMDA receptor trafficking by amyloid-β. *Nat. Neurosci.* **2005**, *8*, 1051–1058. [CrossRef]

75. Takuma, K.; Fang, F.; Zhang, W.; Yan, S.; Fukuzaki, E.; Du, H.; Sosunov, A.; McKhann, G.; Funatsu, Y.; Nakamichi, N. RAGE-mediated signaling contributes to intraneuronal transport of amyloid-β and neuronal dysfunction. *Proc. Natl. Acad. Sci. USA* **2009**, *106*, 20021–20026. [CrossRef] [PubMed]

76. Waschuk, S.A.; Elton, E.A.; Darabie, A.A.; Fraser, P.E.; McLaurin, J. Cellular membrane composition defines Aβ-lipid interactions. *J. Biol. Chem.* **2001**, *276*, 33561–33568. [CrossRef] [PubMed]

77. Kim, S.I.; Yi, J.S.; Ko, Y.G. Amyloid β oligomerization is induced by brain lipid rafts. *J. Cell. Biochem.* **2006**, *99*, 878–889. [CrossRef] [PubMed]

78. Kawarabayashi, T.; Shoji, M.; Younkin, L.H.; Wen-Lang, L.; Dickson, D.W.; Murakami, T.; Matsubara, E.; Abe, K.; Ashe, K.H.; Younkin, S.G. Dimeric amyloid β protein rapidly accumulates in lipid rafts followed by apolipoprotein E and phosphorylated tau accumulation in the Tg2576 mouse model of Alzheimer's disease. *J. Neurosci.* **2004**, *24*, 3801–3809. [CrossRef] [PubMed]

79. LaFerla, F.M.; Green, K.N.; Oddo, S. Intracellular amyloid-β in Alzheimer's disease. *Nat. Rev. Neurosci.* **2007**, *8*, 499–509. [CrossRef] [PubMed]

80. Tackenberg, C.; Nitsch, R.M. The secreted APP ectodomain sAPPα, but not sAPPβ, protects neurons against Aβ oligomer-induced dendritic spine loss and increased tau phosphorylation. *Mol. Brain* **2019**, *12*, 27. [CrossRef] [PubMed]

81. Deng, Y.; Xiong, Z.; Chen, P.; Wei, J.; Chen, S.; Yan, Z. β-Amyloid impairs the regulation of N-methyl-D-aspartate receptors by glycogen synthase kinase 3. *Neurobiol. Aging* **2014**, *35*, 449–459. [CrossRef]

82. Gakhar-Koppole, N.; Hundeshagen, P.; Mandl, C.; Weyer, S.W.; Allinquant, B.; Müller, U.; Ciccolini, F. Activity requires soluble amyloid precursor protein α to promote neurite outgrowth in neural stem cell-derived neurons via activation of the MAPK pathway. *Eur. J. Neurosci.* **2008**, *28*, 871–882. [CrossRef]

83. Fol, R.; Braudeau, J.; Ludewig, S.; Abel, T.; Weyer, S.W.; Roederer, J.-P.; Brod, F.; Audrain, M.; Bemelmans, A.-P.; Buchholz, C.J. Viral gene transfer of APPsα rescues synaptic failure in an Alzheimer's disease mouse model. *Acta Neuropathol.* **2016**, *131*, 247–266. [CrossRef]

84. Mandelkow, E.-M.; Mandelkow, E. Biochemistry and cell biology of tau protein in neurofibrillary degeneration. *Cold Spring Harb. Perspect. Med.* **2012**, *2*, a006247. [CrossRef]

85. Woerman, A.L.; Aoyagi, A.; Patel, S.; Kazmi, S.A.; Lobach, I.; Grinberg, L.T.; McKee, A.C.; Seeley, W.W.; Olson, S.H.; Prusiner, S.B. Tau prions from Alzheimer's disease and chronic traumatic encephalopathy patients propagate in cultured cells. *Proc. Natl. Acad. Sci. USA* **2016**, *113*, E8187–E8196. [CrossRef]

86. Friedrich, M.G.; Skora, A.; Hancock, S.E.; Mitchell, T.W.; Else, P.L.; Truscott, R.J. Tau Is Truncated in Five Regions of the Normal Adult Human Brain. *Int. J. Mol. Sci.* **2021**, *22*, 3521. [CrossRef]

87. Hagestedt, T.; Lichtenberg, B.; Wille, H.; Mandelkow, E.; Mandelkow, E. Tau protein becomes long and stiff upon phosphorylation: Correlation between paracrystalline structure and degree of phosphorylation. *J. Cell Biol.* **1989**, *109*, 1643–1651. [CrossRef]

88. Brion, J.P.; Smith, C.; Couck, A.M.; Gallo, J.M.; Anderton, B.H. Developmental Changes in τ Phosphorylation: Fetal τ Is Transiently Phosphorylated in a Manner Similar to Paired Helical Filament-τ Characteristic of Alzheimer's Disease. *J. Neurochem.* **1993**, *61*, 2071–2080. [CrossRef] [PubMed]

89. Ebashi, M.; Toru, S.; Nakamura, A.; Kamei, S.; Yokota, T.; Hirokawa, K.; Uchihara, T. Detection of AD-specific four repeat tau with deamidated asparagine residue 279-specific fraction purified from 4R tau polyclonal antibody. *Acta Neuropathol.* **2019**, *138*, 163–166. [CrossRef] [PubMed]

90. Reynolds, C.H.; Garwood, C.J.; Wray, S.; Price, C.; Kellie, S.; Perera, T.; Zvelebil, M.; Yang, A.; Sheppard, P.W.; Varndell, I.M. Phosphorylation regulates tau interactions with Src homology 3 domains of phosphatidylinositol 3-kinase, phospholipase Cγ1, Grb2, and Src family kinases. *J. Biol. Chem.* **2008**, *283*, 18177–18186. [CrossRef] [PubMed]

91. Rossi, G.; Dalprà, L.; Crosti, F.; Lissoni, S.; Sciacca, F.L.; Catania, M.; Mangieri, M.; Giaccone, G.; Croci, D.; Tagliavini, F. A new function of microtubule-associated protein tau: Involvement in chromosome stability. *Cell Cycle* **2008**, *7*, 1788–1794. [CrossRef]

92. Wang, J.Z.; Grundke-Iqbal, I.; Iqbal, K. Kinases and phosphatases and tau sites involved in Alzheimer neurofibrillary degeneration. *Eur. J. Neurosci.* **2007**, *25*, 59–68. [CrossRef]

93. Burns, L.H.; Wang, H.-Y. Altered filamin A enables amyloid beta-induced tau hyperphosphorylation and neuroinflammation in Alzheimer's disease. *Neuroimmunol. Neuroinflamm.* **2017**, *4*, 263–271. [CrossRef]

94. Patzke, H.; Tsai, L.-H. Calpain-mediated cleavage of the cyclin-dependent kinase-5 activator p39 to p29. *J. Biol. Chem.* **2002**, *277*, 8054–8060. [CrossRef]

95. Plattner, F.; Angelo, M.; Giese, K.P. The roles of cyclin-dependent kinase 5 and glycogen synthase kinase 3 in tau hyperphosphorylation. *J. Biol. Chem.* **2006**, *281*, 25457–25465. [CrossRef]

96. Hernández, F.; Borrell, J.; Guaza, C.; Avila, J.; Lucas, J.J. Spatial learning deficit in transgenic mice that conditionally over-express GSK-3β in the brain but do not form tau filaments. *J. Neurochem.* **2002**, *83*, 1529–1533. [CrossRef] [PubMed]

97. King, M.E.; Kan, H.-M.; Baas, P.W.; Erisir, A.; Glabe, C.G.; Bloom, G.S. Tau-dependent microtubule disassembly initiated by prefibrillar β-amyloid. *J. Cell Biol.* **2006**, *175*, 541–546. [CrossRef] [PubMed]

98. Wang, J.Z.; Wu, Q.; Smith, A.; Grundke-Iqbal, I.; Iqbal, K. τ is phosphorylated by GSK-3 at several sites found in Alzheimer disease and its biological activity markedly inhibited only after it is prephosphorylated by A-kinase. *FEBS Lett.* **1998**, *436*, 28–34. [CrossRef]

99. Gu, G.J.; Lund, H.; Wu, D.; Blokzijl, A.; Classon, C.; von Euler, G.; Landegren, U.; Sunnemark, D.; Kamali-Moghaddam, M. Role of individual MARK isoforms in phosphorylation of tau at Ser 262 in Alzheimer's disease. *Neuromol. Med.* **2013**, *15*, 458–469. [CrossRef]

100. Guise, S.; Braguer, D.; Carles, G.; Delacourte, A.; Briand, C. Hyperphosphorylation of tau is mediated by ERK activation during anticancer drug-induced apoptosis in neuroblastoma cells. *J. Neurosci. Res.* **2001**, *63*, 257–267. [CrossRef]

101. Li, T.; Paudel, H.K. Glycogen synthase kinase 3β phosphorylates Alzheimer's disease-specific Ser396 of microtubule-associated protein tau by a sequential mechanism. *Biochemistry* **2006**, *45*, 3125–3133. [CrossRef] [PubMed]

102. Cho, J.H.; Johnson, G.V. Primed phosphorylation of tau at Thr231 by glycogen synthase kinase 3β (GSK3β) plays a critical role in regulating tau's ability to bind and stabilize microtubules. *J. Neurochem.* **2004**, *88*, 349–358. [CrossRef]

103. Phiel, C.J.; Wilson, C.A.; Lee, V.M.-Y.; Klein, P.S. GSK-3α regulates production of Alzheimer's disease amyloid-β peptides. *Nature* **2003**, *423*, 435–439. [CrossRef]

104. Sotiropoulos, I.; Catania, C.; Pinto, L.G.; Silva, R.; Pollerberg, G.E.; Takashima, A.; Sousa, N.; Almeida, O.F. Stress acts cumulatively to precipitate Alzheimer's disease-like tau pathology and cognitive deficits. *J. Neurosci.* **2011**, *31*, 7840–7847. [CrossRef]

105. Chow, H.-M.; Guo, D.; Zhou, J.-C.; Zhang, G.-Y.; Li, H.-F.; Herrup, K.; Zhang, J. CDK5 activator protein p25 preferentially binds and activates GSK3β. *Proc. Natl. Acad. Sci. USA* **2014**, *111*, E4887–E4895. [CrossRef]

106. Shen, L.L.; Li, W.W.; Xu, Y.L.; Gao, S.H.; Xu, M.Y.; Bu, X.L.; Liu, Y.H.; Wang, J.; Zhu, J.; Zeng, F. Neurotrophin receptor p75 mediates amyloid β-induced tau pathology. *Neurobiol. Dis.* **2019**, *132*, 104567. [CrossRef]

107. Wang, J.-Z.; Gong, C.-X.; Zaidi, T.; Grundke-Iqbal, I.; Iqbal, K. Dephosphorylation of Alzheimer paired helical filaments by protein phosphatase-2A and− 2B. *J. Biol. Chem.* **1995**, *270*, 4854–4860. [CrossRef]

108. Tanimukai, H.; Grundke-Iqbal, I.; Iqbal, K. Up-regulation of inhibitors of protein phosphatase-2A in Alzheimer's disease. *Am. J. Pathol.* **2005**, *166*, 1761–1771. [CrossRef]

109. Yu, Y.; Run, X.; Liang, Z.; Li, Y.; Liu, F.; Liu, Y.; Iqbal, K.; Grundke-Iqbal, I.; Gong, C.X. Developmental regulation of tau phosphorylation, tau kinases, and tau phosphatases. *J. Neurochem.* **2009**, *108*, 1480–1494. [CrossRef] [PubMed]

110. Bolmont, T.; Clavaguera, F.; Meyer-Luehmann, M.; Herzig, M.C.; Radde, R.; Staufenbiel, M.; Lewis, J.; Hutton, M.; Tolnay, M.; Jucker, M. Induction of tau pathology by intracerebral infusion of amyloid-β-containing brain extract and by amyloid-β deposition in APP× Tau transgenic mice. *Am. J. Pathol.* **2007**, *171*, 2012–2020. [CrossRef] [PubMed]

111. Nisbet, R.M.; Polanco, J.-C.; Ittner, L.M.; Götz, J. Tau aggregation and its interplay with amyloid-β. *Acta Neuropathol.* **2015**, *129*, 207–220. [CrossRef]

112. Penke, B.; Szűcs, M.; Bogár, F. Oligomerization and Conformational Change Turn Monomeric β-Amyloid and Tau Proteins Toxic: Their Role in Alzheimer's Pathogenesis. *Molecules* **2020**, *25*, 1659. [CrossRef]

113. Šimić, G.; Babić Leko, M.; Wray, S.; Harrington, C.; Delalle, I.; Jovanov-Milošević, N.; Bažadona, D.; Buée, L.; De Silva, R.; Di Giovanni, G. Tau protein hyperphosphorylation and aggregation in Alzheimer's disease and other tauopathies, and possible neuroprotective strategies. *Biomolecules* **2016**, *6*, 6. [CrossRef]

114. Alonso, A.d.C.; Grundke-Iqbal, I.; Iqbal, K. Alzheimer's disease hyperphosphorylated tau sequesters normal tau into tangles of filaments and disassembles microtubules. *Nat. Med.* **1996**, *2*, 783–787. [CrossRef]

115. Elie, A.; Prezel, E.; Guérin, C.; Denarier, E.; Ramirez-Rios, S.; Serre, L.; Andrieux, A.; Fourest-Lieuvin, A.; Blanchoin, L.; Arnal, I. Tau co-organizes dynamic microtubule and actin networks. *Sci. Rep.* **2015**, *5*, 9964. [CrossRef] [PubMed]

116. Gómez-Ramos, A.; Díaz-Hernández, M.; Cuadros, R.; Hernández, F.; Avila, J. Extracellular tau is toxic to neuronal cells. *FEBS Lett.* **2006**, *580*, 4842–4850. [CrossRef] [PubMed]

117. Morozova, V.; Cohen, L.S.; Makki, A.E.-H.; Shur, A.; Pilar, G.; ElIdrissi, A.; Alonso, A.D. Normal & Pathological Tau Uptake Mediated by M1/M3 Muscarinic Receptors Promotes Opposite Neuronal Changes. *Front. Cell. Neurosci.* **2019**, *13*, 403. [PubMed]

118. Alonso, A.D.; Cohen, L.S.; Morozova, V. The tau misfolding pathway to dementia. In *Protein Folding Disorders in the Central Nervous System*; Ghiso, J., Rostagno, A., Eds.; World Scientific Publishing: New York, NY, USA, 2017; pp. 83–107.

119. Cripps, D.; Thomas, S.N.; Jeng, Y.; Yang, F.; Davies, P.; Yang, A.J. Alzheimer disease-specific conformation of hyperphosphorylated paired helical filament-Tau is polyubiquitinated through Lys-48, Lys-11, and Lys-6 ubiquitin conjugation. *J. Biol. Chem.* **2006**, *281*, 10825–10838. [CrossRef] [PubMed]

120. Reddy, P.H. Mitochondrial dysfunction in aging and Alzheimer's disease: Strategies to protect neurons. *Antioxid. Redox Signal.* **2007**, *9*, 1647–1658. [CrossRef]

121. Massaad, C.A.; Pautler, R.G.; Klann, E. Mitochondrial superoxide: A key player in Alzheimer's disease. *Aging* **2009**, *1*, 758. [CrossRef]

122. Kincaid, B.; Bossy-Wetzel, E. Forever young: SIRT3 a shield against mitochondrial meltdown, aging, and neurodegeneration. *Front. Aging Neurosci.* **2013**, *5*, 48. [CrossRef] [PubMed]

123. Zhou, Y.; Wang, S.; Li, Y.; Yu, S.; Zhao, Y. SIRT1/PGC-1α signaling promotes mitochondrial functional recovery and reduces apoptosis after intracerebral hemorrhage in rats. *Front. Mol. Neurosci.* **2018**, *10*, 443. [CrossRef]

124. Cheng, Y.; Ren, X.; Gowda, A.S.; Shan, Y.; Zhang, L.; Yuan, Y.; Patel, R.; Wu, H.; Huber-Keener, K.; Yang, J. Interaction of Sirt3 with OGG1 contributes to repair of mitochondrial DNA and protects from apoptotic cell death under oxidative stress. *Cell Death Dis.* **2013**, *4*, e731. [CrossRef] [PubMed]

125. Jęśko, H.; Wencel, P.; Strosznajder, R.P.; Strosznajder, J.B. Sirtuins and their roles in brain aging and neurodegenerative disorders. *Neurochem. Res.* **2017**, *42*, 876–890. [CrossRef] [PubMed]

126. Youle, R.J.; Van Der Bliek, A.M. Mitochondrial fission, fusion, and stress. *Science* **2012**, *337*, 1062–1065. [CrossRef] [PubMed]

127. Swerdlow, R.H. Mitochondria in cybrids containing mtDNA from persons with mitochondriopathies. *J. Neurosci. Res.* **2007**, *85*, 3416–3428. [CrossRef]

128. Hirai, K.; Aliev, G.; Nunomura, A.; Fujioka, H.; Russell, R.L.; Atwood, C.S.; Johnson, A.B.; Kress, Y.; Vinters, H.V.; Tabaton, M. Mitochondrial abnormalities in Alzheimer's disease. *J. Neurosci.* **2001**, *21*, 3017–3023. [CrossRef]

129. Szablewski, L. Glucose transporters in brain: In health and in Alzheimer's disease. *J. Alzheimers Dis.* **2017**, *55*, 1307–1320. [CrossRef]

130. Sonntag, K.-C.; Ryu, W.-I.; Amirault, K.M.; Healy, R.A.; Siegel, A.J.; McPhie, D.L.; Forester, B.; Cohen, B.M. Late-onset Alzheimer's disease is associated with inherent changes in bioenergetics profiles. *Sci. Rep.* **2017**, *7*, 14038. [CrossRef]

131. Manczak, M.; Park, B.S.; Jung, Y.; Reddy, P.H. Differential expression of oxidative phosphorylation genes in patients with Alzheimer's disease. *Neuromol. Med.* **2004**, *5*, 147–162. [CrossRef]

132. Caspersen, C.; Wang, N.; Yao, J.; Sosunov, A.; Chen, X.; Lustbader, J.W.; Xu, H.W.; Stern, D.; McKhann, G.; Du Yan, S. Mitochondrial Aβ: A potential focal point for neuronal metabolic dysfunction in Alzheimer's disease. *FASEB J.* **2005**, *19*, 2040–2041. [CrossRef]

133. Gillardon, F.; Rist, W.; Kussmaul, L.; Vogel, J.; Berg, M.; Danzer, K.; Kraut, N.; Hengerer, B. Proteomic and functional alterations in brain mitochondria from Tg2576 mice occur before amyloid plaque deposition. *Proteomics* **2007**, *7*, 605–616. [CrossRef] [PubMed]

134. Eckert, A.; Hauptmann, S.; Scherping, I.; Rhein, V.; Müller-Spahn, F.; Götz, J.; Müller, W.E. Soluble beta-amyloid leads to mitochondrial defects in amyloid precursor protein and tau transgenic mice. *Neurodegener. Dis.* **2008**, *51*, 57–159. [CrossRef] [PubMed]

135. Eckert, A.; Steiner, B.; Marques, C.; Leutz, S.; Romig, H.; Haass, C.; Müller, W.E. Elevated vulnerability to oxidative stress-induced cell death and activation of caspase-3 by the Swedish amyloid precursor protein mutation. *J. Neurosci. Res.* **2001**, *64*, 183–192. [CrossRef]

136. Marques, C.A.; Keil, U.; Bonert, A.; Steiner, B.; Haass, C.; Müller, W.E.; Eckert, A. Neurotoxic Mechanisms Caused by the Alzheimer's Disease-linked Swedish Amyloid Precursor Protein Mutation oxidative stress, caspases, and the jnk pathway. *J. Biol. Chem.* **2003**, *278*, 28294–28302. [CrossRef] [PubMed]

137. Wang, X.; Su, B.; Siedlak, S.L.; Moreira, P.I.; Fujioka, H.; Wang, Y.; Casadesus, G.; Zhu, X. Amyloid-β overproduction causes abnormal mitochondrial dynamics via differential modulation of mitochondrial fission/fusion proteins. *Proc. Natl. Acad. Sci. USA* **2008**, *105*, 19318–19323. [CrossRef]

138. Cieślik, M.; Czapski, G.A.; Wójtowicz, S.; Wieczorek, I.; Wencel, P.L.; Strosznajder, R.P.; Jaber, V.; Lukiw, W.J.; Strosznajder, J.B. Alterations of Transcription of Genes Coding Anti-oxidative and Mitochondria-Related Proteins in Amyloid β Toxicity: Relevance to Alzheimer's Disease. *Mol. Neurobiol.* **2020**, *57*, 1374–1388. [CrossRef] [PubMed]

139. Singulani, M.P.; Pereira, C.P.M.; Ferreira, A.F.F.; Garcia, P.C.; Ferrari, G.D.; Alberici, L.C.; Britto, L.R. Impairment of PGC-1α-mediated mitochondrial biogenesis precedes mitochondrial dysfunction and Alzheimer's pathology in the 3xTg mouse model of Alzheimer's disease. *Exp. Gerontol.* **2020**, *133*, 110882. [CrossRef]

140. Manczak, M.; Reddy, P.H. Abnormal interaction between the mitochondrial fission protein Drp1 and hyperphosphorylated tau in Alzheimer's disease neurons: Implications for mitochondrial dysfunction and neuronal damage. *Hum. Mol. Genet.* **2012**, *21*, 2538–2547. [CrossRef]

141. Hayashi, T.; Rizzuto, R.; Hajnoczky, G.; Su, T.P. MAM: More than just a housekeeper. *Trends Cell Biol.* **2009**, *19*, 81–88. [CrossRef]

142. Yu, J.; Novgorodov, S.A.; Chudakova, D.; Zhu, H.; Bielawska, A.; Bielawski, J.; Obeid, L.M.; Kindy, M.S.; Gudz, T.I. JNK3 signaling pathway activates ceramide synthase leading to mitochondrial dysfunction. *J. Biol. Chem.* **2007**, *282*, 25940–25949. [CrossRef] [PubMed]

143. Pera, M.; Larrea, D.; Guardia-Laguarta, C.; Montesinos, J.; Velasco, K.R.; Agrawal, R.R.; Xu, Y.; Chan, R.B.; Di Paolo, G.; Mehler, M.F. Increased localization of APP-C99 in mitochondria-associated ER membranes causes mitochondrial dysfunction in Alzheimer disease. *EMBO J.* **2017**, *36*, 3356–3371. [CrossRef] [PubMed]

144. Tillement, L.; Lecanu, L.; Yao, W.; Greeson, J.; Papadopoulos, V. The spirostenol (22R, 25R)-20α-spirost-5-en-3β-yl hexanoate blocks mitochondrial uptake of Aβ in neuronal cells and prevents Aβ-induced impairment of mitochondrial function. *Steroids* **2006**, *71*, 725–735. [CrossRef] [PubMed]

145. Petersen, C.A.H.; Alikhani, N.; Behbahani, H.; Wiehager, B.; Pavlov, P.F.; Alafuzoff, I.; Leinonen, V.; Ito, A.; Winblad, B.; Glaser, E. The amyloid β-peptide is imported into mitochondria via the TOM import machinery and localized to mitochondrial cristae. *Proc. Natl. Acad. Sci. USA* **2008**, *105*, 13145–13150. [CrossRef] [PubMed]

146. Lustbader, J.W.; Cirilli, M.; Lin, C.; Xu, H.W.; Takuma, K.; Wang, N.; Caspersen, C.; Chen, X.; Pollak, S.; Chaney, M. ABAD directly links Aß to mitochondrial toxicity in Alzheimer's Disease. *Science* **2004**, *304*, 448–452. [CrossRef]

147. Cha, M.-Y.; Cho, H.J.; Kim, C.; Jung, Y.O.; Kang, M.J.; Murray, M.E.; Hong, H.S.; Choi, Y.-J.; Choi, H.; Kim, D.K. Mitochondrial ATP synthase activity is impaired by suppressed O-GlcNAcylation in Alzheimer's disease. *Hum. Mol. Genet.* **2015**, *24*, 6492–6504. [CrossRef] [PubMed]

148. Halestrap, A. A pore way to die. *Nature* **2005**, *434*, 578–579. [CrossRef]

149. Jacobson, J.; Duchen, M.R. Mitochondrial oxidative stress and cell death in astrocytes requirement for stored Ca^{2+} and sustained opening of the permeability transition pore. *J. Cell Sci.* **2002**, *115*, 1175–1188. [CrossRef] [PubMed]

150. Du, H.; Guo, L.; Fang, F.; Chen, D.; Sosunov, A.A.; McKhann, G.M.; Yan, Y.; Wang, C.; Zhang, H.; Molkentin, J.D. Cyclophilin D deficiency attenuates mitochondrial and neuronal perturbation and ameliorates learning and memory in Alzheimer's disease. *Nat. Med.* **2008**, *14*, 1097–1105. [CrossRef] [PubMed]

151. Beck, S.J.; Guo, L.; Phensy, A.; Tian, J.; Wang, L.; Tandon, N.; Gauba, E.; Lu, L.; Pascual, J.M.; Kroener, S. Deregulation of mitochondrial F1FO-ATP synthase via OSCP in Alzheimer's disease. *Nat. Commun.* **2016**, *7*, 11483. [CrossRef]

152. Sanz-Blasco, S.; Valero, R.A.; Rodríguez-Crespo, I.; Villalobos, C.; Núñez, L. Mitochondrial Ca^{2+} overload underlies Aβ oligomers neurotoxicity providing an unexpected mechanism of neuroprotection by NSAIDs. *PLoS ONE* **2008**, *3*, e2718. [CrossRef]

153. Alikhani, N.; Ankarcrona, M.; Glaser, E. Mitochondria and Alzheimer's disease: Amyloid-β peptide uptake and degradation by the presequence protease, hPreP. *J.Bioenerg. Biomembr.* **2009**, *41*, 447–451. [CrossRef]

154. Mossmann, D.; Vögtle, F.-N.; Taskin, A.A.; Teixeira, P.F.; Ring, J.; Burkhart, J.M.; Burger, N.; Pinho, C.M.; Tadic, J.; Loreth, D. Amyloid-β peptide induces mitochondrial dysfunction by inhibition of preprotein maturation. *Cell Metab.* **2014**, *20*, 662–669. [CrossRef]

155. Quintana, D.D.; Garcia, J.A.; Anantula, Y.; Rellick, S.L.; Engler-Chiurazzi, E.B.; Sarkar, S.N.; Brown, C.M.; Simpkins, J.W. Amyloid-β Causes Mitochondrial Dysfunction via a Ca^{2+}-Driven Upregulation of Oxidative Phosphorylation and Superoxide Production in Cerebrovascular Endothelial Cells. *J. Alzheimers Dis.* **2020**, *75*, 119–138. [CrossRef] [PubMed]

156. Rao, R.V.; Bredesen, D.E. Misfolded proteins, endoplasmic reticulum stress and neurodegeneration. *Curr. Opin. Cell Biol.* **2004**, *16*, 653–662. [CrossRef]

157. Toyoshima, C. How Ca^{2+}-ATPase pumps ions across the sarcoplasmic reticulum membrane. *Biochim. Biophys. ActaMol. Cell Res.* **2009**, *1793*, 941–946. [CrossRef]

158. Lodish, H.; Kong, N.; Wikström, L. Calcium is required for folding of newly made subunits of the asialoglycoprotein receptor within the endoplasmic reticulum. *J. Biol. Chem.* **1992**, *267*, 12753–12760. [CrossRef]

159. Kaufman, R.J. Stress signaling from the lumen of the endoplasmic reticulum: Coordination of gene transcriptional and translational controls. *Genes Dev.* **1999**, *13*, 1211–1233. [CrossRef] [PubMed]

160. Shen, X.; Zhang, K.; Kaufman, R.J. The unfolded protein response—A stress signaling pathway of the endoplasmic reticulum. *J. Chem. Neuroanat.* **2004**, *28*, 79–92. [CrossRef]

161. Hamos, J.E.; Oblas, B.; Pulaski-Salo, D.; Welch, W.; Bole, D.; Drachman, D.A. Expression of heat shock proteins in Alzheimer's disease. *Neurology* **1991**, *41*, 345. [CrossRef]

162. Görlach, A.; Klappa, P.; Kietzmann, D.T. The endoplasmic reticulum: Folding, calcium homeostasis, signaling, and redox control. *Antioxid. Redox Signal.* **2006**, *8*, 1391–1418. [CrossRef] [PubMed]

163. Leissring, M.A.; Akbari, Y.; Fanger, C.M.; Cahalan, M.D.; Mattson, M.P.; LaFerla, F.M. Capacitative calcium entry deficits and elevated luminal calcium content in mutant presenilin-1 knockin mice. *J. Cell Biol.* **2000**, *49*, 793–798. [CrossRef] [PubMed]

164. Müller, M.; Cheung, K.-H.; Foskett, J.K. Enhanced ROS generation mediated by Alzheimer's disease presenilin regulation of InsP3R Ca^{2+} signaling. *Antioxid. Redox Signal.* **2011**, *14*, 1225–1235. [CrossRef] [PubMed]

165. Ferreira, I.; Bajouco, L.; Mota, S.; Auberson, Y.; Oliveira, C.; Rego, A. Amyloid beta peptide 1–42 disturbs intracellular calcium homeostasis through activation of GluN2B-containing N-methyl-d-aspartate receptors in cortical cultures. *Cell Calcium* **2012**, *51*, 95–106. [CrossRef]

166. Ferreiro, E.; Oliveira, C.R.; Pereira, C. Involvement of endoplasmic reticulum Ca^{2+} release through ryanodine and inositol 1, 4, 5-triphosphate receptors in the neurotoxic effects induced by the amyloid-β peptide. *J. Neurosci. Res.* **2004**, *76*, 872–880. [CrossRef] [PubMed]

167. Hayashi, T.; Su, T.P. Sigma-1 receptor chaperones at the ER-mitochondrion interface regulate Ca^{2+} signaling and cell survival. *Cell* **2007**, *131*, 596–610. [CrossRef] [PubMed]

168. Bravo, R.; Vicencio, J.M.; Parra, V.; Troncoso, R.; Munoz, J.P.; Bui, M.; Quiroga, C.; Rodriguez, A.E.; Verdejo, H.E.; Ferreira, J. Increased ER–mitochondrial coupling promotes mitochondrial respiration and bioenergetics during early phases of ER stress. *J. Cell Biol.* **2011**, *124*, 2143–2152.

169. Jakaria, M.; Azam, S.; Haque, M.E.; Jo, S.-H.; Uddin, M.S.; Kim, I.-S.; Choi, D.-K. Taurine and its analogs in neurological disorders: Focus on therapeutic potential and molecular mechanisms. *Redox Biol.* **2019**, *24*, 101223. [CrossRef]

170. Chafekar, S.M.; Hoozemans, J.J.; Zwart, R.; Baas, F.; Scheper, W. A β 1-42 Induces Mild Endoplasmic Reticulum Stress in an Aggregation State—Dependent Manner. *Antioxid. Redox Signal.* **2007**, *9*, 2245–2254. [CrossRef]

171. Barbero-Camps, E.; Fernández, A.; Baulies, A.; Martinez, L.; Fernández-Checa, J.C.; Colell, A. Endoplasmic reticulum stress mediates amyloid β neurotoxicity via mitochondrial cholesterol trafficking. *Am. J. Pathol.* **2014**, *184*, 2066–2081. [CrossRef]

172. Nishitsuji, K.; Tomiyama, T.; Ishibashi, K.; Ito, K.; Teraoka, R.; Lambert, M.P.; Klein, W.L.; Mori, H. The E693Δ mutation in amyloid precursor protein increases intracellular accumulation of amyloid β oligomers and causes endoplasmic reticulum stress-induced apoptosis in cultured cells. *Am. J. Pathol.* **2009**, *174*, 957–969. [CrossRef]

173. Casas-Tinto, S.; Zhang, Y.; Sanchez-Garcia, J.; Gomez-Velazquez, M.; Rincon-Limas, D.E.; Fernandez-Funez, P. The ER stress factor XBP1s prevents amyloid-β neurotoxicity. *Hum. Mol. Genet.* **2011**, *20*, 2144–2160. [CrossRef]

174. Gourmaud, S.; Paquet, C.; Dumurgier, J.; Pace, C.; Bouras, C.; Gray, F.; Laplanche, J.-L.; Meurs, E.F.; Mouton-Liger, F.; Hugon, J. Increased levels of cerebrospinal fluid JNK3 associated with amyloid pathology: Links to cognitive decline. *J Psychiatry Neurosci.* **2015**, *40*, 151. [CrossRef] [PubMed]

175. Wang, X.; Ron, D. Stress-induced phosphorylation and activation of the transcription factor CHOP (GADD153) by p38 MAP kinase. *Science* **1996**, *272*, 1347–1349. [CrossRef] [PubMed]

176. Yoneda, T.; Imaizumi, K.; Oono, K.; Yui, D.; Gomi, F.; Katayama, T.; Tohyama, M. Activation of caspase-12, an endoplastic reticulum (ER) resident caspase, through tumor necrosis factor receptor-associated factor 2-dependent mechanism in response to the ER stress. *J. Biol. Chem.* **2001**, *276*, 13935–13940. [CrossRef] [PubMed]

177. Nishitoh, H.; Matsuzawa, A.; Tobiume, K.; Saegusa, K.; Takeda, K.; Inoue, K.; Hori, S.; Kakizuka, A.; Ichijo, H. ASK1 is essential for endoplasmic reticulum stress-induced neuronal cell death triggered by expanded polyglutamine repeats. *Genes Dev.* **2002**, *16*, 1345–1355. [CrossRef] [PubMed]

178. Lee, J.H.; Won, S.M.; Suh, J.; Son, S.J.; Moon, G.J.; Park, U.-J.; Gwag, B.J. Induction of the unfolded protein response and cell death pathway in Alzheimer's disease, but not in aged Tg2576 mice. *Exp. Mol. Med.* **2010**, *42*, 386–394. [CrossRef]

179. Duran-Aniotz, C.; Cornejo, V.H.; Espinoza, S.; Ardiles, Á.O.; Medinas, D.B.; Salazar, C.; Foley, A.; Gajardo, I.; Thielen, P.; Iwawaki, T. IRE1 signaling exacerbates Alzheimer's disease pathogenesis. *Acta Neuropathol.* **2017**, *134*, 489–506. [CrossRef]

180. Meusser, B.; Hirsch, C.; Jarosch, E.; Sommer, T. ERAD: The long road to destruction. *Nat. Cell Biol.* **2005**, *7*, 766–772. [CrossRef]

181. Khandelwal, P.J.; Herman, A.M.; Hoe, H.-S.; Rebeck, G.W.; Moussa, C.E.-H. Parkin mediates beclin-dependent autophagic clearance of defective mitochondria and ubiquitinated Aβ in AD models. *Hum. Mol. Genet.* **2011**, *20*, 2091–2102. [CrossRef]

182. Poon, W.W.; Carlos, A.J.; Aguilar, B.L.; Berchtold, N.C.; Kawano, C.K.; Zograbyan, V.; Yaopruke, T.; Shelanski, M.; Cotman, C.W. β-Amyloid (Aβ) oligomers impair brain-derived neurotrophic factor retrograde trafficking by down-regulating ubiquitin C-terminal hydrolase, UCH-L1. *J. Biol. Chem.* **2013**, *288*, 16937–16948. [CrossRef]

183. Saido, T.; Leissring, M.A. Proteolytic degradation of amyloid β-protein. *Cold Spring Harb. Perspect. Med.* **2012**, *2*, a006379. [CrossRef]

184. Salon, M.L.; Pasquini, L.; Moreno, M.B.; Pasquini, J.; Soto, E. Relationship between β-amyloid degradation and the 26S proteasome in neural cells. *Exp. Neurol.* **2003**, *180*, 131–143. [CrossRef]

185. Gallart-Palau, X.; Guo, X.; Serra, A.; Sze, S.K. Alzheimer's disease progression characterized by alterations in the molecular profiles and biogenesis of brain extracellular vesicles. *Alzheimers Res. Ther.* **2020**, *12*, 54. [CrossRef] [PubMed]

186. Zhao, X.; Yang, J. Amyloid-β peptide is a substrate of the human 20S proteasome. *ACS Chem. Neurosci.* **2010**, *1*, 655–660. [CrossRef] [PubMed]

187. Desikan, R.S.; Fan, C.C.; Wang, Y.; Schork, A.J.; Cabral, H.J.; Cupples, L.A.; Thompson, W.K.; Besser, L.; Kukull, W.A.; Holland, D. Genetic assessment of age-associated Alzheimer disease risk: Development and validation of a polygenic hazard score. *PLoS Med.* **2017**, *14*, e1002258. [CrossRef]

188. Zhu, B.; Jiang, L.; Huang, T.; Zhao, Y.; Liu, T.; Zhong, Y.; Li, X.; Campos, A.; Pomeroy, K.; Masliah, E. ER-associated degradation regulates Alzheimer's amyloid pathology and memory function by modulating γ-secretase activity. *Nat. Commun.* **2017**, *8*, 1472. [CrossRef] [PubMed]

189. Kaushik, S.; Cuervo, A.M. Chaperone-mediated autophagy: A unique way to enter the lysosome world. *Trends Cell Biol.* **2012**, *22*, 407–417. [CrossRef]
190. Li, W.; Zhu, J.; Dou, J.; She, H.; Tao, K.; Xu, H.; Yang, Q.; Mao, Z. Phosphorylation of LAMP2A by p38 MAPK couples ER stress to chaperone-mediated autophagy. *Nat. Commun.* **2017**, *8*, 1763. [CrossRef] [PubMed]
191. Mizushima, N. The ATG conjugation systems in autophagy. *Curr. Opin. Cell Biol.* **2020**, *63*, 1–10. [CrossRef]
192. Yim, W.W.Y.; Mizushima, N. Lysosome biology in autophagy. *Cell Discov.* **2020**, *6*, 6. [CrossRef]
193. Nixon, R.A.; Wegiel, J.; Kumar, A.; Yu, W.H.; Peterhoff, C.; Cataldo, A.; Cuervo, A.M. Extensive involvement of autophagy in Alzheimer disease: An immuno-electron microscopy study. *J. Neuropathol. Exp. Neurol.* **2005**, *64*, 113–122. [CrossRef] [PubMed]
194. Yu, W.H.; Cuervo, A.M.; Kumar, A.; Peterhoff, C.M.; Schmidt, S.D.; Lee, J.-H.; Mohan, P.S.; Mercken, M.; Farmery, M.R.; Tjernberg, L.O. Macroautophagy—A novel β-amyloid peptide-generating pathway activated in Alzheimer's disease. *J. Cell Biol.* **2005**, *171*, 87–98. [CrossRef] [PubMed]
195. Fedeli, C.; Filadi, R.; Rossi, A.; Mammucari, C.; Pizzo, P. PSEN2 (presenilin 2) mutants linked to familial Alzheimer disease impair autophagy by altering Ca^{2+} homeostasis. *Autophagy* **2019**, *15*, 2044–2062. [CrossRef]
196. Pickford, F.; Masliah, E.; Britschgi, M.; Lucin, K.; Narasimhan, R.; Jaeger, P.A.; Small, S.; Spencer, B.; Rockenstein, E.; Levine, B. The autophagy-related protein beclin 1 shows reduced expression in early Alzheimer disease and regulates amyloid β accumulation in mice. *J. Clin. Investig.* **2008**, *118*, 2190–2199. [CrossRef]
197. Swaminathan, G.; Zhu, W.; Plowey, E.D. BECN1/Beclin 1 sorts cell-surface APP/amyloid β precursor protein for lysosomal degradation. *Autophagy* **2016**, *12*, 2404–2419. [CrossRef]
198. Ma, J.F.; Huang, Y.; Chen, S.D.; Halliday, G. Immunohistochemical evidence for macroautophagy in neurones and endothelial cells in Alzheimer's disease. *Neuropathol. Appl. Neurobiol.* **2010**, *36*, 312–319. [CrossRef]
199. Omata, Y.; Lim, Y.-M.; Akao, Y.; Tsuda, L. Age-induced reduction of autophagy-related gene expression is associated with onset of Alzheimer's disease. *Am. J. Neurodegener. Dis.* **2014**, *3*, 134. [PubMed]
200. Manczak, M.; Kandimalla, R.; Yin, X.; Reddy, P.H. Hippocampal mutant APP and amyloid beta-induced cognitive decline, dendritic spine loss, defective autophagy, mitophagy and mitochondrial abnormalities in a mouse model of Alzheimer's disease. *Hum. Mol. Genet.* **2018**, *27*, 1332–1342. [CrossRef] [PubMed]
201. Smith, K.T. *Amyloid-β Causes Autophagy Dysfunction by Inhibiting Protein Prenylation*; University of Alberta, Canada: Edmonton, AB, Canada, 2016.
202. Pigino, G.; Morfini, G.; Atagi, Y.; Deshpande, A.; Yu, C.; Jungbauer, L.; LaDu, M.; Busciglio, J.; Brady, S. Disruption of fast axonal transport is a pathogenic mechanism for intraneuronal amyloid beta. *Proc. Natl. Acad. Sci. USA* **2009**, *106*, 5907–5912. [CrossRef] [PubMed]
203. Tammineni, P.; Cai, Q. Defective retrograde transport impairs autophagic clearance in Alzheimer disease neurons. *Autophagy* **2017**, *13*, 982–984. [CrossRef]
204. Aboud, O.; Parcon, P.A.; DeWall, K.M.; Liu, L.; Mrak, R.E.; Griffin, S.T. Aging, Alzheimer's, and APOE genotype influence the expression and neuronal distribution patterns of microtubule motor protein dynactin-P50. *Front. Cell. Neurosci.* **2015**, *9*, 103. [CrossRef]
205. Xiao, T.; Zhang, L.; Huang, Y.; Shi, Y.; Wang, J.; Ji, Q.; Ye, J.; Lin, Y.; Liu, H. Sestrin2 increases in aortas and plasma from aortic dissection patients and alleviates angiotensin II-induced smooth muscle cell apoptosis via the Nrf2 pathway. *Life Sci.* **2019**, *218*, 132–138. [CrossRef]
206. Rojo, A.I.; Pajares, M.; Rada, P.; Nuñez, A.; Nevado-Holgado, A.J.; Killik, R.; Van Leuven, F.; Ribe, E.; Lovestone, S.; Yamamoto, M. NRF2 deficiency replicates transcriptomic changes in Alzheimer's patients and worsens APP and TAU pathology. *Redox Biol.* **2017**, *13*, 444–451. [CrossRef] [PubMed]
207. Bahn, G.; Park, J.-S.; Yun, U.J.; Lee, Y.J.; Choi, Y.; Park, J.S.; Baek, S.H.; Choi, B.Y.; Cho, Y.S.; Kim, H.K. NRF2/ARE pathway negatively regulates BACE1 expression and ameliorates cognitive deficits in mouse Alzheimer's models. *Proc. Natl. Acad. Sci. USA* **2019**, *116*, 12516–12523. [CrossRef] [PubMed]
208. Neely, K.M.; Green, K.N.; LaFerla, F.M. Presenilin is necessary for efficient proteolysis through the autophagy–lysosome system in a γ-secretase-independent manner. *J. Neurosci.* **2011**, *31*, 2781–2791. [CrossRef] [PubMed]
209. Lee, J.-H.; Yu, W.H.; Kumar, A.; Lee, S.; Mohan, P.S.; Peterhoff, C.M.; Wolfe, D.M.; Martinez-Vicente, M.; Massey, A.C.; Sovak, G. Lysosomal proteolysis and autophagy require presenilin 1 and are disrupted by Alzheimer-related PS1 mutations. *Cell* **2010**, *141*, 1146–1158. [CrossRef]
210. Coen, K.; Flannagan, R.S.; Baron, S.; Carraro-Lacroix, L.R.; Wang, D.; Vermeire, W.; Michiels, C.; Munck, S.; Baert, V.; Sugita, S. Lysosomal calcium homeostasis defects, not proton pump defects, cause endo-lysosomal dysfunction in PSEN-deficient cells. *J. Cell Biol.* **2012**, *198*, 23–35. [CrossRef]
211. Ditaranto, K.; Tekirian, T.L.; Yang, A.J. Lysosomal membrane damage in soluble Aβ-mediated cell death in Alzheimer's disease. *Neurobiol. Dis.* **2001**, *8*, 19–31. [CrossRef]
212. Krabbe, G.; Halle, A.; Matyash, V.; Rinnenthal, J.L.; Eom, G.D.; Bernhardt, U.; Miller, K.R.; Prokop, S.; Kettenmann, H.; Heppner, F.L. Functional impairment of microglia coincides with Beta-amyloid deposition in mice with Alzheimer-like pathology. *PLoS ONE* **2013**, *8*, e60921. [CrossRef]
213. Guo, X.; Tang, P.; Chen, L.; Liu, P.; Hou, C.; Zhang, X.; Liu, Y.; Chong, L.; Li, X.; Li, R. Amyloid β-induced redistribution of transcriptional factor EB and lysosomal dysfunction in primary microglial cells. *Front. Aging Neurosci.* **2017**, *9*, 228. [CrossRef]

214. Jiang, Y.; Sato, Y.; Im, E.; Berg, M.; Bordi, M.; Darji, S.; Kumar, A.; Mohan, P.S.; Bandyopadhyay, U.; Diaz, A. Lysosomal dysfunction in Down syndrome is APP-dependent and mediated by APP-βCTF (C99). *J. Neurosci.* **2019**, *39*, 5255–5268. [CrossRef]
215. Soura, V.; Stewart-Parker, M.; Williams, T.L.; Ratnayaka, A.; Atherton, J.; Gorringe, K.; Tuffin, J.; Darwent, E.; Rambaran, R.; Klein, W. Visualization of co-localization in Aβ42-administered neuroblastoma cells reveals lysosome damage and autophagosome accumulation related to cell death. *Biochem. J.* **2012**, *441*, 579–590. [CrossRef] [PubMed]
216. Gowrishankar, S.; Yuan, P.; Wu, Y.; Schrag, M.; Paradise, S.; Grutzendler, J.; De Camilli, P.; Ferguson, S.M. Massive accumulation of luminal protease-deficient axonal lysosomes at Alzheimer's disease amyloid plaques. *Proc. Natl. Acad. Sci. USA* **2015**, *112*, E3699–E3708. [CrossRef] [PubMed]
217. Tan, J.Z.A.; Fourriere, L.; Wang, J.; Perez, F.; Boncompain, G.; Gleeson, P.A. Distinct anterograde trafficking pathways of BACE1 and amyloid precursor protein from the TGN and the regulation of amyloid-β production. *Mol. Biol. Cell.* **2020**, *31*, 27–44. [CrossRef]
218. Toh, W.H.; Tan, J.Z.A.; Zulkefli, K.L.; Houghton, F.J.; Gleeson, P.A. Amyloid precursor protein traffics from the Golgi directly to early endosomes in an Arl5b-and AP4-dependent pathway. *Traffic* **2017**, *18*, 159–175. [CrossRef] [PubMed]
219. Gosavi, P.; Gleeson, P.A. The function of the Golgi ribbon structure–an enduring mystery unfolds! *Bioessays* **2017**, *39*, 1700063. [CrossRef] [PubMed]
220. Wei, J.-H.; Seemann, J. Golgi ribbon disassembly during mitosis, differentiation and disease progression. *Curr. Opin. Cell Biol.* **2017**, *47*, 43–51. [CrossRef]
221. Kim, M.J.; Je, A.R.; Kim, H.-J.; Huh, Y.H.; Kweon, H.S. Coat protein I depletion-associated Golgi fragmentation in an Alzheimer's disease model. *Anim. Cells Syst.* **2015**, *19*, 8–15. [CrossRef]
222. Joshi, G.; Chi, Y.; Huang, Z.; Wang, Y. Aβ-induced Golgi fragmentation in Alzheimer's disease enhances Aβ production. *Proc. Natl. Acad. Sci. USA* **2014**, *111*, E1230–E1239. [CrossRef]
223. Joshi, G.; Bekier, M.I.; Wang, Y. Golgi fragmentation in Alzheimer's disease. *Front. Neurosci.* **2015**, *9*, 340. [CrossRef]
224. Sun, K.-H.; de Pablo, Y.; Vincent, F.; Johnson, E.O.; Chavers, A.K.; Shah, K. Novel genetic tools reveal Cdk5's major role in Golgi fragmentation in Alzheimer's disease. *Mol. Biol. Cell.* **2008**, *19*, 3052–3069. [CrossRef]
225. Jiang, Q.; Wang, L.; Guan, Y.; Xu, H.; Niu, Y.; Han, L.; Wei, Y.-P.; Lin, L.; Chu, J.; Wang, Q. Golgin-84-associated Golgi fragmentation triggers tau hyperphosphorylation by activation of cyclin-dependent kinase-5 and extracellular signal-regulated kinase. *Neurobiol. Aging* **2014**, *35*, 1352–1363. [CrossRef] [PubMed]
226. Liazoghli, D.; Perreault, S.; Micheva, K.D.; Desjardins, M.; Leclerc, N. Fragmentation of the Golgi apparatus induced by the overexpression of wild-type and mutant human tau forms in neurons. *Am. J. Pathol.* **2005**, *166*, 1499–1514. [CrossRef]
227. Cervigni, R.I.; Bonavita, R.; Barretta, M.L.; Spano, D.; Ayala, I.; Nakamura, N.; Corda, D.; Colanzi, A. JNK2 controls fragmentation of the Golgi complex and the G2/M transition through phosphorylation of GRASP65. *J. Cell Biol.* **2015**, *128*, 2249–2260. [CrossRef] [PubMed]
228. Bellouze, S.; Schäfer, M.K.; Buttigieg, D.; Baillat, G.; Rabouille, C.; Haase, G. Golgi fragmentation in pmn mice is due to a defective ARF1/TBCE cross-talk that coordinates COPI vesicle formation and tubulin polymerization. *Hum. Mol. Genet.* **2014**, *23*, 5961–5975. [CrossRef] [PubMed]
229. Gonatas, N.K.; Stieber, A.; Gonatas, J.O. Fragmentation of the Golgi apparatus in neurodegenerative diseases and cell death. *J. Neurol. Sci.* **2006**, *246*, 21–30. [CrossRef]
230. Suga, K.; Saito, A.; Mishima, T.; Akagawa, K. ER and Golgi stresses increase ER–Golgi SNARE Syntaxin5: Implications for organelle stress and βAPP processing. *Neurosci. Lett.* **2015**, *604*, 30–35. [CrossRef]
231. Barucker, C.; Harmeier, A.; Weiske, J.; Fauler, B.; Albring, K.F.; Prokop, S.; Hildebrand, P.; Lurz, R.; Heppner, F.L.; Huber, O. Nuclear translocation uncovers the amyloid peptide Aβ42 as a regulator of gene transcription. *J. Biol. Chem.* **2014**, *289*, 20182–20191. [CrossRef]
232. Maloney, B.; Lahiri, D.K. The Alzheimer's amyloid β-peptide (Aβ) binds a specific DNA Aβ-interacting domain (AβID) in the APP, BACE1, and APOE promoters in a sequence-specific manner: Characterizing a new regulatory motif. *Gene* **2011**, *488*, 1–12. [CrossRef]
233. Bailey, J.A.; Maloney, B.; Ge, Y.-W.; Lahiri, D.K. Functional activity of the novel Alzheimer's amyloid β-peptide interacting domain (AβID) in the APP and BACE1 promoter sequences and implications in activating apoptotic genes and in amyloidogenesis. *Gene* **2011**, *488*, 13–22. [CrossRef]
234. Gezen-Ak, D.; Atasoy, İ.L.; Candaş, E.; Alaylıoğlu, M.; Dursun, E. The transcriptional regulatory properties of amyloid beta 1–42 may include regulation of genes related to neurodegeneration. *Neuromol. Med.* **2018**, *20*, 363–375. [CrossRef]
235. Barucker, C.; Sommer, A.; Beckmann, G.; Eravci, M.; Harmeier, A.; Schipke, C.G.; Brockschnieder, D.; Dyrks, T.; Althoff, V.; Fraser, P.E. Alzheimer amyloid peptide Aβ 42 regulates gene expression of transcription and growth factors. *J. Alzheimers Dis.* **2015**, *44*, 613–624. [CrossRef] [PubMed]
236. Xu, Y.; Kim, H.S.; Joo, Y.; Choi, Y.; Chang, K.; Park, C.; Shin, K.; Kim, S.; Cheon, Y.; Baik, T. Intracellular domains of amyloid precursor-like protein 2 interact with CP2 transcription factor in the nucleus and induce glycogen synthase kinase-3 β expression. *Cell Death Differ.* **2007**, *14*, 79–91. [CrossRef]
237. Zhu, H.; Fu, W.; Mattson, M.P. The catalytic subunit of telomerase protects neurons against amyloid β-peptide-induced apoptosis. *J. Neurochem.* **2000**, *75*, 117–124. [CrossRef] [PubMed]

238. Flanary, B.E.; Sammons, N.W.; Nguyen, C.; Walker, D.; Streit, W.J. Evidence that aging and amyloid promote microglial cell senescence. *Rejuvenation Res.* **2007**, *10*, 61–74. [CrossRef] [PubMed]

239. Jaskelioff, M.; Muller, F.L.; Paik, J.-H.; Thomas, E.; Jiang, S.; Adams, A.C.; Sahin, E.; Kost-Alimova, M.; Protopopov, A.; Cadinanos, J. Telomerase reactivation reverses tissue degeneration in aged telomerase-deficient mice. *Nature* **2011**, *469*, 102–106. [CrossRef]

240. Wang, J.; Zhao, C.; Zhao, A.; Li, M.; Ren, J.; Qu, X. New insights in amyloid beta interactions with human telomerase. *J. Am. Chem. Soc.* **2015**, *137*, 1213–1219. [CrossRef]

241. Collado, M.; Blasco, M.A.; Serrano, M. Cellular senescence in cancer and aging. *Cell* **2007**, *130*, 223–233. [CrossRef]

242. Boccardi, V.; Pelini, L.; Ercolani, S.; Ruggiero, C.; Mecocci, P. From cellular senescence to Alzheimer's disease: The role of telomere shortening. *Ageing Res. Rev.* **2015**, *22*, 1–8. [CrossRef]

243. Fuso, A.; Seminara, L.; Cavallaro, R.A.; D'Anselmi, F.; Scarpa, S. S-adenosylmethionine/homocysteine cycle alterations modify DNA methylation status with consequent deregulation of PS1 and BACE and beta-amyloid production. *Mol. Cell. Neurosci.* **2005**, *28*, 195–204. [CrossRef] [PubMed]

244. Taher, N.; McKenzie, C.; Garrett, R.; Baker, M.; Fox, N.; Isaacs, G.D. Amyloid-β alters the DNA methylation status of cell-fate genes in an Alzheimer's disease model. *J. Alzheimers Dis.* **2014**, *38*, 831–844. [CrossRef]

245. Wang, S.-C.; Oelze, B.; Schumacher, A. Age-specific epigenetic drift in late-onset Alzheimer's disease. *PLoS ONE* **2008**, *3*, e2698. [CrossRef]

246. Chen, K.-L.; Wang, S.S.-S.; Yang, Y.-Y.; Yuan, R.-Y.; Chen, R.-M.; Hu, C.-J. The epigenetic effects of amyloid-β1–40 on global DNA and neprilysin genes in murine cerebral endothelial cells. *Biochem. Biophys. Res. Commun.* **2009**, *378*, 57–61. [CrossRef] [PubMed]

247. Hales, C.M.; Seyfried, N.T.; Dammer, E.B.; Duong, D.; Yi, H.; Gearing, M.; Troncoso, J.C.; Mufson, E.J.; Thambisetty, M.; Levey, A.I. U1 small nuclear ribonucleoproteins (snRNPs) aggregate in Alzheimer's disease due to autosomal dominant genetic mutations and trisomy 21. *Mol. Neurodegener.* **2014**, *9*, 15. [CrossRef] [PubMed]

248. Castellano, J.M.; Deane, R.; Gottesdiener, A.J.; Verghese, P.B.; Stewart, F.R.; West, T.; Paoletti, A.C.; Kasper, T.R.; DeMattos, R.B.; Zlokovic, B.V. Low-density lipoprotein receptor overexpression enhances the rate of brain-to-blood Aβ clearance in a mouse model of β-amyloidosis. *Proc. Natl. Acad. Sci. USA* **2012**, *109*, 15502–15507. [CrossRef]

249. Wilkinson, K.; El Khoury, J. Microglial scavenger receptors and their roles in the pathogenesis of Alzheimer's disease. *Int. J. Alzheimers Dis.* **2012**, *2012*, 489456. [CrossRef]

250. Kanekiyo, T.; Cirrito, J.R.; Liu, C.-C.; Shinohara, M.; Li, J.; Schuler, D.R.; Shinohara, M.; Holtzman, D.M.; Bu, G. Neuronal clearance of amyloid-β by endocytic receptor LRP1. *J. Neurosci.* **2013**, *33*, 19276–19283. [CrossRef] [PubMed]

251. Brandenburg, L.O.; Konrad, M.; Wruck, C.J.; Koch, T.; Lucius, R.; Pufe, T. Functional and physical interactions between formyl-peptide-receptors and scavenger receptor MARCO and their involvement in amyloid beta 1–42-induced signal transduction in glial cells. *J. Neurochem.* **2010**, *113*, 749–760. [CrossRef]

252. Takumi, Y.; Ramírez-León, V.; Laake, P.; Rinvik, E.; Ottersen, O.P. Different modes of expression of AMPA and NMDA receptors in hippocampal synapses. *Nat. Neurosci.* **1999**, *2*, 618–624. [CrossRef]

253. Hayashi, Y.; Shi, S.-H.; Esteban, J.A.; Piccini, A.; Poncer, J.-C.; Malinow, R. Driving AMPA receptors into synapses by LTP and CaMKII: Requirement for GluR1 and PDZ domain interaction. *Science* **2000**, *287*, 2262–2267. [CrossRef] [PubMed]

254. Song, I.; Huganir, R.L. Regulation of AMPA receptors during synaptic plasticity. *Trends Neurosci.* **2002**, *25*, 578–588. [CrossRef]

255. Chan, S.L.; Griffin, W.S.T.; Mattson, M.P. Evidence for caspase-mediated cleavage of AMPA receptor subunits in neuronal apoptosis and Alzheimer's disease. *J. Neurosci. Res.* **1999**, *57*, 315–323. [CrossRef]

256. Szegedi, V.; Juhász, G.; Budai, D.; Penke, B. Divergent effects of Aβ1–42 on ionotropic glutamate receptor-mediated responses in CA1 neurons in vivo. *Brain Res.* **2005**, *1062*, 120–126. [CrossRef] [PubMed]

257. Malenka, R.C.; Bear, M.F. LTP and LTD: An embarrassment of riches. *Neuron* **2004**, *44*, 5–21. [CrossRef] [PubMed]

258. Nieweg, K.; Andreyeva, A.; Van Stegen, B.; Tanriöver, G.; Gottmann, K. Alzheimer's disease-related amyloid-β induces synaptotoxicity in human iPS cell-derived neurons. *Cell Death Dis.* **2015**, *6*, e1709. [CrossRef]

259. Popugaeva, E.; Pchitskaya, E.; Bezprozvanny, I. Dysregulation of neuronal calcium homeostasis in Alzheimer's disease—A therapeutic opportunity? *Biochem. Biophys. Res. Commun.* **2017**, *483*, 998–1004. [CrossRef]

260. Gu, Z.; Liu, W.; Yan, Z. β-Amyloid impairs AMPA receptor trafficking and function by reducing Ca^{2+}/calmodulin-dependent protein kinase II synaptic distribution. *J. Biol. Chem.* **2009**, *284*, 10639–10649. [CrossRef]

261. Zhao, W.-Q.; Santini, F.; Breese, R.; Ross, D.; Zhang, X.D.; Stone, D.J.; Ferrer, M.; Townsend, M.; Wolfe, A.L.; Seager, M.A. Inhibition of calcineurin-mediated endocytosis and α-amino-3-hydroxy-5-methyl-4-isoxazolepropionic acid (AMPA) receptors prevents amyloid β oligomer-induced synaptic disruption. *J. Biol. Chem.* **2010**, *285*, 7619–7632. [CrossRef]

262. Kurup, P.; Zhang, Y.; Xu, J.; Venkitaramani, D.V.; Haroutunian, V.; Greengard, P.; Nairn, A.C.; Lombroso, P.J. Aβ-mediated NMDA receptor endocytosis in Alzheimer's disease involves ubiquitination of the tyrosine phosphatase STEP61. *J. Neurosci.* **2010**, *30*, 5948–5957. [CrossRef]

263. Li, S.; Jin, M.; Koeglsperger, T.; Shepardson, N.E.; Shankar, G.M.; Selkoe, D.J. Soluble Aβ oligomers inhibit long-term potentiation through a mechanism involving excessive activation of extrasynaptic NR2B-containing NMDA receptors. *J. Neurosci.* **2011**, *31*, 6627–6638. [CrossRef]

264. Sheng, M.; Hoogenraad, C.C. The postsynaptic architecture of excitatory synapses: A more quantitative view. *Annu. Rev. Biochem.* **2007**, *76*, 823–847. [CrossRef] [PubMed]

265. Roselli, F.; Tirard, M.; Lu, J.; Hutzler, P.; Lamberti, P.; Livrea, P.; Morabito, M.; Almeida, O. Soluble β-amyloid1-40 induces NMDA-dependent degradation of postsynaptic density-95 at glutamatergic synapses. *J. Neurosci.* **2005**, *25*, 11061–11070. [CrossRef] [PubMed]

266. Lacor, P.N.; Buniel, M.C.; Chang, L.; Fernandez, S.J.; Gong, Y.; Viola, K.L.; Lambert, M.P.; Velasco, P.T.; Bigio, E.H.; Finch, C.E. Synaptic targeting by Alzheimer's-related amyloid β oligomers. *J. Neurosci.* **2004**, *24*, 10191–10200. [CrossRef]

267. Dohler, F.; Sepulveda-Falla, D.; Krasemann, S.; Altmeppen, H.; Schlüter, H.; Hildebrand, D.; Zerr, I.; Matschke, J.; Glatzel, M. High molecular mass assemblies of amyloid-β oligomers bind prion protein in patients with Alzheimer's disease. *Brain* **2014**, *137*, 873–886. [CrossRef] [PubMed]

268. Laurén, J.; Gimbel, D.A.; Nygaard, H.B.; Gilbert, J.W.; Strittmatter, S.M. Cellular prion protein mediates impairment of synaptic plasticity by amyloid-β oligomers. *Nature* **2009**, *457*, 1128–1132. [CrossRef]

269. Rushworth, J.V.; Griffiths, H.H.; Watt, N.T.; Hooper, N.M. Prion protein-mediated toxicity of amyloid-β oligomers requires lipid rafts and the transmembrane LRP1. *J. Biol. Chem.* **2013**, *288*, 8935–8951. [CrossRef]

270. Um, J.W.; Kaufman, A.C.; Kostylev, M.; Heiss, J.K.; Stagi, M.; Takahashi, H.; Kerrisk, M.E.; Vortmeyer, A.; Wisniewski, T.; Koleske, A.J. Metabotropic glutamate receptor 5 is a coreceptor for Alzheimer aβ oligomer bound to cellular prion protein. *Neuron* **2013**, *79*, 887–902. [CrossRef]

271. Larson, M.; Sherman, M.A.; Amar, F.; Nuvolone, M.; Schneider, J.A.; Bennett, D.A.; Aguzzi, A.; Lesné, S.E. The complex PrPc-Fyn couples human oligomeric Aβ with pathological Tau changes in Alzheimer's disease. *J. Neurosci.* **2012**, *32*, 16857–16871. [CrossRef]

272. Taylor, D.R.; Whitehouse, I.J.; Hooper, N.M. Glypican-1 mediates both prion protein lipid raft association and disease isoform formation. *PLoS Pathog.* **2009**, *5*, e1000666. [CrossRef]

273. Khan, G.M.; Tong, M.; Jhun, M.; Arora, K.; Nichols, R.A. β-Amyloid activates presynaptic α7 nicotinic acetylcholine receptors reconstituted into a model nerve cell system: Involvement of lipid rafts. *Eur. J. Neurosci.* **2010**, *31*, 788–796. [CrossRef]

274. Cissé, M.; Halabisky, B.; Harris, J.; Devidze, N.; Dubal, D.B.; Sun, B.; Orr, A.; Lotz, G.; Kim, D.H.; Hamto, P. Reversing EphB2 depletion rescues cognitive functions in Alzheimer model. *Nature* **2011**, *469*, 47–52. [CrossRef]

275. Vargas, L.M.; Leal, N.; Estrada, L.D.; González, A.; Serrano, F.; Araya, K.; Gysling, K.; Inestrosa, N.C.; Pasquale, E.B.; Alvarez, A.R. EphA4 activation of c-Abl mediates synaptic loss and LTP blockade caused by amyloid-β oligomers. *PLoS ONE* **2014**, *9*, e92309.

276. Fu, A.K.; Hung, K.-W.; Huang, H.; Gu, S.; Shen, Y.; Cheng, E.Y.; Ip, F.C.; Huang, X.; Fu, W.-Y.; Ip, N.Y. Blockade of EphA4 signaling ameliorates hippocampal synaptic dysfunctions in mouse models of Alzheimer's disease. *Proc. Natl. Acad. Sci. USA* **2014**, *111*, 9959–9964. [CrossRef]

277. Kumar, R.; Herbert, P.; Warrens, A. An introduction to death receptors in apoptosis. *Int. J. Surg.* **2005**, *3*, 268–277. [CrossRef]

278. Ivins, K.J.; Thornton, P.L.; Rohn, T.T.; Cotman, C.W. Neuronal apoptosis induced by β-amyloid is mediated by caspase-8. *Neurobiol. Dis.* **1999**, *6*, 440–449. [CrossRef]

279. Fossati, S.; Ghiso, J.; Rostagno, A. TRAIL death receptors DR4 and DR5 mediate cerebral microvascular endothelial cell apoptosis induced by oligomeric Alzheimer's A β. *Cell Death Dis.* **2012**, *3*, e321. [CrossRef]

280. Rabizadeh, S.; Bitler, C.M.; Butcher, L.L.; Bredesen, D.E. Expression of the low-affinity nerve growth factor receptor enhances beta-amyloid peptide toxicity. *Proc. Natl. Acad. Sci. USA* **1994**, *91*, 10703–10706. [CrossRef] [PubMed]

281. Yaar, M.; Zhai, S.; Pilch, P.F.; Doyle, S.M.; Eisenhauer, P.B.; Fine, R.E.; Gilchrest, B.A. Binding of beta-amyloid to the p75 neurotrophin receptor induces apoptosis. A possible mechanism for Alzheimer's disease. *J. Clin. Investig.* **1997**, *100*, 2333–2340. [CrossRef] [PubMed]

282. Perini, G.; Della-Bianca, V.; Politi, V.; Della Valle, G.; Dal-Pra, I.; Rossi, F.; Armato, U. Role of p75 neurotrophin receptor in the neurotoxicity by β-amyloid peptides and synergistic effect of inflammatory cytokines. *J. Exp. Med.* **2002**, *195*, 907–918. [CrossRef]

283. Sotthibundhu, A.; Sykes, A.M.; Fox, B.; Underwood, C.K.; Thangnipon, W.; Coulson, E.J. β-amyloid1–42 induces neuronal death through the p75 neurotrophin receptor. *J. Neurosci.* **2008**, *28*, 3941–3946. [CrossRef]

284. Morishima, Y.; Gotoh, Y.; Zieg, J.; Barrett, T.; Takano, H.; Flavell, R.; Davis, R.J.; Shirasaki, Y.; Greenberg, M.E. β-Amyloid induces neuronal apoptosis via a mechanism that involves the c-Jun N-terminal kinase pathway and the induction of Fas ligand. *J. Neurosci.* **2001**, *21*, 7551–7560. [CrossRef]

285. Knowles, J.K.; Rajadas, J.; Nguyen, T.-V.V.; Yang, T.; LeMieux, M.C.; Vander Griend, L.; Ishikawa, C.; Massa, S.M.; Wyss-Coray, T.; Longo, F.M. The p75 neurotrophin receptor promotes amyloid-β (1-42)-induced neuritic dystrophy in vitro and in vivo. *J. Neurosci.* **2009**, *29*, 10627–10637. [CrossRef]

286. Wang, Y.; Tang, H.; Yang, C.; Zhao, H.; Jian, C. Involvement of p75NTR in the effects of Aβ on L-type Ca^{2+} channel in cultured neuronal networks. *Life Sci.* **2020**, *243*, 117293. [CrossRef]

287. Sasaki, N.; Toki, S.; Chowei, H.; Saito, T.; Nakano, N.; Hayashi, Y.; Takeuchi, M.; Makita, Z. Immunohistochemical distribution of the receptor for advanced glycation end products in neurons and astrocytes in Alzheimer's disease. *Brain Res.* **2001**, *888*, 256–262. [CrossRef]

288. Paudel, Y.N.; Angelopoulou, E.; Piperi, C.; Othman, I.; Aamir, K.; Shaikh, M. Impact of HMGB1, RAGE, and TLR4 in Alzheimer's Disease (AD): From Risk Factors to Therapeutic Targeting. *Cells* **2020**, *9*, 383. [CrossRef]

289. Hughes, C.; Choi, M.L.; Yi, J.-H.; Kim, S.-C.; Drews, A.; George-Hyslop, P.S.; Bryant, C.; Gandhi, S.; Cho, K.; Klenerman, D. Beta amyloid aggregates induce sensitised TLR4 signalling causing long-term potentiation deficit and rat neuronal cell death. *Commun. Biol.* **2020**, *3*, 79. [CrossRef]

290. Walters, A.; Phillips, E.; Zheng, R.; Biju, M.; Kuruvilla, T. Evidence for neuroinflammation in Alzheimer's disease. *Prog. Neurol. Psychiatry* **2016**, *20*, 25–31. [CrossRef]

291. Varnum, M.M.; Ikezu, T. The classification of microglial activation phenotypes on neurodegeneration and regeneration in Alzheimer's disease brain. *Arch. Immunol. Ther. Exp.* **2012**, *60*, 251–266. [CrossRef] [PubMed]

292. Mietelska-Porowska, A.; Wojda, U. T lymphocytes and inflammatory mediators in the interplay between brain and blood in Alzheimer's disease: Potential pools of new biomarkers. *J. Immunol. Res.* **2017**, *2017*, 4626540. [CrossRef]

293. Dionisio-Santos, D.A.; Olschowka, J.A.; O'Banion, M.K. Exploiting microglial and peripheral immune cell crosstalk to treat Alzheimer's disease. *J. Neuroinflamm.* **2019**, *16*, 74. [CrossRef] [PubMed]

294. Mcgeer, P.; Mcgeer, E.; Rogers, J.; Sibley, J. Anti-inflammatory drugs and Alzheimer disease. *Lancet* **1990**, *335*, 1037. [CrossRef]

295. Giometto, B.; Argentiero, V.; Sanson, F.; Ongaro, G.; Tavolato, B. Acute-phase proteins in Alzheimer's disease. *Eur. Neurol.* **1988**, *28*, 30–33. [CrossRef] [PubMed]

296. Miguel-Álvarez, M.; Santos-Lozano, A.; Sanchis-Gomar, F.; Fiuza-Luces, C.; Pareja-Galeano, H.; Garatachea, N.; Lucia, A. Non-steroidal anti-inflammatory drugs as a treatment for Alzheimer's disease: A systematic review and meta-analysis of treatment effect. *Drugs Aging* **2015**, *32*, 139–147. [CrossRef] [PubMed]

297. Liu, L.; Chan, C. The role of inflammasome in Alzheimer's disease. *Ageing Res. Rev.* **2014**, *15*, 6–15. [CrossRef]

298. Salter, M.W.; Stevens, B. Microglia emerge as central players in brain disease. *Nat. Med.* **2017**, *23*, 1018. [CrossRef]

299. Yu, Y.; Richard, D.Y. Microglial Aβ receptors in Alzheimer's disease. *Cell. Mol. Neurobiol.* **2015**, *35*, 71–83. [CrossRef] [PubMed]

300. Halle, A.; Hornung, V.; Petzold, G.C.; Stewart, C.R.; Monks, B.G.; Reinheckel, T.; Fitzgerald, K.A.; Latz, E.; Moore, K.J.; Golenbock, D.T. The NALP3 inflammasome is involved in the innate immune response to amyloid-β. *Nat. Immunol.* **2008**, *9*, 857. [CrossRef]

301. Heneka, M.T.; Kummer, M.P.; Stutz, A.; Delekate, A.; Schwartz, S.; Vieira-Saecker, A.; Griep, A.; Axt, D.; Remus, A.; Tzeng, T.C. NLRP3 is activated in Alzheimer's disease and contributes to pathology in APP/PS1 mice. *Nature* **2013**, *493*, 674–678. [CrossRef]

302. Sheedy, F.J.; Grebe, A.; Rayner, K.J.; Kalantari, P.; Ramkhelawon, B.; Carpenter, S.B.; Becker, C.E.; Ediriweera, H.N.; Mullick, A.E.; Golenbock, D.T. CD36 coordinates NLRP3 inflammasome activation by facilitating intracellular nucleation of soluble ligands into particulate ligands in sterile inflammation. *Nat. Immunol.* **2013**, *14*, 812–820. [CrossRef] [PubMed]

303. Calsolaro, V.; Edison, P. Neuroinflammation in Alzheimer's disease: Current evidence and future directions. *Alzheimers Dement.* **2016**, *12*, 719–732. [CrossRef]

304. Sasaki, A. Microglia and brain macrophages: An update. *Neuropathology* **2017**, *37*, 452–464. [CrossRef] [PubMed]

305. Perez, J.; Carrero, I.; Gonzalo, P.; Arevalo-Serrano, J.; Sanz-Anquela, J.; Ortega, J.; Rodriguez, M.; Gonzalo-Ruiz, A. Soluble oligomeric forms of beta-amyloid (Aβ) peptide stimulate Aβ production via astrogliosis in the rat brain. *Exp. Neurol.* **2010**, *223*, 410–421. [CrossRef]

306. Carrero, I.; Gonzalo, M.; Martin, B.; Sanz-Anquela, J.; Arevalo-Serrano, J.; Gonzalo-Ruiz, A. Oligomers of beta-amyloid protein (Aβ1-42) induce the activation of cyclooxygenase-2 in astrocytes via an interaction with interleukin-1beta, tumour necrosis factor-alpha, and a nuclear factor kappa-B mechanism in the rat brain. *Exp. Neurol.* **2012**, *236*, 215–227. [CrossRef] [PubMed]

307. Ringman, J.M.; O'Neill, J.; Geschwind, D.; Medina, L.; Apostolova, L.G.; Rodriguez, Y.; Schaffer, B.; Varpetian, A.; Tseng, B.; Ortiz, F. Diffusion tensor imaging in preclinical and presymptomatic carriers of familial Alzheimer's disease mutations. *Brain* **2007**, *130*, 1767–1776. [CrossRef]

308. Desai, M.K.; Guercio, B.J.; Narrow, W.C.; Bowers, W.J. An Alzheimer's disease-relevant presenilin-1 mutation augments amyloid-beta-induced oligodendrocyte dysfunction. *Glia* **2011**, *59*, 627–640. [CrossRef]

309. Vodovotz, Y.; Lucia, M.S.; Flanders, K.C.; Chesler, L.; Xie, Q.-W.; Smith, T.W.; Weidner, J.; Mumford, R.; Webber, R.; Nathan, C. Inducible nitric oxide synthase in tangle-bearing neurons of patients with Alzheimer's disease. *J. Exp. Med.* **1996**, *184*, 1425–1433. [CrossRef]

310. Limatola, C.; Ransohoff, R.M. Modulating neurotoxicity through CX3CL1/CX3CR1 signaling. *Front. Cell. Neurosci.* **2014**, *8*, 229. [CrossRef] [PubMed]

311. D'Ambrosio, A.L.; Pinsky, D.J.; Connolly, E.S. The role of the complement cascade in ischemia/reperfusion injury: Implications for neuroprotection. *Mol. Med.* **2001**, *7*, 367–382. [CrossRef] [PubMed]

312. Shen, Y.; Yang, L.; Li, R. What does complement do in Alzheimer's disease? Old molecules with new insights. *Transl. Neurodegener.* **2013**, *2*, 21. [CrossRef] [PubMed]

313. Shen, Y.; Meri, S. Yin and Yang: Complement activation and regulation in Alzheimer's disease. *Prog. Neurobiol.* **2003**, *70*, 463–472. [CrossRef]

314. Chibnik, L.B.; Shulman, J.M.; Leurgans, S.E.; Schneider, J.A.; Wilson, R.S.; Tran, D.; Aubin, C.; Buchman, A.S.; Heward, C.B.; Myers, A.J. CR1 is associated with amyloid plaque burden and age-related cognitive decline. *Ann. Neurol.* **2011**, *69*, 560–569. [CrossRef] [PubMed]

315. Azizi, G.; Khannazer, N.; Mirshafiey, A. The potential role of chemokines in Alzheimer's disease pathogenesis. *Am. J. Alzheimers Dis. Other Dement.* **2014**, *29*, 415–425. [CrossRef] [PubMed]

316. Zilka, N.; Kazmerova, Z.; Jadhav, S.; Neradil, P.; Madari, A.; Obetkova, D.; Bugos, O.; Novak, M. Who fans the flames of Alzheimer's disease brains? Misfolded tau on the crossroad of neurodegenerative and inflammatory pathways. *J. Neuroinflamm.* **2012**, *9*, 47. [CrossRef]

317. Westin, K.; Buchhave, P.; Nielsen, H.; Minthon, L.; Janciauskiene, S.; Hansson, O. CCL2 is associated with a faster rate of cognitive decline during early stages of Alzheimer's disease. *PLoS ONE* **2012**, *7*, e30525. [CrossRef] [PubMed]

318. Jiang, H.; Hampel, H.; Prvulovic, D.; Wallin, A.; Blennow, K.; Li, R.; Shen, Y. Elevated CSF levels of TACE activity and soluble TNF receptors in subjects with mild cognitive impairment and patients with Alzheimer's disease. *Mol. Neurodegener.* **2011**, *6*, 69. [CrossRef] [PubMed]

319. Wang, W.-Y.; Tan, M.-S.; Yu, J.-T.; Tan, L. Role of pro-inflammatory cytokines released from microglia in Alzheimer's disease. *Ann. Transl. Med.* **2015**, *3*, 136.

320. Goldgaber, D.; Harris, H.W.; Hla, T.; Maciag, T.; Donnelly, R.J.; Jacobsen, J.S.; Vitek, M.P.; Gajdusek, D.C. Interleukin 1 regulates synthesis of amyloid beta-protein precursor mRNA in human endothelial cells. *Proc. Natl. Acad. Sci. USA* **1989**, *86*, 7606–7610. [CrossRef]

321. Griffin, W.S.T.; Sheng, J.G.; Roberts, G.W.; Mrak, R.E. Interleukin-1 expression in different plaque types in Alzheimer's disease: Significance in plaque evolution. *J. Neuropathol. Exp. Neurol.* **1995**, *54*, 276–281. [CrossRef]

322. Park, S.Y.; Kang, M.J.; Han, J.S. Interleukin-1 beta promotes neuronal differentiation through the Wnt5a/RhoA/JNK pathway in cortical neural precursor cells. *Mol. Brain* **2018**, *11*, 39. [CrossRef]

323. Farrar, W.; Kilian, P.; Ruff, M.; Hill, J.; Pert, C. Visualization and characterization of interleukin 1 receptors in brain. *J. Immunol.* **1987**, *139*, 459–463.

324. Buxbaum, J.D.; Oishi, M.; Chen, H.I.; Pinkas-Kramarski, R.; Jaffe, E.A.; Gandy, S.E.; Greengard, P. Cholinergic agonists and interleukin 1 regulate processing and secretion of the Alzheimer beta/A4 amyloid protein precursor. *Proc. Natl. Acad. Sci. USA* **1992**, *89*, 10075–10078. [CrossRef] [PubMed]

325. Quintanilla, R.A.; Orellana, D.I.; González-Billault, C.; Maccioni, R.B. Interleukin-6 induces Alzheimer-type phosphorylation of tau protein by deregulating the cdk5/p35 pathway. *Exp. Cell Res.* **2004**, *295*, 245–257. [CrossRef] [PubMed]

326. Takeuchi, K.; Smale, S.; Premchand, P.; Maiden, L.; Sherwood, R.; Thjodleifsson, B.; Bjornsson, E.; Bjarnason, I. Prevalence and mechanism of nonsteroidal anti-inflammatory drug–induced clinical relapse in patients with inflammatory bowel disease. *Clin. Gastroenterol. Hepatol.* **2006**, *4*, 196–202. [CrossRef]

327. Perry, R.T.; Collins, J.S.; Wiener, H.; Acton, R.; Go, R.C. The role of TNF and its receptors in Alzheimer's disease. *Neurobiol. Aging* **2001**, *22*, 873–883. [CrossRef]

328. Yamamoto, M.; Kiyota, T.; Horiba, M.; Buescher, J.L.; Walsh, S.M.; Gendelman, H.E.; Ikezu, T. Interferon-γ and tumor necrosis factor-α regulate amyloid-β plaque deposition and β-secretase expression in Swedish mutant APP transgenic mice. *Am. J. Pathol.* **2007**, *170*, 680–692. [CrossRef] [PubMed]

329. Chang, R.; Yee, K.-L.; Sumbria, R.K. Tumor necrosis factor α inhibition for Alzheimer's disease. *J. Cent. Nerv. Syst. Dis.* **2017**, *9*, 1179573517709278. [CrossRef]

330. Grammas, P.; Ovase, R. Cerebrovascular transforming growth factor-β contributes to inflammation in the Alzheimer's disease brain. *Am. J. Pathol.* **2002**, *160*, 1583–1587. [CrossRef]

331. Cao, B.-B.; Zhang, X.-X.; Du, C.-Y.; Liu, Z.; Qiu, Y.-H.; Peng, Y.-P. TGF-β1 Provides Neuroprotection via Inhibition of Microglial Activation in 3-Acetylpyridine-Induced Cerebellar Ataxia Model Rats. *Front. Neurosci.* **2020**, *14*, 187. [CrossRef]

332. Caraci, F.; Spampinato, S.; Sortino, M.A.; Bosco, P.; Battaglia, G.; Bruno, V.; Drago, F.; Nicoletti, F.; Copani, A. Dysfunction of TGF-β1 signaling in Alzheimer's disease: Perspectives for neuroprotection. *Cell Tissue Res.* **2012**, *347*, 291–301. [CrossRef]

333. Lio, D.; Licastro, F.; Scola, L.; Chiappelli, M.; Grimaldi, L.; Crivello, A.; Colonna-Romano, G.; Candore, G.; Franceschi, C.; Caruso, C. Interleukin-10 promoter polymorphism in sporadic Alzheimer's disease. *Genes Immun.* **2003**, *4*, 234–238. [CrossRef]

334. Zhang, Y.; Zhang, J.; Tian, C.; Xiao, Y.; Li, X.; He, C.; Huang, J.; Fan, H. The 1082G/A polymorphism in IL-10 gene is associated with risk of Alzheimer's disease: A meta-analysis. *J. Neurol. Sci.* **2011**, *303*, 133–138. [CrossRef] [PubMed]

335. Ho, L.; Purohit, D.; Haroutunian, V.; Luterman, J.D.; Willis, F.; Naslund, J.; Buxbaum, J.D.; Mohs, R.C.; Aisen, P.S.; Pasinetti, G.M. Neuronal cyclooxygenase 2 expression in the hippocampal formation as a function of the clinical progression of Alzheimer disease. *Arch. Neurol.* **2001**, *58*, 487–492. [CrossRef] [PubMed]

336. Xiang, Z.; Ho, L.; Yemul, S.; Zhao, Z.; Pompl, P.; Kelley, K.; Dang, A.; Qing, W.; Teplow, D.; Pasinetti, G.M. Cyclooxygenase-2 promotes amyloid plaque deposition in a mouse model of Alzheimer's disease neuropathology. *Gene Expr. J. Liver Res.* **2002**, *10*, 271–278. [CrossRef] [PubMed]

337. Xiang, Z.; Ho, L.; Valdellon, J.; Borchelt, D.; Kelley, K.; Spielman, L.; Aisen, P.S.; Pasinetti, G.M. Cyclooxygenase (COX)-2 and cell cycle activity in a transgenic mouse model of Alzheimer's disease neuropathology. *Neurobiol. Aging* **2002**, *23*, 327–334. [CrossRef]

338. Hoozemans, J.; Veerhuis, R.; Janssen, I.; Rozemuller, A.J.; Eikelenboom, P. Interleukin-1β induced cyclooxygenase. *Exp. Gerontol.* **2001**, *2*, 559–570. [CrossRef]

339. Cinelli, M.A.; Do, H.T.; Miley, G.P.; Silverman, R.B. Inducible nitric oxide synthase: Regulation, structure, and inhibition. *Med. Res. Rev.* **2020**, *40*, 158–189. [CrossRef]

340. Asiimwe, N.; Yeo, S.G.; Kim, M.-S.; Jung, J.; Jeong, N.Y. Nitric oxide: Exploring the contextual link with alzheimer's disease. *Oxid. Med. Cell. Longev.* **2016**, *2016*, 7205747. [CrossRef]

341. Saez, T.E.; Pehar, M.; Vargas, M.; Barbeito, L.; Maccioni, R.B. Astrocytic nitric oxide triggers tau hyperphosphorylation in hippocampal neurons. *In Vivo* **2004**, *18*, 275–280.

342. Zhang, Q.; Yu, J.-T.; Zhu, Q.-X.; Zhang, W.; Wu, Z.-C.; Miao, D.; Tan, L. Complement receptor 1 polymorphisms and risk of late-onset Alzheimer's disease. *Brain Res.* **2010**, *1348*, 216–221. [CrossRef]

343. Brubaker, W.D.; Crane, A.; Johansson, J.U.; Yen, K.; Garfinkel, K.; Mastroeni, D.; Asok, P.; Bradt, B.; Sabbagh, M.; Wallace, T.L. Peripheral complement interactions with amyloid β peptide: Erythrocyte clearance mechanisms. *Alzheimers Dement.* **2017**, *13*, 1397–1409. [CrossRef] [PubMed]

344. Ager, R.R.; Fonseca, M.I.; Chu, S.H.; Sanderson, S.D.; Taylor, S.M.; Woodruff, T.M.; Tenner, A.J. Microglial C5aR (CD88) expression correlates with amyloid-β deposition in murine models of Alzheimer's disease. *J. Neurochem.* **2010**, *113*, 389–401. [CrossRef] [PubMed]

345. Yazawa, H.; Yu, Z.X.; Takeda, K.; Le, Y.; Gong, W.; Ferrans, V.J.; Oppenheim, J.J.; Li, C.C.H.; Wang, J.M. β Amyloid peptide (Aβ42) is internalized via the G-protein-coupled receptor FPRL1 and forms fibrillar aggregates in macrophages1. *FASEB J.* **2001**, *15*, 2454–2462. [CrossRef] [PubMed]

346. Tiffany, H.L.; Lavigne, M.C.; Cui, Y.-H.; Wang, J.-M.; Leto, T.L.; Gao, J.-L.; Murphy, P.M. Amyloid-β induces chemotaxis and oxidant stress by acting at formylpeptide receptor 2, a G protein-coupled receptor expressed in phagocytes and brain. *J. Biol. Chem.* **2001**, *276*, 23645–23652. [CrossRef] [PubMed]

347. Paresce, D.M.; Ghosh, R.N.; Maxfield, F.R. Microglial cells internalize aggregates of the Alzheimer's disease amyloid β-protein via a scavenger receptor. *Neuron* **1996**, *17*, 553–565. [CrossRef]

348. Coraci, I.S.; Husemann, J.; Berman, J.W.; Hulette, C.; Dufour, J.H.; Campanella, G.K.; Luster, A.D.; Silverstein, S.C.; El Khoury, J.B. CD36, a class B scavenger receptor, is expressed on microglia in Alzheimer's disease brains and can mediate production of reactive oxygen species in response to β-amyloid fibrils. *Am. J. Pathol.* **2002**, *160*, 101–112. [CrossRef]

349. Stewart, C.R.; Stuart, L.M.; Wilkinson, K.; Van Gils, J.M.; Deng, J.; Halle, A.; Rayner, K.J.; Boyer, L.; Zhong, R.; Frazier, W.A. CD36 ligands promote sterile inflammation through assembly of a Toll-like receptor 4 and 6 heterodimer. *Nat. Immunol.* **2010**, *11*, 155. [CrossRef]

350. Origlia, N.; Bonadonna, C.; Rosellini, A.; Leznik, E.; Arancio, O.; Yan, S.S.; Domenici, L. Microglial receptor for advanced glycation end product-dependent signal pathway drives β-amyloid-induced synaptic depression and long-term depression impairment in entorhinal cortex. *J. Neurosci.* **2010**, *30*, 11414–11425. [CrossRef]

351. Slowik, A.; Merres, J.; Elfgen, A.; Jansen, S.; Mohr, F.; Wruck, C.J.; Pufe, T.; Brandenburg, L.O. Involvement of formyl peptide receptors in receptor for advanced glycation end products (RAGE)-and amyloid beta 1-42-induced signal transduction in glial cells. *Mol. Neurodegener.* **2012**, *7*, 55. [CrossRef]

352. Michaud, J.P.; Hallé, M.; Lampron, A.; Thériault, P.; Préfontaine, P.; Filali, M.; Tribout-Jover, P.; Lanteigne, A.-M.; Jodoin, R.; Cluff, C. Toll-like receptor 4 stimulation with the detoxified ligand monophosphoryl lipid A improves Alzheimer's disease-related pathology. *Proc. Natl. Acad. Sci. USA* **2013**, *10*, 1941–1946. [CrossRef]

353. Reed-Geaghan, E.G.; Savage, J.C.; Hise, A.G.; Landreth, G.E. CD14 and toll-like receptors 2 and 4 are required for fibrillar Aβ-stimulated microglial activation. *J. Neurosci.* **2009**, *29*, 11982–11992. [CrossRef]

354. Doi, Y.; Mizuno, T.; Maki, Y.; Jin, S.; Mizoguchi, H.; Ikeyama, M.; Doi, M.; Michikawa, M.; Takeuchi, H.; Suzumura, A. Microglia activated with the toll-like receptor 9 ligand CpG attenuate oligomeric amyloid β neurotoxicity in in vitro and in vivo models of Alzheimer's disease. *Am. J. Pathol.* **2009**, *175*, 2121–2132. [CrossRef] [PubMed]

355. Lee, S.; Varvel, N.H.; Konerth, M.E.; Xu, G.; Cardona, A.E.; Ransohoff, R.M.; Lamb, B.T. CX3CR1 deficiency alters microglial activation and reduces beta-amyloid deposition in two Alzheimer's disease mouse models. *Am. J. Pathol.* **2010**, *177*, 2549–2562. [CrossRef] [PubMed]

356. Leyns, C.E.; Ulrich, J.D.; Finn, M.B.; Stewart, F.R.; Koscal, L.J.; Serrano, J.R.; Robinson, G.O.; Anderson, E.; Colonna, M.; Holtzman, D.M. TREM2 deficiency attenuates neuroinflammation and protects against neurodegeneration in a mouse model of tauopathy. *Proc. Natl. Acad. Sci. USA* **2017**, *114*, 11524–11529. [CrossRef] [PubMed]

357. Zhao, Y.; Wu, X.; Li, X.; Jiang, L.-L.; Gui, X.; Liu, Y.; Sun, Y.; Zhu, B.; Piña-Crespo, J.C.; Zhang, M. TREM2 is a receptor for β-amyloid that mediates microglial function. *Neuron* **2018**, *97*, 1023–1031. [CrossRef] [PubMed]

358. Griciuc, A.; Serrano-Pozo, A.; Parrado, A.R.; Lesinski, A.N.; Asselin, C.N.; Mullin, K.; Hooli, B.; Choi, S.H.; Hyman, B.T.; Tanzi, R.E. Alzheimer's disease risk gene CD33 inhibits microglial uptake of amyloid beta. *Neuron* **2013**, *78*, 631–643. [CrossRef]

359. Chen, H.; Liu, S.; Ji, L.; Wu, T.; Ji, Y.; Zhou, Y.; Zheng, M.; Zhang, M.; Xu, W.; Huang, G. Folic acid supplementation mitigates Alzheimer's disease by reducing inflammation: A randomized controlled trial. *Mediat. Inflamm.* **2016**, *2016*, 5912146. [CrossRef]

360. Rafiee, S.; Asadollahi, K.; Riazi, G.; Ahmadian, S.; Saboury, A.A. Vitamin B12 inhibits tau fibrillization via binding to cysteine residues of tau. *ACS Chem. Neurosci.* **2017**, *8*, 2676–2682. [CrossRef]

361. Jia, L.; Wang, Y.; Wei, W.; Zhao, W.; Lu, F.; Liu, F. Vitamin B12 inhibits α-synuclein fibrillogenesis and protects against amyloid-induced cytotoxicity. *Food Funct.* **2019**, *10*, 2861–2870. [CrossRef]

362. Huang, L.K.; Chao, S.P.; Hu, C.J. Clinical trials of new drugs for Alzheimer disease. *J. Biomed. Sci.* **2020**, *27*, 18. [CrossRef]

Article

Relationship between Cognitive Dysfunction and Age-Related Variability in Oxidative Markers in Isolated Mitochondria of Alzheimer's Disease Transgenic Mouse Brains

Naoki Yoshida [1], Yugo Kato [2], Hirokatsu Takatsu [3] and Koji Fukui [1,2,*]

1   Molecular Cell Biology Laboratory, Department of Systems Engineering and Science, Graduate School of Engineering and Science, Shibaura Institute of Technology, Fukasaku 307, Minuma-ku, Saitama 337-8570, Japan; mf16068@shibaura-it.ac.jp
2   Molecular Cell Biology Laboratory, Department of Functional Control Systems, Graduate School of Engineering and Science, Shibaura Institute of Technology, Fukasaku 307, Minuma-ku, Saitama 337-8570, Japan; nb19102@shibaura-it.ac.jp
3   Department of Medical Technology, Faculty of Health Sciences, Kyorin University, Shimorenjaku 5-4-1, Mitaka, Tokyo 181-8612, Japan; takatsu@ks.kyorin-u.ac.jp
*   Correspondence: koji@sic.shibaura-it.ac.jp

Abstract: Many neurodegenerative disorders, including Alzheimer's disease (AD), are strongly associated with the accumulation of oxidative damage. Transgenic animal models are commonly used to elucidate the pathogenic mechanism of AD. Beta amyloid (Aβ) and tau hyperphosphorylation are very famous hallmarks of AD and well-studied, but the relationship between mitochondrial dysfunction and the onset and progression of AD requires further elucidation. In this study we used transgenic mice (the strain name is 5xFAD) at three different ages (3, 6, and 20 months old) as an AD model. Cognitive impairment in AD mice occurred in an age-dependent manner. Aβ1-40 expression significantly increased in an age-dependent manner in all brain regions with or without AD, and Aβ1-42 expression in the hippocampus increased at a young age. In a Western blot analysis using isolated mitochondria from three brain regions (cerebral cortex, cerebellum, and hippocampus), NMNAT-3 expression in the hippocampi of aged AD mice was significantly lower than that of young AD mice. SOD-2 expression in the hippocampi of AD mice was lower than for the age-matched controls. However, 3-NT expression in the hippocampi of AD mice was higher than for the age-matched controls. NQO-1 expression in the cerebral cortex of AD mice was higher than for the age-matched controls at every age that we examined. However, hippocampal NQO-1 expression in 6-month-old AD mice was significantly lower than in 3-month-old AD mice. These results indicate that oxidative stress in the hippocampi of AD mice is high compared to other brain regions and may induce mitochondrial dysfunction via oxidative damage. Protection of mitochondria from oxidative damage may be important to maintain cognitive function.

Keywords: Alzheimer's disease; reactive oxygen species; oxidative stress; cognitive impairment; mitochondria

Citation: Yoshida, N.; Kato, Y.; Takatsu, H.; Fukui, K. Relationship between Cognitive Dysfunction and Age-Related Variability in Oxidative Markers in Isolated Mitochondria of Alzheimer's Disease Transgenic Mouse Brains. *Biomedicines* **2022**, *10*, 281. https://doi.org/10.3390/biomedicines10020281

Academic Editors: Kumar Vaibhav, Jun Lu, Meenakshi Ahluwalia and Pankaj Gaur

Received: 17 December 2021
Accepted: 25 January 2022
Published: 26 January 2022

Publisher's Note: MDPI stays neutral with regard to jurisdictional claims in published maps and institutional affiliations.

Copyright: © 2022 by the authors. Licensee MDPI, Basel, Switzerland. This article is an open access article distributed under the terms and conditions of the Creative Commons Attribution (CC BY) license (https://creativecommons.org/licenses/by/4.0/).

1. Introduction

Alzheimer's disease (AD) is a severe irreversible brain disorder. The number of patients suffering from it in Japan is currently higher than 3 million, and the rate of its increase there is the fastest in the world [1]. The most important symptom of AD is dementia via dysfunction of cholinergic neurons [2]; in particular, neuronal cell death occurs frequently in the hippocampal region [3]. Donepezil hydrochloride, an AD therapeutic, acts as an inhibitor of acetylcholine esterase [4], but because treatment with donepezil does not stop AD progression [5], the development of novel therapeutic agents is necessary.

To elucidate the mechanisms of AD onset and progression, researchers have used many experimental models, such as fruit flies [6], Caenorhabditis elegans [7], and monkeys [8].

The most frequently used animal models are transgenic mice, and many kinds have been developed [9]. AD research focuses on two major neuropathological features, namely, the accumulation of amyloid-beta (Aβ) [10] and tau hyperphosphorylation [11], but their relative contributions have not been determined.

Reactive oxygen species (ROS) such as superoxide anion radicals are mainly produced in the mitochondrial electron transport chain [12,13]. Superoxide is reduced to hydroxyl radicals in several steps and attacks cell membranes. Because oxygen is essential for life, ROS production cannot be eliminated in aerobic organisms. The accumulation of oxidative products such as lipid hydroperoxides is associated with many serious diseases such as aging, cancer, cardiovascular disease, arteriosclerosis, and high blood pressure [13,14]. In particular, the brain is rich in polyunsaturated fatty acids and consumes a large volume of oxygen compared to other organs; thus, it is easily affected by ROS. Recently, many reports have noted a relationship between ROS and AD [15]. ROS accelerates Aβ production, and Aβ plaques release ROS, resulting in neuronal cell death [16]. Our previous study found Aβ deposits in the hippocampal CA1 region of rats with vitamin E deficiency, normally aged rats, and young rats exposed to high oxygen [17]. These animals showed significant oxidation, especially in the brain, compared to normal young rats [17].

To prevent AD onset and progression, identification of early changes in the brain is necessary, and many researchers have searched for suitable markers. In the early stage of AD, axonal degenerative effects, such as shrinkage and beading, appear in the brain [18,19]. These phenomena are also seen in other neurodegenerative disorders such as Huntington's disease [20] and multiple sclerosis [21]. In our previous study, treatment with a low concentration of hydrogen peroxide induces axonal degeneration [22]. In this condition, calcium homeostasis is disrupted, and mitochondrial oxidation is induced [23]. These results indicate that mitochondrial dysfunction via oxidation is strongly related to the onset and progression of AD. To further our understanding of this relationship, in this study we used AD-transgenic mice (5xFAD) and isolated mitochondria at different ages to measure ROS-related protein expression.

2. Materials and Methods

2.1. Animals

All animal experiments were approved by the Animal Protection and Ethics Committee of Shibaura Institute of Technology (Approval number #15001), and 5xFAD transgenic mice (#008730, MMRRC034848, B6.Cg-Tg (APPSwFlLon, PSEN1*M146L*L286V) 6799Vas/Mmjax, Alias/5XFAD) from The Jackson Laboratory (Bar Harbor, ME, USA) and self-bred prior to use in these experiments. In the present study, 3-, 6- and 20-month-old AD-transgenic mice were used. C57BL/6 Ncr male mice of the same age were obtained from Sankyo Labo Service Corp. Inc. (Tokyo, Japan) and used as a control group. All mice were maintained in conditions of controlled temperature ($22 \pm 2\ ^{\circ}$C) and a 12 h light/dark cycle, and were provided with free access to food and water. Food consisted of normal diet pellets (Labo MR Stock) purchased from Nosan Corp. (Kanagawa, Japan). Cognition and motor function were assessed using tests as described below. Following assessments, the mice were euthanized, and brain samples (cerebral cortex (Cortex), cerebellum (Cer), and hippocampus (Hip)) were collected for analysis. All other chemical agents were obtained from FUJIFILM Wako Pure Chemical Corp. (Osaka, Japan).

2.2. Behavioral Assessment

2.2.1. Morris Water Maze

Cognitive function was assessed using a Morris water maze apparatus [24,25]. The maze apparatus (140 cm in diameter and 45 cm in height) consists of a pool constructed of acrylic resin. The bottom of the pool was divided into four quadrants by lines and was set up with four different visible marks positioned around the pool. A submerged platform was placed in the center of one quadrant. The water temperature of the pool was maintained at $22 \pm 2\ ^{\circ}$C. Before starting the cognitive performance trials, the animals were acclimated to the pool. by being allowed to swim freely for 60 s in the absence of a platform, and to handling by the experimenter over a 3-day period. The cognitive trials were performed four times per day and continued for five consecutive days. All trials were performed at the same time of day, and were carried out every 3 h (starting at 9:00, 12:00, 15:00, and 18:00). We performed a total of 20 trials per mouse. The platform was maintained in the same location of the pool for all trials. The escape latency (time to reach the goal), swimming distance, swimming speed, and the proportion of time spent swimming in the quadrant containing the platform were measured using the ANY-maze software (ver. 4.98; Stoelting Co., Wood Dale, IL, USA).

2.2.2. Rota-Rod Test

The rota-rod (Muromachi Kikai Co., Ltd., Tokyo, Japan) test was used to assess coordinated movement ability, as described previously with some modifications [26]. The rod was accelerated from 5 to 50 rpm over a duration of 120 s. The latency to the time to fall was measured.

2.3. Mitochondria Isolation

Mitochondria were isolated from the cerebral cortex, cerebellum and hippocampus of each mouse using a self-made buffer (2.5 M sucrose, 5 mM 2-[4-(2-hydroxyethyl)-1-piperazinyl] ethanesulfonic acid (HEPES), 1 mM ethylenediaminetetraacetic acid (EDTA), pH 7.2). After sonication, the samples were centrifuged at $500\times g$ for 10 min at 4 °C. The supernatant was centrifuged at $10,000\times g$ for 10 min at 4 °C. The precipitate was diluted in lysis buffer and mixed with $5\times$ sodium dodecyl sulfate (SDS) buffer.

2.4. Western Blotting

All samples except isolated mitochondria samples were homogenized in lysis buffer and used in Western blotting as described previously [27], with some modifications. Sample lysates were centrifuged, and protein contents were determined using the Bradford assay (Bio-Rad protein assay, #500-0006JA, Bio-Rad Laboratories, Inc., Hercules, CA, USA) according to the manufacturer's protocol. Protein extracts (brain homogenate samples of 10 and 50 μg, respectively) were separated on 10, 12, and 15% SDS-polyacrylamide gels and transferred to polyvinylidene difluoride (PVDF) transfer membranes (Immobilon; Merck KGaA, Darmstadt, Germany). The PVDF membranes were washed and incubated in blocking solution (Tris-HCl-buffered saline, pH 7.6 (TBS), containing 0.1% Tween 20 and 2% non-fat skim milk) for 1 h at room temperature. The membranes were washed in TBS containing 0.1% Tween 20, and then treated with each primary antibody (anti-β-amyloid (1-40), mouse monoclonal antibody (4H308), #GTX17420, GeneTex Inc., Los Angeles, CA, USA; anti-amyloid beta 42 rabbit polyclonal antibody, #PA3-16761 Thermo Fisher Scientific Inc., Waltham, MA, USA; anti-tau [TAU-5] mouse monoclonal antibody, #ab80579, Abcam Inc., Cambridge, UK; anti-tau (phosphor-S262) rabbit polyclonal antibody, #ab131354 Abcam Inc.; anti-brain-derived neurotropic factor (BDNF) (N-20) rabbit polyclonal antibody, #sc-546 SANTA CRUZ BIOTECHNOLOGY (SCBT) Inc., Dallas, TX, USA; anti-nerve growth factor (NGF) (H-20) rabbit polyclonal antibody, #sc-548, SCBT Inc.; anti-tropomyosin receptor kinase A (TrkA) (763) rabbit polyclonal antibody, #sc-118, SCBT Inc.; anti-TrkB (H-181) rabbit polyclonal antibody, #sc-8316, SCBT Inc.; rabbit polyclonal anti-COX-IV mitochondrial loading control antibody, #ab16056, Abcam plc.; mouse monoclonal [2A12] anti-3-nitrotyrosine antibody (3-NT), #ab52309, Abcam plc.; mouse monoclonal anti-cytochrome C [37BA11] antibody, #ab110325, Abcam plc.; goat polyclonal anti-nicotinamide mononucleotide adenylyl transferase (NMNAT-3) antibody, #ab121030, Abcam plc.; rabbit polyclonal anti-nicotine adenine dinucleotide phosphate (NAD(P)H) quinone oxidoreductase-1 (NQO-1) antibody, #bs-2184R, Bioss Antibodies Inc., Woburn, MA, USA; rabbit polyclonal anti-superoxide dismutase-2 (SOD-2) antibody, #bs-1080R, Bioss Antibodies Inc.) overnight at 4 °C. Anti-mouse, -rabbit or -goat IgG horseradish peroxidase-conjugated antibodies (Promega Corp., Madison, WI, USA) were used as secondary antibodies at 1:4000 dilution for 1 h at room temperature. All Western blotting experiments were performed at least three times. All chemiluminescent signals were generated by incubation with the detection reagents (Immobilon; Merck KGaA) according to the manufacturer's protocol. For normalization of the bands for each protein, the membranes were reprobed with anti-α-tubulin rabbit monoclonal antibody (#2125, Cell Signaling Technology Inc., Danvers, MA, USA). The relative intensities were determined using LAS-3000 (FUJIFILM Corp., Tokyo, Japan). Expression ratios were calculated by dividing each protein value by that of α-tubulin using ImageJ software (National Institutes of Health, Bethesda, MD, USA).

2.5. Statistical Analysis

Data are expressed as means $\pm$ standard error (SE). They were analyzed using GraphPad Prism 9.2.0 (GraphPad Software, San Diego, CA, USA); p values of less than 0.05 were considered statistically significant. The detailed statistical methods are described in the individual figure captions.

3. Results

The results of the Morris water maze test are shown in Figure 1. The goal time gradually decreased for all mouse groups. The average goal times on the final trial day for 6- and 20-month-old AD mice were significantly higher than those of age-matched control groups (Figure 1A). The swimming trajectories of the AD mice showed that their swimming distances were remarkably longer than those of age-matched controls (Figure 1B). However, no significant differences in the swimming speeds among all mouse groups were noted (Figure 1C). The ratio of staying time in the platform area was higher for the control mice compared to the age-matched AD mice (Figure 1D). However, no significant differences were found for any mouse groups.

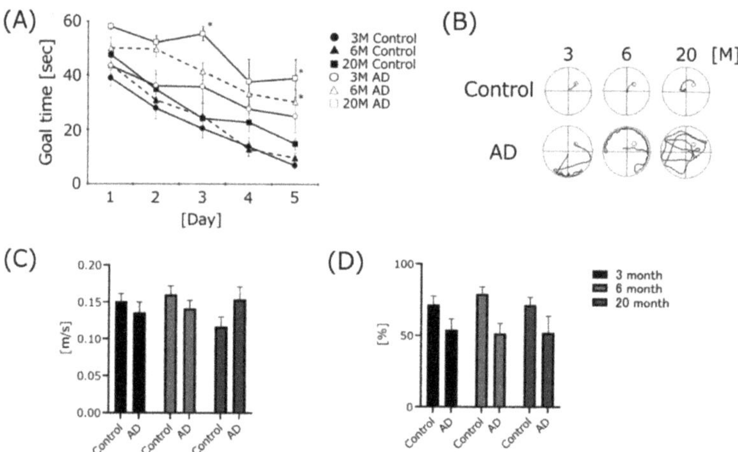

Figure 1. Differences in cognitive function between control and Alzheimer's disease (AD) transgenic mice depending on age. The time to goal (escape latency) in the Morris water maze test is shown in panel (**A**). The swimming trajectory and swimming speed are shown in panels (**B,C**). The ratio of staying time in the platform quadrant is shown in panel (**D**). Three different ages of AD mice (3 M, 6 M, 20 M) were used (3 M AD, $n = 10$; 6 M AD, $n = 10$; 20 M AD, $n = 5$). Age-matched C57BL/6 mice were used as a control group (3 M control, $n = 15$; 6 M control, $n = 15$; 20 M control, $n = 10$). * $p < 0.05$, vs. the age-matched control group. The data are shown as means $\pm$ SE. Statistical analyses of goal time were performed using two-way analysis of variance. Statistical analysis of goal time among all group was performed using the two-way analysis of variance. Statistical analyses of the goal time of each day, swimming speed, and the ratio of staying time in the platform quadrant were performed using the Tukey–Kramer method.

The coordination abilities were measured using a rota-rod apparatus (Figure 2). The time to fall for control and AD mice gradually decreased in an age-dependent manner. The time to fall for young mice did not differ between AD and control mice. However, the time to fall for 6- and 20-month-old control mice tended to decrease compared to the age-matched AD mice.

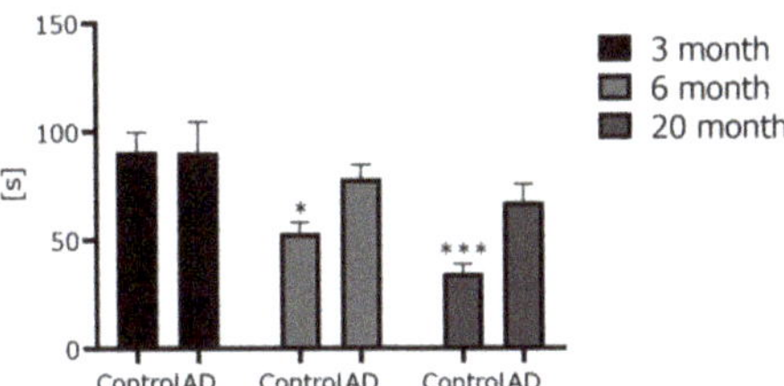

Figure 2. Time to fall in the rota-rod test. Three different ages of AD mice (3 M AD, $n = 8$; 6 M AD, $n = 8$; 20 M AD, $n = 8$) and age-matched C57BL/6 mice (3 M control, $n = 8$; 6 M control, $n = 8$; 20 M control, $n = 5$) were used. * $p < 0.05$ *** $p < 0.001$ vs. 3 M control. The data are shown as means $\pm$ SE. Comparisons were performed using the Tukey–Kramer method.

We measured Aβ protein expression using Western blotting, and measured band intensities for Aβ1-40 expression (Figure 3A). Aβ1-40 protein expression in all regions in AD and control mice gradually increased in an age-dependent manner. A band of high molecular size (fibril formation), representing Aβ1-42 was seen, and the band intensity was calculated (Figure 3B). Aβ1-42 expression in the hippocampus of AD mice dramatically increased compared to age-matched controls. The ratio of the hippocampal score for Aβ1-42 expression in the AD mice was higher than the score in the other two brain regions.

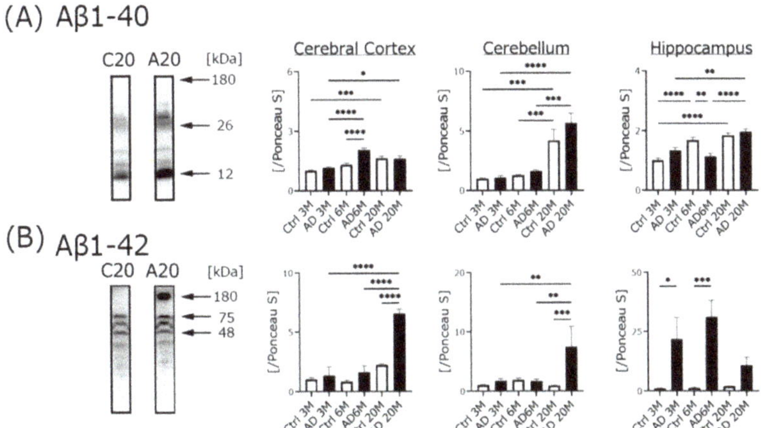

Figure 3. Changes in the levels of Aβ protein according to Western blot and immunohistochemical analyses. Western blotting experiments were performed using three different ages (3, 6, and 20 months) and three different brain regions (cerebral cortex, Cortex; cerebellum, Cer; hippocampus, Hip). Total protein expression in all bands was calculated for Aβ1-40 analysis (**A**). For Aβ1-42 analysis, the high-molecular-weight band (amyloid fibrils) was used for calculation (**B**). The ratio of each protein band intensity to Ponceau S intensity is shown in panels (**A**) and (**B**), with the ratios of each brain region of 3 M control samples set to 1. Black columns show 3 M, 6 M, and 20 M AD transgenic mice (3 M AD, $n = 9$; 6 M AD, $n = 8$; 20 M AD, $n = 5$), and white columns show age-matched C57BL/6 mice (3 M control, $n = 10$; 6 M control, $n = 10$; 20 M control, $n = 10$). Asterisks show significant differences (* $p < 0.05$, ** $p < 0.01$, *** $p < 0.001$, **** $p < 0.0001$). The data are shown as means $\pm$ SE. Comparisons were performed using the Tukey–Kramer method; 6 M AD and control mice were used.

We used 5xFAD mice in this experiment, in which no mutation at the site of tau phosphorylation is present. To analyze the relationship between Aβ accumulation and the ratio of tau protein including the phosphorylated form in AD mice, we analyzed tau and phospho6-tau expression in the three different brain regions with Western blotting (Figure 4).

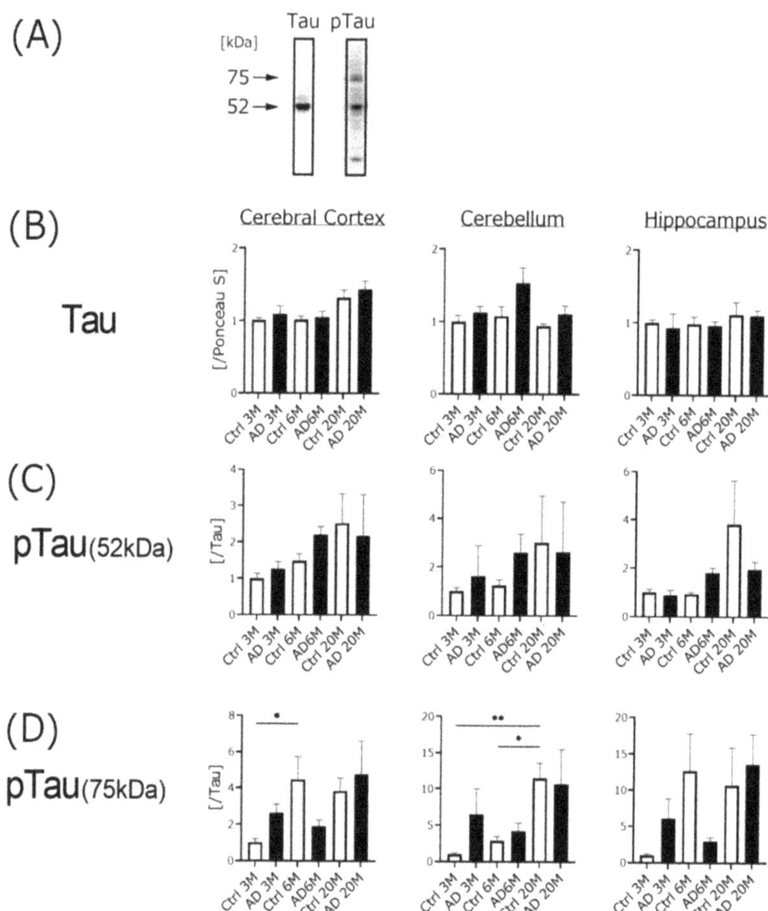

Figure 4. Western blotting analysis of the levels of tau and phospho6-tau expression in the brains of three different ages of AD mice. All experiments were performed using three different brain regions (cerebral cortex, Cortex; cerebellum, Cre; hippocampus, Hip). Western blotting images are shown in panel (**A**). Tau protein expression was calculated and shown in panel (**B**). Phospho-tau protein expression was calculated using two different band sizes (52 (**C**) and 75 kDa (**D**)). The ratio of each phospho-tau band intensity to Ponceau S intensity is shown, with the ratios of each brain region of 3 M control samples set to 1. Black columns show 3 M, 6 M, and 20 M AD mice (3 M AD, $n = 10$; 6 M AD, $n = 10$; 20 M AD, $n = 5$), and white columns show age-matched C57BL/6 mice (3 M control, $n = 10$; 6 M control, $n = 10$; 20 M control, $n = 5$). Asterisks show significant differences (* $p < 0.05$, ** $p < 0.01$). The data are shown as means ± SE. Comparisons were performed using the Tukey–Kramer method.

No differences in tau expression were seen among any samples. Two bands of different sizes representing phospho-tau (52 and 75 kDa) were detected and calculated. Each phospho6-tau expression was nominally increased (but not significant) in AD and control mice in an age-dependent manner. However, phospho6-tau (75 kDa) in the 6-month-old AD mice tended to be lower than that in the 3-month-old AD mice.

To clarify the relationship between cognitive dysfunction in AD mice and neuronal function, we assessed the expression of neurotrophic factors and their receptor proteins (Figure 5). However, the expression ratios of NGF, BDNF, and their receptors did not differ in any brain regions of either mouse group, except for NGF and TrkB expressions in the cortex region, and parts of cerebellum and hippocampus.

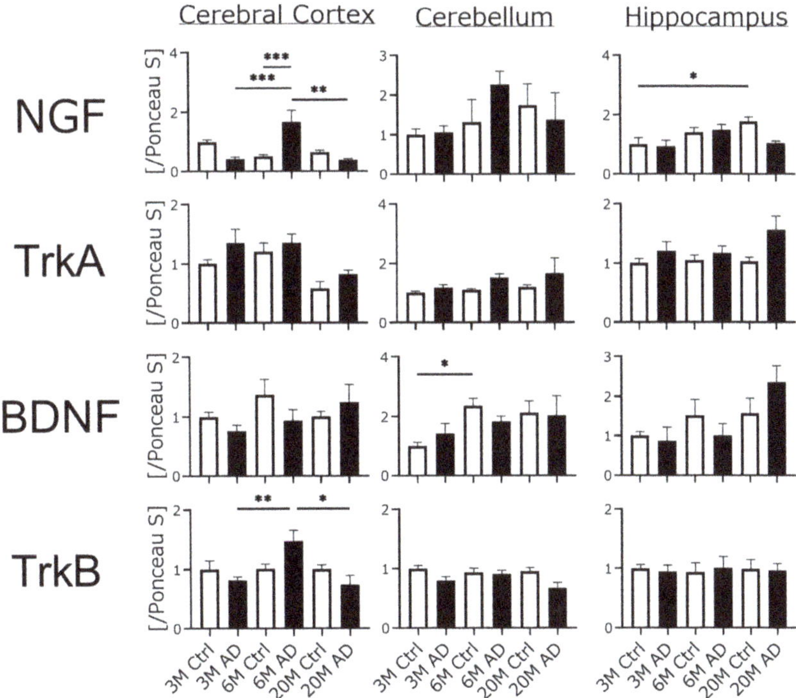

Figure 5. Western blotting analysis of the levels of neurotrophic-factor-related proteins in the brains of three different ages of AD transgenic mice. All experiments were performed using three different brain regions (cerebral cortex, Cortex; cerebellum, Cre; hippocampus, Hip). The ratio of each protein band intensity to Ponceau S intensity is shown, with ratios of each brain regions of 3 M control samples set to 1. Black columns show 3 M, 6 M, and 20 M AD mice (3 M AD, n = 8; 6 M AD, n = 8; 20 M AD, n = 4), and white columns show age-matched C57BL/6 mice (3 M control, n = 8; 6 M control, n = 8; 20 M control, n = 8). Asterisks show significant differences (* $p < 0.05$, ** $p < 0.01$, *** $p < 0.001$). The data are shown as means ± SE. Comparisons were performed using the Tukey–Kramer method.

We isolated mitochondria and measured protein expression using Western blotting. First, we checked the quality of our mitochondria isolation technique using our self-made buffer (Figure 6A). After isolation, we checked COX-IV expression using Western blotting, and found that the COX-IV band was very clear. These results suggest that our self-made buffer and isolation method has the same separation ability as the commercial kit. Hippocampal 3-NT expression in the AD and control mice gradually increased in an age-dependent manner. The 3-NT expression level in the hippocampus of the AD mice was higher than that in the age-matched controls. NQO-1 expression levels in the cerebral cortex of the AD mice were higher than those in the age-matched controls. However, the hippocampal SOD-2 expression level in AD mice was lower than those in the age-matched controls. NMNAT-3 expression tended to be lower in 20-month-old AD mice compared to the age-matched controls of all brain regions.

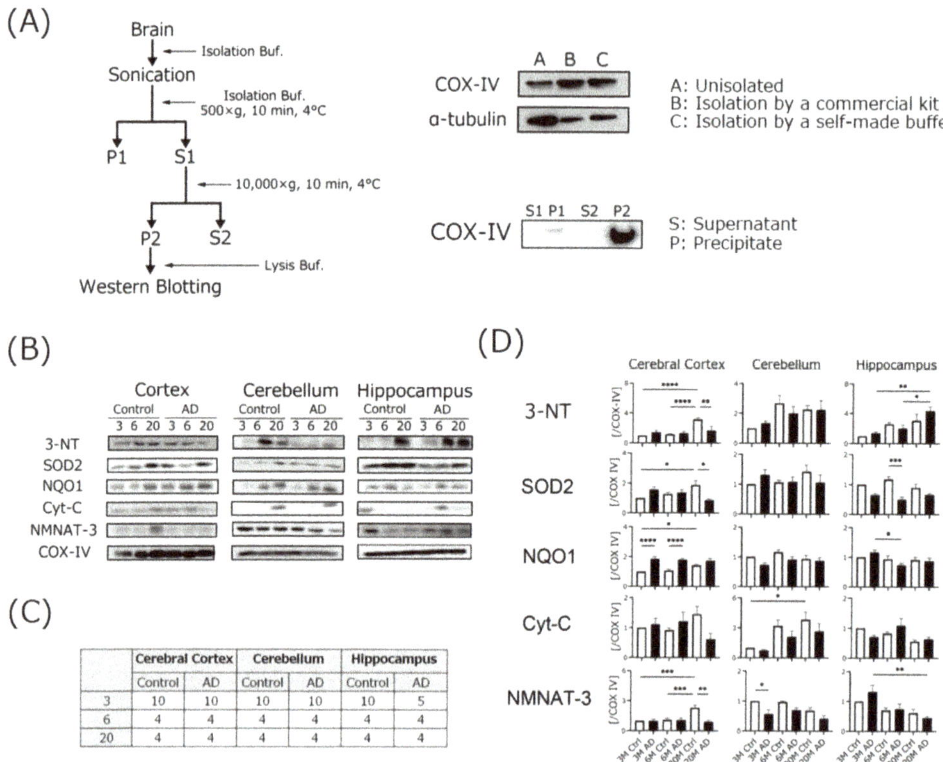

Figure 6. Western blotting analysis of the levels of each protein in isolated mitochondria of brains of three different ages of AD-transgenic mice. All experiments were performed using three different brain regions (cerebral cortex, Cortex; cerebellum, Cer; hippocampus, Hip). The mitochondrial isolation method is shown in panel (**A**). Western blotting images are shown in panel (**B**). The number of each sample is shown in panel (**C**). The ratios of each protein band intensity to COX-IV intensity are shown, with ratios of each brain region in 3 M control samples set to 1 (**D**). Black columns show 3 M, 6 M, and 20 M AD mice. White columns show age-matched control mice. Asterisks show significant differences (* $p < 0.05$, ** $p < 0.01$, *** $p < 0.001$, **** $p < 0.0001$). The data are shown as means $\pm$ SE. Comparisons were performed using the Tukey–Kramer method.

4. Discussion

4.1. AD Transgenic Mice Developed Cognitive Dysfunction, but Swimming Speed Did Not Change

To clarify the relationship between senescence and cognitive function in AD transgenic mice, we assessed their cognitive function using the Morris water maze task at three different ages. The goal time gradually decreased for all mouse groups with or without AD. The goal time of the final trial day gradually decreased in an age-dependent manner in both mouse groups. These results showed that cognitive function was gradually impaired by aging with or without AD and are consistent with our previous study [24]. The goal times on the final trial day for 6- and 20-month-old AD mice were significantly impaired compared to the age-matched controls. The goal time on the final trial day for the 3-month-old AD mice was nominally impaired (but not significantly) compared to the age-matched controls. The swimming distance of the AD mice was longer than that of the age-matched controls. These results showed that the AD mice developed cognitive dysfunction even at a young age, and the difference in the goal time between the controls and the AD mice at the same age gradually widened.

To clarify that these differences were not simply due to motor dysfunction, we measured swimming speed. We found no significant differences among the mouse groups. Additionally, we measured coordination ability using the rota-rod test. The fall time gradually decreased in an age-dependent manner in mice with or without AD. The fall time for the 20-month-old control group was significantly faster than that of the 3-month-old group. These results, using two different apparatuses, showed that AD mice did not induce much motor dysfunction and coordination disorder. Even during normal breeding, we found no particular difference in gait or movement in the AD mice. We found that the cognitive function of the AD mice simply deteriorated at a younger age. However, the fall time in the rota-rod test for the AD mice tended to be nominally longer (but not significantly so) compared to that of the age-matched controls. The reason is likely due to the difference in body weight. The fall time in the rota-rod test tends to be body-weight-dependent. In our previous study, obese mice fell faster than age-matched controls, and a negative correlation between body weight and fall time was observed [28]. In this study, the mean body weights of the AD mice at each age were lower than those of the age-matched controls, despite being fed ad libitum (control 3-month-old (3 M), 22.44 g; control 6-month-old (6 M), 30.8 g; Control 20-month-old (20 M), 29.18 g; AD 3 M, 18.6 g; AD 6 M, 20.4 g; AD 20 M, 24.2 g). The difference in the fall time in the rota-rod test may be due to the difference in weight between the two groups. The reason why the body weight of the AD mice was lower than those of the age-matched controls is unknown.

4.2. Aβ Expression, but Not Phospho-Tau Was Significantly Increased in AD Transgenic Mice

In this experiment, we purchased AD transgenic mice, which are widely used [29,30]. Basic information on the 5xFAD strain is provided on the JAX official website (https: //www.jax.org/strain/008730, accessed on 17 December 2021). This strain overexpresses both human Aβ precursor protein with the Swedish (K670N, M671L), Florida (I716V), and London (V717I) Familial Alzheimer's Disease (FAD) mutations and human presenilin-1 harboring two FAD mutations, M146L and L286V. JAX recommends that this strain be used as an Aβ1-42-induced neurodegeneration model. To elucidate which AD-related proteins affected cognitive dysfunction in the Morris water maze trial, we examined Aβ1-40, -42, tau, and phospho-tau protein expression using Western blotting analysis.

Aβ1-40 expression in the AD mice was increased in all brain regions in an age-dependent manner. However, the rate of the increase in its expression in the AD mice, especially in the 20-month-old mice, was lower than that of Aβ1-42, and the control mice also gradually showed an increase in Aβ1-40 in every brain region in an age-dependent manner. Aβ1-42 expression in the AD mice also increased in an age-dependent manner, except in the hippocampus of the 20-month-old mice. Surprisingly, the increased rate of Aβ1-42 expression in the hippocampus was much higher than that of the other two brain regions in all AD mouse groups. Generally, the toxicity of Aβ1-42 is considered to be stronger than that of Aβ1-40 [31], and Aβ1-42 tends to aggregate [32]. Changes in the ratio of Aβ1-40 and -42 are associated with Aβ toxicity, and increased Aβ1-42 exacerbates the AD condition [33]. We selected a high-molecular-weight band (more than 180 kDa) and used it in the calculations for this study. Our results indicated that Aβ1-42 began to aggregate in the hippocampus of the AD mice from a young age. JAX explains that Aβ1-42 aggregation begins to accumulate in 1.5-month-old mice, which is broadly consistent with our results. Unfortunately, we did not examine the protein expression of both Aβ species in mice younger than 3 months. Because AD mice are self-breeding, they are expensive, time consuming, and very difficult to produce in quantity. We prioritized the creation of a 20-month-old group in this study. In the near future, we intend to conduct a similar study in AD mice younger than 1.5 months to clarify when Aβ1-42 aggregates.

We found no significant difference in tau expression in any of our samples. We calculated phospho-tau expression using two different molecular weight bands, and the expression of the 75-kDa band gradually increased in an age-dependent manner in both the AD and the control mice, except in the 6-month-old AD mice. The increased rate of

75-kDa phospho-tau expression was higher than that of the 52-kDa band, and hippocampal expression of the 75-kDa band was higher than that in the other two brain regions in both the AD and the control mice. These results indicate that cognitive impairment in AD mice is likely induced by Aβ1-42 aggregation in the hippocampus.

4.3. Neurotrophic Factors and Their Receptor Expression Were Unchanged in AD Transgenic Mouse Brains

Neurotrophic factors such as NGF and BDNF play an important role in the maintenance of higher brain function. The functions of neurotrophic factors include axonal elongation, neuronal survival, and neurotransmitter synthesis [34]. To clarify the effect of neurotrophic factor expression in cognitive dysfunction in AD mice, we examined the expression of NGF, BDNF, and their receptors with Western blotting. Contrary to expectation, we found no significant differences in any samples except for parts of the NGF and TrkB expressions in the cerebral cortex. In general, neurotrophic factor secretion declines gradually with age [35], but this study did not show that trend. Cholinergic neurons are severely damaged in AD mice, but NGF-releasing cells may not be significantly affected, and the rate of damage or dysfunction may vary by neuron type. Had we looked at the mRNA, a difference may have been seen or large individual differences may have been apparent. Any underlying causal factors remain unknown.

4.4. Mitochondrial Oxidative Damage Was Increased in AD Transgenic Mouse Brains

ROS attack many tissues and induce oxidative damage and cell death. The "free radical theory of aging" hypothesis was first proposed by Harman in 1956 [36]. The accumulation of oxidative damage is related to the onset or progression of many severe diseases, including neurodegenerative disorders and aging [13]. Oxidation levels in the brain are accelerated in AD mice and cause various symptoms [37]. Lipid hydroperoxides are one golden marker for determining oxidation. Thiobarbituric acid-reactive substances, 4-hydroxynonenal and prostaglandin F2alpha may also be measured. However, these are end products. We wanted to determine the oxidation-related markers that occur further upstream, not in the final product. We thought that there might be significant changes upstream, even if there were no significant differences in the final product. In particular, mitochondrial dysfunction is critical and plays a key role in aging and disease [38], and mitochondrial dysfunction-induced cellular calcium homeostasis deterioration is a major issue [15,39]. In our previous study, treatment with a low concentration of hydrogen peroxide induced neurite degeneration in neuroblastoma cell lines, and mitochondria accumulated in regions of neurite degeneration [40]. Excessive influx of calcium ions induces mitochondrial superoxide generation and membrane oxidation [23]. These results suggest that ROS-induced mitochondrial damage may be closely related to neuronal survival.

To determine the mitochondrial oxidation levels, we measured the expression of proteins using isolated mitochondria. SOD-2 (also known as Mn-SOD), which detoxifies superoxide, is specifically present in mitochondria. SOD-2 mutations are related to the onset of AD, Parkinson's disease, and stroke [41]. SOD-2 expression levels, especially in the hippocampus, were nominally (but not significantly) lower compared to age-matched controls in our experiment. This result indicates that AD mice may have an imbalanced redox state. However, 3-NT expression in the hippocampi of the AD mice was higher than that in the age-matched controls. 3-NT, which is an index of nitric oxide production in cells, is synthesized from peroxynitrite and tyrosine, and peroxynitrite is synthesized from superoxide and nitric oxide [42]. 3-NT is a marker of protein modification and is related to many diseases such as arteriosclerosis and AD [43,44]. These results indicated that nitric oxide production in the hippocampus of AD mice may be relatively high compared to other brain regions of AD mice, as well as in age-matched controls.

NMNAT-3 levels were also attenuated in every region of AD mouse brains in this study. NMNAT-3 catalyzes the formation of nicotinamide adenine dinucleotide (NAD+) synthesis from nicotinamide mononucleotide and ATP, and is localized in the mitochondria [45]. NAD+ is needed for the citric acid cycle and respiratory chain in mitochondria, and the NAD+ level is an index of mitochondrial function [46]. The activation of NMNAT-3 induces mitochondrial function and has an anti-aging effect in NMNAT-3 over expression mouse model [47]. We could not demonstrate a direct interaction between oxidative stress and changes in NMNAT-3 levels in mitochondria. Low NMNAT-3 levels in AD mice indicate that mitochondrial function may be impaired by ROS. Further investigation is needed to clarify this interaction.

NQO-1 is a flavin-adenine-dinucleotide-dependent flavoprotein that serves as an oxidative stress response protein [48]. NQO-1 expression is regulated by a complex of the Kelch-like ECH-associated protein 1 (Keap 1) and the nuclear factor erythroid 2-related factor 2 (Nrf2) system [49]. The induction or knockdown of NQO-1 may be directly related to the reduction or enhancement of the oxidative stress status in cells [50]. NQO-1 expression in the cerebral cortex of the AD mice was higher than that in the age-matched controls at every age that we examined. However, hippocampal NQO-1 expression in 6-month-old AD mice was significantly lower than 3-month-old AD mice. These results suggest that the antioxidant response function, including the Keap1/Nrf2 system, was functioning in the cerebral cortex, albeit with gradual weakening. In the hippocampus, the response of defense mechanisms including the expression of antioxidant enzymes such as SOD-2 to ROS was already weakened, and as a result, oxidation including 3-NT was promoted. Oxidation may result in impaired mitochondrial function, such as by NMNAT-3, leading to mitochondrial dysfunction, and finally, to the development of cognitive dysfunction.

However, all analytical data for Western blotting of isolated mitochondria were normalized by COX-IV. COX-IV is one of the mitochondrial-specific proteins and is a subunit of cytochrome c oxidase in the mitochondrial respiratory chain. For Western blotting using isolated mitochondria, COX-IV, VDAC1/Porin, and heat shock protein 60 (HSP60) may be used as loading controls. However, care should be taken when using these proteins as loading controls. COX-IV may already be damaged in AD transgenic mice [51,52]. It is important to consider the use of multiple loading proteins for more accurate results. In that sense, this result may not be completely correct, so that this should be kept in mind for future experiments.

5. Conclusions

In our experiments, we found that the AD mice showed significant damage from ROS. However, the number of samples of isolated mitochondria from AD mice was limited other than for the outside of the cerebral cortex region. If mitochondria can be protected from ROS, mitochondrial function and cognitive function may be maintained. Future studies should explore compounds that can specifically exert a strong antioxidant capacity in mitochondria.

Author Contributions: Conceptualization, K.F.; data curation, N.Y., Y.K. and K.F.; formal analysis, K.F.; investigation, N.Y., Y.K., H.T. and K.F.; project administration, K.F.; writing, review and editing, K.F.; all other contributions to the research, K.F. All authors have read and agreed to the published version of the manuscript.

Funding: The author(s) received no specific funding for this work.

Institutional Review Board Statement: All animal experiments were approved by the Animal Protection and Ethics Committee of Shibaura Institute of Technology (Approval number #15001).

Informed Consent Statement: Not applicable.

Data Availability Statement: All data generated or analyzed during this study are included in this published article.

Conflicts of Interest: The authors declare no conflict of interest.

References

1. Ohno, S.; Chen, Y.; Sakamaki, H.; Matsumaru, N.; Yoshino, M.; Tsukamoto, K. Humanistic burden among caregivers of patients with Alzheimer's disease or dementia in Japan: A large-scale cross-sectional survey. *J. Med. Econ.* **2021**, *24*, 181–192. [CrossRef] [PubMed]
2. Breijyeh, Z.; Karaman, R. Comprehensive review on Alzheimer's disease: Cause and Treatment. *Molecules* **2020**, *25*, 5789. [CrossRef] [PubMed]
3. Engstrom, A.K.; Walker, A.C.; Moudgal, R.A.; Myrick, D.A.; Kyle, S.M.; Bai, Y.; Rowley, M.J.; Katz, D.J. The inhibition of LSD1 via sequestration contributes to tau-mediated neurodegeneration. *Proc. Natl. Acad. Sci. USA* **2020**, *117*, 29133–29143. [CrossRef] [PubMed]
4. Vaz Miguel Silvestre, S. Alzheimer's disease: Recent treatment strategies. *Eur. J. Pharmacol.* **2020**, *887*, 173554.
5. Alzheimer's Association. 2020 Alzheimer's diseases facts and figures. *Alzheimers Dement.* **2020**, *16*, 391–460. [CrossRef]
6. Wang, X.; Davis, R.L. Early mitochondrial fragmentation and dysfunction in a drosophila model for Alzheimer's disease. *Mol. Neurobiol.* **2021**, *58*, 143–155. [CrossRef]
7. Malar, D.S.; Prasanth, M.I.; Jeyakumar, M.; Balamurugan, K.; Devi, K.P. Vitexin prevents Aβ proteotoxicity in transgenic Caenorhabditis elegans model of Alzheimer's disease by modulating unfolded protein response. *J. Biochem. Mol. Toxicol.* **2021**, *35*, e22632. [CrossRef]
8. Beckman, D.; Chakrabarty, P.; Ott, S.; Dao, A.; Zhou, E.; Janssen, W.G.; Donis-Cox, K.; Muller, S.; Kordower, J.H.; Morrision, J.H. A novel tau-based rhesus monkey model of Alzheimer's pathogenesis. *Alzheimers Dement.* **2021**, *17*, 933–945. [CrossRef]
9. Drummond, E.; Wisniewski, T. Alzheimer's disease: Experimental models and reality. *Acta Neuropathol.* **2017**, *133*, 155–175. [CrossRef]
10. Barage, S.H.; Sonawane, K.D. Amyloid cascade hypothesis: Pathogenesis and therapeutic strategies in Alzheimer's disease. *Neurpeptides* **2015**, *52*, 1–18. [CrossRef]
11. Wisniewski, T.; Boutajangout, A. Tau-based therapeutic approaches for Alzheimer's disease—A mini-review. *Gerontology* **2014**, *60*, 381–385.
12. Zorov, D.B.; Juhaszova, M.; Sollott, S.J. Mitochondrial reactive oxygen species (ROS) and ROS-induced ROS release. *Physiol. Rev.* **2014**, *94*, 909–950. [CrossRef]
13. Davalli, P.; Mitic, T.; Caporali, A.; Lauriola, A.; D'Arca, D. ROS, cell senescence, and novel molecular mechanisms in aging and age-related diseases. *Oxid. Med. Cell. Longev.* **2016**, *2016*, 3565127. [CrossRef]
14. Alfadda, A.A.; Sallam, R.M. Reactive oxygen species in health and disease. *J. Biomed. Biotechnol.* **2012**, *2012*, 936486. [CrossRef]
15. Fang, E.F.; Hou, Y.; Palikaras, K.; Adriaanse, B.A.; Kerr, J.S.; Yang, B.; Lautrup, S.; Hasan-Olive, M.M.; Caponio, D.; Dan, X.; et al. Mitophagy inhibits amyloid-β and tau pathology and reverses cognitive deficits in models of Alzheimer's disease. *Nat. Neurosci.* **2019**, *22*, 401–412. [CrossRef]
16. Cheignon, C.; Tomas, M.; Bonnefont-Rousselot, D.; Faller, P.; Hureau, C.; Collin, F. Oxidative stress and the amyloid beta peptide in Alzheimer's disease. *Redox Biol.* **2018**, *14*, 450–464. [CrossRef] [PubMed]
17. Fukui, K.; Takatsu, H.; Shinkai, T.; Suzuki, S.; Abe, K.; Urano, S. Appearance of amyloid β-like substances and delayed-type apoptosis in rat hippocampus CA1 region through aging and oxidative stress. *J. Alzheimers Dis.* **2005**, *8*, 299–309. [CrossRef] [PubMed]
18. Stokin, G.B.; Lillo, C.; Falzone, T.L.; Brusch, R.G.; Rockenstein, E.; Mount, S.L.; Raman, R.; Davis, P.; Masliah, E.; Williams, D.S. Axonopathy and transport deficits early in the pathogenesis of Alzheimer's disease. *Science* **2005**, *307*, 1282–1288. [CrossRef] [PubMed]
19. Krstic, D.; Knuesel, I. Deciphering the mechanism underlying late-onset Alzheimer disease. *Nat. Rev. Neurol.* **2013**, *9*, 25–34. [CrossRef]
20. Li, H.; Li, S.H.; Yu, Z.X.; Shelbourne, P.; Li, X.J. Huntingtin aggregate-associated axonal degeneration is an early pathological event in Huntington's disease mice. *J. Neurosci.* **2001**, *21*, 8473–8481. [CrossRef]
21. Trapp, B.D.; Peterson, J.; Ransohoff, R.M.; Rudick, R.; Mörk, S.; Bö, L. Axonal transection in the lesions of multiple sclerosis. *N. Engl. J. Med.* **1998**, *338*, 278–285. [CrossRef]
22. Fukui, K.; Ushiki, K.; Takatsu, H.; Koike, T.; Urano, S. Tocotrienols prevent hydrogen peroxide-induced axon and dendrite degeneration in cerebellar granule cells. *Free Radic. Res.* **2012**, *46*, 184–193. [CrossRef] [PubMed]
23. Nakamura, S.; Nakanishi, A.; Okihiro, S.; Urano, S.; Fukui, K. Ionomycin-induced calcium influx induces neurite degeneration in mouse neuroblastoma cells: Analysis of a time-lapse live cell imaging system. *Free Radic. Res.* **2016**, *50*, 1214–1225. [CrossRef]
24. Fukui, K.; Onodera, K.; Shinkai, T.; Suzuki, S.; Urano, S. Impairment of learning and memory in rats caused by oxidative stress and aging and changes in antioxidative defense systems. *Ann. N. Y. Acad. Sci.* **2001**, *928*, 168–175. [CrossRef]
25. Morris, R. Developments of a water-maze procedure for studying spatial learning in the rat. *J. Neurosci. Methods* **1984**, *11*, 47–60. [CrossRef]
26. Kato, Y.; Aoki, Y.; Fukui, K. Tocotrienols influence body weight gain and brain protein expression in long-term high-fat diet-treated mice. *Int. J. Mol. Sci.* **2020**, *21*, 4533. [CrossRef]
27. Fukui, K.; Okihiro, S.; Ohfuchi, Y.; Hashimoto, M.; Kato, Y.; Yoshida, N.; Mochizuki, K.; Tsumoto, H.; Miura, Y. Proteomic study on neurite responses to oxidative stress: Search for differentially expressed proteins in isolated neurites of N1E-115 cells. *J. Clin. Biochem. Nutr.* **2019**, *64*, 36–44. [CrossRef]

28. Kato, Y.; Uchiumi, H.; Usami, R.; Takatsu, H.; Aoki, Y.; Yanai, S.; Endo, S.; Fukui, K. Tocotrienols reach the brain and play roles in the attenuation of body weight gain and improvement of cognitive function in high-fat diet-treated mice. *J. Clin. Biochem. Nutr.* **2021**, *69*, 256–264. [CrossRef]

29. Oakley, H.; Cole, S.L.; Logan, S.; Maus, E.; Shao, P.; Craft, J.; Guillozet-Bongaarts, A.; Ohno, M.; Disterhoft, J.; Van Eldik, L.; et al. Intraneuronal beta-amyloid aggregates, neurodegeneration, and neuron loss in transgenic mice with five familial Alzheimer's disease mutations: Potential factors in amyloid plaque formation. *J. Neurosci.* **2006**, *26*, 10129–10140. [CrossRef]

30. Richard, B.C.; Kurdakova, A.; Baches, S.; Bayer, T.A.; Weggen, S.; Wirths, O. Gene Dosage Dependent Aggravation of the Neurological Phenotype in the 5XFAD Mouse Model of Alzheimer's Disease. *J. Alzheimers Dis.* **2015**, *45*, 1223–1236. [CrossRef]

31. Sanderberg, A.; Luheshi, L.M.; Söllvander, S.; Pereira de Barros, T.; Macao, B.; Knowles, T.P.; Biverstål, H.; Lendel, C.; Ekholm-Petterson, F.; Dubnovitsky, A.; et al. Stabilization of neurotoxic Alzheimer amyloid-beta oligomers by protein engineering. *Proc. Natl. Acad. Sci. USA* **2010**, *107*, 15595–15600. [CrossRef]

32. Jarrett, J.T.; Berger, E.P.; Lansbury, P.T., Jr. The carboxy terminus of the beta amyloid protein is critical for the seeding of amyloid formation: Implications for the pathogenesis of Alzheimer's disease. *Biochemistry* **1993**, *32*, 4693–4697. [CrossRef] [PubMed]

33. Kuperstein, I.; Broersen, K.; Benilova, I.; Rozenski, J.; Jonckheere, W.; Debulpaep, M.; Vandersteen, A.; Segers-Nolten, I.; van der Werf, K.; Subramaniam, V.; et al. Neurotoxicity of Alzheimer's disease $A\beta$ peptides id induced by small changes in the $A\beta42$ to $A\beta40$ ratio. *EMBO J.* **2010**, *29*, 3408–3420. [CrossRef] [PubMed]

34. Guo, L.; Yeh, M.L.; Cuzon Carlson, V.C.; Johnson-Venkatesh, E.M.; Yeh, H.H. Nerve growth factor in the hippocamposeptal system: Evidence for activity-dependent anterograde delivery and modulation of synaptic activity. *J. Neurosci.* **2012**, *32*, 7701–7710. [CrossRef]

35. Lärkfors, L.; Ebendal, T.; Whittemore, S.R.; Persson, H.; Hoffer, B.; Olson, L. Decreased level of nerve growth factor (NGF) and its messenger RNA in the aged rat brain. *Brain Res.* **1987**, *427*, 55–60. [CrossRef]

36. Harman, D. Aging: A theory based on free radical and radiation chemistry. *J. Gerontol.* **1956**, *11*, 298–300. [CrossRef]

37. Pao, P.C.; Patnaik, D.; Watson, L.A.; Gao, F.; Pan, L.; Wang, J.; Adaikkan, C.; Penney, J.; Cam, H.P.; Huang, W.C.; et al. HDAC1 modulates OGC1-initiated oxidative DNA damage repair in the aging brain and Alzheimer's disease. *Nat. Commun.* **2020**, *11*, 2484. [CrossRef]

38. Annesley, S.; Fisher, P.R. Mitochondria in health and disease. *Cells* **2019**, *8*, 680. [CrossRef]

39. Kerr, J.S.; Adriaanse, B.A.; Greig, N.H.; Mattson, M.P.; Cader, M.Z.; Bohr, V.A.; Fang, E.F. Mitophagy and Alzheimer's disease: Cellular and molecular mechanisms. *Trends Neurosci.* **2017**, *40*, 151–166. [CrossRef]

40. Fukui, K.; Masuda, A.; Hosono, A.; SUwabe, R.; Yamashita, K.; Shinkai, T.; Urano, S. Changes in microtubule-related proteins and authohagy in long-term vitamin E-deficient mice. *Free Radic. Res.* **2014**, *48*, 649–658. [CrossRef]

41. Flynn, J.M.; Melov, S. SOD2 in mitochondrial dysfunction and neurodegeneration. *Free Radic. Biol. Med.* **2013**, *62*, 4–12. [CrossRef] [PubMed]

42. Ahsan, H. 3-Nitrotyrosine: A biomarker of nitrogen free radical species modified proteins in systemic autoimmunogenic conditions. *Hum. Immunol.* **2013**, *74*, 1392–1399. [CrossRef] [PubMed]

43. Thomson, L. 3-Nitrotyrosine modified proteins in arteriosclerosis. *Dis. Markers* **2015**, *2015*, 708282. [CrossRef]

44. Butterfield, D.A.; Reed, T.T.; Perluigi, M.; de Marco, C.; Coccia, R.; Keller, J.N.; Markesbery, W.R.; Sultana, R. Elevated levels of 3-nitrotyrosine in brain from subjects with amnestic mild cognitive impairment: Implications for the role of nitration in the progression of Alzheimer's disease. *Brain Res.* **2007**, *1148*, 243–248. [CrossRef]

45. Cambronne, X.A.; Stewart, M.L.; Kim, D.H.; Jones-Brunette, A.M.; Morgan, R.K.; Farrens, D.L.; Cohen, M.S.; Goodman, R.H. NAD+ biosensor reveals multiple sources for mitochondrial NAD+. *Science* **2016**, *352*, 1474–1477. [CrossRef]

46. Imai, S.; Guarente, L. NAD+ and sirtuins in aging and disease. *Trends Cell Biol.* **2014**, *24*, 464–471. [CrossRef]

47. Gulshan, M.; Yaku, K.; Okabe, K.; Mohamood, A.; Sasaki, T.; Yamamoto, M.; Hikosaka, K.; Usui, I.; Kitamura, T.; Tobe, K.; et al. Overexpression of Nmnat3 efficiently increases NAD and NGD levels and ameliorates age-associated insulin resistance. *Aging Cell* **2018**, *17*, e12798. [CrossRef]

48. Chhetri, J.; King, A.E.; Gueven, N. Alzheimer's disease and NQO-1: Is there a link? *Curr. Alzheimer Res.* **2018**, *15*, 56–66. [CrossRef]

49. Dinkova-Kostova, A.; Abramov, A.Y. The emerging role of Nrf2 in mitochondrial function. *Free. Radc. Biol. Med.* **2015**, *88*, 179–188. [CrossRef]

50. Dinkova-Kostova, A.; Talalay, P. NAD(P)H: Quinone acceptor oxidoreductase 1 (NQO) 1, a multifunctional antioxidant enzyme and exceptionally versatile cytoprotector. *Arch. Biochem. Biophys.* **2010**, *501*, 116–123. [CrossRef]

51. Cardoso, S.M.; Proença, M.T.; Santos, S.; Santana, I.; Oliveira, C.R. Cytochrome c oxidase is decreased in Alzheimer's disease platelet. *Neurobiol. Aging* **2004**, *25*, 105–110. [CrossRef]

52. Maurer, I.; Zierz, S.; Möller, H.J. A selective defect of cytochrome c oxidaseis present in brain of Alzheimer disease patients. *Neurobiol. Aging* **2000**, *21*, 455–462. [CrossRef]

Systematic Review

Considerations about Hypoxic Changes in Neuraxis Tissue Injuries and Recovery

Simona Isabelle Stoica [1,2], Coralia Bleotu [3], Vlad Ciobanu [4], Anca Mirela Ionescu [1], Irina Albadi [5,6], Gelu Onose [1,2,*] and Constantin Munteanu [2,7,8,*]

1   Faculty of Medicine, University of Medicine and Pharmacy "Carol Davila" (UMPCD), 020022 Bucharest, Romania; stoica.simona@umfcd.ro (S.I.S.); anca.ionescu@umfcd.ro (A.M.I.)
2   Teaching Emergency Hospital "Bagdasar-Arseni" (TEHBA), 041915 Bucharest, Romania
3   Stefan S. Nicolau Institute of Virology, 030304 Bucharest, Romania; cbleotu@yahoo.com
4   Computer Science Department, Politehnica University of Bucharest (PUB), 060042 Bucharest, Romania; vlad.ciobanu@cs.pub.ro
5   Teaching Emergency County Hospital "Sf. Apostol Andrei", 900591 Constanta, Romania; irina.albadi@yahoo.com
6   Faculty of Medicine, "Ovidius" University of Constanta, 900470 Constanta, Romania
7   Department of Research, Romanian Association of Balneology, 022251 Bucharest, Romania
8   Faculty of Medical Bioengineering, University of Medicine and Pharmacy "Grigore T. Popa", 700115 Iasi, Romania
*   Correspondence: gelu.onose@umfcd.ro (G.O.); constantin.munteanu.biolog@umfiasi.ro (C.M.)

Citation: Stoica, S.I.; Bleotu, C.; Ciobanu, V.; Ionescu, A.M.; Albadi, I.; Onose, G.; Munteanu, C. Considerations about Hypoxic Changes in Neuraxis Tissue Injuries and Recovery. *Biomedicines* **2022**, *10*, 481. https://doi.org/10.3390/ biomedicines10020481

Academic Editor: Kumar Vaibhav

Received: 29 November 2021
Accepted: 13 February 2022
Published: 18 February 2022

Publisher's Note: MDPI stays neutral with regard to jurisdictional claims in published maps and institutional affiliations.

Copyright: © 2022 by the authors. Licensee MDPI, Basel, Switzerland. This article is an open access article distributed under the terms and conditions of the Creative Commons Attribution (CC BY) license (https:// creativecommons.org/licenses/by/ 4.0/).

Abstract: Hypoxia represents the temporary or longer-term decrease or deprivation of oxygen in organs, tissues, and cells after oxygen supply drops or its excessive consumption. Hypoxia can be (para)-physiological—adaptive—or pathological. Thereby, the mechanisms of hypoxia have many implications, such as in adaptive processes of normal cells, but to the survival of neoplastic ones, too. Ischemia differs from hypoxia as it means a transient or permanent interruption or reduction of the blood supply in a given region or tissue and consequently a poor provision with oxygen and energetic substratum-inflammation and oxidative stress damages generating factors. Considering the implications of hypoxia on nerve tissue cells that go through different ischemic processes, in this paper, we will detail the molecular mechanisms by which such structures feel and adapt to hypoxia. We will present the hypoxic mechanisms and changes in the CNS. Also, we aimed to evaluate acute, subacute, and chronic central nervous hypoxic-ischemic changes, hoping to understand better and systematize some neuro-muscular recovery methods necessary to regain individual independence. To establish the link between CNS hypoxia, ischemic-lesional mechanisms, and neuro-motor and related recovery, we performed a systematic literature review following the" Preferred Reporting Items for Systematic Reviews and Meta-Analyses (PRISMA") filtering method by interrogating five international medical renown databases, using, contextually, specific keywords combinations/"syntaxes", with supplementation of the afferent documentation through an amount of freely discovered, also contributive, bibliographic resources. As a result, 45 papers were eligible according to the PRISMA-inspired selection approach, thus covering information on both: intimate/molecular path-physiological specific mechanisms and, respectively, consequent clinical conditions. Such a systematic process is meant to help us construct an article structure skeleton giving a primary objective input about the assembly of the literature background to be approached, summarised, and synthesized. The afferent contextual search (by keywords combination/syntaxes) we have fulfilled considerably reduced the number of obtained articles. We consider this systematic literature review is warranted as hypoxia's mechanisms have opened new perspectives for understanding ischemic changes in the CNS neuraxis tissue/cells, starting at the intracellular level and continuing with experimental research to recover the consequent clinical-functional deficits better.

Keywords: hypoxia; ischemia; neuraxis; hypoxic-ischemic injuries; neural ischemia; neural tissue hypoxic injuries; neuro-recovery; neurorehabilitation

1. Introduction

Without oxygen, many species on Earth would not survive because molecular oxygen is indispensable for biochemical and bioenergetic cellular processes [1,2]. Therefore, hypoxia is defined as the decrease or deprivation of oxygen in organs, tissues, and cells by decreasing oxygen supply (due to damage to the vascular network— in case of ischemia; due to anemia or other lack of oxygen conditions) or increasing oxygen consumption (as in the sudden increase in the rate of cell proliferation) [3,4].

Hypoxia can be physiological (with beneficial effects on the nervous system, respiratory system, cardiovascular system, and different metabolisms) or pathological (neoplasms, rheumatoid arthritis, and atherosclerosis) [5,6]. Although the brain represents only 2% of the body's weight, it is the organ that consumes the most energy, needing at least 20% of the total oxygen to function normally [7]. Furthermore, it seems that the medullary blood supply is similar to the cerebral one, with a blood flow of 5:1 between the white substance and the grey matter, and the mechanisms of cerebral and medullary vascular self-regulation are independent of blood pressure values. In contrast, systemic variations in blood gases alter the medullary vascular flow (without redistributing it) [8,9]. Therefore, reducing the amount of oxygen can be harmful to nerve tissue, leading to neurological disorders, with significant medical and socio-economic implications.

Each cell and tissue has characteristic abilities to adapt to hypoxia conditions by stabilizing HIF alpha and regulating the various genes involved in angiogenesis and oxygen transport [6].

In 2019, the Nobel Prize for Medicine and Physiology was awarded for clarifying the molecular mechanism of cell sensitivity to oxygen, with implications in cell physiology and the pathophysiology of complex processes such as metabolic adaptation, neovascularization, and tumor progression. The mechanisms of molecular biology through which hypoxia influences cellular activity are difficult to understand, their complete elucidation being performed only recently, when the correlations between cellular oxygenation level and hypoxia-inducible factor (HIF), HIF inhibiting factor (FIH), hypoxia regulating element (HRE), Von Hipple-Lindau protein (VHL), proline hydroxylase (PHD) were understood, along with all cellular elements modulated by these factors [10].

Perinatal hypoxic-ischemic lesions produce brain injuries. About 40% of newborns do not survive, and 30% develop permanent neurological disorders (cerebral palsy, visual disturbances, epilepsy, neuro-cognitive delay, learning disorders). Hypoxic-ischemic lesions from perinatal nerve distress cause energy disorders in cell metabolism, leading to cell death through apoptosis, necrosis, and autolysis (autophagic cell death) [11].

Experimentally, it has been observed how prenatal hypoxia affects the migration of embryonic neuroblasts with the subsequent impairment of the development of the CNS in rats [12,13]. Furthermore, experimental ischemia in mice, through MCAO (middle cerebral artery occlusion—model of stroke), is associated with immune cells presence in the meningeal vessels and, respectively, "the mice with MCAO showed an invasion of LysM GFP+ cells into the brain parenchyma especially in the peri-infarct region" [14].

On the other hand, in human adults, vascular disorders (such as stroke) are risk factors for neurodegenerative diseases [15], like Parkinson's disease [16], dementia (and other cognitive impairments). In addition, there are also innate metabolic disorders that affect the CNS's functioning, such as hyperhomocysteinemia (a consequence of impaired homocysteine metabolism or other cofactors involved in its degradation), which is an independent risk factor for stroke [17,18] occurrence [19,20].

Considering the implications of hypoxia on nerve tissue cells that go through different ischemic processes, in this paper, we will detail the molecular mechanisms by which cells feel and adapt to hypoxia. We will present the hypoxic mechanisms and changes in central nervous tissue. We also aim to evaluate acute, subacute, and chronic central nervous hypoxic-ischemic changes in the hope of discovering or improving some methods of neuro-muscular recovery necessary to regain individual independence.

2. Method

The documentation afferent to this paper (about the relationship between the processes of ischemia—hypoxia—recovery in the central nervous system—CNS) relies on both works freely identified in the literature and a rigorous related selection within a systematic literature review, following the "PRISMA" paradigm. Thereby, we interrogated the following medical databases: Elsevier, NCBI/PubMed, NCBI/PMC, PEDro, and—to check whether the (initially) found articles are published in ISI indexed journals—ISI (Institute for Scientific Information—ex Thomson Reuters—currently administered by Clarivate Analytics).

Our search referred to the period 1 January 2016 to 31 December 2020. We considered only works published in English and issued in ISI-indexed journals. Accordingly, we used—contextually—a series of key word combinations/"syntaxes": " Hypoxia" + "nerve tissue" + "lesion(s)", "Hypoxia" + "nerve tissue" + "injury(es)", "Hypoxia" + "nerve tissue" + "recovery", "Hypoxia" + "nerve tissue" + "cellular mechanisms", "Hypoxia" + "CNS" + "lesion(s)", "Hypoxia" + "CNS" + "lesion(s)", "Hypoxia" + "CNS" + "injuries(es)", "Hypoxia" + "CNS" + "injury(es)", "Ischemia" + "nerve tissue" + "lesion(s)", "Ischemia" + "nerve tissue" + "injury(s)", "Ischemia" + "nerve tissue" + "recovery", "ischemia" + "nerve tissue" + "cellular mechanisms", "Ischemia" + "CNS" + "lesion(s)", "Ischemia" + "CNS" + "lesion(s)", "Ischemia" + "CNS" + "injuries(es)", "Ischemia" + "CNS" + "injury(es)" The articles thus found have then been filtered in five steps (without meta-analysis), on the standardized base of the PRISMA inspired selection methodology.

3. Results

The PRISMA standardized methodology for achieving systematic reviews requires specific steps and a high level of strictness, which we have respected, resulting in 45 eligible and contributive works (see Figure 1 and Table S1). Although our systematic literature review has been thoroughly conducted, some bibliographic resources of interest still might have been overlooked. However, considering the contribution mentioned above of the freely discovered publications, hopefully, we have adequately covered the necessary information, as presented in the skeleton structure (see Table 1 at the end).

3.1. The Intimate Mechanisms of Hypoxia

HIF plays a central role in detecting and adapting cells to oxygen by transcriptionally activating genes controlling oxygen homeostasis and metabolic activation (Figure 2) [21]. There are several cytoplasmic HIF isoforms: HIF-1α and HIF-2α (activating the HRE's transcription, without redundant activity) and HIF-3α (the most distant isoform which, in some cases, may encode a polypeptide that inhibits the expression of HRE-dependent genes). HIF-2α is present in tumor cells (such as those in clear-cell renal cell carcinoma associated with von Hippel-Lindau disease), and HIF-1α has been found in normal cells of the human body. However, it appears that activation of the erythropoietin production gene (EPO) is preferentially achieved under the action of HIF-2α (also occurring through activation produced by HIF-1α) [3,22].

HIF-1 is essential for normal development and the response to ischemia/hypoxia, tumor development, energy metabolism, angiogenesis, apoptosis, proliferation, and vasomotor function [23]. HIF 1 binds many genes that contain in their structure hypoxic response elements such as vascular endothelial growth factor (VEGF), glucose transporter 1 (GLUT-1), adenylate kinase 3 (AK-3), aldolase A (ALD-A), phosphoglycerate kinase 1 (PGK-1), 6-phosphofructokinase, liver type (PFK-L), and lactate dehydrogenase A (LDH-A) [24].

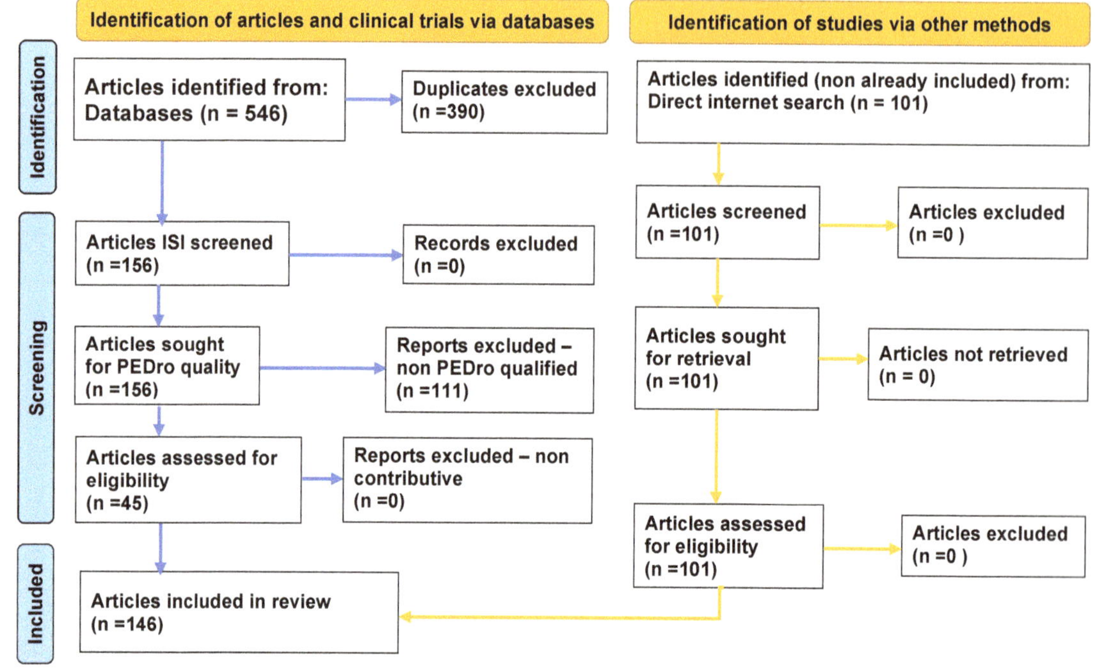

Figure 1. PRISMA flow diagram adapted to our study.

The HIF-DNA complex is a heteromer, the binding being achieved by the subunits α and β. While the expression of the HIF-1β subunit is constitutive, the presence of the HIF-1α-subunit increases exponentially with the decreases of cellular oxygenation below 6%. That means an oxygen partial pressure of 40 mmHg measured at sea level when the average partial pressure of the oxygen in the nervous tissue is 30–48 mmHg [25–27]. The amino-terminal end of HIF-1α is sufficient for dimerization with the HIF-1β subunit and DNA binding [25]. Cellular sensitivity to oxygen is ensured by the activity of the enzyme proline hydroxylase (having 3 isoforms), which, in the presence of iron ions and oxygen, adds two hydroxyl groups to the HIF-1α subunit (to 2 terminal proline residues: Pro-402 and Pro-564) [10,28,29]. At the same time, the oxygen-dependent hydroxylation, performed by FIH to arginine residue of HIF-1α C-terminal transactivation domain (CAD), cancels the HIF-1α interaction with p300, preventing its translational activity [27,30].

Under normal oxygenation, the HIF-1α protein subunit is rapidly degraded by proteosomes. The whole process is mediated by the Von Hipple-Lindau protein factor (pVHL), an E3 ubiquitin recognition and binding component that promotes ubiquitin-dependent proteolysis of this subunit [31]. Under hypoxic conditions, the degradation of the HIF-1α subunit is suppressed, leading to gene transcription activation. At the experimental level, cobalt ions and iron chelators simulate the conditions of cellular hypoxia. In their presence, the binding of HIF-1α to pVHL is prevented [10,32].

HIF-1 is also found in mammalian cell cultures raised under low oxygen pressure conditions and is needed to amplify the transcription of genes that mediate erythropoietin (EPO) production in hypoxic cells [33–35]. Oxygen is essential for mammalian life through oxidative phosphorylation and adenosine triphosphate (ATP) synthesis [36,37].

In response to hypoxia, HIF-1 mediates the activation of anaerobic glycolysis pathways, erythropoietin synthesis pathways (in anemia or individuals living at altitude), and production of VEGF (with new blood vessel formation as in chronic myocardial ischemia). HIF influences genes such as VEGF, EPO by binding to the HRE gene sequence (thus, HRE shows HIF activity) [6]. Molecular mechanisms that mediate the cellular response to hypoxia have been studied regarding erythropoietin production, which controls erythrocytes formation (and increased tissue oxygenation) through specific growth factors. Hypoxia conditions (growth of cell cultures in the atmosphere with 0.1% oxygen, in the presence of cobalt chloride or deferoxamine) determine the binding to the cis end of the EPO3 gene of the trans region of the α subunit of HIF-1. In conditions of deficient oxygenation obtained by treating the cell cultures with deferoxamine or cobalt chloride, the EPO3 gene expression has cell specificity based on HIF-1α-mediated activation [38]. The persistence of HIF-1α in cells grown in hypoxemic environments is due to the binding of cobalt ions and deferoxamine to CAD located at the carboxy-terminal end of HIF-1 [25].

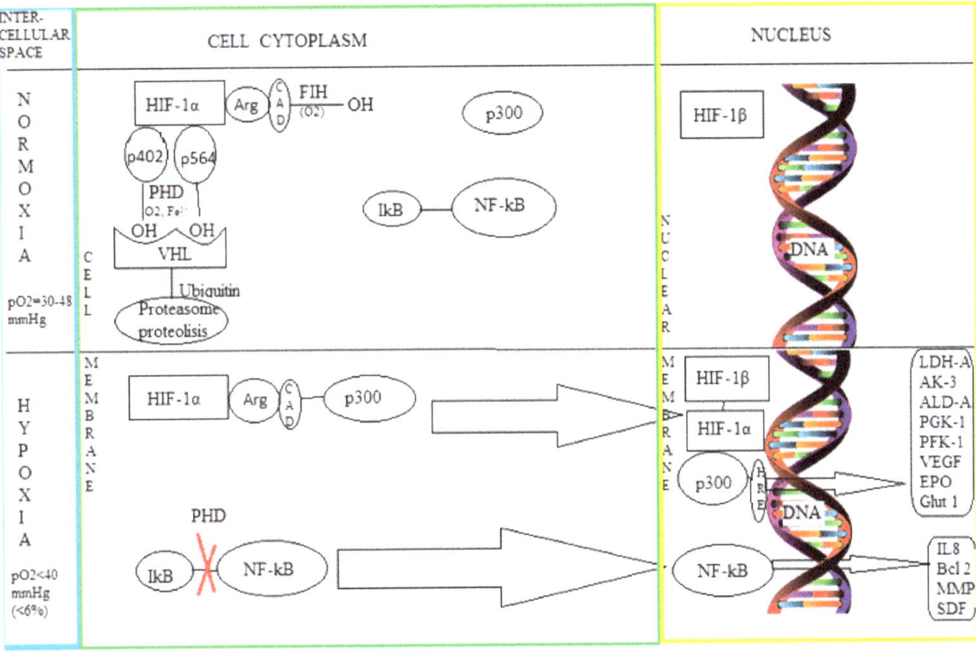

Figure 2. The cellular mechanism of hypoxia (showing how oxygen partial pressure, pO2, influences deoxynucleic acid, DNA, via hypoxia-inducible factor, HIF; with involvement of factor inhibiting HIF, FIH; arginine, ARG; C-terminal transactivation domain, CAD; protein 300, p300 and Von Hipple-Lindau factor, VHL and proline hydroxylase, PHD. We can see how hypoxia response elements (HRE) are activated, as the atypical Nuclear Factor kB, NF-kB, activation pathway, with involvement of inhibitor kB, I-kb. This triggers the genes of lactate dehydrogenase A, LDH-A; adenylate kinase 3, AK-3; aldolase A, ALD-A; phosphoglycerate kinase 1, PGK-1; phosphofructokinase, liver type, PFK-L 6; vascular endothelial growth factor, VEGF; erythropoietin, EPO; glucose transporter 1, Glut-1; interleukin 8, IL 8; B cell lymphoma-2, Bcl-2; matrix metalloproteases, MMP; stromal cell-derived factor 1, SDF).

Nuclear Factor kB (NF-kB) is a major transcription factor under stress, being activated (see Figure 3) by hypoxia or decreased oxygen availability [39]. NF-kB is part of a family of transcription factors composed of RelA, RelB, cRel, NF-kB1 (p105/p50), and NF-kB2 (p100/p52) [40]. These transcription factors are maintained in the cytoplasm in an inactive form by the kB inhibitor family's action (IkB) [41]. The stimulation produced by various stressors leads to the accumulation of NF-kB in the nucleus and its binding to DNA, using some paths classified as canonical, non-canonical, and atypical. The most studied pathway is the canonical (or classical) activation of NF-kB. This one involves the activation of transforming growth factor activating kinase-B (TAK1) and the kB inhibitor complex kinase (IKK), composed of IKKα, IKKβ, and IKKγ or the essential modulator of NF-kB (Nemo) [41]. The non-canonical NF-κB activation pathway consists of NF-κB-induced kinase (NIK) and the IKKγ activating homodimer [42,43]. The atypical NF-kB activation pathway usually does not require the presence of the IKK complex and acts directly on the IkB [41]. NF-κB activation pathways are strongly influenced by changes in ubiquitin components [39].

A ligand for the receptor, such as cytokines or foreign deoxyribonucleic acid (DNA) and ribonucleic acid (RNA), is usually required to activate most NF-κB pathways. Stimulation by damaged DNA activates sensitive NF-kB kinase in the nucleus, followed by cytoplasmic growth of IKK. When activating HIF, the cell uses oxygen sensors such as PHD and FIH. PHD links oxygen sensitivity to NF-kB activation, acting directly on IKK. At the same time, PHD1 was identified as the enzymatic isoform with the highest level of control of IKK activity [39]. Most studies have investigated the activation of NF-κB in response to hypoxia in neoplastic cells and have shown a decrease in apoptosis and increased angiogenesis.

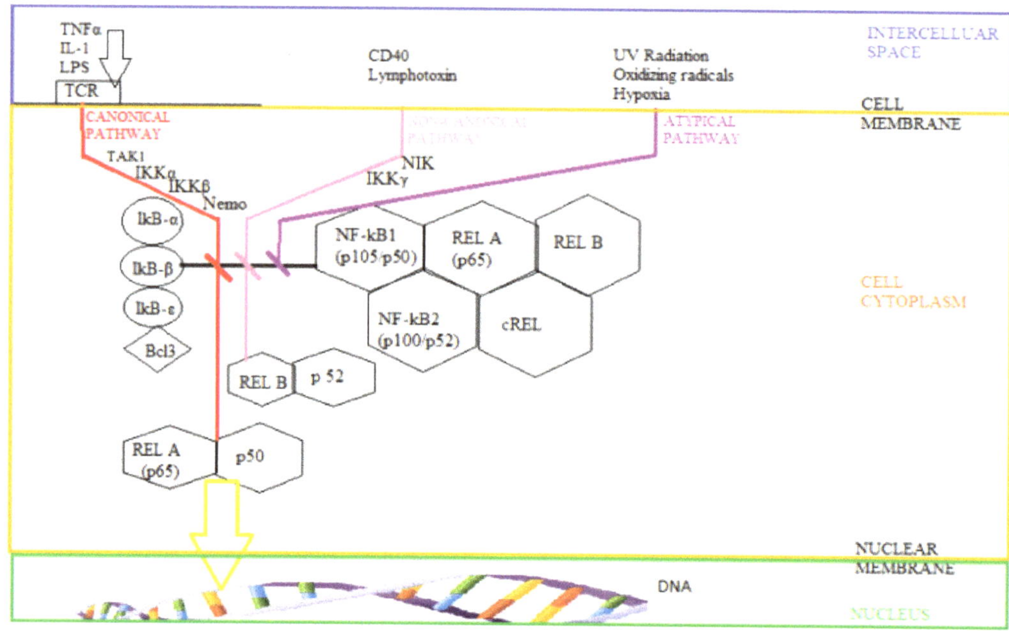

Figure 3. The Nuclear Factor kB (NF-kB) pathways activation: the canonical pathway (triggered by the action of tumor necrosis α, TNFα; interleukin 1, IL-1, and lipopolysaccharides, LPS, on T cell receptor, TCR, with involvement of transforming growth factor activating kinase—B, TAK1; the kB inhibitor complex kinase α, β, IKK α, β; of the essential modulator of NF-kB, IkB-α, β, ε kB inhibitor family, Nemo), the non-canonical pathway (involving NF-κB-induced kinase, NIK, and IKK γ) and the atypical pathway.

However, the situation is different in tissues such as the brain and myocardium, in which NF-kB's role appears to be more complex through the induction and suppression of apoptosis depending on the context [39]. Under hypoxic conditions, NF-kB modulates the expression of numerous proteins involved in controlling apoptosis (such as members of the Bcl-2 family) and the inhibition of programmed cell death [44–46]. In addition, NF-kB increases the expression of IL-8, an essential cytokine in inducing angiogenesis that contributes to the generation of neovascularization in hypoxia [47,48]. NF-kB also induces the expression of proteins involved in motility and adhesion, such as matrix metalloproteases

(MMP) and stromal cell-derived factor 1 (SDF) in tumor, immune and neuronal cells [46,49]. Therefore, the hypoxia-induced NF-kB factor is essential for many cellular responses to this stimulus, (especially) prevention of apoptosis, induction of angiogenesis, and promotion of cell motility [39,50,51].

3.2. The Influence of Hypoxia on the Nervous Tissue

Hypoxia complexly influences the physiology of all types of cells, including the nerve cells, in all stages of nerve tissue development (starting with the neonatal period and ending with that of old age), as well as in various traumatic conditions (spinal cord injury, traumatic brain injury) and non-traumatic conditions (stroke, chronic pain, epilepsy, certain congenital or acquired neurodegenerative diseases). Hypoxia has some pathological consequences on the CNS (CNS) in different post-lesional stages divided into acute (in the first 2 weeks), subacute (in the next 3–11 weeks), early chronic (for the 12–24 next weeks), and chronic (for a more extended period than 24 weeks) [52–54].

Nerve tissue comprises neurons and glial cells, such as microglia or astrocytes [55]. Microglia are derived from erythromyeloid progenitors migrated to the brain during intrauterine life [56]. Microglia have an essential role in defending against microbial aggression and CNS aggression and synaptic budding, neurogenesis, and cerebral homeostasis [57,58]. It has also been observed that, among flavonoids, wogonin and baicalein inhibit microglial inflammatory activity (by decreasing the production of nitric oxide (NO) and the inducible activity of NO synthase and NF-κB activation) [59]. In addition, melatonin is a pineal hormone with an anti-inflammatory effect after stroke (by clearing ROS and inhibiting the inflammatory response) in microglia [60].

Astrocytes are essential in maintaining the integrity of the blood-brain barrier and the perineuronal environment; in the metabolism of glutamate (and decreased excitotoxicity), in the preservation of calcium and potassium homeostasis in the extracellular environment [61]. In addition, astrocytes have a critical homeostatic role in the CNS through neurogenesis, neuroprotection, immunomodulation, and their antioxidant role [62]. It has also been found that HIF1α stimulates the production of peroxisomes (organelles with essential functions in glial cell metabolism) [62].

In the glial cell membranes, Aquaporin 1 (AQP-1) water channel [63] is very little expressed in white and grey matter (in astrocytes ependymal cells and thin fibers in the posterior medullary horns of the normal CNS), act as an ion channel in the secretion of cerebrospinal fluid and is well expressed in choroid plexus cells [64].

At the nuclear level, the High mobility group box 1 (HMGB1) protein is present in all cells, having a role in stabilizing and repairing deoxyribonucleic acid (DNA) [65]. HMGB1 is released from necrotic neurons or is actively secreted from microglia, monocytes/macrophages, and neutrophils, mediating the neuroinflammatory response and contributing to the pathogenesis of ischemic stroke [66]. Extracellular HMGB1 binds to different membrane receptors as toll-like receptors (TLR-2, TRL-4). HMGB1 can be altered by redox reactions with implications for acute and chronic cellular changes in stroke [67,68].

The mechanisms of hypoxia act in a very complex manner on the CNS through all the molecules it involves. In this way, erythropoietin has a neuroprotective effect, engaging in antiapoptotic, anti-inflammatory, neuro-neurovascular remodeling, and stem cell proliferation mechanisms [69].

At the mitochondrial level, hypoxia causes the reprogramming of the functioning of the respiratory chain by changing (transient and reversible) the path of nicotinamide adenine dinucleotides oxidation (NAD) of the substrate (complex I) to the oxidation of succinate (complex II). Activation of the respiratory chain complex II is a vital hypoxia adaptation mechanism required for the production of cellular energy by succinate (for immediate resistance of the body), stabilization of HIF-1α by succinate and initiation of its transcriptional activity for long-term adaptation, and activation of the GPR91 succinate receptor (G protein receptor 91) [70]. This mechanism of activation of the respiratory chain complex II has the following regulatory functions: sensitization and adaptation to the gradual decrease of oxygen in the environment (for the selection of the most efficient ways of oxidation of the substrate in hypoxia), compensation (with the immediate formation of the body's response and resistance to hypoxia), transcriptional (with activation of HIF-1α synthesis and genes that ensure adaptation to hypoxia for long periods), receptor (for mitochondrial participation in the GPR91 succinate receptor-mediated intercellular signaling system) [71–73].

A specific neural pathway for adapting to hypoxia has been discovered, which determines the increase in the biosynthesis of cellular fatty acids with their subsequent esterification. Experimentally, it was observed that the inhibitors of fatty acid synthesis (such as Acetyl-Co-A carboxylase and fatty acid synthetase) increased the ratios of $NADH^{+2}/NAD^+$ and $NADPH^{+2}/NADP^+$ in hypoxic conditions. Inhibition of fatty acid synthetase increased lactic acid levels in normoxia and hypoxia. They show that fatty acids can be proton acceptors from anaerobic glycolysis under hypoxic conditions [74].

3.3. Newborn Hypoxic-Ischemic Encephalopathy

Perinatal hypoxic ischemia is the leading cause of cerebral distress in newborns. Therefore, this condition can have significant individual and social implications [75,76].

Hypoxic-ischemic brain lesions in the newborn are significantly influenced by inflammation, activating neural cells, and local infiltration of circulating leukocytes. An initial inflammatory response is followed by a secondary one (which can last for several days) and a subsequent anti-inflammatory reaction. The pathophysiology of hypoxic-ischemic encephalopathy is based on reduced blood flow and brain oxygenation. Mitochondria tend to hyperpermeability in response to the process of hypoxia-ischemia, and this excessive permeabilization of the mitochondrial membrane is considered a point of irreversibility in the apoptosis following perinatal hypoxic encephalopathy [11,77,78].

Severe hypoxic-ischemic lesions initially cause energy deficiency followed by primary neuronal death, correlated with the decrease in ATP and the increase in lactic acid production intracellular level associated with an increase in the level of reactive oxygen species (ROS). ROS are involved in cell physiological processes but can cause cell damage, inflammation, and oxidative stress [79]. Thus, rapid swelling and cell necrosis occur. Severe hypoxic-ischemic lesions shorten the period of revascularization and cerebral metabolic recovery, leading to secondary lesions (at least 6 h after the initial lesion) with delayed neuronal apoptosis (related to excitotoxicity, oxidative stress, and inflammation) [11,80].

Hypoxic-ischemic lesions of the newborn's brain intensely activate microglial cells, which are involved in secondary energy disorders through the production of proinflammatory cytokines (TNF-α, IL-1b, IL-6, and IL-18), complement system factors, excitotoxic amino acids. On the other hand, microglia are involved in the relief of inflammation and repair processes after suffering from hypoxic-ischemic encephalopathy by phagocytosis processes of cell debris [81,82].

Proinflammatory cytokines and ROS resulting from damage of hypoxic-ischemic neurons [11] may activate reactive astrogliosis and delay local production of proinflammatory cytokines (TNF-α, IL-1a, and b IL-6). It appears that reactive astrogliosis also produces anti-inflammatory cytokines (IL-9, IL-10, and IL-11), promoting tissue healing by activating Toll-like receptor 3 (TLR3). As a result of neonatal hypoxic-ischemic brain damage, T lymphocytes can also enter the CNS releasing micro RNA 210 (miR-210) due to transient focal ischemia [83]. MiR-210 inhibits HIF-1α by performing a negative feedback loop in hypoxic differentiation of T lymphocytes [84]. In other words, acute or chronic suffering in the CNS causes the release of adenosine triphosphate (ATP), which in the intercellular space has an immediate excitotoxic effect by coupling with astrocyte receptors such as P2X7 (ion gate type) and P2Y1 (G protein-coupled receptor) following astrogliosis (which isolates the injured areas) and the synthesis of neurotrophic substances (necessary for neuronal recovery). It seems that purine mechanisms mediate astrogliosis following neurotrauma and local hypoxia/ischemia. In the long run, the effect of nucleotides is to amplify primary lesions, with involvement in chronic pain, epilepsy, and post-traumatic cell death [85].

3.4. Adult Brain Ischemic Vascular Lesions

Stroke (which can have multiple etiologies and is at higher risk of production in diabetic patients or after SARS COV2 infection) is a neurological condition that is becoming more common, with many individual, family, and social implications [86–89].

In adulthood, ischemic vascular lesions in stroke are risk factors for neurodegenerative diseases, and systemic hypoxic episodes increase the production and accumulation of Aβ proteins along with the decreased expression of neprilysin (NEP) [90]. The decrease of NEP, the primary enzyme that degrades Aβ proteins, affects the clearance of this protein. On the other hand, Aβ proteins derived from the transmembrane domain of amyloid precursor protein (APP), together with other active metabolites (including the C-terminal fragment in the intracellular domain of APP), regulate the expression of NEP and other neuronal genes. Some studies have also shown that caspase activation may be necessary for regulating brain NEP in hypoxic and ischemic conditions, and the decrease of the Aβ proteins elimination increases the risk of Alzheimer's dementia [15,91,92]. In the acute phase, post-stroke VEGF increases the permeability of the brain-blood barrier; in the chronic phase, this molecule promotes neurogenesis and cerebral angiogenesis [93]. In patients with hyperhomocysteinemia (a genetic disorder involving cellular hypoxia), neural cells are sensitive to prolonged exposure to the elevated level of homocysteine, due to the adverse effects of the reactive oxygen species and posttranslational changes in proteins (which occur later), and the adaptive CNS (survival) response to sublethal ischemia is also preserved [19].

Primary and secondary lesions occur in traumatic brain injury (TBI). Primary lesions are triggered by endogenous changes such as oxidative stress, glutamate-related excitotoxicity, immune response, disorders of ionic homeostasis, and increased vascular permeability, causing degeneration and neuronal apoptosis [94]. Secondary TBI lesions result from oxidative stress, by excessive accumulation of ROS that causes damage to cellular components (lipids, proteins, DNA), followed by impaired functioning and apoptosis of neuronal cells [95]. For its role in proliferation, signal transduction, and regulation, miRNA can also be considered a TBI biomarker [57,58,83,96–98].

3.5. Adult Spinal Cord Injury and Hypoxia

The dimensions of the modern world are moving at increasing speeds, and the risk of polytrauma (including TBI and SCI) is growing proportionately, with consequences that often are difficult to manage [87,99,100].

After SCI (similar to TBI), primary and secondary lesions appear. The initial injuries are consecutive to the direct traumatic impact (and are accompanied by neuro-vascular lesions), and secondary lesions subsequently occur through vascular dysfunction, inflammation, demyelination, and neuronal morphophysiological impairment [86,101].

In spinal cord injury (SCI) and TBI, the neural stem cells (NSCs) migrate and differentiate around damaged nerve tissue areas. In the case of SCI, the interstitial environment changes its composition due to transient hypoxia, accumulation of potassium ions, calcium, reactive oxygen species, and increase of glutamate activity. All these biochemical changes are not conducive to the survival of locally migrated NSCs. The granulocyte-macrophage colony-stimulating factor (GM-CSF), a cytokine that stimulates the differentiation and proliferation of hematopoietic cells, influences the nervous system's functioning. After SCI, GM-CSF accumulates locally and ensures the survival of dopaminergic neurons, inhibits the formation of glial scars, having a neuroprotective role. Experimentally, GM-CSF may promote in vitro NSC proliferation and in vivo motor recovery in adult mice. Overexpression of the GM-CSF gene protects the NSC by increasing the resistance of these cells to apoptosis induced by hydrogen peroxide in hypoxia, thus ensuring the survival and differentiation of NSC in experimental SCI models [57,102–104].

It has been observed that VHL protein is increased at the medullary level after SCI, reaching a peak level at 3 days post-trauma. The increase of VHL protein was associated with neuronal apoptosis because it was not found in astrocytes and microglia. Decreased VHL levels may reduce glutamate-dependent neuronal apoptosis, but increased expression of VHL protein above a critical limit does not appear to produce further changes in neuronal apoptosis [31].

HIF-1α and VEGF expression in spinal cord injury areas after SCI was associated with the process of angiogenesis and improvement of local microcirculation. The main function of VGEF in the CNS is angiogenesis, which can protect nerve cells against ischemic and mechanical damage to the axons [105–107].

Increasing HIF-1α expression has a protective role in promoting functional recovery after spinal cord injury. Using DMOG-dimethyloxalylglycine treatment, it was obtained a sustained activation of HIF-1α by inhibition of prolyl hydroxylase. DMGO treatment significantly increases HIF-1α expression, inducing molecule stability that decreases apoptotic protein expression and promotes neuronal survival. This treatment also stimulates axonal regeneration by controlling the stability of microtubules in both in vivo and in vitro. On the other hand, DMGO promotes neuronal survival and axonal regeneration by activating the autophagy induced by the HIF-1α/BNIP3 signaling pathway. These experiments sustain the idea that the DMGO molecule can help treat patients with SCI [108]. It seems that amplification of autophagy reduces initial cell death by restricting the function of autophagy-associated genes and modulating the expression of inflammatory cytokines (TNF α, IL 1β) [109].

3.6. Hypoxia and Functional Recovery

Experiments with intermittent exposure to hypoxia have also been performed in relatively short sessions in patients with incomplete deficits after SCI. Respiratory, psychological, and motor function benefits were observed. In addition, following intermittent exposure to hypoxia of rats with SCI, a serotonin-dependent increase of brain-derived neurotrophic factor (BDNF) synthesis was observed in the areas of the anterior medullary horns containing the phrenic nerve nucleus; this could explain the improvement in respiratory function [110–113].

Other potential mechanisms of the beneficial exposure to intermittent hypoxia in SCI could be the increase of cytoglobin, the induction of heat shock protein 70 (HSP70), together with the rise of HIF-1α [114]. Hypoxic exposure (60 cycles of 30 s of intermittent hypoxia with 1.5% atmospheric oxygen, followed by 5 min of normoxia) induced HIF-1α accumulation, following the generation of reactive oxygen species by nicotinamide adenine dinucleotide phosphate (NADPH) oxidase. Also, in rodents exposed to intermittent hypoxia, there was increased gene expression of VEGF and increased growth hormone release [115,116].

This conditioning of neural activity by intermittent hypoxic stimulation is thought to stimulate neuroplasticity of the CNS [117–119]. It has also been found that sleep apnea syndrome (in its mild to moderate forms) is associated with better outcomes rather than intermittent exposure to hypoxia (in patients with incomplete SCI) [120].

It has also been observed that HIF-1α expression is increased during hypoxia or in the ischemia/reperfusion change after SCI, which could suppress the autophagy of neuronal cells. On the other hand, HIF-1α activation produces an anti-inflammatory effect (by decreasing TNF-α, IL-1β, IL-6, and IL-18 in SCI models in rats). In addition, intermittent hypoxia may induce HIF-1α expression, resulting in intermittent activation of autophagy (HIF-1α-dependent during the process of intermittent hypoxia) [121]. HIF-1α can cause additional expression of transcriptional factor p62, which can subsequently release from its binding to the Bcl-2 protein, activating it to participate in the autophagy process [38]. HIF-1α may be decreased in the first 24 h after SCI, just as it can be slightly lowered under normoxic conditions. It was also reported that HIF-1α had a significant increase after the first 24 consecutive hours of an SCI [38].

Erythropoietin is essential for the recovery of cognitive and memory disorders (following ischemic hypoxia) by inducing long-term synaptogenesis and repairing lesions of nerve terminals based on increased expression of synaptic proteins (synapsin 1 and postsynaptic density protein 9, PSD95) of the dendritic marker microtubule-associated protein 2 (MAP-2), of the axonal density and the decrease of a factor associated with the axonal injury, amyloid precursor protein (APP) [122].

Knowing the importance of mild chronic hypoxia in functional recovery by stimulating vascular remodeling in the brain, its implication in spinal vascular remodeling has been investigated after exposure to chronic mild hypoxia (8% O2, for 7 days) [123]. Thus, it was observed how chronic mild hypoxia promotes endothelial proliferation and increased vascularization by increasing angiogenesis and arteriogenesis markers such as rising vascular expression of fibronectin in the extracellular protein matrix, simultaneously with increased endothelial expression of the $\alpha5\beta1$ integrin receptor of fibronectin and increased endothelial expression of the junctional proteins claudin-5, ZO-1 and occluding and astrocyte activation (Halder SK, 2018). It was noted that these changes in exposure to mild chronic hypoxia were more critical in the medullary white matter. Spinal blood vessels appear to have considerable remodeling potential, with $\alpha5\beta1$ integrin essential in promoting endothelial proliferation [19,123].

Research has also been done on the effect of deferoxamine administration in the first 1–2 weeks after SCI in rats. Significant neovascularization was observed in the spinal cord injury [124], demonstrating increased expression of HIF-1α and VEGF. In addition, after deferoxamine treatment, rats with SCI showed a significant improvement in motor deficit, spared nerve tissue area, and electrophysiological conduction. However, all these favorable effects produced after deferoxamine treatment in post-SCI advance were suppressed by treatment with lenvatinib, a VEGF receptor inhibitor, suggesting that deferoxamine's main pharmacological effect in SCI is to promote neovascularization by HIF-1α and VEGF overexpression [125].

It was also found that VEGF production is stimulated by neuropeptide Y (NPY), whose serum levels increase during exercise, hypoxia, cold exposure, tissue injury, ischemia, and hemorrhagic shock [126]. In addition, NPY is an orexigenic hormone whose insulin negatively regulates hypothalamic activity [127,128].

Erythropoietin appears to promote functional recovery after spinal cord injury (SCI). Studies regarding EPO efficiency have been performed both in vitro, on neural stem cells harvested from animals, and in vivo, on rats that have undergone spinal cord contusive patterns. In vivo results showed superior β-tubulin production in erythropoietin-treated neuronal and glial cells. Also, only rats with SCI treated with erythropoietin resumed gait compared to the control group [129]. The positive effects of erythropoietin are also found in recovery after traumatic brain injuries (TBI) [130].

The role of aquaporins in pathogenesis and post-SCI recovery was also studied. In SCI, AQP-1 is 4–8 times better expressed in the traumatic area, probably having a role in the appearance of post-traumatic edema and the formation of spinal cysts. Moreover, maintaining elevated AQP-1 values in the subacute and chronic post-SCI phases, similar to high HIF-1α values, may result in consecutive SCI hypoxia.

Another study highlighted the post-SCI protective role of glutamine synthetase, which metabolizes glutamate to glutamine. Thus, protection against hypoxia-induced excitability has been observed (with inhibition of decreased compound action potentials), probably by blocking gamma-aminobutyric acid A (GABA A) receptors [131].

An interesting fact also refers to the Mediterranean diet based on an abundant consumption of olive oil. Olive oil has anti-inflammatory and immunomodulatory effects on the nervous system via (poly)-phenols, which modulate the activity of NF-κB, HIF-1α, signal transducer, and transcriptional activator 3 (STAT3), and mitogen-activated protein kinases (MAPKs) [132,133].

Natural flavonoids (such as wogonin, baicalein, curcumin, apigenin, quercetin, luteolin) have an anti-inflammatory effect because they have been shown to inhibit the production of IL-6, TNF-α, and IL-1β from the MAPK pathway of the nervous system [59,134,135]. It should also be mentioned that a beneficial effect of cannabinoid receptor agonists on oligodendrocytes (and their precursor cells) has been studied, like apamin (bee venom) [136–140].

Table 1. PRISMA resulting conceptual skeleton structure of the article's organization approach.

1. *The cellular mechanism of hypoxia*

Article	Ref. no	Subject
(Thornton, 2017)	[11]	Hypoxic-ischemic lesions cause energy disorders in cell metabolism, leading to cell death through apoptosis, necrosis and autolysis
(Cai, 2019)	[14]	MCAO mice showed an invasion of immune cells into the brain
(Nowak-Sliwinskaet, 2018)	[22]	HIF-1 is essential for normal development and the response to ischemia/hypoxia, tumor development, energy metabolism, angiogenesis, apoptosis, proliferation, and vasomotor function
(Yuniati, 2019)	[43]	NF-kB modulates the expression of numerous proteins
(Gschwandtner, 2019)	[44]	apoptosis and the inhibition of programmed cell death
(Yang, 2017)	[47]	NF-kB increases the expression of IL-8, inducing angiogenesis that contributes to the generation of neovascularization in hypoxia

2. *The influence of hypoxia on the nervous tissue*

Article	Ref. no	Subject
(Clark, 2019)	[54]	Nerve tissue is made up of neurons and glial cells
(Miller, 2017)	[55]	Microglia are derived from erythromyeloid progenitors
(Greenhalgh, 2018)	[56]	Microglia have an essential role
(Barrett, 2017)	[57]	cerebral homeostasis
(Ginwala, 2019)	[58]	NO synthase and NF-κB activation
(Liu, 2019)	[59]	Melatonin is a pineal hormone with anti-inflammatory effect
(Becerra-Calixto, 2017)	[60]	calcium and potassium homeostasis
(Islinger, 2018)	[61]	HIF1α stimulates the production of peroxisomes
(Gorgulho, 2019)	[63]	High mobility group box 1 (HMGB1) protein
(Kim, 2017)	[64]	neuroinflammatory response, pathogenesis of ischemic stroke

3. *Newborn hypoxic-ischemic encephalopathy*

Article	Ref. no	Subject
(Geisler, 2019)	[75]	reduced blood flow and brain oxygenation
(Rohowetz, 2018)	[76]	Mitochondria tend to hyperpermeabilize
(Weiskirchen, 2016)	[77]	ROS are involved in cell physiological / pathological processes,
(de Faria, 2019)	[80]	phagocytosis processes of cell debris

4. *Adult brain ischemic vascular lesions*

Article	Ref. no	Subject
(Carvajal, 2016)	[85]	The ionotropic glutamate receptor AMPA
(Galicia-Garcia, 2020)	[86]	Stroke (neurological condition) - individuals, family and social
(Pennisi, 2020)	[87]	Stroke neurological condition and SARS-CoV-2
(Shahabipour, 2017)	[88]	Aβ proteins along with the decreased expression of neprilysin
(Tanaka, 2020)	[90]	Aβ proteins - Alzheimer's dementia
(Şekerdağ, 2018)	[91]	acute phase post stroke VEGF increases permeability of BBB
(Morya, 2019)	[92]	Primary and secondary lesions occur in traumatic brain injury
(Ramirez, 2018)	[94]	proliferation, signal transduction
(Iraci, 2016)	[95]	regulation, miRNA - traumatic brain injury
(Ciregia, 2017)	[96]	Traumatic brain injury (TBI) biomarker

Table 1. *Cont.*

5.	*Adult spinal cord injury and hypoxia*	
(Poniatowski, 2017)	[98]	risk of polytrauma
(Lin, 2020)	[99]	primary and secondary lesions
(Kim, 2019)	[102]	Overexpression of the GM-CSF gene protects

6.	*Hypoxia and functional recovery*	
(Miranda, 2019)	[111]	intermittent exposure to hypoxia
(Zhou, 2016)	[117]	hypoxic stimulation is thought to stimulate neuroplasticity
(Ke, 2019)	[119]	intermittent hypoxia may induce HIF-1α expression
(Tan, 2018)	[124]	VEGF production is stimulated by neuropeptide Y (NPY)
(Yung, 2020)	[126]	NPY is an orexigenic hormone, negatively regulated by insulin
(Gaforio, 2019)	[130]	Mediterranean diet based on an abundant consumption of olive oil
(Angeloni, 2017)	[131]	olive oil has anti-inflammatory and immunomodulatory effects
(Libro, 2016)	[132]	Natural flavonoids (wogonin, curcumin, apigenin, quercetin)
(Teleanu, 2019)	[133]	anti-inflammatory effect
(Ilyasov, 2018)	[134]	inhibit the production of IL-6, TNF-α, and IL-1β - MAPK pathway
(Gu, 2020)	[137]	apamin (bee venom)
(Cramer, 2020)	[138]	cannabinoid receptor agonists on oligodendrocytes

4. Discussion and Conclusions

In the last decade, huge scientific research efforts have been deployed to acquire a better understanding of CNS lesions and to significantly improve the clinical-function outcomes, including management, because of their dreadful, multiplane life-threatening, and disabling potential, but at least for now, there are still no medical (of any kind) interventions [17,18,141] able to cure or decisively contribute to their healing [14,140].

Under these circumstances, enhancing the thorough approach of this very complex and challenging pathology domain must be continued without omitting any possible contribution. Serving this purpose, for instance, there are to be found in the literature the beneficial effects in restoring the nervous system through the action of natural plant substances on the mechanisms of hypoxia [140].

It is known that hypoxia induces various adaptive and survival changes for both normal and tumor cells. However, there are essential differences between hypoxic mechanisms in normal cells compared to tumor ones. Therefore, further hypoxia is critical for recovery from ischemic disorders in neuraxis tissue, for instance, by modulating the related involvement of BDNF [141–143], cytoglobin [114], HSP70, VEGF, erythropoietin, fibronectin.

In contrast, in tumoral tissue, due to the uncontrolled cell proliferation and relative low vascularization—ischemia—it results in cell oxygen-starvation—hypoxia. Even if hypoxia inhibits normal cell development, neoplastic cells develop, and hypoxic mechanisms are used for tumor proliferation (although unsystematized tumor development can cause central necrosis and cell death) [144]. All neoplastic ischemic changes, called "pseudo-hypoxia", are essential for angiogenesis, growth, and tumor metabolic adaptation, including thorough the NF-kB way and resistance to treatment. The discovery of links between hypoxia and neoplastic metabolism has led to the development of immuno-oncology (and the synthesis of cytostatics that specifically inhibit specific pathways of pseudo-hypoxia like VEGF inhibitors HIF-2α inhibitors) [145,146].

As mentioned above, the functional recovery of CNS lesions appears to be positively influenced by the activation of normal hypoxia pathways.

Considering all the above, the study regarding hypoxia's mechanisms and ischemia mechanisms targeted to emphasize and synthesize the actual perspectives in understanding ischemic related CNS changes, especially at the intimate level involved in lesions development and or (dialectical, paradoxically) recovery.

Thus, it was observed that HIF-1α and VEGF expression in spinal cord injury areas after SCI is associated with the process of angiogenesis and improvement of local microcirculation. Erythropoietin is essential for recovering cognitive and memory disorders (following ischemic hypoxia) and appears to promote functional recovery after spinal cord injury. Also, after SCI, GM-CSF ensures the survival of dopaminergic neurons inhibits the formation of glial scars, having a neuroprotective role, too. The short intermittent experimental exposure to hypoxia produced respiratory, psychological, and motor function benefits in patients with incomplete deficits after SCI. In addition, deferoxamine treatment of rats with SCI induced a significant improvement in motor deficit, spared nerve tissue area, and electrophysiological conduction.

All these experimental findings justify the need to study the influence of hypoxia in the recovery of CNS disorders.

Supplementary Materials: The following are available online at https://www.mdpi.com/article/10.3390/biomedicines10020481/s1, Table S1. Works selected through a customized (Gelu Onose, 2018) PEDro inspired filtering indirect quality classification.

Author Contributions: All authors had specific but overall equal contributions in achieving this article: conceptualization, S.I.S., G.O., C.B., and C.M.; methodology, S.I.S., G.O., C.B., A.M.I., and C.M.; software, V.C. and C.M.; validation, S.I.S., G.O., C.M., and C.B.; formal analysis, S.I.S., G.O., C.B., C.M., A.M.I., and I.A.; data curation, C.M., S.I.S., and G.O.; writing—original draft preparation, S.I.S., G.O., and C.M.; writing—review and editing, all authors; visualization, S.I.S., G.O., I.A., and C.M.; supervision, G.O., V.C., C.M., and S.I.S. All authors have read and agreed to the published version of the manuscript.

Funding: Financing Project Competitiveness Operational Programme (COP) A1.1.4. ID: P_37_798 MyeloAL-EDiaProT, Contract 149/26.10.2016, (SMIS: 106774), MyeloAL Project.

Institutional Review Board Statement: Not applicable.

Informed Consent Statement: Not applicable.

Data Availability Statement: Not applicable.

Conflicts of Interest: The authors declare that they have no known competing financial interests or personal relationships that could have influenced the work reported in this paper.

References

1. Lee, P.; Chandel, N.S.; Simon, M.C. Cellular adaptation to hypoxia through HIFs and beyond. *Nat. Rev. Mol. Cell Biol.* **2020**, *21*, 268–283. [CrossRef] [PubMed]
2. Hadanny, A.; Efrati, S. The Hyperoxic-Hypoxic Paradox. *Biomolecules* **2020**, *10*, 958. [CrossRef] [PubMed]
3. Hoteteu, M.; Munteanu, C.; Ionescu, E.V.; Almăşan, R.E. Bioactive substances of the Techirghiol therapeutic mud. *Balneo Res. J.* **2018**, *9*, 5–10. [CrossRef]
4. Yang, L.; Roberts, D.; Takhar, M.; Erho, N.; Bibby, B.A.S.; Thiruthaneeswaran, N.; Bhandari, V.; Cheng, W.C.; Haider, S.; McCorry, A.M.B.; et al. Development and Validation of a 28-gene Hypoxia-related Prognostic Signature for Localized Prostate Cancer. *EBioMedicine* **2018**, *31*, 182–189. [CrossRef]
5. Navarrete-Opazo, A.; Mitchell, G.S. Therapeutic potential of intermittent hypoxia: A matter of dose. *Am. J. Physiol. Regul. Integr. Comp. Physiol.* **2014**, *307*, R1181–R1197. [CrossRef]
6. Wu, D.; Yotnda, P. Induction and testing of hypoxia in cell culture. *J. Vis. Exp.* **2011**, *54*, e2899. [CrossRef]
7. Mergenthaler, P.; Lindauer, U.; Dienel, G.A.; Meisel, A. Sugar for the brain: The role of glucose in physiological and pathological brain function. *Trends Neurosci.* **2013**, *36*, 587–597. [CrossRef]
8. Joshi, S.; Ornstein, E.; Young, W.L. Cerebral and Spinal Cord Blood Flow. In *Cottrell and Young's Neuroanesthesia*, 5th ed.; Mosby: Maryland Heights, MI, USA, 2010; pp. 17–59. ISBN 9780323059084. [CrossRef]

9. Marcus, M.L.; Heistad, D.D.; Ehrhardt, J.C.; Abboud, F.M. Regulation of total and regional spinal cord blood flow. *Circ. Res.* **1977**, *41*, 128–134. [CrossRef]

10. Jaakkola, P.; Mole, D.R.; Tian, Y.M.; Wilson, M.I.; Gielbert, J.; Gaskell, S.J.; Von Kriegsheim, A.; Hebestreit, H.F.; Mukherji, M.; Schofield, C.J.; et al. Targeting of HIF-α to the von Hippel-Lindau ubiquitylation complex by O2-regulated prolyl hydroxylation. *Science* **2001**, *292*, 468–472. [CrossRef]

11. Thornton, C.; Leaw, B.; Mallard, C.; Nair, S.; Jinnai, M.; Hagberg, H. Cell death in the developing brain after hypoxia-ischemia. *Front. Cell. Neurosci.* **2017**, *11*, 1–19. [CrossRef]

12. Zhaoa, L.-R.; Willing, A. Enhancing endogenous capacity to repair a stroke-damaged brain: An evolving field for stroke research. *Prog. Neurobiol.* **2018**, *163*, 5–26. [CrossRef] [PubMed]

13. Vasilev, D.S.; Dubrovskaya, N.M.; Tumanova, N.L.; Zhuravin, I.A. Prenatal hypoxia in different periods of embryogenesis differentially affects cell migration, neuronal plasticity, and rat behavior in postnatal ontogenesis. *Front. Neurosci.* **2016**, *10*, 126. [CrossRef] [PubMed]

14. Cai, R.; Pan, C.; Ghasemigharagoz, A.; Todorov, M.I.; Förstera, B.; Zhao, S.; Bhatia, H.S.; Parra-Damas, A.; Mrowka, L.; Theodorou, D.; et al. Panoptic imaging of transparent mice reveals whole-body neuronal projections and skull-meninges connections. *Nat. Neurosci.* **2019**, *22*, 317–327. [CrossRef] [PubMed]

15. Kerridge, C.; Kozlova, D.I.; Nalivaeva, N.N.; Turner, A.J. Hypoxia affects neprilysin expression through caspase activation and an APP intracellular domain-dependent mechanism. *Front. Neurosci.* **2015**, *9*, 426. [CrossRef] [PubMed]

16. Mehanna, R.; Jankovic, J. Movement disorders in cerebrovascular disease. *Lancet Neurol.* **2013**, *12*, 597–608. [CrossRef]

17. Onose, G.; Anghelescu, A.; Blendea, D.; Ciobanu, V.; Daia, C.; Firan, F.C.; Oprea, M.; Spinu, A.; Popescu, C.; Ionescu, A.; et al. Cellular and Molecular Targets for Non-Invasive, Non-Pharmacological Therapeutic/Rehabilitative Interventions in Acute Ischemic Stroke. *Int. J. Mol. Sci.* **2022**, *23*, 907. [CrossRef]

18. Onose, G.; Anghelescu, A.; Blendea, C.D.; Ciobanu, V.; Daia, C.O.; Firan, F.C.; Munteanu, C.; Oprea, M.; Spinu, A.; Popescu, C. Non-invasive, non-pharmacological/bio-technological interventions towards neurorestoration upshot after ischemic stroke, in adults—systematic, synthetic, literature review. *Front. Biosci.* **2021**, *26*, 1204–1239. [CrossRef]

19. Lehotský, J.; Tothová, B.; Kovalská, M.; Dobrota, D.; Benová, A.; Kalenská, D.; Kaplán, P. Role of homocysteine in the ischemic stroke and development of ischemic tolerance. *Front. Neurosci.* **2016**, *10*, 538. [CrossRef]

20. Pang, H.; Fu, Q.; Cao, Q.; Hao, L.; Zong, Z. Sex differences in risk factors for stroke in patients with hypertension and hyperhomocysteinemia. *Sci. Rep.* **2019**, *9*, 14313. [CrossRef]

21. Choudhry, H.; Harris, A.L. Advances in Hypoxia-Inducible Factor Biology. *Cell Metab.* **2018**, *27*, 281–298. [CrossRef]

22. Sköld, M.K.; Marti, H.H.; Lindholm, T.; Lindå, H.; Hammarberg, H.; Risling, M.; Cullheim, S. Induction of HIF1α but not HIF2α in motoneurons after ventral funiculus axotomy—Implication in neuronal survival strategies. *Exp. Neurol.* **2004**, *188*, 20–32. [CrossRef] [PubMed]

23. Nowak-Sliwinska, P.; Alitalo, K.; Allen, E.; Anisimov, A.; Aplin, A.C.; Auerbach, R.; Augustin, H.G.; Bates, D.O.; van Beijnum, J.R.; Bender, R.H.F.; et al. Consensus guidelines for the use and interpretation of angiogenesis assays. *Angiogenesis* **2018**, *21*, 425–532. [CrossRef] [PubMed]

24. Maxwell, P.H.; Wiesener, M.S. Ratcliffe1999. *Nature* **1999**, *459*, 271–275. [CrossRef]

25. Semenza, G.L. HIF-1 and mechanisms of hypoxia sensing. *Curr. Opin. Cell Biol.* **2001**, *13*, 167–171. [CrossRef]

26. Ortiz-Prado, E.; Dunn, J.F.; Vasconez, J.; Castillo, D.; Viscor, G. Partial pressure of oxygen in the human body: A general review. *Am. J. Blood Res.* **2019**, *9*, 1–14. Available online: http://www.ncbi.nlm.nih.gov/pubmed/30899601%0Ahttp://www.pubmedcentral.nih.gov/articlerender.fcgi?artid=PMC6420699 (accessed on 28 November 2021). [PubMed]

27. Burslem, G.M.; Kyle, H.F.; Nelson, A.; Edwards, T.A.; Wilson, A.J. Hypoxia inducible factor (HIF) as a model for studying inhibition of protein-protein interactions. *Chem. Sci.* **2017**, *8*, 4188–4202. [CrossRef] [PubMed]

28. Elvidge, G.P.; Glenny, L.; Appelhoff, R.J.; Ratcliffe, P.J.; Ragoussis, J.; Gleadle, J.M. Concordant regulation of gene expression by hypoxia and 2-oxoglutarate-dependent dioxygenase inhibition: The role of HIF-1α, HIF-2α, and other pathways. *J. Biol. Chem.* **2006**, *281*, 15215–15226. [CrossRef]

29. Appelhoffl, R.J.; Tian, Y.M.; Raval, R.R.; Turley, H.; Harris, A.L.; Pugh, C.W.; Ratcliffe, P.J.; Gleadle, J.M. Differential function of the prolyl hydroxylases PHD1, PHD2, and PHD3 in the regulation of hypoxia-inducible factor. *J. Biol. Chem.* **2004**, *279*, 38458–38465. [CrossRef]

30. Hewitson, K.S.; McNeill, L.A.; Riordan, M.V.; Tian, Y.M.; Bullock, A.N.; Welford, R.W.; Elkins, J.M.; Oldham, N.J.; Bhattacharya, S.; Gleadle, J.M.; et al. Hypoxia-inducible factor (HIF) asparagine hydroxylase is identical to factor inhibiting HIF (FIH) and is related to the cupin structural family. *J. Biol. Chem.* **2002**, *277*, 26351–26355. [CrossRef]

31. Hao, J.; Chen, X.; Fu, T.; Liu, J.; Yu, M.; Han, W.; He, S.; Qian, R.; Zhang, F. The Expression of VHL (Von Hippel-Lindau) After Traumatic Spinal Cord Injury and Its Role in Neuronal Apoptosis. *Neurochem. Res.* **2016**, *41*, 2391–2400. [CrossRef]

32. Chen, F.; Qi, Z.; Luo, Y.; Taylor Hinchliffe, G.D.; Xia, Y.; Ji, X. Non-pharmaceutical therapies for stroke: Mechanisms and clinical implications. *Prog Neurobiol.* **2014**, 246–269. [CrossRef] [PubMed]

33. Bartoszewski, R.; Moszyńska, A.; Serocki, M.; Cabaj, A.; Polten, A.; Ochocka, R.; Dell'Italia, L.; Bartoszewska, S.; Króliczewski, J.; Dąbrowski, M.; et al. Primary endothelial-specific regulation of hypoxiainducible factor (HIF)-1 and HIF-2 and their target gene expression profiles during hypoxia. *FASEB J.* **2019**, *33*, 7929–7941. [CrossRef] [PubMed]

34. Horiguchi, H.; Kayama, F.; Oguma, E.; Willmore, W.G.; Hradecky, P.; Bunn, H.F. Cadmium and platinum suppression of erythropoietin production in cell culture: Clinical implications. *Blood* **2000**, *96*, 3743–3747. [CrossRef] [PubMed]

35. Johansen, J.L.; Sager, T.N.; Lotharius, J.; Witten, L.; Mørk, A.; Egebjerg, J.; Thirstrup, K. HIF prolyl hydroxylase inhibition increases cell viability and potentiates dopamine release in dopaminergic cells. *J. Neurochem.* **2010**, *115*, 209–219. [CrossRef] [PubMed]

36. Mookerjee, S.A.; Gerencser, A.A.; Nicholls, D.G.; Brand, M.D. Quantifying intracellular rates of glycolytic and oxidative ATP production and consumption using extracellular flux measurements. *J. Biol. Chem.* **2017**, *292*, 7189–7207. [CrossRef] [PubMed]

37. Lange, C.A.K.; Bainbridge, J.W.B. Oxygen sensing in retinal health and disease. *Ophthalmologica* **2012**, *227*, 115–131. [CrossRef] [PubMed]

38. Wang, G.L.; Jiang, B.H.; Rue, E.A.; Semenza, G.L. Hypoxia-inducible factor 1 is a basic-helix-loop-helix-PAS heterodimer regulated by cellular O2 tension. *Proc. Natl. Acad. Sci. USA* **1995**, *92*, 5510–5514. [CrossRef]

39. D'ignazio, L.; Rocha, S. Hypoxia induced NF-kB. *Cells* **2016**, *5*, 10. [CrossRef]

40. Simmons, L.J.; Surles-Zeigler, M.C.; Li, Y.; Ford, G.D.; Newman, G.D.; Ford, B.D. Regulation of inflammatory responses by neuregulin-1 in brain ischemia and microglial cells in vitro involves the NF-kappa B pathway. *J. Neuroinflamma.* **2016**, *13*, 237. [CrossRef]

41. Osipo, C.; Golde, T.E.; Osborne, B.A.; Miele, L.A. Off the beaten pathway: The complex cross talk between Notch and NF-κB. *Lab. Investig.* **2008**, *88*, 11–17. [CrossRef]

42. Sun, S.-C. The non-canonical NF-κB pathway in immunity and inflammation. *Nat. Rev. Immunol.* **2017**, *17*, 545–558. [CrossRef]

43. Scott, O.; Roifman, C.M. NF-κB pathway and the Goldilocks principle: Lessons from human disorders of immunity and inflammation. *J. Allergy Clin. Immunol.* **2019**, *143*, 1688–1701. [CrossRef] [PubMed]

44. Yuniati, L.; Scheijen, B.; van der Meer, L.T.; van Leeuwen, F.N. Tumor suppressors BTG1 and BTG2: Beyond growth control. *J. Cell. Physiol.* **2019**, *234*, 5379–5389. [CrossRef]

45. Gschwandtner, M.; Derler, R.; Midwood, K.S. More Than Just Attractive: How CCL2 Influences Myeloid Cell Behavior Beyond Chemotaxis. *Front. Immunol.* **2019**, *10*, 759. [CrossRef] [PubMed]

46. Song, Z.; Zhang, X.; Ye, X.; Feng, C.; Yang, G.; Lu, Y.; Lin, Y.; Dong, C. High expression of stromal cell-derived factor 1 (SDF-1) and NF-κB predicts poor prognosis in cervical cancer. *Med. Sci. Monit.* **2017**, *23*, 151–157. [CrossRef]

47. Feng, W.; Xue, T.; Huang, S.; Shi, Q.; Tang, C.; Cui, G.; Yang, G.; Gong, H.; Guo, H. HIF-1α promotes the migration and invasion of hepatocellular carcinoma cells via the IL-8–NF-κB axis. *Cell. Mol. Biol. Lett.* **2018**, *23*, 26. [CrossRef]

48. Yang, Y.; Liu, L.; Naik, I.; Braunstein, Z.; Zhong, J.; Ren, B. Transcription factor C/EBP homologous protein in health and diseases. *Front. Immunol.* **2017**, *8*, 1612. [CrossRef] [PubMed]

49. Wang, Q.; Liu, N.; Ni, Y.S.; Yang, J.M.; Ma, L.; Lan, X.B.; Wu, J.; Niu, J.G.; Yu, J.Q. TRPM2 in ischemic stroke: Structure, molecular mechanisms, and drug intervention. *Channels* **2021**, *15*, 136–154. [CrossRef] [PubMed]

50. Li, Y.; Xia, J.; Jiang, N.; Xian, Y.; Ju, H.; Wei, Y.; Zhang, X. Corin protects H2O2-induced apoptosis through PI3K/AKT and NF-κB pathway in cardiomyocytes. *Biomed. Pharmacother.* **2018**, *97*, 594–599. [CrossRef]

51. Qiu, X.; Liu, K.; Xiao, L.; Jin, S.; Dong, J.; Teng, X.; Guo, Q.; Chen, Y.; Wu, Y. Alpha-lipoic acid regulates the autophagy of vascular smooth muscle cells in diabetes by elevating hydrogen sulfide level. *Biochim. Biophys. Acta Mol. Basis Dis.* **2018**, *1864*, 3723–3738. [CrossRef]

52. Wu, P.; Zeng, F.; Li, Y.X.; Yu, B.L.; Qiu, L.H.; Qin, W.; Li, J.; Zhou, Y.M.; Liang, F.R. Changes of resting cerebral activities in subacute ischemic stroke patients. *Neural Regen. Res.* **2015**, *10*, 760–765. [CrossRef] [PubMed]

53. Lindsey, H.M.; Wilde, E.A.; Caeyenberghs, K.; Dennis, E.L. Longitudinal Neuroimaging in Pediatric Traumatic Brain Injury: Current State and Consideration of Factors That Influence Recovery. *Front. Neurol.* **2019**, *10*, 1296. [CrossRef] [PubMed]

54. Alizadeh, A.; Dyck, S.M.; Karimi-Abdolrezaee, S. Traumatic spinal cord injury: An overview of pathophysiology, models and acute injury mechanisms. *Front. Neurol.* **2019**, *10*, 282. [CrossRef]

55. Clark, A.R.; Ohlmeyer, M. Protein phosphatase 2A as a therapeutic target in inflammation and neurodegeneration. *Pharmacol. Ther.* **2019**, *201*, 181–201. [CrossRef] [PubMed]

56. Miller, A.D.; Zachary, J.F. Nervous System. *Pathol. Basis Vet. Dis. Expert Consult* **2017**, *14*, 805–907.e1. [CrossRef]

57. Greenhalgh, A.D.; Zarruk, J.G.; Healy, L.M.; Baskar Jesudasan, S.J.; Jhelum, P.; Salmon, C.K.; Formanek, A.; Russo, M.V.; Antel, J.P.; McGavern, D.B.; et al. Peripherally derived macrophages modulate microglial function to reduce inflammation after CNS injury. *PLoS Biol.* **2018**, *16*, e2005264. [CrossRef] [PubMed]

58. Barrett, J.P.; Henry, R.J.; Villapol, S.; Stoica, B.A.; Kumar, A.; Burns, M.P.; Faden, A.I.; Loane, D.J. NOX2 deficiency alters macrophage phenotype through an IL-10/STAT3 dependent mechanism: Implications for traumatic brain injury. *J. Neuroinflamm.* **2017**, *14*, 65. [CrossRef]

59. Ginwala, R.; Bhavsar, R.; Chigbu, D.G.I.; Jain, P.; Khan, Z.K. Potential role of flavonoids in treating chronic inflammatory diseases with a special focus on the anti-inflammatory activity of apigenin. *Antioxidants* **2019**, *8*, 35. [CrossRef]

60. Liu, Z.J.; Ran, Y.Y.; Qie, S.Y.; Gong, W.J.; Gao, F.H.; Ding, Z.T.; Xi, J.N. Melatonin protects against ischemic stroke by modulating microglia/macrophage polarization toward anti-inflammatory phenotype through STAT3 pathway. *CNS Neurosci. Ther.* **2019**, *25*, 1353–1362. [CrossRef]

61. Becerra-Calixto, A.; Cardona-Gómez, G.P. The role of astrocytes in neuroprotection after brain stroke: Potential in cell therapy. *Front. Mol. Neurosci.* **2017**, *10*, 88. [CrossRef]

62. Islinger, M.; Voelkl, A.; Fahimi, H.D.; Schrader, M. The peroxisome: An update on mysteries 2.0. *Histochem. Cell Biol.* **2018**, *150*, 443–471. [CrossRef] [PubMed]

63. MUNTEANU, C.; TEOIBAS-SERBAN, D.; IORDACHE, L.; BALAUREA, M.; BLENDEA, C.-D. Water intake meets the Water from inside the human body—physiological, cultural, and health perspectives—Synthetic and Systematic literature review. *Balneo PRM Res. J.* **2021**, *12*, 196–209. [CrossRef]

64. Mader, S.; Brimberg, L. Aquaporin-4 Water Channel in the Brain and Its Implication for Health and Disease. *Cells* **2019**, *8*, 90. [CrossRef]

65. Gorgulho, C.M.; Romagnoli, G.G.; Bharthi, R.; Lotze, M.T. Johnny on the spot-chronic inflammation is driven by HMGB1. *Front. Immunol.* **2019**, *10*, 1561. [CrossRef] [PubMed]

66. Kim, D.S.; Choi, H.I.; Wang, Y.; Luo, Y.; Hoffer, B.J.; Greig, N.H. A New Treatment Strategy for Parkinson's Disease through the Gut–Brain Axis: The Glucagon-Like Peptide-1 Receptor Pathway. *Cell Transplant.* **2017**, *26*, 1560–1571. [CrossRef]

67. Gou, X.; Ying, J.; Yue, Y.; Qiu, X.; Hu, P.; Qu, Y.; Li, J.; Mu, D. The Roles of High Mobility Group Box 1 in Cerebral Ischemic Injury. *Front. Cell. Neurosci.* **2020**, *14*, 280. [CrossRef]

68. Ye, Y.; Zeng, Z.; Jin, T.; Zhang, H.; Xiong, X.; Gu, L. The role of high mobility group box 1 in ischemic stroke. *Front. Cell. Neurosci.* **2019**, *13*, 127. [CrossRef]

69. Nekoui, A.; Blaise, G. Erythropoietin and Nonhematopoietic Effects. *Am. J. Med. Sci.* **2017**, *353*, 76–81. [CrossRef]

70. Fuhrmann, D.C.; Brüne, B. Mitochondrial composition and function under the control of hypoxia. *Redox Biol.* **2017**, *12*, 208–215. [CrossRef]

71. Lukyanova, L.D.; Kirova, Y.I. Mitochondria-controlled signaling mechanisms of brain protection in hypoxia. *Front. Neurosci.* **2015**, *9*, 320. [CrossRef]

72. Fuhrmann, D.C.; Olesch, C.; Kurrle, N.; Schnütgen, F.; Zukunft, S.; Fleming, I.; Brüne, B. Chronic Hypoxia Enhances β-Oxidation-Dependent Electron Transport via Electron Transferring Flavoproteins. *Cells* **2019**, *8*, 172. [CrossRef] [PubMed]

73. Azuma, K.; Ikeda, K.; Inoue, S. Functional mechanisms of mitochondrial respiratory chain supercomplex assembly factors and their involvement in muscle quality. *Int. J. Mol. Sci.* **2020**, *21*, 3182. [CrossRef] [PubMed]

74. Brose, S.A.; Golovko, S.A.; Golovko, M.Y. Fatty acid biosynthesis inhibition increases reduction potential in neuronal cells under hypoxia. *Front. Neurosci.* **2016**, *10*, 546. [CrossRef] [PubMed]

75. Millar, L.J.; Shi, L.; Hoerder-Suabedissen, A.; Molnár, Z. Neonatal hypoxia ischaemia: Mechanisms, models, and therapeutic challenges. *Front. Cell. Neurosci.* **2017**, *11*, 78. [CrossRef] [PubMed]

76. Rocha-Ferreira, E.; Hristova, M. Plasticity in the neonatal brain following hypoxic-ischaemic injury. *Neural Plast.* **2016**, *2016*, 4901014. [CrossRef] [PubMed]

77. Geisler, J. 2,4 Dinitrophenol as Medicine. *Cells* **2019**, *8*, 280. [CrossRef]

78. Rohowetz, L.J.; Kraus, J.G.; Koulen, P. Reactive oxygen species-mediated damage of retinal neurons: Drug development targets for therapies of chronic neurodegeneration of the retina. *Int. J. Mol. Sci.* **2018**, *19*, 3362. [CrossRef]

79. Weiskirchen, R. Hepatoprotective and anti-fibrotic agents: It's time to take the next step. *Front. Pharmacol.* **2016**, *6*, 303. [CrossRef]

80. Li, B.; Concepcion, K.; Meng, X.; Zhang, L. Brain-immune interactions in perinatal hypoxic-ischemic brain injury. *Prog. Neurobiol.* **2017**, *159*, 50–68. [CrossRef]

81. Bhalala, U.S.; Koehler, R.C.; Kannan, S. Neuroinflammation and neuroimmune dysregulation after acute hypoxic-ischemic injury of developing brain. *Front. Pediatr.* **2015**, *2*, 144. [CrossRef]

82. de Faria, O.; Gonsalvez, D.G.; Nicholson, M.; Xiao, J. Activity-dependent central nervous system myelination throughout life. *J. Neurochem.* **2019**, *148*, 447–461. [CrossRef] [PubMed]

83. Reis, C.; Wang, Y.; Akyol, O.; Ho, W.M.; Applegate, R.; Stier, G.; Martin, R.; Zhang, J.H. What's new in traumatic brain injury: Update on tracking, monitoring and treatment. *Int. J. Mol. Sci.* **2015**, *16*, 11903–11965. [CrossRef] [PubMed]

84. Raichle, M.E. Cerebral blood flow and metabolism. *Outcome Sev. Damage Cent. Nerv. Syst.* **2008**, 85–96. [CrossRef]

85. Franke, H.; Illes, P. Nucleotide signaling in astrogliosis. *Neurosci. Lett.* **2014**, *565*, 14–22. [CrossRef] [PubMed]

86. Nas, K.; Yazmalar, L.; Şah, V.; Aydin, A.; Öneş, K. Rehabilitation of spinal cord injuries. *World J. Orthop.* **2015**, *6*, 8–16. [CrossRef] [PubMed]

87. Carvajal, F.J.; Mattison, H.A.; Cerpa, W. Role of NMDA Receptor-Mediated Glutamatergic Signaling in Chronic and Acute Neuropathologies. *Neural Plast.* **2016**, *2016*, 2701526. [CrossRef]

88. Galicia-Garcia, U.; Benito-Vicente, A.; Jebari, S.; Larrea-Sebal, A.; Siddiqi, H.; Uribe, K.B.; Ostolaza, H.; Martín, C. Pathophysiology of type 2 diabetes mellitus. *Int. J. Mol. Sci.* **2020**, *21*, 6275. [CrossRef]

89. Pennisi, M.; Lanza, G.; Falzone, L.; Fisicaro, F.; Ferri, R.; Bella, R. Sars-cov-2 and the nervous system: From clinical features to molecular mechanisms. *Int. J. Mol. Sci.* **2020**, *21*, 5475. [CrossRef]

90. Shahabipour, F.; Barati, N.; Johnston, T.P.; Derosa, G.; Maffioli, P.; Sahebkar, A. Exosomes: Nanoparticulate tools for RNA interference and drug delivery. *J. Cell. Physiol.* **2017**, *232*, 1660–1668. [CrossRef]

91. Minhas, G.; Mathur, D.; Ragavendrasamy, B.; Sharma, N.K.; Paanu, V.; Anand, A. Hypoxia in CNS pathologies: Emerging role of miRNA-based Neurotherapeutics and yoga based alternative therapies. *Front. Neurosci.* **2017**, *11*, 386. [CrossRef]

92. Tanaka, M.; Toldi, J.; Vécsei, L. Exploring the etiological links behind neurodegenerative diseases: Inflammatory cytokines and bioactive kynurenines. *Int. J. Mol. Sci.* **2020**, *21*, 2431. [CrossRef] [PubMed]

93. Sekerdag, E.; Solaroglu, I.; Gursoy-Ozdemir, Y. Cell Death Mechanisms in Stroke and Novel Molecular and Cellular Treatment Options. *Curr. Neuropharmacol.* **2018**, *16*, 1396–1415. [CrossRef] [PubMed]

94. Morya, E.; Monte-Silva, K.; Bikson, M.; Esmaeilpour, Z.; Biazoli, C.E.; Fonseca, A.; Bocci, T.; Farzan, F.; Chatterjee, R.; Hausdorff, J.M.; et al. Beyond the target area: An integrative view of tDCS-induced motor cortex modulation in patients and athletes. *J. Neuroeng. Rehabil.* **2019**, *16*, 1–29. [CrossRef] [PubMed]

95. Yang, Y.; Wang, H.; Li, L.; Li, X.; Wang, Q.; Ding, H.; Wang, X.; Ye, Z.; Wu, L.; Zhang, X.; et al. Sinomenine provides neuroprotection in model of traumatic brain injury via the Nrf2-ARE pathway. *Front. Neurosci.* **2016**, *10*, 580. [CrossRef]

96. Ramirez, S.H.; Andrews, A.M.; Paul, D.; Pachter, J.S. Extracellular vesicles: Mediators and biomarkers of pathology along CNS barriers. *Fluids Barriers CNS* **2018**, *15*, 19. [CrossRef]

97. Iraci, N.; Leonardi, T.; Gessler, F.; Vega, B.; Pluchino, S. Focus on extracellular vesicles: Physiological role and signalling properties of extracellular membrane vesicles. *Int. J. Mol. Sci.* **2016**, *17*, 171. [CrossRef]

98. Ciregia, F.; Urbani, A.; Palmisano, G. Extracellular vesicles in brain tumors and neurodegenerative diseases. *Front. Mol. Neurosci.* **2017**, *10*, 276. [CrossRef]

99. Falkenberg, L.; Zeckey, C.; Mommsen, P.; Winkelmann, M.; Zelle, B.A.; Panzica, M.; Pape, H.C.; Krettek, C.; Probst, C. Long-term outcome in 324 polytrauma patients: What factors are associated with posttraumatic stress disorder and depressive disorder symptoms? *Eur. J. Med. Res.* **2017**, *22*, 44. [CrossRef]

100. Poniatowski, Ł.A.; Wojdasiewicz, P.; Krawczyk, M.; Szukiewicz, D.; Gasik, R.; Kubaszewski, Ł.; Kurkowska-Jastrzębska, I. Analysis of the Role of CX3CL1 (Fractalkine) and Its Receptor CX3CR1 in Traumatic Brain and Spinal Cord Injury: Insight into Recent Advances in Actions of Neurochemokine Agents. *Mol. Neurobiol.* **2017**, *54*, 2167–2188. [CrossRef]

101. Lin, W.; Stone, S. Unfolded protein response in myelin disorders. *Neural Regen. Res.* **2020**, *15*, 636–645. [CrossRef]

102. Kim, H.J.; Oh, J.S.; An, S.S.; Pennant, W.A.; Gwak, S.J.; Kim, A.N.; Han, P.K.; Yoon, D.H.; Kim, K.N.; Ha, Y. Hypoxia-specific GM-CSF-overexpressing neural stem cells improve graft survival and functional recovery in spinal cord injury. *Gene Ther.* **2012**, *19*, 513–521. [CrossRef] [PubMed]

103. You, Y.; Che, L.; Lee, H.Y.; Lee, H.L.; Yun, Y.; Lee, M.; Oh, J.; Ha, Y. Antiapoptotic effect of highly secreted GMCSF from neuronal cell-specific GMCSF overexpressing neural stem cells in spinal cord injury model. *Spine* **2015**, *40*, E1284–E1291. [CrossRef] [PubMed]

104. Kim, J.Y.; Kim, J.Y.; Kim, J.H.; Jung, H.; Lee, W.T.; Lee, J.E. Restorative mechanism of neural progenitor cells overexpressing arginine decarboxylase genes following ischemic injury. *Exp. Neurobiol.* **2019**, *28*, 85–103. [CrossRef] [PubMed]

105. Li, X.; Lou, X.; Xu, S.; Du, J.; Wu, J. Hypoxia inducible factor-1 (HIF-1α) reduced inflammation in spinal cord injury via miR-380-3p/ NLRP3 by Circ 0001723. *Biol. Res.* **2020**, *53*, 1–14. [CrossRef]

106. Zhong, D.; Cao, Y.; Li, C.J.; Li, M.; Rong, Z.J.; Jiang, L.; Guo, Z.; Lu, H.B.; Hu, J.Z. Highlight article: Neural stem cell-derived exosomes facilitate spinal cord functional recovery after injury by promoting angiogenesis. *Exp. Biol. Med.* **2020**, *245*, 54–65. [CrossRef]

107. Long, H.Q.; Li, G.S.; Cheng, X.; Xu, J.H.; Li, F.B. Role of hypoxia-induced VEGF in blood-spinal cord barrier disruption in chronic spinal cord injury. *Chin. J. Traumatol. Engl. Ed.* **2015**, *18*, 293–295. [CrossRef]

108. Li, Y.; Han, W.; Wu, Y.; Zhou, K.; Zheng, Z.; Wang, H.; Xie, L.; Li, R.; Xu, K.; Liu, Y.; et al. Stabilization of Hypoxia Inducible Factor-1α by Dimethyloxalylglycine Promotes Recovery from Acute Spinal Cord Injury by Inhibiting Neural Apoptosis and Enhancing Axon Regeneration. *J. Neurotrauma* **2019**, *36*, 3394–3409. [CrossRef]

109. Wang, X.; Ma, J.; Fu, Q.; Zhu, L.; Zhang, Z.; Zhang, F.; Lu, N.; Chen, A. Role of hypoxia-inducible factor-1α in autophagic cell death in microglial cells induced by hypoxia. *Mol. Med. Rep.* **2017**, *15*, 2097–2105. [CrossRef]

110. Dale-Nagle, E.A.; Hoffman, M.S.; MacFarlane, P.M.; Satriotomo, I.; Lovett-Barr, M.R.; Vinit, S.; Mitchell, G.S. Spinal plasticity following intermittent hypoxia: Implications for spinal injury. *Ann. N. Y. Acad. Sci.* **2010**, *1198*, 252–259. [CrossRef]

111. Perim, R.R.; Mitchell, G.S. Circulatory control of phrenic motor plasticity. *Respir. Physiol. Neurobiol.* **2019**, *265*, 19–23. [CrossRef]

112. Wen, M.H.; Wu, M.J.; Vinit, S.; Lee, K.Z. Modulation of Serotonin and Adenosine 2A Receptors on Intermittent Hypoxia-Induced Respiratory Recovery following Mid-Cervical Contusion in the Rat. *J. Neurotrauma* **2019**, *36*, 2991–3004. [CrossRef] [PubMed]

113. Miranda, M.; Morici, J.F.; Zanoni, M.B.; Bekinschtein, P. Brain-Derived Neurotrophic Factor: A Key Molecule for Memory in the Healthy and the Pathological Brain. *Front. Cell. Neurosci.* **2019**, *13*, 363. [CrossRef] [PubMed]

114. Astorino, T.A.; Harness, E.T.; White, A.C. Efficacy of acute intermittent hypoxia on physical function and health status in humans with spinal cord injury: A brief review. *Neural Plast.* **2015**, *2015*, 409625. [CrossRef] [PubMed]

115. Naidu, A.; Peters, D.M.; Tan, A.Q.; Barth, S.; Crane, A.; Link, A.; Balakrishnan, S.; Hayes, H.B.; Slocum, C.; Zafonte, R.D.; et al. Daily acute intermittent hypoxia to improve walking function in persons with subacute spinal cord injury: A randomized clinical trial study protocol. *BMC Neurol.* **2020**, *20*, 273. [CrossRef] [PubMed]

116. Gonzalez-Rothi, E.J.; Lee, K.Z.; Dale, E.A.; Reier, P.J.; Mitchell, G.S.; Fuller, D.D. Intermittent hypoxia and neurorehabilitation. *J. Appl. Physiol.* **2015**, *119*, 1455–1465. [CrossRef]

117. Vinit, S.; Lovett-Barr, M.R.; Mitchell, G.S. Intermittent hypoxia induces functional recovery following cervical spinal injury. *Respir. Physiol. Neurobiol.* **2009**, *169*, 210–217. [CrossRef] [PubMed]

118. Dougherty, B.J.; Terada, J.; Springborn, S.R.; Vinit, S.; MacFarlane, P.M.; Mitchell, G.S. Daily acute intermittent hypoxia improves breathing function with acute and chronic spinal injury via distinct mechanisms. *Respir. Physiol. Neurobiol.* **2018**, *256*, 50–57. [CrossRef]

119. Zhou, L.; Chen, P.; Peng, Y.; Ouyang, R. Role of Oxidative Stress in the Neurocognitive Dysfunction of Obstructive Sleep Apnea Syndrome. *Oxidative Med. Cell. Longev.* **2016**, *2016*, 9626831. [CrossRef]
120. Vivodtzev, I.; Tan, A.Q.; Hermann, M.; Jayaraman, A.; Stahl, V.; Rymer, W.Z.E.V.; Mitchell, G.S.; Hayes, H.B.; Trumbower, R.D. Mild to moderate sleep apnea is linked to hypoxia-induced motor recovery after spinal cord injury. *Am. J. Respir. Crit. Care Med.* **2020**, *202*, 887–890. [CrossRef]
121. Ke, P.Y. Diverse functions of autophagy in liver physiology and liver diseases. *Int. J. Mol. Sci.* **2019**, *20*, 300. [CrossRef]
122. Xiong, T.; Yang, X.; Qu, Y.; Chen, H.; Yue, Y.; Wang, H.; Zhao, F.; Li, S.; Zou, R.; Zhang, L.; et al. Erythropoietin induces synaptogenesis and neurite repair after hypoxia ischemia-mediated brain injury in neonatal rats. *Neuroreport* **2019**, *30*, 783–789. [CrossRef] [PubMed]
123. Halder, S.K.; Kant, R.; Milner, R. Chronic mild hypoxia promotes profound vascular remodeling in spinal cord blood vessels, preferentially in white matter, via an α5β1 integrin-mediated mechanism. *Angiogenesis* **2018**, *21*, 251–266. [CrossRef]
124. Munteanu, C. Cell biology considerations in Spinal Cord Injury—Review. *Balneo Res. J.* **2017**, *8*, 136–151. [CrossRef]
125. Tang, G.; Chen, Y.; Chen, J.; Chen, Z.; Jiang, W. Deferoxamine Ameliorates Compressed Spinal Cord Injury by Promoting Neovascularization in Rats. *J. Mol. Neurosci.* **2020**, *70*, 1437–1444. [CrossRef]
126. Tan, C.M.J.; Green, P.; Tapoulal, N.; Lewandowski, A.J.; Leeson, P.; Herring, N. The role of neuropeptide Y in cardiovascular health and disease. *Front. Physiol.* **2018**, *9*, 1281. [CrossRef] [PubMed]
127. Vandini, E.; Ottani, A.; Zaffe, D.; Calevro, A.; Canalini, F.; Cavallini, G.M.; Rossi, R.; Guarini, S.; Giuliani, D. Mechanisms of hydrogen sulfide against the progression of severe Alzheimer's disease in transgenic mice at different ages. *Pharmacology* **2019**, *103*, 93–100. [CrossRef] [PubMed]
128. Yung Justin Hou Ming, G.A. Role of c-Jun N-terminal Kinase (JNK) in Obesity and Type 2 Diabetes. *Cells* **2020**, *9*, 706. [CrossRef] [PubMed]
129. Zhang, H.; Fang, X.; Huang, D.; Luo, Q.; Zheng, M.; Wang, K.; Cao, L.; Yin, Z. Erythropoietin signaling increases neurogenesis and oligodendrogenesis of endogenous neural stem cells following spinal cord injury both in vivo and in vitro. *Mol. Med. Rep.* **2018**, *17*, 264–272. [CrossRef]
130. Ng, S.Y.; Lee, A.Y.W. Traumatic Brain Injuries: Pathophysiology and Potential Therapeutic Targets. *Front. Cell. Neurosci.* **2019**, *13*, 528. [CrossRef]
131. Matsumoto, M.; Ichikawa, T.; Young, W.; Kodama, N. Glutamine synthetase protects the spinal cord against hypoxia-induced and GABAA receptor-activated axonal depressions. *Surg. Neurol.* **2008**, *70*, 122–128. [CrossRef]
132. Gaforio, J.J.; Visioli, F.; Alarcón-De-la-lastra, C.; Castañer, O.; Delgado-Rodríguez, M.; Fitó, M.; Hernández, A.F.; Huertas, J.R.; Martínez-González, M.A.; Menendez, J.A.; et al. Virgin olive oil and health: Summary of the iii international conference on virgin olive oil and health consensus report, JAEN (Spain) 2018. *Nutrients* **2019**, *11*, 2039. [CrossRef]
133. Angeloni, C.; Malaguti, M.; Barbalace, M.C.; Hrelia, S. Bioactivity of olive oil phenols in neuroprotection. *Int. J. Mol. Sci.* **2017**, *18*, 2230. [CrossRef]
134. Libro, R.; Giacoppo, S.; Rajan, T.S.; Bramanti, P.; Mazzon, E. Natural phytochemicals in the treatment and prevention of dementia: An overview. *Molecules* **2016**, *21*, 518. [CrossRef] [PubMed]
135. Teleanu, R.I.; Chircov, C.; Grumezescu, A.M.; Volceanov, A.; Teleanu, D.M. Antioxidant therapies for neuroprotection—A review. *J. Clin. Med.* **2019**, *8*, 1659. [CrossRef] [PubMed]
136. Ilyasov, A.A.; Milligan, C.E.; Pharr, E.P.; Howlett, A.C. The Endocannabinoid System and Oligodendrocytes in Health and Disease. *Front. Neurosci.* **2018**, *12*, 733. [CrossRef] [PubMed]
137. Garcia-Arencibia, M.; Molina-Holgado, E.; Molina-Holgado, F. Effect of endocannabinoid signalling on cell fate: Life, death, differentiation and proliferation of brain cells. *Br. J. Pharmacol.* **2019**, *176*, 1361–1369. [CrossRef] [PubMed]
138. Maccarrone, M.; Guzman, M.; Mackie, K.; Doherty, P.; Sciences, B.; Kingdom, U.; Institutet, K.; Maccarrone, M.; Guzmán, M.; Mackie, K.; et al. Programming of neural cells by (endo)cannabinoids: From physiological rules to emerging therapies. *Nat. Rev. Neurosci.* **2014**, *15*, 786–801. [CrossRef] [PubMed]
139. Gu, H.; Han, S.M.; Park, K.K. Therapeutic effects of apamin as a bee venom component for non-neoplastic disease. *Toxins* **2020**, *12*, 195. [CrossRef]
140. Cramer, S.W.; Chen, C.C. Photodynamic Therapy for the Treatment of Glioblastoma. *Front. Surg.* **2020**, *6*, 81. [CrossRef]
141. Munteanu, C.; Munteanu, D.; Onose, G. Hydrogen sulfide (H2S)—Therapeutic relevance in rehabilitation and balneotherapy Systematic literature review and meta-analysis based on the PRISMA paradig. *Balneo PRM Res. J.* **2021**, *12*, 176–195. [CrossRef]
142. Firan, F.C.; Romila, A.; Onose, G. Current synthesis and systematic review of main effects of calf blood deproteinized medicine (Actovegin®) in ischemic stroke. *Int. J. Mol. Sci.* **2020**, *21*, 3181. [CrossRef] [PubMed]
143. Duan, X.; Yao, G.; Liu, Z.; Cui, R.; Yang, W. Mechanisms of Transcranial Magnetic Stimulation Treating on Post-stroke Depression. *Front. Hum. Neurosci.* **2018**, *12*, 215. [CrossRef] [PubMed]
144. Munteanu, C.; Dogaru, G.; Rotariu, M.; Onose, G. Therapeutic gases used in balneotherapy and rehabilitation medicine—Scientific relevance in the last ten years (2011–2020)—Synthetic literature review. *Balneo PRM Res. J.* **2021**, *12*, 111–122. [CrossRef]
145. Hapke, R.Y.; Haake, S.M. Hypoxia-induced epithelial to mesenchymal transition in cancer. *Cancer Lett.* **2020**, *487*, 10–20. [CrossRef]
146. Murugesan, T.; Rajajeyabalachandran, G.; Kumar, S.; Nagaraju, S.; Jegatheesan, S.K. Targeting HIF-2α as therapy for advanced cancers. *Drug Discov. Today* **2018**, *23*, 1444–1451. [CrossRef] [PubMed]

Article

Fatty Acid-Binding Proteins Aggravate Cerebral Ischemia-Reperfusion Injury in Mice

Qingyun Guo [1], Ichiro Kawahata [1], Tomohide Degawa [1], Yuri Ikeda-Matsuo [2], Meiling Sun [1], Feng Han [3] and Kohji Fukunaga [1,*]

[1] Department of Pharmacology, Graduate School of Pharmaceutical Sciences, Tohoku University, 6-3 Aramaki-Aoba, Aoba-Ku, Sendai 980-8578, Japan; guo.qingyun.r2@dc.tohoku.ac.jp (Q.G.); kawahata@tohoku.ac.jp (I.K.); tomohide.degawa.r2@dc.tohoku.ac.jp (T.D.); reiko.fukunaga.e2@tohoku.ac.jp (M.S.)

[2] Laboratory of Pharmacology, Department of Clinical Pharmacy, Faculty of Pharmaceutical Sciences, Hokuriku University, Kanagawa-Machi, Kanazawa 920-1181, Japan; y-matsuo@hokuriku-u.ac.jp

[3] School of Pharmacy, Nanjing Medical School, Nanjing 211166, China; fenghan169@njmu.edu.cn

* Correspondence: kfukunaga@tohoku.ac.jp; Tel.: +81-22-795-6836

Abstract: Fatty acid-binding proteins (FABPs) regulate the intracellular dynamics of fatty acids, mediate lipid metabolism and participate in signaling processes. However, the therapeutic efficacy of targeting FABPs as novel therapeutic targets for cerebral ischemia is not well established. Previously, we synthesized a novel FABP inhibitor, i.e., FABP ligand 6 [4-(2-(5-(2-chlorophenyl)-1-(4-isopropylphenyl)-1H-pyrazol-3-yl)-4-fluorophenoxy)butanoic acid] (referred to here as MF6). In this study, we analyzed the ability of MF6 to ameliorate transient middle cerebral artery occlusion (tMCAO) and reperfusion-induced injury in mice. A single MF6 administration (3.0 mg/kg, per os) at 0.5 h post-reperfusion effectively reduced brain infarct volumes and neurological deficits. The protein-expression levels of FABP3, FABP5 and FABP7 in the brain gradually increased after tMCAO. Importantly, MF6 significantly suppressed infarct volumes and the elevation of FABP-expression levels at 12 h post-reperfusion. MF6 also inhibited the promotor activity of FABP5 in human neuroblastoma cells (SH-SY5Y). These data suggest that FABPs elevated infarct volumes after ischemic stroke and that inhibiting FABPs ameliorated the ischemic injury. Moreover, MF6 suppressed the inflammation-associated prostaglandin E_2 levels through microsomal prostaglandin E synthase-1 expression in the ischemic hemispheres. Taken together, the results imply that the FABP inhibitor MF6 can potentially serve as a neuroprotective therapeutic for ischemic stroke.

Keywords: ischemia; FABP3; FABP5; FABP7; mPGES-1; PGE_2

Citation: Guo, Q.; Kawahata, I.; Degawa, T.; Ikeda-Matsuo, Y.; Sun, M.; Han, F.; Fukunaga, K. Fatty Acid-Binding Proteins Aggravate Cerebral Ischemia-Reperfusion Injury in Mice. *Biomedicines* **2021**, *9*, 529. https://doi.org/10.3390/biomedicines9050529

Academic Editor: Kumar Vaibhav

Received: 20 March 2021
Accepted: 4 May 2021
Published: 10 May 2021

Publisher's Note: MDPI stays neutral with regard to jurisdictional claims in published maps and institutional affiliations.

Copyright: © 2021 by the authors. Licensee MDPI, Basel, Switzerland. This article is an open access article distributed under the terms and conditions of the Creative Commons Attribution (CC BY) license (https://creativecommons.org/licenses/by/4.0/).

1. Introduction

Ischemic stroke is the most common cerebrovascular disease, accounting for approximately 80% of all strokes and causing a large number of deaths and disabilities worldwide [1,2]. Ischemic stroke is caused by the blockage of the artery supplying blood to the brain, and an insufficient blood supply triggers a series of complex neurochemical responses including excitotoxicity, oxidative stress and inflammation, which ultimately leads to the death and dysfunction of brain cells [3–5]. Even though the administration of intravenous thrombolysis with recombinant tissue plasminogen activator is the only clinically approved drug therapy that can effectively treat ischemic stroke, a short therapeutic window and intracerebral hemorrhagic complications limit its availability [5]. Therefore, alternative neuroprotective drugs are still required for treating acute-phase cerebral ischemia.

Fatty acid-binding proteins (FABPs) are expressed in various tissues in a highly specific manner, where they regulate fatty acid uptake, transport and metabolism [6], and play vital roles in the pathogenesis of many common diseases [7]. Most mammals produce 12 distinct subtypes of FABPs, although humans produce up to 10 [8], three of which are expressed in the brain, including FABP3 (heart-type), FABP5 (epidermal-type) and FABP7 (brain-type) [9,10]. These three FABPs are expressed at specific stages during brain development. For example, FABP3 is not expressed during embryonic period and its expression gradually increases in rodent brains after birth, whereas FABP5 and FABP7 show an opposite pattern and decrease postnatally in rodents [9]. FABP homeostasis is critical for normal brain development and functions at different stages, and FABP imbalances can cause various neurodegenerative and neuropsychiatric disorders. FABP5 and FABP7 mRNA levels were higher in the cortexes of postmortem brains from schizophrenic patients than in those from healthy controls, and similar results were found for FABP7 mRNA in postmortem brains from patients with autism spectrum disorder [9]. Higher FABP3 levels have been observed in the cerebrospinal fluid and serum of patients with Alzheimer's disease, dementia with Lewy bodies or Parkinson's disease [11,12]. FABP3 deficiency prevents nicotine-induced preference behavior [13,14]. The FABP3 deficiency also completely abolished the induction of Parkinson's syndrome after 1-methyl-4-phenyl-1,2,3,6-tetrahydropiridine (MPTP) treatment in mice [15].

Developing specific ligands for each FABP is essential for studying the mechanisms of FABPs in neurological diseases. For example, the FABP4 ligand BMS309403 has been used as a potent and selective biphenyl azole inhibitor to improve ischemia/reperfusion (I/R) injury of the brain [16] and kidneys [17] in mice. Recently, we have developed a series of FABP3 ligands, based on BMS309403 [18]. Among these ligands, we confirmed that MF1 [19] and MF8 [20] could target FABP3 and improve motor deficits and cognitive impairments in a mouse model of MPTP-induced Parkinson's disease. Therefore, we hypothesized that FABP ligands also have neuroprotective effects to rescue neurons from ischemic injury.

In this study, we used a newly synthesized ligand 6 [4-(2-(5-(2-chlorophenyl)-1-(4-isopropylphenyl)-1H-pyrazol-3-yl)-4-fluorophenoxy)butanoic acid] (named here as MF6), which has a high affinity for FABP7 (dissociation constant [Kd] value: 20 ± 9 nM) [21] and has a much weaker affinity for FABP3 (Kd value: 1038 ± 155 nM) and FABP5 (Kd value: 874 ± 66 nM) [22,23], to examine its therapeutic effects on ischemic stroke. We defined the elevation of FABP3, FABP5 and FABP7 expression after I/R injury in mice brain and the neuroprotective effects of FABP inhibitor MF6. Our results provided a new target protein for ischemic stroke therapeutics and demonstrated the potential of FABP ligands as neuroprotective therapeutic drugs.

2. Materials and Methods

2.1. Animals

Male ICR mice (Institute of Cancer Research (ICR) mice, Slc:ICR) (5 weeks old, 25–30 g) were purchased from Japan SLC, Inc. (Shizuoka, Japan). The animals were housed under conditions of constant temperature and humidity, kept on a 12-h light-dark cycle (lights on: 09:00–21:00) and fed ad libitum. All procedures for handling animals complied with the Guide for Care and Use of Laboratory Animals and were approved by the Experimentation Committee of Tohoku University Graduate School of Pharmaceutical Sciences [2019PhLM0-021 (approved date: 1 December 2019) and 2019PhA-024 (approved date: 1 April 2019)].

2.2. Surgical Procedures Used for Establishing Transmit Middle Cerebral Artery Occlusion (Tmcao) and Reperfusion

Mice were randomly assigned to tMCAO and sham groups. The mouse model of tMCAO was generated as previously described [24]. Briefly, adult male ICR mice were anesthetized via intraperitoneal injection with a combination of 0.3 mg/kg body weight medetomidine (Domitol, Meiji Seika Pharma Co., Ltd., Tokyo, Japan), 4.0 mg/kg midazolam (Dormicum, Astellas Pharma Inc., Tokyo, Japan) and 5.0 mg/kg butorphanol (Vetorphale, Meiji Seika Pharma Co., Ltd.). Following this, tMCAO surgical procedure was performed as follows: a silicone-coated 6–0 suture (602356PK10, Doccol Corporation,

Sharon, MA, USA) was inserted from the right external carotid artery to the internal carotid artery, extending to the origin of a middle cerebral artery, for 2 h. After 2 h, the suture was removed, allowing reperfusion to occur. Mice in the sham-operation group underwent the same procedure, except for the suture insertion. Following reperfusion, mice were sacrificed at the indicated time. A homoeothermic heating blanket was used to the maintain core body temperature in each mouse at 37 °C during the I/R operation. Regional cerebral blood flow (rCBF) was monitored by laser-doppler flowmetry (FLOC1, OMEGAWAVE, Tokyo, Japan) to confirm whether the right hemisphere was in an ischemic state. When CBF was reduced by approximately 70–90%, the surgery was considered successful as previously described [25].

2.3. Drug Treatment

MF6 (FABP ligand 6) was synthesized from BMS309403 [23], and its chemical structure is shown in Figure 1A. MF6 was suspended in 0.5% carboxymethyl cellulose (CMC; Wako, Osaka, Japan) and administered per os (p.o.) at different doses (0.5, 1 or 3 mg/kg) just before the experiments were initiated, according to the experimental schedule described in Figure 1B–E. The corresponding control group was orally administered an equivalent volume of 0.5% CMC.

2.4. Infarct Volume Evaluation

After 24 h of reperfusion, the mice were decapitated, and their brains were rapidly removed and cooled to −30 °C for 10 min. The brains were sliced into five sections (2-mm thick), incubated in 1% 2,3,5-triphenyltetrazolium chloride (TTC; Wako, Osaka, Japan) for 20 min at 37 °C and then soaked overnight in 4% paraformaldehyde (PFA; Wako, Osaka, Japan). The infarcted areas appeared white, whereas the non-infarcted regions appeared red. Infarct volumes were measured using ImageJ software and were expressed as a percentage of the total hemisphere [24].

2.5. Neurological Score

Neurological function impairment was also evaluated after 24 h of reperfusion using a neurological deficit grading system with a scale ranging from 0 to 4, as described previously [24]. The following scale was used as a rating system: 0, normal motor function; 1, forelimb flexion when lifted by the tail; 2, circling to the contralateral side when held by the tail on a flat surface, but a normal posture at rest; 3, spontaneous leaning towards to the contralateral side when moving freely; 4, no spontaneous motor activity with an apparent reduction in consciousness. The mice in the sham group exhibited no manifestations of neurological deficits.

2.6. Western Blot Analysis

Following decapitation, a second coronal slice was dissected at 12 h after reperfusion, and regions of the right hemisphere (the ischemic side, ipsilateral) and left hemisphere (the contralateral side) were selected and stored at −80 °C Frozen samples were homogenized in lysis buffer (50 mM Tris–HCl, pH 7.4, 0.5% Triton X-100, 4 mM EGTA, 10 mM EDTA, 1 mM Na_3VO_4, 40 mM $Na_2P_2O_7{\cdot}10H_2O$, 50 mM NaF, 100 nM calyculin A, 50 µg/mL leupeptin, 25 µg/mL pepstatin A, 50 µg/mL trypsin inhibitor and 1 mM dithiothreitol). The samples were then centrifuged at 12,000 rpm for 10 min at 4 °C to remove insoluble material. Protein concentrations were determined using Bradford's assay, and samples were boiled for 3 min at 100 °C with 6× Laemmli's sample buffer [26].

For electrophoresis, equal amounts of proteins were loaded on 15% sodium dodecyl sulfate-polyacrylamide gels and transferred to a polyvinylidene difluoride membrane for 2 h. After blocking with T-TBS solution (50 mM Tris–HCl, pH 7.5, 150 mM NaCl and 0.1% Tween 20) containing 5% fat-free milk powder for 1 h at room temperature, the membranes were incubated overnight at 4 °C with primary antibodies against the proteins of interest. The following working dilutions were used for the indicated monoclonal antibodies, per manufacturer's suggestions: mouse anti-FABP3 (1:1000; Hycult Biotech, HM2016, Uden, NLD), goat anti-FABP5 (1:1000; R&D Systems, AF3077, Minneapolis, MN, USA), goat anti-FABP7 (1:1000; R&D Systems, AF3166, Minneapolis, MN, USA), rabbit anti-mPGES-1 (1:200; Cayman Chemical, 160140, Ann Arbor, MI, USA) and mouse anti-β-actin (1:5000; Sigma, A5441, St Louis, MO, USA). After washing, membranes were incubated with appropriate secondary antibodies diluted in T-TBS solution for 2 h, at room temperature. The membranes were developed using an enhanced chemiluminescence immunoblotting detection system (Amersham Biosciences, NJ, USA) and visualized with a Luminescent Image Analyzer (LAS-4000 mini, Fuji Film, Tokyo, Japan). The densities of the bands were analyzed with ImageJ software (NIH, Bethesda, MA, USA).

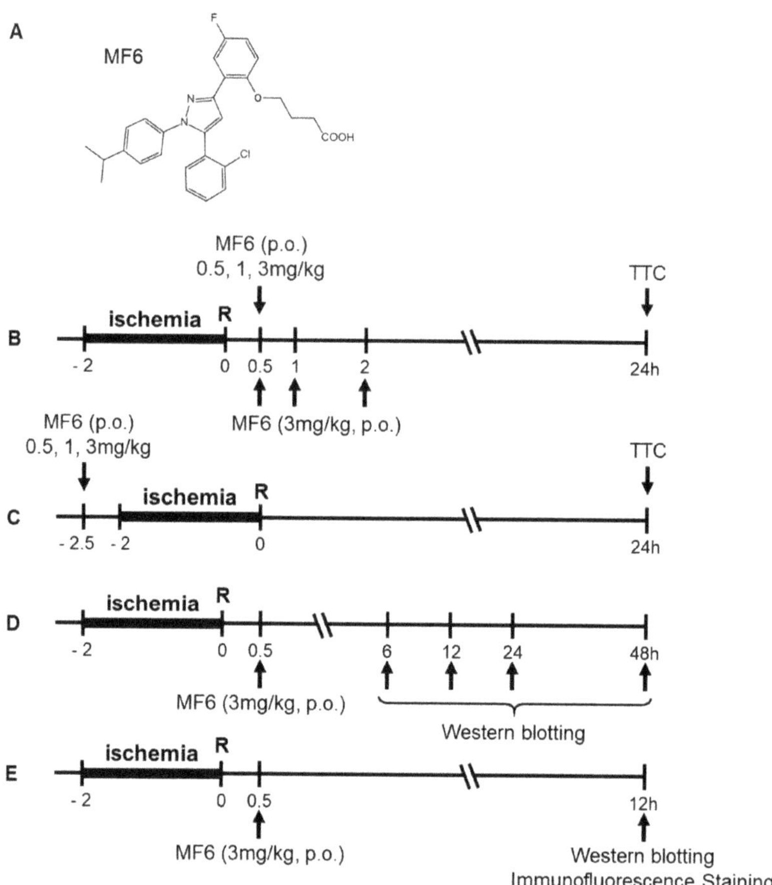

Figure 1. Experimental schedule for MF6 administration (0.5, 1 or 3 mg/kg, p.o.) in I/R mice. (**A**) Chemical structure of MF6. (**B**) ICR mice were subjected to tMCAO for 2 h. Subsequently, they were administered orally (p.o.) different concentrations of MF6 (0.5, 1 or 3 mg/kg) 30 min after reperfusion (for dose–response experiments) or 3 mg/kg MF6 at 0.5, 1 and 2 h after reperfusion (for time–response experiments). TTC staining was performed at 24 h after reperfusion. (**C**) MF6 (0.5, 1 or 3 mg/kg) was administered p.o. to mice 30 min before reperfusion to study the effects of pretreatment. (**D**) MF6 (3 mg/kg) was administered to mice 30 min after reperfusion, and FABP levels were measured after 6, 12, 24 and 48 h. (**E**) I/R mice were administered MF6 (3 mg/kg) 30 min after reperfusion, and after 12 h, the brains were collected for Western blot and immunostaining analyses.

2.7. Immunofluorescence Staining

Mice were anesthetized and transcardially perfused 12 h after reperfusion with ice-cold phosphate-buffered saline (PBS) immediately followed by 4% PFA, as previously described. Mouse brains were removed and fixed in 4% PFA overnight at 4 °C The brain samples were cut into 50 μm coronal sections using a vibratome (Dosaka EM Co. Ltd., Kyoto, Japan). Sections were washed in PBS for 30 min, permeabilized in PBS with 0.1% Triton X-100 for 2 h and blocked in PBS containing 1% BSA and 0.3% Triton X-100 for 1 h at room temperature [26]. The brain samples were then incubated with the following primary antibodies in blocking solution for 3 d at 4 °C: mouse anti-FABP3 (1:500), goat anti-FABP5 (1:500), goat anti-FABP7 (1:500), rabbit anti-mPGES-1 (1:200) and antibodies against cellular markers (1:500). After washing with PBS, the brain sections were incubated with Alexa Flour-conjugated secondary antibodies overnight at 4 °C. After sufficient washing with PBS, the brain sections were mounted on slides with Vectashield (Vector Laboratories, Inc., Burlingame, CA, USA). Immunofluorescent images were analyzed using a confocal laser scanning microscope (Nikon, Tokyo, Japan).

2.8. Measuring MF6 Concentrations in the Blood and Brain

MF6 (3 mg/kg, p.o.) was administered to the mice directly (sham group) or 30 min after reperfusion (I/R group). Blood was collected from the tail vein for measurement at specific time intervals after injection. At 1, 4 and 48 h after injection, the mice were anesthetized and perfused with ice-cold PBS to remove blood to obtain brain samples, which were stored at −80 °C until use. The blood in heparinized tubes was centrifuged at $16,500\times g$ for 10 min at 4 °C, after which the supernatants were collected as plasma. Plasma samples (10 μL) were deproteinized by adding 250 μL of acetonitrile containing ligand 2 (2 ng/mL) as an internal standard, followed by vortexing and sonication. The samples were centrifuged at $16,500\times g$ for 10 min at 4 °C. After centrifugation, the supernatants were evaporated to dryness with a centrifugal concentrator (CC-105, TOMY, Tokyo, Japan), and then the residues were dissolved in 20 μL 80% acetonitrile. For brain sample preparations, 1 mL of acetonitrile was added to 1.5 mL tubes containing brain hemisphere sections, after which the mixtures were homogenized with an ultrasonic homogenizer (SONIFIER Model: 250-Advanced, Branson, CT, USA) for 1 min. Then, the samples were centrifuged at $16,500\times g$ for 10 min at 4 °C. Ten micro-liters of each supernatant were processed following the same procedure used for the plasma samples.

Ultra-performance liquid chromatography (UPLC; Ultimate 3000, Dionex) was performed with an ACQUITY UPLC®BEH C18 column (2.150 mm, 1.7 μm, Waters, Milford, MA, USA) maintained at 40 °C, with a flow rate of 400 μL/min (0–1.0 min, 2.0–3.0 min) or 600 μL/min (1.0–2.0 min). Mobile phase A was composed of water containing 0.1% formic acid, and mobile phase B was composed of acetonitrile containing 0.1% formic acid. The following gradient program was used: 0–1.0 min, 80% B; 1.0–2.0 min, 98% B; 2.0–3.0 min, 80% B. An injection volume of 1 μL was used for analysis.

Mass spectrometry (MS) was performed using a TSQ Vantage mass spectrometer (Thermo Fisher Scientific, Waltham, MA, USA) with an electrospray ionization inter-face. Quantitative analysis was performed in selected-reaction monitoring mode (ligand 6; m/z $[M + H]^+$ 493.2 > 407.2, ligand 2; m/z $[M + H]^+$ 479.1 > 393.1).

2.9. Measuring Brain PGE_2 Concentrations by Performing Enzyme-Linked Immunosorbent Assays (ELISAs)

Prostaglandin E_2 (PGE_2) concentrations in the ipsilateral and contralateral hemispheres (Figure 4A) were determined using a Prostaglandin E_2 ELISA Kit (Cayman Chemical, 514010, Ann Arbor, MI, USA). Samples were collected into liquid nitrogen as described above and weighed. Fifty microliters homogenization buffer (0.1M phosphate, pH 7.4, containing 1 mM EDTA and 10 μM indomethacin) was added to 1 mg of each tissue, the samples were homogenized and centrifugated at $10,000\times g$ for 15 min at 4 °C, and then PGE_2 was extracted from the supernatant and quantitated according to the manufacturer's protocol.

2.10. Cell Culture and Luciferase Reporter Assay

Human neuroblastoma cells (SH-SY5Y) were grown in Dulbecco's modified Eagle's medium (DMEM, Wako) supplemented with 15% fetal bovine serum (FBS, Gibco, CA, USA) and 1% penicillin-streptomycin at 37 °C in a humidified incubator with 5% CO_2/95% air. Human genomic DNA extracted from HEK293 cells was used to amplify the FABP5 promoter fragment (positions -1250/-1) and subcloned into pGL3-Basic-luciferase vector (Promega, Madison, WI, USA). All cloned DNA fragments were confirmed by DNA sequencing. SH-SY5Y cells in 35 mm dishes were transfected with 2 μg of FABP5-pGL3 vector, as well as 50 ng of renilla luciferase plasmid (internal control) for 6 h using lipofectamine LTX and Plus Reagent (Invitrogen, Carlsbad, CA, USA) according to the manufacturer's protocol. After transfection, cells were treated with BSA-AA (arachidonic acid) and MF6 and maintained in D-MEM (1% penicillin-streptomycin) without FBS for 24 h. AA (Sigma, 10931, St Louis, MO, USA) and BSA (Sigma, A7030, fatty acid free) were used to prepare BSA-AA complexes at a 1:5 ratio mixed in binding buffer (10 mM Tris-HCl (pH 8.0), 150 mM NaCl) at 37 °C for 30 min. Firefly luciferase and renilla luciferase activities were measured with the dual-luciferase reporter assay kit (Promega) using a luminometer (Gene Light 55, Microtec, Funabashi, Japan). Relative luciferase activity was expressed as the ratio of firefly luciferase activity to renilla luciferase activity [25].

2.11. Statistical Analysis

The results are presented as box and whisker plots (median, first and third quartile, range), overlaid by dot plot of the raw data. Statistical analysis was conducted using SigmaPlot®version 14 (SYSTAT Software Inc., San Jose, CA, USA). Data from all experiments were analyzed by one-way analysis of variance (ANOVA) followed by Dunnett's test for multiple comparisons (results of concentration gradient and time gradient), or two-way ANOVA followed by Student–Newman–Keuls test for multiple comparisons (MF6 treatment groups). A value of $p < 0.05$ was considered to reflect statistically significant differences.

3. Results

3.1. Pharmacokinetics of MF6 in the Blood and Brain after I/R

We initially determined whether MF6 could penetrate the blood–brain barrier (BBB) as a prerequisite step for evaluating the therapeutic effects of MF6 on cerebral ischemic injury. We measured MF6 concentrations in the plasma and brain tissues of mice after administering MF6 (3 mg/kg). The results revealed that MF6 easily entered the bloodstream. MF6 concentrations in the plasma of sham-operated mice increased rapidly within the first 3 h (15 min: 88.88 ± 25.46 nM; 3 h: 514.95 ± 66.12 nM), reached the maximum value after approximately 4 h (Cmax = 522.18 ± 74.33 nM) and then gradually decreased to 47.14 ± 7.50 nM by 48 h (Figure 2A). However, concentrations of MF6 in the brain did not increase in mice with I/R-induced BBB dysfunction (Figure 2B), indicating that MF6 penetrates to brain even in non-ischemic mice. Moreover, no significant difference was found in MF6 concentrations between the contralateral and ipsilateral portions of ischemic brains (Figure 2B).

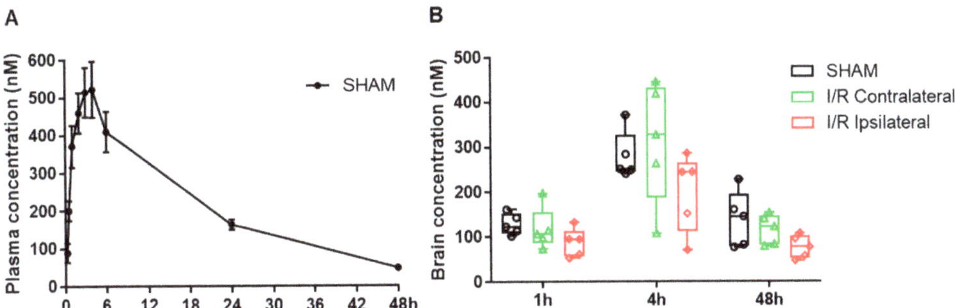

Figure 2. Pharmacokinetics of MF6 in the plasma and brain in sham and I/R mice. (**A**) MF6 plasma concentrations were measured at 0.25, 0.5, 1, 2, 3, 4, 6, 24 and 48 h after MF6 administration in sham mice (n = 5 per group). The data shown in each represent the mean $\pm$ SEM. (**B**) The MF6 concentrations in the contralateral and ipsilateral regions of the brains of sham and I/R mice were measured at 1, 4 and 48 h after MF6 administration (n = 5 per group). No statistical difference was observed among the groups at the same time point by one-way analysis of variance (ANOVA) followed by Dunnett's test.

3.2. MF6 Reduced Infarct Volumes and Ameliorated Neurological Deficits in I/R Mice

We then examined whether MF6 treatment could reduce cerebral ischemic injury. Mice were treated with or without MF6 (0.5, 1 or 3 mg/kg) 30 min after reperfusion. Mice with I/R surgery showed large infarct volumes (70.28$\pm$3.19%) and MF6 treatment signif-icantly reduced infarct volumes in a concentration-dependent manner (1 mg/kg group: 50.13 $\pm$ 5.70%, $p = 0.021$ vs. I/R group; 3 mg/kg group: 43.26 $\pm$ 4.63%, $p < 0.001$ vs. I/R group; Figure 3A,B). Moreover, MF6-treated I/R mice showed significantly decreased neurological deficits compared to those of the I/R group treated without MF6 (1mg/kg group: $p = 0.034$; 3 mg/kg group, $p = 0.014$ vs. I/R group; Figure 3C). Next, we tested the optimal timing of MF6 treatment and confirmed that the time of MF6 administration after reperfusion was crucial for the therapeutic effects, with earlier treatment resulting in better outcomes. Administering MF6 30 min after reperfusion was more effective (46.64 $\pm$ 3.01%, $p = 0.007$ vs. I/R group, 76.66 $\pm$ 3.10%), than administering MF6 1 h or 2 h after reperfusion (60.70$\pm$9.88%, $p = 0.231$; 63.98 $\pm$ 8.80%, $p = 0.403$, respectively; Figure 3D–F).

To confirm the long-lasting protective effects of MF6 on I/R-induced cerebral infarct after the single administration, we evaluated the brain infarct volumes and neurological scores at 7 days after reperfusion. The MF6 administration (3 mg/kg) reduced the cerebral infarction and improved the neurological deficits on day 7 after reperfusion (25.93 $\pm$ 2.09%, $p < 0.001$ vs I/R group, 55.79 $\pm$ 3.32%; Figure 3H,I) and improved the mortality of mice caused by ischemia ($p = 0.0276$ vs. I/R group; Figure 3J). Furthermore, we confirmed the protective effects of MF6 pre-administration in which MF6 was administered once 30 min before ischemia surgery. The MF6 pre-administration significantly reduced the infarct volume (I/R group: 72.88 $\pm$ 6.24%) at 1 mg/kg (47.61 $\pm$ 3.00%, $p = 0.012$ vs. I/R group) and 3 mg/kg (39.31 $\pm$ 5.11%, $p < 0.001$ vs. I/R group) (Supplementary Materials Figure S1). Taken together, MF6 protects brain against I/R injury. Since the effects of 3 mg/kg MF6 treatment were re-productive, we chose the dose of 3 mg/kg as the optimum concentration for the subsequent experiments.

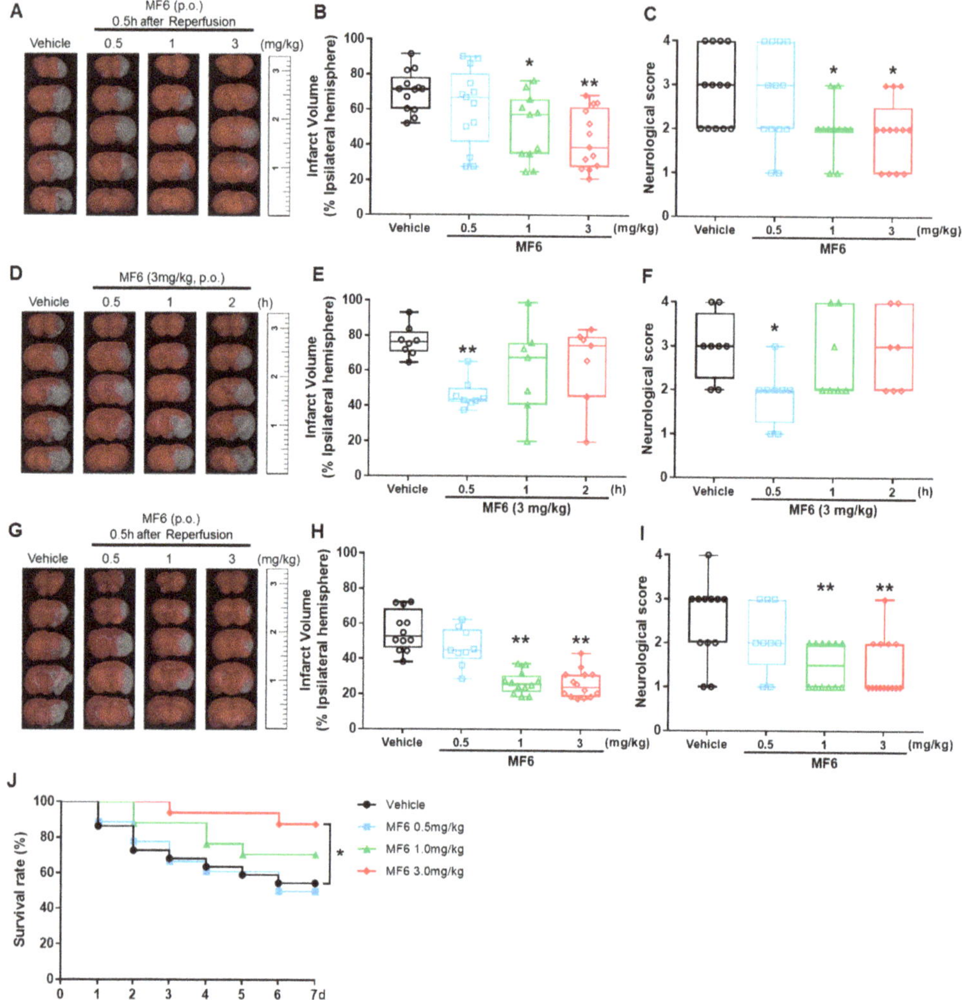

Figure 3. Effects of MF6 post-treatment on I/R injury in mice. Mice were subjected to tMCAO for 2 h. (**A–C**) MF6 was administrated at different concentrations (0.5, 1 or 3 mg/kg) 30 min after reperfusion. Representative images of TTC staining (**A**), quantitative analysis of the infarct volumes (**B**) and neurological deficits (**C**) at 24 h after reperfusion (n = 11–13). (**D–F**) MF6 (3 mg/kg) was administrated at 0.5, 1 or 2 h after reperfusion. Representative images of TTC staining (**D**), quantitative analysis of infarct volumes (**E**) and neurological deficits (**F**) at 24 h after reperfusion (n = 7–8). (**G–J**) MF6 was administrated with different concentrations (0.5, 1, 3 mg/kg) 30 min after reperfusion. Representative images of TTC staining (**G**), quantitative analysis of infarct volume (**H**) and neurological deficits (**I**) at day 7 after reperfusion. The number of living animals in each group every day was recorded, and survival rate (**J**) was calculated (n = 9–14). * $p < 0.05$, ** $p < 0.01$ vs. the I/R-treated group with the CMC group (vehicle). Differences were statistically analyzed using one-way analysis of variance (ANOVA) followed by Dunnett's test.

3.3. I/R Induced FABP3, FABP5 and FABP7 Protein Expression in Mouse Brains

As previous findings indicated that the FABP5 and FABP7 proteins were significantly upregulated in post-ischemic adult monkey brains [27,28], we hypothesized that the FABP3, FABP5 and FABP7 proteins might also be upregulated in ischemic mouse brains. We measured protein-expression levels in the half-brain region including the cortex and striatum (Figure 4A), as previously described [25]. In the right ischemic hemisphere (ipsilateral area), FABP3 significantly increased at 12 h after reperfusion (p = 0.004 vs. sham group) and maintained high protein-expression levels until 48 h (24 h: p = 0.024 vs. sham group; 48 h: p = 0.047 vs. sham group). I/R also significantly upregulated FABP3 expression in the left, non-ischemic hemisphere (contralateral area; 12 h: p = 0.001 vs. sham group; 24 h: p = 0.047 vs. sham group), although the trend was weaker than observed with the ipsilateral areas and declined after 24 h (Figure 4C). Moreover, FABP5 and FABP7 expression were also upregulated in ipsilateral brain tissues (Figure 4D,E), and their expression levels increased faster than that of FABP3 (FABP3: 6 h: p = 0.088 vs. sham group) and had significantly increased at 6 h post-reperfusion (FABP5: p = 0.016 vs. sham group; FABP7: p = 0.048 vs. sham group). Specifically, FABP7 expression did not peak until 24 h. However, the rate of FABP7 upregulation was stronger than that of FABP5 in contralateral brain tissues. All considered, these results show that I/R upregulated the FABP3, FABP5 and FABP7 proteins in a time-dependent manner in the whole brains of mice.

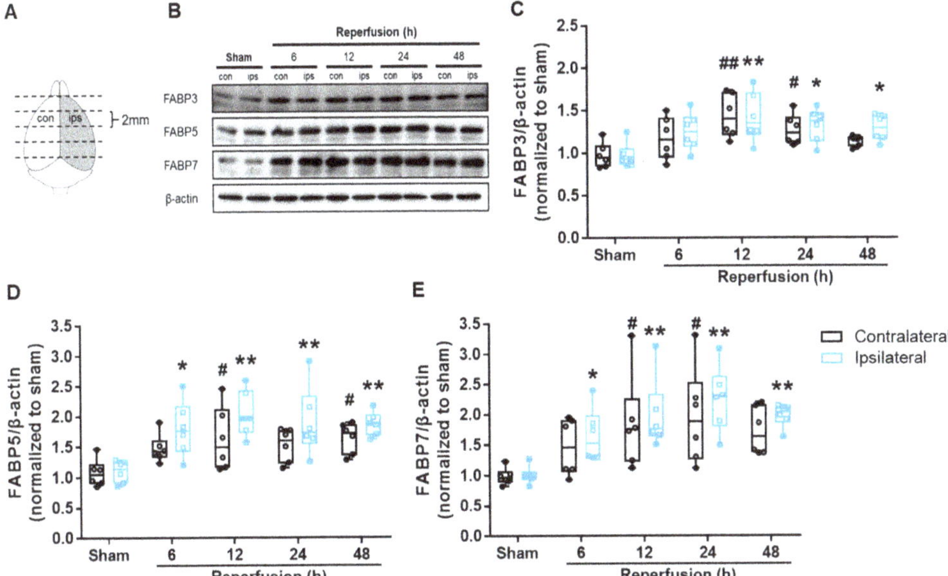

Figure 4. Effects of I/R injury on FABP3, FABP5 and FABP7 expression levels in the brain. (**A**) The locations of samples. Brains were cut into four slices (2-mm thick) from the front of the cortex. The second slices (designated as con and ips slices) of the brain, including the cortex and striatum, were used for Western blot analysis and PGE_2-content analysis in the following experiments. (**B–E**) Mice were subjected to right tMCAO for 2 h. At 6, 12, 24 and 48 h after reperfusion, the second slice of the right brain (ipsilateral) and the left brain (contralateral) areas were collected for Western blot analysis of FABPs levels. (**B**) Representative images of Western blots. Quantitative analyses of FABP3 (**C**), FABP5 (**D**) and FABP7 (**E**) expression levels in contralateral (black) and ipsilateral (blue) areas of the brains. [#] p < 0.05, [##] p < 0.01 vs. the sham group (contralateral); * p < 0.05, ** p < 0.01 vs. the sham group (ipsilateral) (n = 6 per group). Differences were statistically analyzed using one-way analysis of variance (ANOVA) followed by Dunnett's test.

3.4. I/R-Induced FABPs Were Expressed in Specific Cells in the Cortex

We investigated the phenotypes of FABP3-, FABP5- and FABP7-positive cells in the cortical area of the penumbra (Figure 5A) at 12 h after reperfusion. We co-stained the FABP3, FABP5 and FABP7 proteins and cell type-specific markers, which included NeuN for neurons, GFAP for astrocytes, Olig2 for oligodendrocytes and Iba1 for microglia. In non-ischemic mouse cortexes, FABP3 was almost co-expressed with NeuN (Figure 5B1) but not with Iba1 (Figure S2A), indicating that FABP3 was localized to neurons. I/R-induced FABP3 expression also occurred almost in neurons (Figure 5B2). A previous report showed that FABP5 was expressed both in neurons and glial cells [10]. In agreement, we found that FABP5 was expressed in Olig2-positive cells (Figure 5D1), expressed at a slightly lower level in NeuN-positive cells (Figure 5C1) and weaker still in GFAP-positive cells (Figure S3A) in non-ischemic mouse cortexes, but that FABP5 was not co-expressed with Iba1 (Figure S2B). After I/R, FABP5 was significantly expressed in oligodendrocytes (Olig2$^+$ cells, Figure 5D2) and neurons (NeuN$^+$ cells, Figure 5C2), and marginally expressed in astrocytes (GFAP$^+$ cells, Figure S3B). Furthermore, FABP7 was usually expressed in astrocytes (GFAP$^+$ cells, Figure 5E1) and oligodendrocytes (Olig2$^+$ cells, Figure 5F1) in the cortex, and I/R-induced FABP7 expression was mainly observed in astrocytes (Figure 5E2) but was not obviously induced in oligodendrocytes (Figure 5F2). In addition, FABP7 was not co-expressed with Iba1 (Figure S2C).

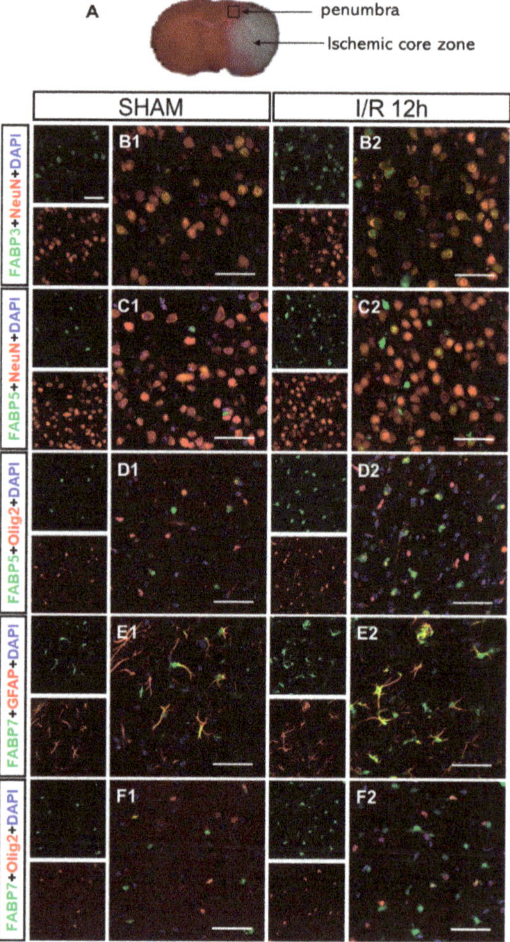

Figure 5. Immunofluorescence of FABP3, FABP5 and FABP7 in the cortexes of sham and I/R mice. (**A**) Representative micro-graphs of immunofluorescence staining of the cortical penumbra region (shown in the black box area) at 12 h after reperfusion. (**B**) Double staining for FABP3 (green) and NeuN (a neuronal marker, red) expression in sham mice (**B1**) and I/R mice (**B2**, ipsilateral). (**C,D**) Double staining for FABP5 (green) and NeuN (red; **C**) or Olig2 (an oligodendrocyte marker, red; **D**) in sham mice and I/R mice (ipsilateral). (**E,F**) Double staining for FABP7 (green) and GFAP (an as-trocyte marker, red; **E**) or Olig2 (red; **F**). Scale bar = 50 μm. The two small images on the left show immunofluorescence for FABPs and cell markers, whereas the larger image on the right is a merged image.

3.5. MF6 Suppressed FABP3, FABP5 and FABP7 Protein Upregulation in I/R Mouse Brains

FABP3, FABP5 and FABP7 expression levels significantly increased at 12 h post-reperfusion, without particularly serious destruction to the structures of brain tissues. Therefore, we next determined whether MF6 could prevent activation of the FABP3, FABP5 and FABP7 proteins at 12 h after reperfusion. MF6 slightly attenuated the increase in FABP3 protein expression ($p = 0.034$ vs. vehicle-treated I/R group) in the ischemic area (ipsilateral), but not in the contralateral area ($p = 0.851$ vs. vehicle-treated I/R group), as shown in Figure 6B. FABP5 and FABP7 expression in ischemic area were significantly suppressed by MF6 treatment (both: $p < 0.001$ vs. vehicle-treated I/R group), and FABP5 expression in the contralateral area was also significantly suppressed ($p = 0.015$ vs. vehicle-treated I/R group), but FABP7 was not ($p = 0.051$ vs. vehicle-treated I/R group) (Figure 6C,D). Moreover, MF6 did not affect FABP3, FABP5 and FABP7 protein levels in non-ischemic mouse brains (sham mice). The present Western blotting data showed that MF6 ameliorated the effects of I/R injury by inhibiting FABP3, FABP5 and FABP7 upregulation.

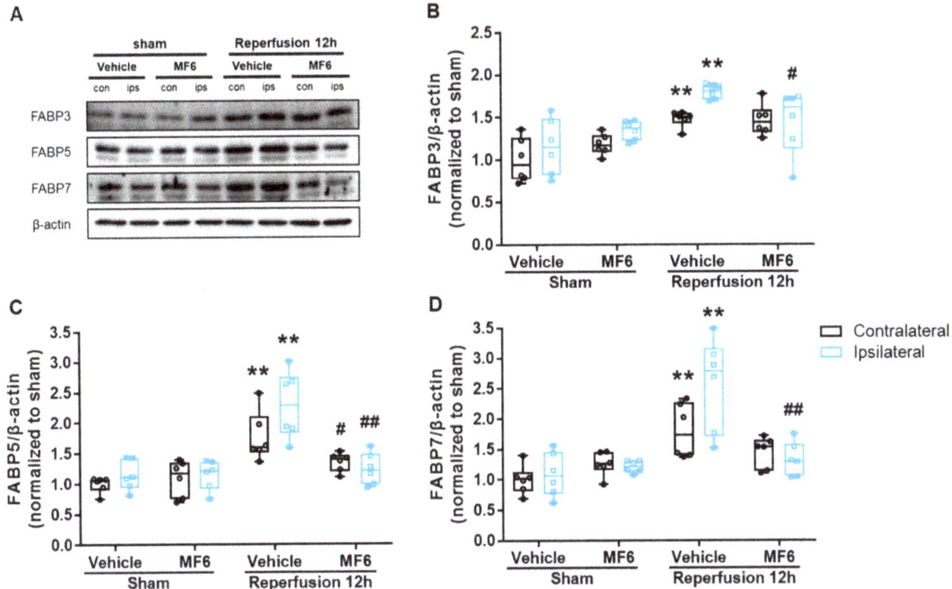

Figure 6. Effects of MF6 administration on FABP3, FABP5 and FABP7 expression levels in I/R mice. Mice were subjected to right tMCAO for 2 h and administered 3 mg/kg MF6 at 30 min after reperfusion. At 12 h after reperfusion, the second slice of the right (ipsilateral) and left (contralateral) brain areas, including the cortex and striatum, were used for Western blot analysis of FABP levels. (**A**) Representative images of Western blots. Quantitative analyses of FABP3 (**B**), FABP5 (**C**) and FABP7 (**D**) proteins expression levels in the contralateral (black) and ipsilateral (blue) regions of the brains. ** $p < 0.01$ vs. the sham-treated group, administered the vehicle (contralateral/ipsilateral); # $p < 0.05$, ## $p < 0.01$ vs. the I/R-treated group, administered the vehicle (contralateral/ipsilateral) (n = 6 per group). Differences were statistically analyzed using two-way analysis of variance (ANOVA) followed by Student–Newman–Keuls test.

3.6. MF6 Prevented AA-Induced FABP5 Upregulation in SH-SY5Y Cells

To elicit the mechanism of MF6 on FABP induction, we focused on FABP5 gene expression because the neuronal expression of FABP5 is pronounced compared to FABP3. Moreover, the stimulation with arachidonic acid (AA) increases the activation of PPARβ/δ in MCF-7 cells [29] and PPARβ/δ agonist, GW0742 up-regulates FABP5 expression in PC3M cells [30]. SH-SY5Y human neuroblastoma cells were used to investigate the effects of AA on FABP5 transcriptional activity. When SH-SY5Y cells were exposed to AA for 24 h, the FABP5 promoter activity increased significantly (30 μM AA, 1.89-fold the control cells level, $p < 0.001$) and AA effects was bell shape. (Figure 7A). MF6 treatment with 1 μM inhibited the increased FABP5 promoter activity by AA ($p = 0.001$ vs. vehicle-treated AA group, Figure 7B). Taken together, MF6 reduces FABP5 protein expression by inhibiting FABP5 transcription activity.

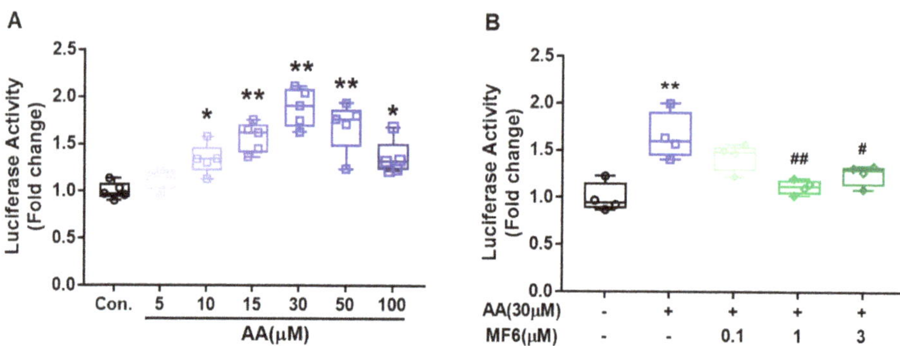

Figure 7. Effects of MF6 administration on AA-induced FABP5 transcriptional activity in SH-SY5Y cells. SH-SY5Y cells were co-transfected with FABP5-pGL3 and Renilla luciferase reporter vectors for 6h. (**A**) After transfection, cells were stimulated with different doses of AA (5 μM, 10 μM, 15 μM, 30 μM, 50 μM or 100 μM) for 24 h (n = 5 per group). (**B**) After transfection, cells were treated with AA/BSA and MF6 (0.1 μM, 1 μM or 3 μM) at the same time for 24 h (n = 4 per group). Promoter activity were represented as firefly/Renilla luciferase activity and were normalized to control cells (without AA treatment). * $p < 0.05$, ** $p < 0.01$ vs. the BSA-treated cells (Con.); [#] $p < 0.05$, [##] $p < 0.01$ vs. the AA and vehicle-treated cells. Differences were statistically analyzed using Student's *t*-test or one-way analysis of variance (ANOVA) followed by Dunnett's test.

3.7. MF6 Suppressed the Microsomal Prostaglandin E Synthase-1 (mPGES-1)–PGE$_2$ Signaling Pathway in I/R Mouse Brains

Post-ischemic inflammation is an important part of the injury mechanism occurring in ischemic stroke. Accumulation of PGE$_2$ and its synthase mPGES-1 aggravates cerebral I/R injury [31]. Therefore, we examined changes of mPGES-1 protein expression and PGE$_2$ levels after reperfusion in mice. Our results showed that I/R significantly increased mPGES-1 protein expression in ischemic and contralateral areas, and both peaked at 24 h after reperfusion ($p < 0.001$ vs. sham group) (Figure 8A). After 12 h of reperfusion, the PGE$_2$ levels in ischemic brains were significantly higher than those of sham-operated brains (ipsilateral: $p < 0.001$ vs. vehicle-treated sham group, contralateral: $p = 0.003$ vs. vehicle-treated sham group) (Figure 8D). Moreover, MF6 treatment significantly reversed I/R-induced increases in mPGES-1 expression ($p = 0.001$ vs. vehicle-treated I/R group) and PGE$_2$ levels ($p = 0.001$ vs. vehicle-treated I/R group) in ipsilateral areas without contralateral area (mPGES-1: $p = 0.435$, PGE$_2$: $p = 0.304$; Figure 8C,D). By immunostaining mPGES-1-positive cells with cell-type specific markers, we observed that mPGES-1 was expressed in NeuN$^+$ neurons but not in GFAP$^+$ astrocytes or Olig2$^+$ oligodendrocytes in the cortexes of sham mice (Figure 8B) and that I/R-induced increases also occurred in neurons.

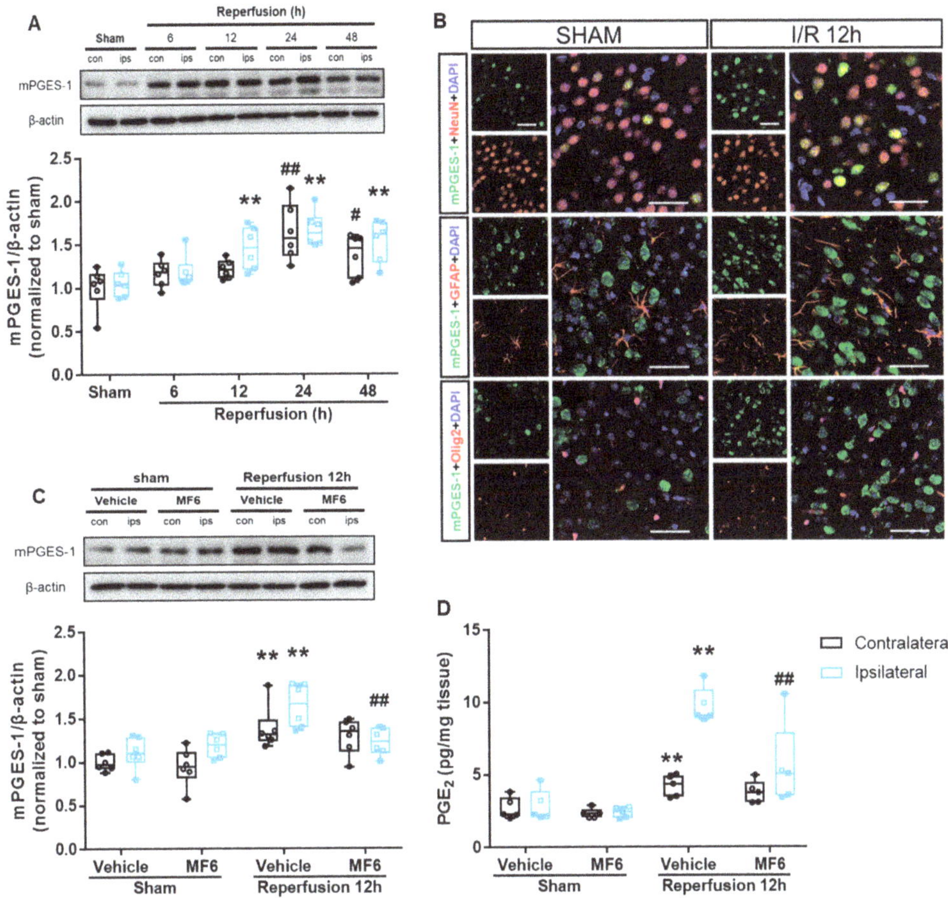

Figure 8. Effects of MF6 administration on mPGES-1 expression and PGE_2 levels in I/R mice. (**A**) Mice were subjected to tMCAO for 2 h. Representative Western blot images and quantitative analyses of mPGES-1 protein expression in contralateral and ipsilateral brain regions at 6, 12, 24 and 48 h after reperfusion. [#] $p < 0.05$, [##] $p < 0.01$ vs. the sham group (contralateral); ** $p < 0.01$ vs. the sham group (ipsilateral) (n = 6 per group). (**B**) Representative micrographs of immunostaining for mPGES-1 (green) and NeuN, GFAP or Olig2 (red) in the cortical penumbra region at 12 h after reperfusion. Scale bar = 50 μm. (**C**) Mice were subjected to tMCAO for 2 h and administered 3 mg/kg MF6 at 30 min after reperfusion. Representative Western blot images and quantitative analyses of mPGES-1 protein expression in contralateral and ipsilateral brain regions at 12 h after reperfusion in I/R mice, treated with MF6 30 min after reperfusion (n = 6 per group). (**D**) PGE_2 levels in contralateral and ipsilateral regions (Figure 4A) of sham and I/R mice treated with or without MF6 (n = 5 per group). ** $p < 0.01$ vs. the sham-treated group, administered the vehicle (contralateral/ipsilateral); [#] $p < 0.05$, [##] $p < 0.01$ vs. the I/R-treated group, administered the vehicle (contralateral/ipsilateral). Differences were statistically analyzed using one-way analysis of variance (ANOVA) followed by Dunnett's test (**A**) or two-way analysis of variance (ANOVA) followed by Student–Newman–Keuls test (**C,D**).

4. Discussion

In this study, we demonstrated that MF6 decreased brain infarction volumes and neurological deficits in mice subjected to tMCAO and reperfusion by inhibiting FABP3, FABP5 and FABP7 expression in the brain. The FABP inhibitor MF6 inhibited the mPGES-1-dependent induction. These results supported our hypothesis that brain FABPs are key molecules in ischemic stroke and that their excessive activation exacerbates brain inflammation through mPGES-1. Therefore, our findings show that MF6 was beneficial for treating ischemic stroke.

In this study, we first discovered that I/R significantly induced FABP3, FABP5 and FABP7 protein expression in the mouse brain, especially that of FABP5 and FABP7. Our results were consistent with several previous findings showing that FABP5 and FABP7 proteins were abundantly expressed in the post-ischemic hippocampus in adult monkeys [28] and that FABP7 gradually increased in the cerebellar cortex of adult monkeys after ischemia, reaching a maximum at day 15 [27]. These results imply that FABP3, FABP5 and FABP7 protein induction appears to exhibit beneficial of detrimental effects for ischemic injury. Moreover, co-staining FABPs in different cell types in the cortical area of the penumbra suggests that increased FABP3, FABP5 and FABP7 expression occurred only in specific cells expressing these proteins before ischemia. FABP3 immunoreactivity increased in neurons, whereas FABP7 was predominantly expressed and increased in astrocytes. In contrast, FABP5 expression increased both in neurons and oligodendrocytes. The localization of FABP3, FABP5 and FABP7 are consistent with previous studies [10,11,32,33].

Under pathological conditions, elevated FABP3 expression level promotes embryonic cancer cell apoptosis [34] and cardiomyocyte apoptosis during myocardial infarction [35]. These studies proposed that FABP3 overexpression reduces mitochondrial activity characterized by lower ATP synthesis and a lower mitochondrial membrane potential, as well as increased reactive oxygen species (ROS) levels and abnormal mitochondrial morphology. FABP3 is also critical for the loss of mitochondrial activity and ROS production during the neurodegenerative process in dopaminergic neurons [36]. In contrast, FABP5 expression is elevated in CA1 neurons (which are apoptotic after ischemia), but was unchanged in DG neurons that remained relatively stable after ischemia in the hippocampus of adult monkeys, suggesting that FABP5 is likely involved in the survival of hippocampal neurons [37].

Furthermore, FABP7 can participate in the proliferation of hippocampal astrocytes after ischemia [37]. However, it remains unclear whether the ischemia-induced enhancement of reactive astrocytes and FABP7 levels elicit beneficial or deleterious effects in neurons [38,39]. Astrocytes expressing FABP7 are crucial for the normal development of dendritic arbors, the formation of and transmission through the excitatory synapses of cortical neurons [40], and even increases the loss of ventral horn neurons in FABP7-knockout mice with spinal cord injury [41]. Conversely, FABP7 overexpression can directly promote a nuclear factor-kappa B (NF-κB)-driven pro-inflammatory response in astrocytes of mice with amyotrophic lateral sclerosis and can ultimately reduce motor neuron survival [42], whereas FABP7 knockdown in the developing brain can increase the proportion of neurons [43,44]. Based on these findings, we hypothesized that FABPs elicit detrimental effects on mitochondrial homeostasis and neuroinflammation, whereas FABP7 is partially required for neuronal differentiation in the developmental stages. However, further studies are required to identify the roles of FABP3, FABP5 and FABP7 after I/R injury.

In addition, we identified important roles for the elevated FABP levels in non-ischemic areas (left hemispheres) after brain ischemia. Previous reports indicated that ischemic lesions cause metabolite changes [45], edema and decreased cerebral blood flow [46] in remote non-ischemic regions. However, in non-ischemic areas, ischemia does not directly cause these changes, and some intermediate incentives are necessary to explain the induction processes. In addition to cerebral ischemia, many findings have also confirmed that in other neurodegenerative disorders (such as Alzheimer's disease and Parkinson's disease), cerebrospinal fluid (CSF) FABP3 levels [47,48] and serum FABP7 levels [49] are elevated in patients, and FABP5 is upregulated in a mouse model of Alzheimer's disease [50]. Since neuroinflammation is a key contributing factor for neurodegenerative diseases, the high levels of pro-inflammatory cytokines, such as tumor necrosis factor-alpha (TNF-α), interleukin (IL)-1β and IL-6 [51–53], are found in the brain and CSF of patients. We speculated that pro-inflammatory cytokines spread from the ischemic core lesion to non-ischemic areas. FABP5 expression is upregulated by lipopolysaccharide (LPS) in human lung epithelial BEAS-2B cells [54] or by TNF-α in human aortic endothelial cells [55]. FABP7 levels are upregulated in spinal cord astrocytes in a mouse model of experimental autoimmune encephalomyelitis [56] and in primary cortical neuronal cultures after exposure to glutamate [57]. Taken together, FABPs trigger the neuronal injury in non-ischemic regions. However, further studies are required to confirm the FABP-induced neuronal injury.

In this study, we administrated MF6 to treat I/R injury in mice, and MF6 effectively reduced infarct volumes and lessened neurological dysfunction, suggesting that it can potentially be used as a therapeutic drug for ischemic injury. Regarding the site of action of MF6, we previously determined that MF6 significantly inhibited FABP3 and FABP5 protein expression and had a high affinity for both proteins [23]. Based on the above results, we envisioned the mechanisms underlying the ameliorative effects of MF6 on an ischemic injury. After administering MF6, it quickly entered the brain through the BBB and associated with FABP3 and FABP5 to different degrees to block their activities, which inhibited FABP-induced signaling pathways such as apoptosis and inflammation [8,34]. Consequently, the degree of cerebral ischemic injury was reduced, and then the weakened injury in turn lowered the stimulation of FABP expression levels.

We recently found that FABP5 causes cell death under oxidative stress in glial cells [58]. When cells were exposed with psychosine, a phospholipase A2 activator, it caused the mitochondria-induced glial death via forming mitochondrial macropores with voltage-dependent anion channel (VDAC-1) and BAX [58]. The MF6 treatment completely abolished the psychosine-induced mitochondrial injury. Even though cell type expressed FABP5 are different between neuronal cells in brain ischemia in the present study and glial cells treated with psychosine by Cheng et al, FABP5-mediated mitochondrial injury in part mediates the neuronal death in the brain ischemia.

FABPs themselves may participate in the FABP upregulation, because MF6 inhibited the FABP5 transcriptional activity induced by AA. AA is known to elevate in during ischemia [59] and it is also a strong ligand of FABP3 and FABP5 [23]. Moreover, AA is an endogenous activator of PPARs [60]. In this context, we speculate that FABPs transport AA from the cytoplasm to the nucleus to promote the activities of PPARs [61].

In addition, there are several the binding sites (PPRE) of PPARs on the promoter region of FABP3, FABP5 and FABP7 genes and the activations of PPARβ/δ and PPARγ mediate the expression of FABP5 [30,62]. Taken together, MF6 may reduce functions of FABPs, thereby inactivating transcriptional activities through PPARs. However, to resolve the precise interaction between PPAR subclasses (α, β/δ and γ) and FABP isoforms in neuronal cells, further extensive studies are required.

As a major inflammatory mediator, PGE$_2$ is involved in the development of injury in several inflammatory neurological diseases, such as stroke and Alzheimer's disease [63,64]. In this study, we confirmed that ischemia induced PGE$_2$ accumulation in the brain, which was caused by the activation of mPGES-1, as previously reported [31]. Since COX-2 level is not changed in the non-ischemic regions [31], the increased mPGES-1 in the contralateral region may count for the increased PGE$_2$ level. Since PPARγ activation enhances FABP5 mRNA expression and increase the levels of mPGES-1 mRNA and PGE$_2$ in porcine trophoblast cells, and a PPARγ antagonist blocked their up-regulation [62], the regulation of the FABP5 and mPGES-1 upregulation may depend on PPARγ. Furthermore, FABP5 inhibition reduces the nuclear transport of NF-κB, thereby decreasing mPGES-1 expression and PGE$_2$ synthesis [65], NF-κB also induces the expression of various pro-inflammatory genes [66] and induces neuronal death in cerebral ischemia [67]. In this context, both PPAR and NF-κB may underly the mechanism of upregulation of mPGES-1–PGE$_2$ in brain ischemia. Moreover, our findings showed that administering MF6 inhibited FABP3 and FABP5 expression and strongly prevented the induction of mPGES-1 expression and PGE$_2$ levels in the brain of I/R mice, indicating that the activation of FABP–mPGES-1–PGE$_2$-signaling pathways promotes ischemic injury. However, it is unclear whether one FABP or FABP3, FABP5 and FABP7 collectively induced mPGES-1 after ischemia. This question requires further investigation, but our data indicated that at least FABP5 was present. Further extensive studies are required to define whether FABP3 or FABP7 is involved in PPAR and NF-κB signaling and whether MF6 treatment inhibits the activity of NF-κB.

Our present results first defined that FABP inhibitor MF6 can inhibit ischemic brain injury by inhibiting the expression and function of activated FABPs after ischemia. Further extensive studies are required to confirm the detrimental functions of FABPs in neuron and glia by using the knockout mice of FABPs. In addition, MF6 has high affinity for FABP7 as compared to FABP3 and FABP5. Since FABP7 is extensively expressed in the astrocytes, the effect of MF6 on the PGE_2 production should be defined in the astrocytes after brain ischemic injury.

In summary, the present study suggested that increased FABP3, FABP5 and FABP7 expression in the brain is a novel biochemical marker of cerebral ischemia and that the FABP inhibitor, MF6 inhibited their expression levels to play a neuroprotective role in cerebral I/R in mice.

Supplementary Materials: The following are available online at https://www.mdpi.com/article/10.3390/biomedicines9050529/s1, Figure S1: Effects of MF6 pre-treatment on ischemia/reperfusion injury in mice, Figure S2: Immunofluorescence of FABP3, FABP5 and FABP7 proteins, as well as Iba1 (a marker of microglia), in the cortexes of sham and ischemia/reperfusion mice, Figure S3: Immunofluorescence of FABP5 with GFAP in the cortex of sham and ischemia/reperfusion mice.

Author Contributions: Q.G.: investigation and original draft writing; I.K. and T.D.: investigation; Y.I.-M., M.S. and F.H.: methodology; K.F.: supervision, review/editing, project administration and funding. All authors have read and agreed to the published version of the manuscript.

Funding: This research was supported by the Strategic Research Program for Brain Sciences from Japan Agency for Medical Research and Development, AMED [grant numbers JP18dm0107071, JP19dm0107071 and JP20dm0107071] and KAKENHI [grant number 19H03406].

Institutional Review Board Statement: The study was conducted according to the guidelines of the Declaration of Helsinki, and approved by the Institutional Animal Care and Use Committee of the Tohoku University Environmental and Safety Committee, Tohoku University, Japan (Approval ID: 2019PhLM0-021 and 2019PhA-024).

Informed Consent Statement: Not applicable.

Data Availability Statement: The data presented in this study are available on request from the corresponding author.

Acknowledgments: We gratefully thank the Kobayashi Foundation for their financial support.

Conflicts of Interest: The authors declare no conflict of interest.

References

1. Donnan, G.A.; Fisher, M.; Macleod, M.; Davis, S.M. Stroke. *Lancet* **2008**, *371*, 1612–1623. [CrossRef]
2. Durukan, A.; Tatlisumak, T. Acute ischemic stroke: Overview of major experimental rodent models, pathophysiology, and therapy of focal cerebral ischemia. *Pharmacol. Biochem. Behav.* **2007**, *87*, 179–197. [CrossRef]
3. Terasaki, Y.; Liu, Y.; Hayakawa, K.; Pham, L.D.; Lo, E.H.; Ji, X.; Arai, K. Mechanisms of Neurovascular Dysfunction in Acute Ischemic Brain. *Curr. Med. Chem.* **2014**, *21*, 2035–2042. [CrossRef] [PubMed]
4. Brouns, R.; De Deyn, P.P. The complexity of neurobiological processes in acute ischemic stroke. *Clin. Neurol. Neurosurg.* **2009**, *111*, 483–495. [CrossRef]
5. Wu, Q.J.; Tymianski, M. Targeting NMDA receptors in stroke: New hope in neuroprotection. *Mol. Brain* **2018**, *11*, 15. [CrossRef]
6. Matsumata, M.; Inada, H.; Osumi, N. Fatty acid binding proteins and the nervous system: Their impact on mental conditions. *Neurosci. Res.* **2016**, *102*, 47–55. [CrossRef]
7. Amiri, M.; Yousefnia, S.; Seyed Forootan, F.; Peymani, M.; Ghaedi, K.; Nasr Esfahani, M.H. Diverse roles of fatty acid binding proteins (FABPs) in development and pathogenesis of cancers. *Gene* **2018**, *676*, 171–183. [CrossRef] [PubMed]
8. Shimamoto, C.; Ohnishi, T.; Maekawa, M.; Watanabe, A.; Ohba, H.; Arai, R.; Iwayama, Y.; Hisano, Y.; Toyota, T.; Toyoshima, M.; et al. Functional characterization of FABP3, 5 and 7 gene variants identified in schizophrenia and autism spectrum disorder and mouse behavioral studies. *Hum. Mol. Genet.* **2015**, *24*, 2409. [CrossRef]
9. Owada, Y. Fatty acid binding protein: Localization and functional significance in the brain. *Tohoku J. Exp. Med.* **2008**, *214*, 213–220. [CrossRef] [PubMed]

10. Liu, R.Z.; Mita, R.; Beaulieu, M.; Gao, Z.; Godbout, R. Fatty acid binding proteins in brain development and disease. *Int. J. Dev. Biol.* **2010**, *54*, 1229–1239. [CrossRef] [PubMed]

11. Chiasserini, D.; Biscetti, L.; Eusebi, P.; Salvadori, N.; Frattini, G.; Simoni, S.; De Roeck, N.; Tambasco, N.; Stoops, E.; Vanderstichele, H.; et al. Differential role of CSF fatty acid binding protein 3, alpha-synuclein, and Alzheimer's disease core biomarkers in Lewy body disorders and Alzheimer's dementia. *Alzheimers Res. Ther.* **2017**, *9*, 52. [CrossRef] [PubMed]

12. Mollenhauer, B.; Steinacker, P.; Bahn, E.; Bibl, M.; Brechlin, P.; Schlossmacher, M.G.; Locascio, J.J.; Wiltfang, J.; Kretzschmar, H.A.; Poser, S.; et al. Serum heart-type fatty acid-binding protein and cerebrospinal fluid tau: Marker candidates for dementia with Lewy bodies. *Neurodegener. Dis.* **2007**, *4*, 366–375. [CrossRef] [PubMed]

13. Jia, W.; Wilar, G.; Kawahata, I.; Cheng, A.; Fukunaga, K. Impaired Acquisition of Nicotine-Induced Conditioned Place Preference in Fatty Acid-Binding Protein 3 Null Mice. *Mol. Neurobiol.* **2021**, *58*, 2030–2045. [CrossRef]

14. Jia, W.; Kawahata, I.; Cheng, A.; Fukunaga, K. The Role of CaMKII and ERK Signaling in Addiction. *Int. J. Mol. Sci.* **2021**, *22*, 3189. [CrossRef]

15. Shioda, N.; Yabuki, Y.; Kobayashi, Y.; Onozato, M.; Owada, Y.; Fukunaga, K. FABP3 protein promotes alpha-synuclein oligomerization associated with 1-methyl-1,2,3,6-tetrahydropiridine-induced neurotoxicity. *J. Biol. Chem.* **2014**, *289*, 18957–18965. [CrossRef]

16. Liao, B.; Geng, L.; Zhang, F.; Shu, L.; Wei, L.; Yeung, P.K.K.; Lam, K.S.L.; Chung, S.K.; Chang, J.; Vanhoutte, P.M.; et al. Adipocyte fatty acid-binding protein exacerbates cerebral ischaemia injury by disrupting the blood-brain barrier. *Eur. Heart J.* **2020**, *41*, 3169–3180. [CrossRef]

17. Shi, M.; Huang, R.S.; Guo, F.; Li, L.Z.; Feng, Y.H.; Wei, Z.J.; Zhou, L.; Ma, L.; Fu, P. Pharmacological inhibition of fatty acid-binding protein 4 (FABP4) protects against renal ischemia-reperfusion injury. *RSC Adv.* **2018**, *8*, 15207–15214. [CrossRef]

18. Beniyama, Y.; Matsuno, K.; Miyachi, H. Structure-guided design, synthesis and in vitro evaluation of a series of pyrazole-based fatty acid binding protein (FABP) 3 ligands. *Bioorg. Med. Chem. Lett.* **2013**, *23*, 1662–1666. [CrossRef] [PubMed]

19. Matsuo, K.; Cheng, A.; Yabuki, Y.; Takahata, I.; Miyachi, H.; Fukunaga, K. Inhibition of MPTP-induced alpha-synuclein oligomerization by fatty acid-binding protein 3 ligand in MPTP-treated mice. *Neuropharmacology* **2019**, *150*, 164–174. [CrossRef] [PubMed]

20. Haga, H.; Yamada, R.; Izumi, H.; Shinoda, Y.; Kawahata, I.; Miyachi, H.; Fukunaga, K. Novel fatty acid-binding protein 3 ligand inhibits dopaminergic neuronal death and improves motor and cognitive impairments in Parkinson's disease model mice. *Pharmacol. Biochem. Behav.* **2020**, *191*, 172891. [CrossRef]

21. Cheng, A.; Wang, Y.; Shinoda, Y.; Kawahata, I.; Yamamoto, T.; Jia, W.; Yamamoto, H.; Mizobata, T.; Kawata, Y.; Fukunaga, K. Fatty acid-binding protein 7 triggers α-synuclein oligomerization in glial cells and oligodendrocytes associated with oxidative stress. *Acta Pharmacol. Sin.* **2021**, *0*, 1–11.

22. Cheng, A.; Shinoda, Y.; Yamamoto, T.; Miyachi, H.; Fukunaga, K. Development of FABP3 ligands that inhibit arachidonic acid-induced alpha-synuclein oligomerization. *Brain Res.* **2019**, *1707*, 190–197. [CrossRef] [PubMed]

23. Shinoda, Y.; Wang, Y.; Yamamoto, T.; Miyachi, H.; Fukunaga, K. Analysis of binding affinity and docking of novel fatty acid-binding protein (FABP) ligands. *J. Pharmacol. Sci.* **2020**, *143*, 264–271. [CrossRef] [PubMed]

24. Sun, M.; Izumi, H.; Shinoda, Y.; Fukunaga, K. Neuroprotective effects of protein tyrosine phosphatase 1B inhibitor on cerebral ischemia/reperfusion in mice. *Brain Res.* **2018**, *1694*, 1–12. [CrossRef]

25. Sun, M.; Shinoda, Y.; Fukunaga, K. KY-226 Protects Blood-brain Barrier Function Through the Akt/FoxO1 Signaling Pathway in Brain Ischemia. *Neuroscience* **2019**, *399*, 89–102. [CrossRef]

26. Yabuki, Y.; Fukunaga, K. Oral administration of glutathione improves memory deficits following transient brain ischemia by reducing brain oxidative stress. *Neuroscience* **2013**, *250*, 394–407. [CrossRef]

27. Boneva, N.B.; Mori, Y.; Kaplamadzhiev, D.B.; Kikuchi, H.; Zhu, H.; Kikuchi, M.; Tonchev, A.B.; Yamashima, T. Differential expression of FABP 3, 5, 7 in infantile and adult monkey cerebellum. *Neurosci. Res.* **2010**, *68*, 94–102. [CrossRef]

28. Boneva, N.B.; Kaplamadzhiev, D.B.; Sahara, S.; Kikuchi, H.; Pyko, I.V.; Kikuchi, M.; Tonchev, A.B.; Yamashima, T. Expression of fatty acid-binding proteins in adult hippocampal neurogenic niche of postischemic monkeys. *Hippocampus* **2011**, *21*, 162–171. [CrossRef]

29. Armstrong, E.H.; Goswami, D.; Griffin, P.R.; Noy, N.; Ortlund, E.A. Structural Basis for Ligand Regulation of the Fatty Acid-binding Protein 5, Peroxisome Proliferator-activated Receptor beta/delta (FABP5-PPAR beta/delta) Signaling Pathway. *J. Biol. Chem.* **2014**, *289*, 14941–14954. [CrossRef]

30. Morgan, E.; Kannan-Thulasiraman, P.; Noy, N. Involvement of Fatty Acid Binding Protein 5 and PPAR beta/delta in Prostate Cancer Cell Growth. *PPAR Res.* **2010**, *2010*, 234629. [CrossRef] [PubMed]

31. Ikeda-Matsuo, Y.; Ota, A.; Fukada, T.; Uematsu, S.; Akira, S.; Sasaki, Y. Microsomal prostaglandin E synthase-1 is a critical factor of stroke-reperfusion injury. *Proc. Natl. Acad. Sci. USA* **2006**, *103*, 11790–11795. [CrossRef]

32. Matsumata, M.; Sakayori, N.; Maekawa, M.; Owada, Y.; Yoshikawa, T.; Osumi, N. The Effects of Fabp7 and Fabp5 on Postnatal Hippocampal Neurogenesis in the Mouse. *Stem Cells* **2012**, *30*, 1532–1543. [CrossRef]

33. Sharifi, K.; Ebrahimi, M.; Kagawa, Y.; Islam, A.; Tuerxun, T.; Yasumoto, Y.; Hara, T.; Yamamoto, Y.; Miyazaki, H.; Tokuda, N.; et al. Differential expression and regulatory roles of FABP5 and FABP7 in oligodendrocyte lineage cells. *Cell Tissue Res.* **2013**, *354*, 683–695. [CrossRef]

34. Song, G.X.; Shen, Y.H.; Liu, Y.Q.; Sun, W.; Miao, L.P.; Zhou, L.J.; Liu, H.L.; Yang, R.; Kong, X.Q.; Cao, K.J.; et al. Overexpression of FABP3 promotes apoptosis through inducing mitochondrial impairment in embryonic cancer cells. *J. Cell. Biochem.* **2012**, *113*, 3701–3708. [CrossRef]

35. Zhuang, L.; Li, C.; Chen, Q.; Jin, Q.; Wu, L.; Lu, L.; Yan, X.; Chen, K. Fatty acid-binding protein 3 contributes to ischemic heart injury by regulating cardiac myocyte apoptosis and MAPK pathways. *Am. J. Physiol. Heart Circ. Physiol.* **2019**, *316*, H971–h984. [CrossRef]

36. Kawahata, I.; Bousset, L.; Melki, R.; Fukunaga, K. Fatty Acid-Binding Protein 3 is Critical for α-Synuclein Uptake and MPP+-Induced Mitochondrial Dysfunction in Cultured Dopaminergic Neurons. *Int. J. Mol. Sci.* **2019**, *20*, 5358. [CrossRef] [PubMed]

37. Ma, D.X.; Zhang, M.M.; Mori, Y.; Yao, C.J.; Larsen, C.P.; Yamashima, T.; Zhou, L.F. Cellular Localization of Epidermal-Type and Brain-Type Fatty Acid-Binding Proteins in Adult Hippocampus and Their Response to Cerebral Ischemia. *Hippocampus* **2010**, *20*, 811–819. [CrossRef] [PubMed]

38. Rossi, D.J.; Brady, J.D.; Mohr, C. Astrocyte metabolism and signaling during brain ischemia. *Nat. Neurosci.* **2007**, *10*, 1377–1386. [CrossRef] [PubMed]

39. Roy-O'Reilly, M.; McCullough, L.D. Astrocytes fuel the fire of lymphocyte toxicity after stroke. *Proc. Natl. Acad. Sci. USA* **2017**, *114*, 425–427. [CrossRef] [PubMed]

40. Ebrahimi, M.; Yamamoto, Y.; Sharifi, K.; Kida, H.; Kagawa, Y.; Yasumoto, Y.; Islam, A.; Miyazaki, H.; Shimamoto, C.; Maekawa, M.; et al. Astrocyte-expressed FABP7 regulates dendritic morphology and excitatory synaptic function of cortical neurons. *Glia* **2016**, *64*, 48–62. [CrossRef]

41. Senbokuya, N.; Yoshioka, H.; Yagi, T.; Owada, Y.; Kinouchi, H. Effects of FABP7 on functional recovery after spinal cord injury in adult mice. *J. Neurosurg. Spine* **2019**, *31*, 291–297. [CrossRef]

42. Killoy, K.M.; Harlan, B.A.; Pehar, M.; Vargas, M.R. FABP7 upregulation induces a neurotoxic phenotype in astrocytes. *Glia* **2020**, *68*, 2693–2704. [CrossRef]

43. Arai, Y.; Funatsu, N.; Numayama-Tsuruta, K.; Nomura, T.; Nakamura, S.; Osumi, N. Role of Fabp7, a downstream gene of Pax6, in the maintenance of neuroepithelial cells during early embryonic development of the rat cortex. *J. Neurosci.* **2005**, *25*, 9752–9761. [CrossRef]

44. Tashiro, R.; Sakayori, N.; Matsumata, M.; Owada, Y.; Wakamatsu, Y.; Osumi, N. Fatty acid binding protein (Fabp7) is involved in the maintenance of neural stem/progenitor cells, survival of neurons and maturation of astrocytes. *Neurosci. Res.* **2011**, *71*, E127. [CrossRef]

45. Ruan, L.; Wang, Y.; Chen, S.C.; Zhao, T.; Huang, Q.; Hu, Z.L.; Xia, N.Z.; Liu, J.J.; Chen, W.J.; Zhang, Y.; et al. Metabolite changes in the ipsilateral and contralateral cerebral hemispheres in rats with middle cerebral artery occlusion. *Neural Regen. Res.* **2017**, *12*, 931–937. [PubMed]

46. Xu, Z.N.S.; Lee, R.J.; Chu, S.S.; Yao, A.N.; Paun, M.K.; Murphy, S.P.; Mourad, P.D. Evidence of Changes in Brain Tissue Stiffness After Ischemic Stroke Derived From Ultrasound-Based Elastography. *J. Ultras Med.* **2013**, *32*, 485–494. [CrossRef] [PubMed]

47. Gangishetti, U.; Christina Howell, J.; Perrin, R.J.; Louneva, N.; Watts, K.D.; Kollhoff, A.; Grossman, M.; Wolk, D.A.; Shaw, L.M.; Morris, J.C.; et al. Non-beta-amyloid/tau cerebrospinal fluid markers inform staging and progression in Alzheimer's disease. *Alzheimers Res. Ther.* **2018**, *10*, 98. [CrossRef]

48. Sepe, F.N.; Chiasserini, D.; Parnetti, L. Role of FABP3 as biomarker in Alzheimer's disease and synucleinopathies. *Future Neurol.* **2018**, *13*, 199–207. [CrossRef]

49. Teunissen, C.E.; Veerhuis, R.; De Vente, J.; Verhey, F.R.J.; Vreeling, F.; van Boxtel, M.P.J.; Glatz, J.F.C.; Pelsers, M.A.L. Brain-specific fatty acid-binding protein is elevated in serum of patients with dementia-related diseases. *Eur. J. Neurol.* **2011**, *18*, 865–871. [CrossRef]

50. Sebastian Monasor, L.; Muller, S.A.; Colombo, A.V.; Tanrioever, G.; Konig, J.; Roth, S.; Liesz, A.; Berghofer, A.; Piechotta, A.; Prestel, M.; et al. Fibrillar Abeta triggers microglial proteome alterations and dysfunction in Alzheimer mouse models. *eLife* **2020**, *9*, e54083. [CrossRef] [PubMed]

51. Chitnis, T.; Weiner, H.L. CNS inflammation and neurodegeneration. *J. Clin. Investig.* **2017**, *127*, 3577–3587. [CrossRef]

52. Wang, W.Y.; Tan, M.S.; Yu, J.T.; Tan, L. Role of pro-inflammatory cytokines released from microglia in Alzheimer's disease. *Ann. Transl. Med.* **2015**, *3*, 136.

53. Zheng, C.; Zhou, X.W.; Wang, J.Z. The dual roles of cytokines in Alzheimer's disease: Update on interleukins, TNF-alpha, TGF-beta and IFN-gamma. *Transl. Neurodegener.* **2016**, *5*, 7. [CrossRef]

54. Rao, D.; Perraud, A.L.; Schmitz, C.; Gally, F. Cigarette smoke inhibits LPS-induced FABP5 expression by preventing c-Jun binding to the FABP5 promoter. *PLoS ONE* **2017**, *12*, e0178021. [CrossRef]

55. Han, Q.A.; Yeung, S.C.; Ip, M.S.M.; Mak, J.C.W. Effects of intermittent hypoxia on A-/E-FABP expression in human aortic endothelial cells. *Int. J. Cardiol.* **2010**, *145*, 396–398. [CrossRef]

56. Kamizato, K.; Sato, S.; Shil, S.K.; Umaru, B.A.; Kagawa, Y.; Yamamoto, Y.; Ogata, M.; Yasumoto, Y.; Okuyama, Y.; Ishii, N.; et al. The role of fatty acid binding protein 7 in spinal cord astrocytes in a mouse model of experimental autoimmune encephalomyelitis. *Neuroscience* **2019**, *409*, 120–129. [CrossRef]

57. MacDougall, G.; Anderton, R.S.; Mastaglia, F.L.; Knuckey, N.W.; Meloni, B.P. Proteomic analysis of cortical neuronal cultures treated with poly-arginine peptide-18 (R18) and exposed to glutamic acid excitotoxicity. *Mol. Brain* **2019**, *12*. [CrossRef]

58. Cheng, A.; Kawahata, I.; Fukunaga, K. Fatty Acid Binding Protein 5 Mediates Cell Death by Psychosine Exposure through Mitochondrial Macropores Formation in Oligodendrocytes. *Biomedicines* **2020**, *8*, 635. [CrossRef] [PubMed]
59. Rink, C.; Khanna, S. Significance of Brain Tissue Oxygenation and the Arachidonic Acid Cascade in Stroke. *Antioxid. Redox Sign.* **2011**, *14*, 1889–1903. [CrossRef] [PubMed]
60. Korbecki, J.; Bobinski, R.; Dutka, M. Self-regulation of the inflammatory response by peroxisome proliferator-activated receptors. *Inflamm. Res.* **2019**, *68*, 443–458. [CrossRef] [PubMed]
61. Yu, S.; Levi, L.; Casadesus, G.; Kunos, G.; Noy, N. Fatty acid-binding protein 5 (FABP5) regulates cognitive function both by decreasing anandamide levels and by activating the nuclear receptor peroxisome proliferator-activated receptor beta/delta (PPARbeta/delta) in the brain. *J. Biol. Chem.* **2014**, *289*, 12748–12758. [CrossRef] [PubMed]
62. Blitek, A.; Szymanska, M. Peroxisome proliferator-activated receptor beta/delta and gamma agonists differentially affect prostaglandin E2 and cytokine synthesis and nutrient transporter expression in porcine trophoblast cells during implantation. *Theriogenology* **2020**, *152*, 36–46. [CrossRef]
63. de Oliveira, A.C.; Candelario-Jalil, E.; Bhatia, H.S.; Lieb, K.; Hull, M.; Fiebich, B.L. Regulation of prostaglandin E2 synthase expression in activated primary rat microglia: Evidence for uncoupled regulation of mPGES-1 and COX-2. *Glia* **2008**, *56*, 844–855. [CrossRef]
64. Ikeda-Matsuo, Y. The Role of mPGES-1 in Inflammatory Brain Diseases. *Biol. Pharm. Bull.* **2017**, *40*, 557–563. [CrossRef] [PubMed]
65. Bogdan, D.; Falcone, J.; Kanjiya, M.P.; Park, S.H.; Carbonetti, G.; Studholme, K.; Gomez, M.; Lu, Y.; Elmes, M.W.; Smietalo, N.; et al. Fatty acid-binding protein 5 controls microsomal prostaglandin E synthase 1 (mPGES-1) induction during inflammation. *J. Biol. Chem.* **2018**, *293*, 5295–5306. [CrossRef]
66. Liu, T.; Zhang, L.; Joo, D.; Sun, S.C. NF-kappaB signaling in inflammation. *Signal Transduct. Target Ther.* **2017**, *2*, 17023. [CrossRef]
67. Ridder, D.A.; Schwaninger, M. NF-kappaB signaling in cerebral ischemia. *Neuroscience* **2009**, *158*, 995–1006. [CrossRef] [PubMed]

Article

Bioactive Flavonoids Icaritin and Icariin Protect against Cerebral Ischemia–Reperfusion-Associated Apoptosis and Extracellular Matrix Accumulation in an Ischemic Stroke Mouse Model

Cheng-Tien Wu [1,2,†], Man-Chih Chen [3], Shing-Hwa Liu [3,4,5,†], Ting-Hua Yang [6], Lin-Hwa Long [7], Siao-Syun Guan [8] and Chang-Mu Chen [7,*]

1 Department of Nutrition, China Medical University, Taichung 406040, Taiwan; ct-wu@mail.cmu.edu.tw
2 Master Program for Food and Drug Safety, China Medical University, Taichung 406040, Taiwan
3 Institute of Toxicology, College of Medicine, National Taiwan University, Taipei 10051, Taiwan; r06447007@ntu.edu.tw (M.-C.C.); shinghwaliu@ntu.edu.tw (S.-H.L.)
4 Department of Medical Research, China Medical University Hospital, China Medical University, Taichung 406040, Taiwan
5 Department of Pediatrics, College of Medicine and Hospital, National Taiwan University, Taipei 10051, Taiwan
6 Department of Otolaryngology, National Taiwan University Hospital, Taipei 10051, Taiwan; thyang37@ntu.edu.tw
7 Division of Neurosurgery, Department of Surgery, College of Medicine and Hospital, National Taiwan University, Taipei 10051, Taiwan; linhua1976@yahoo.com.tw
8 Institute of Nuclear Energy Research, Atomic Energy Council, Taoyuan 32546, Taiwan; ssguan@iner.gov.tw
* Correspondence: cmchen10@ntuh.gov.tw
† These authors contributed equally to this study.

Citation: Wu, C.-T.; Chen, M.-C.; Liu, S.-H.; Yang, T.-H.; Long, L.-H.; Guan, S.-S.; Chen, C.-M. Bioactive Flavonoids Icaritin and Icariin Protect against Cerebral Ischemia–Reperfusion-Associated Apoptosis and Extracellular Matrix Accumulation in an Ischemic Stroke Mouse Model. *Biomedicines* **2021**, *9*, 1719. https://doi.org/10.3390/biomedicines9111719

Academic Editors: Kumar Vaibhav and Pankaj Gaur

Received: 21 October 2021
Accepted: 18 November 2021
Published: 19 November 2021

Publisher's Note: MDPI stays neutral with regard to jurisdictional claims in published maps and institutional affiliations.

Copyright: © 2021 by the authors. Licensee MDPI, Basel, Switzerland. This article is an open access article distributed under the terms and conditions of the Creative Commons Attribution (CC BY) license (https://creativecommons.org/licenses/by/4.0/).

Abstract: Stroke, which is the second leading cause of mortality in the world, is urgently needed to explore the medical strategies for ischemic stroke treatment. Both icariin (ICA) and icaritin (ICT) are the major active flavonoids extracted from *Herba epimedii* that have been regarded as the neuroprotective agents in disease models. In this study, we aimed to investigate and compare the neuroprotective effects of ICA and ICT in a middle cerebral artery occlusion (MCAO) mouse model. Male ICR mice were pretreated with both ICA and ICT, which ameliorated body weight loss, neurological injury, infarct volume, and pathological change in acute ischemic stroke mice. Furthermore, administration of both ICA and ICT could also protect against neuronal cell apoptotic death, oxidative and nitrosative stress, lipid peroxidation, and extracellular matrix (ECM) accumulation in the brains. The neuroprotective effects of ICT are slightly better than that of ICA in acute cerebral ischemic stroke mice. These results suggest that pretreatment with both ICA and ICT improves the neuronal cell apoptosis and responses of oxidative/nitrosative stress and counteracts the ECM accumulation in the brains of acute cerebral ischemic stroke mice. Both ICA and ICT treatment may serve as a useful therapeutic strategy for acute ischemic stroke.

Keywords: extracellular matrix; icariin; icaritin; ischemic stroke; middle cerebral artery occlusion

1. Introduction

According to the statistic report of the World Health Organization (WHO), stroke is the second leading cause of mortality in the world [1]. The health care disbursements spent on stroke have been estimated to be approximately 3% to 4% of total expenditures in many countries [2]. Although the death rates and prevalence of stroke have reduced over time, the medical burden or substantially rehabilitative economic costs for post-stroke care remain likely to become a heavy load of many countries [3]. The latest published Global Burden of Disease Study revealed an estimated 5.5 million people died due to stroke, while

116 million people suffered stroke-associated problems, including aphasia, spasticity, and memory problems worldwide [4]. On the basis of the difference of the etiological process of stroke, approximately 80% of all cases are ischemic stroke, the rest being hemorrhagic stroke [5]. Ischemic stroke occurs when clogged blood vessels occur due to thrombus or atherosclerosis, leading to the interruption of the blood supply to the brain [6]. During ischemic stroke, low levels of oxygen are delivered into the brain to first evoke the hypoxia response signals, and then the reperfusion injury-associated networks, such as glutamate excitation with calcium overload, oxido-nitrosative stress, danger-associated molecular patterns (DAMPS) [7], matricellular protein accumulation [8,9], and proinflammatory induction [10,11], which are aroused after removing or dissolving the clots. The increased number of peri-infarct depolarization during ischemic stroke triggers a larger infarct region [12]. The apoptotic neuronal cell death [13] and accumulated fibrotic proteins [14] are consequently developed to lead the permanent brain damage. The therapeutic effects of traditional thrombolytic agents, including plasminogen activators, for ischemic stroke are limited. Exploring the medical strategies for ischemic stroke prevention or therapy is urgently needed.

Both icariin (ICA) and icaritin (ICT) are the major bioactive flavonoids isolated from the Chinese medicine horny goat weed (also known as Ying Yang Huo and *Herba epimedii*) [15]. ICT, a metabolite of ICA, has been suggested to process several biological activities, including neuroprotection against β-amyloid-induced neurotoxicity [16], immunomodulation [17], and anticancer effects [18]. ICA and ICT have been used to improve memory and learning abilities in experimental Alzheimer's disease models [19,20]. Furthermore, ICT has also been indicated to reveal an anti-inflammatory property and adjust a chloride influx in mouse brain cortical cells [21] or act as an antioxidant agent to prevent neuronal cells against oxidative stress [22]. Zhu et al. [23] and Xiong et al. [24] have shown that ICA treatment can protect the brain from ischemia–reperfusion injury in mice and rats, respectively. Recently, the neuroprotective effects of ICT on focal cerebral ischemic–reperfusion injury in mice have been reported [17]. However, the detailed effects and mechanisms of both ICA and ICT on acute cerebral ischemic stroke still remain to be clarified.

In this study, we aimed to investigate and compare the effects and mechanisms of ICA and ICT on acute ischemic stroke using a middle cerebral artery occlusion (MCAO) mouse model. The neuronal functions, infarct volume, and pathology changes were assessed. The examinations for apoptosis, oxidative/nitrosative stress, and matricellular proteins in the cortex and hippocampus were also performed.

2. Materials and Methods

2.1. Cerebral Ischemia–Reperfusion (I/R) Injury Model

Male ICR mice aged 4 to 5 weeks old were purchased from the Laboratory Animal Center of the College of Medicine, National Taiwan University (Taipei, Taiwan). Mice were housed under controlled temperature as well as photoperiod conditions (12 h light/dark) with food and water freely available. All animal surgical procedures were followed to the approved animal protocol (no. 20190111) and guidelines of the Animal Research Committee of the College of Medicine in National Taiwan University. Briefly, mice ($n = 30$) were randomly divided into five groups, namely, sham, I/R, I/R + edaravone (3 mg/kg; Selleck Chemicals, Houston, TX, USA), I/R + ICA (60 mg/kg), and I/R + ICT (60 mg/kg). Drugs were dissolved in dimethyl sulfoxide (DMSO; Sigma-Aldrich, St. Louis, MO, USA). The dosage of 60 mg/kg for both ICA and ICT was selected according to the previous studies [25,26] and our preliminary test. Edaravone treatment was as a positive control [27]. ICA and ICT were purchased from Sigma-Aldrich (cat. no.: 56601, Purity > 95%; St. Louis, MO, USA) and Cayman Chemical (cat. no.: Cay20236-500, Purity > 98%; Ann Arbor, MI, USA), respectively. Drugs edaravone, ICA, and ICT were given by intraperitoneal injection (i.p.) before the focal cerebral ischemia for 1 h, and then the procedure of MCAO surgery was carried out. Mice were anesthetized by inhalation of isoflurane (Tokyo Chemical Industry Co., Tokyo, Japan), which was mixed with 3% oxygen. The middle incision was operated on the neck, and then the left common carotid artery was separated. A 6-0 nylon thread was inserted from the incision of external carotid artery to the middle cerebral artery for 50 min occlusion. During this process, a Laser Doppler (PeriFlux 4001, Perimed, Stockholm, Sweden) was applied to monitor the cerebral blood flow of MCAO mice as previously described [28,29]. The same procedure was also operated on the sham group without the nylon thread insertion. After 24 h of reperfusion, all mice were euthanized, and necropsy was performed to observe the ischemic stroke signal networks. The rectal temperature of mice was maintained at 37.0 °C with a temperature-control heating pad during and after the MCAO surgery. The dose selection for ICT, ICA, and edaravone was decided according to the previous studies and a preliminary test.

2.2. Neurological Score Assessment

To evaluate the neurological injury, we assessed the modified neurological severity score (mNSS). It was scored on a scale of 0 to 14 (normal score, 0; maximal deficit score, 14), which was detected for the behavior of motor, sensory, reflex, and balance. The evaluated neurological functions for mice have been indicated and modified by Li et al. [30]. The scales were followed to the levels: 10–14, severe injury; 6–10, moderate injury; 1–5, mild injury.

2.3. Determination of Infarct Volume and Histopathological Detection

After assessing the behavior and neuronal functions, we euthanized the mice, and the brains were isolated and sliced into 2 mm thick coronal sections. The tissue sections were stained with 2% 2,3,5-triphenyl tetrazolium chloride (TTC; Sigma-Aldrich, St. Louis, MO, USA) at 37 °C for 20 min. In viable brain tissue, TTC was converted by mitochondria to appear red in color, while the colorless area was considered as an infarct. TTC-stained slices were photographed, and infarct volumes were analyzed by ImageJ software [31]. The infarct areas were summed and divided through the total volume of the slices, which were shown as the percentage of the volume of the contralateral hemisphere.

The brain tissue preparation and histopathological detection were determined as previously described [29]. The paraffin-embedded 4 μm sections were stained with hematoxylin and eosin (H&E; Sigma-Aldrich, St. Louis, MO, USA) for histological examination.

2.4. Terminal Deoxynucleotidyl Transferase (TdT) dUTP Nick end Labeling (TUNEL) Assay

The TUNEL assay was determined by a DeadEnd™ Fluorometric TUNEL System (Promega, Madison, WI, USA) as previously described [30]. The procedure was followed to the manufacturer's instruction to detect the DNA fragments of late apoptotic cells. Briefly, brain slides were deparaffinized at 60 °C for 30 min and then transferred to xylene buffer for washing and then rehydrated through the decreased strength of ethanol/saline buffer with 0.85% NaCl. The sections were fixed with paraformaldehyde for 15 min. The slides were washed with a saline buffer and incubated in a dark humidity chamber at 37 °C in 100 μL of TdT incubation buffer for 1 h. The sections were counterstained with 4′,6-diamidino-2-phenylindole (DAPI; Sigma-Aldrich, St. Louis, MO, USA), slides were mounted, and we detected the fluorescein-12-dUTP-label DNA by using a fluorescence microscope.

2.5. Western Blotting Analysis

The brain tissues of the cortex and hippocampus were collected. The samples were lysed by radioimmunoprecipitation (RIPA; Sigma-Aldrich, St. Louis, MO, USA) buffer and centrifuged at 13,000 rpm for 30 min. After the quantification of protein with a bicinchoninic acid (BCA) protein assay kit (Thermo Fisher Scientific, Waltham, MA, USA), the equal concentrations (10–20 μg) of supernatants with sodium dodecyl sulfate (SDS; Millipore, Burlington, MA, USA) buffer were heated at 95 °C for 10 min. Next, the protein samples were separated by 10–15% SDS-polyacrylamide gel electrophoresis (SDS-PAGE) and blotted onto polyvinylidene difluoride (PVDF) membrane (Millipore, Burlington, MA, USA). The membranes were blocked with 5% skimmed milk, which was dissolved in 0.1%

TBST (50 mM Tris-HCl (pH 7.5), 150 mM NaCl, 0.1% Tween 20) buffer for 1 h. Samples were incubated with primary antibodies for CD31 (#3528, 1:1000 dilution; Cell Signaling, Danvers, MA, USA), Bax (#14796, 1:1000 dilution; Cell Signaling, Danvers, MA, USA), Bcl-2 (#3498, 1:1000 dilution; Cell Signaling, USA), caveolin-1 (#3267, 1:1000 dilution; Cell Signaling, Danvers, MA, USA), cleaved caspase 3 (#9664, 1:1000 dilution, Cell Signaling, Danvers, MA, USA), cleaved PARP (#9548, 1:1000 dilution, Cell Signaling, Danvers, MA, USA), vimentin (#5741, 1:1000 dilution, Cell Signaling, Danvers, MA, USA), eNOS (#32027, 1:1000 dilution, Cell Signaling, USA), fibronectin (#26836, 1:1000 dilution, Cell Signaling, Danvers, MA, USA), iNOS (610431, 1:1000 dilution, BD Biosciences, San Jose, CA, USA), and β-actin (sc-47778, Santa-Cruz, Dallas, TX, USA) overnight at 4 °C. Samples were washed by 1% TBST for 10 min three times and then incubated with HRP-conjugated secondary antibodies for 1 h. Finally, the levels of protein expression were densitometric quantification by ImageJ analysis software [32] and normalized by β-actin.

2.6. Measurement of Lipid Peroxide (Thiobarbituric Acid Reactive Substances, TBARS) Levels

The levels of malondialdehyde (MDA; a product of lipid peroxidation) were measured as previously described [33]. A TBARS Assay Kit (Cayman Chemical, Ann Arbor, MI, USA) for colorimetric measurement of MDA was used. The brain tissues of the cortex and hippocampus were collected. The reaction between thiobarbituric acid (TBA) and MDA in the tissue homogenates was performed. The absorbance at 540 nm was detected by a spectrophotometer.

2.7. Statistical Analysis

The results are presented as the mean ± standard error of the mean of at least three independent experiments. The statistical significance of differences between groups was analyzed by one-way analysis of variance (ANOVA) and followed by Dunnett's post hoc test. When the p-value was less than 0.05, it was considered as a significant difference. Data analysis was performed by the GraphPad Prism software (San Diego, CA, USA).

3. Results

3.1. Both ICT and ICA Ameliorated the Neurological Functions and Brain Pathological Changes in Acute Ischemic Stroke Mice

To evaluate the neuroprotective potency, we pretreated ICA, ICT, and edaravone (a positive control) before MCAO surgery in mice. The chemical structures of ICA and ICT are shown in Figure 1A. We first observed the effects of these drugs on the brain pathological changes and neuronal function loss in MCAO mice. As shown in Figure 1B–E, the bodyweight loss, infarct volume, and the mNSS assessment were significantly increased in MCAO group. After both ICA and ICT treatment, these pathological changes and neurological dysfunctions were significantly reversed. Edaravone treatment significantly reversed the pathological changes and neurological dysfunctions, but not bodyweight loss, in MCAO mice (Figure 1). The histopathological changes of the cerebral cortex and hippocampus in MCAO mice showed a locally extensive neuronal necrosis, neuronal cell loss with cytoplasmic vacuolation of neuropils, and the increased amount of irregularly atrophic neuronal cells as well as the shrunken nucleus (Figure 2). The hemorrhagic dots were also presented in the brain cortex after MCAO surgery (Figure 2). These histopathological changes in the cerebral cortex and hippocampus of MCAO mice could be effectively reversed by both ICA and ICT treatment (Figure 2).

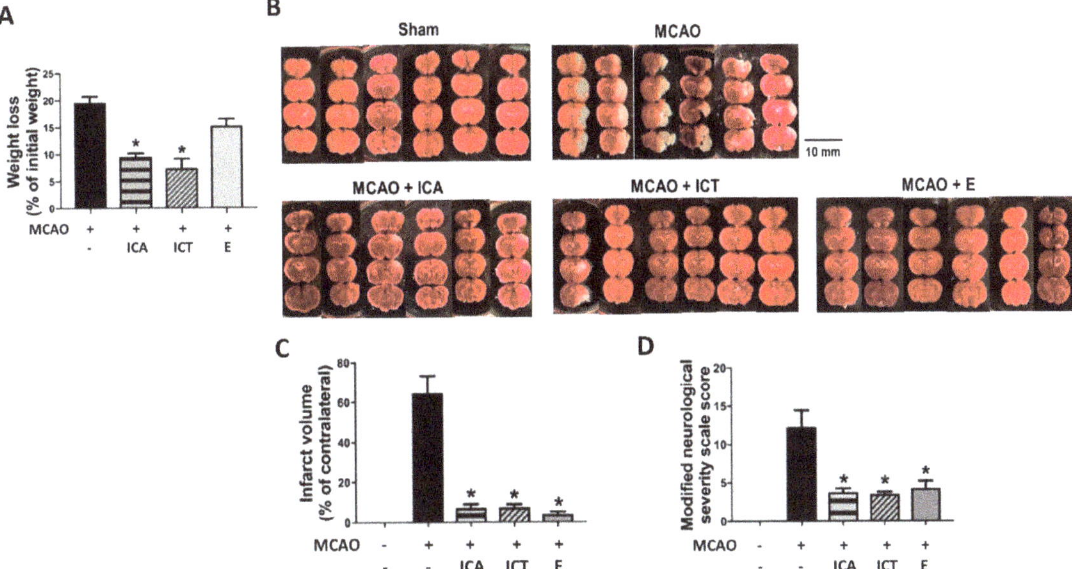

Figure 1. The preventive effects of both icariin (ICA) and icaritin (ICT) on body weight loss, neurological severity, and infarct volume in mice with acute cerebral ischemia–reperfusion. Mice were pretreated with ICA (60 mg/kg), ICT (60 mg/kg), and edaravone (E; 3 mg/kg, as a positive control) before MCAO followed by reperfusion. (**A**) The average body weight loss percentage of mice in each group was calculated. (**B**) Photographs of the mice cerebral infarct areas in each group were shown. (**C**) The quantification of infarct volume is shown. (**D**) The modified neurological severity score (mNSS) in each group was evaluated. Data are presented as mean $\pm$ SD ($n = 6$). * $p < 0.05$ compared to the MCAO group.

3.2. Both ICA and ICT Protected against Neuronal Cell Apoptosis in the Brains of Acute Ischemic Stroke Mice

To evaluate the numbers of apoptotic cells in the ischemic brain, we performed TUNEL staining to detect the fragmented DNA of apoptotic cells. As shown in Figure 3, the TUNEL-positive cells were revealed as a green color while the cell nucleus was stained by DAPI to a blue color. TUNEL-positive cells were clearly observed in the cerebral hippocampus and cortex of MCAO mice, which could be effectively reversed by both ICA and ICT treatment.

We next examined the protein expression for apoptosis-related signaling molecules in the cerebral hippocampus and cortex as determined by Western blot. As shown in Figure 4, the levels of protein expression for cleaved caspase-3, cleaved PARP, and Bax were markedly increased, while the protein expression of Bcl-2 was dramatically decreased in the MCAO group. Nonetheless, pretreatment with both ICA and ICT could effectively reverse the changes of protein expression for these apoptosis-related proteins in MCAO mice (Figure 4).

3.3. Both ICT and ICA Counteracted Oxidative Stress and Nitrosative Stress in the Brains of Acute Ischemic Stroke Mice

We next observed the changes in the levels of protein expression for antioxidant enzymes and nitric oxide synthases and the lipid peroxide generation in acute ischemic stroke mice. As shown in Figure 5A, the levels of protein expression of catalase and SOD-1 were significantly decreased in the cortex and hippocampus of MCAO mice. Both ICT and ICA treatment significantly and conspicuously reversed the decreased protein expression of SOD-1, while partially reversing the decreased catalase protein expression (Figure 5A).

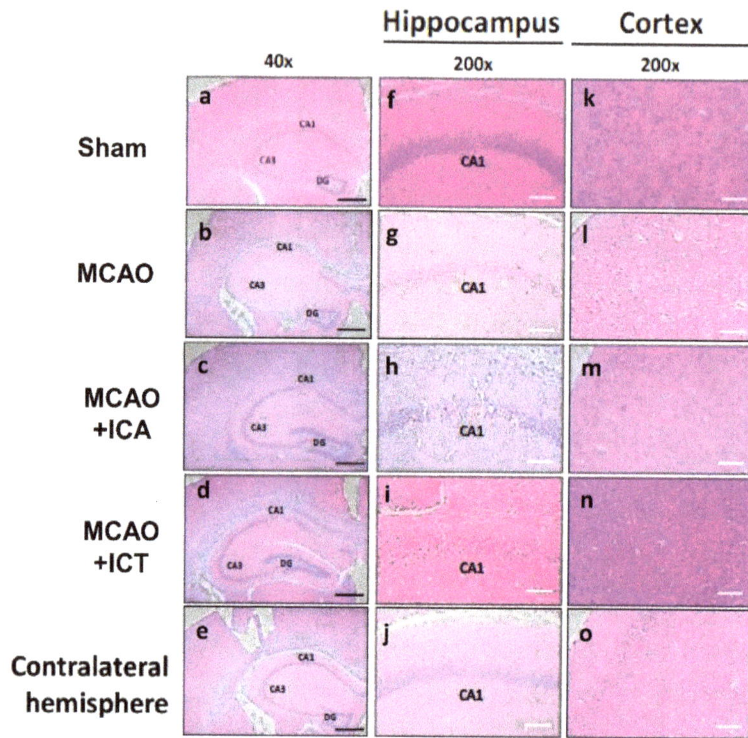

Figure 2. Both ICA and ICT treatment alleviated the histological changes in the left cerebral hemisphere of mice with acute cerebral ischemia–reperfusion. H&E staining was used to detect the histological changes of both cortex and hippocampus under 40× and 200× magnification. The typical ischemia–reperfusion-injured neuronal cells are demonstrated to be irregularly atrophic and eosinophilic cytoplasm as well as a shrunken nucleus with darkly stained pyknotic nuclei (irreversible condensation of chromatin). Black scale bar = 100 μm; white scale bar = 500 μm. Sham (**a,f,k**), MCAO (**b,g,l**), MCAO + ICA (**c,h,m**), MCAO + ICT (**d,i,n**), and contralateral hemisphere (**e,j,o**) are shown. The results are representative from at least four independent experiments.

The activated nitrosative stress is one of the important initial signals to arouse the stroke-induced neuroinflammation and the followed extracellular matrix deposition as well as fibrosis [8]. We next investigated whether both ICA and ICT possess preventive effects on the reduction of nitrosative stress. Caveolin-1 can react with NOSs and inhibit NO synthesis [34]. Caveolin-1 has been found to be reduced after cerebral ischemia–reperfusion injury [35]. As shown in Figure 5A, the protein expression of caveolin-1 was diminished after cerebral ischemia–reperfusion injury, while both eNOS and iNOS protein expression was elevated. Pretreatment with ICT effectively reversed the changes in these protein expression levels in the cortex and hippocampus of MCAO mice (Figure 5A). Moreover, the levels of lipid peroxidation product MDA in the brains of MCAO mice were markedly and significantly increased in MCAO mice, which could be reversed by both ICA and ICT administration (Figure 5B).

3.4. Both ICT and ICA Alleviated the Endothelial–Mesenchymal Transition in the Ischemic Stroke Mice

It has been reported that ischemia–reperfusion may lead to endothelial cell damage and trigger endothelial-to-mesenchymal transition (EndMT) during ischemic acute kidney injury [36]. We next examined the effects of ICA and ICT on endothelial–mesenchymal transition in the brains of MCAO mice. As shown in Figure 6, the levels of protein expression for CD31 (an endothelial cell marker) as well as fibronectin and vimentin (the fibroblast/mesenchymal markers) were significantly decreased and increased, respectively, in the brains of MCAO mice. Both ICA and ICT treatment significantly reversed the increased fibronectin and vimentin protein expression in both cortex and hippocampus tissues of MCAO mice (Figure 6). Both ICA and ICT treatment could not significantly reverse the decreased CD31 protein expression in the cortex tissues of MCAO mice, but ICT treatment significantly reversed the decreased CD31 protein expression in the hippocampus tissues of MCAO mice (Figure 6). These results suggest that both ICA and ICT can inhibit endothelial–mesenchymal transition and extracellular matrix accumulation in the brains of acute ischemic stroke mice.

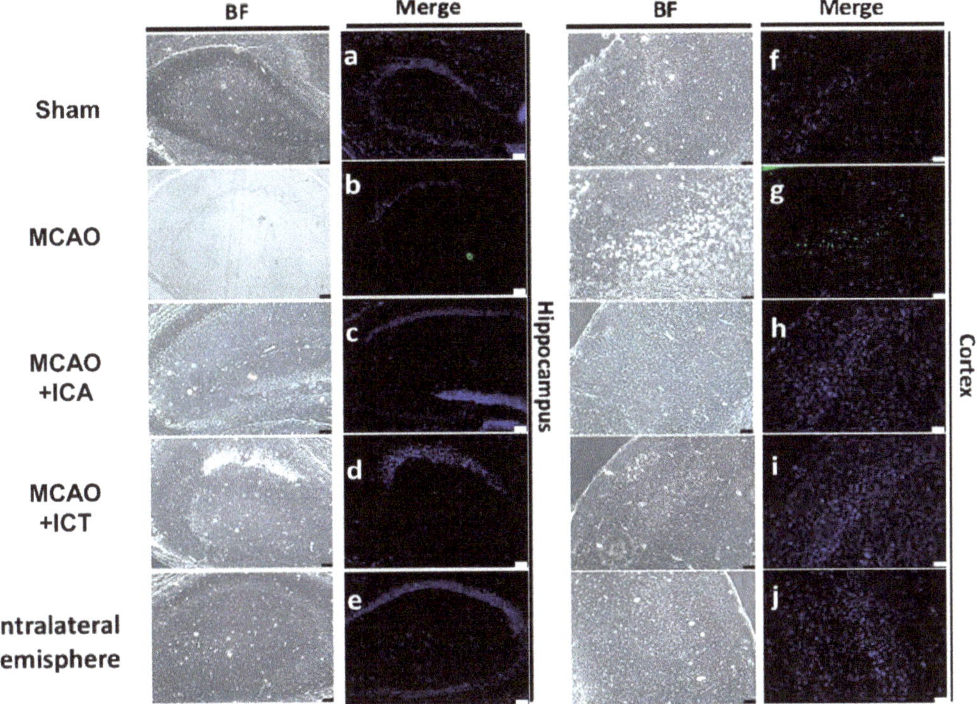

Figure 3. Both ICA and ICT treatment inhibited the neuronal cell apoptosis in the left cerebral hemisphere of mice with acute cerebral ischemia–reperfusion. TUNEL staining was used to detect neuronal apoptotic cells in the hippocampus ((**a**), Sham; (**b**), MCAO; (**c**), MCAO + ICA; (**d**), MCAO + ICT; (**e**), Contralateral hemisphere) and cortex ((**f**), Sham; (**g**), MCAO; (**h**), MCAO + ICA; (**i**), MCAO + ICT; (**j**), Contralateral hemisphere). TUNEL-positive cells were presented as a fluorescent green color, while cell nuclei were displayed as a fluorescent blue color. Scale bar = 75 μm. BF: bright field. The results are representative of at least four independent experiments.

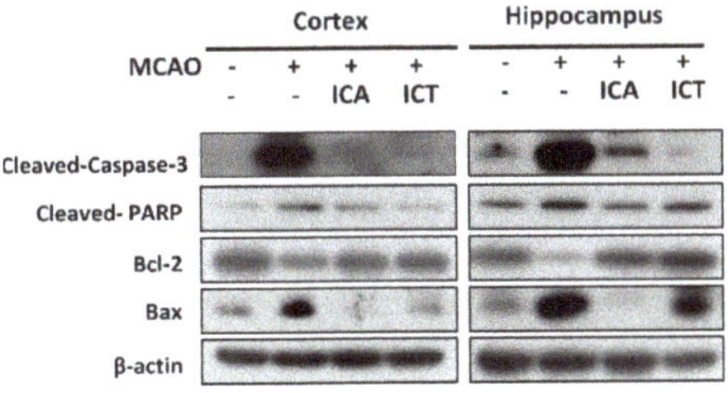

Figure 4. *Cont.*

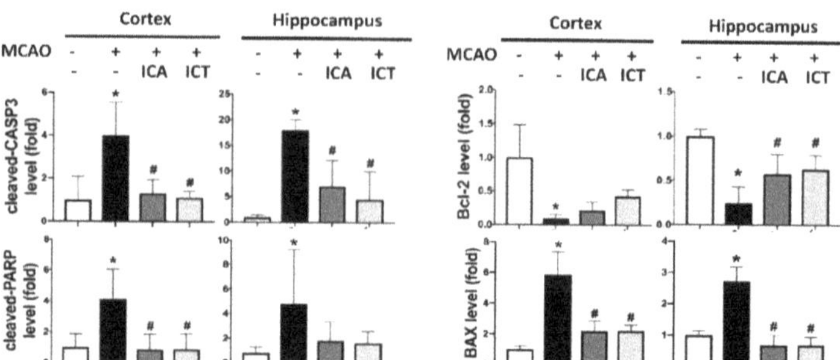

Figure 4. Both ICA and ICT treatment reduced the levels of protein expression of apoptotic markers in the left cerebral hemisphere of mice with acute cerebral ischemia–reperfusion. The levels of protein expression of apoptotic markers (cleaved caspase-3, cleaved PARP, Bcl-2, and Bax) in the hippocampus and cortex areas were determined by Western blot. The quantification of protein expression was determined by densitometry. Data are presented as mean ± SD ($n \geq 4$). * $p < 0.05$ compared to sham group; # $p < 0.05$ compared to MCAO group.

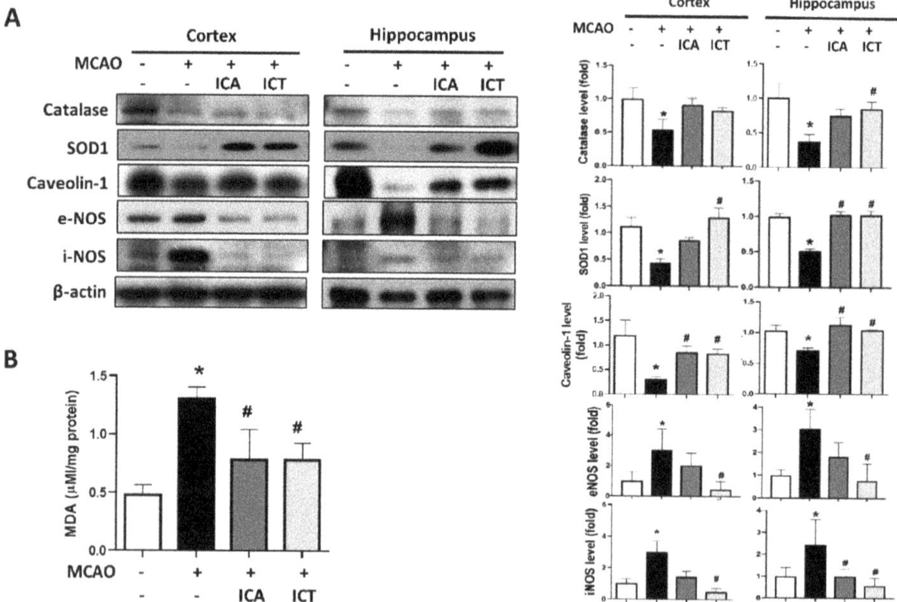

Figure 5. Both ICA and ICT treatment decreased oxidative/nitrosative stress and lipid peroxidation in the left cerebral hemisphere tissues of mice with acute cerebral ischemia–reperfusion. (**A**) The levels of protein expression for antioxidant enzymes (catalase and SOD1) and nitrosative stress-related proteins (caveolin-1, eNOS, and iNOS) in the hippocampus and cortex tissues were determined by Western blot. The quantification of protein expression was determined by densitometry. (**B**) The measurement of brain lipid peroxidation product MDA is shown. Data are presented as mean ± SD ($n \geq 4$). * $p < 0.05$ compared to sham group; # $p < 0.05$ compared to MCAO group.

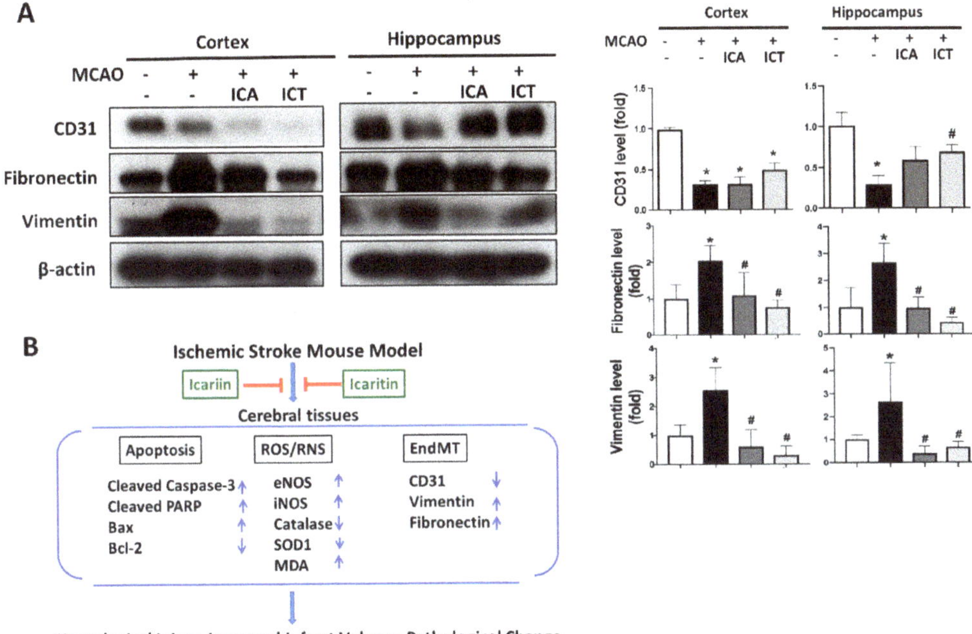

Figure 6. Both ICA and ICT treatment reduced endothelial-to-mesenchymal transition and extracellular matrix accumulation in the left cerebral hemisphere tissues of mice with acute cerebral ischemia–reperfusion. (**A**) The levels of protein expression for CD31 (endothelial marker) and fibronectin and vimentin (fibroblast markers) in the hippocampus and cortex tissues were determined by Western blot. The quantification of protein expression was determined by densitometry. Data are presented as mean $\pm$ SD ($n \geq 4$). * $p < 0.05$ compared to sham group; # $p < 0.05$ compared to MCAO group. (**B**) A schematic summary of our main findings for the effects of both ICA and ICT on the acute cerebral ischemia–reperfusion injury is shown.

4. Discussion

Stroke is a life-threatening morbidity condition that causes long-term disability. Although the first FDA-approved thrombolytic agent recombinant tissue plasminogen activator for stroke has been developed for a decade, the current therapeutic drugs remain problematic and controversial. Researchers have developed the therapeutic strategies for stroke including searching the neuroprotective agents, such as antioxidants, anti-inflammatory agents, and anti-atherosclerosis drugs [37–39]. Both ICA and ICT, the bioactive compounds from *Herba epimedii*, have been revealed to possess the biological properties of antioxidant and anti-atherosclerosis [40,41]. ICA administered by gavage at doses of 50–200 mg/kg/day after reperfusion has been shown to alleviate ischemia reperfusion-induced brain injury in MCAO mice [23]. Xiong et al. [24] have also found that ICA administered by gavage at doses of 10 and 30 mg/kg twice per day for three consecutive days before reperfusion attenuates cerebral ischemia–reperfusion injury. Recently, ICT administered by intraperitoneal injection at a dose of 3 mg/kg/day at before reperfusion has been shown to possess the neuroprotective effects in cerebral I/R mice [17]. In this study, our results revealed that both ICA and ICT administered by intraperitoneal injection at a dose of 60 mg/kg effectively improved brain injury in acute cerebral ischemic stroke mice. The efficacy of ICT seemed to be slightly better than that ICA treatment. In a cell model of Alzheimer's disease, the effects of ICT on decreasing the levels of GSK-3β and phosphorylated Tau have been found to be slightly better than that of ICA [19]. Taken together, these results suggest that both ICA and ICT treatment may potentially possess the advantaged neuroprotective function in stroke.

The progression of ischemic stroke can induce the activation of stress signals, hypoxia, oxygen–glucose deprivation, and oxidative/nitrosative stress, leading to further neuro-inflammation and fibrosis in the brains [8]. As the stroke clot is removed, the reperfusion injury may refer to the reactive oxygen species (ROS)/reactive nitrogen species (RNS) response; protein expression of EMC; and its associated proteins such as matrix metallo-proteinases [42,43], basement membrane changes [44], and inflammation induction [45]. Sun et al. [17] have indicated that ICT pretreatment effectively prevents the neuroinflam-matory response and oxidative damage in the brains of cerebral ischemia–reperfusion mice. On the other hand, cerebral ischemic stroke-induced apoptosis is known to contribute to a significant proportion of neuronal death [13,46]. The overproduction of free radicals, Ca^{2+} overload, and excitotoxicity may be the key molecular events to initiate apoptosis during acute brain ischemia [46]. Both apoptotic and anti-apoptotic proteins have been suggested to be simultaneously over-expressed in the penumbra after ischemic stroke [47]. In the present study, we observed that treatment with both ICA and ICT could also mitigate the pathophysiological changes, TUNEL-positive neuronal cells, imbalance of pro-apoptotic and anti-apoptotic proteins, apoptosis-related signaling molecules, and ROS/RNS-related signaling molecules in the hippocampus and cortex of the MCAO mice. Taken together, these results suggest that both ICA and ICT treatment protects against ischemic stroke injury-associated oxidative/nitrosative stress and apoptosis in the brain.

Fibrosis refers to excess accumulation of fibrous connective tissue during the repair process reacted to specific damage in tissues. EndMT, which endothelial cells lose in their specific phenotype and de-differentiate into cells with mesenchymal phenotype, is a complex process involved in physiologically embryonic development and pathogenesis of human diseases, such as vascular, inflammatory, and fibrotic disorders and cancer [48,49]. EndMT-derived fibroblasts may contribute to the formation of atherosclerotic plaques, leading to the development of cardiac or vascular disorders including stroke [49]. EndMT has been shown to be induced in an acute renal ischemia–reperfusion pig model that the expression of endothelial marker CD31 was significantly decreased, while the expression of fibrotic marker α-SMA was obviously elevated after 24 h reperfusion [50]. Jiang et al. [51] have revealed that vimentin participates in neurotoxicity and microglia activation in cerebral ischemia mice. Fasipe et al. [52] have indicated that the pharmacological targets to the vimentin/VWF (von Willebrand Factor) interaction complex can effectively improve brain injury after ischemic stroke. Furthermore, it has been indicated that the expression of fibronectin is associated with brain edema, hemorrhagic transformation, and poor functional outcome after stroke [53]. Khan et al. [54] have also found that the expression of fibronectin promotes inflammatory injury after ischemic stroke, which prolongs chronic inflammatory conditions. In the present study, we also found that endothelial marker CD31 was decreased, and fibroblastic/mesenchymal markers fibronectin and vimentin were increased in the brains of acute cerebral ischemia–reperfusion mice, which could be effectively reversed by administration of both ICA and ICT. These results suggest that both ICA and ICT administration counteract the EndMT induction and improves ischemic stroke injury in mice.

5. Conclusions

In conclusion, these results demonstrate for the first time that both ICA and ICT pretreatment ameliorate the acute cerebral ischemia–reperfusion injury through the im-provement in apoptotic neuronal cell death, ROS/RNS-induced lipid peroxidation, ECM accumulation, and EndMT-related fibrosis in the mouse brains (Figure 6B). Both icariin and icaritin treatment may serve as a useful therapeutic strategy for acute ischemic stroke.

Author Contributions: Conceptualization, S.-H.L. and C.-M.C.; methodology, M.-C.C.; software, M.-C.C.; validation, M.-C.C, L.-H.L., and C.-T.W.; investigation, C.-T.W.; resources, S.-S.G.; data curation, S.-S.G. and T.-H.Y.; writing—original draft preparation, C.-T.W.; writing—review and editing, S.-H.L. and C.-M.C.; supervision, S.-H.L. and C.-M.C.; project administration, C.-T.W. and S.-H.L.; funding acquisition, S.-H.L. and C.-M.C. All authors have read and agreed to the published version of the manuscript.

Funding: This research was funded by the Ministry of Science and Technology of Taiwan (MOST 110-2314-B-002-033 to C.-M.C.) and the China Medical University (CMU109-MF-83 to C.-T.W.).

Institutional Review Board Statement: The study was conducted according to the guidelines of the Declaration of Helsinki and approved by the Institutional Review Board of the College of Medicine in National Taiwan University (protocol code: 20190111; date of approval: 25 June 2019).

Informed Consent Statement: Not applicable.

Data Availability Statement: The data presented in this study are available from the corresponding author upon reasonable request.

Conflicts of Interest: The authors declare no conflict of interest.

References

1. World Health Organization. Global Health Estimates. Available online: http://www.who.int/healthinfo/global_burden_disease/en/ (accessed on 7 January 2021).
2. Katan, M.; Luft, A. Global Burden of Stroke. *Semin. Neurol.* **2018**, *38*, 208–211. [CrossRef] [PubMed]
3. GBD 2015 Neurological Disorders Collaborator Group. Global, regional, and national burden of neurological disorders during 1990–2015: A systematic analysis for the Global Burden of Disease Study. *Lancet Neurol.* **2017**, *16*, 877–897. [CrossRef]
4. Feigin, V.L.; Nichols, E.; Alam, T.; Bannick, M.S.; Beghi, E.; Blake, N.; Culpepper, W.J.; Dorsey, E.R.; Elbaz, A.; Ellenbogen, R.G.; et al. Global, regional, and national burden of neurological disorders, 1990–2016: A systematic analysis for the Global Burden of Disease Study. *Lancet Neurol.* **2019**, *18*, 459–480. [CrossRef]
5. Thrift, A.; Dewey, H.M.; MacDonell, R.A.; McNeil, J.; Donnan, G.A. Incidence of the Major Stroke Subtypes. *Stroke* **2001**, *32*, 1732–1738. [CrossRef]
6. AHA American Stroke Association Types of Stroke. Available online: https://www.strokeassociation.org/en/about-stroke/types-of-stroke (accessed on 17 July 2021).
7. Agalave, N.M.; Svensson, C.I. Extracellular High-Mobility Group Box 1 Protein (HMGB1) as a Mediator of Persistent Pain. *Mol. Med.* **2014**, *20*, 569–578. [CrossRef]
8. Amruta, N.; Rahman, A.A.; Pinteaux, E.; Bix, G. Neuroinflammation and fibrosis in stroke: The good, the bad and the ugly. *J. Neuroimmunol.* **2020**, *346*, 577318. [CrossRef]
9. Kawakita, F.; Kanamaru, H.; Asada, R.; Suzuki, H. Potential roles of matricellular proteins in stroke. *Exp. Neurol.* **2019**, *322*, 113057. [CrossRef]
10. Nakamura, K.; Shichita, T. Cellular and molecular mechanisms of sterile inflammation in ischaemic stroke. *J. Biochem.* **2019**, *165*, 459–464. [CrossRef]
11. Jian, Z.; Liu, R.; Zhu, X.; Smerin, D.; Zhong, Y.; Gu, L.; Fang, W.; Xiong, X. The Involvement and Therapy Target of Immune Cells After Ischemic Stroke. *Front. Immunol.* **2019**, *10*, 2167. [CrossRef]
12. Mies, G.; Lijima, T.; Hossmann, K.-A. Correlation between periinfarct DC shifts and ischaemic neuronal damage in rat. *NeuroReport* **1993**, *4*, 709–711. [CrossRef]
13. Broughton, B.R.; Reutens, D.C.; Sobey, C.G.; Sims, K.; Politei, J.; Banikazemi, M.; Lee, P. Apoptotic Mechanisms After Cerebral Ischemia. *Stroke* **2009**, *40*, 788–794. [CrossRef]
14. Chen, D.; Li, L.; Wang, Y.; Xu, R.; Peng, S.; Zhou, L.; Deng, Z. Ischemia-reperfusion injury of brain induces endothelial-mesenchymal transition and vascular fibrosis via activating let-7i/TGF-βR1 double-negative feedback loop. *FASEB J.* **2020**, *34*, 7178–7191. [CrossRef]
15. Hwang, E.; Lin, P.; Ngo, H.T.T.; Gao, W.; Wang, Y.-S.; Yu, H.-S.; Yi, T.-H. Icariin and icaritin recover UVB-induced photoaging by stimulating Nrf2/ARE and reducing AP-1 and NF-κB signaling pathways: A comparative study on UVB-irradiated human keratinocytes. *Photochem. Photobiol. Sci.* **2018**, *17*, 1396–1408. [CrossRef] [PubMed]
16. Wang, Z.; Zhang, X.; Wang, H.; Qi, L.; Lou, Y. Neuroprotective effects of icaritin against beta amyloid-induced neurotoxicity in primary cultured rat neuronal cells via estrogen-dependent pathway. *Neuroscience* **2007**, *145*, 911–922. [CrossRef] [PubMed]
17. Sun, C.-H.; Pan, L.-H.; Yang, J.; Yao, J.-C.; Li, B.-B.; Tan, Y.-J.; Zhang, G.-M.; Sun, Y. Protective effect of icaritin on focal cerebral ischemic–reperfusion mice. *Chin. Herb. Med.* **2017**, *10*, 40–45. [CrossRef]
18. Zhao, H.; Guo, Y.; Li, S.; Han, R.; Ying, J.; Zhu, H.; Wang, Y.; Yin, L.; Han, Y.; Sun, L.; et al. A novel anti-cancer agent Icaritin suppresses hepatocellular carcinoma initiation and malignant growth through the IL-6/Jak2/Stat3 pathway. *Oncotarget* **2015**, *6*, 31927–31943. [CrossRef]

19. Li, Y.; Dai, S.; Huang, N.; Wu, J.; Yu, C.; Luo, Y. Icaritin and icariin reduce p-Tau levels in a cell model of Alzheimer's disease by downregulating glycogen synthase kinase 3β. *Biotechnol. Appl. Biochem.* **2021**, in press. [CrossRef]

20. Li, Y.-Y.; Huang, N.-Q.; Feng, F.; Li, Y.; Luo, X.-M.; Tu, L.; Qu, J.-Q.; Xie, Y.-M.; Luo, Y. Icaritin Improves Memory and Learning Ability by Decreasing BACE-1 Expression and the Bax/Bcl-2 Ratio in Senescence-Accelerated Mouse Prone 8 (SAMP8) Mice. *Evid.-Based Complement. Altern. Med.* **2020**, *2020*, 1–10. [CrossRef]

21. Huang, F.; Zhang, Y.; Xu, Y.; Xiang, W.; Xie, C. Analgesic, anti-inflammatory and sedative/hypnotic effects of Icaritin, and its effect on chloride influx in mouse brain cortical cells. *Cell. Mol. Biol.* **2019**, *65*, 99–104. [CrossRef]

22. Xu, Y.; Lu, X.; Zhang, L.; Wang, L.; Zhang, G.; Yao, J.; Sun, C. Icaritin activates Nrf2/Keap1 signaling to protect neuronal cells from oxidative stress. *Chem. Biol. Drug Des.* **2021**, *97*, 111–120. [CrossRef]

23. Zhu, H.-R.; Wang, Z.-Y.; Zhu, X.-L.; Wu, X.-X.; Li, E.-G.; Xu, Y. Icariin protects against brain injury by enhancing SIRT1-dependent PGC-1α Expression in experimental stroke. *Neuropharmacology* **2010**, *59*, 70–76. [CrossRef] [PubMed]

24. Xiong, D.; Deng, Y.; Huang, B.; Yin, C.; Liu, B.; Shi, J.; Gong, Q. Icariin attenuates cerebral ischemia–reperfusion injury through inhibition of inflammatory response mediated by NF-κB, PPARα and PPARγ in rats. *Int. Immunopharmacol.* **2016**, *30*, 157–162. [CrossRef] [PubMed]

25. Liu, D.; Ye, Y.; Xu, L.; Yuan, W.; Zhang, Q. Icariin and mesenchymal stem cells synergistically promote angiogenesis and neurogenesis after cerebral ischemia via PI3K and ERK1/2 pathways. *Biomed. Pharmacother.* **2018**, *108*, 663–669. [CrossRef] [PubMed]

26. Xiong, Y.; Chen, Y.; Huang, X.; Yang, Z.; Zhang, J.; Yu, X.; Fang, J.; Tao, J.; You, K.; Cheng, Z.; et al. Icaritin ameliorates hepatic steatosis via promoting fatty acid β-oxidation and insulin sensitivity. *Life Sci.* **2021**, *268*, 119000. [CrossRef] [PubMed]

27. Watanabe, K.; Tanaka, M.; Yuki, S.; Hirai, M.; Yamamoto, Y. How is edaravone effective against acute ischemic stroke and amyotrophic lateral sclerosis? *J. Clin. Biochem. Nutr.* **2018**, *62*, 20–38. [CrossRef] [PubMed]

28. Chen, C.-M.; Wu, C.-T.; Yang, T.-H.; Liu, S.-H.; Yang, F.-Y. Preventive Effect of Low Intensity Pulsed Ultrasound against Experimental Cerebral Ischemia/Reperfusion Injury via Apoptosis Reduction and Brain-derived Neurotrophic Factor Induction. *Sci. Rep.* **2018**, *8*, 1–11. [CrossRef]

29. Chen, C.-M.; Liu, S.-H.; Lin-Shiau, S. Honokiol, a Neuroprotectant against Mouse Cerebral Ischaemia, Mediated by Preserving Na+, K+-ATPase Activity and Mitochondrial Functions. *Basic Clin. Pharmacol. Toxicol.* **2007**, *101*, 108–116. [CrossRef]

30. Li, Z.; Xiao, G.; Lyu, M.; Wang, Y.; He, S.; Du, H.; Wang, X.; Feng, Y.; Zhu, Y. Shuxuening injection facilitates neurofunctional recovery via down-regulation of G-CSF-mediated granulocyte adhesion and diapedesis pathway in a subacute stroke mouse model. *Biomed. Pharmacother.* **2020**, *127*, 110213. [CrossRef]

31. Schmidt-Kastner, R.; Truettner, J.; Zhao, W.; Belayev, L.; Krieger, C.; Busto, R.; Ginsberg, M.D. Differential changes of bax, caspase-3 and p21 mRNA expression after transient focal brain ischemia in the rat. *Mol. Brain Res.* **2000**, *79*, 88–101. [CrossRef]

32. Schneider, C.A.; Rasband, W.S.; Eliceiri, K.W. NIH Image to ImageJ: 25 years of image analysis. *Nat. Methods* **2012**, *9*, 671–675. [CrossRef]

33. Yen, Y.-P.; Tsai, K.-S.; Chen, Y.-W.; Huang, C.-F.; Yang, R.-S.; Liu, S.-H. Arsenic induces apoptosis in myoblasts through a reactive oxygen species-induced endoplasmic reticulum stress and mitochondrial dysfunction pathway. *Arch. Toxicol.* **2012**, *86*, 923–933. [CrossRef]

34. Kone, B.C. Protein-protein interactions controlling nitric oxide synthases. *Acta Physiol. Scand.* **2000**, *168*, 27–31. [CrossRef] [PubMed]

35. Chen, H.; Chen, X.; Li, W.-T.; Shen, J.-G. Targeting RNS/caveolin-1/MMP signaling cascades to protect against cerebral ischemia-reperfusion injuries: Potential application for drug discovery. *Acta Pharmacol. Sin.* **2018**, *39*, 669–682. [CrossRef] [PubMed]

36. Basile, D. The endothelial cell in ischemic acute kidney injury: Implications for acute and chronic function. *Kidney Int.* **2007**, *72*, 151–156. [CrossRef] [PubMed]

37. Khosravi, A.; Bahonar, A.; Saadatnia, M.; Khorvash, F.; Maracy, M.R. Carotenoids as potential antioxidant agents in stroke prevention: A systematic review. *Int. J. Prev. Med.* **2017**, *8*, 70. [CrossRef]

38. Kikuchi, K.; Uchikado, H.; Morioka, M.; Murai, Y.; Tanaka, E. Clinical Neuroprotective Drugs for Treatment and Prevention of Stroke. *Int. J. Mol. Sci.* **2012**, *13*, 7739–7761. [CrossRef] [PubMed]

39. Kim, J.; Fann, D.Y.-W.; Seet, R.C.S.; Jo, D.-G.; Mattson, M.P.; Arumugam, T.V. Phytochemicals in Ischemic Stroke. *Neuromol. Med.* **2016**, *18*, 283–305. [CrossRef]

40. Hu, Y.; Liu, K.; Yan, M.; Zhang, Y.; Wang, Y.; Ren, L. Effects and mechanisms of icariin on atherosclerosis. *Int. J. Clin. Exp. Med.* **2015**, *8*, 3585–3589.

41. Zhang, Z.-K.; Li, J.; Yan, D.-X.; Leung, W.-N.; Zhang, B.-T. Icaritin Inhibits Collagen Degradation-Related Factors and Facilitates Collagen Accumulation in Atherosclerotic Lesions: A Potential Action for Plaque Stabilization. *Int. J. Mol. Sci.* **2016**, *17*, 169. [CrossRef]

42. Baeten, K.M.; Akassoglou, K. Extracellular matrix and matrix receptors in blood-brain barrier formation and stroke. *Dev. Neurobiol.* **2011**, *71*, 1018–1039. [CrossRef]

43. Nalamolu, K.R.; Chelluboina, B.; Magruder, I.B.; Fru, D.N.; Mohandass, A.; Venkatesh, I.; Klopfenstein, J.D.; Pinson, D.M.; Boini, K.M.; Veeravalli, K.K. Post-stroke mRNA expression profile of MMPs: Effect of genetic deletion of MMP-12. *Stroke Vasc. Neurol.* **2018**, *3*, 153–159. [CrossRef] [PubMed]

44. Kang, M.; Yao, Y. Basement Membrane Changes in Ischemic Stroke. *Stroke* **2020**, *51*, 1344–1352. [CrossRef] [PubMed]

45. Edwards, D.N.; Bix, G.J. The Inflammatory Response After Ischemic Stroke: Targeting β2 and β1 Integrins. *Front. Neurosci.* **2019**, *13*, 540. [CrossRef] [PubMed]
46. Radak, D.; Katsiki, N.; Resanovic, I.; Jovanović, A.; Sudar-Milovanovic, E.; Zafirovic, S.; Mousa, S.; Isenovic, E.R. Apoptosis and Acute Brain Ischemia in Ischemic Stroke. *Curr. Vasc. Pharmacol.* **2017**, *15*, 115–122. [CrossRef]
47. Uzdensky, A.B. Regulation of apoptosis in the ischemic penumbra in the first day post-stroke. *Neural Regen. Res.* **2020**, *15*, 253–254. [CrossRef]
48. Ma, J.; Sanchez-Duffhues, G.; Goumans, M.-J.; Dijke, P.T. TGF-β-Induced Endothelial to Mesenchymal Transition in Disease and Tissue Engineering. *Front. Cell Dev. Biol.* **2020**, *8*, 260. [CrossRef]
49. Piera-Velazquez, S.; Jimenez, S. Endothelial to Mesenchymal Transition: Role in Physiology and in the Pathogenesis of Human Diseases. *Physiol. Rev.* **2019**, *99*, 1281–1324. [CrossRef]
50. Curci, C.; Castellano, G.; Stasi, A.; Divella, C.; Loverre, A.; Gigante, M.; Simone, S.; Cariello, M.; Montinaro, V.; Lucarelli, G.; et al. Endothelial-to-mesenchymal transition and renal fibrosis in ischaemia/reperfusion injury are mediated by complement anaphylatoxins and Akt pathway. *Nephrol. Dial. Transplant.* **2014**, *29*, 799–808. [CrossRef]
51. Jiang, S.X.; Slinn, J.; Aylsworth, A.; Hou, S.T. Vimentin participates in microglia activation and neurotoxicity in cerebral ischemia. *J. Neurochem.* **2012**, *122*, 764–774. [CrossRef]
52. Fasipe, T.A.; Hong, S.-H.; Da, Q.; Valladolid, C.; Lahey, M.T.; Richards, L.M.; Dunn, A.K.; Cruz, M.A.; Marrelli, S.P. Extracellular Vimentin/VWF (von Willebrand Factor) Interaction Contributes to VWF String Formation and Stroke Pathology. *Stroke* **2018**, *49*, 2536–2540. [CrossRef]
53. Wang, L.; Deng, L.; Yuan, R.; Liu, J.; Li, Y.; Liu, M. Association of Matrix Metalloproteinase 9 and Cellular Fibronectin and Outcome in Acute Ischemic Stroke: A Systematic Review and Meta-Analysis. *Front. Neurol.* **2020**, *11*, 523506. [CrossRef] [PubMed]
54. Khan, M.M.; Gandhi, C.; Chauhan, N.; Stevens, J.W.; Motto, D.G.; Lentz, S.R.; Chauhan, A.K. Alternatively-Spliced Extra Domain A of Fibronectin Promotes Acute Inflammation and Brain Injury After Cerebral Ischemia in Mice. *Stroke* **2012**, *43*, 1376–1382. [CrossRef] [PubMed]

Review

Histone Deacetylases and Their Isoform-Specific Inhibitors in Ischemic Stroke

Svetlana Demyanenko *, Valentina Dzreyan and Svetlana Sharifulina

Laboratory of Molecular Neurobiology, Academy of Biology and Biotechnology, Southern Federal University,
pr. Stachki 194/1, Rostov-on-Don 344090, Russia; dzreyan@sfedu.ru (V.D.); svetlana.sharifulina@gmail.com (S.S.)
* Correspondence: demyanenkosvetlana@gmail.com; Tel.: +7-918-5092185; Fax: +7-863-2230837

Abstract: Cerebral ischemia is the second leading cause of death in the world and multimodal stroke therapy is needed. The ischemic stroke generally reduces the gene expression due to suppression of acetylation of histones H3 and H4. Histone deacetylases inhibitors have been shown to be effective in protecting the brain from ischemic damage. Histone deacetylases inhibitors induce neurogenesis and angiogenesis in damaged brain areas promoting functional recovery after cerebral ischemia. However, the role of different histone deacetylases isoforms in the survival and death of brain cells after stroke is still controversial. This review aims to analyze the data on the neuroprotective activity of nonspecific and selective histone deacetylase inhibitors in ischemic stroke.

Keywords: ischemic stroke; epigenetics; histone deacetylase; histone deacetylase inhibitor

Citation: Demyanenko, S.; Dzreyan,
V.; Sharifulina, S. Histone Deacetylases
and Their Isoform-Specific Inhibitors in
Ischemic Stroke. *Biomedicines* **2021**, *9*,
1445. https://doi.org/10.3390/
biomedicines9101445

Academic Editors: Kumar Vaibhav,
Meenakshi Ahluwalia and Pankaj Gaur

Received: 10 September 2021
Accepted: 9 October 2021
Published: 11 October 2021

Publisher's Note: MDPI stays neutral
with regard to jurisdictional claims in
published maps and institutional affiliations.

Copyright: © 2021 by the authors.
Licensee MDPI, Basel, Switzerland.
This article is an open access article
distributed under the terms and
conditions of the Creative Commons
Attribution (CC BY) license (https://
creativecommons.org/licenses/by/
4.0/).

1. Ischemic Stroke Treatment Challenges

Stroke is one of the leading causes of death in the world [1]. About 5 million people die every year. No more than 20% of surviving patients can return to their previous job. 2/3 of strokes occur in people over 65. However, strokes are getting younger. In recent years, about 20% of cerebrovascular accidents have been reported in people aged fewer than 50, and this number is steadily increasing. Thus, the stroke problem is of extreme medical and social importance [2–7].

In ischemic stroke (about 80% of all strokes) the occlusion of cerebral arteries by a thrombus, atherosclerotic plaque, spasm, or abrupt changes in blood pressure sharply decreases or stops the blood supply of the brain tissue [4–6,8]. Cerebral ischemia may be the presenting manifestation of hematological diseases and hematological disorders account for about 1.3% of all causes of acute stroke [9,10].

Stroke is a developing over time multi-stage process. It starts from the primary minor changes and leads to the formation of a penumbra, death or restoration of its cells, and to the irreversible structural damage of brain tissue causing neurodegeneration. A huge number of pathophysiological and biochemical processes in intracellular and intercellular signaling, proteolysis, and regulation of the transcriptional activity of the genome are occurred in between these phases. Each of these stages is the time point of possible application of anti-stroke drugs. Many compounds tested in animal and cell models with the exception of tissue plasminogen activator (tPA) or endovascular thrombectomy [11] reduced apoptotic cell counts, increased infarction size, and improved neurological deficits after stroke [12–14]. However, none of these drugs have been successful in clinical trials. Probably, the complex pathophysiology of cerebral ischemia, the complexity of signaling cascades and differences in the mechanisms of the acute and restorative phases of stroke are the main reasons for failures in translating experimental studies into clinical practice.

Numerous studies searching targets for neuroprotection have highlighted the importance of multimodal stroke therapy. An example of such a strategy, which has been shown to be effective in various models of ischemia, is the inhibition of histone deacetylases

(HDACs) [15–19]. This review analyzes data on the neuroprotective activity of nonspecific HDACs inhibitors (iHDACs) and selective iHDACs.

2. Histone Deacetylases

Cerebral ischemia generally reduces the global level of gene expression due to suppression of acetylation of histones H3 and H4 [20,21]. The antibody microarray study and immunofluorescence microscopy have shown the twofold decrease in the acetylation of lysine 9 in histone H3 (H3K9Ac) in the ischemic penumbra at 1 h after photothrombotic stroke in the rat brain cortex, and more than fourfold decrease at 4 and 24 h. H3K9Ac was shown to localize exclusively in the neuronal, but not astroglial nuclei. These effects could be associated either with downregulation of histone acetyltransferases, or with overexpression of histone deacetylases [22]. Histone acetyltransferases (HATs) acetylate lysine residues in the histone tails. This promotes DNA unfolding and chromatin decondensation that facilitates transcription and protein synthesis. Histone deacetylases remove the acetyl groups from histones. This leads to formation of the transcriptionally inactive heterochromatin, in which gene expression is hindered. To maintain the transcriptionally active state of chromatin, HATs and HDACs should work together [23–25].

Among a large number of cellular non-histone proteins deacetylated by HDACs are transcription factors and co-regulators (e.g., c-MYC, HMG, YY1, EKLF, E2F1, factors GATA, HIF-1α, MyoD, NF-κB, and FoxB3), tumor suppressor proteins (e.g., p53, RUNX3), signaling mediators (e.g., STAT 1 and 3, β-catenin, and SMAD7), steroid receptors (e.g., androgens, estrogen, and SHP), and chaperone proteins and nuclear transport proteins (e.g., α-tubulin, importin-α, cortactin, Ku70, and HSP90) [26–28]. These proteins determine the growth, differentiation, migration, and activity of the protein determining cell survival both in normal conditions and under damage [27]. Thus, deacetylation-dependent signaling pathways play a crucial role in cell homeostasis.

Four classes of HDACs are distinguished in mammals according to their functions, intracellular localization, and expression patterns (Figure 1). Class I includes HDAC1, HDAC2, HDAC3, and HDAC8, class II consists of HDAC4, HDAC5, HDAC6, HDAC7, HDAC9, and HDAC10), and class IV contains only one HDAC11. All of them are zinc-dependent enzymes. Sirtuins use nicotinamide adenine dinucleotide NAD$^+$ as a cofactor. They form the class III histone deacetylases. HDACs are evolutionary conservative [29–32].

Neurons. Acute phase after stroke

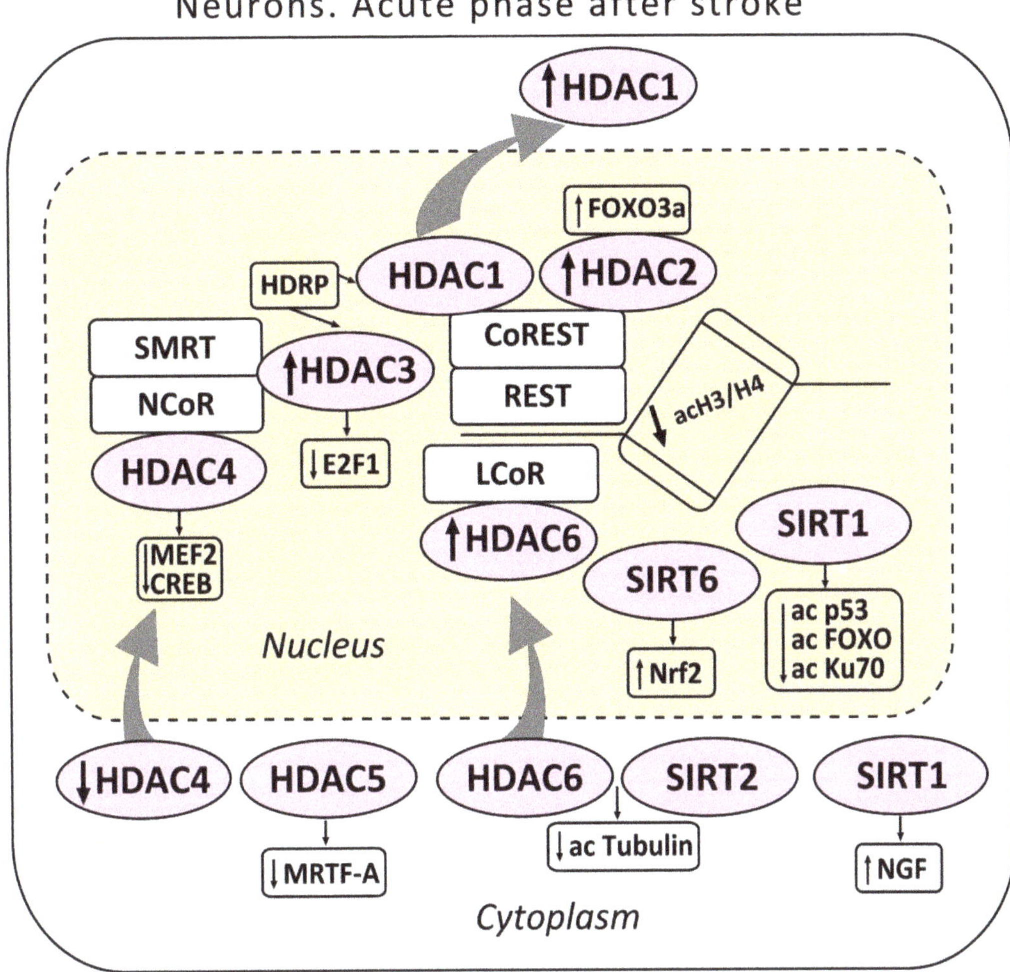

Figure 1. The acute phase after stroke in the neuron. The localization of histone deacetylases (HDAC 1–6) including sirtuins (SIRT 1,2) in the cell is shown. HDAC3 and HDAC1 interact with histone deacetylase-related protein (HDRP), a shortened form of HDAC9. HDRP inhibited the HDAC1/HDAC3 interaction. HDAC1 and HDAC2 are included in the repressor element-1 silencing transcription factor (REST) corepressor 1 (CoREST) complex. HDAC3 in the nuclear receptor co-repressor (NCOR)/silencing mediator for retinoid or thyroid-hormone receptors (SMRT) complex suppresses the production of the pro-apoptotic transcription factor E2F1. HDAC4 suppresses the activity of myocyte enhancer factor 2 (MEF2) and cAMP response element-binding protein (CREB). HDAC4 interacts with the co-repressor complex HDAC3/NCOR. HDAC5 suppressed myocardial transcription factor-A (MRTF-A). One of the cytoplasmic substrates of HDAC6 - α-tubulin is shown. SIRT1 deacetylates p53, FOXO, and Ku70 proteins. The cytoplasmic SIRT1 induces nerve growth factor NGF. SIRT6 stimulates the nuclear factor erythroid 2–related factor (Nrf2).

3. Role of Histone Deacetylases in Cell Damage and Recovery after Cerebral Ischemia

3.1. Class I HDACs

The first class HDACs is widely represented in the brain [31]. HDAC1 localizes both in the neuronal nuclei, and in the cytoplasm, where it deacetylates some cytoplasmic proteins. HDAC2 localizes exclusively in the neuronal nuclei [22,33]. HDAC1 suppresses the production of proteins, which regulate the cell cycle in somatic cells. It also contributes to cell protection against DNA damage [34]. HDAC1 can serve as a molecular switch between neuronal survival and death [35]. HDAC2 regulates apoptosis in the ischemic penumbra [33,36]. HDAC1 and HDAC2 can be included in the multienzyme complexes Sin3, NuRD, CoREST, or NODE that suppress transcription of different sets of target genes [37–39]. The complex CoREST suppresses genes involved in synaptic plasticity and post-stroke recovery [40,41] (Figure 1).

The upregulation of HDAC2, but not HDAC1, in the PTS-induced penumbra was associated with development of apoptosis [22]. Other authors also showed the critical role of HDAC2 in death of neurons in the peri-infarction area after ischemic stroke. The upregulation of HDAC2 observed in the early recovery phase from five to seven post-stroke days reduced survival of neurons and augment neuroinflammation. HDAC2 targeting is apparently a novel therapeutic strategy for stroke recovery [36]. MCAO-induced over-expression of HDAC2 decreased the number of synapses, impaired synaptic plasticity, reduced memory, and deteriorated other cerebral functions [42,43]. Knockdown or knock-out of the HDAC2 gene restored brain functions due to plasticity of the surviving neurons in the peri-infarction zone [43,44].

The overexpression of HDAC1 and HDAC2 in ischemic penumbra neurons and white matter glial cells was observed in the mouse brain during the early regeneration period, 1 week after MCAO. Their levels in the infarct core, oppositely, decreased [31]. Long-term overexpression of HDAC2 and HDAC8 was observed in neurons and astro-cytes at 3–14 days after photothrombotic stroke in the mouse cerebral cortex [45]. Thus, HDAC2 deteriorates synaptic processes, impairs memory, disturbs various cerebral func-tions, and stimulates apoptosis in the ischemic brain.

HDAC3 also deacetylates histones H3 and H4 and some non-histone proteins. It also contributes to regulation of DNA replication and repair. The complex HDAC3/NCOR/SMRT is essential for maintaining chromatin structure and genome stability [46]. Overexpression of HDAC3, HDAC6, and HDAC11 was observed in the ischemic penumbra 3 and 24 h after MCAO, and persisted for a week after reperfusion. The upregulation of HDAC3 and HDAC6 in the mouse cortical neurons was also observed in vitro in the neuronal cell culture. The inhibition of HDAC3 or HDAC6 expression by the short hairpin shRNA increased cell viability. This suggested their involvement in ischemia-induced neurotoxicity [47]. The ischemia-induced neurotoxicity of HDAC3 was demonstrated in other studies [35,48,49]. The neurotoxic effect of HDAC3 was associated with its binding to HDAC1. Actually, the knockdown of HDAC3 suppressed the neurotoxicity of HDAC1, whereas HDAC1 knockdown suppressed the neurotoxicity of HDAC3. HDAC3 and HDAC1 interact with histone deacetylase-related protein (HDRP), a shortened form of HDAC9, whose expression was reduced during neuronal death. The interaction between HDRP and HDAC1, but not HDAC3 protected neurons. HDRP inhibited the HDAC1/HDAC3 interaction and prevented the neurotoxic effect of any of these proteins. This is a possible mechanism of HDAC1-mediated action as a switch "survival/death" in cerebral neurons. HDAC1 interaction with HDRP promotes neuron survival, whereas its interaction with HDAC3 leads to apoptosis [50]. On the other hand, HDAC3 was shown to suppress the production of the pro-apoptotic transcription factor E2F1 in neurons, and thus to contribute to survival of these cells [51].

Another class I histone deacetylase HDAC8 is present mainly in the cytoplasm of neurons and astrocytes of the cerebral cortex, amygdala, hippocampus, and hypothalamus [45,52]. The expression of HDAC8 in the mouse cortical neurons and astrocytes increased significantly during the recovery period, from 3 to 14 days after photothrombotic stroke [43].

3.2. Class II HDACs

The data on the role of HDAC4 in neurodegeneration and neuroprotection are con-tradictory. On one hand, some authors have reported the ability of HDAC4 to maintain neuronal survival [53–55]. However, other authors did not find a dependence of neuronal survival on HDAC4 expression [56,57]. In cultured neurons, HDAC4 rapidly translocates into the nucleus under glutamate release, or decreased K^+ concentration in the medium.

This stimulated cell death [58,59]. The administration of brain-derived neurotrophic factor (BDNF) prevented nuclear translocation of HDAC4 [60]. On the contrary, inhibition or loss of calmodulin-dependent kinase IV (CaMKIV) stimulated HDAC4 accumulation in the neuronal nuclei [59,61]. Nuclear HDAC4 was shown to promote neuronal apoptosis by suppressing the activity of prosurvival transcription factors MEF2 (myocyte enhancer factor 2) and CREB (cAMP response element-binding protein) [58]. Other authors have reported that HDAC4 translocation into the nuclei of neurons, but not astrocytes, did not cause apoptosis in the MCAO-induced ischemic penumbra. Moreover, the nuclear localization of HDAC4 promoted post-stroke brain recovery [62]. The HDAC4 level in the cytoplasmic, but not nuclear fraction of the rat brain cortex decreased at 24 h after photothrombotic stroke [33]. The downregulation of HDAC4 and its relocalization into the neuronal nuclei continued during the recovery period, 2 weeks after stroke [63,64]. In the neuronal nuclei HDAC4 deacetylates histones H3 and H4 and decreases the levels of some prosurvival proteins that finally lead to the neuronal death [58,59,64]. Since HDAC4 was assumed to be inactive against histones, these effects could be mediated by its interaction with other nuclear proteins. Actually, HDAC4 was shown to exhibit deacetylase activity after interacting with the co-repressor complex HDAC3/NCOR [65]. Further studies of HDAC4 interactions with different proteins are needed to understand its role in survival and death of cerebral cells after stroke.

HDAC5, another member of class II histone deacetylases, is involved in neuronal differentiation and axon regeneration in the injured sensory neurons [66,67]. The overexpression of HDAC5 and its nuclear localization was shown to be associated with apoptosis of the cultured neurons from the cerebellar granular layer [68]. After transient MCAO, HDAC5 suppressed the antiapoptotic effect of the transcription factor MRTF-A (myocardial transcription factor-A) in the rat brain neurons [69]. HDAC5 expression in the ischemic penumbra decreased 1, 2, and 14 days after MCAO [47,64]. The downregulation of HDAC5 in the mouse cerebral cortex was observed at 3 days after photothrombotic stroke. However, the number of the apoptotic HDAC5-positive cells did not change [63]. Possibly, the decrease in the level of HDAC5 in cortical neurons was associated with the regeneration processes [70].

HDAC6 belongs to the IIb class of histone deacetylases. It is involved in various cellular processes such as degradation of damaged proteins, cell migration, and intercellular interactions [71]. One of the cytoplasmic substrates of HDAC6 is α-tubulin. Its deacetylation induced destabilization of microtubules in the course of cytoskeleton reorganization and axonal growth during post-stroke regeneration [72]. In the mouse or rat brains HDAC6 presents not only in the cytoplasm, but also in the nuclei of some cortical neurons, but not astrocytes [33,63]. During the first two weeks after the photothrombotic stroke, HDAC6 was upregulated in the neurons not only in the penumbra, but also in the contralateral cerebral cortex, where it appeared in the neuronal nuclei. In the PTS-induced penumbra, HDAC6 co-localized with apoptotic neurons that indicated its involvement in neuronal apoptosis [57,63].

Thus, HDACs are widely represented in the brain. The expression of HDAC1, HDAC2, HDAC3, HDAC4, and HDAC6 increased in the ischemic penumbra. Some of them are located in the neuronal nuclei, some in the cytoplasm, and others—both in the nucleus and cytoplasm. Their functions after ischemic stroke differed. Some HDACs mediate prosurvival processes, whereas others are involved in neurotoxicity. HDAC2 and HDAC6 were apparently involved in apoptosis in the post-ischemic brain.

3.3. Sirtuins

Sirtuins (SIRT) are class III histone deacetylases. The coenzyme nicotinamide adenine dinucleotide (NAD$^+$) makes sirtuins sensitive to metabolic and redox changes [73]. In mammals, seven sirtuins have been identified. Of these, SIRT1 and SIRT6 are localized mainly in the cell nuclei, SIRT7 in the nucleoli, SIRT2 in the cytoplasm, and SIRT3, SIRT4, and SIRT5 are mitochondrial proteins [74]. Sirtuins deacetylate a variety of substrates such as transcription factors, enzymes, and histones. They control diverse biological processes including metabolism, cell growth, aging apoptosis, and autophagy [75]. In the present review we focus on the role of non-mitochondrial SIRT1, SIRT2, and SIRT6 in the brain damage and recovery after ischemic stroke.

SIRT1 content in the brain is higher than in other organs [74]. In the hippocampus it regulates synaptic plasticity and memory. Since SIRT1 deacetylates histones and various transcription factors [72,76], and, also, has the chaperone-like activity [77], its subcellular localization is of significant importance for its functioning. The presence of the nuclear localization signal (NLS) and the nuclear export signal (NES) in the SIRT1 molecule allows it to shuttle from the nucleus to the cytoplasm and back that was assumed to be required for synaptic plasticity and memory formation [78,79]. The subcellular location of SIRT1 changed during brain development and in response to physiological and pathological stimuli [79,80].

SIRT1 mediates neuroprotection after ischemic stroke, traumatic brain injury, and neurodegenerative diseases. It regulates neurogenesis, neurite outgrowth, and gliogenesis, which are involved in postischemic brain regeneration [76,81]. In the SIRT1 knockout mice, MCAO induced greater cerebral infarction than in control animals [82]. On the contrary, mice overexpressing SIRT1 were more resistant to ischemia than control animals [83]. The activation of SIRT1 by resveratrol reduced the MCAO-induced infarction volume [84]. SIRT1 was overexpressed in the ischemic penumbra 7 days after MCAO in the mouse cerebral cortex [82]. Nuclear SIRT1 was reported to prevent apoptosis by deacetylation of proteins p53 [85], FOXO [86], and Ku70 [87]. On the contrary, SIRT1 localized in the cytoplasm enhanced caspase-dependent cell apoptosis [88]. Nevertheless, the translocation of SIRT1 into the cytoplasm was not associated with cell apoptosis in the peri-infarct area at 7 days after photothrombotic stroke in the mouse cerebral cortex. In this case, the cytoplasmic localization of SIRT1 was associated with the upregulation of synaptophysin and GAP-43 that mediate the axon outgrowth and restoration of synaptic connections [89]. The cytoplasmic SIRT1 was shown to enhance the neurite outgrowth that was induced by nerve growth factor NGF. Oppositely, inhibitors of SIRT1 or SIRT1-siRNA significantly reduced this effect [90].

SIRT2 is expressed predominantly in oligodendrocytes and in the myelin-rich regions of the ischemic brain. It was not found in astrocytes, microglia, or neurons [91]. However, other authors reported the presence of SIRT2 in the cytoplasm of neurons, but not astrocytes in the mouse cerebral cortex [89,92]. In the cytoplasm, SIRT2 such as HDAC6 regulates the microtubule dynamics through deacetylation of α-tubulin [93]. SIRT2 and HDAC6 can deacetylate α-tubulin either together [94], or separately [93]. Interestingly, the inhibition of SIRT2 increased acetylation of the microtubular α-tubulin mainly in the perinuclear zone, whereas inhibition of HDAC6 caused the general hyperacetylation of microtubules throughout the cell [95]. Although some studies pointed to the pathological role of SIRT2 after cerebral ischemia [76,96], the functions of SIRT2 in the ischemic brain are possibly more complicated than only pathological or only neuroprotective. Indeed, transient MCAO reduced the expression of SIRT2 and its translocation into the neuronal nuclei that played a neuroprotective role [96]. On the contrary, the overexpression of SIRT2 in the cytoplasm of the cerebellar neurons or in vitro in the differentiated PC12 cell line was shown to induce apoptosis [77,97]. Sirt2 was shown to mediate the myelin-dependent neuronal dysfunction during the early phase after MCAO in the mouse brain. Notably, the dynamics of Sirt2 mRNA and the protein level after ischemia differed [91].

SIRT6 was found in both: cerebral neurons and astrocytes [89,98]. It deacetylates mainly the lysine residues 9 and 56 in histone H3 that possibly represses genes associated with aging [99]. The role of SIRT6 in ischemia is still unclear and controversial. On one hand, SIRT6 protected the brain from postischemic reperfusion injury due to stimulation of transcription factor Nrf2 (nuclear factor-like (erythroid 2)-like 2), which regulates the expression of antioxidant proteins and suppresses oxidative stress [100,101]. SIRT6 was co-expressed with GAP-43, a marker of axon growth and synapse formation, at 14 days after photothrombotic stroke in the mouse cerebral cortex [89]. Whether SIRT6 functions as a part of the multiprotein complex in the postsynaptic membranes [102], or it regulates the neurite growth during the post-stroke recovery phase should be further studied. SIRT6 immunofluorescence was not observed in apoptotic cells in the PTS-induced penumbra [89]. On the other hand, its overexpression in cultured neurons under oxygen and glucose deprivation was associated with necrosis of cortical cells [103].

Thus, sirtuins, SIRT1 and SIRT6, are involved in the postischemic brain regeneration. SIRT1 regulates synaptic plasticity, memory, neuritogenesis, neurogenesis, and gliogenesis. SIRT6 protects neurons and astrocytes from the post ischemic reperfusion injury via stimulation of the transcription factor that regulates the production of antioxidant proteins Nrf2.

The HDAC inhibitors that have been developed to date are capable of inhibiting almost all HDAC isoforms of these four classes with varying degrees of specificity.

4. Pan-Inhibitors of Histone Deacetylases in Cerebral Ischemia

Inhibitors of different histone deacetylases that efficiently protect the brain from ischemic injury belong to two chemical groups: (a) Small carboxylates: valproic acid (VPA), sodium butyrate (SB), and sodium 4-phenylbutyrate (4-PBA), and (b) Hydroxamate-containing compounds: suberoylanilide hydroxamic acid (SAHA, Vorinostat), trichostatin A (TSA), and others. They protect the brain against excitotoxicity, oxidative stress, ER stress, apoptosis, inflammation, and BBB breakdown. They also induce angiogenesis, neurogenesis, and migration of stem cells to the damaged brain regions that improves functional recovery after cerebral ischemia. HDAC inhibitors are the promising neuroprotectors for treating ischemic stroke [15–18].

VPA, a pan-HDAC inhibitor, was shown to reduce brain injury in various stroke models. It improved the functional outcome, and demonstrated the anti-inflammatory activity [104–106]. VPA induces different proteins such as NeuroD, Math1, Ngn1, and p15, which contribute to differentiation of neural precursors in the hippocampus [107]. VPA administration during 7 days after MCAO considerably improved the neurological outcome in rats. This effect was associated with enhanced white matter repair and neurogenesis [108,109]. The VPA treatment increased survival of oligodendrocytes and caused the generation of new oligodendrocytes. These effects were associated with the increased density of myelinated axons in the ischemic boundary around the infarction core. At the molecular level, VPA increased the acetylation of histone H4 and caused overexpression of glutamate transporter 1 (GLT1) in neuroblasts. It also increased the number of new neurons [109]. Prolonged application of VPA during two weeks after MCAO increased the acetylation of histones H3 and H4 in the rat brain, and caused the upregulation of transcription factor HIF-1α (hypoxia-inducible factor-1α) and its downstream pro-angiogenic molecules such as vascular endothelial growth factor (VEGF) and matrix metalloproteinases MMP2 and MMP9. This enhanced the microvessel density and promoted functional recovery [106]. VPA suppressed the nuclear translocation of the NF-κB subunit p65, reduced activity of matrix metalloproteinase MMP9, and restored the BBB integrity that was broken after stroke [110].

Other pan-HDAC inhibitors SB and 4-PBA showed similar neuroprotective activity [111–113]. SB stimulated proliferation of neuronal progenitor cells and neurogenesis in the SVZ and DG zones of the rat brain after MCAO. It increased the levels of acetylated histone H3, neural cell adhesion molecule nestin, glial fibrillary acidic protein, transcription factor CREB (phospho-cAMP response element-binding protein), and brain-derived neurotrophic factor (BDNF) that were reduced after cerebral ischemia [111]. The neuroprotective effect of SB was also associated with inhibition of oxidative stress, reduction of BBB permeability, and anti-inflammatory action [110]. Notably, VPA and SB stimulated neurogenesis in the peri-infarct regions during the post-stroke recovery period [108,109,111]. SB effect on proliferation, differentiation, and migration of the neural precursor cells was mediated by the BDNF-TrkB signaling pathway [114]. Microglia-mediated neuroinflammation is an important component of the stroke-induced brain pathology. As shown in vitro, SB reduced the acetylation of histone H3 that was increased in the activated microglial cells. It also altered the transcription of pro-inflammatory genes Tnf-α, Nos2, Stat1, and Il6. Simultaneously, SB induced the expression of anti-inflammatory genes regulated by the IL10/STAT3 pathway. In the microglia of mice subjected to MCAO, SB reduced the expression of pro-inflammatory proteins TNF-α and NOS2, and stimulated expression of the anti-inflammatory mediator IL10. Therefore, HDAC inhibition by SB turned the microglial activity in the ischemic brain from the pro-inflammatory to the anti-inflammatory mechanism [115].

The application of SAHA protected the rat brain from MCAO-induced ischemia [116]. It prevented the ischemia-induced decrease of the histone H3 acetylation level and reduced the infarct volume. It also increased the levels of neuroprotective proteins Hsp70 and Bcl-2 in both control and ischemic brains [117]. The intraperitoneal administration of SAHA at 12 h after transient MCAO reduced the infarction volume in the mouse brain and improved the post-stroke outcome. It reduced the level of pro-inflammatory cytokines and inhibited the microglia-mediated inflammatory response [118].

TSA increased survival of cultured neuronal cells after oxygen and glucose deprivation. It also decreased the stroke-induced infarct volume in the mice brain. These effects were associated with the activation of antioxidant processes. TSA mediated HDAC inhibition and activated the transcription factor Nrf2, which regulates expression of diverse antioxidant proteins. As a result, heme oxygenase 1, NAD(P)H:quinone oxidoreductase 1, and glutamate-cysteine ligase were overexpressed and mediated neuroprotection in the neuronal culture and in the ischemic brain [106,116].

These HDAC inhibitors are non-selective. The neuroprotective effects of non-selective inhibitors such as TSA, SB, and SAHA have been well reviewed [119–122]. They inhibit a group of HDACs that belong to class I, or II, or both. Some HDACs in these groups are involved in ischemia-induced cell death, whereas others participate in the neuroprotective processes. It is of interest to use more selective inhibitors in order to affect only the pathogenic HDACs in the ischemic brain.

5. Selective Inhibitors of Histone Deacetylases in Cerebral Ischemia

5.1. Inhibition of Class I HDACs

In the work of Lin and colleagues photothrombotic stroke impaired neuronal survival and neuroplasticity in the mouse brain, deteriorated motor functions, and stimulated neuroinflammation [36]. These effects were associated with the upregulation of HDAC2 in the peri-infarct zone from 5 to 7 days after PTS. Since the absolute selective inhibitors of some HDACs are not available, the author used several HDACs inhibitors with the limited anti-HDAC selectivity: MGCD0103 that inhibits HDAC1, HDAC2, and HDAC3; SAHA that inhibits mostly HDAC1 and HDAC2, and TMP269, a class II inhibitor that acts on HDAC4, HDAC5, HDAC7, and HDAC9. The comparison of their effects showed the critical role of HDAC2 inhibition in the restoration of brain function, whereas HDAC2 overexpression exacerbated the stroke-induced functional disorders. HDAC2 inhibition at 5 to 7 post-stroke days promoted survival and neuroplasticity of neurons, suppressed neuroinflammation, recovered motor functions, and improved the stroke outcome. The inhibition of other HDAC isoforms was ineffective [36].

The recent inhibitory analysis also confirmed the participation of HDAC2 in the PTS-induced injury of the mouse cerebral cortex. The administration of α-phenyltroplon that inhibits both HDAC2 and HDAC8 [123], or MI-192, an inhibitor of HDAC2 and HDAC3 [124], suppressed apoptosis of cortical cells in the penumbra and decreased the infarct volume after photothrombotic stroke in mice. These effects contributed to the restoration of cerebral functions. However, the selective HDAC3 inhibitor BRD3308 was ineffective. Hence, HDAC2 was involved in PTS-induced apoptosis in the mouse brain, and its inhibition was beneficial against ischemic stroke [57,123] (Table 1).

Table 1. Effects of selective HDAC inhibitors and activators in animal models of ischemic stroke.

Class I	Inhibitor	Model of stroke	Treatment time	Effective Doses	Effects on Stroke	Citations
HDAC1/2/3	MS-275, entinostat	Mouse, MCAO Mouse, MCAO	Post at 7 h Post at 0h, 24 h and 48 h	20 µg/kg 200 µg/kg 30 mg/kg	↓Infarct volume ↓Neurological deficit ↑Cell survival	[20] [125]
HDAC2/3	MI192	Mouse, PTS	Post for 3 days	40 mg/kg	↓Infarct volume ↓Apoptosis ↑Acetylated α-tubulin ↑GAP43 ↓Neurological deficit	[124]
HDAC2/8	α-phenyl tropolone	Mouse, PTS	Post for 7 days	10 mg/kg	↓Infarct volume ↓Apoptosis	[123]
HDAC6	Tubastatin A	Rats, MCAO Mouse, PTS	Post for 1 or 3 days Post for 4 or 7 days	25 mg/kg 40 mg/kg 25 mg/kg	↓Infarct volume ↓Apoptosis ↑Acetylated α-tubulin ↓Neurological deficit ↑FCF-21 ↑GAP43	[126] [22]
	Tubacin	Mouse, MCAO	Pre for 3 h	5 mg/kg	↓Infarct volume ↑eNOS	[127]
	HPOB	Mouse, PTS	Post for 7 days	10 mg/kg	↓Infarct volume ↓Apoptosis	[89]
SIRT1	EX-527	Rats, MCAO	Post at 6 h, 12 h or 24 h	10 µg	↓Infarct volume ↓Necroptosis ↑Survival	[128]
SIRT1 and SIRT2	Sirtinol	Mouse, MCAO	Post at 48 h	10 mg/kg	↑Infarct volume ↑Apoptosis ↑Acetylated p53 and p65	[82]
SIRT2	AGK2	Mouse, MCAO	Post at 24 h	1 mg/kg	↓Infarct volume ↓Apoptosis ↓Neurological deficit ↓JNK, c-jun ↓AKT/FOXO3a	[128]
Activators						
SIRT1	Resveratrol	Mouse, MCAO Mouse, MCAO Rats, MCAO	Post for 7 days Post for 7 days Post at 24 h	50 mg/kg 6800 µg/kg 30 mg/kg	↓Infarct volume ↓Neurological deficit ↑MMP-2 ↑VEGF ↓Infarct volume ↓Neurological deficit ↓Infarct volume ↓Neurological deficit ↑pAkt ↑pGSK-3β	[129] [20] [130]
	Tetrahydroxystilbene glucoside (TSG)	Mouse, MCAO	Post at 24 h	15 mg/kg 40 mg/kg	↓Infarct volume ↓Apoptosis ↓ROS	[131]
	Icariin	Mouse, MCAO	Post for 1, 3 or 7 days	100 mg/kg 200 mg/kg	↓Infarct volume ↓Brain edema ↓Neurological deficit ↑PGC-1α	[132]
	Activator III	Mouse, MCAO	Post at 48 h	10 mg/kg	↓Infarct volume	[82]
	Alpha-lipoic acid (ALA)	Mouse, MCAO	Post at 24 h	50 mg/kg	↓Infarct volume Brain edema ↓Neurological deficit	[133]

Model of transient middle cerebral artery occlusion (MCAO), photothrombotic stroke (PTS).

It has been shown that in ischemia reperfusion kidney injury in HDAC1 or HDAC2 knockout mice, deletion of HDAC2 but not HDAC1 leads to an ischemic damage decrease [134]. Recently, it was shown that in ischemic reperfusion kidney injury in mice with inducible HDAC1 or HDAC2 knockout, it is the deletion of HDAC2, but not HDAC1, that leads to a decrease in ischemic damage.

Currently, new carbamide based inhibitors of ortho-amino anilides have been developed which show high selectivity for HDAC2 compared to the highly homologous isoform HDAC1. These kinetically selective HDAC2 inhibitors (BRD6688 and BRD4884) increased the acetylation of histones H4K12 and H3K9 in primary mouse hippocampal cell cultures and improved learning and memory in a model of neurodegenerative disease in mice [135].

The selective HDAC3 inhibitor RGFP966 reduced infarction size and alleviated neurological deficits after MCAO by reducing the inflammatory response. RGFP966 enhances STAT1 acetylation and subsequently attenuates STAT1 phosphorylation that may lead to AIM2 inflammasome downregulation by RGFP966 [136]. Inhibition of HDAC3 by RGFP966 was protective against I / R cerebral injury in in vivo and in vitro models of diabetes by modulating oxidative stress, apoptosis, and autophagy that may be mediated by upregulation of Bmal1 [137].

RGFP966 administration mimicked the neuroprotective effect of ischemic tolerance after MCAO causing a decrease in infarction volume and neurological deficits. The mechanism of the protective action of the HDAC3 inhibitor may be based on the weakened recruitment of HDAC3 to the promoter regions after preconditioning potentiating the initiation of transcription of genes including Hspa1a, Bcl2l1, and Prdx2 [138].

The selective HDAC8 inhibitor PCI-34051 has a neuroprotective effect. However, it was found that HDAC8 is not involved in the PCI-34051 mechanism of action [139]. Using BRD3811, an inactive PCI-34051 ortholog, has been shown to exhibit strong neuroprotective properties despite its inability to inhibit HDAC8. Research by Sleiman et al. (2014) [139] showed that the protective effects of small molecules containing hydroxamic acid, such as PCI-34051, are probably not related to direct epigenetic regulation through inhibition of HDAC8, but rather their neuroprotective effect is based on their ability to bind metals exhibiting antioxidant properties.

5.2. Inhibition of Class II HDACs

Selective inhibition of the class IIa HDACs by MC1568 worsened brain recovery. This was associated with the inactivation of CREB and c-Fos and exacerbated neurological deficit [140]. However, recent studies have shown that pharmacological inhibition of class IIa HDAC by MC1568 reduces infarction volume and neurological deficit in rats after tMCAO [138]. The effect was associated with the activation of transcription of the ncx3 gene encoding the plasma membrane Naþ / Ca2þ exchanger 3 (NCX3), which plays a neuroprotective role in stroke.

LMK235, a selective inhibitor of HDAC, did not influence the infarct volume, apoptosis of penumbra cells, and the neurological deficit. This indicated that HDAC4 did not participate in the apoptosis of penumbral cells after photothrombotic stroke [57]. On the other hand, LMK235 dose-dependently inhibited MKK7 transcription and JNK / c-Jun activity, which protected cultured cerebellar granule (CGN) neurons from apoptosis caused by potassium deficiency [141].

HDAC6 inhibitors were reported to reduce ischemia-induced apoptosis in the rat brain [126,142], the retina [143], or the myocardium [144]. They inhibited oxidative stress by decreasing the acetylation of peroxiredoxin 1 [144]. The selective HDAC6 inhibitors tubastatin A or HPOB promoted post-stroke regeneration of the mouse brain. They reduced apoptosis of the penumbral cells, and decreased the infarction volume after photothrombotic stroke. These compounds also stimulated the axon outgrowth, as was evidenced by the overexpression of GAP43 and restoration of acetylation of α-tubulin [57,63]. Tubastatin A also recovered the acetylated state of fibroblast growth factor-21 (FGF-21) that contributed to restoration of brain functions [126].

Another selective HDAC6 inhibitor tubacin increases eNOS expression in vivo improving endothelial function in diabetic db / db mice and significantly reduces ischemic brain damage in a mouse stroke model [127].

Currently, new low molecular weight inhibitors of HDAC6 pyrimidine hydroxyl amide have been developed that are bioavailable in the brain when administered systemically, such as ACY-738 and ACY-775, which have antidepressant effects [145] and are effective in peripheral neuropathy [146]. Their effectiveness in cerebral ischemia has not been studied yet.

These data showed the involvement of different HDACs in some neuropathological processes. In particular, HDAC2 and HDAC6 are involved in the stroke-induced death of neurons in the ischemic brain. Their activity may be associated with deacetylation of histones H3 and H4, following chromatin condensation, and inhibition of protein biosynthesis in the cell. Additionally, some HDACs deacetylate cytoplasmic proteins and modulate their functions. HDAC inhibitors prevent these effects and improve cell viability. Notably, some HDAC inhibitors showed the efficient neuroprotection not only during 3–6 h, but later, up to 5-7 days after ischemic stroke. Such delay may expand the therapeutic window that is very important for practical anti-stroke medicine.

5.3. Sirtuins Activators and Inhibitors

Since the modulation of the sirtuin activity can have a beneficial effect on many diseases, there is growing interest in the development and testing of sirtuin activators and inhibitors [147,148]. Here we consider those compounds that activate or inhibit sirtuins under conditions of cerebral ischemia, discuss the data confirming their effectiveness in cellular and animal models of ischemia.

Almost all sirtuin activators are described only for SIRT1 (Table. 1) The most famous of the SIRT1 activators is resveratrol (3, 5, 4′-trihydroxy-trans-stilbene). It is a natural compound that activates SIRT1 and may help treat or prevent obesity, reduce carcinogenesis, and age-related decline in heart function and neuronal loss. Pharmacological modulation of SIRT1 can have a pronounced effect on the outcome of ischemic stroke [76].

Resveratrol applied to mice before and after reperfusion reduced neurological deficit and infarction volume in the MCAO model by increasing the expression of angiogenic factors MMP-2 and VEGF and the number of microvessels [129]. Preliminary administration of resveratrol to mice for 7 days reduced the expression and activity of MMP-9, increasing the viability of neurons [79]. Recent studies have shown that pre-intraperitoneally administered resveratrol reduced neurological deficits and infarction volume 24 hours after MCAO in male rats (30 mg/kg) [130]. Moreover, the same study showed that resveratrol attenuated neuronal cell death by increasing the phosphorylation level of Akt and GSK-3β [130].

A recent review by Ghazavi H et al (2020) [149] collected evidence that resveratrol may not only affect neuronal function, but also plays an important role in reducing neurotoxicity by altering glial activity by modulating a number of signaling pathways. Numerous studies indicate that resveratrol enhances anti-inflammatory effects and decreases inflammatory cytokines by acting on signaling pathways in microglia such as AMP-activated protein kinase (5′-adenosine monophosphate-activated protein kinase, AMPK), SIRT1, and SOCS1 (suppressor of cytokine signaling Resveratrol increases AMPK activity and inhibits GSK-3β (glycogen synthase kinase 3 beta) activity in astrocytes), which makes ATP available to neurons and reduces reactive oxygen species (ROS). In addition, resveratrol promotes microglial activation and increases survival oligodendrocytes, which can lead to the maintenance of post-stroke brain homeostasis [149].

Intraperitoneal combined administration of the MS-275 histone deacetylase I inhibitor (20 μg / kg) with resveratrol (680 μg / kg) to mice after tMCAO reduced infarction volume and neurological deficits, assessed 48 hours after tMCAO [150]. Then, 24 hours after administration of the drugs, a decrease in the binding of RelA to the Nos2 promoter was observed, which reduced the overexpression of a number of proteins associated with the activation of microglia and macrophages, which ultimately led to a decrease in leukocyte infiltration in the ischemic region [150].

In addition to sirtuins, resveratrol affects many other targets (e.g., kinases and ATP synthase) [151], so the molecular mechanisms found in physiological studies with resveratrol should be interpreted with caution. In addition, resveratrol can also inhibit SIRT1, SIRT3, or SIRT5 depending on the substrate used [152,153].

It should be noted that natural resveratrol suffers from low bioavailability and activity and, like early synthetic SIRT1 activators such as SRT1720, has low selectivity [147]. However, SRT1720 induced mitochondrial biogenesis, increasing the mitochondrial respiration rate and ATP level, reducing ischemia-reperfusion kidney damage [154,155]. SRT1720 accelerated the recovery of mitochondrial function after acute oxidative damage to renal proximal tubule cells [156]. The SRT1720 activator was also effective in ischemia-reperfusion myocardial injury [157]. In a model of brain oxidative stress in vivo, it was shown that SRT1720 has a neuroprotective effect, reducing neurological deficit by inhibiting poly (ADP-ribose) polymerase [158].

In addition to SRT1720, molecules structurally unrelated to resveratrol, such as SRT2104, SRT2379, and SRT3657, have been developed to stimulate the activity of sirtuins, but similar to SRT1720 these SIRT1 activators have not been shown to have a neuroprotective effect in models of ischemic stroke.

Other SIRT1 activators such as tetrahydroxystilbene glucoside (TSG) reduced the effects of OGD and MCAO in cultured neurons and animals, respectively [23]. Icariin is able to increase the expression of SIRT1 and its downstream target PGC-1α, which stimulates mitochondrial activity, both in the MCAO mouse model and in neurons treated with OGD. Incarion improved neurological parameters after MCAO in mice, reduced the size of myocardial infarction and cerebral edema [132]. The neuroprotective effect of incarion was reversed by inhibition of SIRT1.

Treatment of mice with Activator III reduced the infarction volume, while the SIRT1 and SIRT2 inhibitor sirtinol increased ischemic damage both in the model of ischemic stroke [82] and in the model of hemorrhagic stroke [159]. Sirtinol increased acetylation of p53 and nuclear factor κB (p65), which led to the activation of neuronal apoptosis [82,159].

Another SIRT1 activator, citicoline (CDP-choline) was effective for patients with moderate stroke [160]. Treatment with citicoline increased SIRT1 levels in rat brains after tMCAO concurrently with neuroprotection. Sirtinol blocked the decrease in infarction volume caused by citicoline, while resveratrol caused a strong synergistic neuroprotective effect with citicoline [160].

Thus, most studies provide evidence of a neuroprotective effect of SIRT1 activators. Inhibition of neuroprotective SIRT1 usually worsens stroke outcome [76]. On the other hand, the SIRT1 inhibitor nicotinamide can also protect neurons from excitotoxicity and cerebral ischemia [109]. Clinically used selective inhibitor SIRT1 Selisistat, also known as EX-527, reduces the volume of ischemic brain infarction and improves survival, but does not reduce neurological deficits associated with stroke [128]. In addition, the administration of EX-527 effectively increased the expression of metabolic enzymes associated with the regulation of necroptosis [128].

Besides SIRT1, modulation of the activity of other sirtuins can also influence the progress or outcome of cerebral ischemia.

The administration of AGK2, a potent inhibitor of SIRT2, had a neuroprotective effect in the MCAO model; neurological indicators were better compared to the control group [96]. Inhibition of SIRT2 by AGK2 or SIRT2 knockdown reduced cell death caused by hydrogen peroxide by decreasing ROS production in cell culture [97].

However, the use of another SIRT2 inhibitor AK-7 in animal stroke models did not show a neuroprotective effect in the MCAO mouse model [161]. Whereas the administration of other SIRT2 inhibitors AK-1 and AGK2 reduced the infarction volume by suppressing the proapoptotic signaling pathways AKT / FOXO3a and MAPK [162].

Thus, SIRT2 also performs contradictory functions in stroke [163]. Its effect probably depends on the localization of the enzyme [63] as well as on the stage of ischemic stroke. It is known that Sirt2 is able to mediate myelin-dependent neuronal dysfunction at an early stage after ischemic stroke [91].

Modulators for other sirtuins are less studied. To date, little is known about the effectiveness of the pharmacological modulators SIRT3-7 in cerebral ischemia.

6. Conclusions

The brain responses to ischemic stroke are very complex and dynamic. They involve a plethora of molecular processes that occur in different cerebral elements—neurons, glial cells, and blood vessels. The neuroprotective strategies in vivo have failed in human trials [164]. However, the previous presence of a transient ischemic attack is associated—in humans—with a good early outcome in nonlacunar ischemic strokes, thus suggesting a neuroprotective effect of transient ischemic attack possibly by inducing a phenomenon of ischemic tolerance [165]. After numerous unsuccessful searches of effective neuroprotective anti-stroke agents, which used the compounds affecting only one side of this complex process, such as glutamate-mediated excitotoxicity, calcium homeostasis, or different apoptosis stages, it became that the complex approach aimed at several targets in the ischemic cerebral tissue is needed. In addition to brain neuroprotection, it is also important to study and stimulate the processes involved in neurorepair in the adult brain, either from angiogenesis, neurogenesis, or synaptic plasticity, including through endogenous neurorepair phenomena [166].

The inhibitors of different histone deacetylases efficiently protected the animal brains from ischemic injury. They induce angiogenesis, neurogenesis, stem cell migration to the damaged brain regions in order to promote functional recovery after cerebral ischemia. Hence, HDAC inhibitors may be considered as the hopeful neuroprotector agents for treatment of ischemic stroke. The first generation HDAC inhibitors are non-selective. They inhibit not only the HDACs involved in the ischemia-induced cell death, but also those participating in the neuroprotective processes. Possibly, more selective inhibitors that affect specific pro-apoptotic HDACs in the ischemic brain may be promising neuroprotectors for treating ischemic stroke.

Author Contributions: V.D. have made substantial contributions to acquisition of data and drafting the article, S.D. and S.S. have made substantial contributions to all of the following: the conception of review and acquisition of data, drafting the article and revising it critically for important intellectual content. All authors have read and agreed to the published version of the manuscript.

Funding: The work was supported by the Russian Science Foundation, grant #21-15-00188.

Institutional Review Board Statement: Not applicable.

Informed Consent Statement: Not applicable.

Data Availability Statement: Not applicable.

Conflicts of Interest: The authors declare no conflict of interest.

Abbreviations

blood-brain barrier	BBB
middle cerebral artery occlusion	MCAO
photothrombotic stroke	PTS
subventricular zone	SVZ
hippocampal dentate gyrus	DG
reactive oxygen species	ROS
Nitric oxide	NO
endothelial NO synthase	eNOS
neuronal NO synthase	nNOS
iNOS	inducible NO synthase
amyloid precursor protein	APP
MBD	methyl-CpG-binding domain
acetylation of lysine 9 in histone H3	H3K9Ac
Histone acetyltransferase	HAT
HAT1	histone acetyltransferase 1
PCAF	p300/CREB binding protein-associated factor
Histone deacetylase	HDAC
BDNF	brain-derived neurotrophic factor
calmodulin-dependent kinase IV	CaMKIV
MEF2	myocyte enhancer factor 2
CREB	cAMP response element-binding protein

References

1. Meng, H.; Jin, W.; Yu, L.; Xu, S.; Wan, H.; He, Y. Protective effects of polysaccharides on cerebral ischemia: A mini-review of the mechanisms. *Int. J. Biol. Macromol.* **2021**, *169*, 463–472. [CrossRef] [PubMed]
2. Feigin, V.L.; Forouzanfar, M.H.; Krishnamurthi, R.; Mensah, G.A.; Connor, M.; Bennett, D.A.; Moran, A.E.; Sacco, R.L.; Anderson, L.; Truelsen, T.; et al. Global and regional burden of stroke during 1990–2010: Findings from the Global Burden of Disease Study. *Lancet* **2014**, *383*, 245–255. [CrossRef]
3. Wang, H.; Naghavi, M.; Allen, C.; Barber, R.M.; Bhutta, Z.A.; Carter, A.; Casey, D.C.; Charlson, F.J.; Chen, A.Z.; Coates, M.M.; et al. Global, regional, and national life expectancy, all-cause mortality, and cause-specific mortality for 249 causes of death, 1980–2015: A systematic analysis for the Global Burden of Disease Study. *Lancet* **2016**, *388*, 1459–1544. [CrossRef]
4. Hankey, G.J. Stroke. *Lancet* **2017**, *389*, 641–654. [CrossRef]
5. Powers, W.J.; Rabinstein, A.A.; Ackerson, T.; Adeoye, O.M.; Bambakidis, N.C.; Becker, K.; Biller, J.; Brown, M.; Demaerschalk, B.M.; Hoh, B.; et al. 2018 Guidelines for the early management of patients with acute ischemic stroke: A guideline for healthcare professionals from the american heart association/american stroke association. *Stroke* **2018**, *49*, e46–e99. [CrossRef]
6. Powers, W.J. Acute ischemic stroke. *N. Engl. J. Med.* **2020**, *383*, 252–260. [CrossRef]
7. Singh, T.P.; Murphy, S.P.; Weinstein, J.R. Stroke: Basic and clinical. *Adv. Neurobiol.* **2017**, *15*, 281–293. [CrossRef]
8. Pimentel, B.C.; Willeit, J.; Töll, T.; Kiechl, S.; e Melo, T.P.; Canhão, P.; Fonseca, C.; Ferro, J. Etiologic evaluation of ischemic stroke in young adults: A comparative study between two european centers. *J. Stroke Cerebrovasc. Dis.* **2019**, *28*, 1261–1266. [CrossRef]
9. Arboix, A.; Besses, C. Cerebrovascular disease as the initial clinical presentation of haematological disorders. *Eur. Neurol.* **1997**, *37*, 207–211. [CrossRef]
10. Arboix, A.; Jiménez, C.; Massons, J.; Parra, O.; Besses, C. Hematological disorders: A commonly unrecognized cause of acute stroke. *Expert Rev. Hematol.* **2016**, *9*, 891–901. [CrossRef]
11. Leng, T.; Xiong, Z.-G. Treatment for ischemic stroke: From thrombolysis to thrombectomy and remaining challenges. *Brain Circ.* **2019**, *5*, 8–11. [CrossRef]
12. Christophe, B.R.; Mehta, S.H.; Garton, A.L.A.; Sisti, J.; Connolly, E.S. Current and future perspectives on the treatment of cerebral ischemia. *Expert Opin. Pharmacother.* **2017**, *18*, 573–580. [CrossRef] [PubMed]
13. Uzdensky, A.B.; Demyanenko, S.V. Epigenetic mechanisms of ischemic stroke. *Biochem. Suppl. Ser. A Membr. Cell Biol.* **2019**, *13*, 289–300. [CrossRef]
14. Dhir, N.; Medhi, B.; Prakash, A.; Goyal, M.K.; Modi, M.; Mohindra, S. Pre-clinical to clinical translational failures and current status of clinical trials in stroke therapy: A brief review. *Curr. Neuropharmacol.* **2020**, *18*, 596–612. [CrossRef]
15. Shein, N.; Shohami, E. Histone deacetylase inhibitors as therapeutic agents for acute central nervous system injuries. *Mol. Med.* **2011**, *17*, 448–456. [CrossRef]
16. Baltan, S.; Morrison, R.S.; Murphy, S.P. Novel protective effects of histone deacetylase inhibition on stroke and white matter ischemic injury. *Neurotherapy* **2013**, *10*, 798–807. [CrossRef]
17. Fessler, E.; Chibane, F.; Wang, Z.; Chuang, D.-M. Potential roles of HDAC inhibitors in mitigating ischemia-induced brain Damage and facilitating endogenous regeneration and recovery. *Curr. Pharm. Des.* **2013**, *19*, 5105–5120. [CrossRef] [PubMed]
18. Merson, T.D.; Bourne, J.A. Endogenous neurogenesis following ischaemic brain injury: Insights for therapeutic strategies. *Int. J. Biochem. Cell Biol.* **2014**, *56*, 4–19. [CrossRef] [PubMed]
19. Ganai, S.A.; Ramadoss, M.; Mahadevan, V. Histone Deacetylase (HDAC) Inhibitors—Emerging roles in neuronal memory, learning, synaptic plasticity and neural regeneration. *Curr. Neuropharmacol.* **2016**, *14*, 55–71. [CrossRef]
20. Lanzillotta, A.; Pignataro, G.; Branca, C.; Cuomo, O.; Sarnico, I.; Benarese, M.; Annunziato, L.; Spano, P.; Pizzi, M. Targeted acetylation of NF-kappaB/RelA and histones by epigenetic drugs reduces post-ischemic brain injury in mice with an extended therapeutic window. *Neurobiol. Dis.* **2013**, *49*, 177–189. [CrossRef]
21. Jhelum, P.; Karisetty, B.C.; Kumar, A.; Chakravarty, S. Implications of epigenetic mechanisms and their targets in cerebral ischemia models. *Curr. Neuropharmacol.* **2017**, *15*, 815–830. [CrossRef]
22. Demyanenko, S.; Uzdensky, A. Epigenetic alterations induced by photothrombotic stroke in the rat cerebral cortex: Deacetylation of histone h3, Upregulation of Histone Deacetylases and Histone acetyltransferases. *Int. J. Mol. Sci.* **2019**, *20*, 2882. [CrossRef]
23. Wang, Z.; Zang, C.; Cui, K.; Schones, D.E.; Barski, A.; Peng, W.; Zhao, K. Genome-wide mapping of HATs and HDACs reveals distinct functions in active and inactive genes. *Cell* **2009**, *138*, 1019–1031. [CrossRef]
24. Schweizer, S.; Meisel, A.; Märschenz, S. Epigenetic mechanisms in cerebral ischemia. *Br. J. Pharmacol.* **2013**, *33*, 1335–1346. [CrossRef] [PubMed]
25. Hu, Z.; Zhong, B.; Tan, J.; Chen, C.; Lei, Q.; Zeng, L. The emerging role of epigenetics in cerebral ischemia. *Mol. Neurobiol.* **2016**, *54*, 1887–1905. [CrossRef] [PubMed]
26. Glozak, M.A.; Sengupta, N.; Zhang, X.; Seto, E. Acetylation and deacetylation of non-histone proteins. *Gene* **2005**, *363*, 15–23. [CrossRef] [PubMed]
27. Spange, S.; Wagner, T.; Heinzel, T.; Krämer, O.H. Acetylation of non-histone proteins modulates cellular signalling at multiple levels. *Int. J. Biochem. Cell Biol.* **2009**, *41*, 185–198. [CrossRef]
28. Mrakovcic, M.; Kleinheinz, J.; Fröhlich, L.F. p53 at the crossroads between different types of HDAC inhibitor-mediated cancer cell death. *Int. J. Mol. Sci.* **2019**, *20*, 2415. [CrossRef]

29. De Ruijter, A.J.; Van Gennip, A.H.; Caron, H.N.; Kemp, S.; Van Kuilenburg, A.B. Histone deacetylases (HDACs): Characterization of the classical HDAC family. *Biochem. J.* **2003**, *370*, 737–749. [CrossRef]

30. Haberland, M.; Montgomery, R.L.; Olson, E.N. The many roles of histone deacetylases in development and physiology: Implications for disease and therapy. *Nat. Rev. Genet.* **2009**, *10*, 32–42. [CrossRef]

31. Baltan, S.; Bachleda, A.; Morrison, R.S.; Murphy, S.P. Expression of histone deacetylases in cellular compartments of the mouse brain and the effects of ischemia. *Transl. Stroke Res.* **2011**, *2*, 411–423. [CrossRef] [PubMed]

32. Lin, T.-N.; Kao, M.-H. Histone deacetylases in stroke. *Chin. J. Physiol.* **2019**, *62*, 95–107. [CrossRef]

33. Demyanenko, S.V.; Dzreyan, V.A.; Uzdensky, A.B. The expression and localization of histone acetyltransferases HAT1 and PCAF in neurons and astrocytes of the photothrombotic stroke-induced penumbra in the rat brain cortex. *Mol. Neurobiol.* **2020**, *57*, 3219–3227. [CrossRef]

34. Xu, Y.; Wang, Q.; Chen, J.; Ma, Y.; Liu, X. Updating a strategy for histone deacetylases and its inhibitors in the potential treatment of cerebral ischemic stroke. *Dis. Markers* **2020**, *2020*, 1–8. [CrossRef]

35. Bardai, F.H.; Verma, P.; Smith, C.; Rawat, V.; Wang, L.; D'Mello, S.R. Disassociation of histone deacetylase-3 from normal huntingtin underlies mutant huntingtin neurotoxicity. *J. Neurosci.* **2013**, *33*, 11833–11838. [CrossRef]

36. Lin, Y.-H.; Dong, J.; Tang, Y.; Ni, H.-Y.; Zhang, Y.; Su, P.; Liang, H.-Y.; Yao, M.-C.; Yuan, H.-J.; Wang, D.-L.; et al. Opening a New Time Window for Treatment of Stroke by Targeting HDAC2. *J. Neurosci.* **2017**, *37*, 6712–6728. [CrossRef]

37. Montgomery, R.L.; Hsieh, J.; Barbosa, A.C.; Richardson, J.A.; Olson, E.N. Histone deacetylases 1 and 2 control the progression of neural precursors to neurons during brain development. *Proc. Natl. Acad. Sci. USA* **2009**, *106*, 7876–7881. [CrossRef]

38. Kelly, R.D.; Cowley, S. The physiological roles of histone deacetylase (HDAC) 1 and 2: Complex co-stars with multiple leading parts. *Biochem. Soc. Trans.* **2013**, *41*, 741–749. [CrossRef]

39. Dovey, O.M.; Foster, C.T.; Conte, N.; Edwards, S.A.; Edwards, J.M.; Singh, R.; Vassiliou, G.; Bradley, A.; Cowley, S. Histone deacetylase 1 and 2 are essential for normal T-cell development and genomic stability in mice. *Blood* **2013**, *121*, 1335–1344. [CrossRef] [PubMed]

40. Noh, K.-M.; Hwang, J.-Y.; Follenzi, A.; Athanasiadou, R.; Miyawaki, T.; Greally, J.M.; Bennett, M.V.L.; Zukin, R.S. Repressor element-1 silencing transcription factor (REST)-dependent epigenetic remodeling is critical to ischemia-induced neuronal death. *Proc. Natl. Acad. Sci. USA* **2012**, *109*, E962–E971. [CrossRef] [PubMed]

41. Hwang, J.-Y.; Zukin, R.S. REST, a master transcriptional regulator in neurodegenerative disease. *Curr. Opin. Neurobiol.* **2018**, *48*, 193–200. [CrossRef]

42. Guan, J.-S.; Haggarty, S.J.; Giacometti, E.; Dannenberg, J.-H.; Joseph, N.; Gao, J.; Nieland, T.J.F.; Zhou, Y.; Wang, X.; Mazitschek, R.; et al. HDAC2 negatively regulates memory formation and synaptic plasticity. *Nature* **2009**, *459*, 55–60. [CrossRef] [PubMed]

43. Tang, Y.; Lin, Y.; Ni, H.; Dong, J.; Yuan, H.; Zhang, Y.; Liang, H.; Yao, M.; Zhou, Q.; Wu, H.; et al. Inhibiting histone deacetylase 2 (HDAC2) promotes functional recovery from stroke. *J. Am. Hear. Assoc.* **2017**, *6*, 007236. [CrossRef] [PubMed]

44. Lin, Y.; Yao, M.; Wu, H.; Dong, J.; Ni, H.; Kou, X.; Chang, L.; Luo, C.; Zhu, D. HDAC2 (Histone deacetylase 2): A critical factor in environmental enrichment-mediated stroke recovery. *J. Neurochem.* **2020**, *155*, 679–696. [CrossRef]

45. Demyanenko, S.; Neginskaya, M.; Berezhnaya, E. Expression of class I histone deacetylases in ipsilateral and contralateral hemispheres after the focal photothrombotic infarction in the mouse brain. *Transl. Stroke Res.* **2018**, *9*, 471–483. [CrossRef] [PubMed]

46. Bhaskara, S.; Knutson, S.K.; Jiang, G.; Chandrasekharan, M.B.; Wilson, A.J.; Zheng, S.; Yenamandra, A.; Locke, K.; Yuan, J.-L.; Bonine-Summers, A.R.; et al. Hdac3 is essential for the maintenance of chromatin structure and genome stability. *Cancer Cell* **2010**, *18*, 436–447. [CrossRef]

47. Chen, Y.-T.; Zang, X.-F.; Pan, J.; Zhu, X.-L.; Chen, F.; Chen, Z.-B.; Xu, Y. Expression patterns of histone deacetylases in experimental stroke and potential targets for neuroprotection. *Clin. Exp. Pharmacol. Physiol.* **2012**, *39*, 751–758. [CrossRef] [PubMed]

48. Xia, Y.; Wang, J.; Liu, T.-J.; Yung, W.A.; Hunter, T.; Lu, Z. C-jun downregulation by HDAC3-dependent transcriptional repression promotes osmotic stress-induced cell apoptosis. *Mol. Cell* **2007**, *25*, 219–232. [CrossRef]

49. Soriano, F.; Hardingham, G.E. In cortical neurons HDAC3 activity suppresses RD4-dependent SMRT export. *PLoS ONE* **2011**, *6*, e21056. [CrossRef] [PubMed]

50. Bardai, F.H.; Price, V.; Zaayman, M.; Wang, L.; D'Mello, S.R. Histone deacetylase-1 (HDAC1) is a molecular switch between neuronal survival and death. *J. Biol. Chem.* **2012**, *287*, 35444–35453. [CrossRef]

51. Panteleeva, I.; Rouaux, C.; Larmet, Y.; Boutillier, S.; Loeffler, J.-P.; Boutillier, A.-L. HDAC-3 Participates in the repression of e2f-dependent gene transcription in primary differentiated neurons. *Ann. NY Acad. Sci.* **2004**, *1030*, 656–660. [CrossRef] [PubMed]

52. Takase, K.; Oda, S.; Kuroda, M.; Funato, H. Monoaminergic and neuropeptidergic neurons have distinct expression profiles of histone deacetylases. *PLoS ONE* **2013**, *8*, e58473. [CrossRef]

53. Bolger, T.A.; Zhao, X.; Cohen, T.J.; Tsai, C.-C.; Yao, T.-P. The neurodegenerative disease protein Ataxin-1 antagonizes the neuronal survival function of myocyte enhancer factor-2. *J. Biol. Chem.* **2007**, *282*, 29186–29192. [CrossRef] [PubMed]

54. Majdzadeh, N.; Wang, L.; Morrison, B.; Bassel-Duby, R.; Olson, E.N.; D'Mello, S.R. HDAC4 inhibits cell-cycle progression and protects neurons from cell death. *Dev. Neurobiol.* **2008**, *68*, 1076–1092. [CrossRef]

55. Chen, B.; Cepko, C.L. HDAC4 regulates neuronal survival in normal and diseased retinas. *Science* **2009**, *323*, 256–259. [CrossRef]

56. Price, V.; Wang, L.; D'Mello, S.R. Conditional deletion of histone deacetylase-4 in the central nervous system has no major effect on brain architecture or neuronal viability. *J. Neurosci. Res.* **2013**, *91*, 407–415. [CrossRef]

57. Demyanenko, S.; Dzreyan, V.; Uzdensky, A. Overexpression of HDAC6, but not HDAC3 and HDAC4 in the penumbra after photothrombotic stroke in the rat cerebral cortex and the neuroprotective effects of α-phenyl tropolone, HPOB, and sodium valproate. *Brain Res. Bull.* **2020**, *162*, 151–165. [CrossRef]

58. Bolger, T.A. Intracellular trafficking of histone deacetylase 4 regulates neuronal cell death. *J. Neurosci.* **2005**, *25*, 9544–9553. [CrossRef]

59. Yuan, H.; Denton, K.; Liu, L.; Li, X.-J.; Benashski, S.; McCullough, L.; Li, J. Nuclear translocation of histone deacetylase 4 induces neuronal death in stroke. *Neurobiol. Dis.* **2016**, *91*, 182–193. [CrossRef]

60. Koppel, I.; Timmusk, T. Differential regulation of Bdnf expression in cortical neurons by class-selective histone deacetylase inhibitors. *Neuropharmacology* **2013**, *75*, 106–115. [CrossRef]

61. McCullough, L.D.; Tarabishy, S.; Liu, L.; Benashski, S.; Xu, Y.; Ribar, T.; Means, A.; Li, J. Inhibition of calcium/calmodulin-dependent protein kinase kinase β and calcium/calmodulin-dependent protein kinase IV is detrimental in cerebral ischemia. *Stroke* **2013**, *44*, 2559–2566. [CrossRef]

62. Kassis, H.; Shehadah, A.; Chopp, M.; Roberts, C.; Zhang, Z.G. Stroke induces nuclear shuttling of histone deacetylase 4. *Stroke* **2015**, *46*, 1909–1915. [CrossRef]

63. Demyanenko, S.; Berezhnaya, E.; Neginskaya, M.; Rodkin, S.; Dzreyan, V.; Pitinova, M. Class II histone deacetylases in the post-stroke recovery period—expression, cellular, and subcellular localization—Promising targets for neuroprotection. *J. Cell. Biochem.* **2019**, *120*, 19590–19609. [CrossRef]

64. He, M.; Zhang, B.; Wei, X.; Wang, Z.; Fan, B.; Du, P.; Zhang, Y.; Jian, W.; Chen, L.; Wang, L.; et al. HDAC 4/5- HMGB 1 signalling mediated by NADPH oxidase activity contributes to cerebral ischaemia/reperfusion injury. *J. Cell. Mol. Med.* **2013**, *17*, 531–542. [CrossRef]

65. Lee, H.-A.; Song, M.-J.; Seok, Y.-M.; Kang, S.-H.; Kim, S.-Y.; Kim, I. histone deacetylase 3 and 4 complex stimulates the transcriptional activity of the mineralocorticoid receptor. *PLoS ONE* **2015**, *10*, e0136801. [CrossRef]

66. Cho, Y.; Sloutsky, R.; Naegle, K.; Cavalli, V. Injury-induced HDAC5 nuclear export is essential for axon regeneration. *Cell* **2013**, *155*, 894–908. [CrossRef]

67. Schneider, J.W.; Gao, Z.; Li, S.; Farooqi, M.; Tang, T.-S.; Bezprozvanny, I.; Frantz, U.E.; Hsieh, J. Small-molecule activation of neuronal cell fate. *Nat. Chem. Biol.* **2008**, *4*, 408–410. [CrossRef] [PubMed]

68. Wei, J.-Y.; Lu, Q.-N.; Li, W.-M.; He, W. Intracellular translocation of histone deacetylase 5 regulates neuronal cell apoptosis. *Brain Res.* **2015**, *1604*, 15–24. [CrossRef] [PubMed]

69. Li, N.; Yuan, Q.; Cao, X.-L.; Zhang, Y.; Min, Z.-L.; Xu, S.-Q.; Yu, Z.-J.; Cheng, J.; Zhang, C.; Hu, X.-M. Opposite effects of HDAC5 and p300 on MRTF-A-related neuronal apoptosis during ischemia/reperfusion injury in rats. *Cell Death Dis.* **2017**, *8*, e2624. [CrossRef] [PubMed]

70. Gu, X.; Fu, C.; Lin, L.; Liu, S.; Su, X.; Li, A.; Wu, Q.; Jia, C.; Zhang, P.; Chen, L.; et al. miR-124 and miR-9 mediated downregulation of HDAC5 promotes neurite development through activating MEF2C-GPM6A pathway. *J. Cell. Physiol.* **2018**, *233*, 673–687. [CrossRef] [PubMed]

71. Valenzuela-Fernández, A.; Cabrero, J.R.; Serrador, J.M.; Sánchez-Madrid, F. HDAC6: A key regulator of cytoskeleton, cell migration and cell–cell interactions. *Trends Cell Biol.* **2008**, *18*, 291–297. [CrossRef]

72. Zhang, Y.; Li, N.; Caron, C.; Matthias, G.; Hess, D.; Khochbin, S.; Matthias, P. HDAC-6 interacts with and deacetylates tubulin and microtubules in vivo. *EMBO J.* **2003**, *22*, 1168–1179. [CrossRef]

73. Khoury, N.; Koronowski, K.B.; Young, J.I.; Perez-Pinzon, M.A. The NAD+-dependent family of sirtuins in cerebral ischemia and preconditioning. *Antioxid. Redox Signal* **2018**, *28*, 691–710. [CrossRef] [PubMed]

74. Michishita, E.; Park, J.Y.; Burneskis, J.M.; Barrett, J.C.; Horikawa, I. Evolutionarily conserved and nonconserved cellular localizations and functions of human SIRT proteins. *Mol. Biol. Cell* **2005**, *16*, 4623–4635. [CrossRef]

75. Ham, P.B., 3rd; Raju, R. Mitochondrial function in hypoxic ischemic injury and influence of aging. *Prog. Neurobiol.* **2017**, *157*, 92–116. [CrossRef]

76. She, D.T.; Jo, N.-G.; Arumugam, T. Emerging roles of sirtuins in ischemic stroke. *Transl. Stroke Res.* **2017**, *8*, 405–423. [CrossRef]

77. Pfister, J.A.; Ma, C.; Morrison, B.E.; D'Mello, S.R. Opposing effects of sirtuins on neuronal survival: SIRT1-mediated neuroprotection is independent of its deacetylase activity. *PLoS ONE* **2008**, *3*, e4090. [CrossRef]

78. Tanno, M.; Sakamoto, J.; Miura, T.; Shimamoto, K.; Horio, Y. Nucleocytoplasmic shuttling of the NAD+-dependent histone deacetylase SIRT1. *J. Biol. Chem.* **2007**, *282*, 6823–6832. [CrossRef] [PubMed]

79. Gao, D.; Zhang, X.; Jiang, X.; Peng, Y.; Huang, W.; Cheng, G.; Song, L. Resveratrol reduces the elevated level of MMP-9 induced by cerebral ischemia–reperfusion in mice. *Life Sci.* **2006**, *78*, 2564–2570. [CrossRef] [PubMed]

80. Hisahara, S.; Chiba, S.; Matsumoto, H.; Tanno, M.; Yagi, H.; Shimohama, S.; Sato, M.; Horio, Y. Histone deacetylase SIRT1 modulates neuronal differentiation by its nuclear translocation. *Proc. Natl. Acad. Sci. USA* **2008**, *105*, 15599–15604. [CrossRef]

81. Eng, F.; Ewijaya, L.; Etang, B.L. SIRT1 in the brain—connections with aging-associated disorders and lifespan. *Front. Cell. Neurosci.* **2015**, *9*, 64. [CrossRef]

82. Jiménez, M.H.; Hurtado, O.; Cuartero, M.; Ballesteros, I.; Moraga, A.; Pradillo, J.; McBurney, M.W.; Lizasoain, I.; Moro, M.A. Silent information regulator 1 protects the brain against cerebral ischemic damage. *Stroke* **2013**, *44*, 2333–2337. [CrossRef] [PubMed]

83. Hattori, Y.; Okamoto, Y.; Nagatsuka, K.; Takahashi, R.; Kalaria, R.N.; Kinoshita, M.; Ihara, M. SIRT1 attenuates severe ischemic damage by preserving cerebral blood flow. *NeuroReport* **2015**, *26*, 113–117. [CrossRef] [PubMed]

84. Li, Z.; Pang, L.; Fang, F.; Zhang, G.; Zhang, J.; Xie, M.; Wang, L. Resveratrol attenuates brain damage in a rat model of focal cerebral ischemia via up-regulation of hippocampal Bcl-2. *Brain Res.* **2012**, *1450*, 116–124. [CrossRef] [PubMed]

85. Luo, J.; Nikolaev, A.; Imai, S.-I.; Chen, D.; Su, F.; Shiloh, A.; Guarente, L.; Gu, W. Negative control of p53 by Sir2α promotes cell survival under stress. *Cell* **2001**, *107*, 137–148. [CrossRef]

86. Brunet, A.; Sweeney, L.B.; Sturgill, J.F.; Chua, K.F.; Greer, P.L.; Lin, Y.; Tran, H.; Ross, S.E.; Mostoslavsky, R.; Cohen, H.Y.; et al. Stress-dependent regulation of FOXO transcription factors by the SIRT1 deacetylase. *Science* **2004**, *303*, 2011–2015. [CrossRef]

87. Cohen, H.Y.; Lavu, S.; Bitterman, K.J.; Hekking, B.; Imahiyerobo, T.A.; Miller, C.; Frye, R.; Ploegh, H.; Kessler, B.; A. Sinclair, D. Acetylation of the C terminus of Ku70 by CBP and PCAF controls bax-mediated apoptosis. *Mol. Cell* **2004**, *13*, 627–638. [CrossRef]

88. Jin, Q.; Yan, T.; Ge, X.; Sun, C.; Shi, X.; Zhai, Q. Cytoplasm-localized SIRT1 enhances apoptosis. *J. Cell. Physiol.* **2007**, *213*, 88–97. [CrossRef]

89. Demyanenko, S.; Gantsgorn, E.; Rodkin, S.; Sharifulina, S. Localization and expression of Sirtuins 1, 2, 6 and plasticity-related proteins in the recovery period after a photothrombotic stroke in mice. *J. Stroke Cerebrovasc. Dis.* **2020**, *29*, 105152. [CrossRef]

90. Sugino, T.; Maruyama, M.; Tanno, M.; Kuno, A.; Houkin, K.; Horio, Y. Protein deacetylase SIRT1 in the cytoplasm promotes nerve growth factor-induced neurite outgrowth in PC12 cells. *FEBS Lett.* **2010**, *584*, 2821–2826. [CrossRef]

91. Krey, L.; Lühder, F.; Kusch, K.; Czech-Zechmeister, B.; Könnecke, B.; Outeiro, T.F.; Trendelenburg, G. Knockout of Silent Information Regulator 2 (SIRT2) preserves neurological function after experimental stroke in mice. *Br. J. Pharmacol.* **2015**, *35*, 2080–2088. [CrossRef] [PubMed]

92. Maxwell, M.M.; Tomkinson, E.M.; Nobles, J.; Wizeman, J.W.; Amore, A.M.; Quinti, L.; Chopra, V.; Hersch, S.M.; Kazantsev, A.G. The Sirtuin 2 microtubule deacetylase is an abundant neuronal protein that accumulates in the aging CNS. *Hum. Mol. Genet.* **2011**, *20*, 3986–3996. [CrossRef] [PubMed]

93. North, B.J.; Marshall, B.L.; Borra, M.T.; Denu, J.M.; Verdin, E. The human Sir2 ortholog, SIRT2, is an NAD+-dependent tubulin deacetylase. *Mol. Cell* **2003**, *11*, 437–444. [CrossRef]

94. Nahhas, F.; Dryden, S.C.; Abrams, J.; Tainsky, M.A. Mutations in SIRT2 deacetylase which regulate enzymatic activity but not its interaction with HDAC6 and tubulin. *Mol. Cell. Biochem.* **2007**, *303*, 221–230. [CrossRef]

95. Skoge, R.H.; Ziegler, M. SIRT2 inactivation reveals a subset of hyperacetylated perinuclear microtubules inaccessible to HDAC6. *J. Cell Sci.* **2016**, *129*. [CrossRef] [PubMed]

96. Xie, X.Q.; Zhang, P.; Tian, B.; Chen, X.Q. Downregulation of NAD-dependent deacetylase SIRT2 protects mouse brain against ischemic stroke. *Mol. Neurobiol.* **2016**, *54*, 7251–7261. [CrossRef]

97. Nie, H.; Hong, Y.; Lu, X.; Zhang, J.; Chen, H.; Li, Y.; Ma, Y.; Ying, W. SIRT2 mediates oxidative stress-induced apoptosis of differentiated PC12 cells. *NeuroReport* **2014**, *25*, 838–842. [CrossRef]

98. Favero, G.; Rezzani, R.; Rodella, L.F. Sirtuin 6 nuclear localization at cortical brain level of young diabetic mice: An immunohistochemical study. *Acta Histochem.* **2014**, *116*, 272–277. [CrossRef]

99. Tennen, R.I.; Chua, K.F. Chromatin regulation and genome maintenance by mammalian SIRT6. *Trends Biochem. Sci.* **2011**, *36*, 39–46. [CrossRef]

100. Wang, B.; Zhu, X.; Kim, Y.; Li, J.; Huang, S.; Saleem, S.; Li, R.-C.; Xu, Y.; Dore, S.; Cao, W. Histone deacetylase inhibition activates transcription factor Nrf2 and protects against cerebral ischemic damage. *Free. Radic. Biol. Med.* **2012**, *52*, 928–936. [CrossRef]

101. Zhang, W.; Wei, R.; Zhang, L.; Tan, Y.; Qian, C. Sirtuin 6 protects the brain from cerebral ischemia/reperfusion injury through NRF2 activation. *Neuroscience* **2017**, *366*, 95–104. [CrossRef] [PubMed]

102. Cardinale, A.; De Stefano, M.C.; Mollinari, C.; Racaniello, M.; Garaci, E.; Merlo, D. Biochemical characterization of Sirtuin 6 in the brain and its involvement in oxidative stress response. *Neurochem. Res.* **2014**, *40*, 59–69. [CrossRef] [PubMed]

103. Shao, J.; Yang, X.; Liu, T.; Zhang, T.; Xie, Q.R.; Xia, W. Autophagy induction by SIRT6 is involved in oxidative stress-induced neuronal damage. *Protein Cell* **2016**, *7*, 281–290. [CrossRef] [PubMed]

104. Kim, H.J.; Rowe, M.; Ren, M.; Hong, J.-S.; Chen, P.-S.; Chuang, D.-M. Histone deacetylase inhibitors exhibit anti-inflammatory and neuroprotective effects in a rat permanent ischemic model of stroke: Multiple mechanisms of action. *J. Pharmacol. Exp. Ther.* **2007**, *321*, 892–901. [CrossRef] [PubMed]

105. Ren, M.; Leng, Y.; Jeong, M.; Leeds, P.R.; Chuang, D. Valproic acid reduces brain damage induced by transient focal cerebral ischemia in rats: Potential roles of histone deacetylase inhibition and heat shock protein induction. *J. Neurochem.* **2004**, *89*, 1358–1367. [CrossRef]

106. Wang, Z.; Tsai, L.-K.; Munasinghe, J.; Leng, Y.; Fessler, E.B.; Chibane, F.; Leeds, P.; Chuang, D.-M. Chronic valproate treatment enhances postischemic angiogenesis and promotes functional recovery in a rat model of ischemic stroke. *Stroke* **2012**, *43*, 2430–2436. [CrossRef]

107. Yu, I.T.; Park, J.-Y.; Kim, S.H.; Lee, J.-S.; Kim, Y.-S.; Son, H. Valproic acid promotes neuronal differentiation by induction of proneural factors in association with H4 acetylation. *Neuropharmacology* **2009**, *56*, 473–480. [CrossRef]

108. Hao, Y.; Creson, T.; Zhang, L.; Li, P.; Du, F.; Yuan, P.; Gould, T.D.; Manji, H.K.; Chen, G. Mood stabilizer valproate promotes ERK pathway-dependent cortical neuronal growth and neurogenesis. *J. Neurosci.* **2004**, *24*, 6590–6599. [CrossRef]

109. Liu, D.; Gharavi, R.; Pitta, M.; Gleichmann, M.; Mattson, M.P. Nicotinamide prevents NAD+ depletion and protects neurons against excitotoxicity and cerebral ischemia: NAD+ consumption by SIRT1 may endanger energetically compromised neurons. *NeuroMolecular Med.* **2009**, *11*, 28–42. [CrossRef] [PubMed]

110. Park, M.J.; Sohrabji, F. The histone deacetylase inhibitor, sodium butyrate, exhibits neuroprotective effects for ischemic stroke in middle-aged female rats. *J. Neuroinflammation* **2016**, *13*, 1–14. [CrossRef]

111. Kim, H.J.; Leeds, P.; Chuang, D.-M. The HDAC inhibitor, sodium butyrate, stimulates neurogenesis in the ischemic brain. *J. Neurochem.* **2009**, *110*, 1226–1240. [CrossRef] [PubMed]

112. Langley, B.; D'Annibale, M.A.; Suh, K.; Ayoub, I.; Tolhurst, A.; Bastan, B.; Yang, L.; Ko, B.; Fisher, M.; Cho, S.; et al. Pulse inhibition of histone deacetylases induces complete resistance to oxidative death in cortical neurons without toxicity and reveals a role for cytoplasmic p21waf1/cip1 in cell cycle-independent neuroprotection. *J. Neurosci.* **2008**, *28*, 163–176. [CrossRef] [PubMed]

113. Qi, X.; Hosoi, T.; Okuma, Y.; Kaneko, M.; Nomura, Y. Sodium 4-phenylbutyrate protects against cerebral ischemic injury. *Mol. Pharmacol.* **2004**, *66*, 899–908. [CrossRef] [PubMed]

114. Jaworska, J.; Zalewska, T.; Sypecka, J.; Ziemka-Nalecz, M. Effect of the hdac inhibitor, sodium butyrate, on neurogenesis in a rat model of neonatal hypoxia–ischemia: Potential mechanism of action. *Mol. Neurobiol.* **2019**, *56*, 6341–6370. [CrossRef] [PubMed]

115. Patnala, R.; Arumugam, T.; Gupta, N.; Dheen, S.T. HDAC inhibitor sodium butyrate-mediated epigenetic regulation enhances neuroprotective function of microglia during ischemic stroke. *Mol. Neurobiol.* **2016**, *54*, 6391–6411. [CrossRef] [PubMed]

116. Tang, F.; Guo, S.; Liao, H.; Yu, P.; Wang, L.; Song, X.; Chen, J.; Yang, Q. Resveratrol enhances neurite outgrowth and synaptogenesis via sonic hedgehog signaling following oxygen-glucose deprivation/reoxygenation injury. *Cell. Physiol. Biochem.* **2017**, *43*, 852–869. [CrossRef]

117. Faraco, G.; Pancani, T.; Formentini, L.; Mascagni, P.; Fossati, G.; Leoni, F.; Moroni, F.; Chiarugi, A. Pharmacological inhibition of histone deacetylases by suberoylanilide hydroxamic acid specifically alters gene expression and reduces ischemic injury in the mouse brain. *Mol. Pharmacol.* **2006**, *70*, 1876–1884. [CrossRef] [PubMed]

118. Li, S.; Lu, X.; Shao, Q.; Chen, Z.; Huang, Q.; Jiao, Z.; Huang, X.; Yue, M.; Peng, J.; Zhou, X.; et al. Early histone deacetylase inhibition mitigates ischemia/reperfusion brain injury by reducing microglia activation and modulating their phenotype. *Front. Neurol.* **2019**, *10*, 893. [CrossRef] [PubMed]

119. Langley, B.; Brochier, C.; Rivieccio, M.A. Targeting histone deacetylases as a multifaceted approach to treat the diverse outcomes of stroke. *Stroke* **2009**, *40*, 2899–2905. [CrossRef]

120. Aune, S.E.; Herr, D.J.; Kutz, C.J.; Menick, D.R. Histone deacetylases exert class-specific roles in conditioning the brain and heart against acute ischemic injury. *Front. Neurol.* **2015**, *6*, 145. [CrossRef]

121. Pickell, Z.; Williams, A.M.; Alam, H.B.; Hsu, C.H. Histone deacetylase inhibitors: A novel strategy for neuroprotection and cardioprotection following ischemia/reperfusion injury. *J. Am. Heart Assoc.* **2020**, *9*, e016349. [CrossRef]

122. Shukla, S.; Tekwani, B.L. Histone deacetylases inhibitors in neurodegenerative diseases, neuroprotection and neuronal differentiation. *Front. Pharmacol.* **2020**, *11*, 537. [CrossRef]

123. Demyanenko, S.V.; Dzreyan, V.A.; Neginskaya, M.A.; Uzdensky, A.B. Expression of histone deacetylases HDAC1 and HDAC2 and their role in apoptosis in the penumbra induced by photothrombotic stroke. *Mol. Neurobiol.* **2019**, *57*, 226–238. [CrossRef]

124. Demyanenko, S.V.; Nikul, V.V.; Uzdensky, A.B. The neuroprotective effect of the HDAC2/3 inhibitor MI192 on the penumbra after photothrombotic stroke in the mouse brain. *Mol. Neurobiol.* **2019**, *57*, 239–248. [CrossRef]

125. Murphy, S.P.; Lee, R.J.; McClean, M.E.; Pemberton, H.E.; Uo, T.; Morrison, R.S.; Bastian, C.; Baltan, S. MS-275, a Class I histone deacetylase inhibitor, protects the p53-deficient mouse against ischemic injury. *J. Neurochem.* **2013**, *129*, 509–515. [CrossRef] [PubMed]

126. Wang, Z.; Leng, Y.; Wang, J.; Liao, H.-M.; Bergman, J.; Leeds, P.; Kozikowski, A.; Chuang, D.-M. Tubastatin A, an HDAC6 inhibitor, alleviates stroke-induced brain infarction and functional deficits: Potential roles of α-tubulin acetylation and FGF-21 up-regulation. *Sci. Rep.* **2016**, *6*, 19626. [CrossRef] [PubMed]

127. Chen, J.; Zhang, J.; Shaik, N.F.; Yi, B.; Wei, X.; Yang, X.-F.; Naik, U.P.; Summer, R.; Yan, G.; Xu, X.; et al. The histone deacetylase inhibitor tubacin mitigates endothelial dysfunction by up-regulating the expression of endothelial nitric oxide synthase. *J. Biol. Chem.* **2019**, *294*, 19565–19576. [CrossRef] [PubMed]

128. Nikseresht, S.; Khodagholi, F.; Ahmadiani, A. Protective effects of ex-527 on cerebral ischemia–reperfusion injury through necroptosis signaling pathway attenuation. *J. Cell. Physiol.* **2019**, *234*, 1816–1826. [CrossRef]

129. Dong, W.; Li, N.; Gao, D.; Zhen, H.; Zhang, X.; Li, F. Resveratrol attenuates ischemic brain damage in the delayed phase after stroke and induces messenger RNA and protein express for angiogenic factors. *J. Vasc. Surg.* **2008**, *48*, 709–714. [CrossRef]

130. Park, D.-J.; Kang, J.-B.; Shah, F.-A.; Koh, P.-O. Resveratrol modulates the Akt/GSK-3β signaling pathway in a middle cerebral artery occlusion animal model. *Lab. Anim. Res.* **2019**, *35*, 1–8. [CrossRef] [PubMed]

131. Wang, T.; Gu, J.; Wu, P.-F.; Wang, F.; Xiong, Z.; Yang, Y.-J.; Wu, W.-N.; Dong, L.-D.; Chen, J.-G. Protection by tetrahydroxystilbene glucoside against cerebral ischemia: Involvement of JNK, SIRT1, and NF-κB pathways and inhibition of intracellular ROS/RNS generation. *Free. Radic. Biol. Med.* **2009**, *47*, 229–240. [CrossRef]

132. Zhu, H.-R.; Wang, Z.-Y.; Zhu, X.-L.; Wu, X.-X.; Li, E.-G.; Xu, Y. Icariin protects against brain injury by enhancing SIRT1-dependent PGC-1α Expression in experimental stroke. *Neuropharmacology* **2010**, *59*, 70–76. [CrossRef]

133. Fu, B.; Zhang, J.; Zhang, X.; Zhang, C.; Li, Y.; Zhang, Y.; He, T.; Li, P.; Zhu, X.; Zhao, Y.; et al. Alpha-lipoic acid upregulates SIRT1-dependent PGC-1α expression and protects mouse brain against focal ischemia. *Neuroscience* **2014**, *281*, 251–257. [CrossRef] [PubMed]

134. Aufhauser, D.D.; Hernandez, P.; Concors, S.J.; O'Brien, C.; Wang, Z.; Murken, D.R.; Samanta, A.; Beier, U.H.; Krumeich, L.; Bhatti, T.R.; et al. HDAC2 targeting stabilizes the CoREST complex in renal tubular cells and protects against renal ischemia/reperfusion injury. *Sci. Rep.* **2021**, *11*, 1–13. [CrossRef] [PubMed]

135. Wagner, F.F.; Zhang, Y.-L.; Fass, D.M.; Joseph, N.; Gale, J.P.; Weïwer, M.; McCarren, P.; Fisher, S.L.; Kaya, T.; Zhao, W.-N.; et al. Kinetically selective inhibitors of histone deacetylase 2 (HDAC2) as cognition enhancers. *Chem. Sci.* **2015**, *6*, 804–815. [CrossRef] [PubMed]

136. Zhang, M.; Zhao, Q.; Xia, M.; Chen, J.; Chen, Y.; Cao, X.; Liu, Y.; Yuan, Z.; Wang, X.; Xu, Y. The HDAC3 inhibitor RGFP966 ameliorated ischemic brain damage by downregulating the AIM2 inflammasome. *FASEB J.* **2020**, *34*, 648–662. [CrossRef] [PubMed]

137. Zhao, B.; Yuan, Q.; Hou, J.-B.; Xia, Z.-Y.; Zhan, L.-Y.; Li, M.; Jiang, M.; Gao, W.-W.; Liu, L. Inhibition of HDAC3 ameliorates cerebral ischemia reperfusion injury in diabetic mice in vivo and in vitro. *J. Diabetes Res.* **2019**, *2019*, 1–12. [CrossRef] [PubMed]

138. Yang, X.; Wu, Q.; Zhang, L.; Feng, L. Inhibition of histone deacetylase 3 (HDAC3) mediates ischemic preconditioning and protects cortical neurons against ischemia in rats. *Front. Mol. Neurosci.* **2016**, *9*, 131. [CrossRef] [PubMed]

139. Sleiman, S.F.; Olson, D.; Bourassa, M.W.; Karuppagounder, S.S.; Zhang, Y.-L.; Gale, J.; Wagner, F.F.; Basso, M.; Coppola, G.; Pinto, J.T.; et al. Hydroxamic acid-based histone deacetylase (HDAC) inhibitors can mediate neuroprotection independent of HDAC inhibition. *J. Neurosci.* **2014**, *34*, 14328–14337. [CrossRef]

140. Kassis, H.; Shehadah, A.; Li, C.; Zhang, Y.; Cui, Y.; Roberts, C.; Sadry, N.; Liu, X.; Chopp, M.; Zhang, Z.G. Class IIa histone deacetylases affect neuronal remodeling and functional outcome after stroke. *Neurochem. Int.* **2016**, *96*, 24–31. [CrossRef]

141. Wu, L.; Zeng, S.; Cao, Y.; Huang, Z.; Liu, S.; Peng, H.; Zhi, C.; Ma, S.; Hu, K.; Yuan, Z. Inhibition of HDAC4 attenuated JNK/c-jun-dependent neuronal apoptosis and early brain injury following subarachnoid hemorrhage by transcriptionally suppressing MKK7. *Front. Cell. Neurosci.* **2019**, *13*, 468. [CrossRef]

142. Rivieccio, M.A.; Brochier, C.; Willis, D.E.; Walker, B.A.; D'Annibale, M.A.; McLaughlin, K.; Siddiq, A.; Kozikowski, A.P.; Jaffrey, S.; Twiss, J.L.; et al. HDAC6 is a target for protection and regeneration following injury in the nervous system. *Proc. Natl. Acad. Sci. USA* **2009**, *106*, 19599–19604. [CrossRef]

143. Yuan, H.; Li, H.; Yu, P.; Fan, Q.; Zhang, X.; Huang, W.; Shen, J.; Cui, Y.; Zhou, W. Involvement of HDAC6 in ischaemia and reperfusion-induced rat retinal injury. *BMC Ophthalmol.* **2018**, *18*, 300. [CrossRef]

144. Leng, Y.; Wu, Y.; Lei, S.; Zhou, B.; Qiu, Z.; Wang, K.; Xia, Z. Inhibition of HDAC6 activity alleviates myocardial ischemia/reperfusion injury in diabetic rats: Potential role of peroxiredoxin 1 acetylation and redox regulation. *Oxidative Med. Cell. Longev.* **2018**, *2018*, 1–15. [CrossRef] [PubMed]

145. Jochems, J.; Boulden, J.; Lee, B.G.; Blendy, J.A.; Jarpe, M.; Mazitschek, R.; van Duzer, J.H.; Jones, S.; Berton, O. Antidepressant-like properties of Novel HDAC6-selective inhibitors with improved brain bioavailability. *Neuropsychopharmacology* **2014**, *39*, 389–400. [CrossRef] [PubMed]

146. Benoy, V.; Berghe, P.V.; Jarpe, M.; Van Damme, P.; Robberecht, W.; Bosch, L.V.D. Development of improved HDAC6 inhibitors as pharmacological therapy for axonal Charcot–Marie–Tooth disease. *Neurotherapy* **2017**, *14*, 417–428. [CrossRef] [PubMed]

147. Dai, H.; Sinclair, D.; Ellis, J.L.; Steegborn, C. Sirtuin activators and inhibitors: Promises, achievements, and challenges. *Pharmacol. Ther.* **2018**, *188*, 140–154. [CrossRef] [PubMed]

148. Villalba, J.M.; Alcaín, F.J. Sirtuin activators and inhibitors. *BioFactors* **2012**, *38*, 349–359. [CrossRef]

149. Vafaee, F.; Ghazavi, H.; Shirzad, S.; Forouzanfar, F.; SahabNegah, S.; Rad, M.R. The role of resveratrol as a natural modulator in glia activation in experimental models of stroke. *Avicenna J. Phytomed* **2020**, *10*, 557–573.

150. Mota, M.; Porrini, V.; Parrella, E.; Benarese, M.; Bellucci, A.; Rhein, S.; Schwaninger, M.; Pizzi, M. Neuroprotective epi-drugs quench the inflammatory response and microglial/macrophage activation in a mouse model of permanent brain ischemia. *J. Neuroinflammation* **2020**, *17*, 1–16. [CrossRef]

151. Pirola, L.; Fröjdö, S. Resveratrol: One molecule, many targets. *IUBMB Life* **2008**, *60*, 323–332. [CrossRef] [PubMed]

152. Hubbard, B.; Gomes, A.P.; Dai, H.; Li, J.; Case, A.W.; Considine, T.; Riera, T.; Lee, J.; E, S.Y.; Lamming, D.; et al. Evidence for a common mechanism of SIRT1 regulation by allosteric activators. *Science* **2013**, *339*, 1216–1219. [CrossRef] [PubMed]

153. Gertz, M.; Nguyen, G.T.T.; Fischer, F.; Suenkel, B.; Schlicker, C.; Fränzel, B.; Tomaschewski, J.; Aladini, F.; Becker, C.; Wolters, D.; et al. A Molecular mechanism for direct sirtuin activation by resveratrol. *PLoS ONE* **2012**, *7*, e49761. [CrossRef] [PubMed]

154. Khader, A.; Yang, W.-L.; Kuncewitch, M.; Jacob, A.; Prince, J.M.; Asirvatham, J.R.; Nicastro, J.; Coppa, G.F.; Wang, P. Sirtuin 1 activation stimulates mitochondrial biogenesis and attenuates renal injury after ischemia-reperfusion. *Transplantation* **2014**, *98*, 148–156. [CrossRef] [PubMed]

155. Khader, A.; Yang, W.-L.; Godwin, A.; Prince, J.M.; Nicastro, J.M.; Coppa, G.F.; Wang, P. Sirtuin 1 stimulation attenuates ischemic liver injury and enhances mitochondrial recovery and autophagy. *Crit. Care Med.* **2016**, *44*, e651–e663. [CrossRef] [PubMed]

156. Funk, J.A.; Odejinmi, S.; Schnellmann, R.G. SRT1720 Induces mitochondrial biogenesis and rescues mitochondrial function after oxidant injury in renal proximal tubule cells. *J. Pharmacol. Exp. Ther.* **2010**, *333*, 593–601. [CrossRef] [PubMed]

157. Han, Y.; Sun, W.; Ren, D.; Zhang, J.; He, Z.; Fedorova, J.; Sun, X.; Han, F.; Li, J. SIRT1 agonism modulates cardiac NLRP3 inflammasome through pyruvate dehydrogenase during ischemia and reperfusion. *Redox Biol.* **2020**, *34*, 101538. [CrossRef]

158. Gueguen, C.; Palmier, B.; Plotkine, M.; Marchand-Leroux, C.; Besson, V.C. Neurological and histological consequences induced by in vivo cerebral oxidative stress: Evidence for beneficial effects of SRT1720, a Sirtuin 1 activator, and Sirtuin 1-mediated neuroprotective effects of Poly(ADP-ribose) polymerase inhibition. *PLoS ONE* **2014**, *9*, e87367. [CrossRef]

159. Zhang, X.-S.; Wu, Q.; Wu, L.-Y.; Ye, Z.-N.; Jiang, T.-W.; Ling-Yun, W.; Zhuang, Z.; Zhou, M.-L.; Zhang, X.; Hang, C.-H. Sirtuin 1 activation protects against early brain injury after experimental subarachnoid hemorrhage in rats. *Cell Death Dis.* **2016**, *7*, e2416. [CrossRef]
160. Hurtado, O.; Jiménez, M.H.; Zarruk, J.G.; Cuartero, M.; Ballesteros, I.; Camarero, G.; Moraga, A.; Pradillo, J.; Moro, M.A.; Lizasoain, I. Citicoline (CDP-choline) increases Sirtuin1 expression concomitant to neuroprotection in experimental stroke. *J. Neurochem.* **2013**, *126*, 819–826. [CrossRef]
161. Chen, X.; Wales, P.; Quinti, L.; Zuo, F.; Moniot, S.; Hérisson, F.; Rauf, N.A.; Wang, H.; Silverman, R.B.; Ayata, C.; et al. The Sirtuin-2 inhibitor AK7 is neuroprotective in models of Parkinson's disease but not amyotrophic lateral sclerosis and cerebral ischemia. *PLoS ONE* **2015**, *10*, e0116919. [CrossRef] [PubMed]
162. She, D.; Wong, L.J.; Baik, S.-H.; Arumugam, T.V. SIRT2 inhibition confers neuroprotection by downregulation of FOXO3a and MAPK signaling pathways in ischemic stroke. *Mol. Neurobiol.* **2018**, *55*, 9188–9203. [CrossRef] [PubMed]
163. Chen, X.; Lu, W.; Wu, D. Sirtuin 2 (SIRT2): Confusing roles in the pathophysiology of neurological disorders. *Front. Neurosci.* **2021**, *15*, 518. [CrossRef]
164. Dávalos, A.; Alvarez-Sabin, J.; Castillo, J.; Tejedor, E.D.; Ferro, J.; Martinez-Vila, E.; Serena, J.; Segura, T.; Cruz, V.T.; Masjuan, J.; et al. Citicoline in the treatment of acute ischaemic stroke: An international, randomised, multicentre, placebo-controlled study (ICTUS trial). *Lancet* **2012**, *380*, 349–357. [CrossRef]
165. Arboix, A.; Cabeza, N.; Eroles, L.G.; Massons, J.; Oliveres, M.; Targa, C.; Balcells, M. Relevance of transient ischemic attack to early neurological recovery after nonlacunar ischemic stroke. *Cerebrovasc. Dis.* **2004**, *18*, 304–311. [CrossRef]
166. Arboix, A.; Krupinski, J. Angiogenesis, neurogenesis and neuroplasticity in ischemic stroke. *Curr. Cardiol. Rev.* **2010**, *6*, 238–244. [CrossRef]

Article

Mitoquinone Helps Combat the Neurological, Cognitive, and Molecular Consequences of Open Head Traumatic Brain Injury at Chronic Time Point

Muhammad Ali Haidar [1], Zaynab Shakkour [2], Chloe Barsa [1], Maha Tabet [3], Sarin Mekhjian [1], Hala Darwish [1], Mona Goli [4], Deborah Shear [5], Jignesh D. Pandya [5], Yehia Mechref [4], Riyad El Khoury [6,*], Kevin Wang [7,*] and Firas Kobeissy [1,7,*]

[1] Faculty of Biochemistry and Molecular Genetics, American University of Beirut, Beirut 1107 2020, Lebanon; muhammad.a.haidar@gmail.com (M.A.H.); cab20@mail.aub.edu (C.B.); skm12@mail.aub.edu (S.M.); hd30@aub.edu.lb (H.D.)

[2] Department of Pathology & Anatomical Sciences, University of Missouri School of Medicine, Columbia, MO 65212, USA; zist7p@missouri.edu

[3] Centre de Biologie Integrative (CBI), Molecular, Cellular, and Developmental Biology Department (MCD), University of Toulouse, Centre National de la Recherche Scientifique (CNRS), Université Paul Sabatier (UPS), 31062 Toulouse, France; mahatabet.mt@gmail.com

[4] Chemistry and Bioehcmistry Department, Texas Tech University, Lubbock, TX 79409, USA; Mona.Goli@ttu.edu (M.G.); Yehia.Mechref@ttu.edu (Y.M.)

[5] Brain Trauma Neuroprotection (BTN) Branch, Center for Military Psychiatry and Neuroscience, Walter Reed Army Institute of Research, Silver Spring, MD 20910, USA; deborah.a.shear.civ@mail.mil (D.S.); jignesh.d.pandya.civ@mail.mil (J.D.P.)

[6] Neuromuscular Diagnostic Laboratory, Department of Pathology and Laboratory Medicine, American University of Beirut Medical Center, Beirut 1107 2020, Lebanon

[7] Program for Neurotrauma, Neuroproteomics & Biomarkers Research, Departments of Emergency Medicine, Psychiatry, Neuroscience and Chemistry, University of Florida, Gainesville, FL 32611, USA

* Correspondence: re70@aub.edu.lb (R.E.K.); kwang@ufl.edu (K.W.); firasko@gmail.com (F.K.)

Citation: Haidar, M.A.; Shakkour, Z.; Barsa, C.; Tabet, M.; Mekhjian, S.; Darwish, H.; Goli, M.; Shear, D.; Pandya, J.D.; Mechref, Y.; et al. Mitoquinone Helps Combat the Neurological, Cognitive, and Molecular Consequences of Open Head Traumatic Brain Injury at Chronic Time Point. *Biomedicines* **2022**, *10*, 250. https://doi.org/10.3390/biomedicines10020250

Academic Editors: Kumar Vaibhav, Meenakshi Ahluwalia and Pankaj Gaur

Received: 22 December 2021
Accepted: 17 January 2022
Published: 24 January 2022

Publisher's Note: MDPI stays neutral with regard to jurisdictional claims in published maps and institutional affiliations.

Copyright: © 2022 by the authors. Licensee MDPI, Basel, Switzerland. This article is an open access article distributed under the terms and conditions of the Creative Commons Attribution (CC BY) license (https://creativecommons.org/licenses/by/4.0/).

Abstract: Traumatic brain injury (TBI) is a heterogeneous disease in its origin, neuropathology, and prognosis, with no FDA-approved treatments. The pathology of TBI is complicated and not sufficiently understood, which is the reason why more than 30 clinical trials in the past three decades turned out unsuccessful in phase III. The multifaceted pathophysiology of TBI involves a cascade of metabolic and molecular events including inflammation, oxidative stress, excitotoxicity, and mitochondrial dysfunction. In this study, an open head TBI mouse model, induced by controlled cortical impact (CCI), was used to investigate the chronic protective effects of mitoquinone (MitoQ) administration 30 days post-injury. Neurological functions were assessed with the Garcia neuroscore, pole climbing, grip strength, and adhesive removal tests, whereas cognitive and behavioral functions were assessed using the object recognition, Morris water maze, and forced swim tests. As for molecular effects, immunofluorescence staining was conducted to investigate microgliosis, astrocytosis, neuronal cell count, and axonal integrity. The results show that MitoQ enhanced neurological and cognitive functions 30 days post-injury. MitoQ also decreased the activation of astrocytes and microglia, which was accompanied by improved axonal integrity and neuronal cell count in the cortex. Therefore, we conclude that MitoQ has neuroprotective effects in a moderate open head CCI mouse model by decreasing oxidative stress, neuroinflammation, and axonal injury.

Keywords: neurotrauma; oxidative stress; neurodegeneration; neuroinflammation; moderate traumatic brain injury

1. Introduction

Globally, TBI is one of the leading causes of death and long-lasting disability. The incidence of TBI is estimated to be 939 in 10,000 worldwide, with major causes being falls, vehicle accidents, wars, and sports [1]. The mortality rates worldwide are assessed to be between 7% and 23%, 90% of which are in developing countries [2]. TBI studies in the Middle East are limited; however, it was predicted that the incidence in the region is 45 per 100,000 [2]. Survivors suffering from disabilities endure major socioeconomic burdens as well [3]. TBI leads to neurological complications through two major events, dynamic and overlapping, denoted as primary and secondary injuries. Primary injury defines the severity of TBI and therefore represents the major prognostic factor. Damaged neural cells undergo ionic imbalance characterized by an influx of calcium (Ca^{2+}), potassium (K^+), and glutamate [4]. The resultant excitotoxicity leads to an energetic crisis and oxidative stress by increasing reactive oxygen species (ROS) [5]. These can then directly stimulate the release of cytokines and pro-inflammatory factors, contributing to an elevated inflammatory state [6,7]. Following this, secondary injury is triggered, initiating a cascade of metabolic events that include further excitotoxicity, neuroinflammation, disruption of the blood–brain barrier (BBB), and cell death [8]. Notably, mitochondrial dysfunction, characterized by their membrane disruption and fission/fusion imbalance, plays an important role in the pathology of TBI via the excessive production of ROS, eventually leading to apoptotic cell death [9].

Finding a treatment for TBI remains a major challenge as there are currently no FDA-approved drugs with present approaches mainly targeting the symptoms. Depending on the severity of the injury, these can range from anti-seizure drugs and diuretics to coma-inducing drugs. Anti-seizure prophylaxis is administered to prevent the development of epileptic episodes post-TBI; however, due to the shortage of evidence on its effects in the long term, it is usually discontinued a week after injury [10]. Intracranial pressure is also monitored when individuals experience TBI, and in the case of its elevation, pressure is released by medically inducing venous flow and removing cerebrospinal fluid (CSF) from the intracranial regions [11].

Current animal studies on possible therapies have explored multiple candidates such as calcium channel blockers, erythropoietin, and hypothermia. Calcium channel blockers such as nimodipine and ziconotide have been considered as possible agents to enhance mitochondrial function and improve outcomes in patients with spontaneous subarachnoid hemorrhage [12]. However, the former was later revealed to have no effects on mortality and morbidity in TBI patients, and the latter was shown to lead to numerous side effects such as hypotension [13,14]. On the other hand, other suggested treatments target oxidative stress since antioxidants have been shown to ameliorate the pathology of TBI [15,16]. An example of these is the mitochondria-targeted drug mitoquinone (MitoQ). It is synthesized by conjugating a ubiquinone moiety to a triphenylphosphonium cation (TPP^+) [17]. TPP^+ is a lipophilic cation that directs the ubiquinone moiety to the inner mitochondrial membrane (IMM) and enables its accumulation due to the high electrochemical potential [18]. Accumulated MitoQ in the IMM is reduced by complex II into its antioxidative quinol form [19], noting that almost all molecules that enter the mitochondria are adsorbed to the matrix surface of the IMM where they are constantly recycled [20]. MitoQ can easily cross the BBB [21] and has been shown to act by activating the nuclear factor erythroid 2 (NFE2)-related factor 2 (Nrf2) pathway (Nrf2-ARE) [22]. Previous studies from our laboratory (currently under review) have shown that MitoQ improved the outcomes of closed head repetitive mild TBI in mice at acute, subacute, and chronic time points. Our current study aimed at assessing MitoQ's neuroprotective effects on the long-term behavioral, cognitive, and molecular outcomes in a moderate open head TBI mouse model. We hypothesized that improved outcomes are carried out by MitoQ's ability to dampen the state of oxidative stress, decrease cellular injury, and ameliorate neuroinflammation. We were able to show that MitoQ indeed exhibits neuroprotective properties in a moderate controlled cortical impact (CCI) mouse model, with effects translating to the molecular and behavioral/cognitive levels.

2. Materials and Methods

2.1. Animals

Thirty-four C57BL/6 male mice were housed in a controlled environment (12 h light/dark cycles, 22 ± 2 °C). The study was carried out at the Animal Care Facility of the American University of Beirut (AUB) and all animal experiments were performed in compliance with the AUB Institutional Animal Care and Use Committee (IACUC) guidelines with the reference number 17-01-458, 1 February 2019. All animals were handled under pathogen-free conditions and fed chow diet ad libitum. 7–8 weeks old male mice were divided into three groups: Sham, CCI, and CCI + MitoQ. The mice in the CCI and CCI + MitoQ groups were subjected to TBI using a controlled cortical impact machine (CCI). Thirty minutes after TBI, mice in the CCI + MitoQ group were injected intraperitoneally (IP) with MitoQ at a dosage of 8 mg/kg administered three times per week over 30 days, starting at 30 min post-injury. The dose was adopted from a previous study on MitoQ and TBI [22].

Keeping in mind that the test most prone to variability is the Morris water maze, we used it to determine the number of mice required for cognitive and behavioral tests. To achieve 90% power to detect a difference with 95% confidence, 9 animals were required. Therefore, we assigned 11–12 mice per group for behavioral, cognitive, and neurological tests: group 1: Sham (n = 10); group 2: CCI (n = 10); and group 3: CCI + MitoQ (n = 10). Animals were assigned to experimental groups by randomly selecting which animals received the treatment from those that underwent the injury. Animals that showed weak health post-surgery and exhibited any disease condition were not included in the study as per the IACUC regulations. Following sacrifice, four animals per group (n = 4) were used to perform immunofluorescence and another four animals (n = 4) per group for the RT-qPCR.

2.2. Stereotaxic Surgery: Controlled Cortical Impact

Each mouse was anesthetized with a ketamine/xylazine mixture (50 mg/kg and 15 mg/kg, respectively) administered intraperitoneally. Each mouse was then fixed on a stereotaxic frame, and a longitudinal skin incision was made in the middle of the mouse's head by a surgical scalpel to expose the skull. An ointment (Xailin®, Nicox, France) was applied to the eyes to protect vision during surgery. Using a drill and forceps, the part of the skull above the somatosensory area of the parietal lobe was removed to expose the brain. The injury was created using the Leica Impact One Angle Controlled Cortical Impact (CCI) machine. The center of the impactor was placed above the somatosensory area of the parietal cortex of the brain using the following coordinates: +1.0 mm AP, +1.5 mm ML, and −2 mm DV. The duration of impact was kept constant with a dwell time of 1 s, at a velocity of 4 m/s. The depth of the injury was set to 1.5 mm. The tip of the impactor was 1 mm in diameter. The size of the bone flap was 1.7 mm in diameter, which was carefully removed using a manual trephine to expose the dura matter without damaging it. The wound was closed using silk sutures (MERSILK™-W587H, Ethicon, OH, USA). Each mouse was placed on a heating pad to maintain the body's temperature. The CCI + MitoQ group received the first MitoQ injection 30 min following injury. For the animals in the Sham group, drilling was performed to remove confounding factors.

2.3. Mitoquinone Supplementation

Mitoquinone (Focus Biomolecules, 10-1363, MW = 663.64) was prepared by dissolving 25 mg in 10% dimethylsulfoxide (DMSO) (100 µL in 900 µL phosphate-buffered saline (PBS)). Further dilutions using PBS were carried out to obtain working solutions of 1 mg/mL. MitoQ was intraperitoneally (IP) supplemented at a dose of 8 mg/kg three times per week over 30 days, starting at 30 min post-injury. The experimental timeline is presented below in Figure 1A.

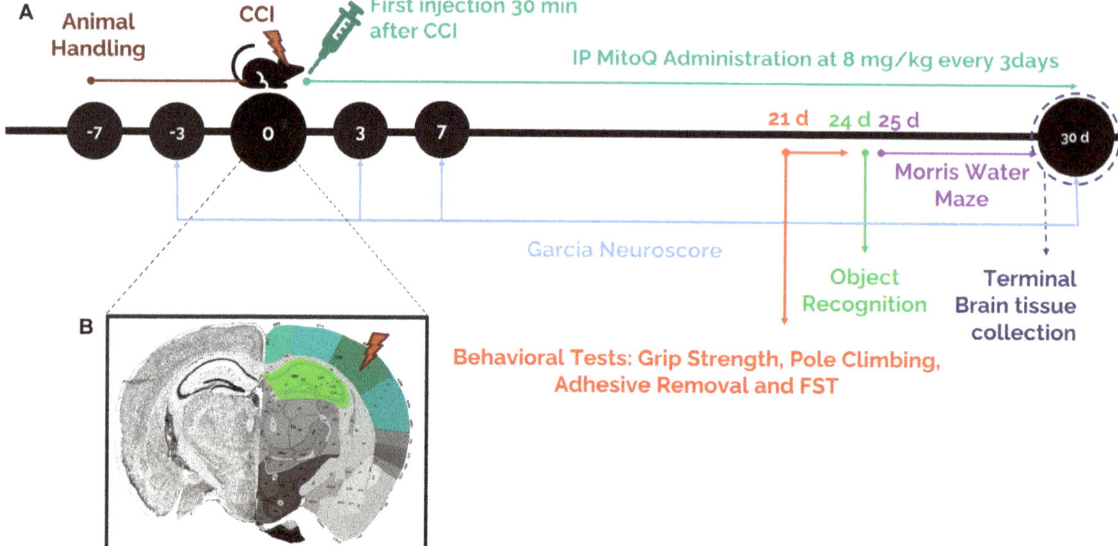

Figure 1. Experimental timeline of the study. (**A**) Mice were divided across three groups: Sham, CCI, and CCI + MitoQ. MitoQ injections were supplemented starting 30 min after the first injury via IP injections at a dose of 8 mg/kg every 3 days for 30 days. Behavioral and cognitive testing commenced after day 21 of the experimental timeline and continued until day 30. (**B**) A coronal section showing the site of injury and the brain regions (green and blue) that were investigated for quantification parameters. The section was obtained from the Allen Mouse Brain Atlas and Allen Reference Atlas—Mouse Brain [23].

2.4. Garcia Neuroscore

The neuroscore was adopted from Garcia et al. to assess the integrity of neurological function via different criteria [24]. Every group was tested 3 days before CCI and then on days 3, 7, and 30 after the injury. The animals underwent six different evaluations on every test day, each being scored from 0 to 3, according to Table 1.

Table 1. Evaluation criteria for Garcia neuroscore.

Criteria	Evaluation
Spontaneous Activity	Ability to approach all four walls of the cage
Limb Symmetry	Limb symmetry when held by the tail
Forepaw Outstretching	Outstretching symmetry of both forelimbs while the hindlimbs are kept in the air
Climbing	Ability to climb into and hang onto the cage
Body Perception	Reaction to stimulus while the mouse is touched on each side of the body with a stick
Vibrissae Touch	Reaction to stimulus while whiskers of the mouse are touched with a stick without entering the visual field

2.5. Grip Strength Test

This grip strength test was performed to assess endurance of motor skills and muscular dysfunction using the 4700-grip strength meter (UGO BASILE®, Gemonio, Italy). Each animal was held by its tail and was allowed to catch a trapeze-shaped metal object with both of its paws. A total of three trials were conducted for each animal. Muscle strength in gram force (gf) was recorded along with the total time of grip and the time the mouse applied the force. For analysis, the average of muscle strength, in gf, in the three trials, was normalized to the mouse weight [25].

2.6. Pole Climbing Test

The pole climbing test was conducted to evaluate motor coordination. The apparatus consisted of a metal pole with a length of 60 cm and a diameter of 1 cm, set perpendicularly in a big cage. The pole was wrapped with tape to facilitate the animal's grip on the apparatus. Animals were habituated to descend on the pole for three trials before the recorded testing. The mice were placed on top of the pole with their heads directed upwards and allowed to descend freely [26]. The time needed for each mouse to reach the bottom of the pole (total time) was recorded. For each animal, a total of three testing trials were conducted on the testing day.

2.7. Adhesive Removal Test

This test was performed by placing a rectangular adhesive tape strip on each mouse's nostril to assess sensorimotor function. Then, the animal was placed back in its cage. Two different values were recorded using two stopwatches: (i) the time needed to make the first contact with the tape, which represents nose sensitivity, and (ii) the time needed to remove the tape, which represents dexterity.

2.8. Forced Swim Test

This test was used to assess despair and depression-like behavior. Animals were placed in a glass cylinder (20 (height) $\times$ 11 cm (diameter)) filled with 25 °C water to a depth of 15 cm for 6 min. The 6 min included a 6 min acclimation period at the beginning of the test, followed immediately by 5 min of testing. Depression was reflected by the time mice spent floating on the surface of the water without moving. Animals were then dried with a towel and returned to their heated cage. The behavior of the mice was recorded using a camera, and immobility time was analyzed later during the last five-minute period of the test. Immobility time was defined as the time a mouse spent floating and only making necessary movements to keep its body balanced and head above the water.

2.9. Morris Water Maze Test

The Morris water maze (MWM) was used to study spatial learning and memory. The apparatus consisted of a circular pool (110 cm diameter, 55 cm depth) half-filled with water and maintained at 22 °C. A 10 cm-diameter platform was placed in the water. Three extra-maze cues were located around the pool to spatially guide the mice. Black and white cues were used to omit any variance in color discrimination among mice. Non-toxic white paint was added to the water surface to make it opaque and ensure accurate tracking. The ANY-maze 5.2 software (Stoelting Co., Wood Dale, IL, USA) was used to track the movement of mice and record the different parameters to be evaluated during the five days of testing.

On the first day, the platform was made visible by mounting a flag for cued trials. Four trials were carried out for each animal where the position of the platform and the starting position of animals changed among trials. On days two through four, which are the acquisition days, the flag was removed, and the platform was fixed in the northeast (NE) quadrant submerged in the water. However, the starting position of the animals varied among trials where three trials were carried out. The maximum time of a trial was set to one minute. If the animal failed to find the platform during this time, it was guided to the platform by the experimenter, where the animals were allowed to sit on the platform for 20 s for memory consolidation. If the animals found the platform before this time, the test was considered completed, and the animals were kept on the platform for five seconds. On day five, or the probe trial day, the platform was removed, and the mice were allowed to swim for 1 min for one trial only. After each trial on all days, the mice were dried with a towel and allowed to rest in a heated cage.

2.10. Novel Object Recognition Test

Novel object recognition depends on the innate capacity of rodents to discriminate a novel object from a familiar object (previously encountered). Mice were placed in the center of an open field at the beginning of each trial and freely explored the open field and objects. At the end of each trial (5 min), mice were removed from the open fields and placed in their home cages next to each testing area for the inter-trial interval (ITI—5 min). The open fields and objects were cleaned with 70% ethanol during the ITI. During Trial 1 (learning trial), the animals explored an open field with two identical objects. During trial 2 (testing trial), one object was kept in the testing field, while the other was replaced by a novel object for exploration. All data were recorded by a video camera that was connected to image analyzer software Any-Maze 5.2 (Stoelting Co., Wood Dale, IL, USA). The zones were located 5 cm around each object on the software, and the software recorded the time the animal spent in each zone.

2.11. RT-qPCR

RNA was extracted using Trizol (T9424-100ML, Sigma Aldrich, St. Louis, MO, USA), and then any contaminating genomic DNA was removed using TURBO DNA-free™ Kit (AM1907, Thermo Fisher Scientific, Waltham, MA, USA) according to the manufacturer's instructions. An amount of 1.5 µg of total RNA was reverse transcribed using the iScript™ cDNA Synthesis Kit (1708890, Bio-Rad, Hercules, CA, USA). Quantitative real-time PCR was applied to 1 µL of the obtained cDNA using the Quantifast® SYBR® Green PCR Master Mix (204054, Qiagen, Hilden, Germany) with 10 µM of each of the reverse and forward primers. Cycling conditions were as follows: 95 °C for 10 min for one cycle, then 95 °C for 10 s followed by 60 °C for 30 s and 72 °C for 30 s for 40 cycles, and, finally, 72 °C for 5 min for one cycle. The primers used are listed in Table 2.

Table 2. Different primer sequences used in RT-qPCR.

Gene	Forward Primer (5′–3′)	Reverse Primer (5′–3′)
mSOD2	GGCCAAGGGAGATGTTACAA	GAACCTTGGACTCCCACA
mCAT	TGAGAAGCCTAAGAACGCAATTC	CCCTTCGCAGCCATGTG
mNrf2	CGAGATATACGCAGGAGAGGTAAGA	GCTCGACAATGTTCTCCAGCTT
mβ-actin	CAGCTGAGAGGGAAATCGTG	CGTTGCCAATAGTGATGACC

2.12. Immunofluorescence Staining

The animals allocated for immunofluorescent staining underwent perfusion right after terminal tissue collection. Prior to surgery, the mice were anesthetized via IP injection of ketamine and xylazine solution (50 mg/kg and 15 mg/kg, respectively). A lateral incision was made through the integument and abdominal wall beneath the rib cage. The diaphragm was then incised, and the pleural cavity was exposed. The sternum was then lifted and any tissue connecting it to the heart was cut. A perfusion needle was introduced into the left ventricle without reaching the aorta, and an incision was made to the animal's right atrium. PBS was firstly perfused into the animals until the fluids were running clear and the liver was clear as well. When this occurred, PBS was switched with PFA until fixation tremors were observed. Brain tissue was placed in PFA for 48 h before being transferred into 30% sucrose solution for further preservation.

In preparation for immunofluorescence staining, brain tissue underwent sectioning using a microtome and was stored as free-floating sections in sodium azide. Free-floating brain sections (40 µm thick) were washed with PBS followed by PBST (0.1% Triton in PBS) and then incubated for 1.5 h in a blocking solution of 10% heat-inactivated fetal bovine serum (FBS) in PBST. Tissues were then incubated overnight at 4 °C with primary antibodies, diluted in 1% FBS solution. The primary antibodies used were anti-GFAP (1:1000 dilution; MCA-5C10, Encor Biotechnology, Gainesville, FL, USA) as an astrocyte marker; anti-NeuN (1:1000 dilution; RPCA-FOX3-AP, Encor Biotechnology) as a marker of mature neurons; anti-Iba-1 (1:1000 dilution; 019-19471, Fujifilm Wako Chemicals, Richmond, WV, USA) as a microglia marker; anti-MBP (1:1000 dilution; MCA-7D2, Encor Biotechnology) as an axonal marker. Then, sections were rinsed in PBST and incubated with the appropriate fluorochrome-conjugated secondary antibody (1:1000 dilution) for 1 h at room temperature, followed by three washes in PBST. The secondaries used were Donkey Anti-Mouse IgG H&L Alexa Fluor® 488 (ab150105, Abcam, MA, USA) and Goat Anti-Rabbit IgG H&L (Alexa Fluor® 568) (ab175696, Abcam, MA, USA). Finally, all sections were counterstained in 1 µg/mL of Hoechst (Sigma Aldrich), diluted in PBS, and mounted using Fluoromount (F4680-25ML, Sigma Aldrich). Microscopic imaging was conducted using a Zeiss LSM 710 confocal microscope. Images were acquired as tile scans with 40× oil objectives and analyzed using the Zeiss ZEN 2009 image analysis software and NIH ImageJ program. Images for the different experimental interventions were acquired under the same laser and microscopic parameters for the purpose of consistency. At least three sections per mouse were used for quantification with different levels to cover the hippocampus and cortex areas. For the hippocampus, quantification was conducted along the CA3 and the dentate gyrus (DG). For the cortex, the entire area around the injury in the somatosensory cortex was captured via tile scanning (Figure 1B). The fields from all three sections were averaged for every animal.

2.13. Statistical Analysis

Statistical analysis was conducted using GraphPad Prism 9.0 (La Jolla, CA, USA). The Kruskal–Wallis test was performed to compare the datasets that contained more than two groups. Then, Dunn's test was used to compare the three groups simultaneously. All results were considered significant at a p-value of < 0.05: *** ($p < 0.001$), ** ($p < 0.01$), * ($p < 0.05$).

3. Results

3.1. MitoQ Improves General Neurological Function by Ameliorating Motor Coordination, Muscle Strength, and Sensorimotor Function

We investigated the general neurological function of TBI mice using the modified Garcia neurological score test. The experiment was first conducted 3 days before the injury to make sure that all animals ($n = 10$ per group) had a similar basal score and no prior neurological dysfunction was present (Figure 2A). As expected, all animals had a perfect score, showing that they were eligible to be included in the study. The test was then repeated at 3, 7, and 30 days post-injury (DPI). On day 3 post-injury, the difference between the CCI + MitoQ and CCI groups was not significant. However, at 7 and 30 days post-injury, the CCI + MitoQ group performed significantly better compared to the CCI group ($p < 0.05$). This first step showed us that MitoQ improved overall neurological performance in mice. However, to be able to further assess this, additional testing was performed, from 21 to 23 DPI, to evaluate specific sensorimotor deficits and observe overall motor coordination.

To start with, the adhesive removal test was performed to assess sensorimotor function by observing sensitivity and dexterity. The CCI + MitoQ group performed significantly better than the CCI group by establishing first contact with the adhesive faster ($p < 0.05$) (Figure 2B,C). This showed improved sensitivity. Additionally, the CCI + MitoQ group required less time to remove the adhesive ($p < 0.05$), indicating better dexterity.

Then, motor function was investigated by exploring motor coordination via the pole climbing test, and muscle strength using the grip strength test. The CCI + MitoQ group was able to perform better by taking less time to descend the pole ($p < 0.05$) (Figure 2E) and revealed increased muscle strength ($p < 0.001$) (Figure 2D) compared to the CCI group. Therefore, MitoQ administration helped improve gross motor function. It is noteworthy here that the data show no significant difference between the CCI and Sham groups in the pole climbing test due to increased variability.

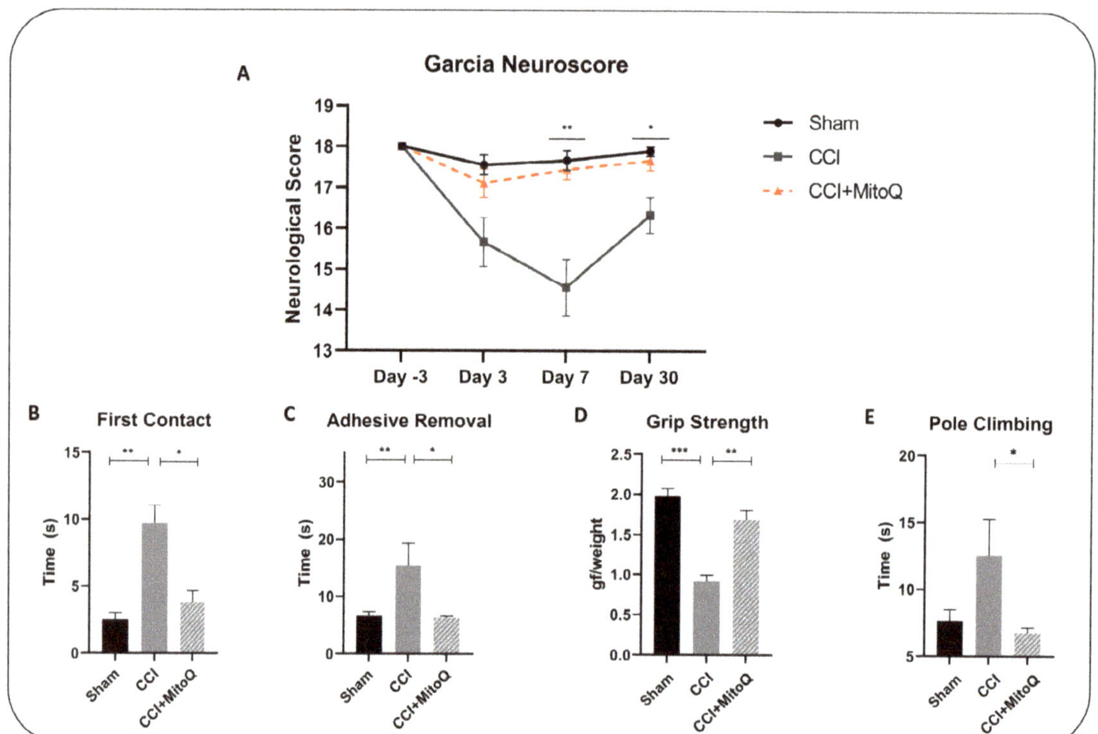

Figure 2. MitoQ ameliorates neurological deficits post-CCI. (**A**) Garcia neuroscore allowed testing for neurological function in mice and showed that MitoQ improved overall neurological performance 7 and 30 days post-CCI. Further testing was conducted to assess sensorimotor function via the adhesive removal test. Mice that received MitoQ took less time to establish contact (**B**) and were able to remove the adhesive faster (**C**) than the CCI group. Additionally, gross motor function was investigated by two tests. (**D**) The grip strength test showed that MitoQ improved overall muscle strength, and the pole climbing test (**E**) showed that CCI + MitoQ had better motor coordination. All tests were executed on Sham ($n = 10$), CCI ($n = 10$), and CCI + MitoQ ($n = 10$). * ($p < 0.05$); ** ($p < 0.01$); *** ($p < 0.001$).

3.2. Learning Deficits and Recognition Memory Dysfunction Are Alleviated by MitoQ

MitoQ's ability to potentially alleviate cognitive deficits that result from CCI was subsequently evaluated using the Morris water maze (MWM) test. No differences in latency to the platform were found among groups on day 1 of the test, which was performed 25 DPI. During the acquisition days of the test, CCI mice took more time to reach the hidden platform as compared to the Sham and CCI + MitoQ groups. This delay was significant on days 2 ($p < 0.05$), 3 ($p < 0.001$), and 4 ($p < 0.05$) of the acquisition days (Figure 3A). The data indicate that CCI results in a cognitive learning deficit reversed upon MitoQ administration. On the probe trial day of the test, there were differences in the latency to reach the target quadrant by the CCI group, indicating that retention deficits are caused by CCI in the model (Figure 3B,C). However, this was shown to be attenuated by MitoQ since the CCI + MitoQ group required less time to reach the target quadrant ($p < 0.05$) and spent more time in it ($p < 0.05$) (Figure 3B,C).

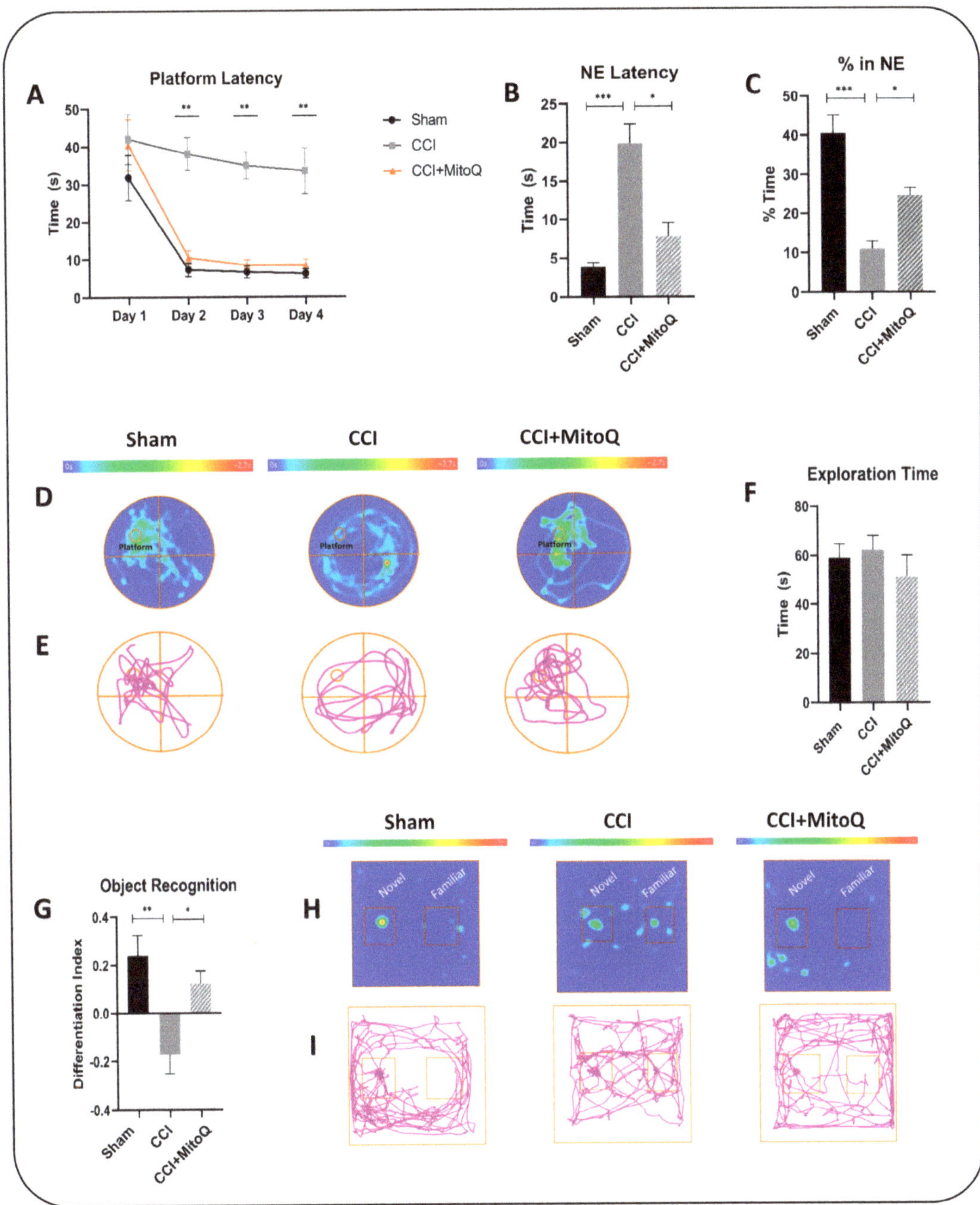

Figure 3. MitoQ improves special learning and recognition memory. Spatial learning and memory in mice from the three groups were evaluated using the Morris water maze. (**A**) shows the time taken for the first arrival at the platform by mice from the three study groups on the acquisition days where the mice learned how to reach the platform based on visual cues. (**B**) shows the time to reach the target quadrant (NE) on the probe trial day. (**C**) shows the percentage of time spent in the target quadrant. (**D**,**E**) show the representative heat maps and trace plots on the probe trial day, respectively. (**H**,**I**) show the representative heat maps and trace plots on the testing day of the object recognition

test. For further investigation of memory acquisition, the novel object recognition test was performed. (**F**) Exploration time for all three groups was recorded, and (**G**) the discrimination index was calculated, showing that recognition memory was enhanced in the CCI + MitoQ group. Data represent the mean ± SEM (n = 10 per group). * ($p < 0.05$), ** ($p < 0.01$), *** ($p < 0.001$).

The novel object recognition (OR) test was conducted for further learning assessment and recognition memory testing. On the testing day, after carrying out habituation and training for the mice in the platform, the total exploration time for all three groups showed no significant difference. This ensured that all animals had an equal chance of exploring the platform and thus ensured the viability of the test (Figure 3F). During the testing trials, the CCI group exhibited no differentiation between familiar and novel objects ($p < 0.01$). This was not the case in the CCI + MitoQ group, which spent more time exploring the novel object, resulting in a positive differentiation index ($p < 0.05$) (Figure 3G).

3.3. MitoQ Decreases Depressive-like Behavior

To understand the effects of MitoQ on depressive-like behavior, the forced swim test allowed quantifying hopelessness in all study groups by measuring the immobility time. The CCI-only group showed an increased immobility time when compared to Sham ($p < 0.001$). Of interest, the CCI + MitoQ group demonstrated a decrease in the immobility time compared to the CCI group ($p < 0.05$) (Figure 4). Therefore, MitoQ administration helped in reducing depressive-like behavior in mice.

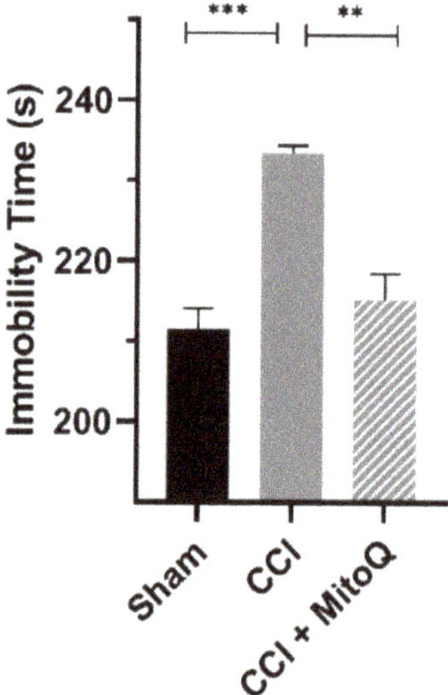

Figure 4. Depressive-like behavior is reduced by MitoQ. Animals were placed in a water container during the forced swim test to measure the immobility time as a sign of hopelessness. The CCI + MitoQ group performed better than the CCI group, with a decreased immobility time. Data represent the mean ± SEM (n = 10 per group). ** ($p < 0.01$), *** ($p < 0.001$).

3.4. MitoQ Upregulates Antioxidative Enzymes

The expression level of Nrf2 and downstream antioxidant enzymes, particularly CAT and SOD2, was quantified by RT-qPCR to draw similarities between previous studies and make sure MitoQ acted on the same axis in this experiment. After collecting ipsilateral cortices 30 DPI, the experiments were performed on total mRNA obtained from all three groups (n = 4). For all three genes of interest, MitoQ treatment resulted in a significant improvement in expression as compared to CCI ($p < 0.05$) (Figure 5). This means that MitoQ increased the expression of antioxidative enzymes via the Nrf2–ARE axis.

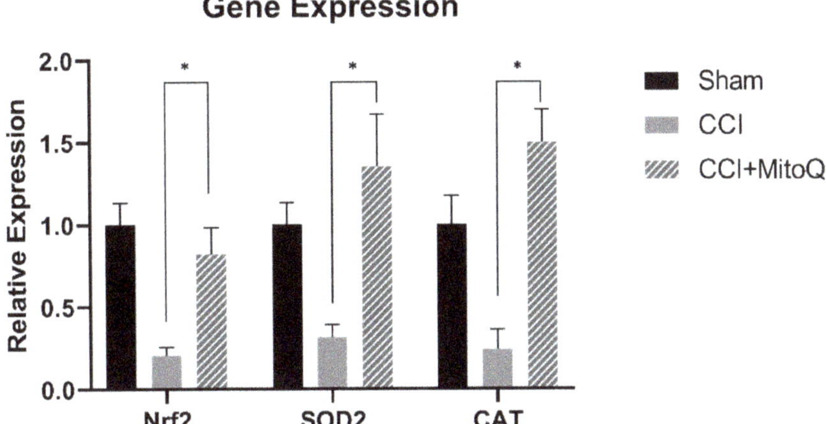

Figure 5. MitoQ improves expression of Nrf2, SOD2, and CAT in the cortex post-CCI. The mRNA levels of Nrf2, CAT, and SOD2 were determined using RT-qPCR in Sham ($n = 4$), CCI ($n = 4$), and CCI + MitoQ ($n = 4$). Data were normalized to β-actin. * ($p < 0.05$).

3.5. MitoQ Diminishes Chronic Activation of CNS Inflammatory Cells, Neuronal Death, and Axonal Damage

We attempted to assess if the decreased state of oxidative stress could be translated into ameliorated cellular recovery post-CCI. For this reason, immunofluorescence staining was performed to examine the activation of glial cells, axonal damage, and neuronal loss ($n = 4$ for every group, three fields were quantified for every n and averaged). Ionized calcium-binding adaptor protein-1 (Iba-1), an actin-binding protein, was used to stain activated microglia using immunofluorescence. The results show that MitoQ administration significantly attenuated the number of activated microglia both in the ipsilateral cortex ($p < 0.05$) and hippocampus ($p < 0.05$) compared to that of the CCI group (Figure 6A,B). Quantifying the number of microglia in both brain regions was not enough without examining the morphology of these cells. The CCI group exhibited amoeboid and elongated microglia with thick extensions, representing an activated phenotype [27] (Figure 6C). This, however, was not the case in the Sham and CCI + MitoQ groups, where microglia retained their ramified shape with extended ramification, indicative of the resting surveillant phenotype. Taken together, these data show that MitoQ reduces microgliosis following CCI, which is indicative of decreased chronic inflammation. Further investigation was conducted by assessing astrogliosis using the astrocytic activation marker GFAP.

Glial fibrillary acidic protein (GFAP) is an intermediate filament protein used as a marker of reactive astrocytes. The intensity of its expression was used as a marker of active astrocytes. GFAP expression was significantly higher in the ipsilateral hippocampus and cortex of the CCI group compared to that of the Sham group ($p < 0.05$). It was noticeable that activated astrocytes congregated around the injury site and were absent in the contralateral side of the cortex of the injured group (Figure 7A). Such a result conforms with the focal nature of the injury. Upon quantification, our data show that this state of increased GFAP expression was significantly improved by MitoQ administration in the cortex ($p < 0.05$) and the hippocampus ($p < 0.05$) (Figure 7B,C). Likewise, the morphology of activated astrocytes was noticeable through Z-stack imaging of the dentate gyrus (DG), where they exhibited abundant arborization exclusively in the CCI group and retained an inactive morphology in the CCI + MitoQ group (Figure 7D).

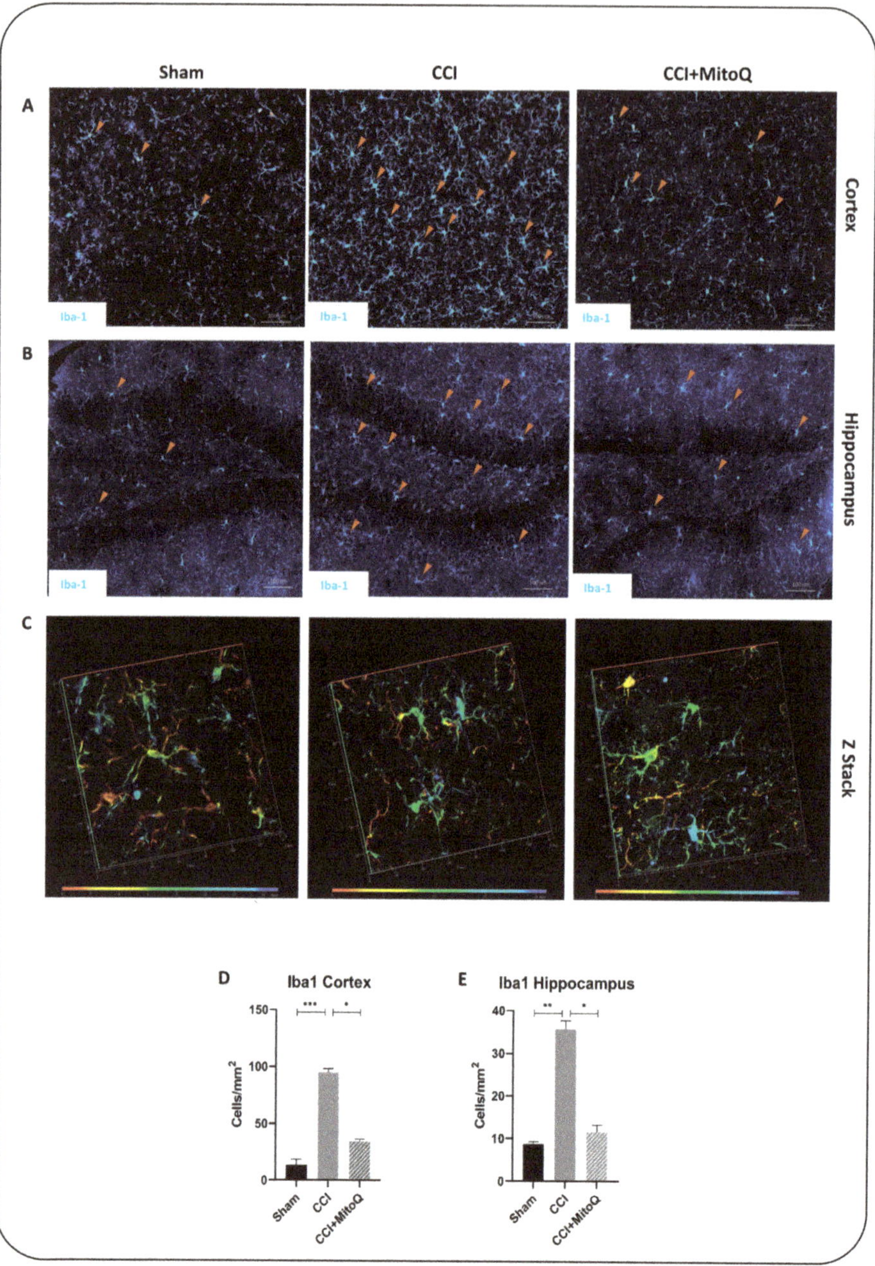

Figure 6. MitoQ decreases microglial activation in the cortex and the hippocampus. Microgliosis was assessed by immunofluorescence staining for Iba-1 (blue/green). (**A**) shows Iba-1-positive cells (arrowheads) across the three experimental groups in the ipsilateral cortex. (**B**) shows the same as (**A**), but in the ipsilateral hippocampus. Scale bar = 100 μm. (**C**) exhibits a depth-coded Z-stack of individual microglial cells to showcase different morphologies across the three groups. (**D**,**E**) are bar graphs showing quantification of Iba-1-positive cells per mm^2. Iba-1-positive cells were counted, and then the obtained number was divided by the area of the selected field in an average of three fields per animal, four animals per group. * ($p < 0.05$); ** ($p < 0.01$); *** ($p < 0.001$).

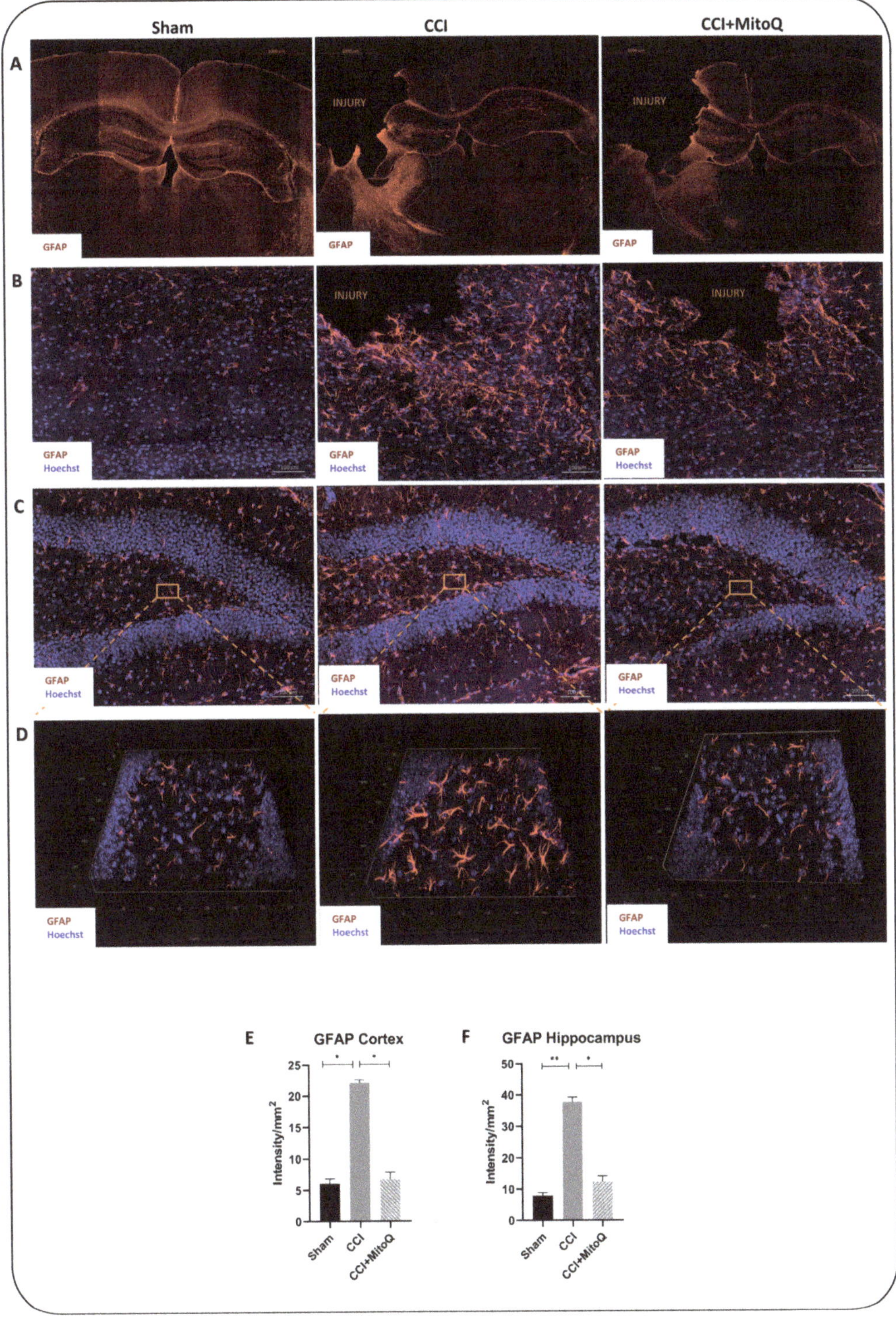

Figure 7. MitoQ ameliorates astrocytosis in the cortex and hippocampus post-CCI. Astrocytosis was

assessed by immunofluorescent staining of GFAP (red). (**A**) shows the assessment of GFAP positivity in the entire brain section of all three groups. (**B**) shows GFAP-positive cells in the cortex and (**C**) in the hippocampus. Hoechst was used as a nuclear counterstain. (**C**) represents Z-stacks of sections in the hippocampal DG from all three groups to show different astrocytic morphologies. Scale bar = 100 µm. (**D**) shows a Z-stack of GFAP-positive cells in the hippocampus across the three groups. (**E,F**) are bar graphs showing the quantification of GFAP intensity per mm^2. GFAP mean fluorescence intensities (MFI) of 3 fields per animal and 4 animals per condition were measured using the NIH ImageJ 1.41 software. Bar graphs display averages $\pm$ SEMs of GFAP MFI/mm^2. * ($p < 0.05$); ** ($p < 0.01$).

Next, we investigated the neuronal cell count to assess cell death. Immunofluorescence staining for NeuN, a nuclear protein found in the nuclei and perinuclear cytoplasm of neurons, was used to assess the number of mature neurons in a semi-quantitative manner [28]. Our results show a significant decrease in the neuronal cell count in CCI as compared to Sham in the cortex and the hippocampus ($p < 0.05$) (Figure 8), with significant improvement in the cortex ($p < 0.05$), but not in the hippocampus, in the CCI + MitoQ group.

To measure axonal integrity, which has been shown to correlate with cognitive and behavioral deficits [29], we investigated myelin basic protein (MBP). Used as a marker for myelination, MBP is responsible for the adhesion of myelin to the cytosolic surfaces of cells [30]. Prominent axonal damage was observed in the cortex of the CCI group, with characteristic axonal shearing and blebbing (Figure 9B). However, in the CCI + MitoQ group, the axons exhibited a more refined integrity similar to that shown in the Sham group. Quantification of the MBP intensity in the cortex showed a significant decrease in the CCI group compared to the Sham group. This was significantly increased upon MitoQ administration (Figure 9D,E). Unlike the cortex, the hippocampus did not exhibit any differences in the integrity of MBP between all three experimental groups, retaining the focal nature of the injury.

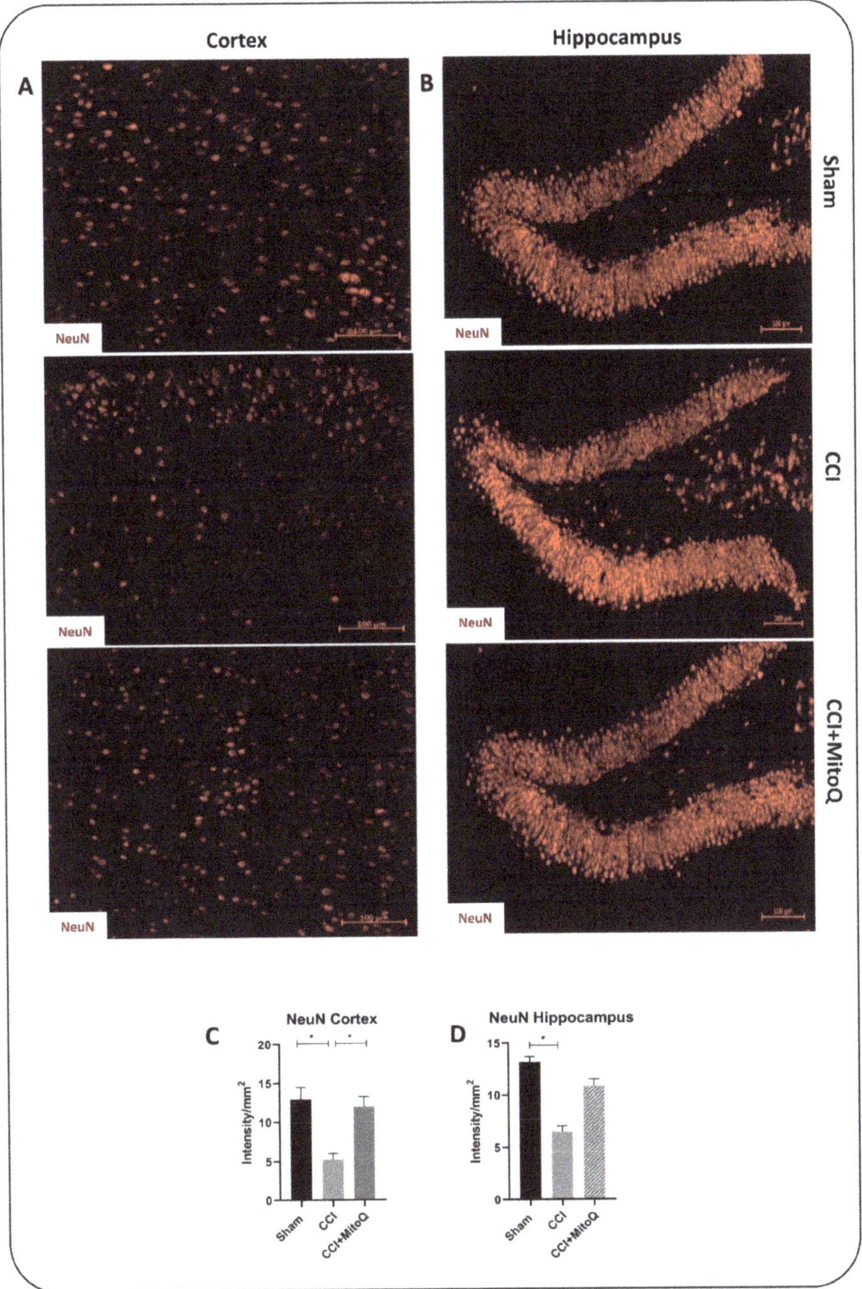

Figure 8. MitoQ decreases neuronal cell loss in the cortex. NeuN staining (red) was performed to assess the number of mature neurons. (**A**) shows representative fluorescent images of NeuN staining in the cortex from the different study groups. NeuN-positive cells were counted, and then the number was divided by the area of the field. (**B**) shows the same as (**A**), but in the hippocampus. GFAP mean fluorescence intensities (MFIs) of 3 fields per animal and 4 animals per condition were measured using the NIH ImageJ software. Scale bar = 100 μm. (**C,D**) are bar graphs displaying average MFIs $\pm$ SEMs of NeuN per mm². * ($p < 0.05$).

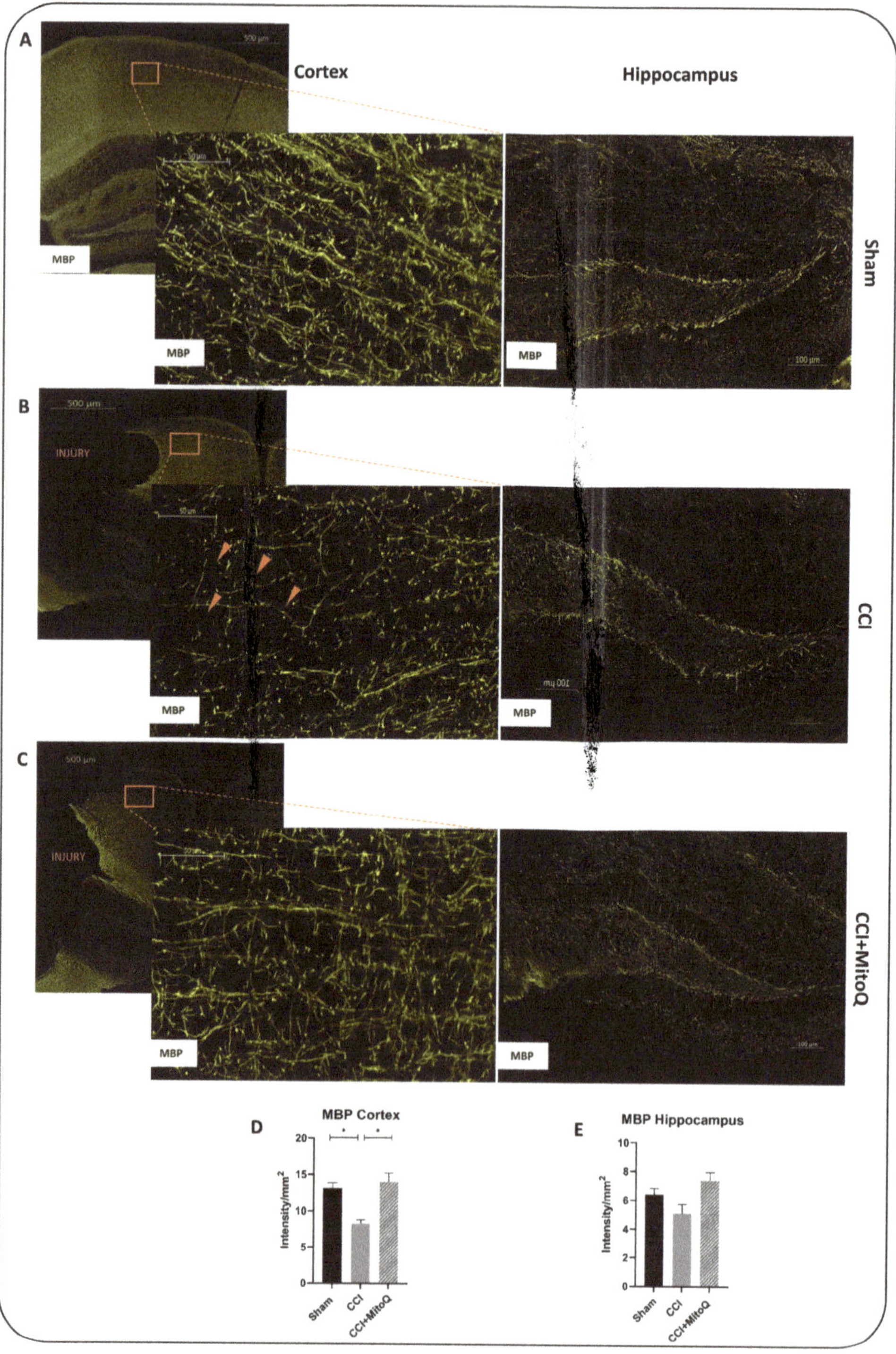

Figure 9. MitoQ decreases axonal shearing in the cortex after CCI. Representative fluorescent images of MBP staining (green) that were used to assess axonal morphology in the cortex and hippocampus for Sham, CCI, and CCI + MitoQ in (**A–C**), respectively. Major axonal blebbing and

shearing are highlighted with arrowheads. Scale bars = 500, 100, 50 μm. (**D,E**) are bar graphs showing the quantification of the MBP intensity per mm^2 in the cortex and hippocampus. MBP mean fluorescence intensities (MFIs) of 3 fields per animal and 4 animals per condition were measured using the NIH ImageJ software. (**D,E**) are bar graphs displaying averages ± SEMs of MBP MFI/mm^2. * ($p < 0.05$).

4. Discussion

TBI involves a plethora of pathological pathways that contribute to its heterogenic nature, making it extremely difficult to understand and treat. This explains the lack of FDA-approved drugs against TBI, exacerbating the need to investigate potential therapeutic approaches. For that purpose, this study was conducted to assess the potential protective effects of MitoQ in a CCI mouse model. Previously, MitoQ was demonstrated to ameliorate pathological effects of TBI by decreasing neuronal apoptosis, improving neurobehavior, and enhancing antioxidative capacity in the cortices of mice subjected to mild TBI (mTBI) [22]. Other studies exploring the neuroprotective properties of MitoQ in animal models of Huntington's disease, Alzheimer's disease (AD), and anterolateral sclerosis (ALS) also showed it can improve memory, enhance motor function, and decrease amyloid-beta (Aβ) accumulation [31–33]. In compliance with this, results from our laboratory further support the claim that MitoQ can decrease neurological and cognitive deficits in mTBI at a chronic time point (paper in publication).

In our study, MitoQ was supplemented at a concentration of 8 mg/kg, adapted from Zhou et al., who investigated the effects of MitoQ in experimental traumatic brain injury for the first time in vivo [22]. During the study, different doses of MitoQ were administered to compare dose-dependent effects. The injured mice that received MitoQ at 8 mg/kg exhibited improved neurological severity scores compared to the vehicle-treated group, as well as decreased edema. There was a lack of any improvement in both parameters when the mice received a dose of 2 mg/kg. This study showed that administering MitoQ at 8 mg/kg was safe and revealed neuroprotective effects in a mouse model of weight drop TBI. Similarly, in another study by Xiao et al., MitoQ was administered in mice at a concentration of 5 mg/kg IP to investigate its effects through the Nrf2 pathway [34]. This concentration yielded significant results, where MitoQ was able to provide protective effects in tubular injury via mitophagy.

Any concerns on the side effects of MitoQ, especially its administration for long periods (30 days in our case), have also been addressed in previous studies. There is a comprehensive body of literature showing this in both clinical and pre-clinical trials. Gane et al. showed that MitoQ helped decrease liver damage in hepatitis C patients with no adverse side effects over 28 days [21]. More recently, Rossmann et al. studied the effect of MitoQ on vascular function in adults. The study was conducted over six weeks and showed no significant adverse side effects following administration [35]. Similarly, Rodriguez-Cuenca et al. showed that long-term administration of MitoQ on wild-type mice for up to 28 weeks was safe and did not act as a pro-oxidant or cause damage over this period [36].

Having said that, the first step in our study was to observe the effects of MitoQ on a macroscale and thus investigate its potential in improving cognitive and neurological functions post-CCI. To start with, general neurological testing was carried out via the modified Garcia neurological score test. This allowed scoring the animals on a scale from 0 to 18 to assess general sensorimotor function, motor skill, and coordination-based functions [24]. In this study, MitoQ was shown to improve overall neurological integrity by resulting in better scores at 7 and 30 DPI, which represented the subacute and chronic phases of injury. With these results, we decided to investigate which aspects of the neurological functions were affected by CCI and which were improved by MitoQ.

The adhesive removal test showed that the current CCI model induced sensorimotor deficits in mice, where they took longer to establish contact with and remove tape that was placed on their noses. However, MitoQ was able to shorten this time, showing that it improved sensorimotor function at a chronic time point. To be more specific, this test provided us with the insight that MitoQ improved both dexterity and sensitivity. On the other hand, during the pole climbing test and grip strength test, the current CCI model displayed decreased muscle strength and adversely affected motor coordination, both of which were improved by MitoQ. Although there was a significant difference between the CCI and CCI + MitoQ groups in the pole climbing test, there was no significance between the CCI and Sham groups. This could either be due to high variability in the test scores or the injury model not being sufficient to induce motor coordination dysfunction. In either case, further investigation is warranted. In addition to this, MitoQ demonstrated a positive impact on depressive-like behavior during the forced swim test, revealing that it helped with the notion of hopelessness. This is especially important as TBI has been linked to many neuropsychiatric sequelae in the long run [37,38].

Moreover, CCI resulted in learning deficits and impairment of recognition memory, as seen in the MWM and OR tests. Firstly, during the former, the CCI group displayed latency to reach the platform on the learning days and a tendency towards worsened memory retention, as seen by the latency to reach the target quadrant and the time spent in it on the probe trial day. During the latter, the CCI group showed no differentiation between the familiar and novel objects, revealing recognition memory dysfunction. These findings align with other studies describing increased latency to reach the hidden platform in mice subjected to TBI [39–41]. However, when animals received MitoQ for 30 days, they exhibited improved learning and memory retention. Their ability to recognize and differentiate novel from familiar objects was also significantly improved, showing enhanced recognition memory. Therefore, we were able to demonstrate two things in this part of the study: (a) the animal model used was sufficient to induce cognitive, neurological, and behavioral dysfunctions, in compliance with the previous literature [42–45]; and (b) MitoQ administration showed exciting potential in improving cognitive dysfunctions in the mice at a chronic time point.

Having seen these data on the macroscale, it was important to assess how MitoQ could affect different pathways involved in TBI pathology at a molecular level. We needed to investigate this in the cortex since it retained the injury, and the hippocampus since it is involved in emotions, memory, and other functions impaired in TBI and the CCI model here [46–48]. The observed alterations in cognitive functions translate to damaged neuronal structures in the cortex and hippocampus [44,49]. Due to synapse disruption and other pathological effects such as chronic inflammation, oxidative stress, and glutamate altercations, circuit alterations have been linked to memory impairment in TBI and other neurological deficits [50–52]. We were able to see in the previous data that MitoQ treatment following CCI resulted in enhanced learning ability and recognition memory, pointing at its potential effects on oxidative stress, chronic inflammation, and neuronal circuitry disruption. This is especially true when referring to a previous study by Zhou et al., which showed MitoQ's potential in decreasing neuronal apoptosis and enhancing Nrf2 nuclear localization to induce antioxidative effects [22]. For this reason, we performed RT-qPCR analysis to observe how effective MitoQ was, in the current animal model, in acting on the previously described Nrf2-ARE pathway [22]. In our experiment, MitoQ significantly increased the expression of the Nrf2 transcription factor and its downstream antioxidative enzymes SOD2 and CAT. This showed us that it worked on the antioxidative pathway involving Nrf2 and thus exerted its antioxidative effect as expected. After examining the effect of MitoQ on antioxidative enzymes, it was important to investigate other parameters that might be related to improved cognition and neurological function. This was conducted via immunofluorescence.

Previous studies have shown that TBI was followed by microglia and astrocyte activation, leading to several neurological and cognitive dysfunctions [53–57]. This is accompanied by neuronal cell loss via activation of the cell death machinery [58–60]. The described CCI model exhibited results that are in line with these previous studies. There was an increased expression of GFAP and Iba-1, showing the presence of chronic activation of astrocytes and microglia, indicative of an active inflammatory state [54,57]. The injury also led to a decreased neuronal count, as examined using the neuronal marker NeuN. This pathology in the cortex and the hippocampus is reflected in the depressive-like state exhibited by the mice, and by the disrupted neurological function and cognition. It must also be highlighted that prominent axonal damage and shearing were observed in the cortex with decreased MBP reactivity due to the breakdown of the protein, indicating disrupted circuitry. However, MitoQ was able to help decrease these effects, where astrocytes and microglia showed dampened activation in the cortico-hippocampal system, and there was reduced neuronal cell loss in the cortex. This decreased activation was apparent in the morphology of both glial cells in the ipsilateral cortices and hippocampi of the mice that received the drug. The used CCI model induced a prominent presence of activated intermediate microglia denoted as bushy microglia, characteristic of their activated morphology [27].

The mitigated neuroinflammatory profile may be related to the antioxidant activity of MitoQ that enhances oxidative enzyme production and therefore prevents excessive production of ROS. Enhanced ROS production is known to directly stimulate the expression of the pro-inflammatory interleukins IL-1α and IL-1β within hours after injury [6,7]. Secreted pro-inflammatory cytokines subsequently contribute to the activation of Ca^{2+}-dependent proteolytic enzymes that activate the cell death machinery and induce a pro-inflammatory state [61]. However, these protective effects could also pertain to unexplored pathways that require further investigation. Therefore, this cycle of oxidative stress, neuroinflammation, and neuronal cell death holds potential for many questions that need to be addressed in MitoQ administration. The proposed mechanisms of MitoQ's effect are summarized in Figure 10.

There are a number of limitations that should be addressed. In this work, only male mice were used, which ignores the female hormonal system that has been shown to have effects in TBI [62]. Additionally, between the animal groups, there was no vehicle-only group because MitoQ was injected in PBS, and previous data from our laboratory showed no difference between the vehicle and Sham groups.

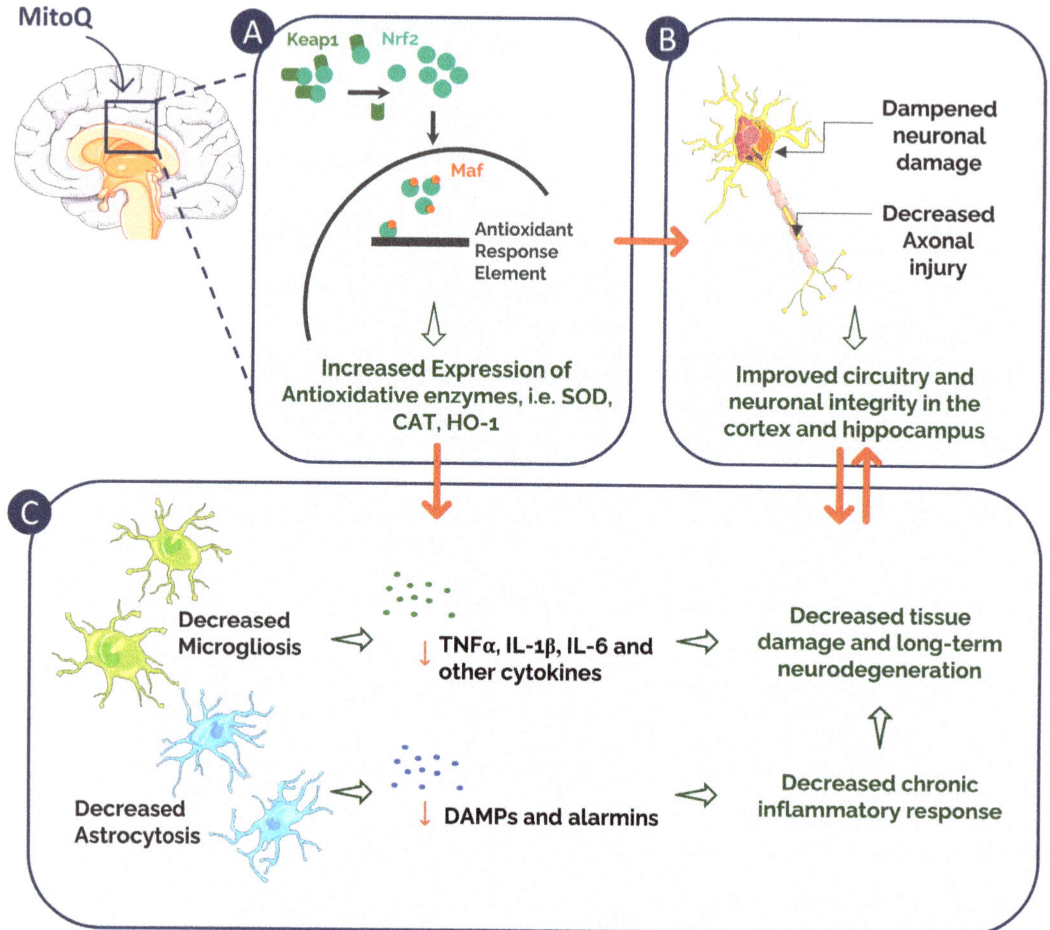

Figure 10. Summary of proposed MitoQ neuroprotective activity post-CCI. When MitoQ is administered, it exerts its effects via several pathways that eventually translate into improved neurological and cognitive functions. (**A**) MitoQ helps in increasing Nr2 expression and translocation into the nucleus by affecting its sequestration by the Kelch-like ECH-associated protein 1 (Keap1) protein. This allows Nrf2 to bind Maf proteins and interact with its antioxidant response element (ARE), leading to increased expression of antioxidative enzymes such as SOD2 and CAT. (**B**) Neurons are prone to increased apoptosis and axonal shearing due to the hostile environment created by the primary and secondary injuries. MitoQ helps decrease neuronal loss and maintains axonal integrity, resulting in improved circuitry in the brain. These effects translate into enhanced cognitive and neurological functions. (**C**) MitoQ helps decrease the activation of both microglia and astrocytes and therefore might help in decreasing chemical signals produced by both cells such as tumor necrosis factor α (TBFα), interleukin 1β (IL-1β), interleukin 6 (IL-6), cytokines, damage-associated molecular proteins (DAMPs), and alarmins. During a chronic inflammatory state, these would lead to damage to brain tissue and neurodegeneration.

5. Conclusions

This study shows that MitoQ administration appears to efficiently reduce the deleterious consequences associated with moderate open head TBI at both the molecular and behavioral/cognitive levels. This paves the way towards identifying a potential therapy for TBI that may improve the quality of life for those affected by it. Moreover, the obtained data provide an additional line of evidence on the potential role of oxidative stress and chronic inflammation in the pathology of brain disorders by mechanisms that need to be unraveled. This also opens more questions on the effects of this supplement on the mitochondrial complexes and neuronal regeneration. This emphasizes two important next steps that need to be addressed. The first is exploring further pathways implicated in the activity of MitoQ in moderate TBI. The second is conducting further assessment of biomarkers of neurodegenerative disorders to check the extent of MitoQ's protective effects at chronic time points.

Author Contributions: Conceptualization, F.K., K.W. and R.E.K.; methodology, M.A.H., Z.S. and C.B.; validation, R.E.K., K.W. and Y.M.; formal analysis, M.A.H. and Z.S.; investigation, M.A.H.; writing—original draft preparation, M.A.H., Z.S., M.T. and S.M.; writing—review and editing, M.G., D.S. and J.D.P.; supervision, F.K. and H.D.; project administration, H.D.; funding acquisition, M.T. and H.D. All authors have read and agreed to the published version of the manuscript.

Funding: This work is partially funded from the American University of Beirut (AUB); Medical Practice Plan (MPP) grant PI: Riyad ElKhoury and Co-PI: Firas Kobeissy.

Institutional Review Board Statement: This study was carried out at the Animal Care Facility of the American University of Beirut (AUB), and all animal experiments were performed in compliance with the AUB Institutional Animal Care and Use Committee (IACUC) guidelines, with the reference number: (17-01-458).

Informed Consent Statement: Not applicable.

Data Availability Statement: The datasets used and/or analyzed during the current study are available from the corresponding author on reasonable request.

Acknowledgments: We would like to thank Yara Yehya and Leila Nasrallah for their technical contribution to the study.

Conflicts of Interest: The authors declare no conflict of interest.

References

1. Dewan, M.C.; Rattani, A.; Gupta, S.; Baticulon, R.E.; Hung, Y.C.; Punchak, M.; Agrawal, A.; Adeleye, A.O.; Shrime, M.G.; Rubiano, A.M.; et al. Estimating the global incidence of traumatic brain injury. *J. Neurosurg.* **2018**, *130*, 1080–1097. [CrossRef] [PubMed]

2. El-Menyar, A.; Mekkodathil, A.; Al-Thani, H.; Consunji, R.; Latifi, R. Incidence, Demographics, and Outcome of Traumatic Brain Injury in The Middle East: A Systematic Review. *World Neurosurg.* **2017**, *107*, 6–21. [CrossRef] [PubMed]

3. Aarabi, B.; Tofighi, B.; Kufera, J.A.; Hadley, J.; Ahn, E.S.; Cooper, C.; Malik, J.M.; Naff, N.J.; Chang, L.; Radley, M.; et al. Predictors of outcome in civilian gunshot wounds to the head. *J. Neurosurg.* **2014**, *120*, 1138–1146. [CrossRef] [PubMed]

4. Schimmel, S.J.; Acosta, S.; Lozano, D. Neuroinflammation in traumatic brain injury: A chronic response to an acute injury. *Brain Circ.* **2017**, *3*, 135–142. [CrossRef]

5. Readnower, R.D.; Chavko, M.; Adeeb, S.; Conroy, M.D.; Pauly, J.R.; McCarron, R.M.; Sullivan, P.G. Increase in blood-brain barrier permeability, oxidative stress, and activated microglia in a rat model of blast-induced traumatic brain injury. *J. Neurosci. Res.* **2010**, *88*, 3530–3539. [CrossRef]

6. Abdul-Muneer, P.M.; Chandra, N.; Haorah, J. Interactions of oxidative stress and neurovascular inflammation in the pathogenesis of traumatic brain injury. *Mol. Neurobiol.* **2015**, *51*, 966–979. [CrossRef]

7. Dalgard, C.L.; Cole, J.T.; Kean, W.S.; Lucky, J.J.; Sukumar, G.; McMullen, D.C.; Pollard, H.B.; Watson, W.D. The cytokine temporal profile in rat cortex after controlled cortical impact. *Front. Mol. Neurosci.* **2012**, *5*, 6. [CrossRef]

8. Masel, B.E.; DeWitt, D.S. Traumatic brain injury: A disease process, not an event. *J. Neurotrauma* **2010**, *27*, 1529–1540. [CrossRef]

9. Fischer, T.D.; Hylin, M.J.; Zhao, J.; Moore, A.N.; Waxham, M.N.; Dash, P.K. Altered Mitochondrial Dynamics and TBI Pathophysiology. *Front. Syst. Neurosci.* **2016**, *10*, 29. [CrossRef]

10. Torbic, H.; Forni, A.A.; Anger, K.E.; Degrado, J.R.; Greenwood, B.C. Use of antiepileptics for seizure prophylaxis after traumatic brain injury. *Am. J. Health-Syst. Pharm. AJHP Off. J. Am. Soc. Health-Syst. Pharm.* **2013**, *70*, 759–766. [CrossRef]

11. Greenberg, M. *Handbook of Neurosurgery*, 6th ed.; Thieme Medical Publishers: New York, NY, USA, 2006; pp. 289–365.

12. Galgano, M.; Toshkezi, G.; Qiu, X.; Russell, T.; Chin, L.; Zhao, L.R. Traumatic Brain Injury: Current Treatment Strategies and Future Endeavors. *Cell Transplant.* **2017**, *26*, 1118–1130. [CrossRef] [PubMed]

13. Xiong, Y.; Mahmood, A.; Chopp, M. Emerging treatments for traumatic brain injury. *Expert Opin. Emerg. Drugs* **2009**, *14*, 67–84. [CrossRef] [PubMed]

14. Vergouwen, M.D.; Vermeulen, M.; Roos, Y.B. Effect of nimodipine on outcome in patients with traumatic subarachnoid haemorrhage: A systematic review. *Lancet Neurol.* **2006**, *5*, 1029–1032. [CrossRef]

15. Deng-Bryant, Y.; Singh, I.N.; Carrico, K.M.; Hall, E.D. Neuroprotective effects of tempol, a catalytic scavenger of peroxynitrite-derived free radicals, in a mouse traumatic brain injury model. *J. Cereb. Blood Flow Metab. Off. J. Int. Soc. Cereb. Blood Flow Metab.* **2008**, *28*, 1114–1126. [CrossRef]

16. Muizelaar, J.P.; Marmarou, A.; Young, H.F.; Choi, S.C.; Wolf, A.; Schneider, R.L.; Kontos, H.A. Improving the outcome of severe head injury with the oxygen radical scavenger polyethylene glycol-conjugated superoxide dismutase: A phase II trial. *J. Neurosurg.* **1993**, *78*, 375–382. [CrossRef]

17. Murphy, M.P.; Smith, R.A. Targeting antioxidants to mitochondria by conjugation to lipophilic cations. *Annu. Rev. Pharmacol. Toxicol.* **2007**, *47*, 629–656. [CrossRef]

18. Smith, R.A.J.; Porteous, C.M.; Gane, A.M.; Murphy, M.P. Delivery of bioactive molecules to mitochondria in vivo. *Proc. Natl. Acad. Sci. USA* **2003**, *100*, 5407–5412. [CrossRef]

19. Maroz, A.; Anderson, R.F.; Smith, R.A.; Murphy, M.P. Reactivity of ubiquinone and ubiquinol with superoxide and the hydroperoxyl radical: Implications for in vivo antioxidant activity. *Free. Radic. Biol. Med.* **2009**, *46*, 105–109. [CrossRef]

20. Smith, R.A.; Murphy, M.P. Animal and human studies with the mitochondria-targeted antioxidant MitoQ. *Ann. N. Y. Acad. Sci.* **2010**, *1201*, 96–103. [CrossRef]

21. Gane, E.J.; Weilert, F.; Orr, D.W.; Keogh, G.F.; Gibson, M.; Lockhart, M.M.; Frampton, C.M.; Taylor, K.M.; Smith, R.A.; Murphy, M.P. The mitochondria-targeted anti-oxidant mitoquinone decreases liver damage in a phase II study of hepatitis C patients. *Liver Int. Off. J. Int. Assoc. Study Liver* **2010**, *30*, 1019–1026. [CrossRef]

22. Zhou, J.; Wang, H.; Shen, R.; Fang, J.; Yang, Y.; Dai, W.; Zhu, Y.; Zhou, M. Mitochondrial-targeted antioxidant MitoQ provides neuroprotection and reduces neuronal apoptosis in experimental traumatic brain injury possibly via the Nrf2-ARE pathway. *Am. J. Transl. Res.* **2018**, *10*, 1887–1899. [PubMed]

23. Sunkin, S.M.; Ng, L.; Lau, C.; Dolbeare, T.; Gilbert, T.L.; Thompson, C.L.; Hawrylycz, M.; Dang, C. Allen Brain Atlas: An integrated spatio-temporal portal for exploring the central nervous system. *Nucleic Acids Res.* **2013**, *41*, D996–D1008. [CrossRef]

24. Garcia, J.H.; Wagner, S.; Liu, K.F.; Hu, X.J. Neurological deficit and extent of neuronal necrosis attributable to middle cerebral artery occlusion in rats. Statistical validation. *Stroke* **1995**, *26*, 627–634. [CrossRef]

25. Smith, J.P.; Hicks, P.S.; Ortiz, L.R.; Martinez, M.J.; Mandler, R.N. Quantitative measurement of muscle strength in the mouse. *J. Neurosci. Methods* **1995**, *62*, 15–19. [CrossRef]

26. Matsuura, K.; Kabuto, H.; Makino, H.; Ogawa, N. Pole test is a useful method for evaluating the mouse movement disorder caused by striatal dopamine depletion. *J. Neurosci. Methods* **1997**, *73*, 45–48. [CrossRef]

27. Fernández-Arjona, M.d.M.; Grondona, J.M.; Granados-Durán, P.; Fernández-Llebrez, P.; López-Ávalos, M.D. Microglia Morphological Categorization in a Rat Model of Neuroinflammation by Hierarchical Cluster and Principal Components Analysis. *Front. Cell. Neurosci.* **2017**, *11*, 235. [CrossRef]

28. Gusel'nikova, V.V.; Korzhevskiy, D.E. NeuN As a Neuronal Nuclear Antigen and Neuron Differentiation Marker. *Acta Nat.* **2015**, *7*, 42–47. [CrossRef]

29. Liu, D.; Wang, Z.; Shu, H.; Zhang, Z. Disrupted white matter integrity is associated with cognitive deficits in patients with amnestic mild cognitive impairment: An atlas-based study. *SAGE Open Med.* **2016**, *4*, 2050312116648812. [CrossRef]

30. Boggs, J.M. Myelin basic protein: A multifunctional protein. *Cell. Mol. Life Sci. CMLS* **2006**, *63*, 1945–1961. [CrossRef]

31. McManus, M.J.; Murphy, M.P.; Franklin, J.L. The mitochondria-targeted antioxidant MitoQ prevents loss of spatial memory retention and early neuropathology in a transgenic mouse model of Alzheimer's disease. *J. Neurosci. Off. J. Soc. Neurosci.* **2011**, *31*, 15703–15715. [CrossRef]

32. Miquel, E.; Cassina, A.; Martínez-Palma, L.; Souza, J.M.; Bolatto, C.; Rodríguez-Bottero, S.; Logan, A.; Smith, R.A.; Murphy, M.P.; Barbeito, L.; et al. Neuroprotective effects of the mitochondria-targeted antioxidant MitoQ in a model of inherited amyotrophic lateral sclerosis. *Free. Radic. Biol. Med.* **2014**, *70*, 204–213. [CrossRef] [PubMed]

33. Pinho, B.R.; Duarte, A.I.; Canas, P.M.; Moreira, P.I.; Murphy, M.P.; Oliveira, J.M.A. The interplay between redox signalling and proteostasis in neurodegeneration: In vivo effects of a mitochondria-targeted antioxidant in Huntington's disease mice. *Free. Radic. Biol. Med.* **2020**, *146*, 372–382. [CrossRef] [PubMed]

34. Xiao, L.; Xu, X.; Zhang, F.; Wang, M.; Xu, Y.; Tang, D.; Wang, J.; Qin, Y.; Liu, Y.; Tang, C.; et al. The mitochondria-targeted antioxidant MitoQ ameliorated tubular injury mediated by mitophagy in diabetic kidney disease via Nrf2/PINK1. *Redox Biol.* **2017**, *11*, 297–311. [CrossRef] [PubMed]

35. Rossman, M.J.; Santos-Parker, J.R.; Steward, C.A.C.; Bispham, N.Z.; Cuevas, L.M.; Rosenberg, H.L.; Woodward, K.A.; Chonchol, M.; Gioscia-Ryan, R.A.; Murphy, M.P.; et al. Chronic Supplementation With a Mitochondrial Antioxidant (MitoQ) Improves Vascular Function in Healthy Older Adults. *Hypertension* **2018**, *71*, 1056–1063. [CrossRef]

36. Rodriguez-Cuenca, S.; Cochemé, H.M.; Logan, A.; Abakumova, I.; Prime, T.A.; Rose, C.; Vidal-Puig, A.; Smith, A.C.; Rubinsztein, D.C.; Fearnley, I.M.; et al. Consequences of long-term oral administration of the mitochondria-targeted antioxidant MitoQ to wild-type mice. *Free. Radic. Biol. Med.* **2010**, *48*, 161–172. [CrossRef]

37. Deb, S.; Lyons, I.; Koutzoukis, C.; Ali, I.; McCarthy, G. Rate of psychiatric illness 1 year after traumatic brain injury. *Am. J. Psychiatry* **1999**, *156*, 374–378. [CrossRef]

38. Rogers, J.M.; Read, C.A. Psychiatric comorbidity following traumatic brain injury. *Brain Inj.* **2007**, *21*, 1321–1333. [CrossRef]

39. Meehan, W.P., 3rd; Zhang, J.; Mannix, R.; Whalen, M.J. Increasing recovery time between injuries improves cognitive outcome after repetitive mild concussive brain injuries in mice. *Neurosurgery* **2012**, *71*, 885–891. [CrossRef]

40. DeFord, S.M.; Wilson, M.S.; Rice, A.C.; Clausen, T.; Rice, L.K.; Barabnova, A.; Bullock, R.; Hamm, R.J. Repeated mild brain injuries result in cognitive impairment in B6C3F1 mice. *J. Neurotrauma* **2002**, *19*, 427–438. [CrossRef]

41. Levine, B.; Kovacevic, N.; Nica, E.I.; Cheung, G.; Gao, F.; Schwartz, M.L.; Black, S.E. The Toronto traumatic brain injury study: Injury severity and quantified MRI. *Neurology* **2008**, *70*, 771–778. [CrossRef]

42. Fox, G.B.; Fan, L.; Levasseur, R.A.; Faden, A.I. Sustained sensory/motor and cognitive deficits with neuronal apoptosis following controlled cortical impact brain injury in the mouse. *J. Neurotrauma* **1998**, *15*, 599–614. [CrossRef] [PubMed]

43. Osier, N.D.; Dixon, C.E. The Controlled Cortical Impact Model: Applications, Considerations for Researchers, and Future Directions. *Front. Neurol.* **2016**, *7*, 134. [CrossRef] [PubMed]

44. Luo, J.; Nguyen, A.; Villeda, S.; Zhang, H.; Ding, Z.; Lindsey, D.; Bieri, G.; Castellano, J.; Beaupre, G.; Wyss-Coray, T. Long-Term Cognitive Impairments and Pathological Alterations in a Mouse Model of Repetitive Mild Traumatic Brain Injury. *Front. Neurol.* **2014**, *5*, 12. [CrossRef] [PubMed]

45. Kinder, H.A.; Baker, E.W.; Howerth, E.W.; Duberstein, K.J.; West, F.D. Controlled Cortical Impact Leads to Cognitive and Motor Function Deficits that Correspond to Cellular Pathology in a Piglet Traumatic Brain Injury Model. *J. Neurotrauma* **2019**, *36*, 2810–2826. [CrossRef]

46. Atkins, C.M. Decoding hippocampal signaling deficits after traumatic brain injury. *Transl. Stroke Res.* **2011**, *2*, 546–555. [CrossRef]

47. Monti, J.M.; Voss, M.W.; Pence, A.; McAuley, E.; Kramer, A.F.; Cohen, N.J. History of mild traumatic brain injury is associated with deficits in relational memory, reduced hippocampal volume, and less neural activity later in life. *Front. Aging Neurosci.* **2013**, *5*, 41. [CrossRef]

48. Morrow, E.L.; Dulas, M.R.; Cohen, N.J.; Duff, M.C. Relational Memory at Short and Long Delays in Individuals with Moderate-Severe Traumatic Brain Injury. *Front. Hum. Neurosci.* **2020**, *14*, 270. [CrossRef]

49. Akamatsu, Y.; Hanafy, K.A. Cell Death and Recovery in Traumatic Brain Injury. *Neurother. J. Am. Soc. Exp. NeuroTherapeutics* **2020**, *17*, 446–456. [CrossRef]

50. Girgis, F.; Pace, J.; Sweet, J.; Miller, J.P. Hippocampal Neurophysiologic Changes after Mild Traumatic Brain Injury and Potential Neuromodulation Treatment Approaches. *Front. Syst. Neurosci.* **2016**, *10*, 8. [CrossRef]

51. Simon, D.W.; McGeachy, M.J.; Bayır, H.; Clark, R.S.B.; Loane, D.J.; Kochanek, P.M. The far-reaching scope of neuroinflammation after traumatic brain injury. *Nat. Rev. Neurol.* **2017**, *13*, 171–191. [CrossRef]

52. Huang, T.-T.; Leu, D.; Zou, Y. Oxidative stress and redox regulation on hippocampal-dependent cognitive functions. *Arch Biochem. Biophys.* **2015**, *576*, 2–7. [CrossRef]

53. Neary, J.T.; Kang, Y.; Tran, M.; Feld, J. Traumatic injury activates protein kinase B/Akt in cultured astrocytes: Role of extracellular ATP and P2 purinergic receptors. *J. Neurotrauma* **2005**, *22*, 491–500. [CrossRef]

54. Burda, J.E.; Bernstein, A.M.; Sofroniew, M.V. Astrocyte roles in traumatic brain injury. *Exp. Neurol.* **2016**, *275 Pt 3*, 305–315. [CrossRef]

55. Fluiter, K.; Opperhuizen, A.L.; Morgan, B.P.; Baas, F.; Ramaglia, V. Inhibition of the membrane attack complex of the complement system reduces secondary neuroaxonal loss and promotes neurologic recovery after traumatic brain injury in mice. *J. Immunol.* **2014**, *192*, 2339–2348. [CrossRef]

56. Morganti, J.M.; Riparip, L.-K.; Rosi, S. Call Off the Dog(ma): M1/M2 Polarization is Concurrent following Traumatic Brain Injury. *PLoS ONE* **2016**, *11*, e0148001. [CrossRef]

57. Loane, D.J.; Kumar, A. Microglia in the TBI brain: The good, the bad, and the dysregulated. *Exp. Neurol.* **2016**, *275*, 316–327. [CrossRef]

58. Clark, R.S.; Chen, J.; Watkins, S.C.; Kochanek, P.M.; Chen, M.; Stetler, R.A.; Loeffert, J.E.; Graham, S.H. Apoptosis-suppressor gene bcl-2 expression after traumatic brain injury in rats. *J. Neurosci. Off. J. Soc. Neurosci.* **1997**, *17*, 9172–9182. [CrossRef]

59. Stoica, B.A.; Faden, A.I. Cell death mechanisms and modulation in traumatic brain injury. *Neurother. J. Am. Soc. Exp. NeuroThera-peutics* **2010**, *7*, 3–12. [CrossRef]

60. Kaya, S.S.; Mahmood, A.; Li, Y.; Yavuz, E.; Göksel, M.; Chopp, M. Apoptosis and expression of p53 response proteins and cyclin D1 after cortical impact in rat brain. *Brain Res.* **1999**, *818*, 23–33. [CrossRef]

61. Buki, A.; Povlishock, J.T. All roads lead to disconnection?—Traumatic axonal injury revisited. *Acta Neurochir.* **2006**, *148*, 181–193. [CrossRef]

62. Wright, D.K.; O'Brien, T.J.; Shultz, S.R.; Mychasiuk, R. Sex matters: Repetitive mild traumatic brain injury in adolescent rats. *Ann. Clin. Transl. Neurol.* **2017**, *4*, 640–654. [CrossRef] [PubMed]

Review

Neuroprotection: Targeting Multiple Pathways by Naturally Occurring Phytochemicals

Andleeb Khan [1,*], Sadaf Jahan [2], Zuha Imtiyaz [3], Saeed Alshahrani [1], Hafiz Antar Makeen [4], Bader Mohammed Alshehri [2], Ajay Kumar [5], Azher Arafah [6] and Muneeb U. Rehman [6]

[1] Department of Pharmacology and Toxicology, College of Pharmacy, Jazan University, Jazan 45142, Saudi Arabia

[2] Medical Laboratories Department, College of Applied Medical Sciences, Majmaah University, Majmaah 15341, Saudi Arabia

[3] Clinical Drug Development, College of Pharmacy, Taipei Medical University, Taipei 11031, Taiwan

[4] Department of Clinical Pharmacy, College of Pharmacy, Jazan University, Jazan 45142, Saudi Arabia

[5] Institute of Nano Science and Technology, Habitat Centre, Phase-10, Sector-64, Mohali 160062, India

[6] Department of Clinical Pharmacy, College of Pharmacy, King Saud University, Riyadh 11451, Saudi Arabia

* Correspondence: drandleebkhan@gmail.com or ankhan@jazanu.edu.sa

Received: 15 June 2020; Accepted: 5 August 2020; Published: 12 August 2020

Abstract: With the increase in the expectancy of the life span of humans, neurodegenerative diseases (NDs) have imposed a considerable burden on the family, society, and nation. In defiance of the breakthroughs in the knowledge of the pathogenesis and underlying mechanisms of various NDs, very little success has been achieved in developing effective therapies. This review draws a bead on the availability of the nutraceuticals to date for various NDs (Alzheimer's disease, Parkinson's disease, Amyotrophic lateral sclerosis, Huntington's disease, vascular cognitive impairment, Prion disease, Spinocerebellar ataxia, Spinal muscular atrophy, Frontotemporal dementia, and Pick's disease) focusing on their various mechanisms of action in various in vivo and in vitro models of NDs. This review is distinctive in its compilation to critically review preclinical and clinical studies of the maximum phytochemicals in amelioration and prevention of almost all kinds of neurodegenerative diseases and address their possible mechanism of action. PubMed, Embase, and Cochrane Library searches were used for preclinical studies, while ClinicalTrials.gov and PubMed were searched for clinical updates. The results from preclinical studies demonstrate the efficacious effects of the phytochemicals in various NDs while clinical reports showing mixed results with promise for phytochemical use as an adjunct to the conventional treatment in various NDs. These studies together suggest that phytochemicals can significantly act upon different mechanisms of disease such as oxidative stress, inflammation, apoptotic pathways, and gene regulation. However, further clinical studies are needed that should include the appropriate biomarkers of NDs and the effect of phytochemicals on them as well as targeting the appropriate population.

Keywords: neurodegenerative diseases; phytochemicals; natural products; neuroprotection

---

1. Introduction

Various conditions affecting nerve cells and the nervous system due to the loss of neurons and their connecting networks are described under the superordinate phrase "Neurodegenerative diseases". They lead to disability due to gradual neuronal death in both the central nervous system (CNS) and the peripheral nervous system (PNS). Several diseases are genetic with few coming being caused due to the exposure to various toxins and chemicals. The main symptoms associated with these disorders are related to movements (ataxia) or mental functioning (dementia) or both causing morbidity and death. This fostered the profound social and economic implications [1]. The most common NDs include cognitive and behavioral disorders, Alzheimer's disease (AD), Parkinson's disease (PD), Amyotrophic lateral sclerosis (ALS), Huntington's disease (HD), Spinocerebellar ataxia (SCA), Spinal muscular atrophy (SMA), Vascular cognitive impairment Prion disease, Frontotemporal disease, Pick's disease, etc. The treatment available for these disorders gives only symptomatic relief to the patient by extending the lifespan to a few years [2]. Still, a lot of research is in progress to find the therapeutic markers in these diseases [3].

There has been an extensive approach to non-pharmacological therapies to combat the ill effects of NDs. Many of them include disease-modifying therapies, non-invasive brain stimulation techniques, physical exercise, adaptive physical activity, complementary and alternative medicine, and several nutraceutical compounds. In a meta-analysis, it was reported that repetitive transcranial magnetic stimulation (rTMS) therapy with Parkinson's patients resulted in mild to moderate improvement of motor activities [4]. In another study, it was reported that both rTMS and transcranial direct current stimulation (tDCS) have shown improvement in the cognitive performance of Alzheimer's patients [5]. Fisicaro and colleagues stated in their review that rTMS can be effectively used as a non-pharmacological tool for various motor and non-motor neurorehabilitation. They suggested that this technique along with other conventional rehabilitative modalities can be effectively used in clinics but still the regime is not clear [6]. Physical exercise has been a promising therapy for motor deficits for years. It could have therapeutic potential for the prevention of mental disorders and neurodegeneration. Extensive research to explain different molecular pathways involved in this therapy is under progress, suggesting multiple pathways may be involved together in neuroprotection [7,8]. The major cornerstones in managing ataxic patients are physiotherapy and kinesitherapy in the present scenario. Studies confirm that customized programs for various exercises of coordination, balance, cognitive skills, and posture maintenance have shown extensive improvement along with the conventional pharmacological therapeutics [9]. Complementary and alternative medicines have always attracted various disease prevention including NDs [10,11]. In addition to these, the various nutraceuticals [12] and phytochemicals multi-target approaches have proved to be a promising adjunct to the present treatment [13].

The large number of pharmacological or biological activities of the phytochemicals have made them appropriate candidates for the treatment of the NDs [14]. They exhibit antioxidant (resveratrol, zingerone), anti-inflammatory property (cineole, thymoquinone), inhibitors of gamma-aminobutyric acid ($GABA_A$) receptors (diterpenes and cyclodepsipeptides), and Monoamine oxidase-B (MAO-B) receptors (selegiline rasagiline), although not proven clinically. Combating disease symptoms by phytochemicals and herbal nutraceutical is known in traditional medicines, and several new research for preventing NDs by phytochemicals are under progress [15]. In this review, our focal point is to discuss the recent advancements in the field of neuroprotection related to the NDs by various phytochemicals and to elicit their potential mechanism of actions. In particular, we discuss phytochemicals used in the prevention of all kinds of NDs and related signaling pathways and mechanism of recovery highlighting the role of phytochemicals. The basic structures of most the common neuroprotective phytochemicals are given in Figure 1.

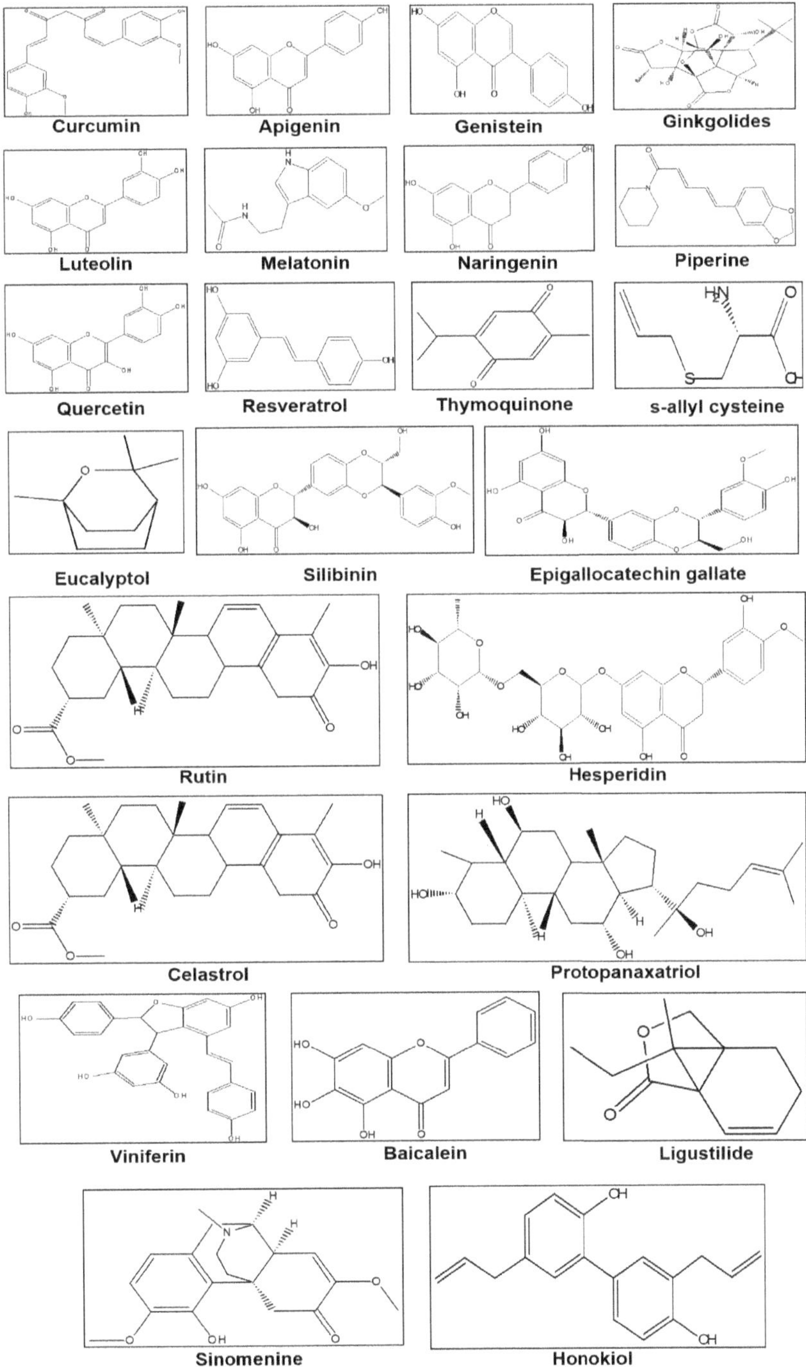

Figure 1. Chemical structures of some common phytochemicals used in the prevention of neurodegenerative diseases (NDs).

2. Methodology

A Medline (PubMed), Cochrane Library and Embase based literature survey was performed by using keywords of "neurodegenerative diseases, prevention, non-pharmacological therapies, and neurodegenerative diseases, phytochemicals, Alzheimer's disease natural products, Parkinson's disease (PD) phytochemicals, Huntington's disease phytoprotection, natural product amyotrophic lateral sclerosis, phytochemicals cerebral ischemia, phytochemicals prion disease, spinocerebellar ataxia prevention by natural products, natural product and spinal muscular atrophy, frontotemporal dementia natural products, pick's disease, and natural products" till June 2020.

Two independent authors (A.K. and S.J.) screened all the titles and abstracts of the retrieved data, disagreements were resolved by the consensus of a third author (Z.I.). Duplicated entries, retracted publications, studies on other diseases or conditions different from NDs or its subtypes, studies without statistical analysis, non-English written papers, publications that are not research studies (i.e., commentaries, letters, editorials, reviews, meta-analysis), and any other article that did not fit within the scope of this review, were excluded. Articles listed in the references were also reviewed in search of more data.

A total of 873 results were retrieved and screened with the above keywords. Out of these 115 publications were selected and eventually used for qualitative analysis (Figure 2) and the same are summarized in Table 1. In more detail, 28 studies work with AD, 17 with PD, 7 with ALS, 8 with HD, 20 with VCI and stroke, 4 with prion, 7 with FTD, 2 with SMA, and 1 with SCA and others on non-pharmacological therapy and neurodegenerative diseases.

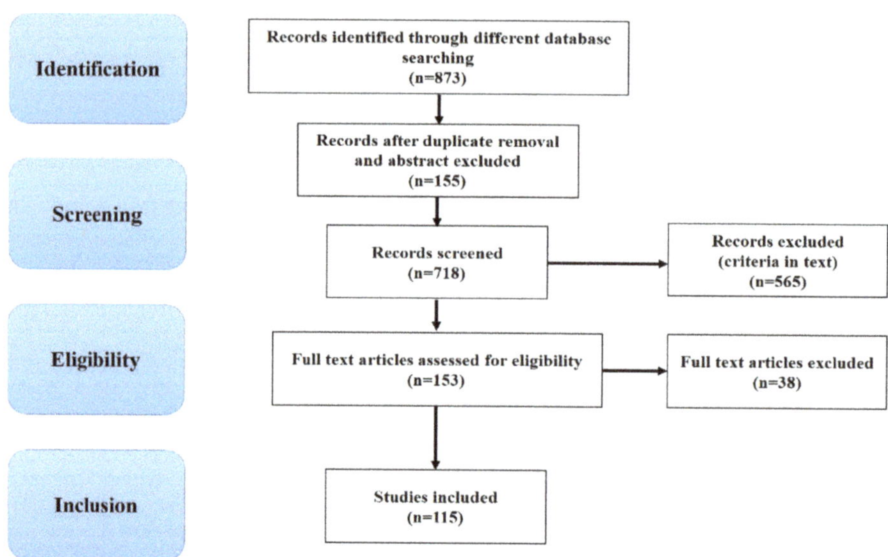

Figure 2. Flow diagram showing the search strategy, the number of records identified, and the included/excluded studies [16].

3. Multifunctional Phytochemicals as Novel Therapeutic Agents for Neurodegenerative Disorders

3.1. Alzheimer's Disease and Related Dementias

AD is a multifactorial, pernicious disease with genetic and environmental factors leading to irreversible cognitive impairment [17]. The main neuropathological features of AD comprises of neuronal loss, amyloid-β plaques, and neurofibrillary tangles in limbic and neocortical regions [18].

Various mechanisms are associated with the progression of the disorder which includes amyloid deposition, tau hyperphosphorylation and aggregation, loss of cholinergic system, oxidative stress, inflammation, apoptosis, glutamatergic excitotoxicity, and decrease in neurogenesis and neurotrophic factors [17,19,20]. In recent years, approved drugs for AD are only modulators of cholinergic and glutamatergic systems, which can only delay the symptoms to some extent [21]. Divergent antioxidant and anti-inflammatory medications are also approved for AD patients [22]. There are recent reports that herbal products and nutraceuticals can act as a multi-target approach as suggested in the traditional system of medicines [23]. Recently extensive research has been conducted which supports their potential in in vitro, in vivo, and clinical studies. In the list of natural compounds, curcumin is considered on the top because of its miraculous qualities. In this section, the mechanistic approach of the curcumin and other important natural compounds will be discussed. Curcumin is a polyphenolic compound and a natural herb which is an integrative part of every Indian kitchen. In literature, many beneficial properties are reported like anti-inflammatory, anti-cancerous, anti-apoptotic, antioxidative, neuroprotective [24]. Apart from that curcumin is also reported as a therapeutic agent to treat rheumatoid disorders, cough, and neurological disorders. Fortunately, it has intensive effects and have multiple sites of action [25]. The relationship between curcumin consumption and a lower prevalence of AD has been observed. In one study, curcumin emulates the improvement of the spatial learning and memory in the rat model and it is reported as an inhibitor of BAC1 in in vitro study. Curcumin has a promising effect in neuroblastoma SHSY5Y cell lines and also has a therapeutic effect on the AD-induced mechanisms [26]. Acrolein exposed HT22 murine hippocampal neuronal cells were protected by curcumin by following the BDNF/TrkB signaling [27]. Orally administrated curcumin downregulates the level of GSK3β and also inhibit the hippocampus Aβ plaques with the improved cognitive impairment and improved maze test in in vivo study [28]. Curcumin effectively counteracted the p25-mediated glial activation and pro-inflammatory chemokines/cytokines production in p25Tg mice. Moreover, this curcumin-mediated suppression of neuro-inflammation reduced the progression of p25-induced tau/amyloid pathology and in turn ameliorated the p25-induced cognitive impairments [29]. Therefore, through the series of research, curcumin is found to be promising phytochemicals against Aβ accumulation and would be a determined remedy against AD. Pilot study with curcumin and *Ginkgo* on 34 possible or probable AD patients was carried out with curcumin and *Ginkgo* extract. It was found that no side effects of curcumin along with that mini-mental examination scores or plasma amyloid beta40 level between 0 and 6 months [30]. AD-related pathology in the brain is well documented, the disease has also been reported to affect the retina, a developmental outgrowth of the brain, which is more accessible for imaging. Early detection of AD, utilizing curcumin fluorescence imaging to highlight Aβ aggregates in the retina. Results show that retinal fluorescence imaging is justified with brain plaque burden from PET scan [31]. Apigenin, a known phytochemical also targets the AD by its antioxidative effects. In plants, Apigenin is found as aglycon and more often in glycoside form. Apigenin inhibits the copper-mediated β-amyloid neurotoxicity (copper-induced) via antioxidative mechanism followed by MAPK signal inactivation in an AD cell model [32]. The Apigenin follows the ERK/CREB/BDNF pathway for neuroprotection [33]. On the list of phytochemical-induced mechanistic pathways, genistin also has an important role in neuroprotection. It is a component of the soya bean meal. The effect of genistin (100 μM) was studied on the in vitro culture of rat neurons [34]. It was observed that the amyloid-beta peptide instigated the damage of mitochondrial membrane potential, nuclear DNA damage, high expression of *Bcl-2* associated X protein (BAX), p53, Caspases and lower expression of B-cell lymphoma-2 (Bcl-2) [35]. The 100 μM dose of genistin was found to be efficient against the deleterious effects of AD and restores the neuronal activity by following the caspase signaling pathways [36]. It is also reported to lower calcium overload and reverse the low fluidity of the neuronal membranes in the AD model. Free radicles were also reduced with the use of genistin and thus it is suggested for the targeting of anti-ROS (reactive oxygen species) pathway and protect the cells from death. Astrocytes are also studied for the efficiency of genistin [37]. Pretreatment of 5 μM concentration of genistin was used against astrocytes and it protects by working on interleukin 1 beta (IL-1β), tumor necrosis factor-α (TNF-α), cyclooxygenase-2 (COX-2), inducible nitric oxide synthase (iNOS) cascade, and reversing the effect of NDs in the AD model [38]. In another study, 50 μM concentration of genistin was effective and initiate IL-1β /TLR4/NF-κB/ IκB-α cascade against neuroinflammation induced by Aβ in the secondary culture of C6 [39]. It can downregulate TNF-α, IL-1β, TLR4, and NF-κB and promote the upregulation of IκB-α. Other phytochemicals are also having miraculous properties against many NDs specifically AD [40]. Genistein protects Aβ-induced neurotoxicity by inducing the PKC signaling pathway, which further regulates the activities of α- and β-secretase and thereby inhibits the formation and toxicity of Aβ [41]. Genistein also protected the hippocampal neurons against injury by calcium/calmodulin dependent protein kinase IV protein levels in AD model [42]. *Ginkgo biloba* is also considered as a neuroprotective agent. Ginkgolide A inhibits Aβ-induced neuronal degradation via JNK phosphorylation and NMDA/AMPA-induced depolarization, thus suppress the activity of plaque [43]. EGb 761 increases the neurogenesis in the adult hippocampus region and CREB phosphorylation in the transgenic mouse model of AD [44]. In one study, *G. biloba* demonstrated protection against a high dose of Bisphenol A (BPA) and played a role in the improvement of cognitive deficits [45]. A flavonoid, luteolin, would be discussed against the AD model as a possible therapeutic agent. In an in vivo study (rat model), it protects against the cognitive dysfunction which was induced by chronic cerebral hypoperfusion [46]. It is also found to be protective against high-fat diet-induced cognitive abnormalities. GSK 3α isoform is involved in the mechanistic signaling. The inactivation of GSK 3α isoform phosphorylates the PS1 which is the catalytic core of the γ-secretase complex causing a decrease in PS1-APP interaction and generation of Aβ. In literature, the role of luteolin against zinc-induced hyperphosphorylation of the protein τ in SH-SY5Y secondary cell culture is reported [47]. Luteolin is suggested as a promising candidate because of its antioxidative properties as well. In the AD animal model, luteolin slows down the escape latency and traveled distance parameters in the morris water maze while increasing the time spent in the target quadrant [48]. Melatonin modulates the proteins (i.e., GSTP1 and CPLX1) that are involved in depression and anxiety. The modulation improves the neuropsychiatric behaviors in the AD model [49]. Melatonin improves anxiety and depression-like behaviors and targets the proteomic changes in the triple transgenic AD mice model [50]. A multicenter, randomized, placebo-controlled clinical trial of 2 dose formulations of oral melatonin coordinated by the National Institute of Aging-funded Alzheimer's Disease Cooperative Study, and it was found that melatonin was effective against AD [51]. Naringenin protects A$\beta_{25\text{-}35}$-caused damage via ER and PI3K/Akt-Mediated Pathways [49,52]. Naringenin also protects AlCl$_3$/D-galactose-induced neurotoxicity in the rat model of AD via inhibition of attenuation of acetylcholinesterase levels and attenuation of oxidative stress [53]. A new study suggested the therapeutic effects of piperine against oxidative insult, neuroinflammation, and neurochemical changes induced by ICV-STZ infusion in mice [54]. In the human brain, microvascular endothelial cells from fibrillar β-amyloid are protected by quercetin [55]. The flavonoid quercetin improves Alzheimer's disease pathology and protects cognitive and emotional function in aged triple transgenic Alzheimer's disease model [56]. in vitro AD model, resveratrol works on oxidative damage via activating mitophagy [57]. Resveratrol is found to be active against the harmful process that occurs in AβPP/PS1 mouse hippocampus and inhibit memory loss [58]. For the AD model, resveratrol is studied in randomized, double-blind, placebo-controlled, phase II trial. Resveratrol was found to be safe, effective, and can protect the BBB integrity with CNS response [59]. Thymoquinone inhibits the Aβ-induced neurotoxicity by blocking the mitochondrial dysfunction and oxidative stress [60]. S-allyl cysteine also targets the oxidative stress-induced cognitive neuronal impairment and provide possible rescue from the diseased [61]. In mice model treated with D-galactose, S-allyl cysteine involves in the alleviation of β-Amyloid, oxidative damage [62]. Eucalyptol ameliorates inflammation induced by A$\beta_{(25\text{-}35)}$ in differentiated PC12 cells by following the NOS-2, COX-2, and NF-κB associated pathways [63]. Silibinin and silymarin administration could recover memory impairment and could decrease Aβ plaque in the brain of APP/PS1 mice [64]. By following the BDNF/TrkB pathway, Silibinin ameliorates anxiety/depression-like behaviors in amyloid β-treated rats [65]. From all the above discussion of the phytochemicals, it is imminent that researchers would be able to hit the specific targets for providing promising therapeutic relief against AD Figure 3. The details of several phytochemicals in AD protection are enlisted in Table 1.

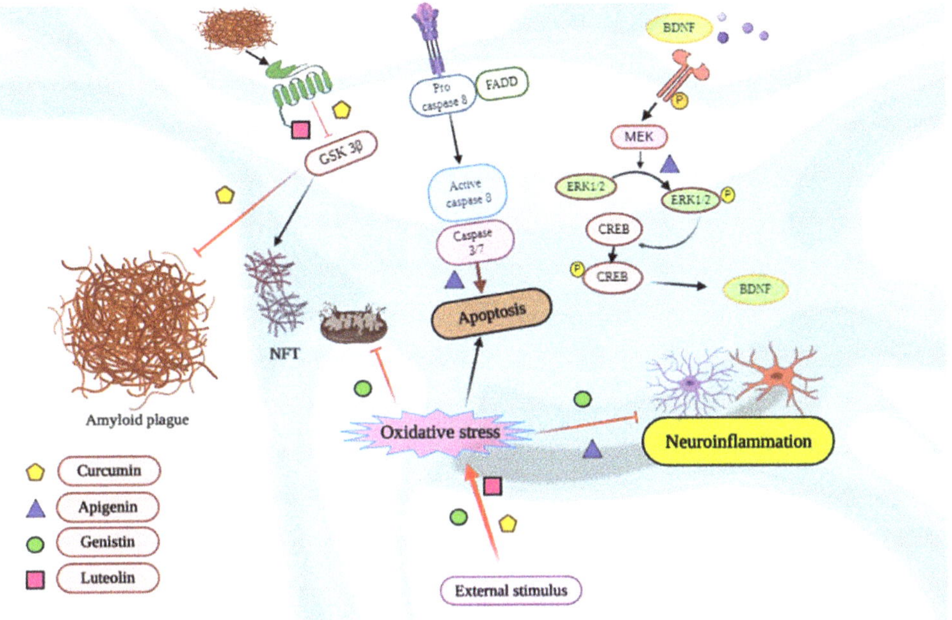

Figure 3. Schematic representation of the several mechanisms associated with Alzheimer's Disease (AD) and their possible prevention by some phytochemicals. The main hallmarks of AD; amyloid plague and NFT are prevented by curcumin and luteolin via GSK 3β pathway. Curcumin, genistin, and luteolin decreases the oxidative stress in the neuron. Neuroinflammation is also inhibited by genistin and apigenin. The pathway of apoptosis, activated in AD is blocked by the use of apigenin. Apigenin also activates BDNF/ERK/CREB pathway leading to neuronal survival.

Table 1. Role of various phytochemicals in different neurodegenerative diseases with possible mechanisms of action with special reference to in vitro, in vivo and clinical studies.

Disease	Plant	Natural Compound	In-Vitro/In-Vivo Models/Human Trial	Dose & Route	Mode of Action	References
Alzheimer's Disease	*Curcuma longa*	Curcumin	Acrolein exposed HT22 murine hippocampal neuronal cells	5 µg/mL/30 min	↓ AD-like pathologies, MDA level, ↑ levels of GSH, SOD, ↓ metalloprotease, APP, β-secretase, ↑BDNF/TrkB signaling	[27]
			p25 Transgenic Mouse model	4 g/kg for 12 weeks; po	↓ tau, amyloid accumulation, ↑ proinflammatory cytokines	[29]
			Pilot study with curcumin and ginkgo on 34 possible or probable AD patients	1 g/day, 4 g/day 120 mg/day ginkgo leaf extract for 6 months	No significant difference in Mini-Mental State Examination scores or plasma Aβ40 levels between 0 and 6 months. No side effect of curcumin	[30]
		Longvida	40 subjects: AD, MCI, healthy	20 g/day for 7 days	Retinal imaging of Aβ plagues differentiated between AD and non-AD subjects	[31]
	Many plants	Apigenin	Copper induced SHSY5Y cells	10.0 mM for 24 h	↑antioxidation, mitochondrion protection, ↑MAPK signaling	[32]
			APP/PS1 double transgenic mouse model	40 mg/kg for 5 day/week	↑behavior, ↓Aβ burden, ↑ERK/CREB/BDNF pathway, ↓oxidative stress	[33]
	Glycine max	Genistein	Rat hippocampal neuronal cells	0.4 µg/mL	↑ α-secretase, ↓ β-secretase, ↑PKC signaling pathway	[41]
			Bilateral hippocampal Aβ_{25-35} injected rat Model	90 mg/kg for 42 days	↓ p-tau, CALM, CAMKK1, p-CAMK4 protein, escape latency	[42]
	Ginkgo biloba	Ginkgolides A or B	Mouse primary cortical neurons	100 or 300 µM/1.5 h	↓ AMPA- and NMDA-induced depolarization, ↓JNK phosphorylation	[43]
			TgAPP/PS1 mice	100 mg/kg for 1 month	↑ cell proliferation in the hippocampus, ↓ Aβ oligomers and ↑ pCREB levels	[44]
	Many plants	Luteolin	ICV-STZ rat model	10 and 20 mg/kg	↑ escape latency, ↑ thickness of CA1 pyramidal layer	[48]
	Many plants	Melatonin	murine N2a neuroblastoma cells	10 µM	↑ cell viability, ↓ cell blebbing and retraction, ↓ LPO, intracellular Ca^{+2}	[49]
			3xTg-AD transgenic mice	10 mg/kg/day in drinking water for 1 month	↑ anxiety and depression like behaviors, ↑ GST and complexin-1	[50]
			157 AD patients	10 mg melatonin	↑ nocturnal total sleep time, ↓ wake after sleep	[51]
	Citrus junos	Naringenin	Aβ_{25-35} treated PC-12 cells	0.4 µM for 2 h	↑ cell viability, ↑ ER-mediated PI3K/Akt signaling pathway, ↓ caspase-3	[52]
			AlCl$_3$+D-gal rat model	50 mg/kg for 2 weeks po	↑ behavioral parameters, ↑ antioxidant enzymes, ↓ LPO, AChE, ↑ ACh levels, 5-HT levels and DA levels, ↓ DOPAC levels	[53]
	Piper nigrum	Piperine	ICV-STZ C57BL/6 mice	10 mg/kg/22 days	↑ behavioral parameters, ↑ antioxidant enzymes, ↓ LPO, ↑ ARG1, CD206, NE, DA, 5-HT and GABA levels and ↓ glutamate level in hippocampus, ↓ CD86 and iNOS,	[54]

Table 1. *Cont.*

Disease	Plant	Natural Compound	In-Vitro/In-Vivo Models/Human Trial	Dose & Route	Mode of Action	References
	Many plants	Quercetin	fAβ 1-40 insulted hBMECs	0.3, 3 and 30 μmol/L	↑ cell viability, ↑ γ-GT and ALP, ↓oxidative stress, ↑ barrier function	[55]
			triple transgenic AD (3xTg-AD) mice	25 mg/kg every 2 days for 3 months	↑ neuronal population, ↓ β-amyloid accumulation, tau accumulation, astrogliosis and microgliosis, ↑behavior parameters	[56]
	Grapes and other plants	Resveratrol	$A\beta_{1-42}$-treated PC12 cells	3 μM	↑ cell viability, ↑ mitophagy, ↓ apoptosis and ROS	[57]
			AβPPswe/PS1dE9mouse model	16 mg/kg/day	↑ behavior, ↓ plague pathology, ↑ p-AMPK/LKB pathway, ↓ SIRT-1, ↑ IL1β and TNF	[58]
			199 patients' mild to moderate dementia due to AD.	500 mg once daily with 500 mg increments every 13 weeks leading to 1000 mg twice daily.	↑ CSF and plasma Aβ 40 in placebo, preserves brain barrier integrity	[59]
	Nigella sativa	Thymoquinone	$A\beta_{25-35}$-treated PC12 cells	4 μM for 24 h	↑ cell viability, ↓ LDH levels, ↓ TBARS, ↑ GSH, antioxidant enzymes, ↓ ROS, AChE, NO levels, ↑ MMP, ↓ iNOS expression	[60]
	Allium sativum	s-allyl cysteine (SAC)	ICV-STZ mice model	30 mg/kg i.p. for 15 days	↑behavior, ↑ GSH, antioxidant enzymes, ↓ LPO, ↑ Bcl2 and ↓ p53 levels	[61]
			D-galactose (DG) mice	1 g/L in drinking water for 7 weeks	↓ Aβ1-40 and Aβ1-42, ↓ APP and BACE1 expression, ↑ PKC activity	[62]
	Many plants	Eucalyptol	$A\beta_{25-35}$-treated PC12 cells	10 μM/24 h	↓LDH levels, ↓ TBARS, ↑ GSH, antioxidant enzymes, ↓ ROS, AChE, NO levels, ↓ cytokines, ↓ NF-κB, COX-2, iNOS, ↑cell viability	[63]
	Silybum marianum	silibinin	APPswe/PS1dE9 double-transgenic mice	100 mg/kg/15 days	↑ behavior in MWM, ↓ amyloid plaque burden, ↓ gut bacterial	[64]
			Aβ1-42 model of rats	25, 50 and 100 mg/kg for 10 days	↓ depression, anxiety behavior, ↓ neuronal damage, ↑ BDNF and TrkB expression, ↑ autophagy	[65]
Parkinson's Disease	Many plants	Quercetin	The mouse dopaminergic MN9D cell line	10 and 30 μM quercetin for 24 h	↑ PKD1 pro-survival signaling, Akt and CREB phosphorylation, BDNF expression, mitochondrial biogenesis	[66]
			6-OHDA model of rats	100, 200, 300 mg/kg for 14 days	↑ motor and non-motor deficits, antioxidant enzyme activities, ↑ neuron density in hippocampus, ↓ AChE	[67]
	Glycine max	Genistein	SH-SY5Y cells overexpressing A53T mutant α-synuclein	20 μM for 24 h	↑ cell viability, ↓ MDA content, ↑ GSH, ATP and Na^+K^+ ATPase levels, ↓ apoptosis	[68]
			Transgenic Drosophila Model expressing human α-synuclein	10, 20, 30, and 40 μM for 24 days	↑ life expectancy, ↓ loss of climbing ability, ↓ oxidative stress, ↑ dopamine content	[69]

Table 1. *Cont.*

Disease	Plant	Natural Compound	In-Vitro/In-Vivo Models/Human Trial	Dose & Route	Mode of Action	References
	Grapefruit and other citrus fruits	Naringin	MPP$^+$-induced Parkinson's disease rat model	80 mg/kg for 4 days	↑ GDNF expression and mTORC1, ↓ TNF-α	[70]
	Curcuma longa	Curcumin	MPP+-induced SH-SY5Y cell	40 μM for 24 h	↑ cell viability, DA, Bcl-xl level, ↓ caspase 3, Bax level and HSP-90 levels	[71]
			Lipopolysaccharide-induced PD rat model	40 mg/kg i.p. for 21 days	↓ GFAP, NFκB, TNF-α, IL-1β, IL-1α, iNOS, oxidative stress, α-synuclein aggregates, apoptotic markers	[72]
	Green tea	Epigallocatechin gallate (EGCG)	Rotenone treated SH-SY5Y cells	25 or 50 μM for 24 h	↓ caspase-3 and apoptosis, ↑ cell viability, SOD	[73]
			MPTP mouse model	25 and 50 mg/kg for 20 days	↑ behavior, TH-positive neurons, ↓ TNF-α and IL-6	[74]
			480 PD patients, randomized, double blind	three dosage groups of green tea polyphenol and one placebo control group	Delay in motor function progression, ↑ cognition, mood and quality of life	ClinicalTrials.gov identifier: NCT00461942
	Grapes and other plants	Resveratrol	6-OHDA treated PC 12 cells	50 μM for 24 h	↑cell viability, MMP, ↓ apoptosis, CXCR4 protein levels	[75]
			MPTP mouse model	50 mg/kg for 3 weeks	↑ TH+ cells and miR-129, ↓ MALAT1, SNCA and apoptosis	[76]
			20 double blind, crossover, randomized, Placebo controlled phase 1 study	BIA 6-512 (trans-resveratrol) 25 mg dose, 50 mg dose, 100 mg dose	To study BIA 6-512 effect on levodopa pharmacokinetics when administered with levodopa/benserazide	ClinicalTrials.gov Identifier: NCT03091543
	Many plants	Rutin	6-OHDA in PC-12 cells	10, 50, and 100 μM for 8 h	↑cell viability, catalase, SOD, GPx, GSH, ↓ MDA	[77]
			6-OHDA-induced PD rat model	25 mg/kg for 21 days	↑ behavior activities, GSH and dependent enzymes, DA and its metabolites, protein carbonyl, ↓ TBARS, H$_2$O$_2$, NO level, TNF-α and IL-1β	[78]
	Citrus trees	Hesperidin	6-OHDA-induced PD mice model	50 mg/kg for 28 days	↑ behavior activities, GSH and dependent enzymes, SOD, Catalase, DA and its metabolites ↓ROS	[79]
	Piper nigrum	Piperine	6-OHDA-induced PD mice model	10 mg/kg for 15 days	↑ behavior activities, GSH and dependent enzymes, SOD, Catalase, ↓ LPO, Caspase-3 and Caspase-9, TNF-α, and IL-1β	[80]

Table 1. *Cont.*

Disease	Plant	Natural Compound	In-Vitro/In-Vivo Models/Human Trial	Dose & Route	Mode of Action	References
Amyotrophic lateral sclerosis	*Panax and Eleutherococcus*	Ginseng root	SOD1-G93A Transgenic Mouse	40, 80 mg/kg	↑ onset to clinical signs and surrogate death	[81]
		Genistein	SOD1-G93A Transgenic Mouse	16 mg/kg 2 times per day	↑ longevity, ↓ ALS symptoms, ↑ motor neurons, ↓ TLR2, TLR4, and NF-kB, p65 levels, IL-1β, IL-6, TNF-α levels	[82]
	Curcuma longa	Brainoil, curcumin supplement	42 randomised ALS patients	600 mg/day for 3 months	↓ disease progression, AOPPs, oxidative damage, ↑ aerobic metabolism	[83]
	Withania somnifera	*Withania somnifera* extract	SOD1-G93A Transgenic Mouse	5 mg of root powder p.o.	↑ longevity, motor performance, Hsp-70, Hsp-60 and Hsp-27, motor neurons, ↓ misfolded SOD1 protein, inflammation	[84]
	Green tea	EGCG	SOD1-G93A Transgenic Mouse	10 mg/kg body	↓ disease onset, ↑ survival, motor neurons, ↓ activated microglia, NF-kB and caspase-9	[85]
	Tripterygium wilfordii and Celastrus regelii	Celastrol	SOD1^{G93A} transfected NSC34 cells	50 nmol/L for 4 h	↑ cell viability, ↓ MDA, ↑ mRNA expressions of GCLC and GST, ERK1/2 and Akt	[86]
	Grapes and other plants	Resveratrol	BM-MSCs derived from ALS patients	1 μM for 12 h	↑ Neuro-progenitor markers, nestin, NF-M, Tuj-1, and Map-2, AMPK/SIRT1 signaling	[87]
Huntington's Disease	*Panax ginseng*	Protopanaxatriol	3-NP-induced HD	5, 10, and 20 mg/kg, po	↑ weight, locomotor activity, antioxidant enzymes, HO-1, NQO1, ↓ ROS	[88]
	Many plants	melatonin	3-NP-induced HD	1 mg/kg for 8 days	↓ LPO, protein carbonyl, ↑ SOD and succinate dehyrogenase	[89]
			20 HD gene carrier subjects	5 mg/day, 30 min before bedtime/month	To improve sleep quality in HD gene carriers	ClinicalTrials.gov Identifier: NCT04421339
	Grapes and other plants	Resveratrol	SH-SY5Y cells hyper-expressing the mutant polyQ Huntingtin (polyQ-Htt) protein	100 μM for 24 h	↑ cell viability, autophagy, autophagy degradation of mutant Huntingtin, ↓ ROS	[90]
			Double blind, randomized controlled 102 early affected HD patients	800 mg/day for 1 year	To evaluate resveratrol effect on caudate volume in HD patients	ClinicalTrials.gov Identifier: NCT02336633
	Green tea	EGCG	Cell culture, yeast model and HD transgenic flies	Different doses	↓ aggregation of mutant htt exon 1 protein, polyQ-mediated htt protein aggregation, cytotoxicity	[91]

Table 1. *Cont.*

Disease	Plant	Natural Compound	In-Vitro/In-Vivo Models/Human Trial	Dose & Route	Mode of Action	References
	Vitis vinifera	trans-(-)-ε-Viniferin	STHdhQ7/Q7, STHdhQ111/Q111, Tet-Off PC12 cells, Neuroblastoma N2a cells and Primary cortical neurons	1 nM, 10nM, 100nM, 1 μM	↓ cell toxicity, oxidative stress, mitochondrial dysfunction, ↑ mitochondrial genesis, SIRT3-dependent AMPK Activation	[92]
	Withania Somnifera	*W. somnifera* root extract	3-NP-induced HD	100 and 200 mg/kg for 14 days	↑ body weight, behavioral activities, SOD, catalase, mitochondrial complex (I, II, III) levels, ↓ LPO, nitrite, LDH	[93]
Vascular Cognitive impairment	*Aster ageratoides*	Aster ageratoides extract (AAE)	PC 12 cells treated with 50 μM glutamate or 2VO/H surgery in Sprague Dawley rats	0, 10, 25, and 50 μg/mL or 50 mg/kg b.w	↓ Memory impairment in vivo, ↓ hippocampal structures, neuronal excitotoxicity in vitro	[94]
		bilobalide	male Sprague Dawley rats (2-vessel occlusion, 2-VO)	2, 4 and 8 mg/kg	↓ Memory impairment, ↓ nuclear condensation, ↑SOD and GSH, ↓ NOS and MDA, TNF-α	[95]
	Ginkgo biloba L	EGb761	bilateral common carotid arteries repeated occlusion in rats and ip injection of sodium nitroprusside	50 mg/kg	↓ Memory impairment, proliferation of neural stem cells in dentate gyrus and subventricular zone	[96]
		Ginkgo biloba extract	90 patients, randomized, double-blind, placebo-controlled trial	60 and 120 mg	↑ CGI scores, no effect on trans cranial Doppler ultrasound, more adverse reactions in Placebo group.	[97]
	Huperzia serrata	Huperzine	randomized, double-blinded, placebo-controlled study with 78 patients with mild to moderate VaD	0.1-mg bid	↑ MMSE, CDR, and ADL scores	[98,99]

Table 1. *Cont.*

Disease	Plant	Natural Compound	In-Vitro/In-Vivo Models/Human Trial	Dose & Route	Mode of Action	References
Stroke (causes secondary neurodegeneration)	Grapes and other plants	Resveratrol	Primary cortical neuron cultures with OGD/R	1, 5, and 20 µmol/L for 24 h	↑ cell viability, neurite outgrowth and synaptogenesis via Shh signaling pathway, Sirt1 activation	[100]
			(MCAO/R) Model	30 mg/kg for 7 days	↑ behavior deficits, ↓ infarction volume, ↑ NeuN+ cells, ↓ TUNEL+ cells	[101]
	Ginkgo biloba	Ginkgolides	NSC line	20, 40 or 60 mg/L	↑ cell viability, process length and cell body area, sizes of NSE, GFAP and SOCS2-positive cells	[102]
			mouse model of myocardial I/R injury	2.5 mL/kg	↑ cardiac function, ↓ LDH and AST, TWEAK expression, infarction volume	[102]
	Curcuma longa	Curcumin	Mouse N2a cells hypoxia model	5, 15, 25, and 35 µmol/L for 24 h	↑ cell viability, mitochondrial disruption, Bcl2 expression, ↓ Tunnel positive cells, Bax and Caspase-3 expression	[103]
			Rat model of global brain ischemia	25 and 50 mg/kg	↑ DA and its metabolites DOPAC and HVANE and 5-HT, cell viability, ↓ COX-2, TNF-α expression	[104]
	Green tea	EGCG	HBMVECs OGD/R model	2 µM for 24 h	↑ cell viability, SOD, migration and tube formation, mRNA expression of VEGF, Bcl2, ↓ apoptosis and autophagy, ROS, LDH, MDA, mRNA expression of Bax and Caspase-3	[105]
			Rat MCAO model	20 mg/kg	↓ infarct volume, TUNEL⁺ cells, NO, MDA, ↑ Behavioral parameters, SOD and GPx	[106]
	Scutellaria baicalensis	Baicalin/baicalein	Rat MCAO model	200 mg/kg, 24 h after reperfusion till 7 days	↑ behavior deficits, microglia/macrophage M2 markers CD206 and CD 163, ↓infarct volume, microglia/ macrophage M1 markers CD86 and CD 16, MAPK and NF-κB	[107]
	Angelica sinensis	Ligustilide	Rat MCAO model	7.5, 15 or 30 mg/kg for 3 days	↑ behavior deficits, ↓ infarct volume, ↑ HSP-70 and MAPK activation	[108]
	Sinomenium acutum	Sinomenine	Mice MCAO model	10 or 20 mg/kg daily for 3 days	↑ behavior deficits, ↓ infarct volume, apoptosis, astrocyte and microglial activation, NLRP3 inflammasome, IL-1β, IL-6, IL-18 and TNF-α generation	[109]
	Magnolia officinalis	Honokiol	Rat cerebral ischemia reperfusion model	0.7–70 µg/kg, 15 min after ischemia	↓ Cerebral edema, p65 level, NO, TNF- α, RANTES/CCL5 levels	[110]
	Zingiber officinale	Zingerone	Rat MCAO model	50, 100 mg/kg at 5 h and 12 h after initiation of surgery	↑ behavior deficits, ↓ infarct volume, LPO, Caspase-3 and -9 Apaf-1, Bax, ↑ GSH and dependent enzymes, Catalase, Bcl-2	[111]

Table 1. *Cont.*

Disease	Plant	Natural Compound	In-Vitro/In-Vivo Models/Human Trial	Dose & Route	Mode of Action	References
	Many plants	Perillyl alcohol	Rat MCAO model	25, 50, 100 mg/kg for 7 days	↑ behavior deficits, ↓ infarct volume, LPO, TNF- α, IL-1β, IL-6, COX-2, iNOS, NF-κB, ↑ GSH and dependent enzymes, Catalase	[112]
	Piper nigrum	Piperine	Rat MCAO model	10 mg/kg for 15 days	↓ infarct volume, ↑ behavior deficits and histopathological findings, ↓ TNF- α, IL-1β, IL-6, COX-2, iNOS, NF-κB	[113]
Prion Disease	*Scutellaria biacalenesis*	Baicalein	PrP 106-126-induced-(SH-SY5Y and PrP 106-126 and SK-N-SH) cells	80 μM	↓ROS production, ↓ mitochondrial dysfunction, ↓ apoptosis, ↓ JNK signalling	[114]
	Grapes and other plants	Resveratrol	PrP 106-126-induced-(SH-SY5Y and PrP 106-126 and SK-N-SH) cells	2–4 μM	↑ autophagy, ↓ mitochondrial dysfunction, ↓ apoptosis	[115]
	Cupressaceous plants	Hinokitiol	PrP 106-126-induced SK-N-SH	8 μM	↑ HIF-1α, ↑ autophagy, ↑ p62/SQSTM1, ↓ apoptosis	[116]
	Many plants	Rutin	PrP induced-HT22 cells	10 μg/mL	↓ ROS and NO, ↓ caspase 3 activity, ↓ caspase 8, FAS, FASL, ↑ BDNF	[117]
Frontotemporal dementia	Many plants	Nicotine	Grn −/− mice	0.6 mg/kg P.I daily for 14 days	↓ CD68, IL-1β, CD11b, ↑ sociability	[118]
	Curcuma longa			250–500 μg/mL	↓ AChE	[119]
	Piper nigrum			250–500 μg/mL	↓ AChE	[119]
	Curcuma longa	Curcumin	Primary astrocytes from NPC +/+ and NPC −/− mice	30 μM	↑ cytosolic Ca^{2+}, ↑ viability	[120]
Pick's disease	Many plants	δ-tocopherol	Human fibroblast and baby hamster kidney cells	40 μM	↓ cholesterol accumulation, ↓ lysosomal size, ↑ intracellular Ca^{2+}, ↑ Ca^{2+} deficiency	[121]
	Many plants	Quercetin	Coca-2 cells and male Wistar rats	100 μM and 5 mg/kg (rats)	↓Cholesterol uptake	[122]
	Many plants	Luteolin	Coca-2 cells and male Wistar rats	100 μM and 5 mg/kg (rats)	↓Cholesterol uptake	[122]
Spinocerebellar ataxia		Melibiose	293 cells and SCA17 transgenic mice	100 nm - 100 μM for 6 days and daily I.P for days (24 mg/kg)	↓ polyQ aggregration. ↓ ROS, ↑ autophagy, ↓ caspase 3	[123,124]

Table 1. *Cont.*

Disease	Plant	Natural Compound	In-Vitro/In-Vivo Models/Human Trial	Dose & Route	Mode of Action	References
Spinal muscular atrophy	*Brucea javanica*	Bruceine D	SMA mice and Δ7 mice	10 to 30 mg/kg i.p once a day for 7 days	Correcting the splice defect in SMN2, ameliorating SMN phenotype defects	[125]
	Tripterygium wilfordii	Triptolide	SMA mice, NSC34 and N18TG2 cells	0.1 mg/kg I.P daily	↑ SMN, Gemin2 and Gemin3 expression levels, ↑ transcription of SMN, ↑ survival rate and ↓ SMA related defects	[126]

↑prevented/loss of/induced/enhanced/improved/increased/upregulated/elicited/promoted/restored/activated. ↓down-regulated/attenuated/decreased/declined/terminated/ blocked/ prevented/ inhibited. 3xTg-AD: Enhanced susceptibility of triple transgenic Alzheimer's disease, 5-HT: 5-hydroxytryptamine, AlCl$_3$+D-gal: Aluminium chloride+d-galactose, ALP:Alkaline phosphatase, AMPA: α-amino-3-hydroxy-5-methylisoxazole-4-propionic acid, APP/PS1:amyloid precursor protein/presenilin 1, APP:Amyloid precursor protein, ARG1:Arginase, Bcl-xl: B-cell lymphoma-extra-large, CALM: Clathrin-assembly lymphoid myeloid leukaemia protein, CAMK4: Calcium/calmodulin-dependent protein kinase type IV, CAMKK1: Calcium/Calmodulin Dependent Protein Kinase Kinase 1, CREB: cAMP response element-binding protein, CXCR4: C-X-C chemokine receptor type 4, DOPAC: 3,4-Dihydroxyphenylacetic acid, GCLC: Glutamate-Cysteine Ligase Catalytic Subunit, GSH: Glutathione, GST: Glutathione S-transferase, hBMECs: Human brain microvascular endothelial cells, Hsp-27: Heat shock protein-27, Hsp-60: Heat shock protein-60, HSP-90: Heat shock protein-90, ICV-STZ: Intracerebroventricular streptozotocin, MALAT1: Metastasis Associated Lung Adenocarcinoma Transcript 1, Map-2: Microtubule associated protein 2, MCAO/R: middle cerebral artery occlusion- reperfusion, MDA: *Malondialdehyde,* MN9D: murine mesencephalon-derived dopaminergic neuronal cell line, mTORC1: mammalian target of rapamycin complex 1, MWM: Morris water maze, NE: Norepinephrine, NF-M: Neurofilament medium polypeptide, NGF: Nerve growth factor, NMDA: N-methyl-D-aspartate, NO: Nitric oxide, p-AMPK/LKB: AMP-activated protein kinase/liver kinase B, PI3K: phosphoinositide 3-kinase, PKC: Protein kinase C, RANTES/CCL5: regulated upon activation, normal T cell expressed and secreted/ Chemokine (C-C motif) ligand 5, SIRT-1: silent mating type information regulation 2 homolog 1, SNCA: Synuclein Alpha, TBARS: Thiobarbituric acid reactive substances, TH[+]: Tyrosine hydroxylase, TLR2: Toll-like receptor 2, TLR4: Toll-like receptor 4, TrkB: Tropomyosin receptor kinase B, Tuj-1: Neuron-specific Class III β-tubulin, TUNEL[+]: terminal deoxynucleotidyl transferase-mediated dUTP nick-end labelling[+], TWEAK: TNF-related weak inducer of apoptosis, γ-GT: Gamma-glutamyl transferase.

3.2. Parkinson's Disease and Parkinsonism

PD is the second most prevalent NDs of the elderly after AD, characterized by dopaminergic neuronal loss in substantia nigra pars compacta region of the brain, the striatum, locus ceruleous, raphe nuclei, nucleus basalis of meynert, and hippocampus. The main sign and symptoms of PD patients are resting tremors, rigidity, bradykinesia, muscular impairment, postural instability, difficulty in walking, soft speech, sleep problems, slow handwriting [127]. The pathological findings are the presence of Lewy Bodies (LBs, aggregation of α-synuclein) with DNA mutations in the brain of PD patients [128], with neuronal cytoplasmic inclusions in the cortex and brain stem [129,130]. The treatment of PD mainly rotates around the enhanced activities of dopaminergic neurons or inhibition of cholinergic effects [131]. Several mechanisms associated with PD accounts for oxidative stress, mitochondrial dysfunction, apoptotic pathway activation etc. [132,133] There is no cure for PD but drugs available in the market increases the life expectancy with relief in the symptoms [131]. Besides, the drug therapy demonstrates many side effects in patients [134,135]. Therefore, researchers have to treat PD at molecular and cellular levels, and use such treatments that have a multifactorial approach. Several scientists have reported plant compounds to be effective in the treatment of PD. After dissecting the oxidative stress role in the pathophysiology of PD, the phytochemicals having antioxidative properties are drawing the attention of researchers [136]. This section deals with the phytochemicals which are generally considered as antioxidative agents. In many fruits and vegetables (apples, broccoli, and onions), a polyphenolic compound is detected named quercetin. It restricts the hippocampal neuronal damage resulting in improved memory and learning (by maze test) in the rat model [137]. In one study, when quercetin was administrated to the PD model, it reversed the neuronal impairment via antioxidative pathways [138]. In PD MitoPark transgenic mouse model, quercetin protects against mitochondrial dysfunction and progressive dopaminergic neurodegeneration [66]. Quercetin also plays a specific role in the upregulation of cognitive effects in the 6-hydroxydopamine (6-OHDA) rat PD model [67]. An isoflavone, genistin is also having several health benefits including nourishment of neurons [139]. It is known as antioxidative, anti-inflammatory, and neuroprotective agent. Genistin (oral dose 10, 50, or 100 mg/kg for one week) acts as a therapeutic agent against LPS-induced cognitive impairment. These dose-dependent treatments of genistin determines to lessen memory deficits, decrease the malondialdehyde levels in the hippocampus region, increases the level of superoxide dismutase, catalase, and glutathione [140]. Moreover, genistin reformed the level of AChE activity in the hippocampus against LPS insult in in vivo model. It targets the modulation of Nrf2/NF-κB/IL-6/TNFα/COX2/iNOS/TLR4/GFAP and thus promote neuronal protection via inhibiting the neuronal inflammation and improve cognitive deficits [140]. Genistein shows the neuroprotection in the A53T mutant α-synuclein overexpressed SH-SY5Y cells [68]. Genistin protects the Transgenic *Drosophila* Model of Parkinson's Disease against oxidative stress [69].

Other phytochemical Naringin was studied and was found to be involved in the neuroprotection by triggering the Nrf2/ARE signaling against 6-OHDA-induced neurotoxicity [141]. In in vitro study (neuroblastoma SH-SY5Y cells), it is reported to show neuronal protection by blocking the c-Jun N-terminal kinase (JNK) phosphorylation and regulation of BAX in the rotenone-induced cell line [142]. Additionally, it also inhibits the caspase-3 and PARP cleavage [143]. Naringin can lead to the anti-inflammatory pathway and provide neuroprotection in PD [70]. In in vitro PD model of SH-SY5Y cells, curcumin imparts a role in neuroprotection against toxic insult via controlling the HSP-90 [71]. Curcumin was studied in Pheochromocytoma Cell Line 12 (PC12) cells against 1-methyl-4-phenylpyridinium (MPP$^+$) toxicity and fixation of mitochondrial membrane potential was found against enhancement of intracellular ROS which leads to the instigation of iNOS, and over-expression of Bcl-2 [144]. Curcumin leads to the restoration of mitochondrial membrane potential (MMP) which causes the up-gradation in Cu/Zn SOD and modulates NF-κB nuclear translocation by inhibition of interleukin-6 (IL-6) and TNF-α [145]. It was further reported that it lowers the iNOS mRNA levels and inhibits the induction of NF-κB. It is found to be effective against oxidative stress via following cytochrome c release or caspase-3 activation. In in vitro study, it is reported to act on the aggregation of synuclein inhibition and aggregation of the PD associated, A53T mutant-synuclein in SH-SY5Y cells in a dose-dependent manner [146]. By inhibiting the α-synuclein aggregation in lipopolysaccharide-induced PD model curcumin plays a role as a promising therapeutic agent [72]. Epigallocatechin-3-gallate (EGCG) is also an important bioactive component with neuroprotective potential. The mode of action of EGCG is on neuroinflammation, oxidative stress, and autophagy. Likewise curcumin, EGCG also ameliorates the mitochondrial restoration by antioxidative properties and associated signaling pathways i.e., via blocking the expression of COX-2, iNOS, and all other enzymes which causes the generation of pro-inflammatory mediators [147]. EGCG protects by antioxidative activity against 6-OHDA administration in SH-SY5Y cells (in vitro dopaminergic model) [148]. EGCG also functions as antagonistic against the NF-κB transcription factor and participates to remove the poisonous accumulation of amyloidogenic α-synuclein in many in vitro models. 1-methyl-4-phenyl-1,2,3,6-tetrahydropyridine (MPTP) treated mice also show improvement when EGCG is administrated orally [149]. EGCG also inhibits the oxidative stress in rotenone-induced in vitro model of SH-SY5Y cells [73]. In the MPTP-induced mouse model of PD EGCG helps in the modulation of peripheral immunity [74]. A natural polyphenolic compound, resveratrol, is involved in the mitochondrial biogenesis by initiating the antioxidative cascade mechanism. It also protects neurotoxic chemicals i.e., 6-OHDA and MPTP in in vitro models [150]. This natural compound shows the promising cure against cell toxicity induced by dopamine (DA) and reverse the neuronal impairment by following the cAMP/PKA pathway and thus inhibit the α-synuclein aggregation [151]. Further study also points out the SIRT1/AMPK/PGC1-α axis as a key neuroprotective pathway and provides a rationale for exploring the therapeutic potential of resveratrol in delaying PD progression [152]. By leading the CXCR4 signaling pathway, resveratrol helps in the restoration of the 6-OHDA-induced damage in the PC 12 cells [75]. Resveratrol attenuates the PD like condition via inhibition of programmed cell death through inducing MALAT1/miR-129/SNCA signaling pathway [76]. A total of 119 participants were randomized to placebo or pure synthetic resveratrol 500 mg orally once daily with a dose-escalation by 500-mg increments every 13 weeks until a final dose of 1000 mg twice daily was reached for the final 13 weeks. Visits occurred at screening, baseline, and every 6 weeks with a total resveratrol exposure of 52 weeks. Compliance with resveratrol and with placebo was confirmed by mass spectrometry of blood samples. Resveratrol may have many other molecular effects, including anti-inflammatory, antioxidant, and anti-Aβ aggregation. 480 PD patients, randomized, double-blind three dosage groups of green tea polyphenol and one placebo control group. Delay in motor function progression, increased cognition, mood, and quality of life was reported [153]. 20 double-blind, crossover, randomized, placebo-controlled phase 1 study BIA 6-512 (trans-resveratrol) 25 mg dose, 50 mg dose, 100 mg dose To study BIA 6-512 effect on levodopa pharmacokinetics when administered with levodopa/benserazide [154]. Rutin, another phytochemical act as an anti-inflammatory agent for neuroprotection which is associated with the involvement of microglia response, cytokines level, and iNOS activation [155]. Rutin facilitates protection in 6-OHDA-induced neurotoxicity in rat pheochromocytoma (PC-12) cells [77]. Rutin also targets the oxidative stress and protects dopaminergic neurons in the animal model of PD [78]. Hesperidin can be a possible promising guide for the designing of inhibitors of mutant SOD1 for therapeutic purposes. Antioxidative and anti-inflammatory properties make hesperidin reliable herbal compound against NDs. ERK/Nrf2 are key signaling molecules for following the neuroprotective pathway [156]. Neurotoxicity induced by 6-OHDA can be suppressed by the use of hesperidin in the mice model of PD [79]. Piperine treatment attenuated rotenone-induced motor deficits, and rescued the loss of dopaminergic neurons in the substantia nigra [157]. Piperine stimulates the anti-apoptotic and anti-inflammatory pathway and shows protection in 6-OHDA-induced PD rat model [80]. PD is known as the second rank on the list of NDs with poor availability of effective treatments still, the studied phytochemicals have given hope for promising and long-lasting therapy. Figure 4. The briefing of phytochemicals and their therapeutic effects on PD are listed in Table 1.

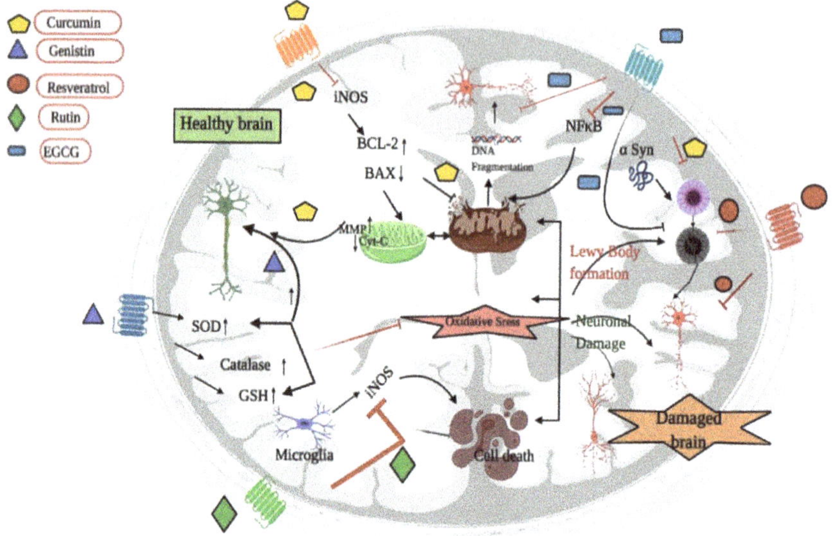

Figure 4. Diagrammatic presentation of Parkinson's disease, cause, and role of phytochemicals to reverse the sign of PD. Mitochondrial dysfunction, oxidative stress, and uncontrolled cell death leads to the accumulation of Lewy bodies and resultant neuronal damage. Phytochemicals function with multiple targets. Curcumin targets the inducible nitric oxide synthase (iNOS) and regulates the ratio of BCL-2/BAX which leads to the restoration of mitochondrial function. Genistin increases the activity of SOD, Catalase, and GSH and protect the neurons from damage ignited by oxidative stress. Resveratrol inhibits the formation of Lewy body (aggregation of α-syn protein-a major cause of PD). Rutin protects the neuron by leading the anti-inflammatory pathway and EGCG inhibits the transcriptional factor NFκB and restores the normal function of mitochondria. EGCG also plays a role to inhibit DNA fragmentation. All over the activities of phytochemicals escape neurons from damage.

3.3. Amyotrophic Lateral Sclerosis

ALS also called motor neuron disease (MND) or Lou Gehrig's disease is another NDs that is heterogeneous and both upper and lower motor neurons are degenerated causing motor symptoms. The initial symptoms vary between individuals, some exhibiting spinal-onset disease with muscle weakness of limbs, and others present with bulbar-onset disease in which they have dysarthria and dysphagia [158]. The preponderance of patients face death within 3–5 years after the first sign of disease due to respiratory failure [159]. The cause of this diseases is unknown but few cases are linked to familial history with gene mutations as a major contributor [160] and associated mechanisms of oxidative stress [161], calcium toxicity [162], inflammation [163], chronic viral infections [164], and excitotoxicity [165]. Other symptoms are spasticity, sialorrhoea, pain, muscular cramps, liver problems, deep venous thrombosis, mood alterations, respiratory insufficiency, fever, vomiting, fatigue, and nausea. In the case of ALS, the ability of the brain to convert the toxic radicals into non-toxic substances gets compromised because of mutations in *SOD1* [166]. Although, this kind of mutation does not lead to the progression of the disease but the mutation makes the *SOD1* more prone to the misfolding which causes the accumulation of protein and thus involves the reduction in the activity of motor neurons. The treatment of ALS is often symptomatic with Riluzole (Rilutek) and Edaravone (Radicava) as the only drugs approved by the FDA. These drugs can only increase the life expectancy by three to six months. Much of the efforts of physicians and other health workers is to make the quality of life (QOL) better for ALS patients. Several studies on natural plants and their active compounds are

In progress to combat the dreadful symptoms of this disease. The molecular mechanisms and potential therapeutic role of the plant compounds have to be studied and implemented.

Few phytochemicals and their possible known mechanism of actions will be discussed in this section. Ginseng root shows the protective effect in *SOD-1* (G93A) transgenic mice [81]. An isoflavone, genistin can easily cross the blood–brain barrier (BBB) and attributes its role by anti-inflammatory activity via extracellular-signal-regulated kinase (ERK) and microtubule-associated protein kinase (MAPK) signaling in the microglia which leads to the neuronal inflammation. The neuroprotective role of genistin was detected in the SOD1-G93A mice [82]. Ginseng roots inhibit the onset and increase the survival of affected cells delays the onset and prolongs the survival of affected cells [167]. The most promising phytochemical, curcumin has a protective role in ALS too. It acts for mitochondrial biogenesis leading the AMP/PGC-1α and/or Nfr2 pathway [83]. The clinical studies show the effectiveness of curcumin for slowing down the progress of the disease. Curcumin helps in the improvement of aerobic metabolism and oxidative stress. The antioxidative mechanism is responsible for leading the protective effect against ALS [83]. A studied phytochemical named withaferin A shows an anti-inflammatory effect by working as antagonistic of NF-κB and is found to be effective for the improved motor activity in transgenic mice model [84]. ALS mouse model generated by the point mutation in SOD1 was studied and it was observed that by the withaferin treatment there was a late progression of the disease and more survival time in comparison to the non-treated ALS mice. The root extract of *Withania* was also tested against mutated transgenic mice (TDP-43A315T mice) and manifested with removal of abruptly cytoplasmic motor neurons, reduction in inflammatory markers, and improvement in cognitive performance [168]. It is reported that ALS is following numerous pathways therefore, combined therapy could be more efficient for targeting more than one signaling pathway within a single attempt. EGCG has a power of antioxidation, hampered activation of caspase-3, NF-κB and thus reduces microglial activation resulting in decreased neuroinflammation and increased neuronal survival [169]. EGCG stimulates the iNOS/NF-κB/ Caspase-3 and shows the increased number of neuronal cells, lower microglial activation in transgenic mice model [85]. One more phytochemical named celastrol also reveals the anti-inflammatory, antioxidative property. Celastrol exhibits an antioxidative effect in SOD1G93ANSC34 cells by leading MEK/ERK and PI3K/Akt signaling pathways [86] Celastrol protects the H_2O_2-induced oxidative stress model by activating the antioxidative pathway in ALS model [86]. The bran oil treatment helps in the significant enhancement of antioxidative enzymes and suppress the effect of ROS and malondialdehyde (MDA) in rotenone-induced ALS model [170].

ALS study was also conducted in bone marrow-derived mesenchymal stem cells (MSCs) by activating the SIRT1/AMPK pathway and it is noted a significant decrease in the ALSMSCs compared to normal healthy control originated BM-MSCs by using resveratrol. Neuro-progenitor markers were increased in resveratrol treated ALS-MSCs [87]. All these phytochemicals along with the ones enumerated in the table could be promising therapeutic agents which can be the best alternative of commercial drugs with negligible side effects Table 1.

3.4. Huntington's Disease

HD is an annihilative, inherited, and familial illness, with progressive loss of brain and muscle functions. This is a single-gene disorder in which there is programmed degeneration of neurons within different regions of the brain. In wild type gene, there are 6-35 CAG repeats in exon 1 of the *Huntingtin (Htt)* gene on chromosome number 4 whereas, in the affected persons, this repeat increases to more than 36 [171]. As a result of this, there is an accumulation of Huntingtin protein in the neuron causing its death. Almost 95% of GABAergic medium spiny neurons (MSNs) projecting to substantia nigra and globus pallidus are lost causing atrophy of cortex, thalamus, and hypothalamic nuclei. The symptoms include concentration problems, short-term memory, tumbling, lack of focus, clumsiness, depression weight loss, feeding problems, difficulty in speech, uncontrolled face movements, itching etc. [172]. There is a lot of research going on in the field of HD prevention and treatment. Many experimental models have been experimentally validated to test the efficacy of different drugs. Hitherto, there is no cure for the disease, drugs available only give symptomatic relief. There have always been natural compounds linked to various diseases and a lot of them have shown promising effects in preclinical studies.

One of the major factors for the progression of HD is the abrupt function of mitochondria [173]. The activity of complex II of the respiratory chain gets reduced in the affected brain region. Protopanaxtriol limits the overproduction of free radicals and attenuates the expression of heat shock protein 70 (Hsp-70), along with the restoration of SOD activity inhibiting the formation of free radicals. Protopanaxtriol enhances the Nrf2 entry to the nucleus and increases the expression of Heme oxygenase-1 (HO-1) and NAD(P)H dehydrogenase [quinone]1(NQO1) [88]. Melatonin also has been shown to reduce neural damage in PD, AD as well as in ischemia-reperfusion injury against d-aminolevulinic acid and a variety of other neural toxins. In the case of mitochondrial injury, induced by 3-nitropropionic acid there was a restoration by the treatment with melatonin by leading the antioxidative pathway [88,174]. Melatonin also rescues the HD animal model from the oxidative stress induced by 3-nitropropionic acid [89]. Resveratrol attenuates the expression of ATG4 and permits the lipidation of LC3 and promotes the mortification of the polyQ-Htt accumulation and saves the neuronal cells against dopamine-induced toxicity [90]. EGCG inhibits the accumulation of misfolded proteins and oligomerization of mutant Htt exon 1 *protein* in vitro, which indicates the early interference of EGCG in the misfolded protein accumulation [175]. In another study, EGCG reduced the cytotoxicity and polyQ-mediated htt protein aggregation in a yeast model of HD [176]. EGCG attenuates the adverse effect of misfolded huntingtin protein in early-stage in HD model [91].

One more phytochemical, trans-ε-Viniferin (viniferin), was also found to be effective against mutant *Htt*-induced depletion of SIRT3 and thus save the cells from a mutant protein [177]. Viniferin also plays a crucial role in reducing the level of free radicals and thus participate in the protection of MMP in the cells having mutant *Htt*. The presence of mutant *Htt* indicates the lower deacetylase activity of SIRT3 which leads to decreased cellular NAD levels with the biogenesis of mitochondrial cells [178]. Viniferin enhances the mitochondrial biogenesis by activation of AMP-activated kinase [92]. It was reported that systemic 3-NP administration significantly enhances the level of lipid peroxidation (LPO), nitrite and lactate dehydrogenase (LDH), downregulate the antioxidant enzyme (superoxide dismutase and catalase) levels, and inhibit the synthesis of ATP by blocking the activity of mitochondrial complex in the different regions (striatum and cortex) of the brain [179]. Figure 5. The root extract of *Withania somnifera* is used against 3-NP-induced gait abnormalities, mitochondrial dysregulation, and oxidative stress, in striatum and cortex of the rat brain, in vivo model [93]. The detailed functions of phytochemicals are well explained in Table 1.

3.5. Vascular Cognitive Impairment

Vascular cognitive impairment is distinguished by a specific cognitive profile involving preserved memory with impairments in attentional and executive functioning [180]. Vascular dementia should be broadened to recognize the important part cerebrovascular disease plays in several cognitive disorders, including the hereditary vascular dementia, multi-infarct dementia, post-stroke dementia, subcortical ischemic vascular disease and dementia, mild cognitive impairment, and degenerative dementias (including Alzheimer's disease, frontotemporal dementia, and dementia with Lewy bodies). Alzheimer's disease (AD) and vascular dementia (VD) share key pathologies including oxidative damage, oral supplement of phytochemical medicines, which are well-known for their antioxidant properties, can be a viable therapy for both types of dementia. In this study, the therapeutic potential of the *Aster ageratoides* extract (AAE), was found to be effective in experimental rat models of AD and VD. These results provided evidence that AAE supplements can exert anti-AD and -VD efficacies and suggested that AAE might be used as an edible phytotherapeutic for the two major types of dementia [94]. Another study suggested that bilobalide (BB) protected against learning and memory impairments, neuronal apoptosis, and oxidative stress in a rat model of AD induced by $A\beta_{25-35}$ peptide. Additionally, the inhibition of TNF-α and $A\beta_{1-40}$ expression is also involved in the action mechanism of

BB in this experimental model. The results indicated that BB has protective effects on Aβ$_{25-35}$-induced learning and memory impairments and neuronal oxidative damage [95]. *Ginkgo biloba* extract (EGb 761) is widely used to treat cerebral disorders. EGb761 increases proliferation of neural stem cells in the subventricular zone and dentate gyrus, and significantly improves learning and memory in rats with vascular dementia. Studies reported the effects on endogenous NSCs in rat models of vascular dementia and NSCs in the mouse cochlear. *G. biloba* seemed to slow down the cognitive deterioration in patients with VCI, but the effect was shown in only one of the four neuropsychological tests administered [96]. In a randomized, double-blind, placebo-controlled clinical trial *G. biloba* standardized extract showed effectiveness against vascular cognitive impairment [97]. Huperzine A is a new type of *Lycopodium* alkaloid monomer isolated from *Huperzia serrata*. It inhibits acetylcholinesterase activity and increases levels of acetylcholine in the brain thereby improving cognitive function in patients with dementia [98]. A total of 92 participants were included in the two studies, with 46 in the Hup A group and 46 in the control group. The number of patients in the individual studies ranged from 14 to 78, and the duration of trial ranged from 12 to 24 weeks [99].

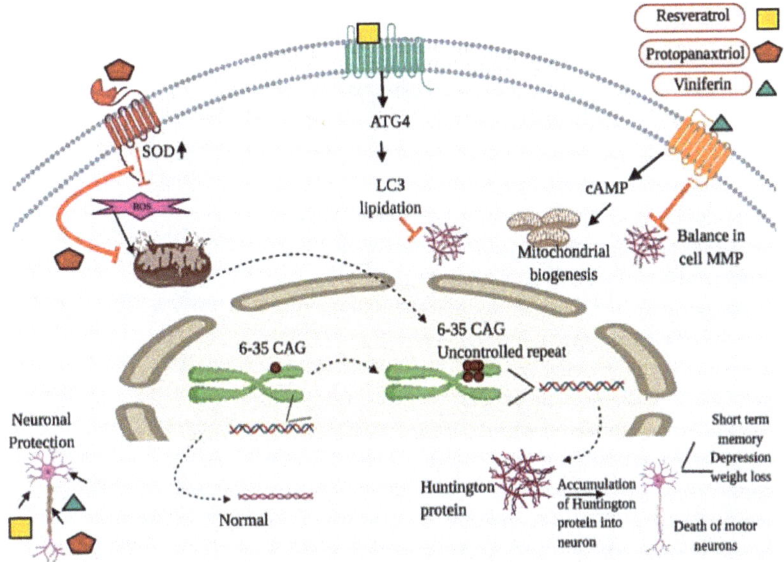

Figure 5. Diagrammatic presentation of Huntington Disease progression and protection of neurons with phytochemicals: 6-35 CAG repeats lead to the formation of Huntingtin protein which consequently accumulates in the neurons and damage motor neurons. In the early stage, damaged motor neurons result in short term memory while on a later stage, this condition changes into depression, weight loss, and many more. The studied phytochemicals generally work against free radicals and restrict the accumulation of Huntingtin protein in the cells. This inhibition further leads to neuronal survival. Resveratrol works on LC3 lipidation while protopanaxtriol and viniferin leads to the antioxidative pathway and restore the function of mitochondria to protect the neurons and stop further accumulation of Huntingtin protein in neurons.

Stroke is itself not a neurodegenerative disease but can lead to a secondary neurodegeneration. It has been known to be a major cause of VCI. It is the third major cause of death in brain disorders worldwide [181]. There is a sudden cessation of transient or permanent reduction of the blood flow to a particular area in the brain leading to deficiency of oxygen, glucose, and energy. The main region affected in the case of Ischemia is cornu ammonis 1 (CA1) of the hippocampus [182]. There is the death of pyramidal neurons of the CA1 region, delayed to 4 to 5 days of ischemia, hence the name is "delayed neuronal death" (DND). It is followed by a series of pathological processes working together including excitotoxicity, oxidative stress, apoptosis, inflammation, gliosis of astrocytes, and microglia in the CA1 region after ischemia/reperfusion. There is a huge burden of the depletion of the endogenous antioxidants and loss of mitochondrial function engendering neuronal death via intrinsic programmed cell death causing behavioral and histological alterations [183]. All this is followed by the breakdown of BBB causing edema. Edema activates the secretion of proinflammatory cytokines like TNF-α, IL-1β, and IL-6 by activated immune cells [184]. The oxidative stress pathway along with the inflammatory and apoptotic pathway causes further loss of neurons and brain tissues expanding the infarct area [184]. The treatment of this dreaded disease may lie in the science of phytochemicals. Many studies showing prevention against stroke have been extensively published, some of them are discussed below.

Resveratrol has been used in one of the studies where it is reported that pretreatment of resveratrol improved the neuronal functions by lowering the chances of ischemic injury in in vivo model. Additionally, it also inhibits axonal deterioration and promotes neuronal growth and synaptic formation. In in vitro study, it is found to be effective against oxygen-glucose deprivation/reoxygenation (OGD/R) injury and promoted neural stem cell (NSC) proliferation [100]. SIRT 1 is involved in the Shh signaling arbitrated the effects of resveratrol on the OGD/R injury neuronal model [100]. Its administration leads to neuronal survival, reticent the apoptosis, and synaptic formations with the upregulation of SIRT1. In rat stroke model, resveratrol slows down the ischemic injury and upgraded the neuronal functions by Sonic Hedgehog Signalling [101]. Ginkgolide B stimulated the proliferation and differentiation of neural stem cells in the in vivo models of NSCs in the rat brain following cerebral ischemia, thereby increasing the proportion of neurons to improve neurological outcomes [102]. Ginkgolide B is a potential neuro-protectant and in a study, the cellular and molecular mechanisms were dissected out by following the BDNF, EGF, and Suppressor of cytokine signaling 2 (SOCS2) [185]. Another phytochemical, curcumin, extensively studied shows beneficial effect in case of stroke too. The role of inflammation is the key factor in raising the problem of the ischemic stroke. Curcumin has shown the neuroprotective effect in the in vivo model of brain ischemia by showing anti-inflammatory properties [186]. It has also shown protection against ischemic injury in N2a cells and mouse brain with stroke [103]. Curcumin plays a significant role in neuronal survival and proliferation in the model of global brain ischemia [104]. EGCG, another phytochemical, also can remove the free radicals by increasing the antioxidative enzyme level and also downregulating the MMP-9. It inhibits the neuronal damage in transient focal cerebral ischemia instigated by middle cerebral artery occlusion (MCAO) in mice. It follows the mTOR-AMPK pathway in stressed ER and impaired the oxidative stress responses and enhance autophagy-dependent survival. It also promotes cell multiplication, differentiation, neovascularization, tube formation, and cell homing by associating the vascular endothelial growth factor (VEGF), regulation of BAX, Bcl-2, LC3B, caspase-3, mTOR and Beclin-1 [187]. It has also protected HBMVECs from Ischemia/Reperfusion injury by ameliorating apoptosis and autophagy and promoting neurovascularization [105]. EGCG has a protective effect on rat brain injury induced by MCAO, possibly by modulating the PI3K/Akt/eNOS signaling pathway [106]. Baicalein also has shown a promising effect in the subacute phase of cerebral Ischemia/reperfusion (I/R) injury in a rat model of ischemia-induced by occlusion of the middle cerebral artery (MCA). Furthermore, it is showed that baicalein inhibits the expression of NF-κB by reducing the IκBα phosphorylation and nuclear translocation of NF-κB/p65, this cascade of signaling molecules is associated with the down-regulation of the pro-inflammatory factors IL-6, IL-18, and TNF-α [107]. Additionally, baicalein inhibits the phosphorylation of ERK, p38, and JNK. These key molecules were involved with the modulation of microglia/macrophage M1/M2 polarization. Studies show the antioxidative activity by decreasing the expression of caspase-3 along with increasing the Bcl-2/Bax ratio, thus protecting the neurons. Baicalein follows the PI3K/Akt/mTOR signaling pathway via decreasing the LC3-II/ LC3-I ratio [188]. Figure 6. *Ligusticum chuanxiong* is used as a therapy for cerebrovascular and neuronal diseases. It works against OGD-reoxy-induced injury and can induce protective HSP-70 expression via the activation of the MAPK pathway [108]. Sinomenine, a natural phytochemical shows the neuroprotection by behaving as an anti-inflammatory and anti-apoptotic agent in a mouse model of middle cerebral artery occlusion (MCAO) and astrocytes/microglia treated with oxygen-glucose deprivation (OGD). It was observed that OGD downregulates the p-AMPK and at the same time activates the inflammasome NLRP3 in mixed glial culture (in vitro model) [109,189]. Honokiol, the main biphenyl neolignan, also shows the antioxidative effects, anti-inflammatory, anti-apoptotic activity. It helps to reduce the cerebral infarction [190]. Honokiol attenuates the inflammatory reaction during cerebral ischemia-reperfusion via inhibiting the by NF-κB activation and cytokine production of glial cells [110]. Zingerone targets the oxidative stress and provide protections in focal transient ischemic rats model [111]. Zingerone (4-(4-hydroxy-3-methoxyphenyl)-2-butanone) is an innocuous, easily available, and economically cost-effective compound with promising therapeutic effects. Having anti-inflammatory, antispasmodic, antioxidative properties, zingerone protects against oxidative stress and damage. It leads to the antioxidative pathway for protection, thus help against ischemic insult [111]. Perillyl alcohol ameliorates the condition of ischemia-reperfusion injury via inhibition of the oxidative damage by following the NF-κB, COX-2, and NOS-2 associated pathway in in vivo Rat model [112]. It is also reported that piperine vanquishes the cerebral ischemia–reperfusion-induced inflammation anti-inflammatory pathway i.e., NOS-2, COX-2, and NF-κB [113]. The list of phytochemicals and the associated mechanisms are explained in Table 1.

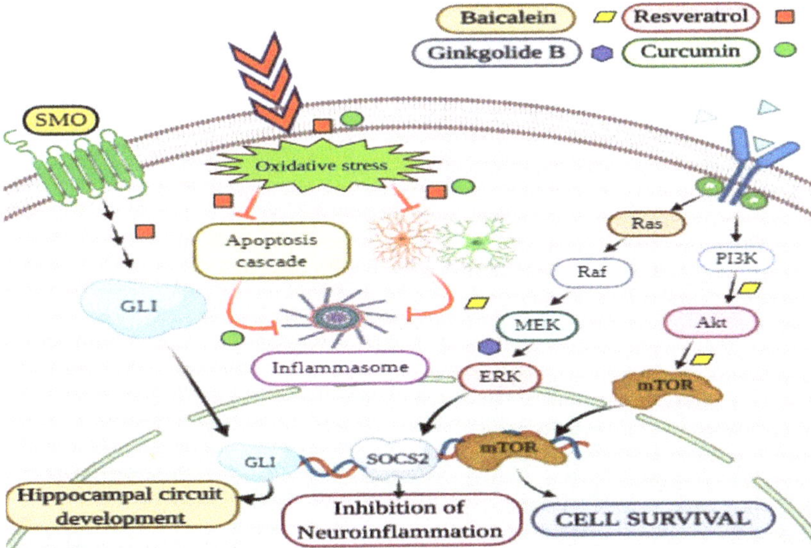

Figure 6. Overall pathogenesis of neuronal loss in ischemic stroke and their prevention by phytochemicals. Oxidative stress, found in the neurons of the ischemic brain is prevented by resveratrol and curcumin. These phytochemicals also prevent the inflammation and apoptosis induced by oxidative stress. Baicalein prevents the formation of inflammasome reducing inflammation and edema. Resveratrol activates SMO/GLI pathway leading to the development of hippocampal neuronal circuit. It also activates the PI3K/Akt/mTOR pathway leading to cell survival. Another phytochemical ginkgolide B acts on MEK/ERK/SOCS2 pathway causing inhibition of inflammation in neurons.

3.6. Prion Disease

Prion disease is also known as transmissible spongiform encephalopathies (TSE) and is the only naturally occurring infectious protein misfolding disorder. Human prion disease is caused by the unnatural conversion of normal prion protein (PrP^c) into an abnormal form of a protein called prions (PrP^sc), it stands for pertinacious infectious particles [191]. PrP^c is a surface glycoprotein expressing in multiple tissues, however, it is highly expressed in the CNS [192]. The onset and progression of the prions disease are explained by the "protein only-hypothesis", wherein it is understood how a protein self-replicates without a nucleic acid. It claims that PrP^c itself is the infectious agent [193]. During the progression of prion disease, the infectious protein PrP^{sc} aggregates and becomes resistant to proteinase digestion. This is followed by PrP^{sc}-seeded conversion of PrP^c into PrP^{sc}, therefore, resulting in the high accumulation of PrP^{sc} in an individual, which leads to neurodegeneration and the clinical manifestation of the disease [194]. This disease is classified as rare both in the progression and manifestation [195].

Several studies have been done to identify the potent agents for drug development for Prion disease. In a recent study, synthetic human prion protein (PrP 106-126) was used as an inducer to manifest prion diseases like signaling. Baicalein from *Scutellaria biacalenesis* attenuated ROS production and also inhibited ROS-induced mitochondrial dysfunction in PrP 106-126-induced-SH-SY5Y. JNK plays an important role in PrP 106-126-induced neuronal apoptosis in SK-N-SH cells, however, baicalein inhibited the JNK signaling in SK-N-SH cells [114]. Resveratrol, a flavonoid with several bioactivities is also known to prevent the PrP 106-126-induced neurotoxicity in SH-SY5Y and SK-N-SH cells by activating autophagy. Thus preventing the mitochondrial dysfunction, apoptosis, and neuronal cell death [115]. Autophagy is a cellular mechanism opted by several phytochemicals to protect against PrP-106-126-induced neurotoxicity, hinokitiol is a monoterpene that activates autophagy and it inhibits apoptosis through stabilization of hypoxia-inducible factor (HIF)-1α, which is important for the regulation of oxygen homeostasis [116]. Rutin has been found to prevent the PrP 106-126-induced neurotoxicity in mouse hippocampal cell line (HT22). Their results showed that rutin was able to increase the expression levels of neurotrophic factors like BDNF, glial cell-derived neurotrophic factor (GDNF), and nerve growth factor (NGF). They also observed the inhibition in apoptosis through suppression of FASL, FAS, caspase 8, and by inhibiting the activity of caspase 3 [117]. Table 1 enlist various neuroprotectant phytochemicals used in the case of Prion disease.

3.7. Frontotemporal Dementia (FTD)

FTD is a term that encompasses a broad group of heterogeneous progressive neurological syndromes. These are non-Alzheimer dementias characterized by the atrophy in the frontal and temporal lobe [196]. As frontal and temporal lobes are highly affected in this disease, the functions controlled by these areas are the outcome of FTD such as behavior and language. Some people with FTD have drastic behavioral and personality changes that make them socially inappropriate and impulsive whereas some people lose the ability to use language [197]. Some other common symptoms include lack of inhibition, apathy and judgment, repetitive compulsive behavior, decline in personal hygiene, eating the inedible objects, and loss of sensitivity towards another person's feelings [198]. FTD can be categorized into three common variant syndromes; behavioral variant FTD (bvFTD), non-fluent/agrammatic variant primary progressive aphasia (nfvPPA), semantic variant PPA (svPPA) [199]. A subtype of FTD is Pick's disease (PiD) which is a neurodegenerative condition that results in irreversible dementia where an excessive amount of tau inclusions known as Pick bodies is a key pathological feature [200]. Pick bodies are round argyrophilic neuronal inclusions. PiD causes cortical atrophy affecting frontal and temporal poles including limbic systems, neocortex, and dentate granular cells of the hippocampus. Such neuronal degenerations result in behavioral changes/alterations and personality disorders [201]. The early signs for PiD are apragmatism and mutism. Some studies refer to PiD as Niemann-Pick disease (NPD) based on the name of the scientist who reported the first case [200].

Not many phytochemicals have been explored for their effect on FTD. Nicotine, a plant alkaloid has been found to have a correctional effect on the behavior of $Grn^{-/-}$ (progranulin gene) mice. Haploinsufficiency of progranulin levels is the result of any mutation in the Grn gene, which is one of the cases of FTD [118]. Mice with insufficient amounts of progranulin secrete inflammatory cytokines at elevated levels and develop microgliosis [202,203]. This study reported that daily intervention of nicotine reversed the sociability and improved the behavioral change in the $Grn^{-/-}$ mice [202]. A traditional Chinese herbal medicine known as Yi-Gan San also known as Yokukansan in Japan containing a mixture of seven dried herbs *Atractylodes lancea, Cnidium officinale, Uncaria rhynchophylla, Angelica acutiloba, Poria cocos, Bupleurum falcatum*, and *Glycyrrhiza uralensis* in a ratio of 4:3:3:3:4:2:1.5 [204] was found to alleviate behavioral changes in patients with FTD [205]. Extracts of *Piper nigrum* (fruit) and *Curcuma longa* (rhizome) also inhibited the AChE significantly which is considered one of the key targets for therapeutic strategies against dementia [119]. The nanoformulation of curcumin can be a promising agent against Niemann-Pick disease type C astrocytes [120]. Curcumin has been reported to have several medicinal properties, wherein it was observed that nanoformulations of curcumin were able to increase cytosolic calcium ion levels and cell viability in astrocytes [120]. Another compound δ-tocopherol was found to decreased cholesterol levels and also lipid accumulation, it also reduced lysosomal enlargement. It was also reported to increase intracellular Ca^{2+} levels [121]. A study by Nekohashi et al., reported that quercetin and luteolin significantly decreased the cholesterol levels as cholesterol [122] as cholesterol is one of the key characteristics of PiD. The details of phytochemicals are given in Table 1.

3.8. Spinocerebellar Ataxia (SCA)

SCA is a term used to refer to hereditary (autosomal dominant) ataxias, these are neurodegenerative conditions affecting the cerebellum (movement controlling) part of the brain and the spinal cord [206]. It comprises a large group of neurodegeneration characterized by cerebellar ataxia with oculomotor dysfunction, dysarthria, also affecting the brain stem. The resulting ataxia is an outcome of atrophy in the cerebellum [207]. All types of SCA are associated with the loss of coordination in eye and hand movements including speech. SCA usually has an adult-onset and the progression is gradual, therefore worsening with time [208]. To date, 44 types of SCA have been identified and classified as SCA1 through SCA44. Many SCAs are an outcome of CAG nucleotide repeat expansion encoding polyglutamine, resulting in the involvement of toxic polyglutamine protein (polyQ). The most recent genetic cause of SCA44 reported in 2017 [209].

The focus on nutraceuticals against this disease is not that much, however, trehalose is a natural compound abundantly found in invertebrates, plants, and microorganisms. Studies have shown that it has a neuroprotective effect exhibited through interaction with protein folding. Analog of trehalose, melibiose has been reported the combat neurotoxicity/neurodegeneration by upregulating autophagy [123]. Melibiose significantly inhibited the aggregation of polyQ decreased ROS levels. It also displayed potential in decreasing SCA17TBP/Q_{79} and significant inhibition to Purkinje cell aggregation in SCA17 transgenic mice [124] Table 1.

3.9. Spinal Muscular Atrophy (SMA)

SMA is a genetic disorder characterized by the atrophy in skeletal muscle. The onset of SMA takes place due to the loss of motor neurons that regulate muscle movement. It results in weakness and makes the movement or any physical activity challenges, this condition worsens with time [210]. This disease is caused by the inadequate production of survival motor neuron (SMN) which is mainly expressed by gene SMN1. SMN2 is also responsible for the production of SMN but relatively far less. [211]. Mutations in SMN1 such as alter or loss results in observed in patients with SMA [212]. The presence of additional third or more copies of the SMN2 gene partially compensates for the damage caused due to the SMN1 gene and milder symptoms are seen in patients [213]. It mainly affects infants and restrict physical movement. SMA is categorized into five types (0-IV) based on the age of the patient and the severity of the disorder [214].

Since SMA is a genetic disorder, the plant products that can reverse or correct the damage are to be identified. *Brucea javanica* and its major compound as found to be bruceine D were found to have SMN2 splicing-correcting property. This study showed that they corrected the splice defect and also increased the activity of SMN2 significantly. *B. javanica* was able to ameliorate the muscle defects [125].

Triptolide is another phytochemical that showed a significant effect in combating SMA in SMA-like mice. The results showed that triptolide isolated from *Tripterygium wilfordii* elevated the production of SMN protein and transcription of SMN2 both in vivo and in vitro experimental models. Hsu et al. concluded that triptolide is a modulator of SMN expression. Triptolide enhances the expression of transcript and protein levels of survival motor neurons in human SMA fibroblasts and improves survival in SMA-like mice [126] Table 1.

4. Conclusions

This review focuses on the use of phytochemicals as a preventive approach to neurodegenerative disorders as they are cost-effective, readily applicable, and easily accessible compounds with a promise in future use. On the other hand, there have always been limitations to the clinical use of these phytochemicals. The main issue focuses on the bioavailability and digestibility of these natural compounds in the body of the patients. Blood–brain barrier is another hurdle in the use of phytochemicals for problems related to the nervous system. Moreover, it is time-consuming therapy for effectiveness against the disease.

Natural products possess variable neuroprotective properties, which can be exploited to target ND therapeutically. Phytochemicals have multi-targeted applications with numerous advantages over currently targeted drugs on the market and have limited adverse reactions. They can also help in developing food supplements and tonics from the edible plants possessing neuroprotective activity. However, given the complexities and heterogeneity of phytochemicals, a cautious approach is necessary to define herbal mixtures and compositions of different compounds for treatment purposes. Keeping this in mind, more preclinical and clinical studies are being conducted. Finally, these plants and other remaining body of studies can very well be executed in not only designing safer drugs to combat the NDs but also preventing the onset. Consequently, this review will abet the researchers for further extensive study in this field.

Author Contributions: Conceptualization: A.K. (Andleeb Khan), M.U.R.; writing—Original draft preparation: A.K. (Andleeb Khan), S.J., and Z.I.; writing—Review and editing: S.A., H.A.M., B.M.A., A.K. (Ajay Kumar), A.A. and M.U.R. All authors have read and agreed to the published version of the manuscript.

Funding: This research received no external funding.

Conflicts of Interest: The authors declare no conflict of interest. The funders had no role in the design of the study; in the collection, analyses, or interpretation of data; in the writing of the manuscript, or in the decision to publish the results.

References

1. Wynford-Thomas, R.; Robertson, N.P. The economic burden of chronic neurological disease. *J. Neurol.* **2017**, *264*, 2345–2347. [CrossRef]
2. Chen, X.; Pan, W. The Treatment Strategies for Neurodegenerative Diseases by Integrative Medicine. *Integr. Med. Int.* **2015**, *1*, 223–225. [CrossRef]
3. Mizuno, Y. Recent Research Progress in and Future Perspective on Treatment of Parkinson's Disease. *Integr. Med. Int.* **2014**, *1*, 67–79. [CrossRef]
4. Shukla, A.W.; Shuster, J.J.; Chung, J.W.; Vaillancourt, D.E.; Patten, C.; Ostrem, J.; Okun, M.S. Repetitive Transcranial Magnetic Stimulation (rTMS) Therapy in Parkinson Disease: A Meta-Analysis. *PM R* **2015**, *8*, 356–366. [CrossRef]
5. Bordet, R.; Ihl, R.; Korczyn, A.D.; Lanza, G.; Jansa, J.; Hoerr, R.; Guekht, A. Towards the concept of disease-modifier in post-stroke or vascular cognitive impairment: A consensus report. *BMC Med.* **2017**, *15*, 107. [CrossRef]
6. Fisicaro, F.; Lanza, G.; Grasso, A.A.; Pennisi, G.; Bella, R.; Paulus, W.; Pennisi, M. Repetitive transcranial magnetic stimulation in stroke rehabilitation: Review of the current evidence and pitfalls. *Ther. Adv. Neurol. Disord.* **2019**, *12*. [CrossRef]

7. Liu, Y.; Yan, T.; Chu, J.M.-T.; Chen, Y.; Dunnett, S.; Ho, Y.-S.; Wong, G.T.C.; Chang, R.C.C. The beneficial effects of physical exercise in the brain and related pathophysiological mechanisms in neurodegenerative diseases. *Lab. Investig.* **2019**, *99*, 943–957. [CrossRef]

8. Lanza, G.; Pino, M.; Fisicaro, F.; Vagli, C.; Cantone, M.; Pennisi, M.; Bella, R.; Bellomo, M. Motor activity and Becker's muscular dystrophy: Lights and shadows. *Physician Sportsmed.* **2019**, *48*, 151–160. [CrossRef]

9. Lanza, G.; Casabona, J.A.; Bellomo, M.; Cantone, M.; Fisicaro, F.; Bella, R.; Pennisi, G.; Bramanti, P.; Pennisi, M.; Bramanti, A. Update on intensive motor training in spinocerebellar ataxia: Time to move a step forward? *J. Int. Med Res.* **2019**. [CrossRef]

10. Lanza, G.; Centonze, S.S.; Destro, G.; Vella, V.; Bellomo, M.; Pennisi, M.; Bella, R.; Ciavardelli, D. Shiatsu as an adjuvant therapy for depression in patients with Alzheimer's disease: A pilot study. *Complement. Ther. Med.* **2018**, *38*, 74–78. [CrossRef]

11. Prasansuklab, A.; Brimson, J.M.; Tencomnao, T. Potential Thai medicinal plants for neurodegenerative diseases: A review focusing on the anti-glutamate toxicity effect. *J. Tradit. Complement. Med.* **2020**. [CrossRef] [PubMed]

12. Pennisi, M.; Lanza, G.; Cantone, M.; D'Amico, E.; Fisicaro, F.; Puglisi, V.; Vinciguerra, L.; Bella, R.; Vicari, E.; Malaguarnera, G. Acetyl-L-Carnitine in Dementia and Other Cognitive Disorders: A Critical Update. *Nutrients* **2020**, *12*, 1389. [CrossRef] [PubMed]

13. Harvey, A.L.; Cree, I.A. High-Throughput Screening of Natural Products for Cancer Therapy. *Planta Medica* **2010**, *76*, 1080–1086. [CrossRef] [PubMed]

14. Harvey, A.L.; Clark, R.L.; Mackay, S.P.; Johnston, B.F. Current strategies for drug discovery through natural products. *Expert Opin. Drug Discov.* **2010**, *5*, 559–568. [CrossRef]

15. Venkatesan, R.; Ji, E.; Kim, S.Y. Phytochemicals That Regulate Neurodegenerative Disease by Targeting Neurotrophins: A Comprehensive Review. *BioMed Res. Int.* **2015**, *2015*, 1–22. [CrossRef]

16. Moher, D.; Liberati, A.; Tetzlaff, J.; Altman, D.G.; PRISMA Group. Preferred reporting items for systematic reviews and meta-analyses: The PRISMA statement. *PLoS Med.* **2009**, *6*, e1000097. [CrossRef]

17. Farkhondeh, T.; Samarghandian, S.; Pourbagher-Shahri, A.M.; Sedaghat, M. The impact of curcumin and its modified formulations on Alzheimer's disease. *J. Cell. Physiol.* **2019**, *234*, 16953–16965. [CrossRef]

18. Macdonald, I.R.; Rockwood, K.; Martin, E.; Darvesh, S. Cholinesterase Inhibition in Alzheimer's Disease: Is Specificity the Answer? *J. Alzheimer's Dis.* **2014**, *42*, 379–384. [CrossRef]

19. Webber, K.M.; Raina, A.K.; Marlatt, M.W.; Zhu, X.; Prat, M.I.; Morelli, L.; Casadesus, G.; Perry, G.; Smith, M.A. The cell cycle in Alzheimer disease: A unique target for neuropharmacology. *Mech. Ageing Dev.* **2005**, *126*, 1019–1025. [CrossRef]

20. Ganguli, M.; Chandra, V.; Kamboh, M.I.; Johnston, J.M.; Dodge, H.H.; Thelma, B.K.; Juyal, R.C.; Pandav, R.; Belle, S.H.; DeKosky, S.T. Apolipoprotein E Polymorphism and Alzheimer Disease. *Arch. Neurol.* **2000**, *57*, 824–830. [CrossRef]

21. Kumar, A.; Singh, A. A review on Alzheimer's disease pathophysiology and its management: An update. *Pharmacol. Rep.* **2015**, *67*, 195–203. [CrossRef] [PubMed]

22. Oghabian, Z.; Mehrpour, O. Treatment of Aluminium Phosphide Poisoning with a Combination of Intravenous Glucagon, Digoxin and Antioxidant Agents. *Sultan Qaboos Univ. Med J.* **2016**, *16*, e352. [CrossRef] [PubMed]

23. Khani, A.H.; Sahragard, A.; Namdari, A.; Zarshenas, M.M. Botanical Sources for Alzheimer's: A Review on Reports from Traditional Persian Medicine. *Am. J. Alzheimer's Dis. Other Dement.* **2017**, *32*, 429–437. [CrossRef]

24. Balasubramanian, S.; Roselin, P.; Singh, K.K.; John, Z.; Saxena, S.N. Postharvest processing and benefits of black pepper, coriander, cinnamon, fenugreek, and turmeric spices. *Crit. Rev. Food Sci. Nutr.* **2016**, *56*, 1585–1607. [CrossRef]

25. Lin, J.-K. The Molecular Targets and Therapeutic Uses of Curcumin in Health and Disease. In *Molecular Targets of Curcumin*; Aggarwal, B.B., Surh, Y.J., Shishodia, S., Eds.; Springer: Boston, MA, USA, 2007; Volume 595, pp. 227–243. [CrossRef]

26. Liu, Z.; Li, T.; Yang, D.; Smith, W.W. Curcumin protects against rotenone-induced neurotoxicity in cell and drosophila models of Parkinson's disease. *Adv. Park. Dis.* **2013**, *2*, 18–27. [CrossRef]

27. Shi, L.-Y.; Zhang, L.; Li, H.; Liu, T.-L.; Lai, J.-C.; Wu, Z.-B.; Qin, J. Protective effects of curcumin on acrolein-induced neurotoxicity in HT22 mouse hippocampal cells. *Pharmacol. Rep.* **2018**, *70*, 1040–1046. [CrossRef]

28. Maan, G.; Sikdar, B.; Kumar, A.; Shukla, R.; Mishra, A. Role of Flavonoids in Neurodegenerative Diseases: Limitations and Future Perspectives. *Curr. Top. Med. Chem.* **2020**, *20*, 1169–1194. [CrossRef]

29. Sundaram, J.R.; Poore, C.P.; Bin Sulaimee, N.H.; Pareek, T.; Cheong, W.F.; Wenk, M.R.; Pant, H.C.; Frautschy, S.A.; Low, C.-M.; Kesavapany, S. Curcumin Ameliorates Neuroinflammation, Neurodegeneration, and Memory Deficits in p25 Transgenic Mouse Model that Bears Hallmarks of Alzheimer's Disease. *J. Alzheimer's Dis.* **2017**, *60*, 1429–1442. [CrossRef]

30. Baum, L.; Lam, C.W.K.; Cheung, S.K.-K.; Kwok, T.; Lui, V.; Tsoh, J.; Lam, L.; Leung, V.; Hui, E.; Ng, C.; et al. Six-Month Randomized, Placebo-Controlled, Double-Blind, Pilot Clinical Trial of Curcumin in Patients with Alzheimer Disease. *J. Clin. Psychopharmacol.* **2008**, *28*, 110–113. [CrossRef]

31. Frost, S.; Kanagasingam, Y.; Macaulay, L.; Koronyo-Hamaoui, M.; Koronyo, Y.; Biggs, D.; Verdooner, S.; Black, K.; Taddei, K.; Shah, T.; et al. O3-13-01: Retinal amyloid fluorescence imaging predicts cerebral amyloid burden and alzheimer's disease. *Alzheimer's Dement.* **2014**, *10*, 234–235. [CrossRef]

32. Zhao, L.; Wang, J.-L.; Wang, Y.-R.; Fa, X.-Z. Apigenin attenuates copper-mediated β-amyloid neurotoxicity through antioxidation, mitochondrion protection and MAPK signal inactivation in an AD cell model. *Brain Res.* **2013**, *1492*, 33–45. [CrossRef] [PubMed]

33. Zhao, L.; Wang, J.-L.; Liu, R.; Li, X.-X.; Li, J.-F.; Zhang, L. Neuroprotective, Anti-Amyloidogenic and Neurotrophic Effects of Apigenin in an Alzheimer's Disease Mouse Model. *Molecules* **2013**, *18*, 9949–9965. [CrossRef] [PubMed]

34. Wiegand, H.; Wagner, A.E.; Saadatmandi, C.B.; Kruse, H.-P.; Kulling, S.; Rimbach, G. Effect of dietary genistein on Phase II and antioxidant enzymes in rat liver. *Cancer Genom. Proteom.* **2009**, *6*, 85–92.

35. Marchenko, N.D.; Zaika, A.; Moll, U.M. Death signal-induced localization of p53 protein to mitochondria a potential role in apoptotic signaling. *J. Biol. Chem.* **2000**, *275*, 16202–16212. [CrossRef] [PubMed]

36. Scalbert, A.; Manach, C.; Morand, C.; Rémésy, C.; Jimenez, L. Dietary Polyphenols and the Prevention of Diseases. *Crit. Rev. Food Sci. Nutr.* **2005**, *45*, 287–306. [CrossRef] [PubMed]

37. Das, S.; Mahapatra, P.; Kumari, P.; Kushwaha, P.P.; Singh, P.; Kumar, S. Phytochemical as Hope for the Treatment of Hepatic and Neuronal Disorders. *Phytochemistry* **2018**, *2*, 289–314. [CrossRef]

38. Farooqui, T.; Farooqui, A.A. *Role of the Mediterranean Diet in the Brain and Neurodegenerative Diseases*; Academic Press: Salt Lake, UT, USA, 2017; p. 484. [CrossRef]

39. Ong, W.-Y.; Farooqui, T.; Ho, C.F.-Y.; Ng, Y.-K.; Farooqui, A.A. Use of Phytochemicals against Neuroinflammation. *Neuroprotective Effects Phytochem. Neurol. Disord.* **2017**, 1–41. [CrossRef]

40. Lee, T.-H.; Jung, M.; Bang, M.-H.; Chung, D.K.; Kim, J.-Y. Inhibitory effects of a spinasterol glycoside on lipopolysaccharide-induced production of nitric oxide and proinflammatory cytokines via down-regulating MAP kinase pathways and NF-κB activation in RAW264.7 macrophage cells. *Int. Immunopharmacol* **2012**, *13*, 264–270. [CrossRef]

41. Liao, W.; Jin, G.; Zhao, M.; Yang, H. The Effect of Genistein on the Content and Activity of α- and β-Secretase and Protein Kinase C in Aβ-Injured Hippocampal Neurons. *Basic Clin. Pharmacol. Toxicol.* **2012**, *112*, 182–185. [CrossRef]

42. Ye, S.; Wang, T.T.; Cai, B.; Wang, Y.; Li, J.; Zhan, J.X.; Shen, G.M. Genistein protects hippocampal neurons against injury by regulating calcium/calmodulin dependent protein kinase IV protein levels in Alzheimer's disease model rats. *Neural Regen. Res.* **2017**, *12*, 1479–1484. [CrossRef]

43. Kuo, L.-C.; Song, Y.-Q.; Yao, C.-A.; Cheng, I.H.; Chien, C.-T.; Lee, G.-C.; Yang, W.-C.; Lin, Y. Ginkgolide A. Prevents the Amyloid-β-Induced Depolarization of Cortical Neurons. *J. Agric. Food Chem.* **2018**, *67*, 81–89. [CrossRef] [PubMed]

44. Tchantchou, F.; Xu, Y.; Wu, Y.; Christen, Y.; Luo, Y. EGb 761 enhances adult hippocampal neurogenesis and phosphorylation of CREB in transgenic mouse model of Alzheimer's disease. *FASEB J.* **2007**, *21*, 2400–2408. [CrossRef] [PubMed]

45. El Tabaa, M.M.; Sokkar, S.S.; Ramadan, E.S.; El Salam, I.Z.A.; Zaid, A.M. Neuroprotective role of Ginkgo biloba against cognitive deficits associated with Bisphenol A exposure: An animal model study. *Neurochem. Int.* **2017**, *108*, 199–212. [CrossRef] [PubMed]

46. Xu, B.; Li, X.-X.; He, G.-R.; Hu, J.-J.; Mu, X.; Tian, S.; Du, G.-H. Luteolin promotes long-term potentiation and improves cognitive functions in chronic cerebral hypoperfused rats. *Eur. J. Pharmacol.* **2010**, *627*, 99–105. [CrossRef] [PubMed]

47. Cavallaro, V.; Baier, C.J.; Murray, M.; Estévez-Braun, A.; Murray, A.P. Neuroprotective effects of Flaveria bidentis and Lippia salsa extracts on SH-SY5Y cells. *S. Afr. J. Bot.* **2018**, *119*, 318–324. [CrossRef]

48. Wang, H.; Wang, H.; Cheng, H.; Che, Z. Ameliorating effect of luteolin on memory impairment in an Alzheimer's disease model. *Mol. Med. Rep.* **2016**, *13*, 4215–4220. [CrossRef]

49. Pappolla, M.A.; Sos, M.; Omar, R.A.; Bick, R.J.; Hickson-Bick, D.L.M.; Reiter, R.J.; Efthimiopoulos, S.; Robakis, N.K. Melatonin Prevents Death of Neuroblastoma Cells Exposed to the Alzheimer Amyloid Peptide. *J. Neurosci.* **1997**, *17*, 1683–1690. [CrossRef]

50. Nie, L.; Wei, G.; Peng, S.; Qu, Z.; Yang, Y.; Yang, Q.; Huang, X.; Liu, J.; Zhuang, Z.; Yang, X. Melatonin ameliorates anxiety and depression-like behaviors and modulates proteomic changes in triple transgenic mice of Alzheimer's disease. *BioFactors* **2017**, *43*, 593–611. [CrossRef]

51. Singer, C.; Tractenberg, R.E.; Kaye, J.A.; Schafer, K.; Gamst, A.; Grundman, M.; Thomas, R.G.; Thal, L.J. A Multicenter, Placebo-controlled Trial of Melatonin for Sleep Disturbance in Alzheimer's Disease. *Sleep* **2003**, *26*, 893–901. [CrossRef]

52. Zhang, N.; Hu, Z.; Zhang, Z.; Liu, G.; Wang, Y.; Ren, Y.; Wu, X.; Geng, F. Protective Role of Naringenin Against Aβ25-35-Caused Damage via ER and PI3K/Akt-Mediated Pathways. *Cell. Mol. Neurobiol.* **2017**, *38*, 549–557. [CrossRef]

53. Haider, S.; Liaquat, L.; Ahmad, S.; Batool, Z.; Ali Siddiqui, R.; Tabassum, S.; Shahzad, S.; Rafiq, S.; Naz, N. Naringenin protects AlCl3/D-galactose induced neurotoxicity in rat model of AD via attenuation of acetylcholinesterase levels and inhibition of oxidative stress. *PLoS ONE* **2020**, *15*, e0227631. [CrossRef] [PubMed]

54. Wang, C.; Cai, Z.; Wang, W.; Wei, M.; Kou, D.; Li, T.; Yang, Z.; Guo, H.; Le, W.; Li, S. Piperine attenuates cognitive impairment in an experimental mouse model of sporadic Alzheimer's disease. *J. Nutr. Biochem.* **2019**, *70*, 147–155. [CrossRef] [PubMed]

55. Li, Y.; Zhou, S.; Li, J.; Sun, Y.; Hasimu, H.; Liu, R.; Zhang, T. Quercetin protects human brain microvascular endothelial cells from fibrillar β-amyloid1–40-induced toxicity. *Acta Pharm. Sin. B* **2015**, *5*, 47–54. [CrossRef] [PubMed]

56. Sabogal-Guáqueta, A.M.; Manco, J.I.M.; Ramírez-Pineda, J.R.; Lamprea-Rodriguez, M.; Osorio, E.; Cardona-Gómez, G.P. The flavonoid quercetin ameliorates Alzheimer's disease pathology and protects cognitive and emotional function in aged triple transgenic Alzheimer's disease model mice. *Neuropharmacology* **2015**, *93*, 134–145. [CrossRef] [PubMed]

57. Wang, H.; Jiang, T.; Li, W.; Gao, N.; Zhang, T. Resveratrol attenuates oxidative damage through activating mitophagy in an in vitro model of Alzheimer's disease. *Toxicol. Lett.* **2018**, *282*, 100–108. [CrossRef]

58. Porquet, D.; Griñán-Ferré, C.; Ferrer, I.; Camins, A.; Sanfeliu, C.; Del Valle, J.; Pallàs, M. Neuroprotective Role of Trans-Resveratrol in a Murine Model of Familial Alzheimer's Disease. *J. Alzheimer's Dis.* **2014**, *42*, 1209–1220. [CrossRef]

59. Sawda, C.; Moussa, C.; Turner, R.S. Resveratrol for Alzheimer's disease. *Ann. NY Acad. Sci.* **2017**, *1403*, 142–149. [CrossRef]

60. Khan, A.; Vaibhav, K.; Javed, H.; Khan, M.M.; Tabassum, R.; Ahmed, E.; Srivastava, P.; Khuwaja, G.; Islam, F.; Siddiqui, M.S.; et al. Attenuation of Aβ-induced neurotoxicity by thymoquinone via inhibition of mitochondrial dysfunction and oxidative stress. *Mol. Cell. Biochem.* **2012**, *369*, 55–65. [CrossRef]

61. Javed, H.; Khan, M.M.; Khan, A.; Vaibhav, K.; Ahmad, A.; Khuwaja, G.; Ahmed, E.; Raza, S.S.; Ashafaq, M.; Tabassum, R.; et al. S-allyl cysteine attenuates oxidative stress associated cognitive impairment and neurodegeneration in mouse model of streptozotocin-induced experimental dementia of Alzheimer's type. *Brain Res.* **2011**, *1389*, 133–142. [CrossRef]

62. Tsai, S.-J.; Chiu, C.P.; Yang, H.T.; Yin, M.C. s-Allyl cysteine, s-ethyl cysteine, and s-propyl cysteine alleviate β-amyloid, glycative, and oxidative injury in brain of mice treated by D-galactose. *J. Agric. Food Chem.* **2011**, *59*, 6319–6326. [CrossRef]

63. Khan, A.; Vaibhav, K.; Javed, H.; Tabassum, R.; Ahmed, E.; Khan, M.M.; Khan, M.B.; Shrivastava, P.; Islam, F.; Siddiqui, M.S.; et al. 1,8-Cineole (Eucalyptol) Mitigates Inflammation in Amyloid Beta Toxicated PC12 Cells: Relevance to Alzheimer's Disease. *Neurochem. Res.* **2013**, *39*, 344–352. [CrossRef] [PubMed]

64. Shen, L.; Liu, L.; Li, X.-Y.; Ji, H.-F. Regulation of gut microbiota in Alzheimer's disease mice by silibinin and silymarin and their pharmacological implications. *Appl. Microbiol. Biotechnol.* **2019**, *103*, 7141–7149. [CrossRef] [PubMed]

65. Song, X.; Liu, B.; Cui, L.; Zhou, B.; Liu, W.; Xu, F.; Hayashi, T.; Hattori, S.; Ushiki-Kaku, Y.; Tashiro, S.-I.; et al. Silibinin ameliorates anxiety/depression-like behaviors in amyloid β-treated rats by upregulating BDNF/TrkB pathway and attenuating autophagy in hippocampus. *Physiol. Behav.* **2017**, *179*, 487–493. [CrossRef] [PubMed]

66. Ay, M.; Luo, J.; Langley, M.; Jin, H.; Anantharam, V.; Kanthasamy, A.; Kanthasamy, A. Molecular mechanisms underlying protective effects of quercetin against mitochondrial dysfunction and progressive dopaminergic neurodegeneration in cell culture and MitoPark transgenic mouse models of Parkinson's Disease. *J. Neurochem.* **2017**, *141*, 766–782. [CrossRef] [PubMed]

67. Sriraksa, N.; Wattanathorn, J.; Muchimapura, S.; Tiamkao, S.; Brown, K.; Chaisiwamongkol, K. Cognitive-Enhancing Effect of Quercetin in a Rat Model of Parkinson's Disease Induced by 6-Hydroxydopamine. *Evid. Based Complement. Altern. Med.* **2011**, *2012*. [CrossRef] [PubMed]

68. Zhang, S.; Liu, Z.-L.; Wu, H.-C.; Hu, Q.-L.; Zhang, S.-J.; Wang, Y.-M.; Jin, Z.-K.; Lv, L.-F.; Wu, H.-L.; Cheng, O.-M. Neuroprotective effects of genistein on SH-SY5Y cells overexpressing A53T mutant α-synuclein. *Neural Regen. Res.* **2018**, *13*, 1375–1383. [CrossRef]

69. Siddique, Y.H.; Naz, F.; Jyoti, S.; Ali, F. Rahul Effect of Genistein on the Transgenic Drosophila Model of Parkinson's Disease. *J. Diet. Suppl.* **2018**, *16*, 550–563. [CrossRef]

70. Jung, U.J.; Kim, S.R. Effects of naringin, a flavanone glycoside in grapefruits and citrus fruits, on the nigrostriatal dopaminergic projection in the adult brain. *Neural Regen. Res.* **2014**, *9*, 1514–1517. [CrossRef]

71. Sang, Q.; Liu, X.; Wang, L.; Qi, L.; Sun, W.; Wang, W.; Sun, Y.; Zhang, H. Curcumin Protects an SH-SY5Y Cell Model of Parkinson's Disease Against Toxic Injury by Regulating HSP90. *Cell. Physiol. Biochem.* **2018**, *51*, 681–691. [CrossRef]

72. Sharma, N.; Nehru, B. Curcumin affords neuroprotection and inhibits α-synuclein aggregation in lipopolysaccharide-induced Parkinson's disease model. *Inflammopharmacology* **2017**, *26*, 349–360. [CrossRef]

73. Chung, W.-G.; Miranda, C.L.; Maier, C.S. Epigallocatechin gallate (EGCG) potentiates the cytotoxicity of rotenone in neuroblastoma SH-SY5Y cells. *Brain Res.* **2007**, *1176*, 133–142. [CrossRef] [PubMed]

74. Zhou, T.; Zhu, M.; Liang, Z. (-)-Epigallocatechin-3-gallate modulates peripheral immunity in the MPTP-induced mouse model of Parkinson's disease. *Mol. Med. Rep.* **2018**, *17*, 4883–4888. [CrossRef]

75. Zhang, J.; Fan, W.; Wang, H.; Bao, L.; Li, G.; Li, T.; Song, S.; Li, H.; Hao, J.; Sun, J. Resveratrol Protects PC12 Cell against 6-OHDA Damage via CXCR4 Signaling Pathway. *Evid. Based Complement. Altern. Med.* **2015**, *2015*, 1–12. [CrossRef] [PubMed]

76. Xia, D.; Sui, R.; Zhang, Z. Administration of resveratrol improved Parkinson's disease-like phenotype by suppressing apoptosis of neurons via modulating the MALAT1/miR-129/SNCA signaling pathway. *J. Cell. Biochem.* **2019**, *120*, 4942–4951. [CrossRef] [PubMed]

77. Magalingam, K.B.; Radhakrishnan, A.; Haleagrahara, N. Protective effects of quercetin glycosides, rutin, and isoquercetrin against 6-hydroxydopamine (6-OHDA)-induced neurotoxicity in rat pheochromocytoma (PC-12) cells. *Int. J. Immunopathol. Pharmacol.* **2015**, *29*, 30–39. [CrossRef] [PubMed]

78. Khan, M.M.; Raza, S.S.; Javed, H.; Ahmad, A.; Khan, A.; Islam, F.; Safhi, M.M.; Islam, F. Rutin Protects Dopaminergic Neurons from Oxidative Stress in an Animal Model of Parkinson's Disease. *Neurotox. Res.* **2011**, *22*, 1–15. [CrossRef] [PubMed]

79. Antunes, M.S.; Goes, A.T.; Boeira, S.P.; Prigol, M.; Jesse, C.R. Protective effect of hesperidin in a model of Parkinson's disease induced by 6-hydroxydopamine in aged mice. *Nutrition* **2014**, *30*, 1415–1422. [CrossRef]

80. Shrivastava, P.; Vaibhav, K.; Tabassum, R.; Khan, A.; Ishrat, T.; Khan, M.M.; Ahmad, A.; Islam, F.; Safhi, M.M.; Islam, F. Anti-apoptotic and Anti-inflammatory effect of Piperine on 6-OHDA induced Parkinson's Rat model. *J. Nutr. Biochem.* **2013**, *24*, 680–687. [CrossRef]

81. Jiang, F.; DeSilva, S.; Turnbull, J. Beneficial effect of ginseng root in SOD-1 (G93A) transgenic mice. *J. Neurol. Sci.* **2000**, *180*, 52–54. [CrossRef]

82. Zhao, Z.; Fu, J.; Li, S.; Li, Z. Neuroprotective Effects of Genistein in a SOD1-G93A Transgenic Mouse Model of Amyotrophic Lateral Sclerosis. *J. Neuroimmune Pharm.* **2019**, *14*, 688–696. [CrossRef]

83. Chico, L.; Caldarazzo-Ienco, E.; Bisordi, C.; LoGerfo, A.; Petrozzi, L.; Petrucci, A.; Mancuso, M.; Siciliano, G. Amyotrophic Lateral Sclerosis and Oxidative Stress: A Double-Blind Therapeutic Trial After Curcumin Supplementation. *CNS Neurol. Disord. Drug Targets* **2018**, *17*, 767–779. [CrossRef] [PubMed]

84. Dutta, K.; Patel, P.; Julien, J.-P. Protective effects of Withania somnifera extract in SOD1G93A mouse model of amyotrophic lateral sclerosis. *Exp. Neurol.* **2018**, *309*, 193–204. [CrossRef] [PubMed]

85. Xu, Z.; Chen, S.; Li, X.; Luo, G.; Li, L.; Le, W. Neuroprotective Effects of (-)-Epigallocatechin-3-gallate in a Transgenic Mouse Model of Amyotrophic Lateral Sclerosis. *Neurochem. Res.* **2006**, *31*, 1263–1269. [CrossRef] [PubMed]

86. Li, Y.-H.; Liu, S.-B.; Zhang, H.-Y.; Zhou, F.-H.; Liu, Y.-X.; Lu, Q.; Yang, L. Antioxidant effects of celastrol against hydrogen peroxide-induced oxidative stress in the cell model of amyotrophic lateral sclerosis. *Sheng Li Xue Bao Acta Physiol. Sin.* **2017**, *69*, 751–758.

87. Yun, Y.C.; Jeong, S.-G.; Kim, S.H.; Cho, G.-W. Reduced sirtuin 1/adenosine monophosphate-activated protein kinase in amyotrophic lateral sclerosis patient-derived mesenchymal stem cells can be restored by resveratrol. *J. Tissue Eng. Regen. Med.* **2019**, *13*, 110–115. [CrossRef] [PubMed]

88. Gao, Y.; Chu, S.-F.; Li, J.-P.; Zhang, Z.; Yan, J.-Q.; Wen, Z.-L.; Xia, C.-Y.; Mou, Z.; Wang, Z.-Z.; He, W.-B.; et al. Protopanaxtriol protects against 3-nitropropionic acid-induced oxidative stress in a rat model of Huntington's disease. *Acta Pharmacol. Sin.* **2015**, *36*, 311–322. [CrossRef] [PubMed]

89. Túnez, I.; Montilla, P.; Muñoz, M.D.C.; Feijóo, M.; Salcedo, M. Protective effect of melatonin on 3-nitropropionic acid-induced oxidative stress in synaptosomes in an animal model of Huntington's disease. *J. Pineal Res.* **2004**, *37*, 252–256. [CrossRef]

90. Vidoni, C.; Secomandi, E.; Castiglioni, A.; Melone, M.A.B.; Isidoro, C. Resveratrol protects neuronal-like cells expressing mutant Huntingtin from dopamine toxicity by rescuing ATG4-mediated autophagosome formation. *Neurochem. Int.* **2018**, *117*, 174–187. [CrossRef]

91. Ehrnhoefer, D.E.; Duennwald, M.; Markovic, P.; Wacker, J.L.; Engemann, S.; Roark, M.; Legieiter, J.; Marsh, J.L.; Thompson, L.M.; Lindquist, S.; et al. Green tea (−)-epigallocatechin-gallate modulates early events in huntingtin misfolding and reduces toxicity in Huntington's disease models. *Hum. Mol. Genet.* **2006**, *15*, 2743–2751. [CrossRef]

92. Fu, J.; Jin, J.; Cichewicz, R.H.; Hageman, S.A.; Ellis, T.K.; Xiang, L.; Peng, Q.; Jiang, M.; Arbez, N.; Hotaling, K.; et al. Trans-(−)-ε-Viniferin increases mitochondrial sirtuin 3 (SIRT3), activates AMP-activated protein kinase (AMPK), and protects cells in models of Huntington Disease. *J. Biol. Chem.* **2012**, *287*, 24460–24472. [CrossRef]

93. Kumar, P.; Kumar, A. Possible Neuroprotective Effect of Withania somnifera Root Extract Against 3-Nitropropionic Acid-Induced Behavioral, Biochemical, and Mitochondrial Dysfunction in an Animal Model of Huntington's Disease. *J. Med. Food* **2009**, *12*, 591–600. [CrossRef] [PubMed]

94. Jeong, J.H.; Lee, S.E.; Lee, J.H.; Kim, H.D.; Seo, K.-H.; Kim, D.H.; Han, S.Y. Aster ageratoides Turcz. extract attenuates Alzheimer's disease-associated cognitive deficits and vascular dementia-associated neuronal death. *Anat. Cell Boil.* **2020**, *53*, 216–227. [CrossRef] [PubMed]

95. Li, W.-Z.; Wu, W.-Y.; Huang, H.; Yin, Y.-Y.; Wu, Y.-Y. Protective effect of bilobalide on learning and memory impairment in rats with vascular dementia. *Mol. Med. Rep.* **2013**, *8*, 935–941. [CrossRef] [PubMed]

96. Wang, J.; Chen, W.; Wang, Y. A ginkgo biloba extract promotes proliferation of endogenous neural stem cells in vascular dementia rats. *Neural Regen. Res.* **2013**, *8*, 1655–1662. [PubMed]

97. Demarin, V.; Kes, V.B.; Trkanjec, Z.; Budišić, M.; Pasić, M.B.; Črnac, P.; Budinčević, H. Efficacy and safety of Ginkgo biloba standardized extract in the treatment of vascular cognitive impairment: A randomized, double-blind, placebo-controlled clinical trial. *Neuropsychiatr. Dis. Treat.* **2017**, *13*, 483–490. [CrossRef] [PubMed]

98. Xu, Z.-Q.; Liang, X.-M.; Wu, J.; Zhang, Y.-F.; Zhu, C.-X.; Jiang, X.-J. Treatment with Huperzine A Improves Cognition in Vascular Dementia Patients. *Cell Biophys.* **2011**, *62*, 55–58. [CrossRef]

99. Xing, S.-H.; Zhu, C.-X.; Zhang, R.; An, L. Huperzine A in the Treatment of Alzheimer's Disease and Vascular Dementia: A Meta-Analysis. *Evid. Based Complement. Altern. Med.* **2014**, *2014*. [CrossRef]

100. Tang, F.; Guo, S.; Liao, H.; Yu, P.; Wang, L.; Song, X.; Chen, J.; Yang, Q. Resveratrol Enhances Neurite Outgrowth and Synaptogenesis Via Sonic Hedgehog Signaling Following Oxygen-Glucose Deprivation/Reoxygenation Injury. *Cell. Physiol. Biochem.* **2017**, *43*, 852–869. [CrossRef]

101. Shen, C.; Cheng, W.; Yu, P.; Wang, L.; Zhou, L.; Zeng, L.; Yang, Q. Resveratrol pretreatment attenuates injury and promotes proliferation of neural stem cells following oxygen-glucose deprivation/reoxygenation by upregulating the expression of Nrf2, HO-1 and NQO1 in vitro. *Mol. Med. Rep.* **2016**, *14*, 3646–3654. [CrossRef]

102. Zheng, J.-S.; Zheng, P.-D.; Mungur, R.; Zhou, H.-J.; Hassan, M.; Jiang, S.-N. Ginkgolide B promotes the proliferation and differentiation of neural stem cells following cerebral ischemia/reperfusion injury, both in vivo and in vitro. *Neural Regen. Res.* **2018**, *13*, 1204–1211. [CrossRef]

103. Xie, C.J.; Gu, A.P.; Cai, J.; Wu, Y.; Chen, R.C. Curcumin protects neural cells against ischemic injury in N2a cells and mouse brain with ischemic stroke. *Brain Behav.* **2018**, *8*, e00921. [CrossRef] [PubMed]

104. De Alcântara, G.F.T.; Simões-Neto, E.; Da Cruz, G.M.P.; Nobre, M.E.P.; Neves, K.R.T.; De Andrade, G.M.; Brito, G.; Viana, G.S.D.B. Curcumin reverses neurochemical, histological and immuno-histochemical alterations in the model of global brain ischemia. *J. Tradit. Complement. Med.* **2016**, *7*, 14–23. [CrossRef] [PubMed]

105. Fu, B.; Zeng, Q.; Zhang, Z.; Qian, M.; Chen, J.; Dong, W.-L.; Li, M. Epicatechin Gallate Protects HBMVECs from Ischemia/Reperfusion Injury through Ameliorating Apoptosis and Autophagy and Promoting Neovascularization. *Oxidative Med. Cell. Longev.* **2019**, *2019*, 7824684. [CrossRef]

106. Park, D.J.; Kang, J.B.; Koh, P.O. Epigallocatechin gallate alleviates neuronal cell damage against focal cerebral ischemia in rats. *J. Vet. Med. Sci.* **2020**, *82*, 639–645. [CrossRef]

107. Yang, S.; Wang, H.; Yang, Y.; Wang, R.; Wang, Y.; Wu, C.; Du, G.-H. Baicalein administered in the subacute phase ameliorates ischemia-reperfusion-induced brain injury by reducing neuroinflammation and neuronal damage. *Biomed. Pharm.* **2019**, *117*, 109102. [CrossRef] [PubMed]

108. Yu, J.; Jiang, Z.; Ning, L.; Zhao, Z.; Yang, N.; Chen, L.; Ma, H.; Li, L.; Fu, Y.; Zhu, H.; et al. Protective HSP70 Induction by Z-Ligustilide against Oxygen–Glucose Deprivation Injury via Activation of the MAPK Pathway but Not of HSF1. *Boil. Pharm. Bull.* **2015**, *38*, 1564–1572. [CrossRef] [PubMed]

109. Qiu, J.; Wang, M.; Zhang, J.; Cai, Q.; Lü, D.; Li, Y.; Dong, Y.; Zhao, T.; Chen, H. The neuroprotection of Sinomenine against ischemic stroke in mice by suppressing NLRP3 inflammasome via AMPK signaling. *Int. Immunopharmacol.* **2016**, *40*, 492–500. [CrossRef] [PubMed]

110. Zhang, P.; Liu, X.; Zhu, Y.; Chen, S.; Zhou, D.; Wang, Y.-Y. Honokiol inhibits the inflammatory reaction during cerebral ischemia reperfusion by suppressing NF-κB activation and cytokine production of glial cells. *Neurosci. Lett.* **2013**, *534*, 123–127. [CrossRef]

111. Vaibhav, K.; Shrivastava, P.; Tabassum, R.; Khan, A.; Javed, H.; Ahmed, E.; Islam, F.; Safhi, M.M.; Islam, F. Delayed administration of zingerone mitigates the behavioral and histological alteration via repression of oxidative stress and intrinsic programmed cell death in focal transient ischemic rats. *Pharmacol. Biochem. Behav.* **2013**, *113*, 53–62. [CrossRef]

112. Tabassum, R.; Vaibhav, K.; Shrivastava, P.; Khan, A.; Ahmed, M.E.; Ashafaq, M.; Khan, M.B.; Islam, F.; Safhi, M.M.; Islam, F. Perillyl alcohol improves functional and histological outcomes against ischemia–reperfusion injury by attenuation of oxidative stress and repression of COX-2, NOS-2 and NF-κB in middle cerebral artery occlusion rats. *Eur. J. Pharmacol.* **2015**, *747*, 190–199. [CrossRef]

113. Vaibhav, K.; Shrivastava, P.; Javed, H.; Khan, A.; Ahmed, E.; Tabassum, R.; Khan, M.M.; Khuwaja, G.; Islam, F.; Siddiqui, M.S.; et al. Piperine suppresses cerebral ischemia–reperfusion-induced inflammation through the repression of COX-2, NOS-2, and NF-κB in middle cerebral artery occlusion rat model. *Mol. Cell. Biochem.* **2012**, *367*, 73–84. [CrossRef] [PubMed]

114. Moon, J.-H.; Park, S.-Y. Baicalein prevents human prion protein-induced neuronal cell death by regulating JNK activation. *Int. J. Mol. Med.* **2014**, *35*, 439–445. [CrossRef] [PubMed]

115. Jeong, J.-K.; Moon, M.-H.; Bae, B.-C.; Lee, Y.-J.; Seol, J.-W.; Kang, H.-S.; Kim, J.-S.; Kang, S.-J.; Park, S.-Y. Autophagy induced by resveratrol prevents human prion protein-mediated neurotoxicity. *Neurosci. Res.* **2012**, *73*, 99–105. [CrossRef] [PubMed]

116. Moon, J.-H.; Lee, J.-H.; Lee, Y.-J.; Park, S.-Y. Hinokitiol protects primary neuron cells against prion peptide-induced toxicity via autophagy flux regulated by hypoxia inducing factor-1. *Oncotarget* **2016**, *7*, 29944–29957. [CrossRef] [PubMed]

117. Na, J.-Y.; Kim, S.; Song, K.; Kwon, J. Rutin Alleviates Prion Peptide-Induced Cell Death Through Inhibiting Apoptotic Pathway Activation in Dopaminergic Neuronal Cells. *Cell. Mol. Neurobiol.* **2014**, *34*, 1071–1079. [CrossRef]

118. Minami, S.S.; Min, S.-W.; Krabbe, G.; Wang, C.; Zhou, Y.; Asgarov, R.; Li, Y.; Martens, L.H.; Elia, L.P.; Ward, M.E.; et al. Progranulin protects against amyloid β deposition and toxicity in Alzheimer's disease mouse models. *Nat. Med.* **2014**, *20*, 1157–1164. [CrossRef]

119. Murata, K.; Matsumura, S.; Yoshioka, Y.; Ueno, Y.; Matsuda, H. Screening of β-secretase and acetylcholinesterase inhibitors from plant resources. *J. Nat. Med.* **2014**, *69*, 123–129. [CrossRef]

120. Maguire, E.; Haslett, L.J.; Welton, J.L.; Lloyd-Evans, E.; Goike, J.; Clark, E.H.; Knifton, H.R.; Shrestha, R.; Wager, K.; Webb, R.; et al. Effects of curcumin nanoformulations on cellular function in Niemann-Pick disease type C astrocytes. *bioRxiv* **2017**, 135830. [CrossRef]

121. Xu, M.; Liu, K.; Swaroop, M.; Porter, F.D.; Sidhu, R.; Finkes, S.; Ory, D.S.; Marugan, J.J.; Xiao, J.; Southall, N.T.; et al. δ-Tocopherol Reduces Lipid Accumulation in Niemann-Pick Type C1 and Wolman Cholesterol Storage Disorders. *J. Boil. Chem.* **2012**, *287*, 39349–39360. [CrossRef]

122. Nekohashi, M.; Ogawa, M.; Ogihara, T.; Nakazawa, K.; Kato, H.; Misaka, T.; Abe, K.; Kobayashi, S. Luteolin and Quercetin Affect the Cholesterol Absorption Mediated by Epithelial Cholesterol Transporter Niemann–Pick C1-Like 1 in Caco-2 Cells and Rats. *PLoS ONE* **2014**, *9*, e97901. [CrossRef]

123. Chen, C.-M.; Lin, C.-H.; Wu, Y.R.; Yen, C.-Y.; Huang, Y.-T.; Lin, J.-L.; Lin, C.-Y.; Chen, W.-L.; Chao, C.-Y.; Lee-Chen, G.-J.; et al. Lactulose and Melibiose Inhibit α-Synuclein Aggregation and Up-Regulate Autophagy to Reduce Neuronal Vulnerability. *Cells* **2020**, *9*, 1230. [CrossRef] [PubMed]

124. Lee, G.-C.; Lin, C.-H.; Tao, Y.-C.; Yang, J.-M.; Hsu, K.-C.; Huang, Y.-J.; Huang, S.-H.; Kung, P.-J.; Chen, W.-L.; Wang, C.-M.; et al. The potential of lactulose and melibiose, two novel trehalase-indigestible and autophagy-inducing disaccharides, for polyQ-mediated neurodegenerative disease treatment. *NeuroToxicology* **2015**, *48*, 120–130. [CrossRef] [PubMed]

125. Baek, J.; Jeong, H.; Ham, Y.; Jo, Y.H.; Choi, M.; Kang, M.; Son, B.; Choi, S.; Ryu, H.W.; Kim, J.; et al. Improvement of spinal muscular atrophy via correction of the SMN2 splicing defect by Brucea javanica (L.) Merr. extract and Bruceine, D. *Phytomedicine* **2019**, *65*, 153089. [CrossRef] [PubMed]

126. Hsu, Y.-Y.; Jong, Y.-J.; Tsai, H.-H.; Tseng, Y.-T.; An, L.-M.; Lo, Y.-C. Triptolide increases transcript and protein levels of survival motor neurons in human SMA fibroblasts and improves survival in SMA-like mice. *Br. J. Pharmacol.* **2012**, *166*, 1114–1126. [CrossRef] [PubMed]

127. Leverenz, J.B.; Quinn, J.F.; Zabetian, C.; Zhang, J.; Montine, K.S.; Montine, T.J. Cognitive impairment and dementia in patients with Parkinson disease. *Curr. Top. Med. Chem.* **2009**, *9*, 903–912.

128. Zecca, L.; Shima, T.; Stroppolo, A.; Goj, C.; Battiston, G.; Gerbasi, R.; Sarna, T.; Swartz, H. Interaction of neuromelanin and iron in substantia nigra and other areas of human brain. *Neuroscience* **1996**, *73*, 407–415. [CrossRef]

129. Hoang, Q.Q. Pathway for Parkinson disease. *Proc. Natl. Acad. Sci. USA* **2014**, *111*, 2402–2403. [CrossRef]

130. Gibb, W.R.; Lees, A.J. The relevance of the Lewy body to the pathogenesis of idiopathic Parkinson's disease. *J. Neurol. Neurosurg. Psychiatry* **1988**, *51*, 745–752. [CrossRef]

131. Albin, R.L. Parkinson's Disease: Background, Diagnosis, and Initial Management. *Clin. Geriatr. Med.* **2006**, *22*, 735–751. [CrossRef]

132. Dauer, W.; Przedborski, S. Parkinson's disease: Mechanisms and models. *Neuron* **2003**, *39*, 889–909. [CrossRef]

133. Yang, Y.X.; Wood, N.W.; Latchman, D.S. Molecular basis of Parkinson's disease. *NeuroReport* **2009**, *20*, 150–156. [CrossRef] [PubMed]

134. Reichmann, H. Modern treatment in Parkinson's disease, a personal approach. *J. Neural Transm.* **2015**, *123*, 73–80. [CrossRef] [PubMed]

135. Peschanski, M.; Defer, G.; Nguyen, J.P.; Ricolfi, F.; Monfort, J.C.; Hantraye, P.; Jeny, R.; Degos, J.D.; Cesaro, P.; Rémy, P.; et al. Bilateral motor improvement and alteration of L-dopa effect in two patients with Parkinson's disease following intrastriatal transplantation of foetal ventral mesencephalon. *Brain* **1994**, *117*, 487–499. [CrossRef] [PubMed]

136. Renaud, J.; Nabavi, S.F.; Daglia, M.; Martinoli, M.-G. Epigallocatechin-3-Gallate, a Promising Molecule for Parkinson's Disease? *Rejuvenation Res.* **2015**, *18*, 257–269. [CrossRef]

137. Haleagrahara, N.; Siew, C.J.; Mitra, N.K.; Kumari, M. Neuroprotective effect of bioflavonoid quercetin in 6-hydroxydopamine-induced oxidative stress biomarkers in the rat striatum. *Neurosci. Lett.* **2011**, *500*, 139–143. [CrossRef]

138. El-Horany, H.E.; El-Latif, R.N.A.; Elbatsh, M.M.; Emam, M.N. Ameliorative Effect of Quercetin on Neurochemical and Behavioral Deficits in Rotenone Rat Model of Parkinson's Disease: Modulating Autophagy (Quercetin on Experimental Parkinson's Disease). *J. Biochem. Mol. Toxicol.* **2016**, *30*, 360–369. [CrossRef]

139. Zhu, J.T.T.; Choi, R.C.Y.; Chu, G.K.Y.; Cheung, A.W.H.; Gao, Q.T.; Li, J.; Jiang, Z.Y.; Dong, T.T.X.; Tsim, K.W.K. Flavonoids Possess Neuroprotective Effects on Cultured Pheochromocytoma PC12 Cells: A Comparison

of Different Flavonoids in Activating Estrogenic Effect and in Preventing β-Amyloid-Induced Cell Death. *J. Agric. Food Chem.* **2007**, *55*, 2438–2445. [CrossRef]

140. Mirahmadi, S.-M.-S.; Shahmohammadi, A.; Rousta, A.-M.; Azadi, M.-R.; Fahanik-Babaei, J.; Baluchnejadmojarad, T.; Roghani, M. Soy isoflavone genistein attenuates lipopolysaccharide-induced cognitive impairments in the rat via exerting anti-oxidative and anti-inflammatory effects. *Cytokine* **2018**, *104*, 151–159. [CrossRef]

141. Lou, H.; Jing, X.; Wei, X.; Shi, H.; Ren, D.; Zhang, X. Naringenin protects against 6-OHDA-induced neurotoxicity via activation of the Nrf2/ARE signaling pathway. *Neuropharmacology* **2014**, *79*, 380–388. [CrossRef]

142. Kim, H.-J.; Song, J.Y.; Park, H.J.; Park, H.-K.; Yun, D.H.; Chung, J.-H. Naringin Protects against Rotenone-induced Apoptosis in Human Neuroblastoma SH-SY5Y Cells. *Korean J. Physiol. Pharmacol.* **2009**, *13*, 281–285. [CrossRef]

143. Shen, S.-C.; Ko, C.H.; Tseng, S.-W.; Tsai, S.-H.; Chen, Y.-C. Structurally related antitumor effects of flavanones in vitro and in vivo: Involvement of caspase 3 activation, p21 gene expression, and reactive oxygen species production. *Toxicol. Appl. Pharmacol.* **2004**, *197*, 84–95. [CrossRef]

144. Lee, W.-H.; Loo, C.-Y.; Bebawy, M.; Luk, F.; Mason, R.S.; Rohanizadeh, R. Curcumin and its Derivatives: Their Application in Neuropharmacology and Neuroscience in the 21st Century. *Curr. Neuropharmacol.* **2013**, *11*, 338–378. [CrossRef]

145. Lee, H.S.; Jung, K.K.; Cho, J.Y.; Rhee, M.H.; Hong, S.; Kwon, M.; Kim, S.H.; Kang, S.Y. Neuroprotective effect of curcumin is mainly mediated by blockade of microglial cell activation. *Die Pharm.* **2007**, *62*, 937–942.

146. Wang, M.S.; Boddapati, S.; Emadi, S.; Sierks, M.R. Curcumin reduces α-synuclein induced cytotoxicity in Parkinson's disease cell model. *BMC Neurosci.* **2010**, *11*, 57. [CrossRef] [PubMed]

147. Dragicevic, N.; Smith, A.; Lin, X.; Yuan, F.; Copes, N.; Delic, V.; Tan, J.; Cao, C.; Shytle, R.D.; Bradshaw, P.C. Green Tea Epigallocatechin-3-Gallate (EGCG) and Other Flavonoids Reduce Alzheimer's Amyloid-Induced Mitochondrial Dysfunction. *J. Alzheimer's Dis.* **2011**, *26*, 507–521. [CrossRef] [PubMed]

148. Chao, J.; Lau, W.K.-W.; Huie, M.J.; Ho, Y.-S.; Yu, M.-S.; Lai, C.S.-W.; Wang, M.; Yuen, W.-H.; Lam, W.H.; Chan, T.H.; et al. A pro-drug of the green tea polyphenol (−)-epigallocatechin-3-gallate (EGCG) prevents differentiated SH-SY5Y cells from toxicity induced by 6-hydroxydopamine. *Neurosci. Lett.* **2010**, *469*, 360–364. [CrossRef]

149. Leaver, K.R.; Allbutt, H.N.; Creber, N.J.; Kassiou, M.; Henderson, J.M. Oral pre-treatment with epigallocatechin gallate in 6-OHDA lesioned rats produces subtle symptomatic relief but not neuroprotection. *Brain Res. Bull.* **2009**, *80*, 397–402. [CrossRef]

150. Segura-Aguilar, J.; Kostrzewa, R.M. Neurotoxin Mechanisms and Processes Relevant to Parkinson's Disease: An Update. *Neurotox. Res.* **2015**, *27*, 328–354. [CrossRef]

151. Hedya, S.A.; Safar, M.M.; Bahgat, A.K. Cilostazol Mediated Nurr1 and Autophagy Enhancement: Neuroprotective Activity in Rat Rotenone PD Model. *Mol. Neurobiol.* **2018**, *55*, 7579–7587. [CrossRef]

152. Ferretta, A.; Gaballo, A.; Tanzarella, P.; Piccoli, C.; Capitanio, N.; Nico, B.; Annese, T.; Di Paola, M.; Dell'Aquila, C.; De Mari, M.; et al. Effect of resveratrol on mitochondrial function: Implications in parkin-associated familiar Parkinson's disease. *Biochim. Biophys. Acta Mol. Basis Dis.* **2014**, *1842*, 902–915. [CrossRef] [PubMed]

153. Quik, M.; Perez, X.A.; Bordia, T. Nicotine as a potential neuroprotective agent for Parkinson's disease. *Mov. Disord.* **2012**, *27*, 947–957. [CrossRef] [PubMed]

154. Hendouei, F.; Moghaddam, H.S.; Mohammadi, M.R.; Taslimi, N.; Rezaei, F.; Akhondzadeh, S. Resveratrol as adjunctive therapy in treatment of irritability in children with autism: A double-blind and placebo-controlled randomized trial. *J. Clin. Pharm. Ther.* **2019**, *45*, 324–334. [CrossRef] [PubMed]

155. Korkmaz, A.; Kolankaya, D. Inhibiting inducible nitric oxide synthase with rutin reduces renal ischemia/reperfusion injury. *Can. J. Surg.* **2013**, *56*, 6–14. [CrossRef] [PubMed]

156. Roohbakhsh, A.; Parhiz, H.; Soltani, F.; Rezaee, R.; Iranshahi, M. Neuropharmacological properties and pharmacokinetics of the citrus flavonoids hesperidin and hesperetin—A mini-review. *Life Sci.* **2014**, *113*, 1–6. [CrossRef] [PubMed]

157. Liu, J.; Chen, M.; Wang, X.; Wang, Y.; Duan, C.; Gao, G.; Lu, L.; Wu, X.; Wang, X.; Yang, H. Piperine induces autophagy by enhancing protein phosphotase 2A activity in a rotenone-induced Parkinson's disease model. *Oncotarget* **2016**, *7*, 60823–60843. [CrossRef]

158. Hardiman, O.; Al-Chalabi, A.; Chio, A.; Corr, E.M.; Logroscino, G.; Robberecht, W.; Shaw, P.J.; Simmons, Z.; van den Berg, L.H. Erratum: Amyotrophic lateral sclerosis. *Nat. Rev. Dis. Primers* **2017**, *3*, 1–19.

159. Boillée, S.; Yamanaka, K.; Lobsiger, C.S.; Copeland, N.G.; Jenkins, N.A.; Kassiotis, G.; Kollias, G.; Cleveland, D.W. Onset and Progression in Inherited ALS Determined by Motor Neurons and Microglia. *Science* **2006**, *312*, 1389–1392. [CrossRef]

160. Al-Chalabi, A.; Berg, L.H.V.D.; Veldink, J. Gene discovery in amyotrophic lateral sclerosis: Implications for clinical management. *Nat. Rev. Neurol.* **2016**, *13*, 96–104. [CrossRef]

161. Simpson, E.P.; Yen, A.A.; Appel, S.H. Oxidative Stress: A common denominator in the pathogenesis of amyotrophic lateral sclerosis. *Curr. Opin. Rheumatol.* **2003**, *15*, 730–736. [CrossRef]

162. Grosskreutz, J.; Bosch, L.V.D.; Keller, B.U. Calcium dysregulation in amyotrophic lateral sclerosis. *Cell Calcium* **2010**, *47*, 165–174. [CrossRef]

163. McGeer, P.L.; McGeer, E.G. Inflammatory processes in amyotrophic lateral sclerosis. *Muscle Nerve* **2002**, *26*, 459–470. [CrossRef] [PubMed]

164. Swash, M.; Schwartz, M.S. What do we really know about amyotrophic lateral sclerosis? *J. Neurol. Sci.* **1992**, *113*, 4–16. [CrossRef]

165. Niebroj-Dobosz, I.; Janik, P. Amino acids acting as transmitters in amyotrophic lateral sclerosis (ALS). *Acta Neurol. Scand.* **1999**, *100*, 6–11. [CrossRef] [PubMed]

166. Semmler, S.; Gagné, M.; Garg, P.; Pickles, S.R.; Baudouin, C.; Hamon-Keromen, E.; Destroismaisons, L.; Khalfallah, Y.; Chaineau, M.; Caron, E.; et al. TNF receptor-associated factor 6 interacts with ALS-linked misfolded superoxide dismutase 1 and promotes aggregation. *J. Boil. Chem.* **2020**, *295*, 3808–3825. [CrossRef] [PubMed]

167. McGeer, E.G.; McGeer, P.L. Pharmacologic Approaches to the Treatment of Amyotrophic Lateral Sclerosis. *BioDrugs* **2005**, *19*, 31–37. [CrossRef] [PubMed]

168. Dutta, K.; Patel, P.; Rahimian, R.; Phaneuf, D.; Julien, J.P. Withania somnifera Reverses Transactive Response DNA Binding Protein 43 Proteinopathy in a Mouse Model of Amyotrophic Lateral Sclerosis/Frontotemporal Lobar Degeneration. *Neurother. J. Am. Soc. Exp. NeuroTher.* **2017**, *14*, 447–462. [CrossRef]

169. Ivanenkov, Y.A.; Balakin, K.V.; Lavrovsky, Y. Small Molecule Inhibitors of NF-B and JAK/STAT Signal Transduction Pathways as Promising Anti-Inflammatory Therapeutics. *Mini Rev. Med. Chem.* **2011**, *11*, 55–78. [CrossRef]

170. Zhang, C.; Liang, W.; Wang, H.; Yang, Y.; Wang, T.; Wang, S.; Wang, X.; Wang, Y.; Feng, H. γ-Oryzanol mitigates oxidative stress and prevents mutant SOD1-Related neurotoxicity in Drosophila and cell models of amyotrophic lateral sclerosis. *Neuropharmacology* **2019**, *160*, 107777. [CrossRef]

171. Szlachcic, W.J.; Switonski, P.M.; Krzyzosiak, W.J.; Figlerowicz, M.; Figiel, M. Huntington disease iPSCs show early molecular changes in intracellular signaling, the expression of oxidative stress proteins and the p53 pathway. *Dis. Model. Mech.* **2015**, *8*, 1047–1057. [CrossRef]

172. Manoharan, S.; Guillemin, G.J.; Abiramasundari, R.S.; Essa, M.M.; Akbar, M. The Role of Reactive Oxygen Species in the Pathogenesis of Alzheimer's Disease, Parkinson's Disease, and Huntington's Disease: A Mini Review. *Oxidative Med. Cell. Longev.* **2016**, *2016*, 1–15. [CrossRef]

173. Friedlander, R.M. Apoptosis and Caspases in Neurodegenerative Diseases. *N. Engl. J. Med.* **2003**, *348*, 1365–1375. [CrossRef] [PubMed]

174. Suwanjang, W.; Govitrapong, P.; Chetsawang, B.; Phansuwan-Pujito, P. The protective effect of melatonin on methamphetamine-induced calpain-dependent death pathway in human neuroblastoma SH-SY5Y cultured cells. *J. Pineal Res.* **2010**, *48*, 94–101. [CrossRef] [PubMed]

175. Andreasen, M.; Lorenzen, N.; Otzen, D. Interactions between misfolded protein oligomers and membranes: A central topic in neurodegenerative diseases? *Biochim. Biophys. Acta Biomembr.* **2015**, *1848*, 1897–1907. [CrossRef] [PubMed]

176. Mason, R.P.; Giorgini, F. Modeling Huntington disease in yeast: Perspectives and future directions. *Prion* **2011**, *5*, 269–276. [CrossRef] [PubMed]

177. Moumné, L.; Betuing, S.; Caboche, J. Multiple Aspects of Gene Dysregulation in Huntington's Disease. *Front. Neurol.* **2013**, *4*, 127. [CrossRef]

178. Szegő, É.M.; Outeiro, T.F.; Kazantsev, A.G. Sirtuins in Brain and Neurodegenerative Disease. *Introd. Rev. Sirtuins Biol. Aging Dis.* **2018**, 175–195. [CrossRef]

179. Túnez, I.; Tasset, I.; La Cruz, V.P.-D.; Santamaría, A. 3-Nitropropionic Acid as a Tool to Study the Mechanisms Involved in Huntington's Disease: Past, Present and Future. *Molecules* **2010**, *15*, 878–916. [CrossRef]

180. O'Brien, J.; Erkinjuntti, T.; Reisberg, B.; Román, G.; Sawada, T.; Pantoni, L.; Bowler, J.V.; Ballard, C.; DeCarli, C.; Gorelick, P.B.; et al. Vascular cognitive impairment. *Lancet Neurol.* **2003**, *2*, 89–98. [CrossRef]

181. Hicks, A.; Jolkkonen, J. Challenges and possibilities of intravascular cell therapy in stroke. *Acta Neurobiol. Exp.* **2009**, *69*, 1–11.

182. Kirino, T. Delayed neuronal death in the gerbil hippocampus following ischemia. *Brain Res.* **1982**, *239*, 57–69. [CrossRef]

183. Abas, F.; Alkan, T.; Goren, B.; Taskapilioglu, O.; Sarandol, E.; Tolunay, S. Neuroprotective effects of postconditioning on lipid peroxidation and apoptosis after focal cerebral ischemia/reperfusion injury in rats. *Turk. Neurosurg.* **2010**, *20*, 1–8. [PubMed]

184. Jin, W.-N.; Shi, S.X.-Y.; Li, Z.; Li, M.; Wood, K.; Gonzales, R.J.; Liu, Q. Depletion of microglia exacerbates postischemic inflammation and brain injury. *Br. J. Pharmacol.* **2017**, *37*, 2224–2236. [CrossRef] [PubMed]

185. Xiao, G.; Lyu, M.; Wang, Y.; He, S.; Liu, X.; Ni, J.; Li, L.; Fan, G.; Han, J.; Gao, X.; et al. Ginkgo Flavonol Glycosides or Ginkgolides Tend to Differentially Protect Myocardial or Cerebral Ischemia–Reperfusion Injury via Regulation of TWEAK-Fn14 Signaling in Heart and Brain. *Front. Pharmacol.* **2019**, *10*, 735. [CrossRef] [PubMed]

186. Jiang, J.; Wang, W.; Sun, Y.J.; Hu, M.; Li, F.; Zhu, D. Neuroprotective effect of curcumin on focal cerebral ischemic rats by preventing blood–brain barrier damage. *Eur. J. Pharmacol.* **2007**, *561*, 54–62. [CrossRef] [PubMed]

187. Nan, W.; Zhonghang, X.; Keyan, C.; Tongtong, L.; Wanshu, G.; Xu, Z.-X. Epigallocatechin-3-Gallate Reduces Neuronal Apoptosis in Rats after Middle Cerebral Artery Occlusion Injury via PI3K/AKT/eNOS Signaling Pathway. *BioMed Res. Int.* **2018**, *2018*. [CrossRef]

188. Aryal, P.; Kim, K.; Park, P.-H.; Ham, S.; Cho, J.; Song, K. Baicalein induces autophagic cell death through AMPK/ULK1 activation and downregulation of mTORC1 complex components in human cancer cells. *FEBS J.* **2014**, *281*, 4644–4658. [CrossRef]

189. Wu, P.-F.; Zhang, Z.; Wang, F.; Chen, J. Natural compounds from traditional medicinal herbs in the treatment of cerebral ischemia/reperfusion injury. *Acta Pharmacol. Sin.* **2010**, *31*, 1523–1531. [CrossRef]

190. Xu, H.; Tang, W.; Du, G.-H.; Kokudo, N. Targeting apoptosis pathways in cancer with magnolol and honokiol, bioactive constituents of the bark of Magnolia officinalis. *Drug Discov. Ther.* **2011**, *5*, 202–210. [CrossRef]

191. Barron, R. Infectious prions and proteinopathies. *Prion* **2017**, *11*, 40–47. [CrossRef]

192. Marín-Moreno, A.; Fernández-Borges, N.; Espinosa, J.C.; Andréoletti, O.; Torres, J.M. Transmission and Replication of Prions. *Prog. Mol. Biol. Transl. Sci.* **2017**, *150*, 181–201. [CrossRef]

193. Ma, J.; Ma, J. Prion disease and the 'protein-only hypothesis'. *Essays Biochem.* **2014**, *56*, 181–191. [CrossRef] [PubMed]

194. Ma, Y.; Ma, J. Immunotherapy against Prion Disease. *Pathogens* **2020**, *9*, 216. [CrossRef] [PubMed]

195. Baiardi, S.; Rossi, M.; Capellari, S.; Parchi, P. Recent advances in the histo-molecular pathology of human prion disease. *Brain Pathol.* **2019**, *29*, 278–300. [CrossRef] [PubMed]

196. Bott, N.; Radke, A.; Stephens, M.L.; Kramer, J.H. Frontotemporal dementia: Diagnosis, deficits and management. *Neurodegener. Dis. Manag.* **2014**, *4*, 439–454. [CrossRef]

197. Kelley, R.E.; El-Khoury, R. Frontotemporal Dementia. *Neurol. Clin.* **2016**, *34*, 171–181. [CrossRef]

198. Mohandas, E.; Rajmohan, V. Frontotemporal dementia: An updated overview. *Indian J. Psychiatry* **2009**, *51*, S65–S69.

199. Miller, B.L. Frontotemporal Dementia. *Front. Dement.* **2014**, *35*, 339–374. [CrossRef]

200. Kertesz, A.; Muñoz, D. Pick's Disease, Frontotemporal Dementia, and Pick Complex. *Arch. Neurol.* **1998**, *55*, 302–304. [CrossRef]

201. Frederick, J. Pick disease: A brief overview. *Arch. Pathol. Lab. Med.* **2006**, *130*, 1063–1066.

202. Minami, S.S.; Shen, V.; Le, D.; Krabbe, G.; Asgarov, R.; Liberty, P.-C.; Lee, C.-H.; Li, J.; Donnelly-Roberts, D.; Gan, L. Reducing inflammation and rescuing FTD-related behavioral deficits in progranulin-deficient mice with α7 nicotinic acetylcholine receptor agonists. *Biochem. Pharmacol.* **2015**, *97*, 454–462. [CrossRef]

203. Yin, F.; Banerjee, R.; Thomas, B.; Zhou, P.; Qian, L.; Jia, T.; Ma, X.; Ma, Y.; Iadecola, C.; Beal, M.F.; et al. Exaggerated inflammation, impaired host defense, and neuropathology in progranulin-deficient mice. *J. Exp. Med.* **2009**, *207*, 117–128. [CrossRef] [PubMed]

204. De Caires, S.; Steenkamp, V. Use of Yokukansan (TJ-54) in the treatment of neurological disorders: A review. *Phytother. Res.* **2010**, *24*, 1265–1270. [CrossRef] [PubMed]
205. Kimura, T.; Hayashida, H.; Furukawa, H.; Takamatsu, J. Pilot study of pharmacological treatment for frontotemporal dementia: Effect of Yokukansan on behavioral symptoms. *Psychiatry Clin. Neurosci.* **2010**, *64*, 207–210. [CrossRef] [PubMed]
206. Klockgether, T.; Mariotti, C.; Paulson, H.L. Spinocerebellar Ataxia. *Nat. Rev. Dis. Primers* **2019**, *5*, 24. [CrossRef] [PubMed]
207. Schöls, L.; Bauer, P.; Schmidt, T.; Schulte, T.; Riess, O. Autosomal dominant cerebellar ataxias: Clinical features, genetics, and pathogenesis. *Lancet Neurol.* **2004**, *3*, 291–304. [CrossRef]
208. Groth, C.L.; Berman, B.D. Spinocerebellar Ataxia 27: A Review and Characterization of an Evolving Phenotype. *Tremor Other Hyperkinetic Mov.* **2018**, *8*, 534. [CrossRef]
209. Sullivan, R.; Yau, W.Y.; O'Connor, E.; Houlden, H. Spinocerebellar ataxia: An update. *J. Neurol.* **2018**, *266*, 533–544. [CrossRef]
210. Kolb, S.J.; Kissel, J.T. Spinal Muscular Atrophy. *Arch. Neurol.* **2011**, *68*, 979. [CrossRef]
211. Farrar, M.A.; Kiernan, M.C. The Genetics of Spinal Muscular Atrophy: Progress and Challenges. *Neurotherapeutics* **2014**, *12*, 290–302. [CrossRef]
212. Rao, V.K.; Kapp, D.; Schroth, M. Gene Therapy for Spinal Muscular Atrophy: An Emerging Treatment Option for a Devastating Disease. *J. Manag. Care Spéc. Pharm.* **2018**, *24*, S3–S16. [CrossRef]
213. Singh, R.N.; Singh, N.N. Mechanism of Splicing Regulation of Spinal Muscular Atrophy Genes. *Adv. Neurobiol.* **2018**, *20*, 31–61. [CrossRef]
214. Fuller, H.R.; Gillingwater, T.H.; Wishart, T.M. Commonality amid diversity: Multi-study proteomic identification of conserved disease mechanisms in spinal muscular atrophy. *Neuromuscul Disord.* **2016**, *26*, 560–569. [CrossRef] [PubMed]

 © 2020 by the authors. Licensee MDPI, Basel, Switzerland. This article is an open access article distributed under the terms and conditions of the Creative Commons Attribution (CC BY) license (http://creativecommons.org/licenses/by/4.0/).

Systematic Review

Effects of Treadmill Exercise on Neural Mitochondrial Functions in Parkinson's Disease: A Systematic Review of Animal Studies

Nguyen Thanh Nhu [1,2], Yu-Jung Cheng [2] and Shin-Da Lee [2,3,4,*]

1   Faculty of Medicine, Can Tho University of Medicine and Pharmacy, Can Tho 94117, Vietnam; ntnhu@ctump.edu.vn
2   Department of Physical Therapy, Graduate Institute of Rehabilitation Science, China Medical University, Taichung 41354, Taiwan; chengyu@mail.cmu.edu.tw
3   School of Rehabilitation Medicine, Weifang Medical University, Weifang 261053, China
4   Department of Physical Therapy, Asia University, Taichung 41354, Taiwan
*   Correspondence: shinda@mail.cmu.edu.tw; Tel.: +886-4-22053366 (ext. 7300)

Abstract: This systematic review sought to determine the effects of treadmill exercise on the neural mitochondrial respiratory deficiency and neural mitochondrial quality-control dysregulation in Parkinson's disease. PubMed, Web of Science, and EMBASE databases were searched through March 2020. The English-published animal studies that mentioned the effects of treadmill exercise on neural mitochondria in Parkinson's disease were included. The CAMARADES checklist was used to assess the methodological quality of the studies. Ten controlled trials were included (median CAMARADES score = 5.7/10) with various treadmill exercise durations (1–18 weeks). Seven studies analyzed the neural mitochondrial respiration, showing that treadmill training attenuated complex I deficits, cytochrome c release, ATP depletion, and complexes II–V abnormalities in Parkinson's disease. Nine studies analyzed the neural mitochondrial quality-control, reporting that treadmill exercise improved mitochondrial biogenesis, mitochondrial fusion, and mitophagy in Parkinson's disease. The review findings supported the hypothesis that treadmill training could attenuate both neural mitochondrial respiratory deficiency and neural mitochondrial quality-control dysregulation in Parkinson's disease, suggesting that treadmill training might slow down the progression of Parkinson's disease.

Keywords: treadmill exercise; Parkinson's disease; neural mitochondrial functions

Citation: Nhu, N.T.; Cheng, Y.-J.; Lee, S.-D. Effects of Treadmill Exercise on Neural Mitochondrial Functions in Parkinson's Disease: A Systematic Review of Animal Studies. *Biomedicines* **2021**, *9*, 1011. https://doi.org/10.3390/biomedicines9081011

Academic Editor: Kumar Vaibhav

Received: 16 July 2021
Accepted: 12 August 2021
Published: 13 August 2021

Publisher's Note: MDPI stays neutral with regard to jurisdictional claims in published maps and institutional affiliations.

Copyright: © 2021 by the authors. Licensee MDPI, Basel, Switzerland. This article is an open access article distributed under the terms and conditions of the Creative Commons Attribution (CC BY) license (https://creativecommons.org/licenses/by/4.0/).

1. Introduction

Parkinson's disease (PD) is the second most common neurodegenerative disorder, causing a considerable number of disabilities globally [1,2]. The underlying mechanisms of PD are unclear, making it difficult to find efficient targeted therapies [3]. The current treatment of PD addresses symptomatic improvement only because none of the available treatment strategies have been confirmed to slow down PD progression [2].

Recently, neural mitochondrial respiratory deficiency has emerged as a central hallmark of Parkinsonian etiology [4,5]. In PD, the electron transport system of the neural mitochondria is impaired, mainly characterized by mitochondrial complex I deficit, cytochrome *c* release, and ATP depletion [6]. Impaired mitochondrial respiration in the brain increases oxidative stress and neuron loss, thereby augmenting PD progression [4].

Studies have also linked neural mitochondrial respiratory deficiency to neural mitochondrial quality-control dysregulation in PD [7,8]. In the physiological condition, neural mitochondrial quality-control involves a balance among biogenesis, dynamics (fusion/fission), and mitophagy (autophagy of mitochondria) [7]. Mitochondrial biogenesis produces the new mitochondria and mitochondrial content, accompanied by mitochondrial fusion to maintain a healthy mitochondrial network [9]. Meanwhile, mitochondrial fission segregates the damaged mitochondria and provides those for the mitophagy process, preventing the accumulation of dysfunctional mitochondria in the brain [10].

In PD, the biogenesis regulators and import machinery of neural mitochondria are reduced, leading to inhibition of mitochondrial biogenesis [11]. Additionally, PD has been confirmed to induce an imbalance of neural mitochondrial dynamics (fusion/fission) [4]. Moreover, the mitophagy process has been proven to be impaired with reductions of lysosomal activities in PD [12]. Disorders of biogenesis, fusion/fission, and mitophagy reduce the quantity and quality of neural mitochondria, leading to a neural mitochondrial respiratory deficiency in PD [4,7].

Treadmill exercise (TE) has been widely applied in PD rehabilitation [13,14]. Previous evidence indicated that TE training improved both the symptoms and the quality of life in PD patients [13]. In addition, a previous study showed that TE training improved gait functions by modulating neural mitochondrial dynamics in a PD rats model [15]. Another study reported that TE training reduced both neuron loss and behavioral disorders by improving mitochondrial respiration in a PD mouse model [16]. Those data suggested that TE training not only improves symptoms but also delays PD progression by attenuating PD-induced neural mitochondrial damage. However, the various approaches of the individual studies make it difficult to comprehensively understand the effects of TE on neural mitochondria in PD. Although several systematic reviews have been carried out to summarize the neuroprotective effects of TE training on PD, none of them specifically analyzed neural mitochondrial respiratory deficiency and neural mitochondrial quality-control dysregulation.

Therefore, we conducted a systematic review of animal studies to summarize the effects of TE training on mitochondrial functions in PD, focusing on the following objectives: (1) The effects of TE training on neural mitochondrial respiratory deficiency in PD; and (2) The effects of TE training on neural mitochondrial quality-control dysregulation in PD, including neural mitochondrial biogenesis, neural mitochondrial dynamics (fusion/fission), and neural mitophagy.

2. Materials and Methods

2.1. Protocol and Registration

The protocol of this systematic review was registered on PROSPERO with the registration number: CRD42020164122. We followed "the Preferred Reporting Items for Systematic Reviews and Meta-Analyses (PRISMA) checklist" [17] and "the PRISMA for abstract checklist" [18] to conduct and report this systematic review.

2.2. Eligibility Criteria

Types of study designs: controlled-trial animal studies with separate experimental groups. The studies were English publications without any restriction of publication date. Protocol articles, case reports, reviews, and conference abstracts were excluded from this systematic review.

Types of animal models: animal models of PD. Sex, age, and species of the subjects were not restricted. Studies were excluded if they did not provide sufficient data about the animal species or the PD induction (model types and timing).

Types of intervention: treadmill exercise (TE) training without any restriction of the protocol. The information about timing, duration, and frequency must be provided. Studies that evaluated the effects of TE in combination with the other therapies were excluded.

Type of comparators: studies that at least reported a comparison among a normal group, a sedentary PD group, and a TE-trained PD group. The normal control group must not be treated with any other therapeutic methods.

Type of outcomes: for the effects of TE training on neural mitochondrial respiratory deficiency in PD (the first objective), our outcomes were the components of the electron transport system (complexes I–V, cytochrome c, coenzyme Q10) and ATP production. For the effects of TE training on neural mitochondrial quality-control dysregulation in PD (the second objective), our outcomes included neural mitochondrial biogenesis, neural mitochondrial dynamics (fusion/fission), and neural mitophagy. The neural mitochondrial biogenesis outcome was measured through biogenesis regulators and translocase factors. The neural mitochondrial dynamic outcome was measured through fusion factors and fission factors. The neural mitophagy outcome was measured through dysfunctional mitochondria detectors, autophagosomal factors, and lysosomal factors.

2.3. Information Sources and Search Strategy

Relevant studies were identified by keywords searching on PubMed, Web of Science and EMBASE databases through March 2020, with a combination of the terms: (*"Parkinson" OR "Parkinsonism" OR "Parkinsonian") AND "treadmill" AND ("mitochondria" OR "mitochondrial" OR "mitochondrion" OR "mitophagy" OR "ATP" OR "SIRT" OR "AMPK" OR "PGC-1α" OR "TFAM" OR "NRF" OR "mito-fusion" OR "mito-fission"*). We also reviewed references of the included studies to find other eligible papers. Briefly, the titles and abstracts of studies were screened to exclude duplications and irrelevant studies that did not mention the relevant information of Parkinson's disease and exercise in their abstracts. Then, full texts were assessed and read to see if they met the eligibility criteria. Two independent assessors conducted the process of study selection. When a disagreement occurred, two assessors discussed with a third consultant to make the final decision.

2.4. Data Collection Process

Two independent reviewers extracted the data (including study characteristics and outcomes) by reading the text, graphs, and tables of the included studies. If the data were not available, we contacted the corresponding authors to request it. For study characteristics, we extracted the data of the first author's name, published year, PD model (type, species, and sex), and TE protocol (timing, frequency, duration, and speed). For the outcome extraction, we extracted the data of neural mitochondrial respiration, neural mitochondrial biogenesis, neural mitochondrial dynamics (fusion/fission), and neural mitophagy.

2.5. Study Quality Evaluation

The study quality was evaluated by using the "Collaborative Approach to Meta-Analysis and Review of Animal Data from Experimental Studies" (CAMARADES) checklist with ten items [19]. Two authors independently evaluated and filled in the predesigned datasheets of the CAMARADES checklist, then disagreements were resolved by discussing among the two evaluators and the third consultant.

2.6. Data Synthesis and Presentation

The results of the search strategy are shown in the PRISMA flowchart and the narrative synthesis. Text and tables were used to present the study characteristics and the outcomes. For presentation of study characteristics, the summaries of the PD models, TE protocols, and types of outcomes are provided. For presentation of the outcomes, the effects of TE on neural mitochondrial respiratory deficiency and mitochondrial quality-control dysregulation in PD are described in the comparison among the normal control group, the PD group, and the TE-trained PD group.

3. Results

3.1. Search Results

A total of 76 articles were found from PubMed (n = 15), Web of Science (n = 28), and EMBASE (n = 33) (Figure 1). After title and abstract screening, we removed 50 records: 31 duplications and 19 irrelevant studies. After full-text reviewing, we excluded 15 studies, including 1 erratum paper, 2 reviews, 4 conference abstracts, and 1 ex-vivo study. In addition, we excluded 2 studies that did not use TE, 4 studies that did not mention our outcomes, and 1 study that was missing data. There were two publications that came from one group of authors, conducting the same protocol and reporting the same outcome in one year. We considered them as one study, therefore reported one study and excluded the other. Besides, the studies that came from the same group of authors and conducting the same protocol, but reported different outcomes were considered as separated publications to discuss. Finally, ten publications were included in the current systematic review. No additional articles were found by reading the reference lists of the included studies.

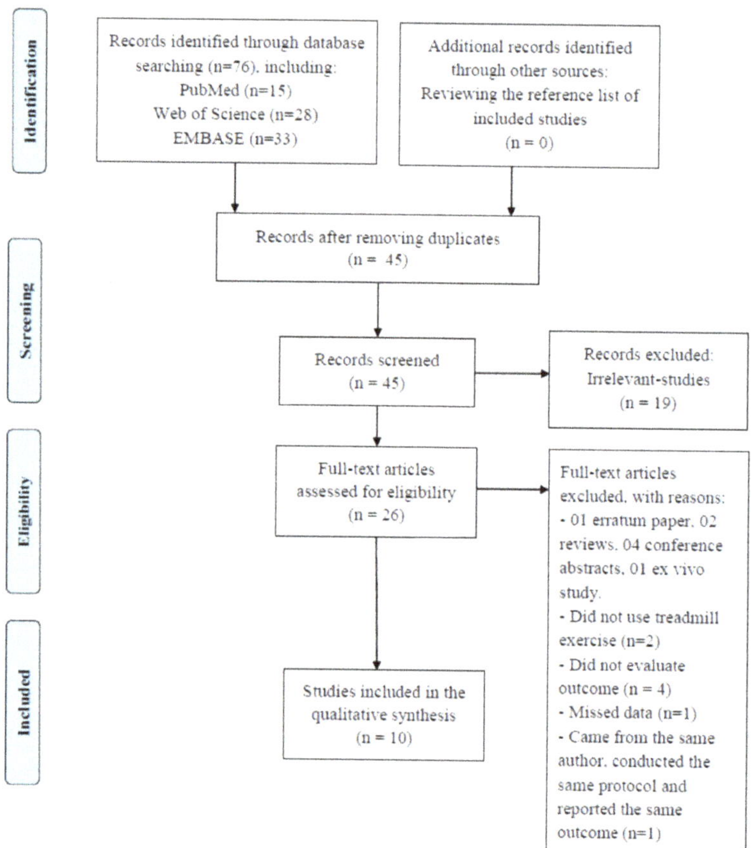

Figure 1. PRISMA flow chart for the selected protocol.

3.2. Study Characteristics

Types of PD models: rats and mice were used with two models, including the 1-methyl-4-phenyl-1,2,3,6-tetrahydropyridine (MPTP) model (n = 6) and the 6-hydroxydopamine (6-ODHA) model (n = 4). Eight studies used young animals (7–8 weeks old), whereas two studies used middle-aged animals (6–10 months old) (Table 1).

Table 1. Characteristics and outcomes of the included studies.

Study	Model	Treadmill Exercise	Brain's Tissue	Outcomes			
				Mitochondrial Respiratory Function	Mitochondrial Biogenesis	Mitochondrial Dynamic	Mitophagy
Koo and Cho, 2017 [20]	MPTP model on male mice (7-wk-old), induced by 25 mg/kg MPTP i.p, twice/wk for 5 wks	After PD induction. Duration: 40–60 min/day, 5 days/wk, 8 wks. Speed: 10–12 min/m.	Substantia Nigra Striatum	Compared to normal, protein levels of complex IV and cytochrome *c* were reduced in PD group. TE training enhanced those levels in PD.	Compared to normal, protein levels of SIRT1, PGC-1α, NRF-1, and TFAM were reduced in PD group. TE training enhanced those levels in PD.		Compared to normal, the protein levels of p62, beclin-1 levels, and LC3II/I ratio were increased in PD group. TE training reduced p62 levels, but unchanged beclin-1 and LC3II/I levels in PD.
Koo et al., 2017a [21]	MPTP model on male mice (7-wk-old) induced by 25 mg/kg MPTP i.p twice/wk for 5 wks	After PD induction. Duration: 40–60 min/day, 5 days/wk, 8 wks. Speed: 10–12 min/m.	Substantia Nigra Striatum	Compared to normal, protein levels of complex IV were reduced in PD group. TE training enhanced those levels in PD.	Compared to normal, protein levels of TOM-40, TOM-20, TIM-23, and mtHSP70 were reduced in PD group. TE training enhanced those levels in PD.		
Rezaee et al., 2019 [22]	6-OHDA model on male rats (8-wk-old) induced by 2 µg/µL injected to the right medial forebrain bundle	Before PD induction. Duration: 25–50 min/day, 5 days/wk, 16 wks. Speed: 15–21 m/min.	Striatum		Compared to normal, both mRNA expression and protein levels of AMPK and PGC-1α were reduced in PD group. Both mRNA and protein levels of SIRT1 and TFAM were increased in PD group. All of those levels were enhanced by TE in PD group.		

Table 1. *Cont.*

Study	Model	Treadmill Exercise	Brain's Tissue	Outcomes			
				Mitochondrial Respiratory Function	Mitochondrial Biogenesis	Mitochondrial Dynamic	Mitophagy
Chuang et al., 2017 [15]	6-OHDA model on female rats (8-wk-old) induced by 15 μg/μL injected to the ascending mesostriatal pathway	After PD induction. Duration: 30 min/day, 7 days/wk, 4 wks. Speed: 15 m/min.	Substantia Nigra Striatum	Compared to normal, protein levels of complex I was reduced, whereas complex II, III, IV protein levels were increased in PD. TE increased complex I levels and reduced complex II, III, IV levels in PD. Complex V protein levels were unchanged among three groups.	Compared to normal, protein levels of TOM-20 were reduced in PD group. TE training enhanced TOM-20 levels in PD in Substantia Nigra. TOM-20 level was unchanged among three groups in striatum.	Compared to normal, protein levels of OPA-1, MFN-2, and Drp1 were reduced in PD group. TE training enhanced those levels in PD.	Compared to normal, the protein levels of PINK1 were increased in PD group. TE training reduced those levels in PD
Jang et al., 2018 [23]	MPTP model on male mice (7-wk-old) induced by 25 mg/kg MPTP i.p daily, 7 days	After PD induction. Duration: 60 min/day, 5 days/wk, 6 wks. Speed: 12 m/min.	Substantia Nigra	Compared to normal, protein levels of complex II and V were reduced in PD group. TE training enhanced complex II and V levels in PD. Complex I, III, IV protein levels were unchanged among all groups.	Compared to normal, protein levels of TFAM, NRF-1, and SIRT3 were reduced in PD group. TE training enhanced those levels in PD. PGC-1α protein level was unchanged among three groups.	Compared to normal, protein levels of OPA-1, MFN-2, and p-Drp1^{Ser637} were reduced in PD group. TE training enhanced those levels in PD.	
Tuon et al., 2015 [16]	6-OHDA model on male mice (8-wk-old) induced by 2 μg/μL injected to the striatum	Before PD induction. Duration: 50 min/day, 3–4 days/wk, 8 wks Speed: 13–17 m/min.	Striatum Hippocampus		Compared to normal, protein levels of SIRT1 were reduced in PD group. TE training enhanced those levels in PD.		

Table 1. *Cont.*

Study	Model	Treadmill Exercise	Brain's Tissue	Outcomes			
				Mitochondrial Respiratory Function	Mitochondrial Biogenesis	Mitochondrial Dynamic	Mitophagy
Patki and Lau, 2011 [24]	MPTP model on male mice (6–10 month-old) induced by 15 mg/kg MPTP, 10 doses, s.c., 5 wks	Before and after PD induction. Duration: 40 min/day, 5 days/wk, 18 wks. Speed: 15 m/min.	Striatum	Compared to normal, cytochrome c protein level in mitochondria was reduced in PD group. TE training enhanced those levels in PD.	Compared to normal, the mRNA levels of TFAM, PGC-1α were increased in PD group. TE training reduced those levels in PD to normal.		
Lau et al., 2011 [25]	MPTP model on male mice (6–10 month-old) induced by 15 mg/kg MPTP, s.c., 10 doses, 5 wks	Before and after PD induction. Duration: 40 min/day, 5 days/wk, 18 wks. Speed: 15 m/min.	Substantia Nigra Striatum	Compared to normal, the mitochondrial respiration stage 3–4, as well as ATP production, were reduced in PD group. TE training enhanced those levels in PD			
Ferreira et al., 2020 [26]	6-OHDA model on male mice (2–3 month-old) induced by 6 μg/μL injected to the striatum	After PD induction Duration: 40 min/day, 3 days/wk, 1 or 4 wks Speed: 10 m/min.	Substantia Nigra Striatum	Compared to normal, the protein level of complex I was reduced in PD group. 4-week TE training enhanced complex I levels in PD. Complex II-V in substantia nigra and complex I-V levels in striatum were unchanged among all groups.	Compared to normal, the levels of PGC-1α, NRF-1, and TFAM were reduced in PD. Those levels significantly increased to normal after 4 weeks training in Substantia Nigra. Those levels in striatum were unchanged among all groups.		

Table 1. *Cont.*

Study	Model	Treadmill Exercise	Brain's Tissue	Outcomes			
				Mitochondrial Respiratory Function	Mitochondrial Biogenesis	Mitochondrial Dynamic	Mitophagy
Hwang et al., 2018 [27]	MPTP model on male mice (8 week-old) induced by 25 mg/kg MPTP, i.p, 10 doses, 5 wks	After PD induction. Duration: 20 min/day, 5 days/wk, 8 wks. Speed: 15 m/min.	Substantia Nigra				Compared to normal, the protein levels of PINK1, Parkin, p62, and LC3II/I ratio were increased in PD group. TE training reduced the levels of PINK1, parkin, and p62, but unchanged LC3II/I ratio in PD. Compared to normal, protein levels of LAMP2 and Cathepsin L were reduced in PD group. TE enhanced those levels in PD.

Note: TE = treadmill exercise; PD = Parkinson's disease; MPTP = 1-methyl-4-phenyl-1,2,3,6-tetrahydropyridine; 6-OHDA = 6-Hydroxydopamine; COX-I, COX-IV = Cytochrome c oxidase subunit I & IV; PGC-1α = peroxisome proliferator-activated receptor gamma coactivator 1-alpha; NRF-1,2 = nuclear respiratory factor 1 and 2; TFAM = mitochondrial transcription factor A; SIRT3 = sirtuin-3; AMPK = AMP-activated protein kinase; SIRT1 = sirtuin-1; TOM = translocase of outer membrane; TIM = translocase of inner membrane; mtHSP70 = mitochondrial heat shock protein 70; PINK1 = PTEN-induced kinase-1; OPA1 = Dynamin-like 120 kDa protein; MFN = Mitofusin-2; Dpr1 = Dynamin-related protein-1; LAMP2 = Lysosome-associated membrane protein 2; i.p = intraperitoneal injection; s.c = subcutaneous injection.

Type of TE training protocol: there were three types of TE protocol: preventive training (n = 2), treatment training (n = 6), and both preventive and treatment training (n = 2). The exercise duration ranged from 1 to 18 weeks. TE training was conducted 3–7 days/week, 20–60 min for each session. The speed of the TE ranged between 10 m/min to 21 m/min (Table 1).

Type of outcome: the included studies analyzed substantia nigra (n = 7), striatum (n = 7), and hippocampus (n = 1). For the electron transport system outcome, six studies reported the expression of neural mitochondrial complex I–V and cytochrome *c*. One study reported the ATP production outcome. For the mitochondrial biogenesis outcome, six studies evaluated the levels of neural mitochondrial biogenesis regulators, including sirtuin-3 (SIRT3), sirtuin-1 (SIRT1), AMP-activated protein kinase (AMPK), peroxisome proliferator-activated receptor gamma coactivator 1-alpha (PGC-1α), Nuclear respiratory factor 1 and 2 (NRF-1,2), and mitochondrial transcription factor A (TFAM). Two studies reported the alterations of import machinery, including translocase of the outer membrane 20 and 40 (TOM-20, TOM-40), translocase of the inner membrane-23 (TIM-23), and mitochondrial heat shock protein (mtHSP70). For the mitochondrial dynamic outcome, two studies reported fusion proteins, i.e., dynamin-like 120 kDa protein (OPA-1) and mitofusin-1,2 (MFN-1,2), as well as fission proteins, i.e., dynamin-related protein-1 (Drp-1) and the phosphorylation at Ser637 of Drp-1 (Drp-1^{Ser637}). For the mitophagy outcome, three studies reported the dysfunctional mitochondria detector proteins, i.e., PTEN-induced kinase-1 (PINK1), parkin, and p62; autophagosomal proteins, i.e., beclin-1 and microtubule-associated protein 1A/1B-light chain 3 (LC3II); as well as lysosomal proteins, i.e., lysosome-associated membrane proteins 2 (LAMP2) and cathepsin L (Table 1).

3.3. Outcome Summary

3.3.1. Effects of TE Training on Neural Mitochondrial Respiratory Deficiency in PD

Three studies analyzed the effects of TE training on mitochondrial complex I in PD [15,23,26]. Two of those showed that the protein levels of complex I were reduced in PD compared to normal, whereas TE training enhanced their levels in PD [15,26]. However, the other study observed that the protein levels of complex I were similar among the normal control group, the PD group, and the TE-trained PD group [23].

Two studies observed that the protein levels of cytochrome *c* in neural mitochondria were reduced in PD compared to normal, whereas TE training increased those levels in PD [20,24]. One study reported that ATP production was reduced in PD compared to normal, whereas TE training enhanced ATP production in PD [25].

Five authors accessed the expression of complexes II, III, IV, and V, reporting different results [15,20,21,23,26]. One study reported that the protein levels of complexes II, III, IV, and V were unchanged among the normal control group, the sedentary PD group, and the TE-trained PD group [26]. The other study showed that the protein levels of complex II and complex V were reduced in PD compared to normal and those levels were restored by TE training in PD, whereas the protein levels of complex III and complex IV were unchanged among the normal control group, the PD group, and the TE-trained PD group [23]. Another study observed that TE training reduced the overexpression of complex II, complex III, and complex IV protein levels in PD, whereas complex V protein levels were unchanged among the normal control group, the PD group, and the TE trained-PD group [15]. Two studies showed that the protein levels of complex IV were reduced in PD compared to normal, whereas TE training increased those levels in PD [20,21].

3.3.2. Effects of TE Training on Neural Mitochondrial Biogenesis in PD

Six publications analyzed TE effects on biogenesis regulators of neural mitochondria in PD [16,20,22–24,26]. Four of those showed that the protein levels of biogenesis regulators, including SIRT3 [23], SIRT1 [16,20], PGC-1α [20,26], NRF-1,2 [20,23,26], and TFAM [20,23,26] were reduced in PD compared to normal, whereas TE training increased those levels in PD. In the other study that analyzed mRNA and protein levels of biogenesis regulators, they observed reduced levels of two biogenesis regulators (AMPK and PGC-1α) along with increased levels of two others (SIRT1 and TFAM) in PD compared to normal [22]. However, in this study, all of those levels were enhanced by TE training in PD [22]. On the contrary, another study showed that the mRNA levels of biogenesis regulators (PGC-1α and TFAM) were increased in PD compared to normal, and those levels were reduced by TE training in PD [24].

Two studies analyzed the effects of TE on translocase factors of neural mitochondria in PD [15,21]. One study reported that the protein levels of translocase proteins (TOM-20, TOM-40, TIM-23, and mtHSP70) were reduced in PD compared to normal, whereas TE training increased those levels in PD [21]. The other study showed that the level of translocase protein (TOM-20) in the substantia nigra was reduced in PD and recovered by TE training, but its level in the striatum was similar among the normal control group, the PD group and the TE-trained PD group [15].

3.3.3. Effects of TE Training on Neural Mitochondrial Dynamics in PD

Two studies analyzed the effects of TE training on mitochondrial fusion and fission proteins [15,23]. They reported that the protein levels of fusion proteins (OPA1, MFN2) were reduced in PD compared to normal, whereas TE training enhanced those levels in PD [15,23]. Regarding neural mitochondrial fission, one of two studies showed that the fission protein (Drp-1) was reduced in PD compared to normal, whereas TE training enhanced those levels in PD [15]. However, the other study showed that the anti-fission protein level (p-Drp1^{Ser637}) was reduced in PD compared to normal, whereas TE training enhanced those levels in PD [23].

3.3.4. Effects of TE Training on Neural Mitophagy in PD

Three studies analyzed the effects of TE training on neural mitophagy in PD [15,20,27]. Those studies showed that the levels of mitophagy detector proteins, including PINK1 [15,27], parkin [27], and p62 [20,27] were increased in PD compared to normal, whereas TE training reduced those levels in PD. Two of those studies showed that the levels of autophagosomal proteins, including beclin-1 [20] and LC3 II/I [20,27] were increased in PD compared to normal, whereas TE training had no effect on their levels in PD. One of those studies reported that the levels of lysosomal proteins (LAMP2 and cathepsin L) were reduced in PD compared to normal, whereas TE training enhanced those levels in PD [27].

3.4. Study Quality Evaluation

All of the studies were published in peer-reviewed journals (item 1), providing a statement of compliance with regulatory requirements (item 9). All of the studies used validated models of PD (item 7). However, 100% of the studies did not mention allocation concealment (item 4), blinded assessment (item 5), or sample size calculation (item 8). Five papers (50%) did not clearly explain the anesthetics process (item 6). Four studies (40%) did not report the randomization of allocation (item 3). Three studies (30%) did not provide temperature control (item 2). Two studies (20%) did not provide conflicts of interest statements (item 10). Together, according to the CAMARADES checklist, the median quality score was 5.7/10 (Table 2).

Table 2. The quality of studies basing-on the CAMARADES checklist.

Study	CAMARADES Checklist of Study Quality										Total
	1	2	3	4	5	6	7	8	9	10	
Koo and Cho, 2017 [20]	√	√	√				√		√	√	6
Koo et al., 2017a [21]	√	√	√				√		√	√	6
Rezaee et al., 2019 [22]	√	√	√			√	√		√	√	7
Chuang et al., 2017 [15]	√	√	√			√	√		√	√	7
Jang et al., 2018 [23]	√	√	√				√		√	√	6
Tuon et al., 2015 [16]	√	√				√	√		√	√	7
Patki and Lau, 2011 [24]	√					√	√		√		4
Lau et al., 2011 [25]	√					√	√		√		4
Ferreira et al., 2020 [26]	√		√				√		√	√	5
Hwang et al., 2018 [27]	√	√					√		√	√	5

Note: (1) Publication in peer-reviewed journal, (2) Statement of control of temperature. (3) Randomization of treatment or control, (4) Allocation concealment, (5) Blinded assessment of outcome, (6) Avoidance of anesthetics with marked intrinsic properties, (7) Use of animals with PD, (8) sample size calculation (9) Statement of compliance with regulatory requirements, (10) Statement regarding possible conflict of interest.

4. Discussion

4.1. Summary of Evidence

Our review findings are synthesized as follows: (1) Treadmill training attenuated neural mitochondrial respiratory deficiency in Parkinson's disease, supported by the evidence that treadmill training normalized the levels of complexes I–V, cytochrome c, and ATP production in the Parkinsonian brain. (2) Treadmill training optimized neural mitochondrial biogenesis in Parkinson's disease, supported by the evidence that treadmill training increased or normalized the levels of biogenesis regulators (SIRT3, SIRT1, AMPK, PGC-1α, NRF-1,2, and TFAM) and import machinery (TOM-20, TOM-40, TIM-23, and mtHSP70) in the Parkinsonian brain. (3) Treadmill training enhanced the neural mitochondrial fusion in Parkinson's disease, supported by the evidence that treadmill training increased mitochondrial fusion factors (OPA-1 and MFN-2) in the Parkinsonian brain. (4) Treadmill training repaired the impairment of mitophagy in Parkinson's disease, supported by the evidence that treadmill training reduced the levels of dysfunctional mitochondria detectors (PINK1, parkin, and p62) and increased the levels of lysosomal factors (LAMP2 and cathepsin L) in the Parkinsonian brain. Taking these findings with the previously hypothesized pathophysiology of Parkinson's disease together, we drew a hypothesized figure (Figure 2), which suggests that treadmill training could counteract the neurodegeneration of Parkinson's disease in both the neural mitochondrial respiratory system and neural mitochondrial quality-control. Our review findings implied that treadmill training might provide therapeutic effects to slow down the progression of Parkinson's disease.

As mentioned, the neurodegeneration of Parkinson's disease on the neural mitochondrial respiratory system is characterized by a complex I deficit, cytochrome c release, and ATP depletion [6]. The included studies showed that treadmill training enhanced the complex I level, cytochrome c concentration, and ATP production in Parkinsonian neural mitochondria [15,20,24–26], suggesting that treadmill training could attenuate the neural mitochondrial respiratory deficiency in Parkinson's disease. Supportively, previous evidence showed that treadmill training increased brain-derived neurotrophic factor (BDNF), which activated neural mitochondrial complex I in Parkinson's disease [28,29]. Of note, a mitochondrial complex I deficit induces oxidative stress and cytochrome c release, promoting neural apoptosis in Parkinson's disease [4,7,30]. Treadmill training appeared to enhance BDNF and complex I level to reduce the progressive development of Parkinson's disease.

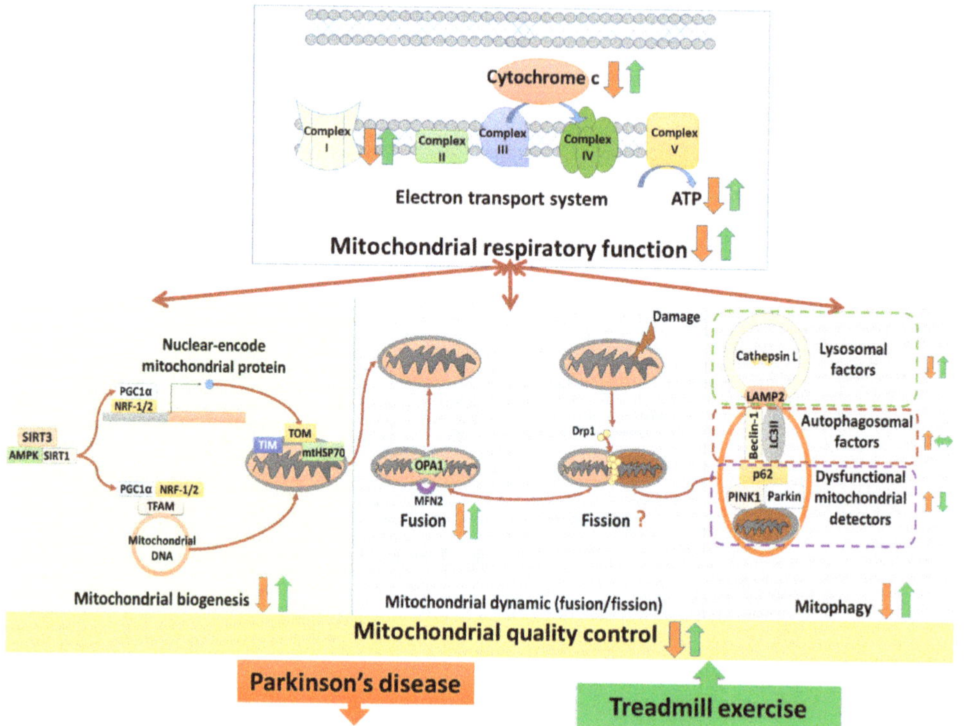

Figure 2. The hypothesized figure. The figure shows the structure of the electron transport system and the cycle of mitochondrial quality-control. The electron transport system includes complexes I–V and cytochrome *c*, taking the primary responsibility for producing ATP (adenosine triphosphate) in neurons. Neural mitochondrial quality-control is the balance among mitochondrial biogenesis, mitochondrial dynamics (fusion/fission), and mitophagy (autophagy of mitochondria). Mitochondrial biogenesis produces the new mitochondria and mitochondrial content, controlled by biogenesis regulators (e.g., SIRT3, SIRT1, AMPK, PGC-1α, NRF-1,2, and TFAM). SIRT3/SIRT1/AMPK activates PGC-1α, then PGC-1α binds with NRF-1,2 in both the nucleus and mitochondria. In the nucleus, PGC-1α and NRF-1,2 promote the production of nuclear-encoded mitochondrial proteins. Nuclear-encoded mitochondrial proteins are imported into mitochondria, which involves the import machinery (e.g., TOM, TIM, and mtHSP70). In the mitochondria, PGC-1α and NRF-1,2 bind with TFAM to activate replication, transcription, and translation of mitochondrial DNA. Mitochondrial fusion merged mitochondria to the large mitochondrion, regulated by OPA-1 in the inner membrane and MFN-2 in the outer membrane. When the mitochondria are damaged, Drp-1 promotes mitochondrial fission to segregate and provide dysfunctional mitochondria for the mitophagy process to destroy. In the mitophagy process, the overexpression of detectors (e.g., PINK1, parkin, p62) on the dysfunctional mitochondrial membrane recruits autophagosomal factors (e.g., beclin-1 and LC3II) to form an autophagosome. Supported by LAMP2, the autophagosome fuses with lysosomes, destroying the dysfunctional mitochondria by enzymes (e.g., cathepsin L). In Parkinson's disease, the evidence shows that complex I, cytochrome *c*, and ATP production in the mitochondria are reduced. Moreover, the mitochondria quality-control is dysregulated in Parkinson's diseases, characterized by biogenesis reduction, fusion/fission imbalance, and mitophagy reduction. The included studies suggested that treadmill training activated complex I, cytochrome *c*, and ATP production. Additionally, treadmill training was shown to optimize the levels of mitochondrial biogenesis regulators (SIRT3, SIRT1, AMPK, PGC-1α, NRF-1,2, and TFAM), translocase factors (TOM-20, TOM-40, TIM-23, and mtHSP70), fusion proteins (OPA-1 and MFN-2), and lysosomal factors (LAMP2 and cathepsin L) as well as reducing dysfunctional mitochondrial detectors (PINK1, parkin, and p62). These data imply that treadmill training could attenuate neural mitochondrial respiratory deficiencies and neural mitochondrial quality-control dysregulation in Parkinson's disease.

In the current included studies, the abnormal expression of complex II, complex III, complex IV, and complex V in Parkinson's disease were varied [15,20,21,23,26]. Consistent with those data, previous studies showed that the alterations of complexes II–V in Parkinson's disease were diverse, which seemed to be remodeled in response to the complex I deficits [5,11,30]. The current included studies showed that treadmill training normalized the levels of neural mitochondrial complexes II–V, suggesting that treadmill training prevented the abnormal remodeling of complexes II–V in Parkinson's disease. In addition to Parkinson's disease, a previous study showed that 12-week treadmill training attenuated the deficits of neural mitochondrial complexes I–IV in the brain of a Huntington's disease mice model [31]. Another study reported that 36-week treadmill training normalized the protein levels of complexes I–V in the cerebral cortex and hippocampus of 40-week-old rats compared to 5-week-old rats [32]. Therefore, we further hypothesized that treadmill training could repair the impairment of the electron transport system in neurodegenerative disorders.

In the physiologic condition, mitochondrial biogenesis is regulated to optimize the neural energy status and guarantee neuronal survival [9]. In Parkinson's disease, α-synuclein binds with genes to inhibit the expression of mitochondrial biogenesis regulators in the brain [33]. The current included studies showed that treadmill training increased or normalized the levels of biogenesis regulators (SIRT3, SIRT1, AMPK, PGC-1α, NRF-1,2, and TFAM), as well as the import machinery (TOM-20, TOM-40, TIM-23, and mtHSP70) [16,20,22–24,26], suggesting that treadmill training could optimize neural mitochondria biogenesis to attenuate the neural energy deficits in Parkinson's disease.

One theory explained that exercise increased oxygen consumption, promoting metabolism challenge and mitochondrial adaptive responses, and thus, the recovery of neural mitochondrial biogenesis [34]. Another researcher suggested that treadmill training may reduce α-synuclein accumulation in Parkinson's disease, thereby increasing the translation of biogenesis regulators and the import of nuclear-encoded mitochondrial proteins into mitochondria [20]. Therefore, our review findings suggested that treadmill training could improve neural mitochondrial biogenesis in Parkinson's disease. Additionally, one study showed that 12-week treadmill training also increased the protein levels of biogenesis regulators (SIRT1, PGC-1α, and TFAM) along with the reduction of beta-amyloid accumulation in the brain of Alzheimer's disease rats model [35]. Another study reported that 12-week treadmill training enhanced neural mitochondrial biogenesis in a Huntington's disease mice model, as evidenced by increases in the ratio of mitochondrial DNA/nucleus DNA in the brain [31]. Overall, we further hypothesized that treadmill training could improve neural mitochondrial biogenesis in neurodegenerative disorders.

In the current review, we found that treadmill training activated neural mitochondrial fusion in Parkinson's disease, as evidenced by increases of mitochondrial fusion factors (OPA-1 and MNF-2) after treadmill training in the Parkinsonian brain [15,23]. Supportively, a previous study showed that 12-week treadmill training increased the levels of fusion proteins (OPA-1 and MNF-2) in the hippocampus of an Alzheimer's disease mouse model [36]. The improvement of neural mitochondrial fusion could reduce the genome mutations and protect the healthy neural mitochondria against the neurodegenerative progression.

Regarding neural mitochondria fission, two included studies reported opposite findings [15,23]. One study showed that neural mitochondrial fission was increased in PD and reduced by TE training [23]. In contrast, another showed that neural mitochondrial fission was reduced in PD and restored to normal by TE [15]. One possible explanation for such a discrepancy is that neural mitochondrial fission is controlled by neural optimization and homeostasis, which are different among Parkinson's disease models [23]. Supportively, a previous study reported that 12-week treadmill exercise normalized the protein levels of fission factors (Drp-1) in the hippocampus of Alzheimer's disease mice model [36]. However, due to the lack of studies, the effect of treadmill training on neural mitochondrial fission was not provided by the current systematic review.

In addition, our review found that treadmill training restored the mitophagy process in Parkinson's disease. In the included studies, treadmill training reduced mitophagy detectors (PINK1, parkin, and p62) but there was no change in autophagosomal factors (beclin-1 and LC3II) in Parkinson's disease [15,20,27], suggesting that treadmill training reduced the accumulation of dysfunctional mitochondria as well as maintained autophagosome flux in Parkinson's disease. Moreover, treadmill training has been shown to upregulate lysosomal proteins (LAMP2 and cathepsin L) in Parkinson's disease, implying that treadmill training increased the activities of the lysosome to fuse with the autophagosome and destroy dysfunctional mitochondria in neural cells [27]. The underlying mechanisms of those benefits may be associated with treadmill training-reduced α-synuclein aggregation in the Parkinsonian brain [20,27]. In addition to Parkinson's disease, a previous study reported that 12-week treadmill training also enhanced mitophagy in an Alzheimer's disease mice model, as evidenced by reductions in the levels of dysfunctional mitochondrial detectors (PINK1 and p62) [37]. Thus, we hypothesized that treadmill training prevents the progressive development of neurodegenerative disorders partially through improving neural mitophagy.

4.2. Study Quality Evaluation

In the current systematic review, we used the CAMARADES checklist to evaluate the methodological quality of the included studies strictly. Because the included studies conducted treadmill training on animals, it is not possible to allocate them in a blinded manner. As is common in animal studies, none of the included studies calculated sample size or used blinded outcome measurements. However, most of the included studies randomly allocated experimental groups and declared no conflicts of interest, suggesting that those studies had no or only minor selection bias and reporting bias. Furthermore, all of the included studies were published in peer-reviewed journals—credible resources. Although there were two publications that were from one group of authors and conducted the similar protocol, they conducted with different sample sizes and reported different aspects of neural mitochondrial functions [24,25]. Similarly, the other two publications were from another group of authors and had the similar protocol, but provided different outcomes [20,21]. Therefore, the data analysis issues (or reporting bias) in those studies might be considered as moderate concern. Taken together, the reviewed findings from the included studies are reasonable. However, due to a lack of data and some concerns mentioned above, further studies need to be conducted to support our review findings.

4.3. Limitations

There were several limitations in our systematic review. First, we only collected the English-published articles with full text, and therefore, we did not include the evidence from non-English articles, conference articles, unpublished articles, or locally published articles. Second, most of the included studies used young animals (n = 8), and all of the included studies used toxin models, but not genetic models. Furthermore, eight of ten studies used male animal models, but not female animal models. All those biases mentioned above limited the generalizability of our review findings to the entire Parkinson's disease population. Third, the studies reviewed herein used treadmill training with various duration (1–18 weeks) and intensities (10–21 m/min). Most of the included studies (n = 9) did not compare durations or intensities of treadmill training on neural mitochondrial functions in Parkinson's disease. Therefore, the current systematic review cannot provide evidence for the optimal protocol of treadmill training in Parkinson's disease. Finally, it should be noted that neural mitochondrial dysfunction in Parkinson's disease involves many factors such as oxidative stress, neuro-inflammation, α-synuclein accumulation, and calcium flux [4]. Our systematic review reported the therapeutic effects of treadmill training on neural mitochondrial functions in Parkinson's disease but cannot provide cause-effects of why treadmill training could attenuate neural mitochondrial respiratory deficiency and neural mitochondrial quality-control dysregulation in Parkinson's disease.

4.4. The Implications for Future Research

In the studies reviewed here, treadmill training appeared to improve both symptoms and neural mitochondrial function in Parkinson's disease animal models. An included study showed that, alongside the enhancement of mitochondrial functions, stride length and swing speed in Parkinsonian rats were enhanced after one month of treadmill training [15]. The other included study reported that, along with neural mitochondrial functions improvement, rest tremor and rigidity in Parkinsonian mice were reduced after 6 weeks of treadmill training [23]. Those results suggested that there may be a possible correlation between symptoms and neural mitochondrial function in Parkinson's disease. Supportively, in clinical observations, a previous randomized controlled clinical trial showed that treadmill training improved symptoms and quality of life in Parkinson's disease patients [38]. Moreover, a previous systematic review showed that treadmill training improved the cognitive function of Parkinson's disease patients [39]. Thus, we hypothesized that treadmill training might reduce the motor and cognitive impairments in Parkinson's disease patients partially through improving neural mitochondrial functions. Further clinical studies should address the correlation between symptoms and mitochondrial function in Parkinson's disease patients with treadmill training.

As mentioned, there were two included studies showed that treadmill exercise training before the onset of PD might also maintain neural mitochondrial functions in animal models of PD [16,22]. Consistently, a previous study provided that 8 weeks of treadmill training could enhance neural mitochondrial biogenesis regulator (PGC-1α protein) as well as neural mitochondrial DNA in the hippocampus of the healthy aged mice (21-month-old) [40]. Another study showed that long-term treadmill training (36 weeks) could increase the protein expressions of neural mitochondrial electron transport chain (complex I, III, and IV) as well as neural mitochondrial biogenesis (SIRT-1, PGC-1α, AMPK) in the hippocampus of 42-week-old healthy rats [32]. Those findings implied that treadmill training might control both neural mitochondrial respiratory functions and neural mitochondrial quality-control to prevent or attenuate the damage of neurodegeneration, including PD. Supportively, regarding clinical evidence, a previous systematic review and meta-analysis has been shown that exercise training could reduce the risk of PD [41]. Therefore, people with the high-risk of neurodegeneration or the neurodegenerative patients should devote themselves to exercise training with treadmill exercise as one choice.

The animal models of PD used in the included studies, including MPTP model and 6-ODHA model, have been proven to be the reliable models and are commonly used to mimic PD in animals [42]. In MPTP model, once entering brain blood barrier, MPTP directly inhibit neural mitochondrial complex I, thereby reduce neural mitochondrial respiration and damage neural mitochondrial functions [42,43]. MPTP has been proven to induce oxidative stress (the accumulation of MPP+ could produce reactive oxygen species) and neuroinflammation (MPTP could activate macrophages, microglia) [43]. In 6-ODHA model, previous evidence proved that 6-ODHA not only generate reactive oxygen species, but also interact with mitochondrial complex I and complex IV to inhibit neural mitochondrial respiration [42]. It should be noted that, in PD, both neural oxidative stress and neuroinflammation could further promote neural mitochondrial dysfunction [44,45]. Alongside the improvement of neural mitochondrial functions, the results of the included studies showed that treadmill training could enhance the levels of antioxidants (e.g., superoxide dismutase) in both neural cytoplasm and neural mitochondria in MPTP mice [20,25] as well as reduce the levels of pro-inflammatory cytokines, including TNF-α, IFN-γ, and IL-1β level in the striatum and hippocampus of 6-ODHA mice [16]. Supportively, a previous study showed that 14-day treadmill training reduced neural oxidative stress in PD, as evidenced by the decreased levels of lipid peroxidation in the striatum of 6-ODHA rats [46]. Another study provided that 4-week treadmill training could attenuate neuroinflammation via the inactivation of microglia in MPTP mice [47]. Those findings might suggest that treadmill exercise could protect neural mitochondrial functions due partially to attenuating the damage of neural oxidative stress and neuroinflammation on neural mitochondria in PD. This issue needs to be addressed in the further studies to clarify the complicated interdependences among the mechanisms of neurodegenerative disorders, including PD.

Epidemiological statistics showed that men have higher risks and proportion to suffer from Parkinson's disease than women in all ages [48]. This phenomenon might be also due partially to the gender-specific characteristics of neural mitochondria. Evidence indicated that the activities of neural mitochondrial electron transport system (complex I–V) and neural mitochondrial capacities in males was lower than that in females [49]. Moreover, the dysregulation of genes associated to neural mitochondrial respiration (neural oxidative phosphorylation) has been proven to be stronger in males when compared to females in PD, suggesting that neural mitochondria in males are more sensitive to the damage of several Parkinsonian mechanisms (e.g., neuroinflammation, neural oxidative stress) than that in females [48]. Supportively, another study provided that estradiol (a hormone in females) could restore the number of hippocampal mitochondria in aged rats (24-month-old), implying that estradiol could maintain neural mitochondrial biogenesis in females [50]. Basing on the differences in neural mitochondrial functions between two genders mentioned here, we might hypothesize that the effects of the same protocol of treadmill training in neural mitochondrial respiratory function as well as neural mitochondrial quality-control in two genders might be different. Therefore, in order to further understand the mitochondria-related effects of treadmill exercise as well as to find the suitable protocol of treadmill training for each gender in PD, further studies are required to evaluate and compare the benefits of treadmill exercise on neural mitochondrial functions in both males and females, regarding the underlying mechanisms (e.g., genes, hormones).

Because treadmill exercise is widely applied for symptomatic improvement in patients with Parkinson's disease [38], the highlighted issue for future clinical trials is how to determine the optimal protocol of treadmill training, which could slow down the progression and improve symptoms in Parkinson's disease patients. To establish this issue, it is necessary to have an efficient method to measure the change of Parkinson's disease mechanisms during and after treadmill training. Recently, the evidence has suggested that 31 phosphorus-magnetic resonance spectroscopy (^{31}P-MRS) is a promised approach for the evaluation of mitochondrial function in Parkinson's disease patients in both the rest-stage and moving-stage [5]. Although the application of ^{31}P-MRS on Parkinson's disease needs to be further investigated, future studies should determine the optimal protocol of treadmill training on neural mitochondrial function in Parkinson's disease by carrying out TE training on Parkinson's disease patients with a ^{31}P-MRS monitor.

5. Conclusions

Our systematic review summarized the recent evidence from animal studies to determine the effects of treadmill training on the neural mitochondrial respiratory deficiency and neural mitochondrial quality-control dysregulation in Parkinson's disease. From the included studies with various Parkinson's disease models and treadmill training protocols, both preventive treadmill training and treatment training were shown to positively affect neural mitochondria in the Parkinsonian brain. Overall, the review found that treadmill training could attenuate the abnormalities of neural mitochondrial complexes I–V, cytochrome *c*, and ATP production, as well as improve the neural mitochondrial biogenesis, neural mitochondrial fusion, and neural mitophagy in Parkinson's disease. Our systematic review suggested that treadmill training might attenuate the neurodegeneration of Parkinson's disease on neural mitochondria, leading to prevention of or delaying the development of Parkinson's disease.

Further interdisciplinary studies are required to investigate the effects of treadmill training on the neural mitochondrial respiratory system, biogenesis, dynamics, and mitophagy in both genetic models and toxin models of Parkinson's disease. Additionally, clinical studies should clarify the possible therapeutic applications through different exercise interventions into neural mitochondrial dysfunction in Parkinson's disease.

Author Contributions: N.T.N. and S.-D.L. contributed to the conceptualization. N.T.N. and S.-D.L. contributed to the methodology. N.T.N., Y.-J.C. and S.-D.L. contributed to the collection, synthesis, and interpretation of data. N.T.N. drafted the manuscript. N.T.N. and S.-D.L. edited and revised the manuscript. All authors have read and agreed to the published version of the manuscript.

Funding: This study was supported by the Ministry of Science and Technology (MOST 107-2314-B-468-002-MY3) and Weifang Medical University. The funders were not associated with design, data searching, data collection, synthesis, and publication decisions.

Institutional Review Board Statement: Not applicable.

Informed Consent Statement: Not applicable.

Data Availability Statement: No new data were created or analyzed in this study. Data sharing is not applicable to this article.

Conflicts of Interest: The authors declare no conflict of interest.

References

1. Sveinbjornsdottir, S. The clinical symptoms of Parkinson's disease. *J. Neurochem.* **2016**, *139*, 318–324. [CrossRef]
2. Kalia, L.V.; Lang, A.E. Parkinson's disease. *Lancet* **2015**, *386*, 896–912. [CrossRef]
3. Cacabelos, R. Parkinson's disease: From pathogenesis to pharmacogenomics. *Int. J. Mol. Sci.* **2017**, *18*, 551. [CrossRef]
4. Chen, C.; Turnbull, D.M.; Reeve, A.K. Mitochondrial dysfunction in Parkinson's disease—Cause or consequence? *Biology* **2019**, *8*, 38. [CrossRef]
5. Dossi, G.; Squarcina, L.; Rango, M. In Vivo Mitochondrial Function in Idiopathic and Genetic Parkinson's Disease. *Metabolites* **2020**, *10*, 19. [CrossRef] [PubMed]
6. Moon, H.E.; Paek, S.H. Mitochondrial dysfunction in Parkinson's disease. *Exp. Neurobiol.* **2015**, *24*, 103–116. [CrossRef] [PubMed]
7. Banerjee, R.; Starkov, A.A.; Beal, M.F.; Thomas, B. Mitochondrial dysfunction in the limelight of Parkinson's disease pathogenesis. *Biochim. Biophys. Acta* **2009**, *1792*, 651–663. [CrossRef]
8. Meng, H.; Yan, W.-Y.; Lei, Y.-H.; Wan, Z.; Hou, Y.-Y.; Sun, L.-K.; Zhou, J.-P. SIRT3 regulation of mitochondrial quality control in neurodegenerative diseases. *Front. Aging Neurosci.* **2019**, *11*, 313. [CrossRef]
9. Diaz, F.; Moraes, C.T. Mitochondrial biogenesis and turnover. *Cell Calcium.* **2008**, *44*, 24–35. [CrossRef] [PubMed]
10. Lodish, H.; Berk, A.; Kaiser, C.A.; Krieger, M.; Scott, M.P.; Bretscher, A.; Ploegh, H.; Matsudaira, P. *Molecular Cell Biology*, 6th ed.; W. H. Freeman and Company: New York, NY, USA, 2008.
11. Perier, C.; Vila, M. Mitochondrial biology and Parkinson's disease. *Cold Spring Harb. Perspect. Med.* **2012**, *4*, a009332. [CrossRef]
12. Liu, J.; Liu, W.; Li, R.; Yang, H. Mitophagy in Parkinson's disease: From pathogenesis to treatment. *Cells* **2019**, *8*, 712. [CrossRef]
13. Mehrholz, J.; Kugler, J.; Storch, A.; Pohl, M.; Elsner, B.; Hirsch, K. *Treadmill Training for Patients with Parkinson's Disease*; John Wiley & Sons: Hoboken, NJ, USA, 2015.
14. Wang, R.; Tian, H.; Guo, D.; Tian, Q.; Yao, T.; Kong, X. Impacts of exercise intervention on various diseases in rats. *J. Sport Health Sci.* **2020**, *9*, 211–227. [CrossRef]
15. Chuang, C.-S.; Chang, J.-C.; Cheng, F.-C.; Liu, K.-H.; Suc, H.-L.; Liu, C.-S. Modulation of mitochondrial dynamics by treadmill training to improve gait and mitochondrial deficiency in a rat model of Parkinson's disease. *Life Sci.* **2017**, *191*, 236–244. [CrossRef]
16. Tuon, T.; Souza, P.S.; Santos, M.F.; Pereira, F.T.; Pedroso, G.S.; Luciano, T.F.; Souza, C.T.D.; Dutra, R.C.; Silveira, P.C.L.; Pinho, R.A. Physical training regulates mitochondrial parameters and neuroinflammatory mechanisms in an experimental model of Parkinson's disease. *Oxidative Med. Cell. Longev.* **2015**, *2015*, 261809. [CrossRef]
17. Moher, D.; Liberati, A.; Tetzlaff, J.; Altman, D.G.; The PRISMA Group. Preferred reporting items for systematic reviews and meta-analyses: The PRISMA statement. *PLoS Med.* **2009**, *6*, e1000097. [CrossRef]
18. Beller, E.M.; Glasziou, P.P.; Altman, D.G.; Hopewell, S.; Bastian, H.; Chalmers, I.; Gøtzsche, P.C.; Lasserson, T.; Tovey, D. PRISMA for abstracts: Reporting systematic reviews in journal and conference abstracts. *PLoS Med.* **2013**, *10*, e1001419. [CrossRef]
19. Auboire, L.; Sennoga, C.A.; Hyvelin, J.-M.; Ossant, F.; Escoffre, J.-M.; Tranquart, F.; Bouakaz, A. Quality assessment of the studies using the collaborative approach to meta-analysis and review of Animal Data from Experimental Studies (CAMARADES) checklist items. *PLoS ONE* **2018**. [CrossRef]
20. Koo, J.-H.; Cho, J.-Y. Treadmill exercise attenuates α-synuclein levels by promoting mitochondrial function and autophagy possibly via SIRT1 in the chronic MPTP/P-induced mouse model of Parkinson's disease. *Neurotox Res.* **2017**, *32*, 473–486. [CrossRef]
21. Koo, J.-H.; Cho, J.-Y.; Lee, U.-B. Treadmill exercise alleviates motor deficits and improves mitochondrial import machinery in an MPTP-induced mouse model of Parkinson's disease. *Exp. Gerontol.* **2017**, *89*, 20–29. [CrossRef] [PubMed]
22. Rezaee, Z.; Marandi, S.M.; Alaei, H.; Esfarjani, F.; Feyzollahzadeh, S. Effects of preventive treadmill exercise on the recovery of metabolic and mitochondrial factors in the 6-hydroxydopamine rat model of Parkinson's disease. *Neurotox. Res.* **2019**, *35*, 908–917. [CrossRef] [PubMed]

23. Jang, Y.; Kwon, I.; Song, W.; Cosio-Lima, L.M.; Taylor, S.; Lee, Y. Modulation of mitochondrial phenotypes by endurance exercise contributes to neuroprotection against a MPTP-induced animal model of PD. *Life Sci.* **2018**, *209*, 455–465. [CrossRef]
24. Patki, G.; Lau, Y.-S. Impact of exercise on mitochondrial transcription factor expression and damage in the striatum of a chronic mouse model of Parkinson's disease. *Neurosci. Lett.* **2011**, *505*, 268–272. [CrossRef]
25. Lau, Y.-S.; Patki, G.; Das-Panja, K.; Le, W.-D.; Ahmad, S.O. Neuroprotective effects and mechanisms of exercise in a chronic mouse model of Parkinson's disease with moderate neurodegeneration. *Eur. J. Neurosci.* **2011**, *33*, 1264–1274. [CrossRef] [PubMed]
26. Ferreira, A.F.F.; Binda, K.H.; Singulani, M.P.; Pereira, C.P.M.; Ferrari, G.D.; Alberici, L.C.; Real, C.C.; Britto, L.R. Physical exercise protects against mitochondria alterations in the 6-hidroxydopamine rat model of Parkinson's disease. *Behav. Brain Res.* **2020**, *387*, 11260. [CrossRef]
27. Hwang, D.; Koo, J.; Kwon, K.; Choi, D.; Shin, S.; Jeong, J.; Um, H.; Cho, J. Neuroprotective effect of treadmill exercise possibly via regulation of lysosomal degradation molecules in mice with pharmacologically induced Parkinson's disease. *J. Physiol. Sci.* **2018**, *68*, 707–716. [CrossRef]
28. Markham, A.; Cameron, I.; Franklin, P.; Spedding, M. BDNF increases rat brain mitochondrial respiratory couplingat complex I, but not complex II. *Eur. J. Neurosci.* **2004**, *20*, 1189–1196. [CrossRef]
29. Marques-Aleixo, I.S.; Oliveira, P.J.; Moreira, P.I.; Magalhães, J.; Ascensão, A.N. Physical exercise as a possible strategy for brain protection: Evidence from mitochondrial-mediated mechanisms. *Prog. Neurobiol.* **2012**, *99*, 149–162. [CrossRef] [PubMed]
30. Moreno-Lastres, D.; Fontanesi, F.; García-Consuegra, I.; Martín, M.A.; Arenas, J.; Barrientos, A.; Ugalde, C. Mitochondrial complex I plays an essential role in human respirasome assembly. *Cell Metab.* **2012**, *15*, 324–335. [CrossRef] [PubMed]
31. Caldwell, C.C.; Petzinger, G.M.; Jakowec, M.W.; Caden, E. Treadmill exercise rescues mitochondrial function and motor behavior in the CAG140 knock-in mouse model of Huntington's disease. *Chem. Biol. Interact.* **2020**, *315*, 108907. [CrossRef] [PubMed]
32. Bayod, S.; Del Valle, J.; Canudas, A.M.; Lalanza, J.F.; Sanchez-Roige, S.; Camins, A.; Escorihuela, R.M.; Pallàs, M. Long-term treadmill exercise induces neuroprotective molecular changes in rat brain. *J. Appl. Physiol.* **2011**, *111*, 1380–1390. [CrossRef] [PubMed]
33. Park, J.-S.; Davis, R.L.; Sue, C.M. Mitochondrial dysfunction in Parkinson's disease: New mechanistic insights and therapeutic perspectives. *Curr. Neurol. Neurosci. Rep.* **2018**, *18*, 21. [CrossRef]
34. Radak, Z.; Zhao, Z.; Koltai, E.; Ohno, H.; Atalay, M. Oxygen consumption and usage during physical exercise: The balance between oxidative stress and ROS-dependent adaptive signaling. *Antioxid. Redox Signal.* **2013**, *18*, 1208–1246. [CrossRef] [PubMed]
35. Koo, J.-H.; Kang, E.-B.; Oh, Y.-S.; Yang, D.-S.; Cho, J.-Y. Treadmill exercise decreases amyloid-β burden possibly via activation of SIRT-1 signaling in a mouse model of Alzheimer's disease. *Exp. Neurol.* **2017**, *288*, 142–152. [CrossRef]
36. Yan, Q.-W.; Zhao, N.; Xia, J.; Li, B.-X.; Yin, L.-Y. Effects of treadmill exercise on mitochondrial fusion and fission in the hippocampus of APP/PS1 mice. *Neurosci. Lett.* **2019**, *701*, 84–91. [CrossRef] [PubMed]
37. Zhao, N.; Yan, Q.; Xia, J.; Zhang, X.; Li, B.; Yin, L.; Xu, B. Treadmill exercise attenuates aβinduced mitochondrial dysfunction and enhances mitophagy activity in APP/PS1 transgenic mice. *Neurochem. Res.* **2020**, *45*, 1202–1214. [CrossRef]
38. Arfa-Fatollahkhani, P.; Cherati, A.S.; Habibi, S.A.H.; Shahidi, G.A.; Sohrabi, A.; Zamani, B. Effects of treadmill training on the balance, functional capacity and quality of life in Parkinson's disease: A randomized clinical trial. *J. Complement. Integr. Med.* **2019**, *17*, 17. [CrossRef]
39. Da Silva, F.C.; da Rosa Iop, R.; de Oliveira, L.C.; Boll, A.M.; de Alvarenga, J.G.S.; Filho, P.J.B.G.; de Melo, L.M.A.B.; Xavier, A.J.; da Silva, R. Effects of physical exercise programs on cognitive function in Parkinson's disease patients: A systematic review of randomized controlled trials of the last 10 years. *PLoS ONE* **2018**, *13*, e0193113. [CrossRef]
40. Lezi, E.; Burns, J.M.; Swerdlow, R.H. Effect of high-intensity exercise on aged mouse brain mitochondria, neurogenesis, and inflammation. *Neurobiol. Aging* **2014**, *35*, 2574–2583. [CrossRef]
41. Fang, X.; Han, D.; Cheng, Q.; Zhang, P.; Zhao, C.; Min, J.; Wang, F. Association of levels of physical activity with risk of Parkinson's disease. *JAMA Netw. Open* **2018**, *1*, e182421. [CrossRef]
42. Duty, S.; Jenner, P. Animal models of Parkinson's disease: A source of novel treatments and clues to the cause of the disease. *Br. J. Pharmacol.* **2011**, *164*, 1357–1391. [CrossRef]
43. Meredith, G.E.; Rademacher, D.J. MPTP mouse models of Parkinson's disease: An update. *J. Parkinson's Dis.* **2011**, *1*, 19–33. [CrossRef] [PubMed]
44. Stepien, K.M.; Heaton, R.; Rankin, S.; Murphy, A.; Bentley, J.; Sexton, D.; Hargreaves, I.P. Evidence of oxidative stress and secondary mitochondrial dysfunction in metabolic and non-metabolic disorders. *J. Clin. Med.* **2017**, *6*, 71. [CrossRef]
45. Jellinger, K.A. Basic mechanisms of neurodegeneration: A critical update. *J. Cell. Mol. Med.* **2010**. [CrossRef] [PubMed]
46. Da Costa, R.O.; Gadelha-Filho, C.V.J.; da Costa, A.E.M.; Feitosa, M.L.; de Araújo, D.P.; de Lucena, J.D.; de Aquino, P.E.A.; Lima, F.A.V.; Neves, K.R.T.; de Barros Viana, G.S. The treadmill exercise protects against dopaminergic neuron loss and brain oxidative stress in parkinsonian rats. *Oxidative Med. Cell. Longev.* **2017**, *2017*, 2138169. [CrossRef]
47. Sung, Y.-H.; Kim, S.-C.; Hong, H.-P.; Park, C.-Y.; Shin, M.-S.; Kim, C.-J.; Seo, J.-H.; Kim, D.-Y.; Kim, D.-J.; Cho, H.-J. Treadmill exercise ameliorates dopaminergic neuronal loss through suppressing microglial activation in Parkinson's disease mice. *Life Sci.* **2012**, *91*, 1309–1316. [CrossRef]

48. Ullah, M.F.; Ahmad, A.; Bhat, S.H.; Abu-Duhier, F.M.; Barreto, G.E.; Ashraf, G.M. Impact of sex differences and gender specificity on behavioral characteristics and pathophysiology of neurodegenerative disorders. *Neurosci. Biobehav. Rev.* **2019**, *102*, 95–105. [CrossRef]
49. Cerri, S.; Mus, L.; Blandini, F. Parkinson's disease in women and men: What's the difference? *J. Parkinson's Dis.* **2019**, *9*, 501–515. [CrossRef] [PubMed]
50. Waters, E.M.; Mazid, S.; Dodos, M.; Puri, R.; Janssen, W.G.; Morrison, J.H.; McEwen, B.S.; Milner, T.A. Effects of estrogen and aging on synaptic morphology and distribution of phosphorylated Tyr1472 NR2B in the female rat hippocampus. *Neurobiol. Aging* **2019**, *73*, 200–210. [CrossRef] [PubMed]

MDPI

St. Alban-Anlage 66

4052 Basel

Switzerland

Tel. +41 61 683 77 34

Fax +41 61 302 89 18

www.mdpi.com

Biomedicines Editorial Office

E-mail: biomedicines@mdpi.com

www.mdpi.com/journal/biomedicines

www.ingramcontent.com/pod-product-compliance
Lightning Source LLC
LaVergne TN
LVHW071617170726
843515LV00009B/2413